ENVIRONMENTAL
Science and Technology

ENVIRONMENTAL
Science and Technology

Stanley E. Manahan
Department of Chemistry
University of Missouri
Columbia, Missouri

Library of Congress Cataloging-in-Publication Data

Manahan, Stanley E.
Environmental science and technology / Stanley E. Manahan
p. cm.
Includes bibliographical references and index.
ISBN 1-56670-213-5
1. Environmental sciences. 2. Ecological engineering. I. Title.
GE105.M36 1997
628—dc21

97-306
CIP

International Standard Book Number 1-56670-213-5
Library of Congress Card Number 97-306
Printed in the United States of America 1 2 3 4 5 6 7 8 9 0
Printed on acid-free paper

PREFACE

Environmental Science and Technology is designed to provide an overview of the environment in its broadest sense. As is the case with other works dealing with environmental science and ecology, this book recognizes the four traditional "spheres" of the environment: (1) the hydrosphere (water), (2) the atmosphere (air), (3) the geosphere (earth), and (4) the biosphere (life). In addition, however, it incorporates a fifth sphere, which might be called the "anthrosphere," dealing with human activities and technology. In so doing, *Environmental Science and Technology* acknowledges that, for better or for worse, technology is intimately intertwined with the other environmental spheres. Humans will continue to utilize resources, manufacture goods, practice agriculture, and engage in other activities that have profound effects on Earth. The challenge, then, is to practice technology in a manner that minimizes environmental disruption and that, through technology constructively directed toward environmental improvement, is a positive influence on Earth's environment. This is the basic theme of this book.

As shown from the Table of Contents, *Environmental Science and Technology* is divided into six major areas. The first of these consists of introductory chapters that relate technology to environmental science and outline basic chemistry, biology, biochemistry, and environmental chemistry. The remaining five units are organized according to the five major environmental spheres consisting of the hydrosphere, atmosphere, geosphere, biosphere, and "technology." Throughout the book, relationships are shown among these five major environmental spheres. For example, as part of the coverage of the hydrosphere, it is explained how the geosphere with which water is in contact influences the environmental chemistry of the water, how aquatic life largely determines aquatic chemical phenomena, and how technological means are used to treat water and reduce pollution.

As with other books by the author, this book is written to serve as both a trade book and a textbook. Courses in environmental science are widely taught at the college level and have substantial enrollments. *Environmental Science and Technology* will fill a specific niche in that market because of its incorporation of technology with the other areas of the environment commonly covered in such courses.

AUTHOR

Stanley E. Manahan is Professor of Chemistry at the University of Missouri–Columbia, where he has been on the faculty since 1965. He received his A.B. in chemistry from Emporia State University in 1960 and his Ph.D. in analytical chemistry from the University of Kansas in 1965. Since 1968 his primary research and professional activities have been in environmental chemistry and have included the development of methods for chemical analysis of pollutant species, environmental aspects of coal conversion processes, development of coal products useful for pollutant control, hazardous waste treatment, and toxicological chemistry. He teaches courses on environmental chemistry, hazardous wastes, toxicological chemistry, and analytical chemistry and has lectured on these topics throughout the U.S. as an American Chemical Society Local Section tour speaker. He is also President of ChemChar Research, Inc., a firm working on the development of non-incinerative thermochemical and electrothermochemical treatment of mixed hazardous substances containing refractory organic compounds and heavy metals.

Professor Manahan has written books on environmental chemistry (*Environmental Chemistry*, 6th ed., Stanley E. Manahan, CRC Press/Lewis Publishers, 1994); general and environmental chemistry (*Fundamentals of Environmental Chemistry*, Stanley E. Manahan, CRC Press/Lewis Publishers, 1993); hazardous wastes (*Hazardous Waste Chemistry, Toxicology and Treatment*, 1990, Lewis Publishers, Inc.); toxicological chemistry (*Toxicological Chemistry*, 2nd ed., 1992, Lewis Publishers, Inc.); applied chemistry; and quantitative chemical analysis. He has been the author or co-author of approximately 80 research articles.

ACKNOWLEDGMENTS

The author gratefully acknowledges the assistance of the staff of CRC Press in producing this book. The editor, Joel Stein, was very helpful in recognizing the need for a book with the approach of this one and in moving the project along. The two project editors who worked on the project, Joan Moscrop and Mimi Williams of CRC Press, were exceptionally helpful and diligent in their editing. It has been a pleasure working with both of them. Barbara Glunn of the CRC Marketing Department continues to do an outstanding job of bringing this work to the attention of potential readers who might have a need for it. Matt Braunel of the University of Missouri Department of Chemistry was very helpful in reading final copy of the manuscript and noting several corrections.

Feedback from individuals who have used this book is very valuable, and their comments regarding the basic premise and approach used in writing the book would be very much appreciated. The author may be contacted at 123 Chemistry Building, University of Missouri, Columbia, MO 65211 or through his e-mail address, which is chemstan@showme.missouri.edu.

TABLE OF CONTENTS

INTRODUCTORY CHAPTERS

1 ENVIRONMENTAL SCIENCE AND THE TECHNOLOGICAL CONNECTION

1.1. WHAT IS ENVIRONMENTAL SCIENCE?

Environmental science in its broadest sense is the science of the complex interactions that occur among the terrestrial, atmospheric, aquatic, and living environments. It includes all the disciplines, such as chemistry, biology, ecology, sociology, and government, that affect or describe these interactions. For the purposes of this book, environmental science will be defined as *the study of the earth, air, water, and living environments and the effects of technology thereon.* To a significant degree, environmental science has evolved from investigations of the ways by which, and places in which living organisms carry out their life cycles. This is the discipline of **natural history**, which in recent times has evolved into **ecology**, the study of environmental factors that affect organisms and of how organisms interact with these factors and with each other.[1]

For better or for worse, the environment in which all humans must live has been affected irrreversibly by technology. Therefore, technology is considered strongly in this book in terms of how it affects the environment and in the ways by which, applied intelligently by those knowledgeable of environmental science, it can serve, rather than damage, this Earth upon which all living beings depend for their welfare and existence.

The physical environments—air, water, and earth—are tied closely with living systems, including humans. These four environmental spheres have strong mutual interactions with technology. These tie-ins are illustrated in Figure 1.1, which in a sense summarizes and outlines the theme of the rest of this book.

A key aspect of environmental science is the "interrelatedness" of things, the influence that one thing, action, or change may have on another. One of the ways in which this interrelatedness is most strongly expressed is through feedback in environmental systems. **Positive feedback** occurs when a change or action tends to amplify itself. An example of detrimental positive feedback is provided by soil erosion set off, for example, by plowing virgin prairie land in hilly terrain. Topsoil is removed by running water causing gullies to form. As these form, they can induce even larger erosive features. As topsoil is lost, it is harder to grow vegetative cover, making the soil even more prone to erosion. **Negative feedback** occurs when a system automatically adjusts to minimize an influence. Left to their own devices, rivers and their floodplains tend to develop in ways that compensate for increased river flow and minimize environmental harm from flooding.

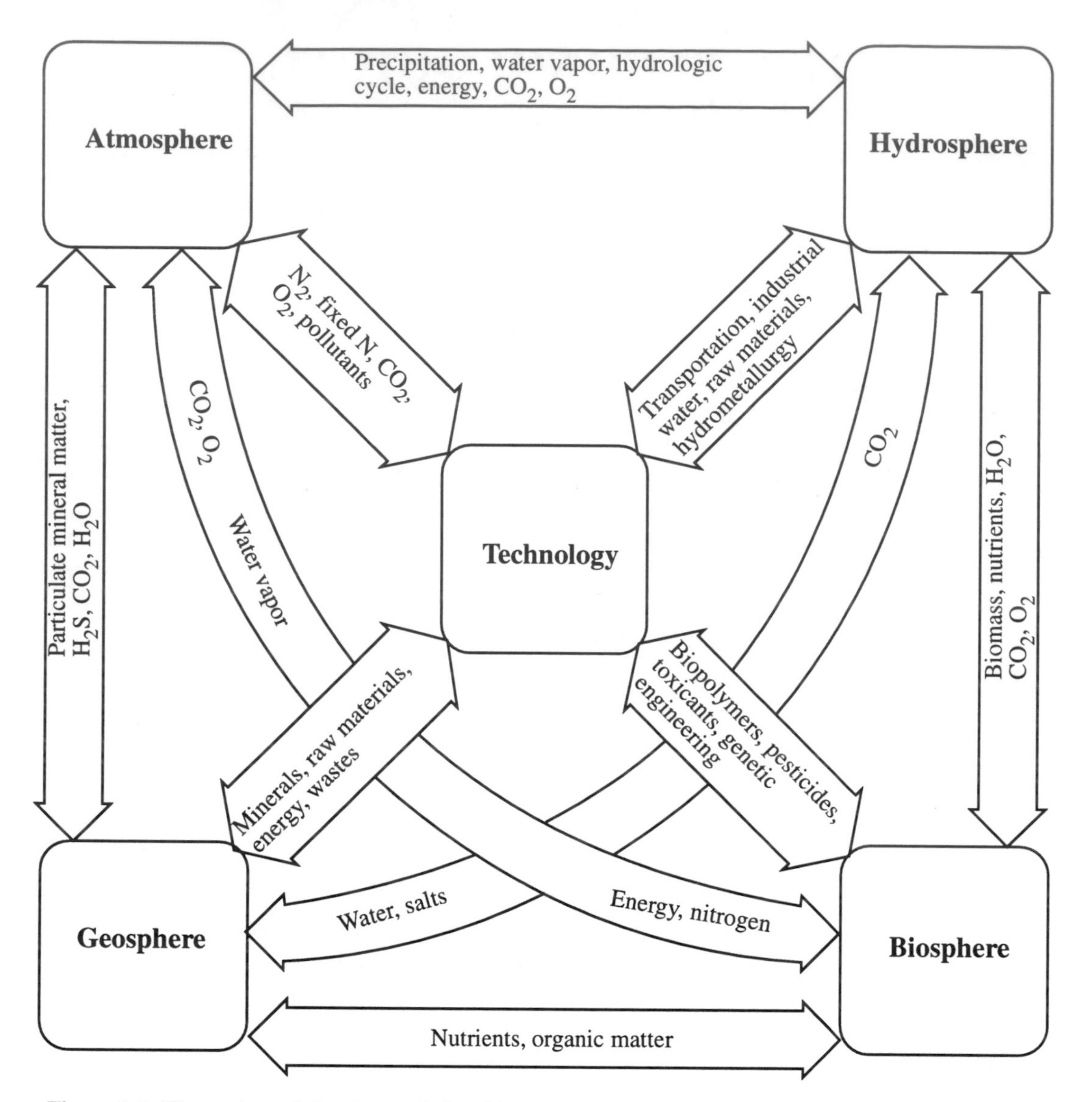

Figure 1.1. Illustration of the close relationships among the air, water, and earth environments with each other and with living systems, as well as the tie-in with technology.

Key Concepts and Definitions

There are numerous key concepts and definitions that are used in discussing the environmental sciences. A few of these that will be helpful in understanding this chapter are summarized here.

Environmental science may be divided among the study of the atmosphere, the hydrosphere, the geosphere, and the biosphere. The **atmosphere** is the thin layer of gases that covers Earth's surface. In addition to its role as a reservoir of gases, the atmosphere moderates Earth's temperature, absorbs energy and damaging ultraviolet radiation from the sun, transports energy away from equatorial regions, and serves as a pathway for vapor-phase movement of water in the hydrologic cycle. The **hydrosphere** contains Earth's water. Over 97% of Earth's water is in oceans, and most of the remaining fresh water is in the form of ice. Therefore, only a relatively small percentage of the total water on Earth is actually involved with terrestrial, atmospheric, and biological processes. Exclusive of seawater, the water that circulates through environmental processes and cycles occurs in the atmosphere, underground as groundwater, and as surface water in streams, rivers, lakes, ponds, and reservoirs. The **geosphere** consists of the solid earth, including soil, which supports most plant life. The part of the geosphere that is directly involved with environmental processes through contact with the atmosphere, the hydrosphere, and living things is the solid **lithosphere**, varying from 50 to 100 km in thickness, and particularly the outer skin of the lithosphere composed largely of lighter silicate-based minerals and called the **crust**. All living entities on Earth compose the **biosphere**. Living organisms and the aspects of the environment pertaining directly to them are called **biotic**, and other portions of the environment are **abiotic**.

There are strong interactions among living organisms and the various spheres of the abiotic environment. To a large extent these are best described by cycles of matter that involve biological, chemical, and geological processes and phenomena. Such cycles are called **biogeochemical cycles**.

There is not space at this point to define all the terms and describe all the concepts essential to the understanding of environmental science. Several of these that should be noted here are the following:

- **Matter**: Substance, that which occupies space and has mass.
- **Chemistry**: The science of matter—the study of the composition, structure, and properties of matter and the changes that matter undergoes.
- **Energy**: The capacity to do work, such as by causing a body of matter to move. The rate at which energy is transferred or moved, that is, energy per unit time, is called **power**.
- **Resources**: Matter of specific kinds and energy needed by humans for their well-being or existence. **Renewable resources** are those that are replenished naturally within a reasonable time span.
- **Climate**: The overall, long-term characteristics of weather, including temperature, precipitation, storms, and wind patterns in an area.
- **Pollutant**: A substance present in greater than natural concentration as a result of human activity and having a net detrimental effect upon its environment or upon something of value in that environment. Every pollutant originates from a **source**. A **receptor** is anything that is affected by a pollutant. A **sink** is a long-time repository of a pollutant.
- **Biological community**: The total of all living organisms inhabiting a specified area. A biological community and the environmental conditions that characterize it are termed a **biome**. A group of organisms of the same species in a biological community is called a **population**. The role played by a biological population in its surroundings, including the pattern by which it uses available resources, is called its **ecological niche**.

- **Productivity**: Rate of production of biomass per unit time per unit area by organisms called **producers**, which use energy (usually from photosynthesis) to produce biological matter from inorganic substances.
- **Limiting factors**: Environmental factors, such as nutrients, that limit the abundance, growth, or distribution of an organism, or that determine whether it even exists in a particular environment.
- **Ecology**: The study of the interactions of organisms with their environment and with each other.

1.2. ENVIRONMENTAL SCIENCE, TECHNOLOGY, AND SOCIETY

Modern societies and economic systems are driven largely by material wants and needs. The more industrially and economically advanced societies, such as those in Western Europe, the United States, and Japan, are characterized by a high level of consumption and the accompanying quest for money and the spending of it for material goods. Countries with developing economies are similarly driven, aspiring to the same standards of living as are enjoyed by citizens of richer nations. Residents of the poorest nations, such as a number of countries in Africa, Latin America, and parts of Asia, are driven by a more fundamental need—the acquisition of enough food, water, fuel, and other basic necessities to ensure their survival.

Given these conditions and human attitudes, what is the role of environmental science in modern societies, and how is it related to technology and human needs and aspirations? Does environmental science even have a place in the richer societies that can afford the necessities of life and many luxuries as well? Should environmental factors even be considered in subsistence societies in which getting the most basic necessities for human survival is the overwhelming concern of most of the population? Can environmental science coexist with modern technology to the benefit of humankind? The answer to all of these questions is a resounding, "yes!"

Consider first the modern industrialized societies. The material benefits enjoyed by people in these societies cannot be sustained for long without coming to terms with the environment upon which the resources, energy, and quality surroundings enjoyed by the people depend. Depletion of resources of scarce metals and strategic minerals will force prices upward so that the possessions required for the so-called "good life"—luxury automobiles, modern single-family dwellings, and other amenities—can no longer be purchased. Increased income levels enable people to move farther from population centers to enjoy less urbanized surroundings. The same conditions that draw people to the more rural areas deteriorate because of the presence of more people. These factors in turn exacerbate social problems and tensions. For example, there is a tendency for the upper economic strata of society to use their influence to ensure conditions under which their material benefits will be sustained or enhanced, even if this means increased poverty in the lower economic strata and a diminished middle class. A large fraction of the poorer population "tunes out" and becomes involved with self-destructive behavior that makes their situation even worse. The environment in which people find themselves certainly influences their behavior and well-being.

The question is commonly asked whether or not developing and poor, undeveloped societies can afford to consider environment. The answer is that they cannot afford to do otherwise. In its broadest sense a good environment is one that is favorable for the existence of life in a flourishing, steady-state condition. Ruthless exploitation of the environment—farming of marginal land without proper conservation practices, poorly considered damming of rivers to provide hydroelectric power, and intense exploitation of mineral and energy resources—will result in environmental deterioration that is eventually counterproductive to the material standard of living that such measures are designed to enhance.

Whereas ill-considered exploitation of the environment is definitely counterproductive, the opposite approach of "going back to nature" and shunning all technological development is wildly impractical. It simply is not going to happen. Therefore, it is essential to realize that technology *will be employed* to meet human needs, that these needs can only be met on the basis of a healthy environment suitable for supporting humankind, and that societies must learn to live in harmony with the environment and with technology.

1.3. WATER

Water, with a deceptively simple chemical formula of H_2O, is a vitally important substance in all parts of the environment. Water covers about 70% of Earth's surface. It occurs in all spheres of the environment—in the oceans as a vast reservoir of saltwater, on land as surface water in lakes and rivers, underground as groundwater, in the atmosphere as water vapor, and in the polar icecaps as solid ice. Water is an essential part of all living systems and is the medium from which life evolved and in which life exists.

Energy and matter are carried through various spheres of the environment by water. Water leaches soluble constituents from mineral matter and carries them to the ocean or leaves them as mineral deposits some distance from their sources. Water carries plant nutrients from soil into the bodies of plants by way of plant roots. Solar energy absorbed in the evaporation of ocean water is carried as latent heat and released inland. The accompanying release of latent heat provides the energy that powers massive storms.

The quantity of water in various forms on or beneath Earth's surface and in the atmosphere is enormous, amounting to about 1.4 billion cubic kilometers of liquid water. By far the largest portion of this water, about 97.6%, occurs in the oceans. The next largest amount, about 2%, is held as ice and snow, largely in the polar and Greenland icecaps. Groundwater held in underground aquifers amounts to 0.28% (to a depth of 1 km). The rest of Earth's water, only about 0.25% of the total, is found in decreasing order of abundance in saline and freshwater lakes and reservoirs; as soil moisture; as water held in living organisms, as vapor, droplets, and miniscule ice crystals in the atmosphere; in swamps and marshes; and in rivers and streams.

Water is obviously an important topic in environmental sciences, and it is discussed in some detail in later chapters of this book. Chapter 5 covers characteristics of water and bodies of water; Chapter 6 discusses the environmental chemistry, biology, and microbiology of water; Chapter 7 deals with water pollution; and Chapter 8 covers water treatment.

1.4. AIR AND THE ATMOSPHERE

The atmosphere is a protective blanket that nurtures life on the Earth and protects it from the hostile environment of outer space. The atmosphere is the source of carbon dioxide for plant photosynthesis and of oxygen for respiration. It provides the nitrogen that nitrogen-fixing bacteria and ammonia-manufacturing industrial plants use to produce chemically bound nitrogen, an essential component of life molecules. As a basic part of the hydrologic cycle (Chapter 5, Figure 5.1) the atmosphere transports water from the oceans to land, thus acting as the condenser in a vast solar-powered still. The atmosphere serves a vital protective function. It absorbs most of the cosmic rays from outer space and protects organisms from their effects. It also absorbs most of the electromagnetic radiation from the sun, allowing transmission of significant amounts of radiation only in the near-ultraviolet, visible, and near-infrared radiation regions. The atmosphere filters out damaging ultraviolet radiation that would otherwise be very harmful to living organisms. Furthermore, because it reabsorbs much of the infrared radiation by which absorbed solar energy is re-emitted to space, the atmosphere stabilizes the Earth's temperature, preventing the great temperature extremes of planets and moons lacking substantial atmospheres.

Atmospheric science deals with the movement of air masses in the atmosphere, atmospheric heat balance, and atmospheric chemical composition and reactions. **Meteorology** is the study of the movement of air masses as well as physical forces in the atmosphere such as heat, wind, and transitions of water, primarily liquid to vapor, or *vice versa*. Short-term variations in the state of the atmosphere, including temperature, clouds, winds, humidity, horizontal visibility, type and quantity of precipitation, and atmospheric pressure, are described as **weather**. Long-term conditions of weather referred to regionally or even globally are classified as **climate**. The nature of climate and potential human effects on it are vitally important for environmental protection.

Atmospheric science is covered in this book in Chapters 9–12. Chapter 9 discusses meteorology and characteristics of the atmosphere. Atmospheric chemistry is discussed specifically in Chapter 10, and air pollution and its control in Chapter 11. Some effects on the atmosphere, such as global warming by the emission of "greenhouse gas" carbon dioxide from fossil fuel combustion or injection into the atmosphere of massive amounts of sunlight-blocking particulate matter by the impacts of asteroids, have the potential to cause massive harm to the environment and are discussed in Chapter 12 under the topic of "The Endangered Global Atmosphere."

1.5. EARTH

The **geosphere**, or solid Earth, discussed in general in Chapter 13, is that part of the Earth upon which humans live and from which they extract most of their food, minerals, and fuels. The earth is divided into layers, including the solid iron-rich inner core, molten outer core, mantle, and crust. Environmental science is most concerned with the **lithosphere**, which consists of the outer mantle and the **crust**. The latter is the earth's outer skin that is accessible to humans. It is extremely thin compared to the diameter of the earth, ranging from 5 to 40 km thick.

Geology is the science of the geosphere. As such it pertains mostly to the solid mineral portions of Earth's crust. But it must also consider water, which is involved in weathering rocks and in producing mineral formations; the atmosphere and climate, which have profound effects on the geosphere and interchange matter and energy with it; and living systems, which largely exist on the geosphere and in turn have significant effects on it. Geological science uses chemistry to explain the nature and behavior of geological materials, physics to explain their mechanical behavior, and biology to explain the mutual interactions between the geosphere and the biosphere.[2] Modern technology, for example, the ability to move massive quantities of dirt and rock around, has a profound influence on the geosphere.

Most of the solid Earth crust consists of rocks. Rocks are composed of minerals, where a **mineral** is a naturally occurring inorganic solid with a definite internal structure and chemical composition. A **rock** is a mass of pure mineral or an aggregate of two or more minerals. The most important part of the geosphere for life on earth is **soil** formed by the disintegrative weathering action of physical, geochemical, and biological processes on rock. It is the medium upon which plants grow, and virtually all terrestrial organisms depend upon it for their existence. The productivity of soil is strongly affected by environmental conditions and pollutants. Because of the importance of soil, all of Chapter 14 is devoted to it. Chapter 15 covers environmental geology and pollution of the geosphere. Mineral resources are discussed in Chapter 16.

1.6. LIFE

Biology is the science of life. It is based on biologically synthesized chemical species that fall into the categories of proteins, lipids (fats, oils, some hormones), carbohydrates, and nucleic acids that largely constitute living organisms and enable them to function. Many of theses substances exist as large molecules, called *macromolecules*. As living beings, the ultimate concern of humans with their environment is the interaction of the environment with life. Therefore, biological science is a key component of environmental science.

The focal point of biochemistry and biochemical aspects of toxicants is the **cell**, the basic building block of living systems where most life processes are carried out. Individual cells are so small that they cannot be seen without a microscope. Bacteria, yeasts, and some algae consist of single cells. However, most living things are made up of many cells. In a more complicated organism the cells have different functions. Liver cells, muscle cells, brain cells, and skin cells in the human body are quite different from each other and do different things.

Consideration of biology from the perspective of cells and macromolecules is done at a microscopic level. At the other end of the scale, environmental scientists may regard the biosphere from the standpoint of **populations** of species interacting with one another in a **biological community**. As noted in Section 1.1, **ecology** is the study of environmental factors that affect organisms and of how organisms interact with these factors and with each other. It is discussed in greater detail in Section 1.7 and in Chapter 17.

The role of life in environmental science is discussed in Chapters 17–20. Chapter 17 addresses biological communities and ecology. Biologically mediated transformations and biodegradation of substances are covered in Chapter 18. Living systems interact with the geosphere, hydrosphere, and atmosphere largely through biogeochemical cycles, the topic of Chapter 19. The effects on living beings of toxic substances, many of which are environmental pollutants, are addressed in Chapter 20, "Toxicology and Toxicological Chemistry."

1.7. ECOLOGY

Ecology is the science that deals with the relationships between living organisms with their physical environment and with each other.[3] Ecology can be approached from the viewpoints of (1) the environment and the demands it places on the organisms in it or (2) organisms and how they adapt to their environmental conditions. An **ecosystem** consists of an assembly of mutually interacting organisms and their environment in which materials are interchanged in a largely cyclical manner. An ecosystem has physical, chemical, and biological components along with energy sources and pathways of energy and materials interchange. The environment in which a particular organism lives is called its **habitat**. The role of an organism in a habitat is called its **niche**.

For the study of ecology it is often convenient to divide the environment into four broad categories. The **terrestrial environment** is based on land and consists of **biomes**, such as grasslands, one of several kinds of forests, savannas, or deserts. The **freshwater environment** can be further subdivided between *standing-water habitats* (lakes, reservoirs) and *running-water habitats* (streams, rivers). The oceanic **marine environment** is characterized by saltwater and may be divided broadly into the shallow waters of the continental shelf composing the **neritic zone** and the deeper waters of the ocean that constitute the **oceanic region**. An environment in which two or more kinds of organisms exist together to their mutual benefit is termed a **symbiotic environment**.

A particularly important factor in describing ecosystems is that of **populations** consisting of numbers of a specific species occupying a specific habitat. Populations may be stable, or they may grow exponentially as a **population explosion**. A population explosion that is unchecked results in resource depletion, waste accumulation, and predation culminating in an abrupt decline called a **population crash**. **Behavior** in areas such as heirarchies, territoriality, social stress, and feeding patterns plays a strong role in determining the fates of populations.

Two major subdivisions of modern ecology are **ecosystem ecology**, which views ecosystems as large units, and **population ecology**, which attempts to explain ecosystem behavior from the properties of individual units. In practice, the two approaches are usually merged. **Descriptive ecology** describes the types and nature of organisms and their environment, emphasizing structures of ecosystems and communities and dispersions and structures of populations. **Functional ecology** explains how things work in an ecosystem, including how populations respond to environmental alteration and how matter and energy move through ecosystems.

An understanding of ecology is essential in the management of modern industrialized societies in ways that are compatible with environmental preservation and enhancement. The branch of ecology that deals with predicting the impacts of technology and development and making recommendations such that these activities will have minimum adverse impacts, or even positive impacts, on ecosystems may be termed **applied ecology**.

Though not identical to environmental science, ecology is an integral part of it. Therefore, ecological concepts are used directly or indirectly throughout this book, especially in Chapters 17–19 dealing with biological communities, biotransformations and biodegradation, and biogeochemical cycles.

1.8. MATTER AND CYCLES OF MATTER

Biogeochemical cycles describe the circulation of matter, particularly plant and animal nutrients, through ecosystems. These cycles are ultimately powered by solar energy, fine-tuned and directed by energy expended by organisms. In a sense, the solar-energy-powered hydrologic cycle (Sections 1.3 and 5.2) acts as an endless conveyer belt to move materials essential for life through ecosystems.

Most biogeochemical cycles can be described as elemental cycles involving nutrient elements such as carbon, oxygen, nitrogen, sulfur, and phosphorus. Many are **gaseous cycles** in which the element in question spends part of the cycle in the atmosphere—O_2 for oxygen, N_2 for nitrogen, CO_2 for carbon. Others, notably the phosphorus cycle, do not have a gaseous component and are called **sedimentary cycles**. All sedimentary cycles involve **salt solutions** or **soil solutions** (see Section 14.2) that contain dissolved substances leached from weathered minerals, that may be deposited as mineral formations, or they may be taken up by organisms as nutrients. The sulfur cycle, which may have H_2S or SO_2 in the gaseous phase or minerals ($CaSO_4 \bullet 2H_2O$) in the solid phase, is a combination of gaseous and sedimentary cycles.

Carbon, the basic building block of life molecules, is circulated through the **carbon cycle**, shown in a simplified version in Figure 1.2. (Readers familiar with fundamentals of chemistry will understand the chemical formulas, such as CO_2, shown in the figure; those who are not, may want to come back to it after reading about chemistry in Chapter 2.) This cycle shows that carbon may be present as gaseous atmospheric CO_2, dissolved in groundwater as HCO_3^- or molecular $CO_2(aq)$, in underlying rock strata as limestone ($CaCO_3$), and as organic matter, represented in a simplified manner as $\{CH_2O\}$. Photosynthesis fixes inorganic carbon as biological carbon, which is a consituent of all life molecules.

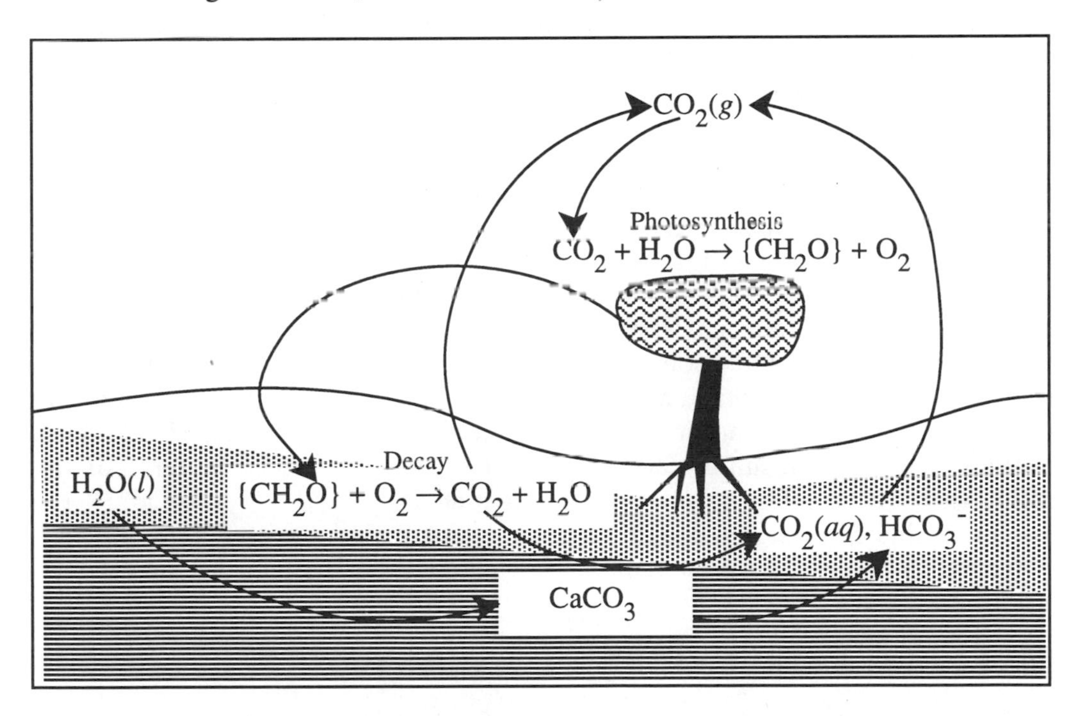

Figure 1.2. The carbon cycle. Mineral carbon is held in a reservoir of limestone, $CaCO_3$, from which it may be leached into a mineral solution as dissolved hydrogen carbonate ion, HCO_3^- formed when dissolved $CO_2(aq)$ reacts with $CaCO_3$. In the atmosphere carbon is present as carbon dioxide, CO_2. Atmospheric carbon dioxide is fixed as organic matter by photosynthesis, and organic carbon is released as CO_2 by microbial decay of organic matter.

An important aspect of the carbon cycle is that it is the cycle by which energy is transferred to biological systems. Organic, or biological carbon, $\{CH_2O\}$, is an energy-rich molecule that can react biochemically with molecular oxygen, O_2, to regenerate carbon dioxide and produce energy. This can occur in an organism as shown by the "decay" reaction in Figure 1.2, or it may take place as combustion, such as when wood is burned.

The **oxygen cycle** involves the interchange of oxygen between the elemental form of gaseous O_2 in the atmosphere and chemically bound O in CO_2, H_2O, and organic matter. Elemental oxygen becomes chemically bound by various energy-yielding processes, particularly combustion and metabolic processes in organisms. It is released in photosynthesis. Photosynthetic release of oxygen and consumption of oxygen in the utilization of organic matter ("decay") are shown in the carbon cycle in Figure 1.2.

Nitrogen, though constituting much less of biomass than carbon or oxygen, is an essential constituent of proteins. The atmosphere is 78% by volume elemental nitrogen, N_2, and constitutes an inexhaustible reservoir of this essential element. The N_2 molecule is very stable so that breaking it down to atoms that can be incorporated in inorganic and organic chemical forms of nitrogen is the limiting step in the **nitrogen cycle**. This does occur by highly energetic processes in lightning discharges such that nitrogen becomes chemically combined with hydrogen or oxygen as ammonia or nitrogen oxides. Elemental nitrogen is also incorporated into chemically bound forms or **fixed** by biochemical processes mediated by micro-organisms. The biological nitrogen is returned to the inorganic form during the decay of biomass by a process called **mineralization**.

The phosphorus cycle is crucial because phosphorus is usually the limiting nutrient in ecosystems. There are no common stable gaseous forms of phosphorus, so the phosphorus cycle is strictly sedimentary. In the geosphere phosphorus is held largely in poorly soluble minerals, such as hydroxyapatite, a calcium salt. Soluble phosphorus from these minerals and other sources, such as fertilizers, is taken up by plants and incorporated into the nucleic acids of biomass (see Section 1.6). Mineralization of biomass by microbial decay returns phosphorus to the salt solution from which it may precipitate as mineral matter.

The sulfur cycle is relatively complex in that it involves several gaseous species, poorly soluble minerals, and several species in solution. It is involved with the oxygen cycle in that sulfur combines with oxygen to form gaseous sulfur dioxide, SO_2, an atmospheric pollutant, and soluble sulfate ion, SO_4^{2-}. Among the significant species involved in the sulfur cycle are gaseous hydrogen sulfide, H_2S; mineral sulfides, such as PbS; sulfuric acid, H_2SO_4, the main constituent of acid rain; and biologically bound sulfur in sulfur-containing proteins.

From this discussion it should be obvious that material cycles, often based on elemental cycles, are very important in the environment. These cycles are discussed further as "Biogeochemical Cycles" in Chapter 19.

1.9. ENERGY AND CYCLES OF ENERGY

Biogeochemical cycles and virtually all other processes on Earth are driven by energy from the sun. The sun acts as a so-called blackbody radiator with an effective surface temperature of 5780 K (Celsius degrees above absolute zero).[4] It transmits energy to Earth as electromagnetic radiation (see below). The maximum energy flux of the incoming solar energy is at a wavelength of about 500 nanometers, which is in the visible region of the spectrum. A 1 square meter area perpendicular to the line of solar flux at the top of the atmosphere receives energy at a rate of 1,340 watts, sufficient, for example, to power an electric iron.[5] This is called the **solar flux** (see Chapter 9, Figure 9.3).

Energy in natural systems is transferred by **heat**, which is the form of energy that flows between two bodies as a result of their difference in temperature, or by **work**, which is a transfer of energy that does not depend upon a temperature difference, as governed by the laws of **thermodynamics**. The **first law of thermodynamics** states that, although energy may be transferred or transformed, it is conserved and is not lost. Chemical energy in the food ingested by organisms is converted by **metabolic processes** to work or heat that can be utilized by the organisms, but there is no net gain or loss of energy overall. The **second law of thermodynamics** describes the tendency toward disorder in natural systems. It demonstrates that each time energy is transformed, some is lost in the sense that it cannot be utilized for work, so only a fraction of the energy that organisms derive from metabolizing food can be converted to work; the rest is dissipated as heat.

Light and Electromagnetic Radiation

Electromagnetic radiation, particularly light, is of utmost importance in considering energy in environmental systems. Therefore, the following important points related to electromagnetic radiation should be noted:

- Energy can be carried through space at the speed of light, 3.00×10^8 meters per second (m/s) in a vacuum, by **electromagnetic radiation**, which includes visible light, ultraviolet radiation, infrared radiation, microwaves, and radio waves.
- Electromagnetic radiation has a **wave character**. The waves move at the speed of light, c, and have characteristics of **wavelength** (λ), amplitude, and **frequency** (ν, Greek "nu") as illustrated below:

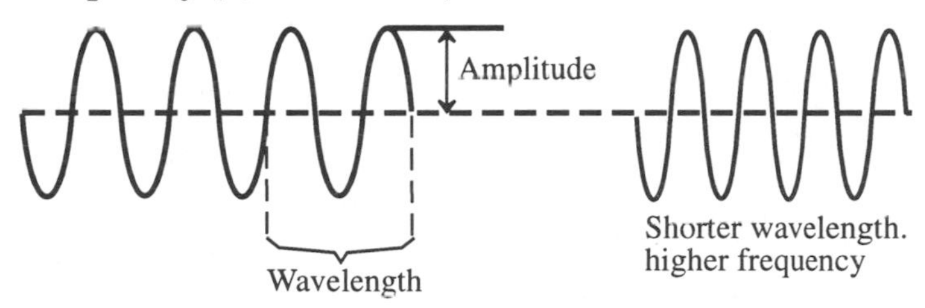

- The wavelength is the distance required for one complete cycle and the frequency is the number of cycles per unit time. They are related by the following equation:

$$\nu\lambda = c$$

where ν is in units of cycles per second (s^{-1}, a unit called the **hertz**, Hz) and λ is in meters (m).

- In addition to behaving as a wave, electromagnetic radiation also has characteristics of particles.
- The dual wave/particle nature of electromagnetic radiation is the basis of the **quantum theory** of electromagnetic radiation, which states that radiant energy may be absorbed or emitted only in discrete packets called **quanta** or **photons**. The energy, E, of each photon is given by

$$E = h\nu$$

where h is Planck's constant, 6.63×10^{-34} J-s (joule x second).

- From the preceding, it is seen that *the energy of a photon is higher when the frequency of the associated wave is higher* (and the wavelength shorter).

Energy Flow and Photosynthesis

Whereas materials are recycled through ecosystems, the flow of useful energy may be viewed as essentially a one-way process. Incoming solar energy can be regarded as high-grade energy because it can cause useful reactions to occur, the most important of which in living systems is photosynthesis. As shown in Figure 1.3, solar energy captured by green plants energizes chlorophyll, which in turn powers metabolic processes that produce carbohydrates from water and carbon dioxide. These carbohydrates represent stored chemical energy that can be converted to heat and work by metabolic reactions with oxygen in organisms. Ultimately, most of the energy is converted to low-grade heat, which is eventually re-radiated away from Earth by infrared radiation.

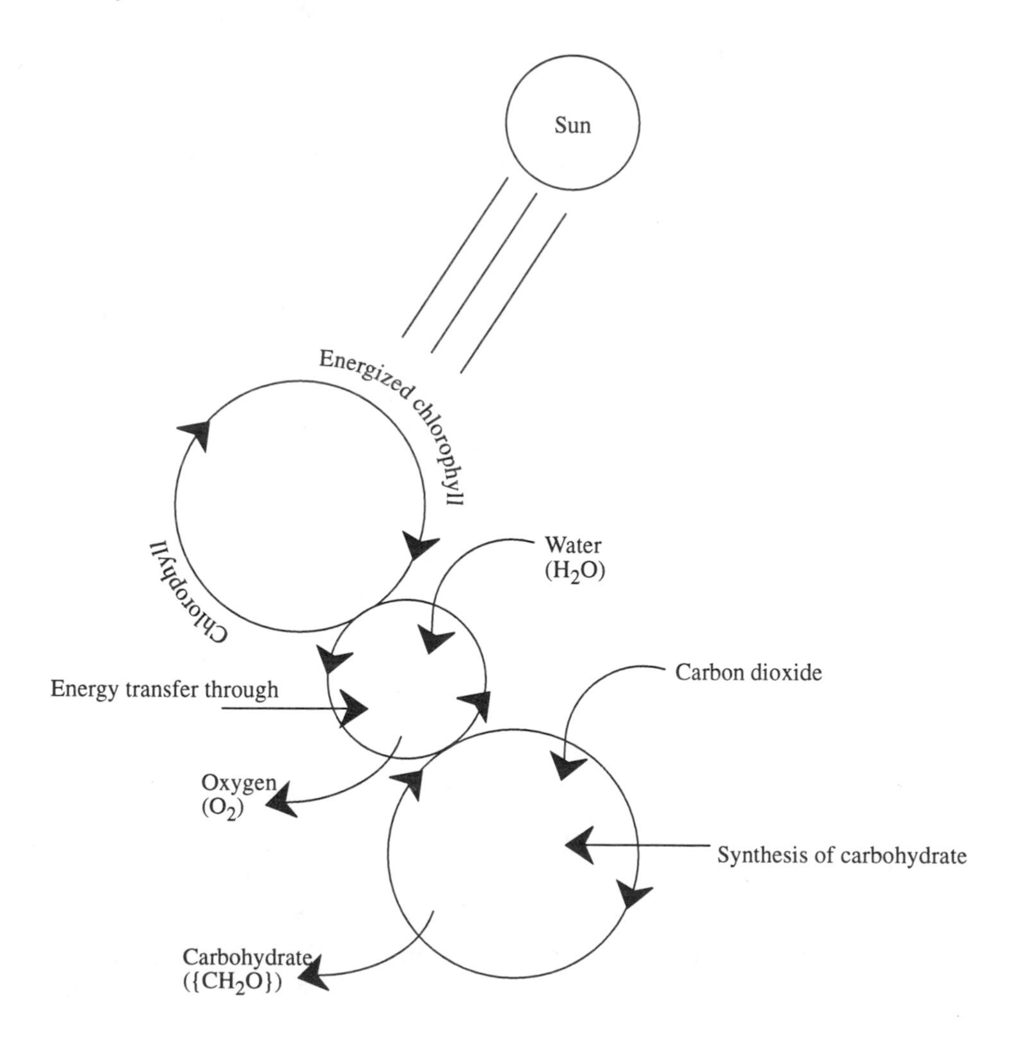

Figure 1.3. Energy conversion and transfer by photosynthesis.

Energy Utilization

During the last two centuries the human impact on energy utilization and conversion has been enormous and has resulted in many of the environmental problems now facing humankind. This time period has seen a transition from the almost exclusive use of energy captured by photosynthesis and utilized as biomass (food to provide muscle power, wood for heat) to the use of fossil fuels for about 90 percent, and nuclear energy for about 5 percent of all energy employed commerically. Fossil fuel consumption is divided primarily among petroleum, natural gas, and coal. These sources of energy are limited and their pollution potential is high. The mining of coal and the extraction of petroleum is environmentally disruptive, the combustion of high-sulfur coal releases acidic sulfur dioxide to the atmosphere, and all fossil fuels produce carbon dioxide, a greenhouse gas. Therefore, it will be necessary to move toward the utilization of alternate energy sources, particularly those that are renewable and that do not pose unacceptable risks to global climate through emission of carbon dioxide. Prominent among these is solar energy, and biomass will to a degree come back as an energy source. The study of energy utilization is crucial in the environmental sciences, and it is discussed in greater detail in Chapter 22 of this book, "Energy and Resource Utilization."

1.10. HUMAN IMPACT AND POLLUTION

The demands of increasing population coupled with the desire of most people for a higher material standard of living are resulting in worldwide pollution on a massive scale. Environmental pollution can be divided among the categories of water, air, and land pollution. All three of these areas are linked. For example, acid gases emitted to the atmosphere can be converted to strong acids by atmospheric chemical processes, fall to the earth as acid rain, and pollute water with acidity. Improperly discarded hazardous wastes can leach into groundwater that is eventually released as polluted water into streams.

Water Pollution

Pollution of surface water and groundwater are discussed in some detail in Chapter 7, "Water Pollution." Throughout history, the quality of drinking water has been a factor in determining human welfare. Waterborne diseases in drinking water have decimated the populations of whole cities. Unwholesome water polluted by natural sources has caused great hardship for people forced to drink it or use it for irrigation. Although waterborne diseases have in general been well controlled in industrialized nations, they are prevalent, and even growing worse in poorer countries, especially where population pressures have overtaxed the resources available to provide safe drinking water and to treat wastewater. Currently, toxic chemicals pose the greatest threat to the safety of water supplies in industrialized nations.

Since World War II there has been a tremendous growth in the manufacture and use of synthetic chemicals. Many of the chemicals have contaminated water supplies. Two examples are insecticide and herbicide runoff from agricultural land, and industrial discharge into surface waters. Another serious problem is the threat to groundwater from waste chemical dumps and landfills, storage lagoons, treating ponds, and other facilities.

It is clear that water pollution should be a concern of every citizen. Understanding the sources, interactions, and effects of water pollutants is essential for controlling pollutants in an environmentally safe and economically acceptable manner. Above all, an understanding of water pollution and its control depends upon a basic knowledge of aquatic environmental science.

Air Pollution

Inorganic air pollutants consist of many kinds of substances. Many solid and liquid substances may become particulate air contaminants. Another important class of inorganic air pollutants consists of oxides of carbon, sulfur, and nitrogen. Carbon monoxide is a directly toxic material that is fatal at relatively small doses. Carbon dioxide is a natural and essential constituent of the atmosphere, and it is required for plants to use during photosynthesis. However, CO_2 may turn out to be the most deadly air pollutant of all because of its potential as a greenhouse gas that might cause devastating global warming. Oxides of sulfur and nitrogen are acid-forming gases that can cause acid precipitation. Ammonia, hydrogen chloride, and hydrogen sulfide, are also inorganic air pollutants.

A number of gaseous inorganic pollutants enter the atmosphere as the result of human activities. Those added in the greatest quantities are carbon monoxide (CO), sulfur dioxide (SO_2), nitric oxide (NO) and nitrogen dioxide (NO_2). Other inorganic pollutant gases include ammonia, (NH_3), nitrous oxide (N_2O) hydrogen sulfide (H_2S), elemental chlorine (Cl_2), hydrogen chloride (HCl), and hydrogen fluoride (HF). Substantial quantities of some of these gases are added to the atmosphere each year by human activities. Globally, atmospheric emissions of carbon monoxide, sulfur oxides, and nitrogen are of the order of one to several hundred million tons per year.

Organic pollutants are common atmospheric contaminants that may have a strong effect upon atmospheric quality. Such pollutants may come from both natural and artificial sources. In some cases contaminants from both kinds of sources interact to produce a pollution effect. This occurs, for example, when terpene hydrocarbons evolved from citrus and conifer trees interact with nitrogen oxides from automobiles to produce photochemical smog.

The effects of organic pollutants in the atmosphere may be divided into two major categories. The first consists of **direct effects**, such as cancer caused by exposure to vinyl chloride. The second is the formation of **secondary pollutants**, especially photochemical smog. In the case of pollutant hydrocarbons in the atmosphere, the latter is the more important effect. In some localized situations, particularly the workplace, direct effects of organic air pollutants may be equally important.

Inorganic and organic air pollutants are discussed in some detail in Chapter 11, "Air Pollution and its Control." Some air pollutants, particularly those that may result in irreversible global warming or destruction of the protective stratsopheric ozone layer, are of a magnitude that they have the potential to threaten life on earth. These are discussed in Chapter 12, "The Endangered Global Atmosphere."

Pollution of the Geosphere and Hazardous Wastes

The most serious kind of pollutant that is likely to contaminate the geosphere, particularly soil, consists of hazardous wastes. A simple definition of a **hazardous**

waste is that it is a hazardous substance that has been discarded, abandoned, neglected, released, or designated as a waste material, or one that may interact with other substances to be hazardous. In a simple sense a hazardous waste is a material that has been left where it may cause harm if encountered.

Humans have always been exposed to hazardous substances going back to prehistoric times when they inhaled noxious volcanic gases or succumbed to carbon monoxide from inadequately vented fires in cave dwellings sealed too well against Ice-Age cold. As the production of dyes and other organic chemicals developed from the coal tar industry in Germany during the 1800s, pollution and poisoning from coal tar by-products was observed. By around 1900 the quantity and variety of chemical wastes produced each year was increasing sharply with the addition of wastes such as spent steel and iron pickling liquor, lead battery wastes, chromic wastes, petroleum refinery wastes, radium wastes, and fluoride wastes from aluminum ore refining. As the century progressed into the World War II era, the wastes and hazardous by-products of manufacturing increased markedly from sources such as chlorinated solvents manufacture, pesticides synthesis, polymers manufacture, plastics, paints, and wood preservatives. The Love Canal affair of the 1970s and 1980s brought hazardous wastes to the public attention as a major political issue in the U.S. Starting around 1940, large quantities of at least 80 different waste chemicals were dumped into this old abandoned canal in Niagara Falls, New York. Serious problems developed that required massive remedial action, and by 1993 state and federal governments had spent well over $100 million to clean up the site and relocate residents.

The geosphere and land provide sites for human habitation, reservoirs of drinking water in the form of groundwater, and soil upon which most food used by humans and animals is grown. Obviously, environmental protection of the geosphere is of great importance. Aspects of this matter are discussed in Chapter 15, "Environmental Geology and Geospheric Pollution"; Chapter 23, "Environmental Impacts, Pollution Control, and Waste Minimization"; and Chapter 24, "Waste Treatment and Disposal."

1.11. TECHNOLOGY AND THE PROBLEMS IT POSES

Modern technology has provided the means for massive alteration of the environment and pollution of the environment. As discussed in the following section, technology, intelligently applied with a strong environmental awareness, also provides the means for dealing with problems of environmental pollution and degradation.

Some of the major ways in which modern technology has contributed to environmental alteration and pollution are the following:

- Agricultural practices that have resulted in intensive cultivation of land, drainage of wetlands, irrigation of arid lands, and application of herbicides and insecticides.
- Manufacturing of huge quantities of industrial products that consumes vast amounts of raw materials and produces large quantities of air pollutants, water pollutants, and hazardous waste byproducts.
- Extraction and production of minerals and other raw materials with accompanying environmental disruption and pollution.

- Energy production and utilization with environmental effects that include disruption of soil by strip mining, pollution of water by release of saltwater from petroleum production, and emission of air pollutants, such as acid-rain-forming sulfur dioxide.
- Modern transportation practices, particularly reliance on the automobile, that cause scarring of land surfaces from road construction, emission of air pollutants, and greatly increased demands for fossil fuel resources.

Important aspects of the impact of technology on the environment are discussed in detail in Chapter 21, "Technology, Manufacturing, and Transportation"; Chapter 22, "Energy and Resource Utilization"; Chapter 23, "Environmental Impacts, Pollution Control, and Waste Minimization"; and Chapter 24, "Waste Treatment and Disposal."

1.12. SOLUTIONS OFFERED BY SCIENCE AND TECHNOLOGY

Technology based on a firm foundation of environmental science can be very effectively applied to the solution of environmental problems. One important example of this is the redesign of basic manufacturing processes to minimize raw material consumption, energy use, and waste production. Consider a generalized manufacturing process shown in Figure 1.4. With proper design such a process can be made environmentally relatively more acceptable. In some cases raw materials and energy sources can be chosen in ways that minimize environmental impact. If the process involves manufacture of a chemical, it may be possible to completely alter the reactions used so that the process is much more environmentally friendly. Raw materials and water may be recycled to the maximum extent possible. Best

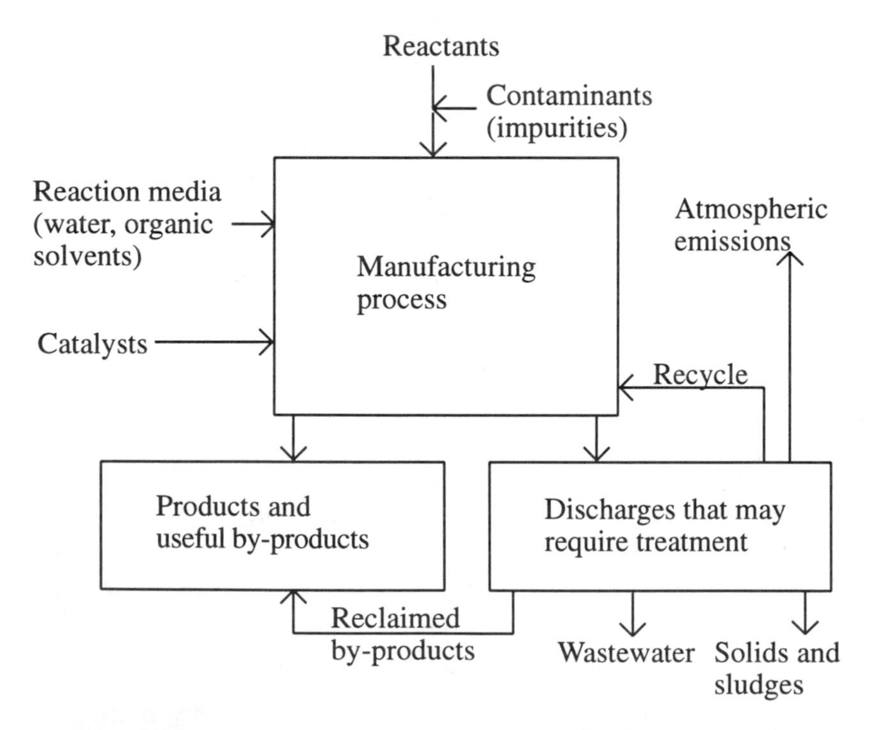

Figure 1.4. A manufacturing process viewed from the standpoint of minimization of environmental impact.

available technologies may be employed to minimize air, water, and solid waste emissions. Among the ways in which technology can be applied to minimize environmental impact are the following:

- Use of state-of-the-art computerized control to achieve maximum energy efficiency, maximum utilization of raw materials, and minimum production of pollutant byproducts
- Use of materials that minimize pollution problems, for example, heat-resistant materials that enable use of high temperatures for efficient thermal processes
- Application of processes and materials that enable maximum materials recycling and minimum waste product production, for example, advanced membrane processes for wastewater treatment to enable water recycling.
- Application of advanced biotechnologies, such as in the biological treatment of wastes
- Use of best available catalysts for efficient synthesis
- Use of lasers for precision machining and processing to minimize waste production

The applications of modern technology to environmental improvement are addressed in several chapters of this book. An overview of technology and how it relates to environmental science is presented in Chapter 4, "Technology and Engineering in Environmental Science." Technological approaches to the minimization and treatment of water and air pollutants are presented in Chapter 8, "Water Treatment," and in Chapter 11, "Air Pollution and Its Control," respectively. Environmentally sound technologies for manufacturing and transportation are discussed in Chapter 21, "Technology, Manufacturing, and Transportation." Technologically based approaches to efficient mineral and energy extraction and utilization are discussed in Chapter 16, "Mineral Resources," and in Chapter 22, Energy and Resource Utilization." The applications of sound technologies to hazardous waste minimization and treatment are given in Chapter 23, "Wastes from the Anthrosphere," and in Chapter 24, "Waste Minimization, Treatment, and Disposal."

1.13. UNITS

Already in this chapter quantities have been expressed in several **units** of various kinds. A basic knowledge of units and how to use them is essential for any of the sciences. Important units are summarized here to enable better understanding of material covered later in the book.

The **metric** system has long been the standard system for scientific measurement and is the one most commonly used in this book. It uses multiples of 10 to designate units that differ by orders of magnitude from a basic unit. The multiples most commonly used are listed in Table 1.1.

Units of Mass

Mass expresses the degree to which an object resists a change in its state of rest or motion and is proportional to the amount of matter in the object. **Weight** is the gravitational force acting upon an object and is proportional to mass. The **gram** (g) is the fundamental unit of mass in the metric system; there are 453.6 g per pound. Although the gram is a convenient unit for many laboratory-scale operations, other units that are multiples of the gram are often more useful for expressing mass. The names of these are obtained by affixing the appropriate prefixes from Table 1.1 to "gram." Human body mass is expressed in kilograms (1 kg = 2.2046 pounds). Global burdens of atmospheric pollutants may be given in units of teragrams, each equal to 1 x 10^{12} grams. Significant quantities of toxic water pollutants may be measured in micrograms (1 x 10^{-6} grams). Large-scale industrial commodities are marketed in units of megagrams (Mg), also known as a metric ton, or tonne, which is somewhat larger (2205 lb) than the 2000-lb short ton still used in the United States.

Table 1.1. Prefixes Commonly Used to Designate Multiples of Units

Prefix	Basic unit is multliplied by	Abbreviation
Mega	1 000 000 (10^6)	M
Kilo	1 000 (10^3)	k
Hecto	100 (10^2)	h
Deka	10 (10)	da
Deci	0.1 (10^{-1})	d
Centi	0.01 (10^{-2})	c
Milli	0.001 (10^{-3})	m
Micro	0.000 001 (10^{-6})	μ
Nano	0.000 000 001 (10^{-9})	n
Pico	0.000 000 000 001 (10^{-12})	p

Units of Length

Length in the metric system is expressed in units based upon the **meter**, m; a meter is 39.37 inches long, slightly longer than a yard. A kilometer (km) is equal to 1000 m and, like the mile, is used to measure relatively great distances. A centimeter (cm) equal to 0.01 m is often convenient to designate lengths such as the dimensions of laboratory instruments. There are 2.540 cm per inch, and the cm is employed to express lengths that would be given in inches in the English system. The micrometer (μm) is about the same length as that of a typical bacterial cell. The μm is also used to express wavelengths of infrared radiation by which Earth re-radiates solar energy back to outer space. The nanometer (nm), equal to 10^{-9} m, is a convenient unit for the wavelength of visible light, which ranges from 400 to 800 nm.

Units of Volume

The basic metric unit of **volume** is the **liter**. This volume is defined in terms of metric units of length. A liter is the volume of a decimeter cubed, that is, 1 L = 1 dm^3 (a decimeter is 0.1 meter, about 4 inches), and is equal to 1.057 quarts. A milliliter (mL) is the same volume as a cubic centimeter, cm^3.

Units of Temperature

In science, temperatures may be expressed in metric units of **Celsius degrees**, °C. On this scale, water freezes at 0°C and boils at 100°C. The most fundamental temperature scale is the **Kelvin** or **absolute** scale, for which zero is the lowest attainable temperature. A unit of temperature on this scale is equal to a Celsius degree, but it is called a **Kelvin**, not a degree, and is designated as K, not °K. The value of absolute zero on the Kelvin scale is -273.15°C, so that the Kelvin temperature is always a number 273.15 (usually rounded to 273) higher than the Celsius temperature. Thus water boils at 373 K and freezes at 273 K.

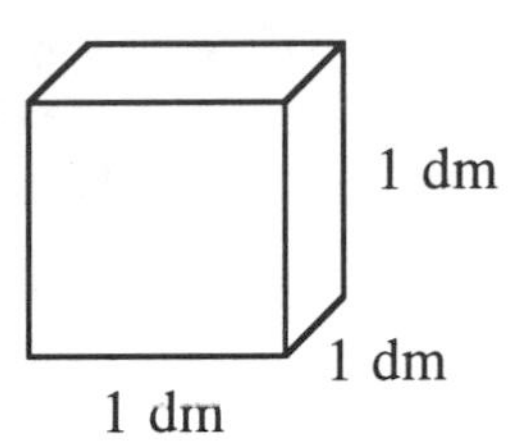

Figure 1.5. A cube that is 1 decimeter to the side has a volume of 1 liter.

Units of Pressure

Pressure is force per unit area and can be expressed in a number of different units, including the **atmosphere** (atm), which is the average pressure exerted by air at sea level, or the pascal (Pa), usually expressed in kilopascal (1 kPa = 1000 Pa, and 101.3 kPa = 1 atm). Based on pressure required to hold up a column of mercury in a mercury barometer, presssure can be given as **millimeters of mercury** (mm Hg), where 1 mm of mercury is a unit called the **torr** and 760 torr equal 1 atm.

CHAPTER SUMMARY

The chapter summary below is presented in a programmed format to review the main points covered in this chapter. It is used most effectively by filling in the blanks, referring back to the chapter as necessary. The correct answers are given at the end of the summary.

Environmental science is [1]____ ____. When a change or action tends to amplify itself it is called [2]______________________, and when a system automatically adjusts to minimize an influence it is called [3]______ __________. A key way in which the atmosphere is related to the hydrosphere is through the [4]__________________ cycle. Earth's crust is that part of the geosphere consisting of [5]____ ____.

All living entities on Earth compose the [6] ____________. Substance, that which occupies space and has mass is called [7] __________________, the science of which is [8] __________________. The capacity to do work is called [9]__________. A pollutant is a substance that is present in [10] ____________ natural concentration as a result of [11]______________ and having a net [12] ______________ upon its environment or upon something of value in that environment. The total of all living organisms inhabiting a specified area is called a [13]____________________. The study of the interactions of organisms with their environment and with each other constitutes the science of [14]__________. About [15]________% of Earth's fresh water occurs in the oceans. The atmosphere

serves a vital protective function in that it [16]______________________________ ______________________________. Because it re-absorbs much of the infrared radiation by which absorbed solar energy is re-emitted to space, the atmosphere [17]______________________________.
A naturally occurring inorganic solid in the geosphere with a definite internal structure and chemical composition is called a [18]____________________. The basic building blocks of living systems of microscopic size where most life processes are carried out are the [19]______________. An assembly of mutually interacting organisms and their environment in which materials are interchanged in a largely cyclical manner constitutes [20]____________________. The environment in which a particular organism lives is called its [21]______________, and the role of an organism therein is called its [22]______________. Numbers of a specific species occupying a specific habitat constitute a [23]______________. The circulation of matter, particularly plant and animal nutrients, through ecosystems is described by [24]______________________________, which may be either [25]________________in which the element in question spends part of the time in the atmosphere, or [26]________________ that do not have a gaseous component. The gaseous component of the carbon cycle consists of [27]__________ __________, which is converted to biological carbon by [28]________________. ______________. The role of photosynthesis in the oxygen cycle is to [29]________ ______________________________.
When elemental nitrogen is incorporated into biomass by biochemical processes mediated by microorganisms it is said to be [30]________________. One reason that the phosphorus cycle is so important is that phosphorus is usually the [31]________ ____________________ in ecosystems. A relatively complex cycle that involves several gaseous species, poorly soluble minerals, and several species in solution is the [32]______________________________. Energy from the sun reaches Earth as electromagnetic radiation with a maximum energy flux at about [33]______________.
The [34]________________, states that radiant energy may be absorbed or emitted only in discrete packets called [35]________________. Solar energy captured by green plants energizes chlorophyll, which in turn powers metabolic processes that produce [36]______________________________ from water and carbon dioxide.
The reason that CO_2 may turn out to be the "most deadly air pollutant of all" is because [37]______________________________ ______________________________.
The effects of organic pollutants in the atmosphere may be divided into two major categories of [38]______________________________. In manufacturing designed to minimize environmental impact, processes and materials can be used that enable maximum [39]______________________________ and minimum [40]______________________________. The base unit of mass in the metric system is the [41]____________________, that of length is the [42]____________________, and that of volume is the [43]____________.

Answers

[1] the study of the earth, air, water, and living environments and the effects of technology thereon

2 positive feedback

3 negative feedback

4 hydrologic

5 the outer skin of the lithosphere composed largely of lighter silicate-based minerals

6 biosphere

7 matter

8 chemistry

9 energy

10 greater than

11 human activity

12 detrimental effect

13 biological community

14 ecology

15 97.6

16 absorbs most of the cosmic rays from outer space and protects organisms from their effects

17 stabilizes the Earth's temperature

18 mineral

19 cells

20 an ecosystem

21 habitat

22 niche

23 population

24 biogeochemical cycles

25 gaseous cycles

26 sedimentary cycles

27 atmospheric carbon dioxide

28 photosynthesis

29 release elemental oxygen

30 fixed

31 limiting nutrient

32 sulfur cycle

33 500 nanometers

34 quantum theory

35 quanta or photons

36 carbohydrates

37 of its potential as a greenhouse gas that might cause devastating global warming

38 direct effects and secondary pollutants

39 materials recycling

40 waste product production

41 gram

42 meter

43 liter

QUESTIONS AND PROBLEMS

1. Classify each of the following as positive feedback or negative feedback and explain your answer. Are there circumstances under which both may occur together?

   (A) Humans living in impoverished circumstances bear more children with the hope of being supported in their old age.

   (B) The population of deer increases in an area leading to an increase in the population of their natural predators.

   (C) Ebola virus is so deadly that all its victims die before they have a chance to spread the disease.

   (D) Depletion of hot water in a household hot water tank leads the person taking a shower to turn off the cold water tap in order to keep the water at a comfortable temperature.

2. Coal as an energy source was produced from large masses of plants many millions of years ago. In what sense is coal a renewable resource? Why, in practical terms, is coal a nonrenewable resource?

3. What is the distinction between a mineral and a rock?

4. What is the distinction between a gaseous cycle and a sedimentary cycle? Give examples of each.

5. The formula $\{CH_2O\}$ is used to represent a broad range of materials of environmental significance. What are these materials, and how are they formed?

6. In what important respect involving a biological process are the carbon and oxygen cycles "opposites"?

7. Describe the nature of the energy which has the maximum flux for transmission from the sun to Earth.

8. In what respect does energy utilization, particularly as it has developed in about the last century, have an enormous potential for influencing the environment?

9. Come up with a plausible explanation of why ultraviolet radiation with a wavelength of around 300 nm is more likely to cause chemical (photochemical) processes to occur than visible radiation around 500 nm.

10. Inhalation of pollutant sulfur dioxide from the atmosphere can aggravate respiratory problems and exposure to this atmospheric pollutant may kill plants. In addition, sulfur dioxide can be oxidized by atmospheric chemical processes to sulfuric acid, the main ingredient of acid rain. Explain how these phenomena relate to direct effects of atmospheric pollutants and to secondary pollutants.

11. What is the environmental significance of Love Canal in the United States?

12. In this chapter, "agricultural practices that have resulted in intensive cultivation of land, drainage of wetlands, irrigation of arid lands, and application of herbicides and insecticides" were cited as "some of the major ways in which modern technology has contributed to environmental alteration and pollution." Suggest ways in which these practices, suitably modified and used, could actually be employed for environmental preservation or improvement.

13. Also cited in this chapter as contributing to environmental degradation were "modern transportation practices, particularly reliance on the automobile, that cause scarring of land surfaces from road construction, emission of air pollutants, and greatly increased demands for fossil fuel resources." Suggest ways in which transportation systems can be modified in ways that result in environmental improvement compared to present practices.

14. What are the ways in which each of the following might reduce pollution?

    (A) Computerized control

    (B) Heat-resistant materials that enable use of high temperatures

    (C) Advanced membrane processes

    (D) Advanced biotechnologies

    (E) Lasers

15. Express and justify appropriate metric units for the

    (A) Mass of a serving of ice cream,

    (B) Size of a socket wrench to fit a bolt,

    (C) Volume of a dose of liquid medicine,

    (D) Size of a bacterial cell about 1/10,000 as large as your toe.

16. Explain why a "first and ten" in football would be closer to a "first and nine" in most parts of the world in which the metric system is used.

17. In what sense is the liter defined in terms of length?

18. Although the millimeter is a unit of length, "millimeters of mercury" are used to express pressure. Explain how this can be done.

19. In what sense is 0 on the Kelvin (absolute) temperature scale a more fundamental value than 0 on the Celsius scale?

20. Why is "incoming solar energy" regarded as high-grade energy?

LITERATURE CITED

[1] Cunningham, William P. and Barbara Woodworth Saigo, *Environmental Science: A Global Concern* , 4th Edition, William C Brown Publishers, Dubuque, IA, 1996.

[2] Montgomery, Carla W., *Environmental Geology*, 5th ed., Wm. C. Brown Publishers, Dubuque, IA, 1997.

[3] Beeby, A. N., and A. Brennan, *First Ecology*, Chapman & Hall, New York, NY,1997.

[4] Graedel, T. E. and Paul J. Crutzen, *Atmospheric Change, An Earth System Perspective*, W. H. Freeman and Company, New York, NY, 1993.

[5] Manahan, Stanley E., *Environmental Chemistry*, 6th ed., CRC Press/Lewis Publishers, Boca Raton, FL, 1994.

SUPPLEMENTARY REFERENCES

Arms, Karen, *Environmental Science*, 2nd ed., HBJ College and School Division, Saddle Brook, NJ, 1994.

Atchia, Michael and Shawna Tropp, Eds., *Environmental Management: Issues and Solutions*, Wiley, New York, NY, 1995

Botkin, Daniel B., *Environmental Science: Earth as a Living Planet*, Wiley, New York, NY, 1995.

Carter, Howard and Gillian Irvine, *Environmental Crime*, Cameron May, New York, NY, 1995.

Easterbrook, Gregg, *A Moment on the Earth: The Coming Age of Environmental Optimism*, Viking Penguin, Bergenfield, NJ, 1995.

Field, Barry C., *Environmental Economics: An Introduction*, 2nd ed., McGraw-Hill, New York, NY, 1996.

Harper, Charles L., *Environment and Society: Human Perspectives on Environmental Issues*, Prentice Hall, New York, NY, 1995.

Henry, J. Glynn and Gary W. Heinke, *Environmental Science and Engineering*, 2nd Ed., Prentice Hall, New York, NY, 1995.

Jackson, Andrew R. and Julie M. Jackson, *Environmental Science: The Natural Environment and Human Impact*, Longman, New York, NY,1996.

Jørgensen, S. E., B. Halling-Sørensen, and S. N. Nielsen, *Handbook of Environmental and Ecological Modeling*, CRC Press/Lewis Publishers, Boca Raton, FL, 1996.

McKibben, Bill, *Hope, Human and Wild: True Stories of Living Lightly on the Earth*, Little Brown & Company, Boston, MA, 1995

Miller, G. Tyler, Jr., *Environmental Science: Working with the Earth*, 5th ed., Wadsworth Publishing Co., Belmont, CA, 1997.

Miller, G. Tyler, Jr., *Living in the Environment*, 10th ed., Wadsworth Publishing Co., Belmont, CA, 1998.

Molak, Vlasta, *Fundamentals of Risk Analysis and Risk Management*, CRC Press/Lewis Publishers, Boca Raton, FL, 1997.

Myers, Norman and Jennifer Kent, *Environmental Exodus: An Emergent Crisis in the Global Arena*, Climate Institute, Washington, DC, 1995.

Pepper, Ian L., Charles P. Gerba, and Mark L. Brusseau, *Pollution Science*, Academic Press Textbooks, San Diego, CA, 1996.

Plant, Glen, *Environmental Protection and the Law of War: A Fifth Geneva Convention on the Protection of the Environment in Time of Armed Conflict*, Wiley, New York, NY, 1994.

Rodgers, William H., *Environmental Law*, West Publishing Co., New York, NY, 1994.

Stone, Christopher D., *Should Trees Have Standing?*, Oceana, Dobbs Ferry, NY, 1996.

Sullivan, Thomas, F. P., Ed., *Environmental Law Handbook*, 14th ed., Government Institutes, Inc., Rockville, MD, 1997.

Tower, Edward, *Environmental and Natural Resource Economics*, Eno River Press, New York, NY, 1995.

2 CHEMISTRY AND ENVIRONMENTAL CHEMISTRY

2.1. INTRODUCTION

Chemistry is the science of matter, where **matter** is *anything that has mass and occupies space*. The most fundamental kind of matter consists of the **elements**. Of these, **metals**, such as copper or silver, are generally solid, shiny in appearance, electrically conducting, and malleable. **Nonmetals** often have a dull appearance, are not at all malleable, and may exist as gases (atmospheric oxygen), liquids (bromine), or solids (sulfur). **Organic substances** consist of virtually all compounds, such as wood, that contain carbon. All other substances are **inorganic substances**.

This chapter discusses the bare essentials of chemistry needed to understand the material in the rest of the book. In the first section general aspects of chemistry are covered with a presentation of the most fundamental information required to understand some of the language of this science. The second portion of the chapter covers organic chemistry, and the chapter ends with an outline of environmental chemistry. Biochemistry is discussed in Chapter 3. For more details the reader is referred to a book that discusses these three areas of chemistry.[1]

2.2. ATOMS, THE BUILDING BLOCKS OF MATTER

As discussed in this chapter, all matter is composed of only about a hundred fundamental kinds of matter called **elements**. Each element is made up of very small entities called **atoms**. Atoms, in turn, are composed of the following very small **subatomic particles**. The two subatomic particles that are located in the small, dense **nucleus** in the center of the atom are the proton, designated p, p^+, or +, having a charge of +1 and a mass of 1.007 atomic mass units (u, defined as exactly 1/12 the mass of the most common kind of atom of carbon, carbon-12) and the neutron, n, charge 0, mass 1.009 u. Moving rapidly around the nucleus of the atom and forming a cloud of negative charge that composes essentially all of the volume of the atom is the electron, designated e, e^-, or -, charge -1, mass 0.0005 u. The neutron and proton, each with a mass of essentially 1 u, are said to have a **mass number** of 1, and the much lighter electron is said to have a mass number of 0. The charges and mass numbers of the three basic subatomic particles can be denoted by the following three symbols: ${}^{1}_{1}p$, ${}^{1}_{0}n$, and ${}^{0}_{-1}e$.

Figure 2.1 represents subatomic particles in two different atoms. On the left is the lightest atom that contains all three subatomic particles, an atom of deuterium, a rare form of hydrogen. On the right is an atom of the most common form of carbon, carbon-12. As is the case with all atoms, the heavy protons and neutrons are contained in a very small nucleus and the much lighter electrons move around the nucleus forming a cloud of negative charge.

Atoms and Elements

Each atom of a particular element has the same number of protons in its nucleus. This is the **atomic number** of the element. Each element has a name and a **chemical symbol**, such as carbon, C; potassium, K (for its Latin name kalium); or cadmium, Cd. Although atoms of the same element are chemically identical, atoms of most elements consist of two or more **isotopes** that have different numbers of neutrons in their nuclei. The **mass number** commonly used to denote isotopes and subatomic particles (see above) *is the sum of the number of protons and neutrons in the nucleus of the isotope*. Each element has an **atomic mass** (atomic weight), the average mass of all atoms of the element, which can be expressed in *atomic mass units, u*, also used to express masses of individual atoms, molecules (aggregates of atoms), and subatomic particles.

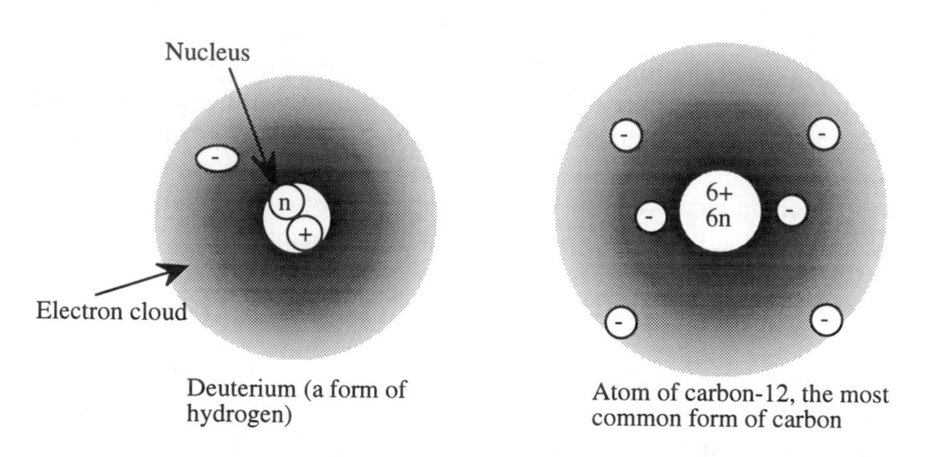

Figure 2.1. On the left is a representation of a deuterium atom. The nucleus contains one proton (+) and one neutron (n). The electron (-) is in constant, rapid motion around the nucleus forming a cloud of negative electrical charge, the density of which drops off with increasing distance from the nucleus. On the right is a carbon atom showing 6 protons and 6 neutrons in its nucleus surrounded by a cloud of negative charge composed of 6 electrons.

Important Elements

Some elements are more commonly encountered than others. It is useful to be aware of the names, symbols, and properties of some of the most significant elements without having to look them up. Most of the "more important common elements" are among the first 20 elements, which will be discussed next. Several other elements the names and symbols of which should be memorized are copper, Cu, atomic number 29, atomic mass 63.54; iodine, I, atomic number 53, atomic mass 126.904; lead, Pb, atomic number 82, atomic mass 207.19; mercury, Hg, atomic number 80, atomic mass 200.59; silver, Ag, atomic number 47, atomic mass 107.87; and zinc, Zn, atomic number 30, atomic mass 65.37.

2.3. ELEMENTS AND THE PERIODIC TABLE

When elements are considered in order of increasing atomic number, it is observed that their properties are repeated in a periodic manner. For example, elements with atomic numbers 2, 10, and 18 are gases that do not undergo chemical reactions and consist of individual atoms, whereas those with atomic numbers larger by one—3, 11, and 19—are unstable, highly reactive metals. The **periodic table** is a very useful listing of elements by symbol, atomic number, and atomic mass in a manner that reflects this recurring behavior. The periodic table gets its name from the fact that the properties of elements are repeated periodically in **periods** that go from left to right across a horizontal row of elements. The table is arranged such that an element has properties similar to those of other elements above or below it in the table. Elements with similar chemical properties are called **groups** of elements and are contained in vertical columns in the periodic table. In this section, a periodic table will be developed for the first 20 elements. A complete periodic table showing all the more than 100 known elements is given on the inside back cover of this book.

Development of the 20-Element Periodic Table

The atom of the first element in the periodic table, **hydrogen**, is the simplest of all, having only one positively charged proton in its nucleus, which is surrounded by a cloud of negative charge formed by only one electron. By far the most abundant kind of hydrogen atom has no neutrons in its nucleus, so its mass number is 1. A small fraction of hydrogen atoms also have 1 neutron (deuterium, see Figure 2.1), and fewer still have 2 neutrons (radioactive tritium). The three different forms of elemental hydrogen are *isotopes* of hydrogen that all have the same number of protons, but different numbers of neutrons in their nuclei; they may be designated as $^{1}_{1}H$, $^{2}_{1}H$, and $^{3}_{1}H$, The average mass of all hydrogen atoms is 1.0079 u (atomic mass units), so hydrogen's *atomic mass* is 1.0079. Hydrogen's box in the periodic table designates its atomic number (1), symbol (H), and atomic mass (1.0079).

Showing Electrons in Atoms

Electrons in chemical symbols and formulas are shown with **electron-dot symbols** or **Lewis symbols**, which use dots around the symbol of an element to show *outer electrons*. These *valence electrons* are the ones that may become involved in chemical bonds. The hydrogen atom has only one electron, and its Lewis symbol is

Lewis symbols for atoms of some other elements and *Lewis formulas* for molecules of compounds are given later in this chapter.

Electron Configurations and the Periodic Table

Electron configurations, which determine chemical behavior, describe how electrons in atoms occupy distinct **energy levels** and are contained in **electron shells**. Each shell can hold a maximum number of electrons. An atom with a **filled**

electron shell has little or no tendency to lose, gain, or share electrons; elements composed of such atoms exist as chemically unreactive gas-phase atoms and are called **noble gases**. The lightest element with filled electron shells is helium, He, atomic number 2. All helium atoms contain 2 protons and 2 electrons. Virtually all helium atoms contain 2 neutrons in their nuclei, and the atomic mass of helium is 4.00260. The two electrons in the filled electron shell of the helium atom are shown by the Lewis symbol illustrated in Figure 2.2.

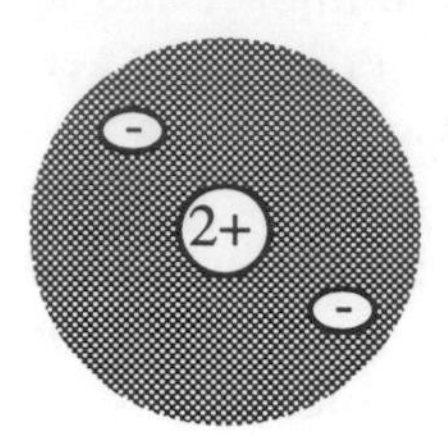

He:

A helium atom has a filled electron shell containing 2 electrons.

It can be represented by the Lewis symbol above.

Figure 2.2. Two representations of the helium atom having a filled electron shell.

The third element in the periodic table is the metal lithium (Li), atomic number 3, atomic mass 6.941. The most abundant lithium isotope has 4 neutrons in its nucleus and a less common isotope has only 3 neutrons. The lithium atom has three electrons. As shown in Figure 2.3, lithium has both **inner electrons**—in this case 2 contained in an **inner shell**—as in the immediately preceding noble gas helium, and an **outer electron** that is farther from, and less strongly attracted to, the nucleus. The outer electron is said to be in the atom's **outer shell**. The inner electrons are, on the average, closer to the nucleus than is the outer electron, are very difficult to remove from the atom, and do not become involved in chemical bonds. Lithium's outer electron is relatively easy to remove from the atom, which is what happens when ionic bonds involving Li^+ ion are formed (see Section 2.4).

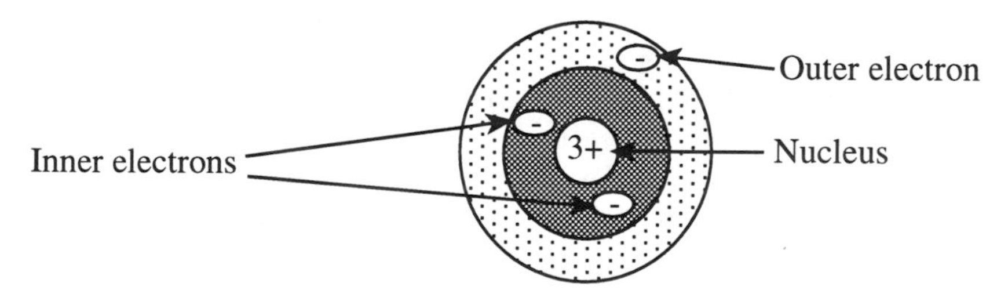

Figure 2.3. An atom of lithium, Li, has 2 inner electrons and 1 outer electron. The latter can be lost to another atom to produce the Li^+ ion, which is present in ionic compounds (see Section 2.4).

In atoms such as lithium that have both outer and inner electrons, the Lewis symbol shows only the outer electrons. Therefore, the Lewis symbol of lithium is

Li·

The second period of the table contains elements with atomic numbers 3 (lithium) through 10. Those other than lithium are the following, listed by atomic number: 4, beryllium, Be; 5, boron, B; 6, carbon, C; 7, nitrogen, N; 8, oxygen, O; 9, fluorine, F; and 10, neon, Ne. These elements are shown in the abbreviated periodic table in Figure 2.4 with Lewis symbols to designate their electron configurations. The last element in the second period under discussion is **neon**. As shown by its Lewis symbol,

the neon atom has 8 outer electrons. These 8 electrons constitute a *filled electron shell*, just as the 2 electrons in helium give it a filled electron shell. Because of this "satisfied" outer shell, the neon atom has no tendency to acquire, give away, or share electrons. Therefore, neon is a *noble gas*, like helium, and consists of individual neon atoms.

First period →	1 H · 1.0							2 He : 4.0
Second period →	3 Li · 6.9	4 Be : 9.0	5 B 10.8	6 C 12.0	7 N 14.0	8 O 16.0	9 F 19.0	10 Ne 20.1
Third period →	11 Na · 23.0	12 Mg : 24.3	13 Al 27.0	14 Si 28.1	15 P 31.0	16 S 32.1	17 Cl 35.5	18 Ar 39.9
Fourth period →	19 K · 39.1	20 Ca : 40.1						

Figure 2.4. Abbreviated 20-element version of the periodic table showing Lewis symbols of the elements. The bottom number in each entry is the atomic mass of the element.

Stability of the Neon Noble Gas Electron Octet

Like neon, all other atoms with 8 outer electrons are chemically unreactive noble gases. In addition to neon, these are argon (atomic number 18), krypton (atomic number 36), xenon (atomic number 54), and radon (atomic number 86). Each of the noble gases may be represented by the Lewis symbol

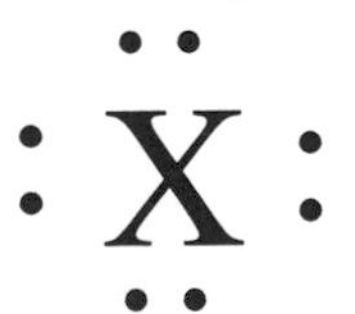

where X is the chemical symbol of the noble gas. It is seen that these atoms each have 8 outer electrons, a group known as an **octet** of electrons. In many cases atoms that do not have an octet of outer electrons acquire one by losing, gaining, or sharing electrons in chemical combination with other atoms; that is, they acquire a **noble gas outer electron configuration**. For all noble gases except helium, which has only 2 electrons, the noble gas outer electron configuration consists of eight outer electrons. The tendency of elements to acquire an 8-electron outer electron configuration is very useful in predicting the nature of chemical bonding and the formulas of compounds that result and is called the **octet rule**.

Completion of the Abbreviated Periodic Table

The abbreviated version of the periodic table can be finished with elements 11 through 20. The atomic numbers, names, and symbols of these elements are 11, sodium, Na; 12, magnesium, Mg; 13, aluminum, Al; 14, silicon, Si; 15, phosphorus, P; 16, sulfur, S; 17, chlorine, Cl; 18, argon, Ar; 19, potassium, K; and 20, calcium, Ca. An abbreviated periodic table with these elements in place is shown in Figure 2.4. This table shows the Lewis symbols of the elements to emphasize their orderly variation across periods and similarity in groups of the periodic table. Note that, with the exception of He at the top of the last group, the configuration of dots for the Lewis symbols in each group (vertical column) are identical.

2.4. CHEMICAL BONDS AND COMPOUNDS

Most atoms are joined by **chemical bonds** to other atoms. Therefore, elemental hydrogen exists as **molecules**, each consisting of 2 H atoms (Figure 2.5) and denoted by the **chemical formula**, H_2. The H atoms in the H_2 molecule are held together by a **covalent bond** made up of 2 electrons, each contributed by one of the H atoms, and shared between the atoms. This can be shown by **electron-dot formulas** or **Lewis formulas** for the H_2 molecule illustrated in the lower portion of Figure 2.5.

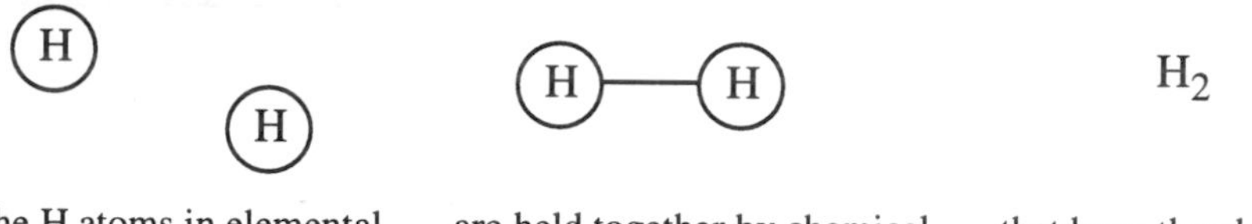

The H atoms in elemental hydrogen | are held together by chemical bonds in molecules | that have the chemical formula H_2.

The bonding of H atoms to form H_2 molecules may also be represented by Lewis symbols (for atoms) and Lewis formulas (for molecules) as shown below. The two dots, :, in the H_2 molecule represent two shared electrons in the covalent bond holding the H molecule together

$$H\cdot \rightleftarrows \cdot H \longrightarrow H:H$$

Figure 2.5. Formation of a molecule of H_2 showing covalent bonds and Lewis formula of the hydrogen molecule.

Chemical Compounds

Most substances consist of two or more elements joined by chemical bonds. Hydrogen atoms combine with oxygen atoms to form molecules in which 2 H atoms are bonded to 1 O atom in a substance with a chemical formula of H_2O (water, Figure 2.6). A **chemical compound** is a substance composed of atoms of two or more different elements bonded together. In the *chemical formula* for water the subscript 2 indicates that there are 2 H atoms per O atom. (The absence of a subscript after the O denotes the presence of just 1 O atom in the molecule.) Each of the chemical bonds holding a hydrogen atom to the oxygen atom in the water molecule is formed by two electrons shared between the hydrogen and oxygen atoms.

Covalent Bonds

Figure 2.7 shows covalent bonding in the formation of hydrogen compounds of C and N. In the case of carbon, 4 H atoms are combined with one C atom having 4 valence electrons, sharing electrons such that in the CH_4 product each H atom has 2 electrons and the C atom has 8 outer shell electrons (a stable octet), all in 4 shared pairs. To form NH_3, only 3 H atoms are required to share their electrons with an atom of N having 5 outer shell electrons, leading to a compound in which the N atom has 8 outer shell electrons, 6 of which are shared with H. The *shared pairs* of electrons, :, in the C-H and N-H bonds shown comprise *covalent bonds*.

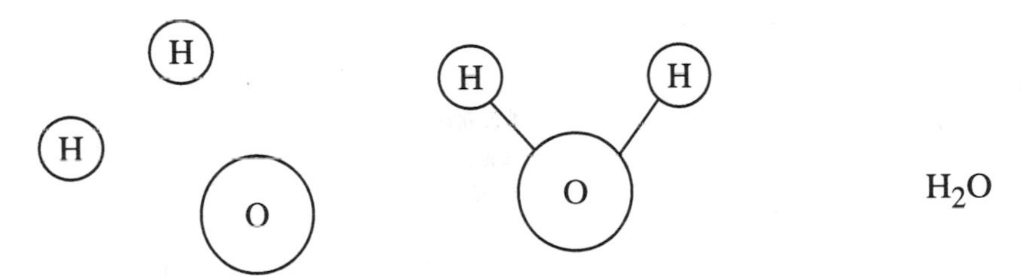

Hydrogen atoms and oxygen atoms bond together | to form molecules in which 2 H atoms are attached to 1 O atom. | The chemical formula of the product compound, water is H_2O.

Figure 2.6. A molecule of water, H_2O, formed from 2 H atoms and 1 O atom held together by chemical bonds.

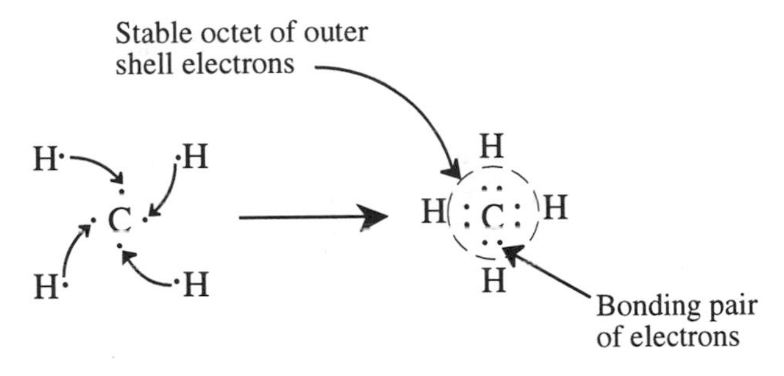

Each of 4 H atoms shares a pair of electrons with a C atom to form a molecule of methane CH_4.

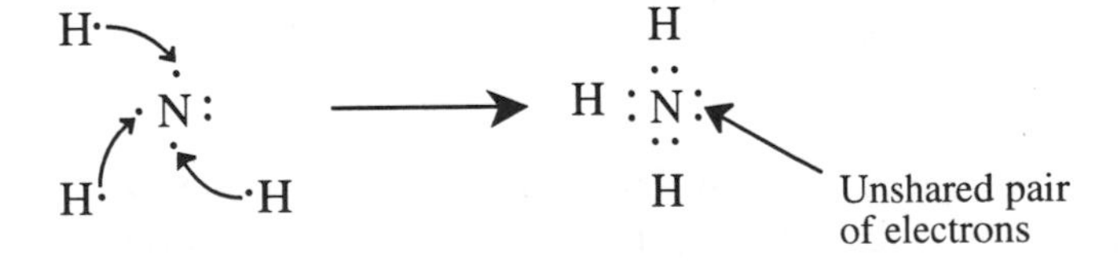

Each of 3 H atoms shares a pair of electrons with an atom of N to form a molecule of ammonia, NH_3.

Figure 2.7. Formation of stable outer electron shells by covalent bonding in compounds.

Lewis symbols and formulas can be used to show the atoms bound together in compounds and the types of bonds in each. The outer shell **valence electrons** are represented as dots, or each pair of valence electrons in a chemical bond is shown as a dash. The electrons in the Lewis formulas of the molecules above are shown in pairs reflecting the fact that electrons tend to be paired in groups of two in both atoms and molecules. They are said to occupy **orbitals**, each of which can contain a maximum of two electrons. The electrons in orbitals have spins in opposite directions. This is an important concept in explaining the behavior of electrons in atoms and molecules.

Ionic Bonds

As shown for the formation of ionic magnesium oxide in Figure 2.8, the transfer of electrons from one atom to another produces charged species called **ions** consisting of positively charged **cations** and negatively charged **anions**. Ions in solids are held together by the attractive forces between the oppositely charged ions (**ionic bonds**) in a **crystalline lattice**.

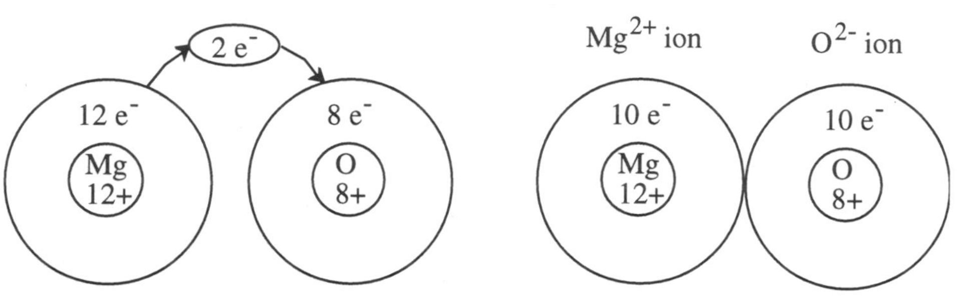

The transfer of two electrons from an atom of Mg to an O atom

yields an ion of Mg^{2+} and one of O^{2+} in the compound MgO.

Figure 2.8. Ionic bonds are formed by the transfer of electrons and the mutual attraction of oppositely charged ions in a crystalline lattice.

Many ions are charged groups of atoms bonded together covalently. A common example of such an ion is the ammonium ion, NH_4^+,

```
    H  +              H
   ..                 | +
H :N: H            H—N—H
   ..                 |
    H                 H
```

Ammonium ion, NH_4^+. Shared pairs of electrons in the chemical bonds are shown as pairs of dots in the Lewis structure on the left and as dashed lines in the structure on the right.

composed of 4 hydrogen atoms covalently bonded to a single nitrogen (N) atom and having a net electrical charge of +1 for the whole cation, as shown above.

Chemical Formulas of Compounds

Chemical formulas, such as H_2 for hydrogen and H_2O for water, contain a lot of information. This is illustrated by the moderately complicated example of calcium phosphate, $Ca_3(PO_4)_2$, Figure 2.9, a compound that contains both covalent and ionic bonds. Each formula unit ("molecule") of this compound is composed of 3 Ca^{2+} ions, each with a +2 charge, and 2 phosphate ions, PO_4^{3-}, each of which has a -3 charge. The four oxygen atoms and the phosphorus atom in the phosphate ion are held together with covalent bonds. The calcium and phosphate ions are bonded together ionically in a lattice composed of these two kinds of ions.

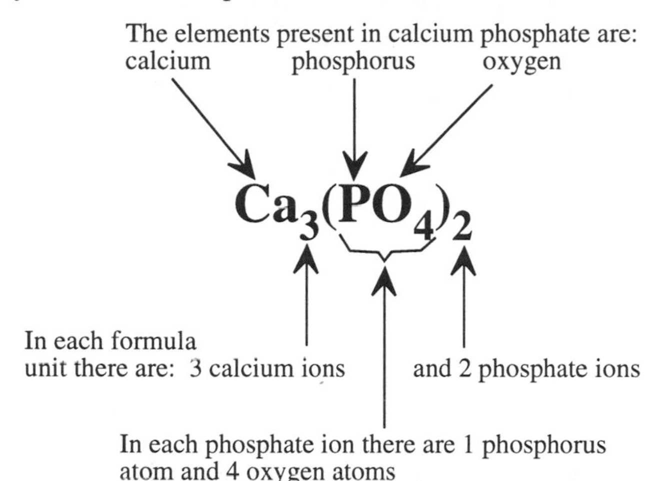

Figure 2.9. Summary of information contained in the chemical formula of calcium phosphate.

2.5. QUANTITY OF MATTER: THE MOLE

The **molecular mass** (formerly called molecular weight) of a compound is calculated by multiplying the atomic mass of each element by the relative number of atoms of the element, then adding all the values obtained for each element in the compound. The molecular mass of NH_3 is 14.0 + 3 x 1.0 = 17.0, where 14.0 is the atomic mass of the single N atom and 1.0 is the atomic mass of each of the 3 H atoms. The molecular mass of ethylene, C_2H_4, is 2 x 12.0 + 4 x 1.0 = 28.0 (the atomic mass of C is 12).

The Mole

The mole, used to express quantity of substance, is defined in terms of specific entities, such as atoms of Ar, molecules of H_2O, or Na^+ and Cl^- ions, each pair of which composes a "molecule" of NaCl. A **mole** is defined as *the quantity of substance that contains the same number of specified entities as there are atoms of C in exactly 0.012 kg (12 g) of carbon-12*. To specify the mass of a substance equivalent to its number of moles, simply state the atomic mass (of an element) or the molecular mass (of a compound) and affix "grams" to it as shown by the examples below:

- *A mole of argon, which always exists as individual Ar atoms:* The atomic mass of Ar is 40.0. Therefore, exactly one mole of Ar is 40.0 grams of argon.
- *A mole of molecular elemental hydrogen, H_2:* The atomic mass of H is 1.0, the molecular mass of H_2 is, therefore, 2.0, and a mole of H_2 contains 2.0 g of H_2.

- *A mole of methane, CH_4:* The atomic mass of H is 1.0 and that of C is 12.0, so the molecular mass of CH_4 is 16.0. Therefore, a mole of methane has a mass of 16.0 g.

Avogadro's number (6.02×10^{23}) *is the number of specified entities in a mole of substance.* The "specified entities" may consist of atoms or molecules or they may be groups of ions that make up the smallest possible unit of an ionic compound, such as two Na^+ ions and one S^{2-} ion in Na_2S.

2.6. CHEMICAL REACTIONS AND EQUATIONS

Chemical reactions occur when substances are changed to other substances. An example is the chemical reaction of hydrogen and oxygen written as the **chemical equation**,

$$2H_2 + O_2 \rightarrow 2H_2O \tag{2.6.1}$$

in which the arrow is read as "yields" and separates the hydrogen and oxygen **reactants** from the water **product**. All correctly written chemical equations are **balanced**, having *the same number of each kind of atom on both sides of the equation*. The equation above is balanced because of the following: On the left there are 2 H_2 *molecules* each containing 2 H *atoms* for a total of 4 H atoms, and there is 1 O_2 *molecule* containing 2 O *atoms* for a total of 2 O atoms. On the right, there are 2 H_2O *molecules* each containing 2 H *atoms* and 1 O atom for a total of 4 H atoms and 2 O atoms.

Some substances enable chemical reactions to occur, but are not themselves consumed in the reactions. These materials are called **catalysts**. Catalysts are very important in chemical manufacturing. Catalysts in automotive exhaust systems enable destruction of pollutant exhaust gases. Special biological catalysts called *enzymes* (see Section 3.15) enable biochemical processes to occur.

2.7. PHYSICAL PROPERTIES AND STATES OF MATTER

Physical properties of matter are those that can be measured without altering the chemical composition of the matter. Three physical properties important in describing and identifying particular kinds of matter are density, color, and solubility. **Density** (d) is defined as mass per unit volume and is expressed by the formula

$$d = \frac{\text{mass}}{\text{volume}} \tag{2.7.1}$$

The densities of liquids and solids are normally given in units of grams per cubic centimeter (g/cm^3, the same as grams per milliliter, g/mL), whereas the densities of much lighter gases are given in units of grams per liter (g/L). **Color** is one of the more useful properties for identifying substances without doing any chemical or physical tests. As examples, a violet vapor is characteristic of iodine, and a characteristic yellow/brown color in water may be indicative of organically bound iron. **Solubility** refers to the degree to which a substance dissolves in a liquid, such as water. It is discussed along with solutions later in this section.

States of Matter

Figure 2.10 illustrates the three **states of matter** in which matter may exist. **Solids** have a definite shape and volume. **Liquids** have an indefinite shape and take on the shape of the container in which they are contained. Solids and liquids are not significantly compressible, which means that a specific quantity of a substance has a definite volume and cannot be squeezed into a significantly smaller volume. **Gases** take on both the shape and volume of their containers. A quantity of gas may be compressed to a very small volume and will expand to occupy the volume of any container into which it is introduced.

Everyone is familiar with the three states of matter for water. These are (1) gas, such as water vapor in a humid atmosphere, steam; (2) liquid, such as water in a lake, groundwater; (3) solid, such as ice in polar ice caps, snow in snowpack. Changes in matter from one phase to another are very important in the environment, such as water vapor changing from the gas phase to liquid forming clouds or precipitation. Water is desalinated by producing water vapor from sea water, leaving the solid salt behind, and recondensing the pure water vapor as a salt-free liquid. Some organic pollutants are extracted from water for chemical analysis by transferring them from the water to another organic phase that is immiscible with water.

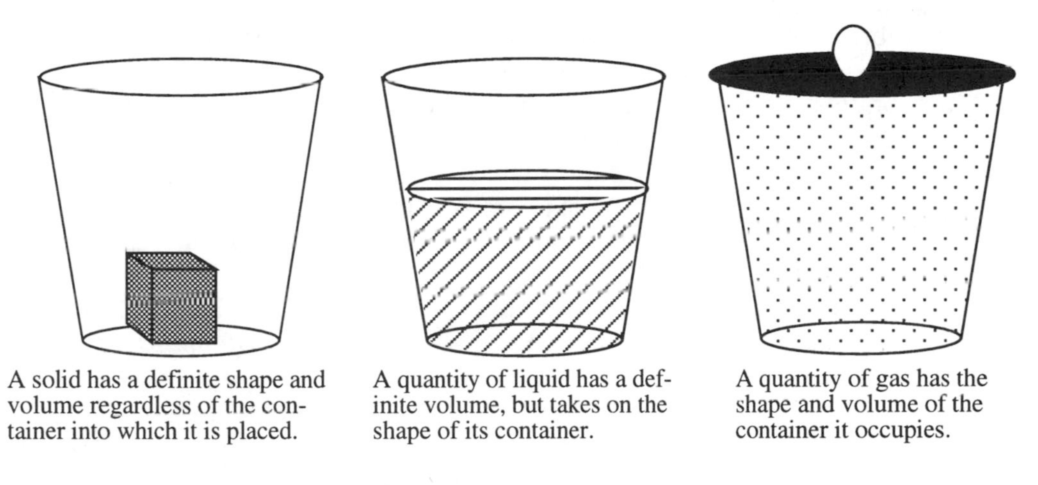

A solid has a definite shape and volume regardless of the container into which it is placed.

A quantity of liquid has a definite volume, but takes on the shape of its container.

A quantity of gas has the shape and volume of the container it occupies.

Figure 2.10. Representations of the three states of matter.

Gases and The Gas Laws

The atmosphere is composed of a mixture of gases, the most abundant of which are nitrogen, oxygen, argon, carbon dioxide, and water vapor. Gases and the atmosphere are addressed in more detail in Chapter 9. A quantity of gas takes on the shape and volume of the container in which it is held. The reason for this behavior is that gas molecules move independently and at random, bouncing off each other as they do so. Gas molecules colliding with container walls exert **pressure**. The rapid, constant motion of gas molecules explains the phenomenon of gas **diffusion** in which gases move large distances from their sources. This occurs, for example, with water vapor in the atmosphere. Evaporated gasoline can diffuse some distance from its source such that, if it contacts a flame, a fire or explosion may result.

The **general gas law** explains the relationships among the quantity of gas in numbers of moles (n) (see Section 2.5), volume (V), temperature (T), and pressure (P) in the form of the **ideal gas equation**:

$$V = \frac{RnT}{P} \text{ or } PV = nRT \quad (2.7.2)$$

For calculations involving volume in liters and pressures in atmospheres, the value of R is 0.0821 L-atm/deg-mol.

A temperature of 0°C (273.15 K) and a pressure of 1 atmosphere (atm) have been chosen as **standard temperature and pressure (STP)**. *At STP the volume of 1 mole of ideal gas is 22.4 L,* a volume called the **molar volume of a gas**. The ideal gas law can be used to calculate changes in volume or other parameters resulting from changes in P, V, n, and T. As an example, calculate the volume of 0.333 moles of gas at 300 K under a pressure of 0.950 atm:

$$V = \frac{nRT}{P} = \frac{0.333 \text{ mol x } 0.0821 \text{ L atm/K mol x } 300 \text{ K}}{0.950 \text{ atm}} = 8.63 \text{ L} \quad (2.7.3)$$

Liquids and Solutions

A given quantity of liquid occupies a fixed volume, but, because of the free movement of molecules relative to each other, it takes on the shape of that portion of the container that it occupies. Molecules of liquids that are energetic enough to escape the attractive forces of the other molecules in the mass of liquid enter the gas phase, a phenomenon called **evaporation**; the opposite process is called **condensation**. Equilibrium between these two processes results in a steady-state level of vapor above a liquid called its **vapor pressure** .

Many liquids act as **solvents** to form **solutions** in which quantities of gases, solids, or other liquids are **dissolved** as **solutes**. The maximum degree to which a solute dissolves in a liquid is its **solubility**. A solution that is at equilibrium with excess solute so that it contains the maximum amount of solute that it can dissolve is called a **saturated solution**. One that can still dissolve more solute is called an **unsaturated solution**.

For the chemist, the most useful way to express the concentration of a solution is in terms of the **molar concentration, M**, *the number of moles of solute dissolved per liter of solution*:

$$M = \frac{\text{moles of solute}}{\text{number of liters of solution}} \quad (2.7.4)$$

Example: Exactly 34.0 g of ammonia, NH_3, were dissolved in water and the solution was made up to a volume of exactly 0.500 L. What was the molar concentration, M, of ammonia in the resulting solution?

Answer: The molar mass of NH_3 is 17.0 g/mole. Therefore

$$\text{Number of moles of } NH_3 = \frac{34.0 \text{ g}}{17.0 \text{ g/mole}} = 2.00 \text{ mol} \quad (2.7.5)$$

$$M = \frac{2.00 \text{ mol}}{0.500 \text{ L}} = 4.00 \text{ mol/L} \quad (2.7.6)$$

Solids

The **solid state** is the most organized form of matter in that the atoms, molecules, and ions in it are in essentially fixed relative positions and are highly attracted to each other. Therefore, solids have a definite shape, maintain a constant volume, are virtually noncompressible under pressure, and expand and contract only slightly with changes in temperature. Because of the strong attraction of the atoms, molecules, and ions of solids for each other, solids do not enter the vapor phase readily at all; the phenomenon by which this happens to a limited extent is called **sublimation**.

2.8. THERMAL PROPERTIES

The **melting point** of a pure substance is the temperature at which the substance changes from a solid to a liquid. At the melting temperature pure solid and pure liquid composed of the substance may be present together in a state of equilibrium. Boiling occurs when a liquid is heated to a temperature such that bubbles of vapor of the substance are evolved. When the surface of a pure liquid substance is in contact with the pure vapor of the substance at 1 atm pressure, boiling occurs at a temperature called the **normal boiling point**.

As the temperature of a substance is raised, energy must be put into it to enable the molecules of the substance to move more rapidly relative to each other and to overcome the attractive forces between them. The **specific heat** of a substance is defined as the amount of heat energy required to raise the temperature of a gram of substance by 1 degree Celsius. The **heat of vaporization** required to convert liquid water to vapor is 2,260 J/g (2.26 kJ/g) for water boiling at 100°C at 1 atm pressure. This amount of heat energy is about 540 times that required to raise the temperature of 1 gram of liquid water by 1°C. When water vapor condenses, similar enormous amounts of heat energy called **heat of condensation** are released. **Heat of fusion** is the quantity of heat taken up in converting a unit mass of solid entirely to liquid at a constant temperature. The heat of fusion of water is 330 J/g for ice melting at 0°C and is 80 times the specific heat of water.

2.9. ACIDS, BASES, AND SALTS

Almost all inorganic compounds and many organic compounds can be classified as acids, bases, or salts, and they are discussed in this section. Acids, bases, and salts are very important in life processes, in the environment, and as industrial chemicals.

An **acid** is a substance that produces hydrogen ions, H^+, in water. (Actually, in water, H^+ ion is associated with water molecules in clusters such as the **hydronium ion**, H_3O^+, but for simplicity in this book, H^+ in water will be shown simply as H^+.) For example, HCl in water is entirely in the form of H^+ ions and Cl^- ions. These two ions in water form hydrochloric acid. Toxic hydrocyanic acid, HCN, also produces hydrogen ions in water:

$$HCN \rightarrow H^+ + CN^- \qquad (2.9.1)$$

When the HCN molecule comes apart, it is said to **ionize**, and the process is called **ionization**. At all but extremely low HCN concentrations, only a small percentage of the acid molecules release H^+ ion. Therefore, HCN is said to be a *weak acid*, a term that will be defined further in this section.

Bases

A **base** is a substance that produces hydroxide ion, OH^-, and/or accepts H^+. Many bases consist of metal ions and hydroxide ions. For example, solid sodium hydroxide dissolves in water

$$NaOH(s) \rightarrow Na^+(aq) + OH^-(aq) \quad (2.9.2)$$

to yield a solution containing OH^- ions. (In this equation (*s*) denotes a solid and (*aq*) stands for a substance dissolved in water; (*g*) is used to designate a gas in equations involving gases.) When ammonia gas is bubbled into water, a limited number of the molecules of NH_3 remove hydrogen ion from water and produce ammonium ion, NH_4^+, and hydroxide ion as shown by the following reaction:

$$NH_3 + H_2O \rightarrow NH_4^+ + OH^- \quad (2.9.3)$$

Salts

Whenever an acid and a base are brought together, water is always a product, leaving a negative ion from the acid and a positive ion from the base:

$$\underbrace{H^+ + Cl^-}_{\text{hydrochloric acid}} + \underbrace{Na^+ + OH^-}_{\text{sodium hydroxide}} \rightarrow \underbrace{Na^+ + Cl^-}_{\text{sodium chloride}} + \underbrace{H_2O}_{\text{water}} \quad (2.9.4)$$

Sodium chloride is a **salt**, a compound composed of a positively charged *cation* other than H^+ and a negatively charged *anion* other than OH^-.

Dissociation of Acids and Bases In Water

The reactions discussed above have shown that acids and bases dissociate, or ionize, in water to produce ions. There is a great difference in how much various acids and bases ionize. Some, like HCl or NaOH, are completely dissociated in water. Because of this hydrochloric acid is called a **strong acid**. Sodium hydroxide is a **strong base**. Partially ionized acids and base, such as hydrocyanic acid and ammonia, mentioned above are **weak acids** and **weak bases**. Some common strong acids are sulfuric acid (H_2SO_4) and nitric acid (HNO_3); weak acids include acetic acid ($HC_2H_3O_2$, of which only one of the four H's can form H^+ ion) and hypochlorous acid (HClO). The percentage of weak acid molecules that are ionized depends upon the concentration of the acid. The lower the concentration, the higher the percentage of ionized molecules.

Hydrogen Ion Concentration and pH

Hydrogen ion concentration, commonly denoted as $[H^+]$ and expressed in moles/liter, M, is a very important characteristic of some solutions. For example, the value of $[H^+]$ in human blood must stay within relatively narrow ranges, or the person will become ill or even die. Fortunately, there are mixtures of chemicals that keep the H^+ concentration of a solution relatively constant. Reasonable quantities of acid or base added to such solutions do not cause large changes in H^+ concentration. Solutions that resist changes in $[H^+]$ are called **buffers**.

Because of the fact that water, itself, produces both H^+ and hydroxide ion,

$$H_2O \leftrightarrows H^+ + OH^- \quad (2.9.5)$$

there is always some H^+ and some OH^- in any solution. (The reverse arrow shows that H^+ and OH^- ions recombine to give H_2O molecules.) Of course, in an acid solution the concentration of OH^- is always very low, and in a solution of base the concentration of H^+ is very low. If the value of either $[H^+]$ or $[OH^-]$ in moles per liter (M) is known, the value of the other can be calculated from the following relationship:

$$[H^+][OH^-] = 1.00 \times 10^{-14} = K_w \quad \text{(at 25°C)} \tag{2.9.6}$$

For example, in a solution of 0.100 M HCl in which $[H^+] = 0.100$ M,

$$[OH^-] = \frac{K_W}{[H^+]} = \frac{1.00 \times 10^{-14}}{0.100} = 1.00 \times 10^{-13}\text{ M} \tag{2.9.7}$$

Molar concentrations of hydrogen ion, $[H^+]$, range over many orders of magnitude and are conveniently expressed by pH defined as

$$\text{pH} = -\log[H^+] \tag{2.9.8}$$

In absolutely pure water $[H^+] = [OH^-]$, the value of $[H^+]$ is exactly 1×10^{-7} mole/L at 25°C, the pH is 7.00, and the solution is **neutral** (neither acidic nor basic). **Acidic** solutions have pH values of less than 7, and **basic** solutions have pH values of greater than 7. When the H^+ ion concentration is 1 times 10 to a power (the superscript number, such as -2, -7, etc.), the pH is simply the negative value of that power. Thus, when $[H^+]$ is 1×10^{-3}, the pH is 3; when $[H^+]$ is 1×10^{-4}, the pH is 4. For a solution with a hydrogen ion concentration between 1×10^{-4} and 1×10^{-3}, such as 3.16×10^{-4}, the pH is obviously going to be between 3 and 4. The pH is calculated very easily on an electronic calculator by entering 3.16×10^{-4} on the keyboard and pressing the "log" button, which gives -3.50, so the pH is 3.50.

2.10. ORGANIC CHEMISTRY

Most carbon-containing compounds are **organic chemicals** and are addressed by the subject of **organic chemistry**. Organic chemistry is a vast, diverse, discipline because of the enormous number of organic compounds that exist as a consequence of the versatile bonding capabilities of carbon. Such diversity is due to the ability of carbon atoms to bond to each other through single (2 shared electrons) bonds, double (4 shared electrons) bonds, and triple (6 shared electrons) bonds, in a limitless variety of straight chains, branched chains, and rings. All organic compounds, of course, contain carbon. Virtually all also contain hydrogen and have at least one C–H bond.

Among organic chemicals are included the majority of important industrial compounds, synthetic polymers, agricultural chemicals, biological materials, and most substances that are of concern because of their toxicities and other hazards. Pollution of the water, air, and soil environments by organic chemicals is an area of significant concern.

Molecular Geometry in Organic Chemistry

The three-dimensional shape of a molecule, called its **molecular geometry**, is particularly important in organic chemistry. This is because its molecular geometry determines in part the properties of an organic molecule, particularly its interactions with biological systems. Shapes of molecules are represented in drawings by lines of normal, uniform thickness for bonds in the plane of the paper, and with broken lines for bonds extending away from, and heavy lines for bonds extending toward the viewer. These conventions are shown by the example of dichloromethane, CH_2Cl_2, an important organochloride solvent and extractant, illustrated in Figure 2.11.

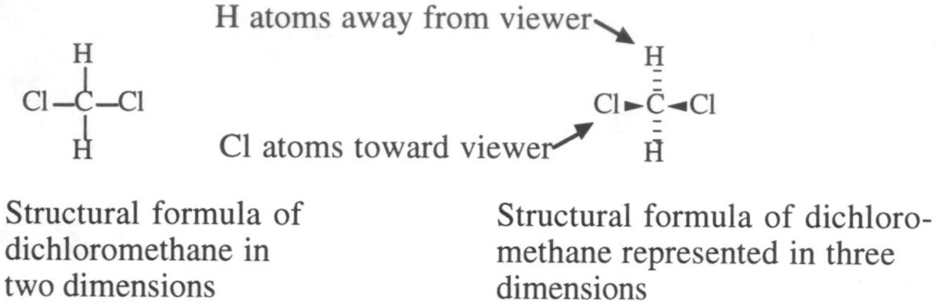

H atoms away from viewer

Cl atoms toward viewer

Structural formula of dichloromethane in two dimensions

Structural formula of dichloromethane represented in three dimensions

Figure 2.11. Structural formulas of dichloromethane, CH_2Cl_2; the formula on the right provides a three-dimensional representation.

2.11. HYDROCARBONS

The simplest and most easily understood organic compounds are **hydrocarbons**, which contain only hydrogen and carbon. As shown in Figure 2.12, the major types of hydrocarbons are alkanes, alkenes, alkynes, and aryl compounds. In the structures shown, C=C represents a **double bond** in which four electrons are shared, and C≡C is a **triple bond** in which six electrons are shared.

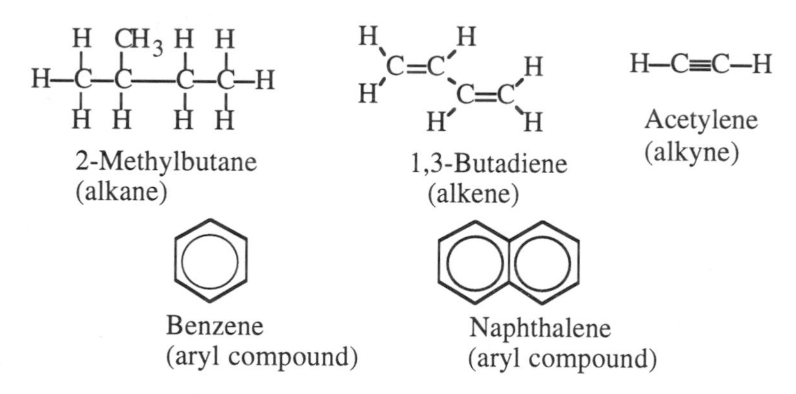

Figure 2.12. Examples of major types of hydrocarbons.

Alkanes

Alkanes, also called **paraffins** or **aliphatic hydrocarbons**, are hydrocarbons in which the C atoms are joined by single covalent bonds (sigma bonds) consisting of two shared electrons (see Section 2.4 and methane in Figure 2.7). Some examples of alkanes are shown in Figure 2.13. The three major kinds of alkanes are, respectively, **straight-chain alkanes**, **branched-chain alkanes**, and **cycloalkanes**. In one of the molecules shown in Figure 2.13, all of the carbon atoms are in a straight chain, and in two they are in branched chains, whereas in a fourth molecule 6 of the carbon atoms are in a ring.

Formulas of Alkanes

Formulas of organic compounds present information at several different levels of sophistication. **Molecular formulas**, such as that of octane (C_8H_{18}), give the number of each kind of atom in a molecule of a compound. As shown in Figure 2.13, however, the molecular formula of C_8H_{18} may apply to several alkanes, each one of which has unique chemical, physical, and toxicological properties. These different compounds are designated by **structural formulas** showing the order in which the atoms in a molecule are arranged. Compounds that have the same molecular, but different structural, formulas are called **structural isomers**. Of the compounds shown in Figure 2.13, *n*-octane, 2,5-dimethylhexane, and 2-methyl-3-ethylpentane are structural isomers, all having the formula C_8H_{18}, whereas 1,4-dimethylcyclohexane is not a structural isomer of the other three compounds because its molecular formula is C_8H_{16}.

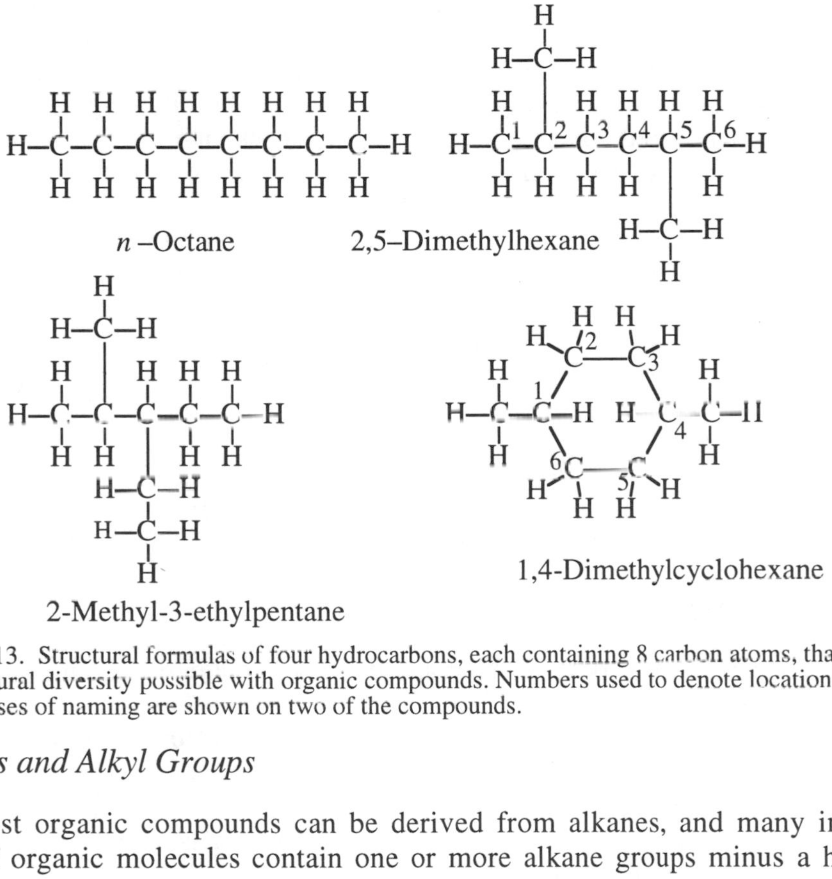

n-Octane

2,5-Dimethylhexane

2-Methyl-3-ethylpentane

1,4-Dimethylcyclohexane

Figure 2.13. Structural formulas of four hydrocarbons, each containing 8 carbon atoms, that illustrate the structural diversity possible with organic compounds. Numbers used to denote locations of atoms for purposes of naming are shown on two of the compounds.

Alkanes and Alkyl Groups

Most organic compounds can be derived from alkanes, and many important parts of organic molecules contain one or more alkane groups minus a hydrogen atom bonded as substituents onto the basic organic molecule. As a consequence, the names of many organic compounds are based upon alkanes. Two important substituent groups derived from alkanes are the methyl group, $-CH_3$, (derived from methane, CH_4) and the ethyl group, $-C_2H_5$, (derived from ethane, C_2H_6).

Names of Alkanes and Organic Nomenclature

Systematic names, from which the structures of organic molecules can be deduced, have been assigned to all known organic compounds. The more common organic compounds, including many toxic and hazardous organic sustances, likewise have **common names** with no structural implications. To provide some idea of how organic compounds are named, consider the alkanes shown in Figure 2.13. The fact that *n*-octane has no side chains is denoted by "*n*", that it has 8 carbon atoms by "oct," and that it is an alkane by "ane." The names of compounds with branched chains or atoms other than H or C attached make use of numbers that stand for positions on the longest continuous chain of carbon atoms in the molecule. This convention is illustrated by the second compound in Figure 2.13. It gets the hexane part of the name from the fact that it is an alkane with 6 carbon atoms in its longest continuous chain ("hex" stands for 6). However, it has a methyl group (CH_3) attached on the second carbon atom of the chain and another on the fifth. Hence the full systematic name of the compound is 2,5-dimethylhexane, where "di" indicates two methyl groups. In the case of 2-methyl-3-ethylpentane, the longest continuous chain of carbon atoms contains 5 carbon atoms, denoted by pentane; a methyl group is attached to the second carbon atom, and an ethyl group, C_2H_5, on the third carbon atom. The last compound shown in the figure has 6 carbon atoms in a ring, indicated by the prefix "cyclo," so it is a cyclohexane compound. Furthermore, the carbon in the ring to which one of the methyl groups is attached is designated by "1" and another methyl group is attached to the fourth carbon atom around the ring. Therefore, the full name of the compound is 1,4-dimethylcyclohexane.

The basic rules used in naming simple alkanes are the following: (1) The name of the compound is based upon the longest continuous chain of carbon atoms, (2) the carbon atoms in the longest continous chain are numbered sequentially from one end, (3) all groups attached to the longest continuous chain are designated by the number of the carbon atom to which they are attached and by the name of the substituent group, (4) a prefix is used to denote multiple substitutions by the same kind of group, and (5) the complete name is assigned such that it denotes the longest continuous chain of carbon atoms and the name and location on this chain of each substituent group.

Reactions of Alkanes

Alkanes are relatively unreactive. At elevated temperatures they readily burn with molecular oxygen in air as shown by the following reaction of propane:

$$C_3H_8 + 5O_2 \rightarrow 3CO_2 + 4H_2O + \text{heat} \qquad (2.11.1)$$

Common alkanes are highly flammable and the more volatile lower molecular mass alkanes form explosive mixtures with air.

In addition to combustion, alkanes undergo **substitution reactions** in which one or more H atoms on an alkane are replaced by atoms of another element. The most common such reaction is the replacement of H by chlorine, to yield **organochlorine** compounds. For example, methane reacts with chlorine to give chloromethane, as shown below:

$$Cl_2 + CH_4 \rightarrow CH_3Cl + HCl \qquad (2.11.2)$$

Alkenes and Alkynes

Alkenes (olefins) are hydrocarbons that have double bonds consisting of 4 shared electrons. The simplest and most widely manufactured alkene is ethene (ethylene),

Ethylene (ethene): $H_2C=CH_2$

used for the production of polyethylene polymer. Another example of an important alkene is 1,3-butadiene (Figure 2.12), widely used in the manufacture of polymers, particularly synthetic rubber.

Acetylene (Figure 2.12) is an **alkyne**, a class of hydrocarbons characterized by carbon-carbon triple bonds consisting of 6 shared electrons. Highly flammable, dangerously explosive acetylene is used in large quantities as a chemical raw material and fuel for oxyacetylene torches.

Addition Reactions

The double and triple bonds in alkenes and alkynes have "extra" electrons capable of forming additional bonds. Therefore, the carbon atoms attached to these bonds can add atoms without losing any atoms already bonded to them, and the multiple bonds are said to be **unsaturated**. Therefore, alkenes and alkynes both undergo **addition reactions** in which pairs of atoms are added across unsaturated bonds as shown in the hydrogenation reaction of ethylene with hydrogen to give ethane:

$$H_2C=CH_2 + H\text{–}H \rightarrow H_3C\text{–}CH_3 \quad (2.11.3)$$

Addition reactions, which are not possible with alkanes, add to the chemical and metabolic versatility of compounds containing unsaturated bonds and constitute a factor contributing to their generally higher toxicities. Addition reactions make unsaturated compounds much more chemically reactive, more hazardous to handle in industrial processes, and more active in atmospheric chemical processes, such as smog formation (see Chapter 11).

Alkenes and *Cis-trans* Isomerism

As shown by the two simple compounds in Figure 2.14, the two carbon atoms connected by a double bond in alkenes cannot rotate relative to each other. For this reason, another kind of isomerism, called ***cis-trans*** isomerism as illustrated in Figure 2.14, is possible for alkenes.

Aryl Hydrocarbons

Benzene (Figure 2.15) is the simplest of a large class of **aryl (aromatic)** hydrocarbons. Many important aryl compounds have substituent groups containing atoms of elements other than hydrogen and carbon and are called **aryl compounds** or **aromatic compounds**. Most aryl compounds discussed in this book contain 6-carbon-atom benzene rings as shown for benzene, C_6H_6, in Figure 2.15. The atoms in aryl compounds are held together in part by particularly stable bonds that contain delocalized clouds of so-called π (pi, pronounced "pie") electrons. In an oversimplified sense the structure of benzene can be visualized as resonating between the two equivalent structures shown on the left in Figure 2.15 by the shifting of electrons in chemical bonds. This structure can be shown more simply and accurately by a hexagon with a circle in it.

Cis-2-butene, both CH_3 groups on the same side of the double bond

Trans-2-butene, both CH_3 groups on opposite sides of the double bond

Figure 2.14. *Cis* and *trans* isomers of the alkene, 2-butene, C_4H_8. *Cis-trans* isomers have different parts of the molecule oriented differently in space, although these parts occur in the same order. In the case of *cis*-2-butene, the two CH_3 (methyl) groups attached to the C=C carbon atoms are on the same side of the molecule, whereas in *trans*-2-butene they are on opposite sides.

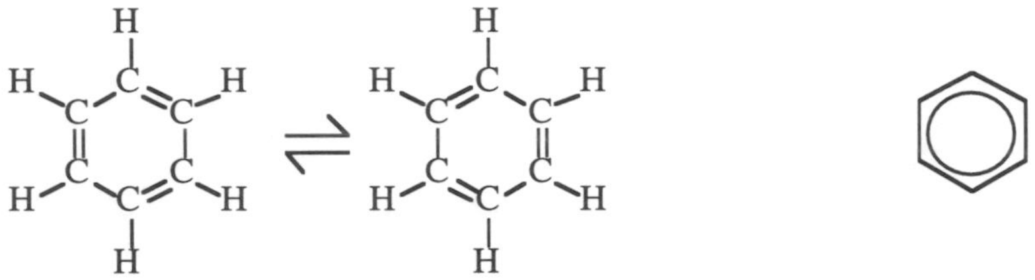

Figure 2.15. Representation of the aryl benzene molecule with two resonance structures (left) and, more accurately, as a hexagon with a circle in it (right). Unless shown by symbols of other atoms, it is understood that a C atom is at each corner and that one H atom is bonded to each C atom.

Many toxic substances, environmental pollutants, and hazardous waste compounds, such as benzene, toluene, naphthalene, and chlorinated phenols, are aryl compounds (see Figure 2.16). As shown in Figure 2.16, some aryl compounds, such as naphthalene and the polycyclic aromatic compound, benzo(a)pyrene, contain fused rings.

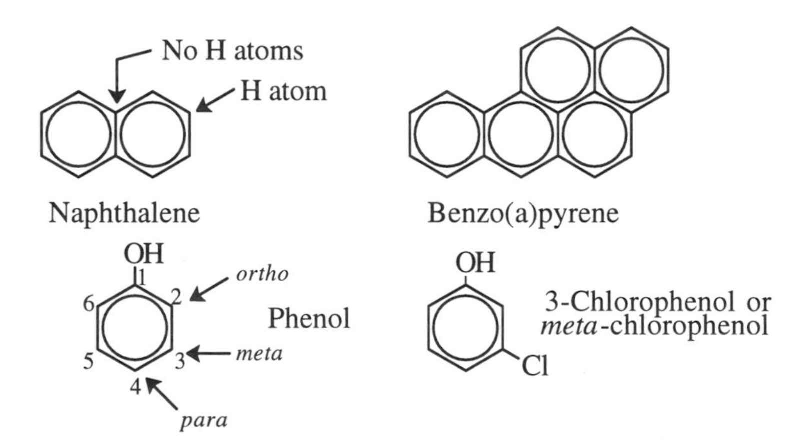

Figure 2.16. Aryl compounds containing fused rings (top) and showing the numbering of carbon atoms for purposes of nomenclature.

Benzo(a)pyrene is the most studied of the polycyclic aryl hydrocarbons (PAHs), which are characterized by condensed ring systems ("chicken wire" structures). These compounds are formed by the incomplete combustion of other hydrocarbons. Some PAH compounds, including benzo(a)pyrene, are of toxicological concern because they are precursors to cancer-causing metabolites.

2.12. FUNCTIONAL GROUPS AND CLASSES OF COMPOUNDS

The presence of elements other than hydrogen and carbon in organic molecules greatly increases the diversity of their chemical behavior. **Functional groups** consist of specific bonding configurations of atoms in organic molecules. Most functional groups contain at least one element other than carbon or hydrogen, although two carbon atoms joined by a double bond (alkenes) or triple bond (alkynes) are likewise considered to be functional groups. Table 2.1 shows some of the major functional groups that determine the nature of organic compounds.

Table 2.1. Examples of Some Important Functional Groups

Type of functional group	Example compound	Structural formula of functional group[a]
Alkene (olefin)	Propene (propylene)	$H_2C{=}CH{-}CH_3$ (C=C outlined)
Alkyne	Acetylene	$H{-}C{\equiv}C{-}H$ (C≡C outlined)
Alcohol (-OH attached to alkyl group)	2-Propanol	$CH_3{-}CH(OH){-}CH_3$ (C–OH outlined)
Phenol (-OH attached to aryl group)	Phenol	$C_6H_5{-}OH$ (OH outlined)
Ketone (When $-\overset{O}{\overset{\|\|}{C}}-H$ group is on an end carbon, compound is an aldehyde.)	Acetone	$CH_3{-}CO{-}CH_3$ (C=O outlined)
Amine	Methylamine	$H_3C{-}NH_2$ (NH_2 outlined)
Nitro compounds	Nitromethane	$H_3C{-}NO_2$ (NO_2 outlined)
Sulfonic acids	Benzenesulfonic acid	$C_6H_5{-}SO_2{-}OH$ (SO_2OH outlined)
Organohalides	1,1–Dichloroethane	$Cl{-}CH_2{-}CH_2{-}Cl$ (Cl outlined)

[a] Functional group outlined by dashed line.

Nonhydrocarbon organic compounds may be classified according to the elements other than carbon and hydrogen in them. Important types of such compounds are organooxygen compounds, organonitrogen compounds, organohalides, organosulfur compounds, and organophosphorus compounds.

2.13. ENVIRONMENTAL CHEMISTRY

As shown graphically by the interchange of chemical species among various environmental spheres in Figure 2.17, **environmental chemistry** is *the study of the sources, reactions, transport, effects, and fates of chemical species in the water, air, terrestrial and living environments and the effects of human activities thereon.*[2] This definition is outlined briefly here and in greater detail in later chapters.

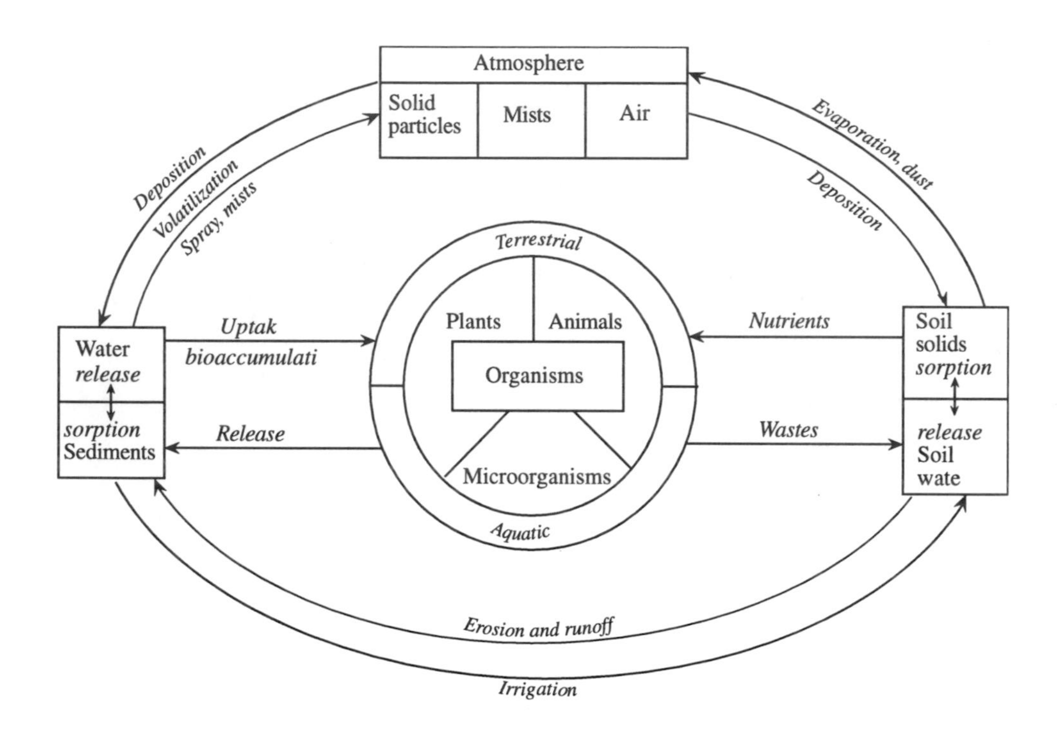

Figure 2.17. Interchange of environmental chemical species among the atmosphere, hydrosphere, geosphere, and biosphere. Human activities (the anthrosphere) have a strong influence on the various processes shown.

2.14. AQUATIC CHEMISTRY

Water has a number of unique properties that are essential to life, due largely to its molecular structure and bonding properties discussed in more detail in Chapter 5, Section 5.3. Among the special characteristics of water are the fact that it is an excellent solvent, it has a temperature/density relationship that results in bodies of water becoming stratified in layers, it is transparent, and it has extraordinary capacity to absorb, retain, and release heat per unit mass of ice, liquid water, or water vapor.

Figure 2.18 summarizes the more important aspects of **aquatic chemistry** as it applies to environmental chemistry, and this topic is covered in greater detail in

Chapter 6. As shown in this figure, a number of chemical phenomena occur in water. Many aquatic chemical processes are influenced by the action of algae and bacteria in water. For example, it is shown that algal photosynthesis fixes inorganic carbon from HCO_3^- ion in the form of biomass (represented as $\{CH_2O\}$), in a process that also produces carbonate ion, CO_3^{2-}. Carbonate undergoes an acid-base reaction to produce OH^- ion and raise the pH, or it reacts with Ca^{2+} ion to precipitate solid $CaCO_3$. Most of the many oxidation-reduction reactions that occur in water are mediated (catalyzed) by bacteria. For example, bacteria convert inorganic nitrogen largely to ammonium ion, NH_4^+, in the oxygen-deficient (anaerobic) lower layers of a body of water. Near the surface, which is aerobic because O_2 is available, bacteria convert inorganic nitrogen to nitrate ion, NO_3^-. Metals in water may be bound to organic chelating agents, such as pollutant nitrilotriacetic acid (NTA) or naturally occurring fulvic acids produced by decay of plant matter. Gases are exchanged with the atmosphere, and various solutes are exchanged between water and sediments in bodies of water.

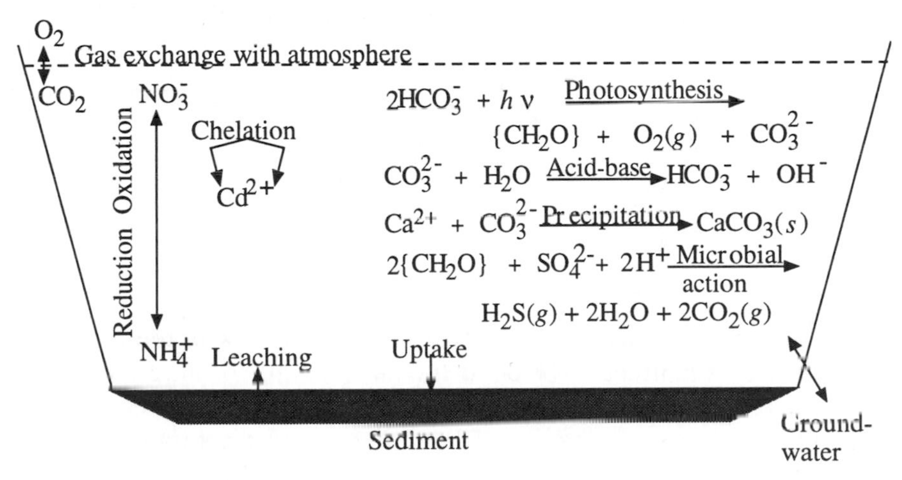

Figure 2.18. Major aquatic chemical processes.

Oxidation-Reduction

Oxidation-reduction (redox) reactions in water involve the transfer of electrons between chemical species. In natural water, wastewater, and soil, most significant oxidation-reduction reactions are carried out by bacteria (Section 3.3).

The relative oxidation-reduction tendencies of a chemical system depend upon the **activity of the electron** e^-. When the electron activity is relatively high, chemical species, including water, tend to accept electrons,

$$2H_2O + 2e^- \leftarrow\rightarrow H_2(g) + 2OH^- \qquad (2.14.1)$$

and are said to be **reduced**. When the electron activity is relatively low, the medium is **oxidizing**, and chemical species such as H_2O may be **oxidized** by the loss of electrons:

$$2H_2O \leftarrow\rightarrow O_2(g) + 4H^+ + 4e^- \qquad (2.14.2)$$

The relative tendency toward oxidation or reduction may be expressed by the electrode potential, E, which is more positive in an oxidizing medium and more negative in a reducing medium. Oxidation-reduction phenomena are discussed in greater detail in Chapter 6, Section 6.10.

Complexation and Chelation

Metal ions in water are always bonded to water molecules in the form of hydrated ions represented by the general formula, $M(H_2O)_x^{n+}$, from which the H_2O is often omitted for simplicity. Other species may be present that bond to the metal ion more strongly than does water. For example, cadmium ion dissolved in water, Cd^{2+}, reacts with cyanide ion, CN^-, as follows:

$$Cd^{2+} + CN^- \rightarrow CdCN^+ \qquad (2.14.3)$$

The product of the reaction is called a **complex** (or complex ion) and the cyanide ion is called a **ligand**. Some (usually organic) ligands can bond with a metal ion in two or more places, forming particularly stable complexes. One such ligand is the nitrilotriacetate (NTA) ligand, which has the following formula:

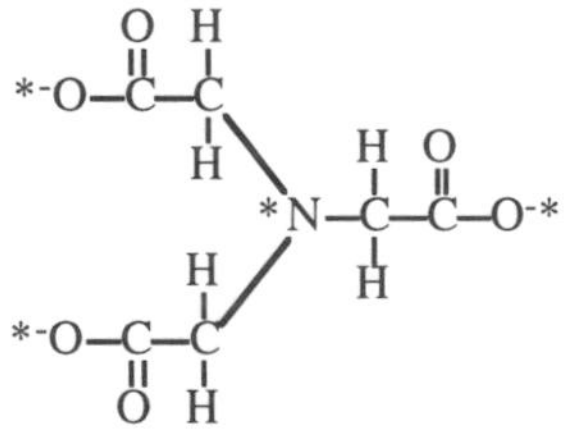

```
         O  H
         ‖  |
  *-O—C—C
            |  \
            H   \      H  O
                 \     |  ‖
                 *N—C—C—O-*
                 /     |
            H   /      H
            |  /
  *-O—C—C
         ‖  |
         O  H
```

This ion has four binding sites, each marked with an asterisk in the preceding illustration, which may simultaneously bond to a metal ion, forming a structure with three rings. Such a species is known as a **chelate**, and NTA is a **chelating** agent. The most important class of complexing agents that occur naturally are the **humic substances**, degradation-resistant materials formed during the decomposition of vegetation and found in water, sediments, and soil.

In addition to metal complexes and chelates, another major type of environmentally important metal species consists of **organometallic compounds**. These differ from complexes and chelates in that the organic portion is bonded to the metal by a carbon-metal bond and the organic ligand is frequently not capable of existing as a stable separate species.

Complexation, chelation, and organometallic compound formation have strong effects upon metals in the environment. For example, complexation with negatively charged ligands may convert a soluble metal species from a cation, which is readily bound and immobilized by ion exchange processes in soil, to an anion, such as $Ni(CN)_4^{2-}$, that is not strongly held by soil. On the other hand, some chelating agents are used for the treatment of heavy metal poisoning, and insoluble chelating agents, such as chelating resins, can be used to remove metals from waste streams.

Water Interactions with Other Phases

Most of the important chemical phenomena associated with water do not occur in solution, but rather through interaction of solutes in water with other phases. Such interactions may involve exchange of solute species between water and sediments, gas exchange between water and the atmosphere, and effects of organic surface films. Substances dissolve in water from other phases, and gases are evolved and solids precipitated as the result of chemical and biochemical phenomena in water.

Sediments are repositories of a wide variety of chemical species and the site of many chemical and biochemical processes. Sediments are sinks for many hazardous organic compounds and heavy metal salts that have gotten into water.

Colloids, which consist of very small particles ranging from 0.001 micrometer (μm) to 1 μm in diameter, have a strong influence on aquatic chemistry. Colloids have very high surface-to-volume ratios, so that they can be very active physically, chemically, and biologically. Colloids may be very difficult to remove from water during water treatment.

Water Pollutants

Natural waters are afflicted with a wide variety of inorganic, organic, and biological pollutants. In some cases, such as that of highly toxic cadmium, a pollutant is directly toxic at a relatively low level. In other cases the pollutant itself is not toxic, but its presence results in conditions detrimental to water quality. For example, biodegradable organic matter in water is often not toxic, but the consumption of oxygen during its degradation prevents the water from supporting fish life. Some contaminants, such as NaCl, are normal constituents of water at low levels, but harmful pollutants at higher levels.

Water Treatment

The treatment of water can be considered under the two major categories of (1) treatment before use and (2) treatment of contaminated water after it has passed through a municipal water system or industrial process. In both cases, consideration must be given to potential contamination by pollutants.

Several operations may be employed to treat water prior to use. Aeration is used to drive off odorous gases, such as H_2S, and to oxidize soluble Fe^{2+} and Mn^{2+} ions to insoluble forms. Lime is added to remove dissolved calcium (water hardness). $Al_2(SO_4)_3$ forms a sticky precipitate of $Al(OH)_3$, which causes very fine particles to settle. Various filtration and settling processes are employed to treat water. Chlorine, Cl_2, is added to kill bacteria.

Municipal wastewater may be subjected to primary, secondary, or advanced water treatment. **Primary** water treatment consists of settling and skimming operations that remove grit, grease, and physical objects from water. **Secondary** water treatment is designed to take out biochemical oxygen demand, BOD. This is normally accomplished by introducing air and microorganisms such that waste biomass in the water, $\{CH_2O\}$, is removed by aerobic respiration:

$$\{CH_2O\} + O_2 \rightarrow CO_2 + H_2O \qquad (2.14.4)$$

2.15. THE GEOSPHERE AND SOIL

The **geosphere**, or solid Earth, is that part of the Earth upon which humans live and from which they extract most of their food, minerals, and fuels. Once thought to have an almost unlimited buffering capacity against the perturbations of humankind, the geosphere is now known to be rather fragile and subject to harm by human activities, such as mining, acid rain, erosion from poor cultivation practices, and disposal of hazardous wastes. It may be readily seen that the preservation of the geosphere in a form suitable for human habitation is one of the greatest challenges facing humankind. Additional aspects of the environmental chemistry of the geosphere and soil are discussed in Chapters 13-16.

Soil

Soil consists of a large variety of material composing the uppermost layer of the earth's crust upon which plants grow. In addition to solids, soil contains air and water. Typically, soil solids consist of about 95% mineral matter and 5% organic material, although the proportions vary widely. Soils are formed by the weathering (physical and chemical disintegration) of parent rocks as the result of interactive geological, hydrological, and biological processes. Soils are porous and are vertically stratified into **horizons** through the action of water, organisms, and weathering processes. Soils are open systems that undergo continual exchange of matter and energy with the atmosphere, hydrosphere, and biosphere. The most active and important part of soil is **topsoil**, the layer in which plants are rooted and in which most biological activity occurs.

2.16. THE ATMOSPHERE AND ATMOSPHERIC CHEMISTRY

The **atmosphere** consists of the thin layer of mixed gases covering the earth's surface. Exclusive of water, atmospheric air is 78.1% (by volume) nitrogen, 21.0% oxygen, 0.9% argon, and 0.03% carbon dioxide. Normally, air contains 1-3% water vapor by volume. In addition, air contains a large variety of trace level gases at levels below 0.002%, including neon, helium, methane, krypton, nitrous oxide, hydrogen, xenon, sulfur dioxide, ozone, nitrogen dioxide, ammonia, and carbon monoxide.

The atmosphere is divided into several layers on the basis of temperature. Of these, the most significant are the troposphere extending in altitude from the earth's surface to approximately 11 kilometers (km) and the stratosphere from about 11 km to approximately 50 km. The temperature of the troposphere ranges from an average of 15°C at sea level to an average of -56°C at its upper boundary. The average temperature of the stratosphere increases from -56°C at its boundary with the troposphere to -2°C at its upper boundary. The reason for this increase is absorption of solar ultraviolet energy by ozone (O_3) in the stratosphere.

Various aspects of the environmental chemistry of the atmosphere are discussed in Chapters 9 and 12. The most significant feature of atmospheric chemistry is the occurrence of **photochemical reactions** resulting from the absorption by molecules of light photons, designated $h\nu$. (The energy, E, of a photon of visible or ultraviolet light is given by the equation, $E = h\nu$, where h is Planck's constant and ν is the frequency of light, which is inversely proportional to its wavelength (see Section 1.9.). Ultraviolet radiation has a higher frequency than visible light and is, therefore, more energetic and more likely to break chemical bonds in molecules that absorb it.) One of the most significant photochemical reactions is the one responsible for the presence of ozone in the troposphere (see above), which is initiated when O_2 absorbs highly energetic ultraviolet radiation in the wavelength ranges of 135-176 nanometers (nm) and 240-260 nm in the stratosphere:

$$O_2 + h\nu \rightarrow O + O \quad (2.16.1)$$

The oxygen atoms produced by the photochemical dissociation of O_2 react with oxygen molecules to produce ozone, O_3,

$$O + O_2 + M \rightarrow O_3 + M \tag{2.16.2}$$

where M is a third body, such as a molecule of N_2, which absorbs excess energy from the reaction. The ozone that is formed is very effective in absorbing ultraviolet radiation in the 220-330 nm wavelength range, which causes the temperature increase observed in the stratosphere. The ozone serves as a very valuable filter to remove ultraviolet radiation from the sun's rays. If this radiation reached the earth's surface, it would cause skin cancer and other damage to living organisms.

Gaseous Oxides in the Atmosphere

Oxides of carbon, sulfur, and nitrogen are important constituents of the atmosphere and are pollutants at higher levels. Of these, carbon dioxide, CO_2, is the most abundant. It is a natural atmospheric constituent, and it is required for plant growth. However, the level of carbon dioxide in the atmosphere, now at about 350 parts per million (ppm) by volume, is increasing by about 1 ppm per year. As discussed in Chapter 12, this increase in atmospheric CO_2 may well cause general atmospheric warming—the "greenhouse effect," with potentially very serious consequences for the global atmosphere and for life on earth. Though not a global threat, carbon monoxide, CO, can be a serious health threat because it prevents blood from transporting oxygen to body tissues.

The two most serious nitrogen oxide air pollutants are nitric oxide, NO, and nitrogen dioxide, NO_2, collectively denoted as "NO_x." These tend to enter the atmosphere as NO, and photochemical processes in the atmosphere tend to convert NO to NO_2. Further reactions can result in the formation of corrosive nitrate salts or nitric acid, HNO_3. Nitrogen dioxide is particularly significant in atmospheric chemistry because of its photochemical dissociation by light with a wavelength less than 430 nm to produce highly reactive O atoms. This is the first step in the formation of photochemical smog (see below). Sulfur dioxide, SO_2, is a reaction product of the combustion of sulfur-containing fuels, such as high-sulfur coal. Part of this sulfur dioxide is converted in the atmosphere to sulfuric acid, H_2SO_4, normally the predominant contributor to acid precipitation.

Hydrocarbons and Photochemical Smog

The most abundant hydrocarbon in the atmosphere is methane, chemical formula CH_4. This gas is released from underground sources as natural gas and produced by the fermentation of organic matter. Methane is one of the least reactive atmospheric hydrocarbons and is produced by diffuse sources, so that its participation in the formation of pollutant photochemical reaction products is minimal. The most significant atmospheric pollutant hydrocarbons are the reactive ones produced as automobile exhaust emissions. In the presence of NO, under conditions of temperature inversion (see Chapter 11), low humidity, and sunlight, these hydrocarbons produce undesirable **photochemical smog** manifested by the presence of visibility-obscuring particulate matter, oxidants such as ozone, and noxious organic species such as aldehydes.

Particulate Matter

Particles ranging from aggregates of a few molecules to pieces of dust readily visible to the naked eye are commonly found in the atmosphere. Some of these particles, such as sea salt formed by the evaporation of water from droplets of sea spray, are natural and even beneficial atmospheric constituents. Very small particles called **condensation nuclei** serve as bodies for atmospheric water vapor to condense upon and are essential for the formation of precipitation.

Colloidal-sized particles in the atmosphere are called **aerosols**. Those formed by grinding up bulk matter are known as **dispersion aerosols**, whereas particles formed from chemical reactions of gases are **condensation aerosols**; the latter tend to be smaller. Smaller particles are in general the most harmful because they have a greater tendency to scatter light and are the most respirable (tendency to be inhaled into the lungs).

Much of the mineral particulate matter in a polluted atmosphere is in the form of oxides and other compounds produced during the combustion of high-ash fossil fuel. Smaller particles of **fly ash** enter furnace flues and are efficiently collected in a properly equipped stack system. However, some fly ash escapes through the stack and enters the atmosphere. Unfortunately, the fly ash thus released tends to consist of smaller particles that do the most damage to human health, plants, and visibility.

CHAPTER SUMMARY

The chapter summary below is presented in a programmed format to review the main points covered in this chapter. It is used most effectively by filling in the blanks, referring back to the chapter as necessary. The correct answers are given at the end of the summary.

Chemistry is the science of [1] ________________________________, which is defined as [2] ____
________________________________. All matter is composed of only about a hundred fundamental kinds of matter called [3] ____________________________, each made of very small [4] ________________. The number of protons in the nucleus of each atom of an element is its [5] ____________________ and the average mass of all the atoms of an element is its [6] ____________________. The Lewis symbol for carbon, ·Ċ·, shows that [7] ____.

Chemically, an element denoted by the Lewis symbol, :Ẍ:, is noted for [8] ________ ____.

In the Lewis formula,

H:Ö:H

The two pairs of dots outlined by the oval dashed lines represent [9] ____________ ____.

A substance composed of atoms of two or more different elements bonded together is called a [10] ________________________________. In the chemical formula K_2SO_4, the subscripts 2 and 4 denote that each formula unit of the chemical compound contains, respectively, [11]______________________________ ______________________________. A quantity of a substance that contains the same number of specified entities as there are atoms of C in exactly 12 g of carbon-12 is known as a [12]______________ of the substance, and the number of such entities is [13]______________________. All correctly written chemical equations have [14] ______________________ ______________________________________. Properties of matter are that can be measured without altering the chemical composition of the matter are called [15]______________ ______________. Liquids constitute a [16]______________ of matter in which the liquids have [17]______________________________________ ______________________________________.

The ideal gas equation is expressed as [18]______________________. Liquids may act as solvents to form [19]______________________ in which quantities of materials are [20]______________. The relationship (moles of solute)/(number of liters of solution) expresses [21] ______________________. The amount of heat energy required to raise the temperature of a gram of substance by 1 degree Celsius is called the [22]______________________ of the substance. A substance that produces H^+ ion in water is called an [23]______________. The reaction $HCN \rightarrow H^+ + CN^-$ shows the [24]______________________ of HCN. A base is a substance that [25]______________________________________. Whenever an acid and a base are brought together, the two products are [26]__________ ______________________________. The mathematical expression for pH is [27]______________________. Organic chemistry addresses most compounds containing the element [28]______________. Hydrocarbons contain only [29]______________. The bond represented as C=C represents a [30]______________________ found in hydrocarbons known as [31]______________________. The name of the compound

```
        CH3
   H    |   H  H  H
   |    |   |  |  |
H--C----C---C--C--C--H
   |    |      |  |
   H    H   |  H  H
            |
           C2H5
```

is based on pentane, a 5-carbon chain, and the name of the compound is [32]__________ ______________________. A general type of reaction that alkenes can undergo that alkanes cannot consist of [33]______________________. The structures,

```
H3C         CH3        H3C         H
   \       /              \       /
    C=====C                C=====C
   /       \              /       \
  H         H            H         CH3
```

illustrate [34]______________________________. A class of hydrocarbons represented by benzene, C_6H_6, is composed of [35]______________________ hydrocarbons. Specific bonding configurations of atoms in organic molecules often containing atoms other than carbon or hydrogen are known as [36]______________________. Environmental chemistry may be defined as[37]______________________________ ______________________________________ ______________________________________.

Major types of aquatic chemical processes are [38]______________________
____
____

Reduced species in water are associated with [39]__________ electron activity and oxidized species with [40]__________ electron activity. The product of the reaction $Cd^{2+} + CN^{-} \rightarrow CdCN^{+}$ is a [41]__________________ and CN^{-} is known as a [42]______________. Compounds in which an organic (hydrocarbon) entity is bound to a metal atom through a carbon atom are called [43]________________ compounds. Very small particles ranging from 0.001 micrometer (µm) to 1 µm in diameter often found in water are known as [44]______________. Secondary water treatment is designed to remove [45]____________________. Soils are formed by [46]______________________________________
____
____.

The process $O_2 + h\nu \rightarrow O + O$ is an example of a [47]________________
______________________. The formula of an atmospheric oxide that is increasing in concentration by about 1 ppm per year is [48]____________, and it is feared that this increase may result in [49]________________________. In the presence of NO, under conditions of temperature inversion, low humidity, and sunlight, reactive hydrocarbons produce [50]__________________ manifested by [51]______________________________________
____. Colloidal-sized particles in the atmosphere are called [52]________________.

Answers

[1] matter

[2] anything that has mass and occupies space

[3] elements

[4] atoms

[5] atomic number

[6] atomic mass

[7] the carbon atom has four outer-shell electrons

[8] being chemically unreactive, a noble gas

[9] covalent chemical bonds between hydrogen and oxygen atoms

[10] chemical compound

[11] 2 K^{+} ions and 4 O atoms

[12] mole

[13] Avogadro's number , 6.02×10^{23}

[14] the same number of each kind of atom on both sides of the equation

[15] physical properties

[16] state

17 an indefinite shape and take on the shape of the container in which they are contained

18 PV = nRT

19 solutions

20 dissolved

21 molar concentration, M

22 specific heat

23 acid

24 dissociation or ionization

25 produces hydroxide ion, OH^-, and/or accepts H^+

26 water and a salt

27 $pH = -\log[H^+]$

28 carbon

29 hydrogen and carbon

30 double bond

31 alkenes

32 2-methyl-3-ethylpentane

33 addition reactions

34 *cis* and *trans* isomers

35 aryl

36 functional groups

37 the study of the sources, reactions, transport, effects, and fates of chemical species in the water, air, terrestrial, and living environments

38 acid-base, precipitation, oxidation-reduction, microbial, photosynthesis, gas exchange, leaching

39 high

40 low

41 complex or complex ion

42 ligand

43 organometallic

44 colloid

45 biochemical oxygen demand, BOD

46 the weathering (physical and chemical disintegration) of parent rocks as the result of interactive geological, hydrological, and biological processes

47 photochemical reaction

48 CO_2

49 global warming

50 photochemical smog

[51] visibility-obscuring particulate matter, oxidants such as ozone, and noxious organic species such as aldehydes

[52] aerosols.

QUESTIONS AND PROBLEMS

1. In terms of subatomic particles in an atom, what defines atomic number?
2. What is indicated by dots numbered from 1 to 8 around the symbol for an element? What is special about 8 dots?
3. Explain how covalent bonds differ from ionic bonds.
4. Consider all the information given by the chemical formula of calcium phosphate in Figure 2.9. What is shown by the chemical formula of aluminum sulfate, $Al_2(SO_4)_3$?
5. Explain in which sense the mole is a more fundamental unit for quantity of matter in chemical reactions than is mass.
6. Consider the quantity (moles) of a gas and the conditions that it is under. What is the equation expressing the quantities and conditions, and what is the meaning of each term?
7. What is happening when a small percentage of hydrogen molecules undergo the reaction $H_2S \rightarrow H^+ + HS^-$ in water?
8. Which element is contained in all organic compounds?
9. What is meant by the molecular geometry of organic compounds? How is it related to isomers?
10. What is an addition reaction? What kinds of hydrocarbons undergo substitution, but not addition, reactions?
11. In terms of functional groups, what distinguishes alcohols from phenols? Ketones from aldehydes?
12. Restate the definition of environmental chemistry in terms of the **five** spheres of the environment shown in Figure 1.1.
13. Which part of the biosphere that lives in water is especially influential on aquatic chemistry? How is that illustrated in Figure 2.18?
14. Why is cyanide ion simply a ligand, whereas the NTA anion is a special case of a ligand called a ______________________.
15. In what sense do sediments tend to be sinks in aquatic chemistry?
16. What distinguishes soil from other parts of the geosphere?
17. In the atmosphere a general type of reaction occurs that is of little importance in other environmental spheres. What kind of reaction is it, and why is it important in the atmosphere?
18. Distinguish between dispersion and condensation aerosols. Why do condensation aerosols tend to be smaller?

LITERATURE CITED

1. Manahan, Stanley E., *Fundamentals of Environmental Chemistry*, CRC Press/Lewis Publishers, Boca Raton, FL, 1993.

2. Manahan, Stanley E., *Environmental Chemistry*, 6th ed., CRC Press/Lewis Publishers, Boca Raton, FL, 1994.

SUPPLEMENTARY REFERENCES

Andrews, J. E., *Environmental Chemistry*, Blackwell Science Publishers, Cambridge, MA, 1996.

Burke, Robert, *Hazardous Materials Chemistry for Emergency Responders*, CRC Press/Lewis Publishers, Boca Raton, FL, 1997.

Csuros, Maria, Ed., *Environmental Sampling and Analysis Laboratory Manual*, CRC Press/Lewis Publishers, Boca Raton, FL, 1997.

Erickson, Mitchell D., *Analytical Chemistry of PCBs*, 2nd ed, CRC Press/Lewis Publishers, Boca Raton, FL, 1997.

Hess, Kathleen, *Environmental Sampling for Unknowns*, CRC Press/Lewis Publishers, Boca Raton, FL, 1996.

Keith, Lawrence H., *EPA's Sampling and Analysis Methods Database, Version 2.0*, CRC Press/Lewis Publishers, Boca Raton, FL, 1996.

Patnaik, Pradyot, *Handbook of Environmental Analysis*, CRC Press/Lewis Publishers, Boca Raton, FL, 1997.

Schlesinger, W. H., *Biogeochemistry*, Academic Press, San Diego, CA, 1991.

Spiro, Thomas G. and William M. Stigliani, *Chemistry of the Environment*, Prentice Hall, Upper Saddle River, NJ, 1996.

LITERATURE CITED

1. Manahan, Stanley E., *Fundamentals of Environmental Chemistry*, CRC Press/Lewis Publishers, Boca Raton, FL, 1993.

2. Manahan, Stanley E., *Environmental Chemistry*, 6th ed., CRC Press/Lewis Publishers, Boca Raton, FL, 1994.

SUPPLEMENTARY REFERENCES

[illegible], Blackwell Science, Cambridge, [illegible]

3 BIOLOGY AND BIOCHEMISTRY

3.1. INTRODUCTION

Chapter 1 introduced and defined the topic of *environmental science and technology*. Chapter 2 presented the minimal background in chemistry and organic chemistry required to understand the material in this book, along with a section on environmental chemistry. The present chapter first discusses biology, the science of life. Then, building on that material and the fundamentals of chemistry presented in Chapter 2, it outlines the chemical basis of life—biochemistry.

Chapters 2 and 3 present the basic chemical and biological science, including general chemistry, organic chemistry, environmental chemistry, biology, and biochemistry needed to understand the rest of the material in the book. It will be assumed throughout the rest of the book that readers have a rudimentary knowledge of these topics from Chapters 2 and 3 and from their outside study and reading.

3.2. BIOLOGY

Biology is the science of life and of living organisms. A **living organism** is constructed of one or more small units called *cells* and has the following characteristics: (1) It is composed in part of large characteristic macromolecules containing carbon, hydrogen, oxygen, and nitrogen along with phosphorus and sulfur. (2) It is capable of **metabolism**; that is, it mediates chemical processes by which it utilizes energy and synthesizes new materials needed for its structure and function. (3) It regulates itself. (4) It interacts with its environment. (5) It reproduces itself.

Biology developed from the fascination that humans have always had with living things. In its earliest stages of development biology was largely confined to descriptions of plants and animals, concentrating largely on their physical forms. **Natural history**, the study of where and the manner in which organisms carry out life cycles, became an integral part of biology. Modern biology has been revolutionized at an ever accelerating pace during the last several decades. This great change in biology dates from the recognition by James Watson and Francis Crick in the early 1950s of the structure of DNA, the molecule of heredity that determines the replication and function of living organisms. Modern techniques of molecular genetics that study DNA and related RNA have revolutionized humankind's understanding of life and the relationships among life forms. For example, it has shown that some species that give the superficial appearance of being closely related are not, whereas others that do not look much alike are really quite close in the scheme of evolution.

The unit of paramount importance in living organisms is the **cell**, the fundamental unit of life consisting of a small body of the order of micrometers in size. There are two basic kinds of cells. Single-celled bacteria are composed of **prokaryotic cells** (see Section 3.3). In a sense, prokaryotic cells are relatively simpler and more primitive than the **eukaryotic cells** which comprise all organisms other than bacteria. A very simple representation of a cell is shown in Figure 3.1.

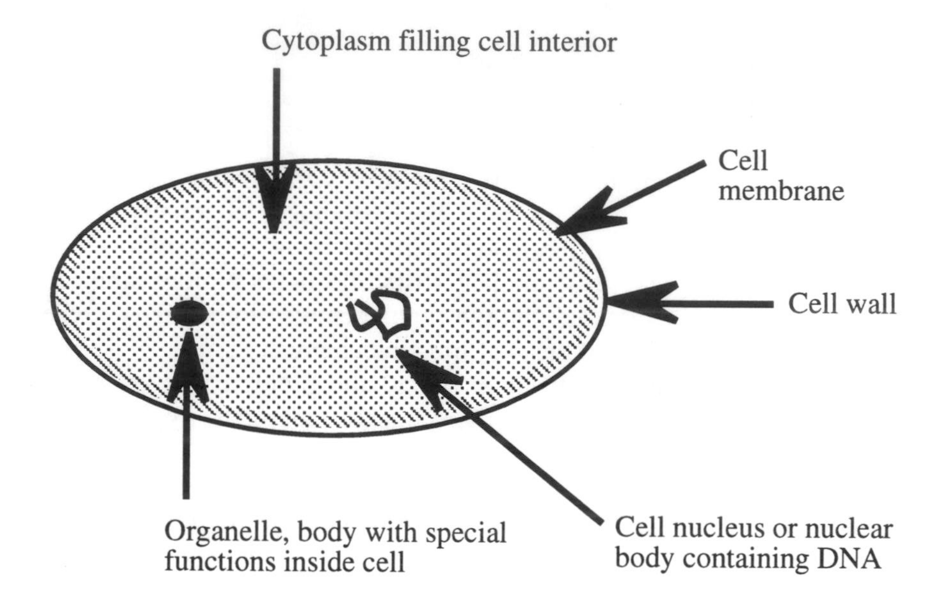

Figure 3.1. Basic features of cells making up living organisms.

Cells are divided into two major categories depending upon whether or not they have a nucleus: **eukaryotic** cells have a nucleus and **prokaryotic** cells do not. Prokaryotic cells are found predominately in single-celled organisms such as bacteria. Eukaryotic cells occur in multicelled plants and animals — higher life forms. The major features of eukaryotic cells are the following:

- **Cell membrane**, which encloses the cell and regulates the passage of ions, nutrients, lipid-soluble ("fat-soluble") substances, metabolic products, toxicants, and toxicant metabolites into and out of the cell interior because of its varying **permeability** for different substances. The cell membrane protects the contents of the cell from undesirable outside influences.
- **Cell nucleus**, which acts as a sort of "control center" of the cell. It contains the genetic directions the cell needs to reproduce itself. The key substance in the nucleus is the nucleic acid **deoxyribonucleic** acid (**DNA**).
- **Cytoplasm**, which fills the interior of the cell not occupied by the nucleus.
- **Mitochondria**, "powerhouses" that mediate energy conversion and utilization in the cell.
- **Ribosomes**, which participate in protein synthesis.

- **Cell walls** of plant cells. These are strong structures composed mostly of cellulose that provide stiffness and strength.
- **Vacuoles** inside plant cells that often contain materials dissolved in water.
- **Chloroplasts** in plant cells that are sites for photosynthesis (the chemical process that uses energy from sunlight to convert carbon dioxide and water to organic matter).

Types of Organisms

Organisms cover a vast range of size, complexity, and type. It is convenient to divide organisms into five major **kingdoms** (Figure 3.2); many authorities believe that it should be at least six and that other subdivisions are likely. These kingdoms are the following:

- **Monera**. Single-celled bacteria made up of prokaryotic cells.
- **Protista.** Single-celled organisms composed of eukaryotic cells. Various members display characteristics of fungi, plants, and animals.
- **Fungi.** Eukaryotic organisms that live off biomass from other sources, which they digest by excreting digestive enzymes and absorbing the breakdown products resulting from enzyme action on biomass.
- **Plantae**. Multicellular organisms capable of utilizing solar energy to convert inorganic carbon from CO_2 to biomass.
- **Animalia**. Multicellular organisms that live off biomass from other organisms, which they ingest.

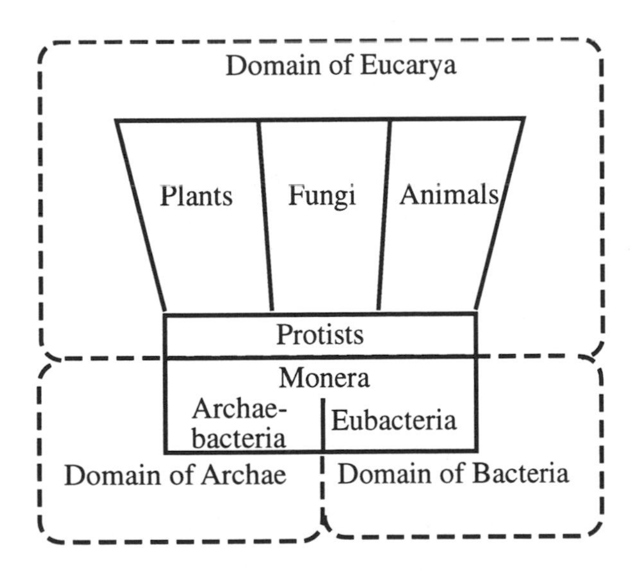

Figure 3.2. Representation of the classification of life forms into five (or six) kingdoms and three domains.

Each of these kingdoms is important in environmental science. They are discussed individually below.

A crucial distinction between organisms is whether or not they can make their own food or must rely for food on biomass produced by other organisms. Those that can synthesize food are called **autotrophs**. Most autotrophs, such as plants, mediate **photosynthesis**; a few other organisms, the **chemoautotrophs**, mediate energy-yielding inorganic chemical reactions. Organisms that must rely on other organisms for food are called **heterotrophs**. Furthermore, most organisms require molecular oxygen for their metabolic processes and are labelled **aerobic**, whereas some kinds of bacteria can use other kinds of oxidants and are called **anaerobic**; those that can do both are **facultative**.

3.3. MONERA

All monera are bacteria and are the only organisms that are composed of prokaryotic cells. Such cells lack a true nucleus, the function of which is served instead by a strand of DNA, the basic molecule of heredity discussed in Section 3.16. They do not possess the smaller internal cellular bodies that characterize eukaryotic cells—*organelles*, such as lysosomes and mitochondria—but the cytoplasm that fills the cells contains large numbers of small structures called **ribosomes**, in which proteins are synthesized. The *plasma membrane* that lines the inner part of the cell wall is the site of the metabolic processes by which energy is extracted from food matter or carbon fixed by photosynthesis. The prokaryotic cells of monera differ subtly from those of other kingdoms in other ways, such as the composition of cell walls.

Monera fossils have been found in rocks dating back 3.5 billion years, and they are the oldest life forms that have persisted to the present time. Monera are subdivided into the two major categories of *archaebacteria* and the more common and recently evolved *eubacteria*. Recent studies of molecular biology suggest that archaebacteria and eubacteria are fundamentally different and belong to two of only three different *domains* of living organisms, the third of which consists of all organisms other than monera. Regardless of which classification system is correct, archaebacteria and eubacteria are both of utmost importance in the environment.

Some monera are *autotrophic*, which means that they mediate photosynthesis or act as catalysts in energy-yielding inorganic chemical processes in order to obtain their food and energy. Most monera are *heterotrophic* in that they rely on other organisms for food. In the latter category are the **saprobes**, which play a vital role in the environment by degrading dead biomass. Many monera, such as the bacteria that thrive in the intestines of animals, live in close association with other organisms and are called **symbionts**. Those that cause illness or other harmful effects to their hosts are called **parasites**.

Archaebacteria

Archaebacteria, named because of their existence from earliest times, exist in extreme environments that probably reflect conditions under which they evolved in the early years of Earth's existence. The *methanogens* live under oxygen-free conditions and produce methane, CH_4. They are found, therefore, in lake sediments, swamp muck, and anaerobic digestors of sewage treatment plants in which sewage sludge degrades in the absence of oxygen. The *thermoacidophilic group* lives in water under outrageously severe conditions of high temperature and high acidity.

Such conditions are encountered in sulfurous thermal vents, such as those in Yellowstone National Park in the U.S. *Halophilic* archaebacteria live under conditions of extreme salinity, such as some landlocked saline bodies of water.

Eubacteria

Eubacteria comprise a wide range of monera with a range of characteristics. Their cells may be shaped as spheres (**cocci**), rods (**bacilli**), or spirals (**spirilla**). Sometimes called *true bacteria*, the evolution of these organisms can be traced into 11 separate evolutionary lines. The most important of these are summarized below.

Purple bacteria, the largest and most diverse group of eubacteria, contains autotrophic members that mediate photosynthesis, though with a type of chlorophyll different from that used by plants. Nonphotosynthetic heterotrophic members of this category include intestinal *Escherichia coli*, nitrogen-fixing *Rhizobium* bacteria, and some pathogenic bacteria.

Cyanobacteria, formerly mislabelled "blue-green algae," carry out photosynthesis using chlorophyll *a*, the same type used by plants. Some cyanobacteria convert atmospheric N_2 to biologically bound nitrogen, so they are *nitrogen-fixing* organisms. Because of their ability to synthesize biomass and release O_2 by photosynthesis, and to fix atmospheric nitrogen, cyanobacteria are very self-sufficient, widespread, and important in the environment. Ancient members of this evolutionary group are believed to have been the first organisms to release O_2, thus resulting in the massive change of Earth's atmosphere from a chemically reducing to an oxidizing one, and were probably the precursors to chloroplasts, the organelles in plant and photosynthetic protista cells in which photosynthesis is carried out. Cyanobacteria contribute bad taste and odor to water.

Other major groups of eubacteria include the following: **Gram-positive bacteria** are named for their response to Gram stain used to stain bacterial cells and include a diversity of kinds, such as *Lactobacillus* that produces yogurt, *Chlostridium*, which produces botulism toxin, and *Staphylococcus*, which causes toxic shock syndrome. **Chlamydia** live inside other cells and include members that cause a majority of sexually transmitted diseases. Spiral-shaped **spirochetes** include pathogens that cause diseases such as syphilis and Lyme disease. Other general categories of eubacteria are **plantomyces**, **bacteroides**, **green sulfur bacteria**, **deinococcus** (some so resistant to ionizing radiation that they cause problems in water subjected to such radiation from nuclear reactors), **green nonsulfur bacteria**, and **thermotoga**.

Bacteria in Aquatic Chemistry

In Chapter 2 it was noted that bacteria were very important in mediating aquatic chemical reactions. Having discussed the nature of bacteria in this section, it is now possible to expand on their role in aquatic chemistry. Bacteria mediate oxidation-reduction reactions to extract the energy that they need for their own growth and reproduction. The most common bacterially mediated reaction in water and soil is the oxidation of organic matter (represented by the simplified formula $\{CH_2O\}$),

$$O_2 + \{CH_2O\} \rightarrow CO_2(g) + H_2O \qquad (3.3.1)$$

a process called **aerobic respiration**. It provides the means for degrading organic wastes, such as sewage. If this reaction occurs at a rate faster than processes for replenishing oxygen in a body of water, the level of dissolved oxygen may become so diminished that the water no longer supports fish life. Some other important oxidation-reduction reactions carried out by bacteria in water are given in Table 3.1.

Table 3.1. Some Oxidation-Reduction Reactions Mediated by Bacteria

Reaction	Significance
$2\{CH_2O\} \rightarrow CH_4 + CO_2$	Fermentation reaction, atmospheric methane source
$SO_4^{2-} + 2\{CH_2O\} + 2H^+ \rightarrow H_2S(g) + 2CO_2 + 2H_2O$	Sulfate reduction, source of atmospheric H_2S
$2FeS_2 + 2H_2O + 7O_2 \rightarrow 4H^+ + 4SO_4^{2-} + 2Fe^{2+}$	Production of acid mine water
$4NO_3^- + 5\{CH_2O\} + 4H^+ \rightarrow 2N_2 + 5CO_2 + 7H_2O$	Denitrification (conversion of fixed nitrogen back to atmospheric N_2)

Bacteria interact with environmental chemicals in a number of ways. Some very toxic materials impede, or even totally stop, bacterial action. A number of organic wastes are partially or completely degraded by bacteria. In a few cases, however, the organic products are even more toxic than the original pollutants.

3.4. PROTISTA

Protista are single-celled organisms composed of eukaryotic cells, rather than the prokaryotic cells of bacteria. The protista form a large variety of cells with many complex and specialized structural features. Of the protista, the **protozoans** act like animals, the **algal protists** resemble plants, and others resemble fungi. Figure 3.3 shows some of the major features typical of protozoa as illustrated by photosynthetic *Euglena*.

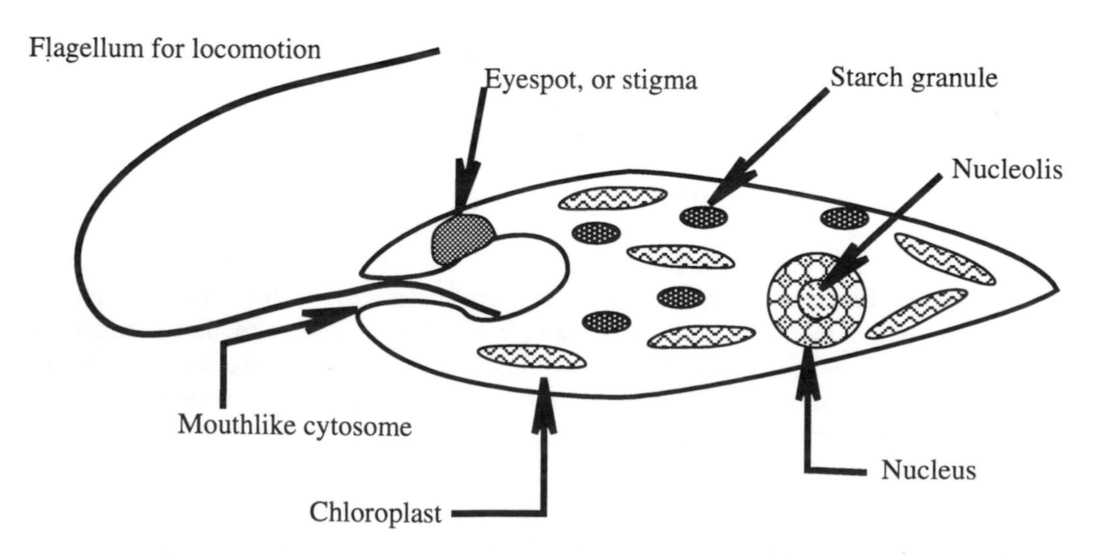

Figure 3.3. Representation of *Euglena* showing some of the many features of single-celled protista.

Protozoans

Heterotrophic protozoans may exist independently, in symbiotic relationships, or as parasites. Pathogenic protozoans cause some of the more serious diseases of humans and animals, such as malaria or amebic dysentery. Among the members of this group are relatively primitive *Mastigophora* that propel themselves with flagella; *amebas* and related organisms with cells which move in a characteristic way that involves "oozing" of the cell contents; parasitic *sporozoans*, including the type that causes malaria; and complicated, highly developed *ciliates*, which move by the action of hairlike cilia on their bodies.

Algal Protists

Algal protists are capable of photosynthesis and are plantlike in their behavior. **Euglenophyta** are interesting because they can live heterotrophically and can generate food by photosynthesis. The **pyrrophyta** are complex, largely marine protists, which include toxin-releasing members capable of producing *red tides* that can cause massive fish kills. The most abundant of the **chrysophyta** are the *diatoms*, which are enclosed in double shells composed of mineral matter. Others in this group are *golden-brown algae* and *yellow-green algae*. **Chlorophyta** contain chlorophyll and some other substances characteristic of higher plants, of which they are probably the ancestors. They exist predominantly in fresh water, though some live in seawater and some on land. They are the algal component of lichens, a mutualistic association of algae and fungi that grow on rocks. Other algal protists include *brown algae*, which form giant marine kepts, and *red algae* in seaweeds.

Protists That Resemble Fungi

Fungilike protists are heterotrophic organisms that consititute so-called *slime molds*. They include *Myxomycota*, *Acrasiomycota*, and *Oomycota*. *Oomycota* are agents that cause some rusts, mildews, and water molds. One of their members caused the catastrophic potato blight that afflicted Ireland in 1848, causing devastating famine.

3.5. FUNGI

As noted previously, **fungi** are heterotrophic organisms that digest food by excreting digestive enzymes and absorbing the breakdown products resulting from enzyme action on biomass. Except for the single-celled yeasts, most are multicellular and grow in branched strands called *hyphae*. The divisions between cells in hyphae are often unclear, and each cell is defined by its own nucleus.

There are four major divisions of fungi. ***Zygomycota*** live in soil and get their food from decaying organic matter. They form rugged *zygospores* that produce viable organisms when conditions are right for growth. These fungi are environmentally very important as primary *decomposers* of dead biomass. ***Ascomycota*** include several important kinds of fungi, including yeasts and mildew-forming fungi. One member of this group is *ergot*, which infects grain and produces toxins that can be very harmful to animals and humans. The fungi that humans are usually most aware of are the ***Basidiomycota***, notable for their production of external spore-producing bodies. These bodies include puffballs, toadstools, and mushrooms. Miscellaneous fungi are included in a category called ***Deuteromycota***.

Some of the more environmentally significant fungi are those that occur in **lichens**, a symbiotic combination of fungi and algae. The fungal part anchors the lichen to rock and produces metabolic products that leach and weather the rock, releasing nutrients needed by the algae. The algae in turn fix carbon dioxide to produce carbohydrates, which serve as the food source for algae. Lichens are of great importance in the weathering processes by which exposed parent rocks are converted to soil. The fungal component is usually a basidiomycote (in warm climates) or an ascomycote (in cold climates).

Another important symbiotic relationship of fungi is their participation in ***mycorrhizae***, an association of fungi and plant roots. Fungi growing on the roots send out hyphae that absorb water and nutrients required by the plant, thereby effectively increasing root area enormously. The plants in turn provide carbohydrate as a food source for the fungi, along with some other nutrients. It is because of mycorrhizae on their roots that the largest organisms ever to exist on Earth, the giant sequoia trees of California, can grow in relatively dry, nutrient-deficient soil to heights almost equal to the length of a football field and to masses of several thousand tons with root systems smaller than those of trees a fraction of their size.

The greatest significance of fungi in the environment is their role as *decomposers* that break down biomass, mostly from plants, and release inorganic components, a process called *mineralization*. Thus fungi are an essential component of the materials cycles that characterize all viable ecosystems. Fungi are very useful to humans, providing antibiotics, which attack disease-causing pathogenic bacteria, acting as yeasts to ferment sugar to ethanol, and playing a role in the manufacture of some foods, such as Roquefort cheese. Fungal infections of plants and animals can be very harmful. Fungi cause plant rusts, and byproducts of fungi growing on grain may consist of highly toxic aflatoxins. Human afflictions caused by fungi include ringworm, athlete's foot, and vaginal yeast infections.

3.6. PLANTAE

Plants are multicellular organisms that act as nature's great producers by generating biomass (shown by the simplified chemical formula $\{CH_2O\}$) photosynthetically according to the overall process,

$$CO_2 + H_2O + h\nu \rightarrow \{CH_2O\} + O_2 \tag{3.6.1}$$

where $h\nu$ represents the energy of a quantum of solar energy. The amount of biomass produced each year photosynthetically amounts to many billions of tons and would fill a freight train wrapped multiple times around the Earth. Humans and virtually all other organisms are dependent for their existence on plant photosynthesis. Starting as single-celled marine protists, plants moved to land and developed a fascinating variety of structures and physiological responses to cope with a vast variety of terrestrial conditions. Plants range in complexity from simple, single-celled algae to rooted plants with leaves, stems, and complex flowering bodies.

Algae

Despite their abundance and environmental importance, particularly in freshwater aquatic environments, algae are difficult to define. For this book, **algae** will be regarded as photosynthetic aquatic organisms that do not have roots and do not produce seeds, including members from both protists and plants. Algae lack vascular systems (ductlike structures for transmitting fluids) possessed by higher plants. There are three major categories of algae in the plant kingdom. The red algae, or *Rhodophyta*, may grow as single cells, in colonies, or as multicellular plants, often exhibiting filamentous or fanlike structures. Their pigments that absorb light used in photosynthesis may be red or purple. The brown algae, or *Phaeophyta*, include the largest algae, particularly giant kelps. The green algae or *Chlorophyta*, are closest to land plants and probably gave rise to them. The chlorophyll in green algae is the same as that of land plants.

Nonvascular Plants

Nonvascular plants, or *Bryophytes*, include liverworts, hornworts, and mosses. The rigidity of their tissues enables them to grow upright, with the competitive advantages for access to sunlight that such growth allows. They possess rhizoids, rootlike structures by which they are anchored to soil and through which they draw water and nutrients.

Vascular Plants

Vascular plants are plants other than algae and other nonvascular plants, named because of their vascular tissue used for support and for transport of water, nutrients, and metabolic products. Roots, stems, and leaves are characteristic of vascular plants. Most of these plants produce seeds, except for the **seedless vascular plants**, including ferns. The abundant and successful **seed plants**, which most people think of when defining plants, are generally characterized by well-defined vascular systems, true roots, leaves, and stems. Prominent among this group are *Coniferophyta*, including pine trees and firs. These plants usually have needlelike leaves and release naked seeds from cones. The **flowering seed plants** produce flowers and, after the flowers have bloomed, fruits and seeds. They are divided into the two main categories of **monocots**, such as corn or daffodils, characterized by a single seed leaf and leaves that have parallel veins, and **dicots**, (marigolds, tomatoes) with two-leaved seedlings and leaves with veins in a netted shape.

3.7. ANIMALIA

Animals are multicellular organisms, most of which move voluntarily, can respond to stimuli by means of nerve cells, and have active means of seeking food, which they usually digest internally. Animals are heterotrophic; those that eat plants are *herbivores*, and those that prey on other animals are *carnivores*. Animals were descended from "animal-like" single-celled protozoa. A special category of animal consists of sponges, simple animals that are permanently attached to some kind of stationary support. Beyond sponges, animals may be radially symmetrical (jellyfish) or bilaterally symmetrical (spiders, humans). The animal kingdom is divided into the two major categories of *invertebrates* (those without a central spinal nerve cord or backbones) and a much smaller category of *vertebrates*, including humans, dogs, and most other animals with which humans are familiar.

Among the invertebrates are sponges (mentioned above), **cnidaria** (hydras, jellyfish), **platyhelminthes** (flatworms), **nematoda** (roundworms), **mollusca** (clams, slugs), **annelida** (earthworms, leeches), and **echinodermata** (sea stars, sea cucumbers). A phylum of animals deserving special mention because of its widespread distribution, persistence, and more than one million species consists of the **arthropoda**, including insects, spiders, ticks, scorpions, lobsters, and shrimp. Several special features mark the majority of invertebrate animals and are standard features of most successful animals species, including the more advanced chordates discussed below. These features, which are crucial to the success of animal species, are the following: (1) Development of symmetrical right and left halves, a feature called **radial symmetry**. (2) **Segmentation** into separate body units and parts with different functions, such as the thorax, abdomen, legs, and wings. (3) A head, consisting of a centralized "control center" that detects, processes, and responds to stimuli, such as light, sound, and chemicals (odor); this feature is called **cephalization**. (4) A **gut** consisting of a tube traversing a significant portion of the body into which food is taken and from which waste products are excreted. (5) Presence of a **coelom**, a fluid-filled space in which the gut and other organs are suspended.

Chordates

Making up only about 5 percent of animal species, **chordates**, have a central nerve cord enclosed in, or supported by, a protective structure. In **vertebrate** animals, including humans, the former is the spinal cord and the latter is the backbone; in some animals the central nerve cord is protected by a flexible rod called a *notochord*. Added to the five key features described above of radial symmetry, segmentation, cephalization, gut, and coelom, the central nerve cord, notochord, gill slits, tail, and muscle blocks unique to them have given the chordates some extraordinary capabilities and advantages. The most primitive groups of chordates are those without vertebrae and possessing only a notochord to support the central nerve cord—the tunicates and cephalochordata. The **vertebrata** chordates, which have segmented backbones, are divided among **fishes**, **amphibians**, **reptiles**, **birds**, and **mammals**.

3.8. CONTROL IN ORGANISMS

Regulation and **control** of processes in organisms is crucial to their reproduction, growth, function, and survival. There are two major ways in which regulation and control are accomplished. The first of these is through **molecular messengers** consisting of specific chemical species that are transported in the organism to regulate processes. The second way is through the **nervous system** of animals. These regulatory mechanisms are discussed briefly in this section.

Molecular Messengers in Animals

Molecular messengers in organisms consist largely of hormones and some other small molecules of which the following is generally true:

- There is a remarkable similarity in molecular messengers among all living things, suggesting a common evolutionary origin.
- Molecular messengers produced in one part or organ of a living thing are often transported to remote locations to carry out their mission.
- Molecular messengers either cause or prevent change in an organism's physiology.

- Molecular messengers generally act by binding to **receptor proteins** on the surface or inside of a cell.

The general action of molecular messengers is illustrated in Figure 3.4. A stimulus, often an environmental change that tends to perturb the organism's crucial internal environment—its **homeostasis**—is detected by a **regulator cell**. This cell is stimulated to produce molecular messengers that are carried through a fluid medium, such as the blood stream, to a **target cell**. A protein on or in the receptor cell with a shape complementary to the molecular messenger binds with the messenger and causes the cell to act in a desired way, such as by synthesizing a protein needed to counteract the environmental influence threatening homeostasis.

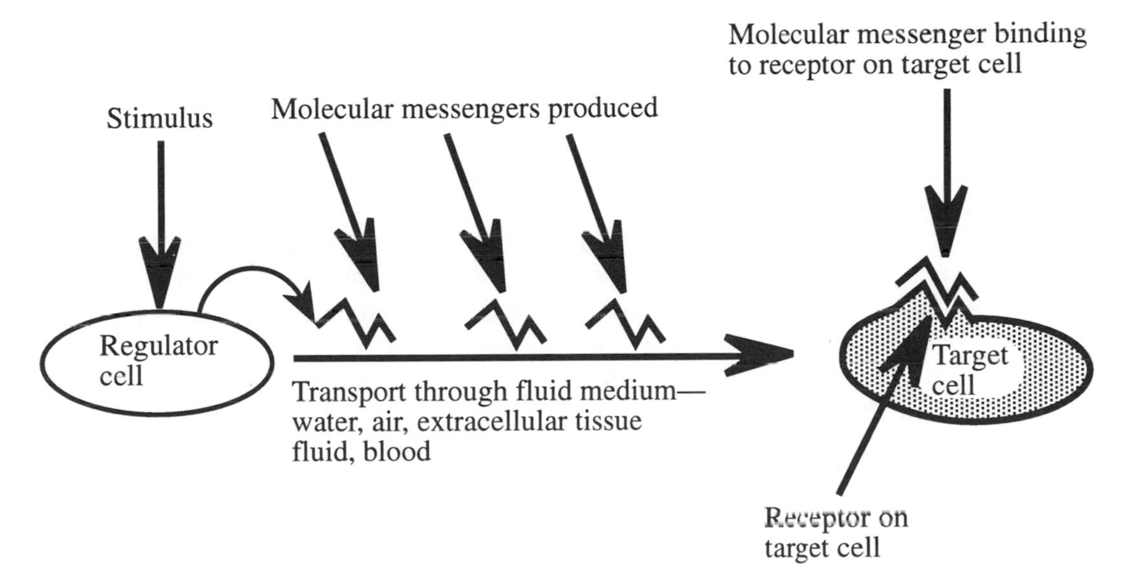

Figure 3.4. General mechanism by which molecular messengers function.

In animals molecular messengers are usually produced by special organs called **glands. Endocrine glands**, such as the adrenal, pituitary, and thyroid glands, release molecular messengers into the fluid external to the cells from which they may be carried to target cells. Harm to endocrine glands, such as damage by radioactive iodine to the thyroid, may be among the most harmful effects of environmental toxicants.

There are five major types of molecular messengers in animals: (1) **Paracrine hormones** that act on target cells adjacent to the cells that secrete the hormones; (2) **neurotransmitters** that are secreted by a regulator nerve cell directly to a target cell; (3) **neurohormones** secreted by nerve cells that may be transported to nonadjacent cells; (4) **true hormones** secreted by endocrine gland cells and carried some distance, usually by the blood stream, to target cells; and (5) **pheromones** that are carried from one organism to another. It is beyond the scope of this short section to discuss molecular messengers and hormones in any detail. Some important examples include growth hormone secreted by the anterior pituitary gland that stimulates body growth; parathyroid hormone secreted by the parathyroid gland that stimulates calcium uptake into the blood from bones and from the digestive tract; insulin from the pancreas that stimulates glucose uptake from blood; and testosterone from the testis that stimulates male development and behavior.

Control Functions of the Nervous System

Control processes in animals provide a mechanism by which they respond to alterations called **stimuli** in the animal's internal and external environments. Normally, stimuli result in **responses** by which the organism makes adjustments to maintain homeostasis (defined above). The use of molecular messengers associated with animal endocrine systems was discussed above as constituting a major part of the control mechanism of animals. The other major aspect of these processes is the **nervous system**, by which stimuli are received, processed, and translated to **nerve impulses**, which are transmitted through the nervous system to initiate necessary physiological changes. All vertebrates and most invertebrates have nervous systems. More advanced animals have a **central nervous system** (CNS) consisting of the brain and spinal cord, that is capable of receiving, processing, and sending nerve impulses in an integrated and sophisticated fashion. The endocrine and nervous systems are similar in several important ways, particularly in that both employ molecular messengers for their function. However, the nervous system tends to act very quickly with short-lasting effects and the endocrine system acts more slowly with longer lasting effects.

The nervous system is composed of a network of nerve cells joined together, somewhat analogous to a telephone system. The scope of nervous system communication covers a broad range, from communication within an individual nerve cell through communication between adjacent nerve cells, communication between different parts (organs) of the body, and communication between individuals. The fundamental function of the nervous system is to receive information, process and integrate it, and pass along information that causes a change in physiology.

The fundamental units of the nervous systems are specialized cells that transmit nerve impulses called **neurons**, surrounded by and supported physically and physiologically by cells called **glial cells**. The main body of a neuron is called the **soma**, attached to which are branched **dendrites** that pick up and transmit stimuli from the surrounding environment or from other nerve cells to the soma. Nerve signals are passed away from the nerve cell through a long structure attached to the soma called the **axon**, at the end of which are knoblike structures called **boutons** that act to transmit signals between cells. The junction through which nerve impulses are transmitted between cells is called the **synapse**. Within a neuron, information is received by dendrites, processed by the soma, and passed on for action by the axon and boutons. A bundle of axons and associated cells constitute a **nerve**.

Neurons pass along bits of information by impulses called **nerve impulses**. These impulses are in a sense electrochemical, in that they involve the transfer of electrical charge by way of ionic chemical species that flow into and out of the cell. It is beyond the scope of this book to discuss the nature of these impulses. Basically, however, they depend upon a difference in the electrical charge between the interior of the neuron and the extracellular fluid surrounding it. This is accomplished by the differential passage of K^+ and Na^+ ions through the cell membrane as controlled by proteins that regulate the passage of these ions. At rest, the inside of the neuron is slightly negative and the outside relatively more positive. When a nerve impulse occurs, sodium ions are allowed to leak into the cell, neutralizing the potential difference and actually causing a charge reversal. The membrane then becomes impermeable to sodium ions again, potassium ions are pumped out of the cell interior, and the resting potential is restored.

Control and Molecular Messengers in Plants

Unlike animals, plants cannot move to accommodate environmental stress and stimuli, such as light or temperature. Instead, they react by growth. For example a plant rooted in partial shade will grow toward the source of sunlight. As is the case with animals, plants respond by way of molecular messengers, most of which can be classified as **growth hormones**. Plant growth hormones act in often complex ways that may involve several hormones performing in concert. In addition to stimulating cell division and growth, they can inhibit growth and promote maturation. They act on cells to cause the cells to divide or to reproduce.

There are five major categories of plant hormones: (1) Auxins, (2) gibberellins, (3) abscisic acid, (4) cytokinins, and (5) ethylene. Of these the simplest is **ethene** (ethylene), chemical formula,

$$\begin{matrix} H & & & & H \\ & \diagdown & & \diagup & \\ & & C{=}C & & \\ & \diagup & & \diagdown & \\ H & & & & H \end{matrix}$$

which promotes the processes associated with maturation. These include the ripening and dropping of fruit and the detachment of leaves and flowers. Ethylene is transported through the air as a gas. **Auxins** are transported through the plant vascular system downward from the tip of the plant shoot. Auxins augment and orient root and shoot growth, but inhibit growth of lateral buds. They act in the opposite way from ethylene in preventing leaves and fruits from dropping. **Gibberellins** move both upward and downward in the plant vascular system. These substances are involved in growth and lengthening of leaves, stems, branches, and fruits; germination of grass seeds, flowering, and fertilization. **Abscisic acid**, which moves over short distances in leaves and fruits, inhibits growth by acting counter to growth hormones. This substance causes dormancy to occur and maintains it. **Cytokinins** are growth and cell division promoters transferred upward from roots by the plant vasculary system. These substances also inhibit aging of leaves (leaf senescence).

3.9. NUTRITION AND METABOLISM

Cells and the organisms that they constitute depend for their existence on a continual flow of energy and materials (water, nutrients, salts, waste products). **Metabolism**, the ways in which organisms process and convert energy and materials, includes *photosynthesis* by which plants capture solar energy and convert it to chemical energy; *digestion* by which complex molecules are broken down to simpler forms that the organism can utilize; *respiration* by which molecules used as "fuel" are oxidized so that the organism can extract needed energy; and *synthesis*, the process of building up complex life molecules from simpler ones. **Nutrients** are substances that organisms must obtain from their surroundings to provide needed energy and structural raw material. The process by which nutrients are obtained is called **nutrition**. This section briefly discusses nutrition and metabolism.

Metabolism

Figure 3.5 summarizes the essential features of metabolism. Virtually all aspects of metabolism are mediated by enzymes, which are discussed in greater detail in Section 3.15. The two major divisions of metabolism are **catabolism**, degradative metabolism, which breaks macromolecules down to their small monomeric constituents, and **anabolism**, synthetic metabolism in which small molecules are assembled into large ones. The former releases energy, part of which is expended as waste heat and the remainder of which is retained as energy that can be used by cellular processes. Useable energy released by catabolism and carried by molecules with high-energy bonds is utilized by energy-consuming anabolic processes to build macromolecules required for cell function and reproduction. The substance most clearly identified with energy transfer processes in living organisms is **adenosine triphosphate, ATP**, a molecule composed of ribose sugar, nitrogen-containing adenine, and phosphate groups. This high-energy molecule can move from one cell to another, where it hydrolyzes to lower-energy **adenosine diphosphate, ADP**, so that the energy released is taken up by the second cell. Much of the activity involved in metabolism has to do with using energy from an energy source to synthesize ATP from ADP, then releasing energy from the ATP after it has moved to where it is needed.

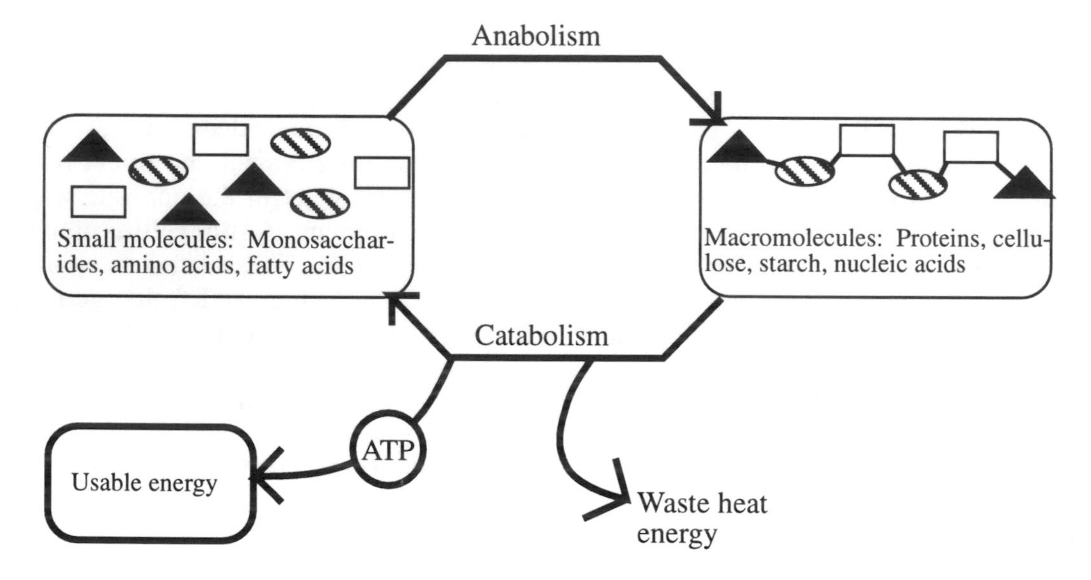

Figure 3.5. Metabolism and energy production.

The process by which fuel molecules react with oxygen and are broken down to carbon dioxide and water to release energy is called **respiration**; more properly, with oxygen as the oxidant, it is called *aerobic respiration*. The overall biochemical reaction for aerobic respiration may be written as,

$$C_6H_{12}O_6 + 6O_2 \rightarrow 6CO_2 + 6H_2O + \text{energy} \quad (3.9.1)$$

To a large extent, respiration occurs in special organelles in cells called **mitochondria**. **Fermentation**, like respiration, is a process by which energy is extracted from food. It differs from respiration in not having a so-called electron transport chain, and yields less energy as a consequence. Yeasts produce ethanol from sugars by respiration:

$$C_6H_{12}O_6 \rightarrow 2CO_2 + 2C_2H_5OH \quad (3.9.2)$$

Photosynthesis

The metabolic process upon which virtually all others depend is **photosynthesis**. Photosynthesis traps energy from the sun and converts it to energetic carbohydrate molecules (glucose, $C_6H_{12}O_6$). It also incorporates water and carbon dioxide into biomass. The basic chemical reaction for photosynthesis may be written as,

$$6CO_2 + 6H_2O + h\nu \rightarrow C_6H_{12}O_6 + 6O_2 \qquad (3.9.3)$$

where $h\nu$ represents the energy of a photon of light. Photosynthesis can be divided into two main phases. In the first of these, **light-dependent** reactions in light absorbing pigments are utilized to boost electrons to higher energy levels; this energy is subsequently used to generate high-energy molecules, such as ATP mentioned above. The high-energy molecules generated in light-dependent reactions are used to make glucose in **light-independent (dark) reactions**. The net result is the generation of glucose, which other organisms can use for their metabolic needs.

Photosynthesis is in a sense the mirror image of aerobic respiration discussed above. Photosynthesis captures nonchemical energy from sunlight in the form of chemical energy, and respiration enables conversion of chemical energy to work and heat energy. Furthermore, both occur in specialized organelles in cells—respiration in mitochondria, and photosynthesis in chloroplasts (see below).

The key structures in photosynthesis are special organelles in leaves called **chloroplasts**. Both light-dependent and light-independent reactions occur in chloroplasts. In a sense, chloroplasts are solar cells that produce a carbohydrate fuel product upon which all nonphotosynthetic organisms depend for their existence.

Nutrition

The chemical nutrients needed by plants are generally relatively simple materials that can be divided into the two main categories of macronutrients and micronutrients. **Macronutrients** are those elements that occur in substantial levels in plant materials or in fluids in the plant. The elements generally recognized as essential macronutrients for plants are carbon, hydrogen, oxygen, nitrogen, phosphorus, potassium, calcium, magnesium, and sulfur. Carbon, hydrogen, and oxygen are obtained from the atmosphere and water. The other essential macronutrients must be obtained from soil. Of these, nitrogen, phosphorus, and potassium are the most likely to be lacking and are commonly added to soil as fertilizers.

Plant **micronutrients** are elements that are essential only at very low levels. Boron, chlorine, copper, iron, manganese, molybdenum (for fixing atmospheric nitrogen into chemically-bound N), sodium, vanadium, and zinc are considered essential plant micronutrients. Most of these elements function as components of essential enzymes. Manganese, iron, chlorine, zinc, and vanadium may be involved in photosynthesis.

Nutrients required by heterotrophic animals are more complex than those needed by autotrophic plants. The macronutrients required by animals are carbohydrates, proteins, and lipids, types of biological materials discussed in Sections 3.12–3.14. Proteins are used largely as structural matter, such as in muscles, whereas carbohydrates and lipid fats and oils are primarily utilized for energy production. Animals also require micronutrients in the form of small-molecule vitamins and minerals, such as iron, phosphorus, and selenium.

3.10. REPRODUCTION AND HEREDITY

Reproduction, the process by which organisms produce new members of their species, is a driving force in life. The simplest type of reproduction is **asexual** in which a single cell splits into two identical cells. More advanced multicelled organisms usually undergo **sexual reproduction**, in which egg cells from the female unite with sperm cells from the male to form a new cell that grows into a new individual of the species.

Reproduction is a crucial aspect of the environmental sciences for several reasons. One of these is that the drive to reproduce, essential for survival, particularly in a hostile world where most individuals met with early and untimely demise, is a major threat to the environment because of the pressures of overpopulation. Secondly, a number of environmental conditions and toxicants threaten reproduction of some species under some circumstances. Finally, environmental toxicants may cause undesirable alterations in heredity.

Living organisms are characterized by specific physical, chemical, and behavioral traits. The most specific of these are those associated with the chemical nature of DNA, which directs reproduction. The development of specific traits is guided by **genes**, which occur on molecules of DNA. Changes in DNA structure in its transfer from parent to offspring result in **mutations**. The very small fraction of mutations that are potentially beneficial can give rise to desirable heritable characteristics in offspring through the process of *natural selection*.

3.11. BIOCHEMISTRY

Biochemistry is that branch of chemistry that deals with the chemical properties, composition, and biologically mediated processes of complex substances in living systems. Biochemistry is discussed in the remainder of this chapter, based on the background of general and organic chemistry from Chapter 2 and biology discussed above. Biochemical processes not only are profoundly influenced by chemical species in the environment, they largely determine the nature of these species, their degradation, and even their syntheses, particularly in the aquatic and soil environments.

Biomolecules

The biomolecules that constitute matter in living organisms are often polymers with molecular masses of the order of a million or even larger. As discussed later in this chapter, they may be divided into the categories of carbohydrates, proteins, lipids, and nucleic acids. Proteins and nucleic acids consist of macromolecules, lipids are usually relatively small molecules, and carbohydrates range from relatively small sugar molecules to high molecular mass **macromolecules** such as those in cellulose.

3.12. PROTEINS

Proteins are nitrogen-containing organic compounds, which are the basic units of life systems. **Simple proteins** contain only amino acids, whereas **conjugated proteins** also contain groups other than amino acids, such as carbohydrates or lipids. Cytoplasm, the jellylike liquid filling the interior of cells is made up largely of protein. Enzymes, which act as catalysts of life reactions, are specialized proteins; they are discussed later in the chapter. **Amino acids** joined together in huge macromolecular chains to make up proteins are organic compounds that contain the carboxylic acid group, $-CO_2H$, and the amino group, $-NH_2$. Proteins are macromolecular polymers of amino acids containing from approximately forty to several thousand amino acid groups joined by peptide linkages (discussed later in this section and illustrated in Figure 3.6). Smaller molecule amino acid polymers, containing only about ten to about forty amino acids per molecule, are called **polypeptides**.

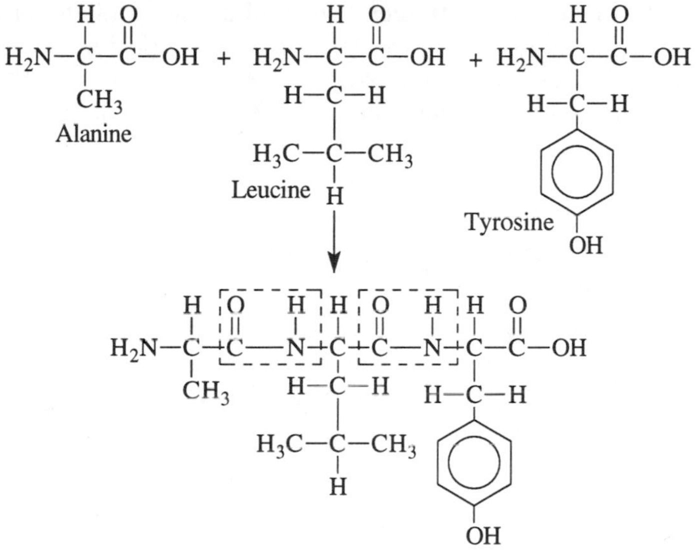

Alanine

Leucine

Tyrosine

Figure 3.6. Condensation of alanine, leucine, and tyrosine to form a tripeptide consisting of three amino acids joined by peptide linkages (outlined by dashed lines).

The 20 common natural amino acids in proteins may all be represented by the general formula,

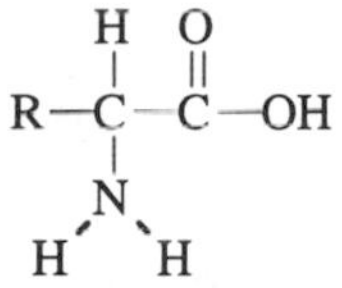

shown here with uncharged $-NH_2$ and $-CO_2H$ groups, which have a strong tendency to exchange H^+ ion and produce the **zwitterion** form as shown for glycine, below. In the general structure for amino acids above, the $-NH_2$ group is always bonded to the "alpha" carbon next to the $-CO_2H$ group, so natural amino acids are alpha amino acids. Other groups, designated as "R," are attached to the basic alpha amino acid structure. The R groups may be as simple as an atom of H found in glycine,

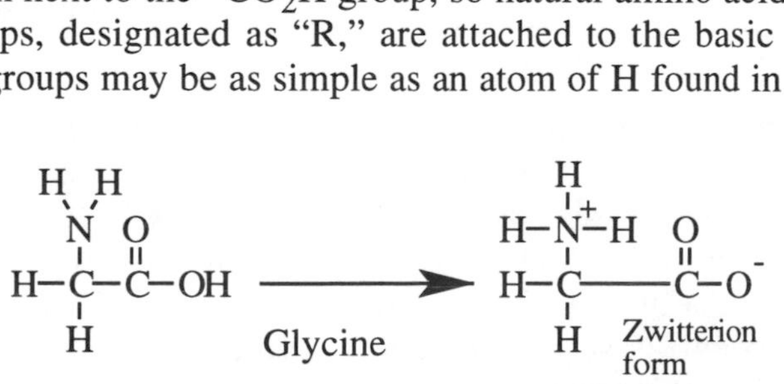

Glycine

Zwitterion form

or, they may be as complicated as the structure below, which constitutes the R group in tryptophan:

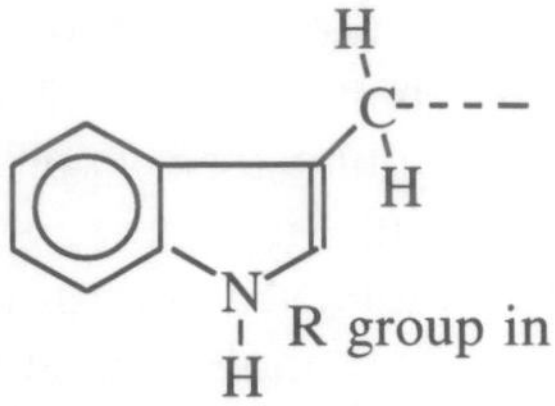

R group in tryptophan

The formation of **peptide linkages** that bind amino acids together in proteins is a condensation process involving the loss of water. Consider as an example the condensation of alanine, leucine, and tyrosine shown in Figure 3.6. When these three amino acids join together, two water molecules are eliminated. The product is a *tri*peptide since there are three amino acids involved. The amino acids in proteins are linked as shown for this tripeptide, except that many more monomeric amino acid groups are involved.

Proteins may be divided into several types that have widely varying functions. These are given in Table 3.2.

Table 3.2. Major Types of Proteins

Protein type	Example	Function and characteristics
Nutrient	Casein (milk protein)	Food
Storage	Ferritin	Storage of iron in animal tissues
Structural	Collagen (tendons) keratin (hair)	Structural and protective components in organisms
Contractile	Actin, myosin in muscle tissue	Strong, fibrous proteins that can contract and cause movement to occur
Transport	Hemoglobin	Transport inorganic and organic species in cell membranes and blood, between organs
Defense	- - -	Antibodies produced by the immune system that act against foreign agents, such as viruses
Regulatory	Insulin, human growth hormone	Regulate biochemical processes such as sugar metabolism or growth by binding to sites inside cells or on cell membranes
Enzymes	Acetylcholinesterase	Catalysts of biochemical reactions (see Section 3.15)

Protein Structure

The order of amino acids in protein molecules, and the resulting three-dimensional structures that form, provide an enormous variety of possibilites for **protein structure**. This is what makes life so diverse. Proteins have primary, secondary, tertiary, and quaternary structures. The structures of protein molecules determine the behavior of proteins in crucial areas such as the processes by which the body's immune system recognizes substances that are foreign to the body. Proteinaceous enzymes depend upon their structures for the very specific functions of the enzymes.

The order of amino acids in the protein molecule determines its **primary structure**. **Secondary protein structures** result from the folding of polypeptide protein chains to produce a maximum number of hydrogen bonds between peptide linkages:

```
 \                   /
  C=O- - -H- - -N
  |   ↑         ↑   |
  | Hydrogen bonds  |
  |   ↓         ↓   |
  N- - -H- - -O=C
 /                   \
```

Illustration of hydrogen bonds between N and O atoms in peptide linkages, which constitute protein secondary structures

Secondary structure is influenced by amino acid side chains. Small R groups enable protein molecules to be hydrogen-bonded together in a parallel arrangement. With larger R groups the molecules tend to take a spiral form. Such a spiral is known as an **alpha-helix.**

Tertiary structures are formed by the twisting of alpha-helices into specific shapes. They are produced and held in place by the interactions of amino side chains on the amino acid residues constituting the protein macromolecules. Tertiary protein structure is very important in the processes by which enzymes identify specific proteins and other molecules upon which they act. It is also involved with the action of antibodies in blood which recognize foreign proteins by their shape and react to them. This is basically the mechanism by which immunity to a disease is developed so that antibodies in blood recognize specific proteins from viruses or bacteria and reject them.

Two or more protein molecules consisting of separate polypeptide chains may be further attracted to each other to produce a **quaternary structure**.

Fibrous and Globular Proteins

Some proteins are **fibrous proteins**, which occur in skin, hair, wool, feathers, silk, muscles, and tendons. The molecules in these proteins are long and threadlike and are laid out parallel in bundles. This structure makes such proteins strong so that they can perform functions that require relatively high strengths. Fibrous proteins are quite tough and they do not dissolve in water.

Aside from fibrous protein, the other major type of protein form is the **globular protein**. These proteins are in the shape of balls and oblongs. Compared to fibrous proteins, globular proteins are relatively soluble in water. A typical, and very important globular protein is hemoglobin. Hemoglobin is the protein that occurs in red blood cells and that functions to carry oxygen in blood. Enzymes are generally globular proteins.

Denaturation of Proteins

Secondary, tertiary, and quaternary protein structures are easily changed by a process called **denaturation**. These changes can be quite damaging. Heating, exposure to acids or bases, and even violent physical action can cause denaturation to occur. The albumin protein in egg white is denatured by heating so that it forms a semisolid mass. Almost the same thing is accomplished by the violent physical action of an egg beater in the preparation of meringue. Heavy metal poisons such as lead and cadmium change the structures of proteins by binding to functional groups on the protein surface.

3.13. CARBOHYDRATES

Carbohydrates have the approximate empirical formula CH_2O and include a diverse range of substances composed of simple sugars such as glucose:

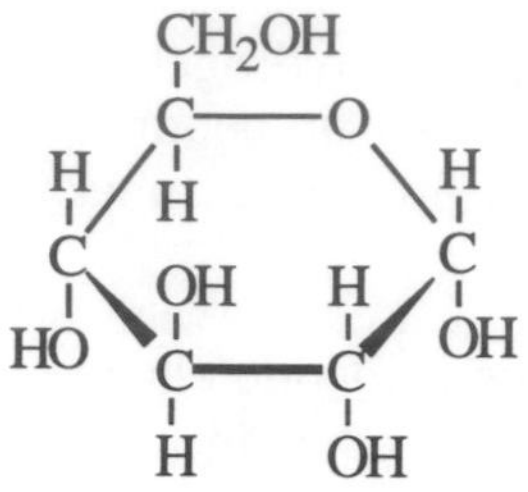

Glucose molecule

High-molecular-mass **polysaccharides**, such as starch and glycogen ("animal starch"), are biopolymers of simple sugars.

When photosynthesis occurs in a plant cell, the energy from sunlight is converted to chemical energy in a carbohydrate, $C_6H_{12}O_6$. This carbohydrate may be transferred to some other part of the plant for use as an energy source. It may be converted to a water-insoluble carbohydrate for storage until it is needed for energy. Or it may be incorporated with cell wall material and become part of the structure of the plant. If the plant is eaten by an animal, the carbohydrate is used for energy by the animal.

The simplest carbohydrates are the **monosaccharides**. These are also called **simple sugars**; or, because they have 6 carbon atoms, *hex*oses. Glucose (structural formula shown above) is the most common simple sugar involved in cell processes. Other simple sugars with the same formula but somewhat different structures are fructose, mannose, and galactose. These must be changed to glucose before they can be used in a cell. Because of its use for energy in body processes, glucose is found in the blood. Normal levels are from 65 to 110 mg glucose per 100 ml of blood. Higher levels may indicate diabetes.

Units of two monosaccharides make up several very important sugars known as **disaccharides**. When two molecules of monosaccharides join together to form a disaccharide,

$$C_6H_{12}O_6 + C_6H_{12}O_6 \rightarrow C_{12}H_{22}O_{11} + H_2O \quad (3.13.1)$$

a molecule of water is lost. (Recall that proteins are also formed from smaller amino acid molecules by condensation reactions involving the loss of water molecules.) Disaccharides include sucrose (cane sugar used as a sweetener), lactose (milk sugar), and maltose (a product of the breakdown of starch).

Polysaccharides are polymers of many simple sugar units. One of the most important polysaccharides is **starch**, which is produced by plants for food storage. Animals produce a related material called **glycogen**. The chemical formula of starch is $(C_6H_{10}O_5)_n$, where n may represent a number as high as several hundreds What this means is that the very large starch molecule consists of many units of $C_6H_{10}O_5$ joined together. For example, if n is 100, there are 6 times 100 carbon atoms, 10 times 100 hydrogen atoms, and 5 times 100 oxygen atoms in the molecule. Its chemical formula is $C_{600}H_{1000}O_{500}$. The atoms in a starch molecule are actually present as linked rings represented by the structure shown in Figure 3.7. Starch is readily digested by animals and occurs in many foods, such as bread and cereals.

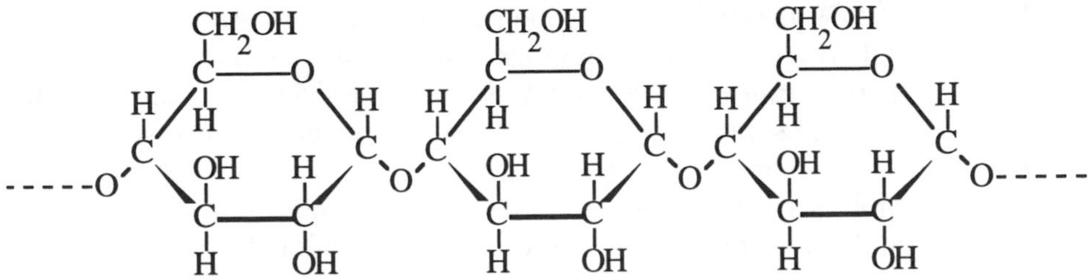

Figure 3.7. Part of a starch molecule showing units of $C_6H_{10}O_5$ condensed together.

Cellulose is a polysaccharide that is also made up of $C_6H_{10}O_5$ units. Molecules of cellulose are huge, with molecular masses of around 400,000. The cellulose structure (Figure 3.8) is similar to that of starch. Cellulose is produced by plants and forms the structural material of plant cell walls. Wood is about 60% cellulose, and cotton contains over 90% of this material. Fibers of cellulose are extracted from wood and pressed together to make paper.

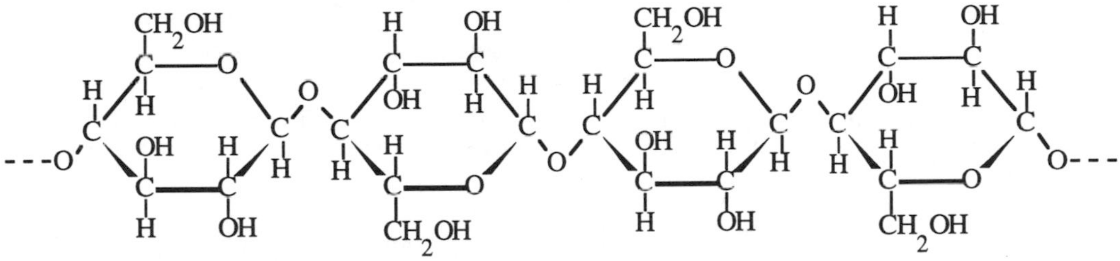

Figure 3.8. Part of the structure of cellulose.

Humans and most other animals cannot digest cellulose. Ruminant animals (cattle, sheep, goats, moose) have bacteria in their stomachs that break down cellulose into products that can be used by the animal. Chemical processes are available to convert cellulose to simple sugars by the reaction

$$\underset{\text{cellulose}}{(C_6H_{10}O_5)n} + nH_2O \rightarrow \underset{\text{glucose}}{nC_6H_{12}O_6} \tag{3.13.2}$$

where n may be 2000-3000. This involves a hydrolysis reaction in which the linkages between units of $C_6H_{10}O_5$ are broken and a molecule of H_2O is added at each linkage. Large amounts of cellulose from wood, sugar cane, and agricultural products go to waste each year. The hydrolysis of cellulose enables these products to be converted to sugars, which can be fed to animals.

Carbohydrate groups are attached to protein molecules in a special class of materials called **glycoproteins**. Collagen is a crucial glycoprotein that provides structural integrity to body parts. It is a major constituent of skin, bones, tendons, and cartilage.

3.14. LIPIDS

Whereas carbohydrates and proteins are characterized predominately by the monomers (monosaccharides and amino acids) of which they are composed, **lipids** are defined by their physical characteristic of organophilicity. The most common lipids are fats and oils composed of **triglycerides** formed from the alcohol glycerol, $CH_2(OH)CH(OH)CH_2(OH)$, and a long-chain fatty acid such as stearic acid, $CH_3(CH_2)_{16}C(O)OH$ (Figure 3.9). Many other biological materials, including waxes, cholesterol, and some vitamins and hormones, are classified as lipids. Common foods, such as butter and salad oils, are lipids. The longer-chain fatty acids, such as stearic acid, are also organic-soluble and are classified as lipids. Constituting a diverse group of biomolecules, lipids are defined as substances that can be extracted from plant or animal matter by organic solvents, such as chloroform or toluene.

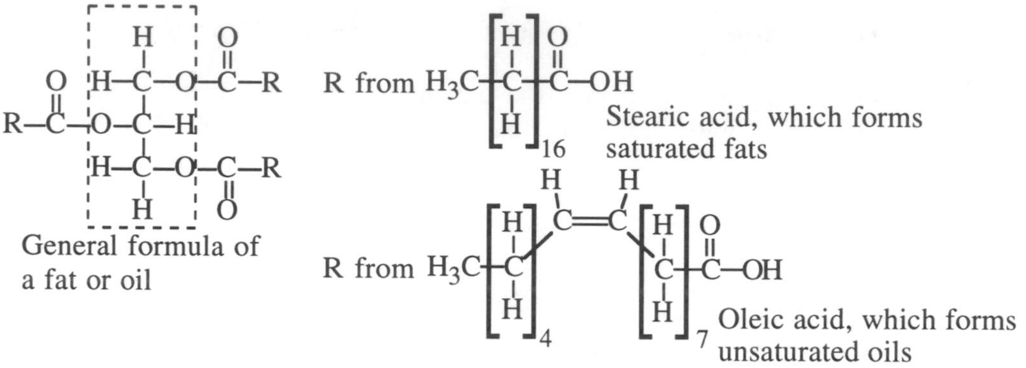

General formula of a fat or oil

R from $H_3C-[CH_2]_{16}-C(O)-OH$ — Stearic acid, which forms saturated fats

R from $H_3C-[CH_2]_4-CH{=}CH-[CH_2]_7-C(O)-OH$ — Oleic acid, which forms unsaturated oils

Figure 3.9. General formula of triglycerides, which make up fats and oils. The glycerol alcohol group is outlined by the dashed line, and the R group is from a fatty acid and is a hydrocarbon chain, such as $-(CH_2)_{16}CH_3$.

An important class of lipids consists of **phosphoglycerides** (glycerophosphatides). These compounds may be regarded as triglyderides in which one of the acids bonded to glycerol is orthophosphoric acid. These lipids are especially important because they are essential constituents of cell membranes consisting of bilayers in which the hydrophilic phosphate "heads" of the molecules are on the outside of the membrane and the hydrophobic "tails" of the molecules are on the inside.

Waxes are also esters of fatty acids. However, the alcohol in a wax is not glycerol and is often a very long chain alcohol. For example, one of the main compounds in beeswax is myricyl palmitate,

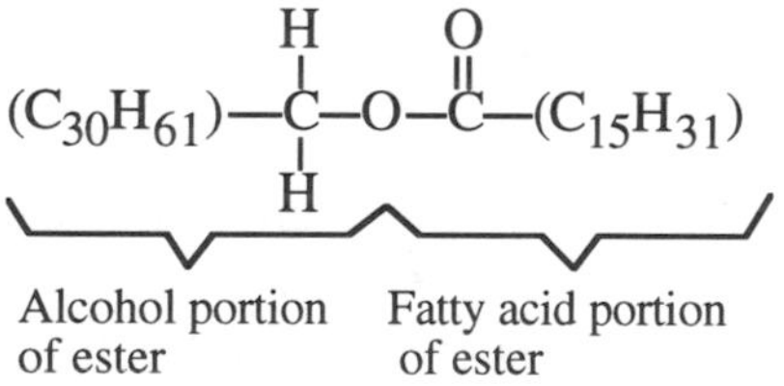

$$(C_{30}H_{61})-CH_2-O-C(O)-(C_{15}H_{31})$$

Alcohol portion of ester   Fatty acid portion of ester

in which the alcohol portion of the ester has a very long hydrocarbon chain. Waxes are produced by both plants and animals, largely as protective coatings. Waxes are found in a number of common products, such as lanolin.

Steroids are lipids found in living systems which all have the ring system shown in Figure 3.10 for cholesterol. Steroids occur in **bile salts**, which are produced by the liver and then secreted into the intestines. Their breakdown products give feces its characteristic color. Bile salts act upon fats in the intestine. They suspend very tiny fat droplets in the form of colloidal emulsions. This enables the fats to be broken down chemically and digested. Some steroids are hormones discussed as molecular messengers in Section 3.8.

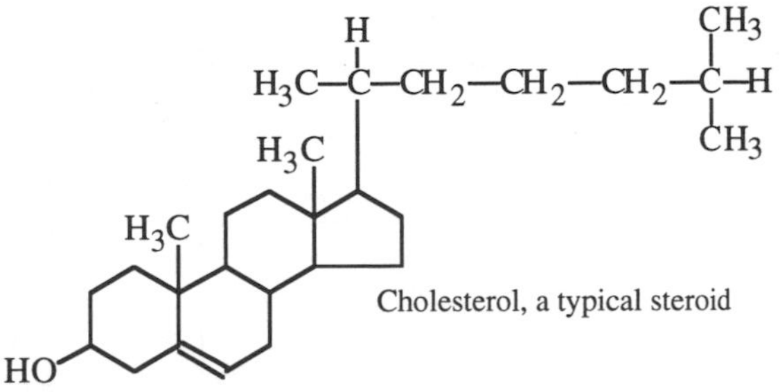

Cholesterol, a typical steroid

Figure 3.10. Steroids are characterized by the ring structure shown above for cholesterol.

3.15. ENZYMES

Catalysts (Section 2.6) are substances that speed up a chemical reaction without themselves being consumed in the reaction. The most sophisticated catalysts of all are **enzymes** found in living systems. They bring about reactions that could not be performed at all, or only with great difficulty, without them. In addition to speeding up reactions by as much as ten- to a hundred-million-fold, enzymes are extremely selective in the reactions that they promote.

Nature and Action of Enzymes

Enzymes are proteinaceous substances with highly specific structures that interact with particular substances or classes of substances called **substrates**. After they enable biochemical reactions to occur, enzymes are regenerated intact to take part in additional reactions. The extremely high specificity with which enzymes interact with substrates results from their "lock and key" action based upon their unique shapes as illustrated in Figure 3.11. This illustration shows that an enzyme "recognizes" a particular substrate by its molecular structure and binds to it to produce an **enzyme-substrate complex**. This complex then breaks apart to form one or more products different from the original enzyme, regenerating the unchanged enzyme, which is then available to catalyze additional reactions. The basic process for an enzyme reaction is, therefore,

$$\text{enzyme + substrate} \leftrightarrows \text{enzyme-substrate complex} \leftrightarrows \text{enzyme + product} \quad (3.15.1)$$

Several important things should be noted about this reaction. As shown in Figure 3.11, an enzyme acts on a specific substrate to form an enzyme-substrate complex because of the fit between their structures. As a result, something happens to the substrate molecule. For example, it might be split in two at a particular location. Then the enzyme-substrate complex comes apart, yielding the enzyme and products. Unchanged in the reaction, the enzyme is now free to react again. Note that the arrows in the formula for enzyme reaction point both ways. This means that the reaction is **reversible**. An enzyme-substrate complex can simply revert back to the enzyme and the substrate. The products of an enzymatic reaction can react with the enzyme to reform the enzyme-substrate complex again. It, in turn, may again form the enzyme and the substrate. Therefore, the same enzyme may act to cause a reaction to go either way.

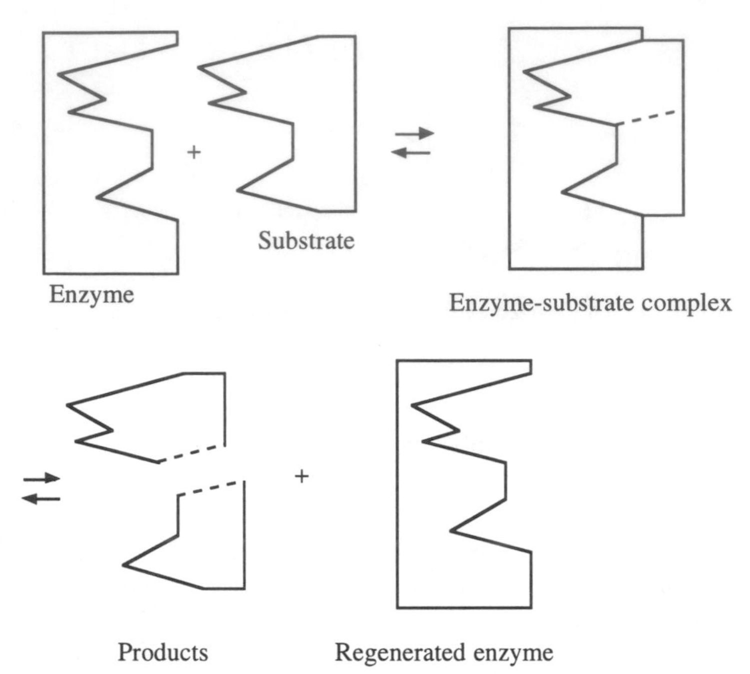

Figure 3.11. Representation of the "lock-and-key" mode of enzyme action, which enables the very high specificity of enzyme-catalyzed reactions.

Some enzymes cannot function by themselves. In order to work, they must first be attached to **coenzymes**. Coenzymes normally are not protein materials. Some of the vitamins are important coenzymes.

Types and Names of Enzymes

Enzymes are named for what they do. For example, the enzyme given off by the stomach, which splits proteins as part of the digestion process, is called *gastric proteinase*. The "gastric" part of the name refers to the enzyme's origin in the stomach. The "proteinase" denotes that it splits up protein molecules. The common name for this enzyme is pepsin. Similarly, the enzyme produced by the pancreas that breaks down fats (lipids) is called *pancreatic lipase*. Its common name is steapsin. In general, lipase enzymes cause lipid triglycerides to dissociate and form glycerol and fatty acids.

The enzymes mentioned above are **hydrolyzing enzymes**, which bring about the breakdown of high-molecular-mass biological compounds by the addition of water. This is one of the most important types of the reactions involved in digestion. The three main classes of energy-yielding foods that animals eat are carbohydrates, proteins, and fats. Recall that the higher carbohydrates humans digest are largely disaccharides (sucrose, or table sugar) and polysaccharides (starch) and that these are formed by the joining together of units of simple sugars, $C_6H_{12}O_6$, with the elimination of an H_2O molecule at the linkage where they join. Proteins are formed by the condensation of amino acids, again with the elimination of a water molecule at each linkage. Fats are esters that are produced when glycerol and fatty acids link together. A water molecule is lost for each of these linkages when a protein, fat, or carbohydrate is synthesized. In order for these substances to be used as a food source, the reverse process must occur to break down large, complicated molecules of protein, fat, or carbohydrate to simple, soluble substances that can penetrate a cell membrane and take part in chemical processes in the cell. This reverse process is accomplished by hydrolyzing enzymes.

Biological compounds with long chains of carbon atoms are broken down into molecules with shorter chains by the breaking of carbon-carbon bonds. This commonly occurs by the elimination of —CO_2H groups from carboxylic acids. For example, *pyruvic decarboxylase* enzyme acts upon pyruvic acid,

$$\underset{\text{Pyruvic acid}}{\text{H—}\overset{\text{H}}{\underset{\text{H}}{\text{C}}}\text{—}\overset{\text{O}}{\overset{\|}{\text{C}}}\text{—}\overset{\text{O}}{\overset{\|}{\text{C}}}\text{—OH}} \xrightarrow[\text{decarboxylase}]{\text{Pyruvate}} \underset{\text{Acetaldehyde}}{\text{H—}\overset{\text{H}}{\underset{\text{H}}{\text{C}}}\text{—}\overset{\text{O}}{\overset{\|}{\text{C}}}\text{—H}} + CO_2 \tag{3.15.2}$$

to split off CO_2 and produce a compound with one less carbon. It is by such one- or two-carbon reactions that long-chain compounds are eventually degraded to CO_2 in the body, or that long-chain hydrocarbons undergo biodegradation by the action of microorganims in the water and soil environments.

Oxidation and reduction are the predominant reactions for the exchange of energy in living systems. Cellular respiration, a crucial energy-yielding process in living systems, is an oxidation reaction in which a carbohydrate, $C_6H_{12}O_6$, is broken down to carbon dioxide and water with the release of energy.

$$C_6H_{12}O_6 + 6O_2 \rightarrow 6CO_2 + 6H_2O + \text{energy} \tag{3.15.3}$$

Actually, such an overall reaction occurs in living systems by a complicated series of individual steps. Some of these steps involve oxidation. The enzymes that bring about oxidation in the presence of free O_2 are called **oxidases**. In general, biological oxidation-reduction reactions are catalyzed by **oxidoreductase enzymes**.

In addition to the types of enzymes discussed above, there are many enzymes that perform miscellaneous duties in living systems. Typical of these are **isomerases**, which form isomers of particular compounds. For example, there are several simple sugars with the formula $C_6H_{12}O_6$. However, only glucose can be used directly for cell processes. The other isomers are converted to glucose by the action of isomerases. **Transferase enzymes** move chemical groups from one molecule to another, **lyase enzymes** remove chemical groups without hydrolysis and participate in the formation of C=C bonds or addition of species to such bonds, and **ligase enzyme**s work in conjunction with ATP (adenosine triphosphate, a high-energy molecule that plays a crucial role in energy-yielding, glucose-oxidizing metabolic processes) to link molecules together with the formation of bonds such as carbon-carbon or carbon-sulfur bonds.

Enzyme action may be affected by many different things. Enzymes require a certain hydrogen ion concentration to function best. For example, gastric proteinase requires the acid environment of the stomach to work well. When it passes into the much less acidic intestines, it stops working. This prevents damage to the intestinal walls, which would occur if the enzyme tried to digest them. Temperature is critical. Not surprisingly, the enzymes in the human body work best at around 98.6°F (37°C), which is the normal body temperature. Heating these enzymes to around 140°F permanently destroys them. Some bacteria that thrive in hot springs have enzymes that work best at relatively high temperatures. Other "cold-seeking" bacteria have enzymes adapted to near the freezing point of water.

A common mechanism of toxicity is the alteration or destruction of enzymes by toxic agents—as examples, cyanide, heavy metals, or organic compounds, such as insecticidal parathion. An enzyme that has been destroyed obviously cannot perform its designated function, whereas one that has been altered may either not function at all or may act improperly. Toxicants can affect enzymes in several ways. Parathion, for example, bonds covalently to the nerve enzyme acetylcholinesterase, which can then no longer serve to stop nerve impulses. Heavy metals tend to bind to sulfur atoms in enzymes (such as sulfur from the amino acid cysteine),

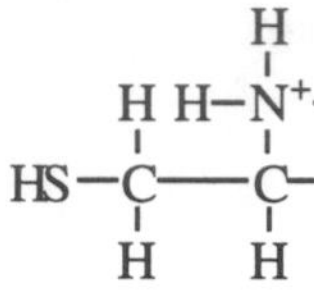

```
          H
          |
    H  H—N⁺-H  O
    |     |     ‖
HS—C——C———C—O⁻
    |     |
    H     H  Cysteine
```

thereby altering the shape and function of the enzyme. Enzymes are denatured by some poisons causing them to "unravel" so that the enzyme no longer has its crucial specific shape.

3.16. NUCLEIC ACIDS

The "essence of life" is contained in **deoxyribonucleic acid (DNA)**, a biopolymer with a molecular mass of 6–16 million atomic mass units (u) mentioned in Section 3.10, and **ribonucleic acid (RNA)**, a biopolymer with a molecular mass of 20–40 thousand u. DNA stays in the cell nucleus, whereas RNA functions in the cell cytoplasm. These substances, which are known collectively as **nucleic acids**, store and pass on genetic information that controls reproduction and protein synthesis.

The monomeric constituents of nucleic acids are pyrimidine or purine nitrogen-containing bases, two sugars, and phosphate (Figure 3.12). DNA molecules are made up of the nitrogen-containing bases adenine, guanine, cytosine, and thymine; phosphoric acid (H_3PO_4); and the simple sugar 2-deoxy-β-D-ribofuranose (commonly called deoxyribose). RNA molecules are composed of the nitrogen-containing bases adenine, guanine, cytosine, and uracil; phosphoric acid (H_3PO_4); and the simple sugar β-D-ribofuranose (commonly called ribose).

Nucleic acid polymers are composed of repeating units of **nucleotides**, each consisting of a nitrogen-containing base, phosphate, and one of the two simple sugars mentioned above, as shown by the example in Figure 3.13. In the nucleic acid the phosphate negative charges are neutralized by metal cations (such as Mg^{2+}) or positively charged proteins (histones).

Both DNA and RNA are high-molecular-mass biomolecules. DNA has a "double helix," figured out in 1953 by an American scientist, James D. Watson, and Francis Crick, a British scientist. This scientific milestone earned them the Nobel prize in 1962. DNA has a so-called double α-helix structure of oppositely wound polymeric strands held together by hydrogen bonds between opposing pyrimidine and purine groups. DNA has both a primary and a secondary structure, the former due to the sequence of nucleotides in the individual strands of DNA and the latter from the α-helix interaction of the two strands. In the secondary structure of DNA, only cytosine can be opposite guanine and only thymine can be opposite adenine. The two strands of DNA are **complementary** (Figure 3.14) so that a particular portion of one strand fits like a key in a lock with the corresponding portion of another strand. When pulled apart, each makes a new complementary strand, forming two copies of the original double helix. This occurs during cell reproduction.

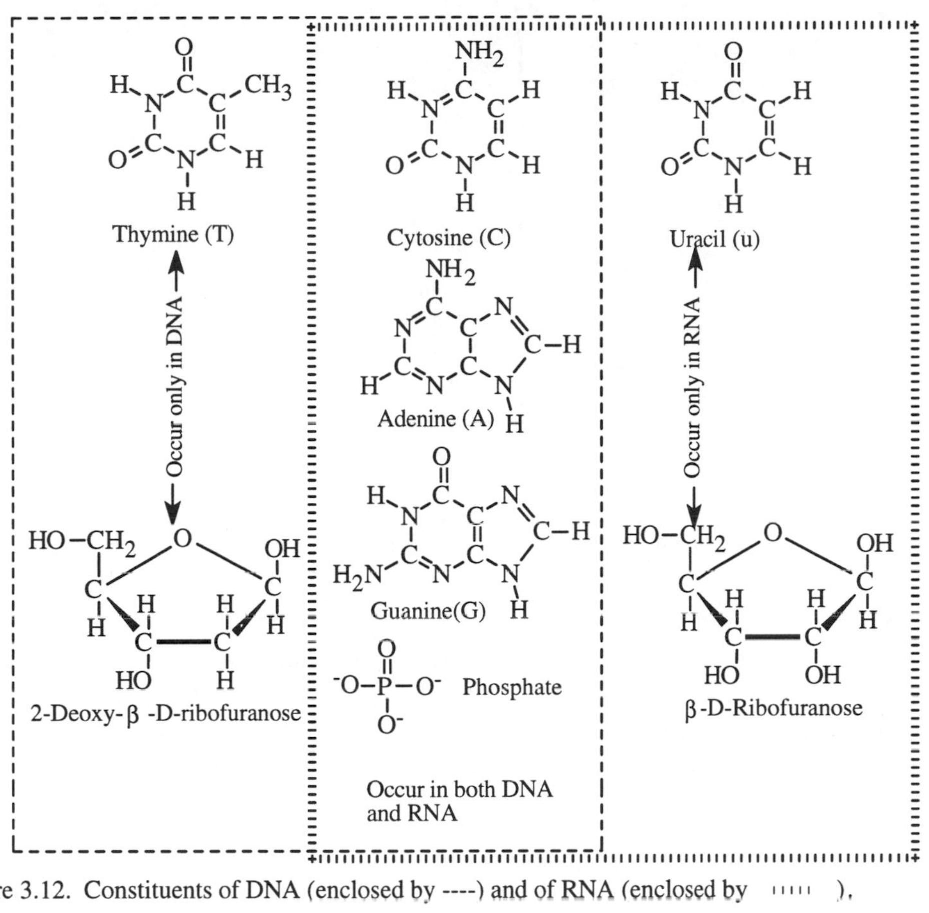

Figure 3.12. Constituents of DNA (enclosed by ----) and of RNA (enclosed by ıııı).

Bond to preceding nucleotide unit

Bond to next nucleotide unit

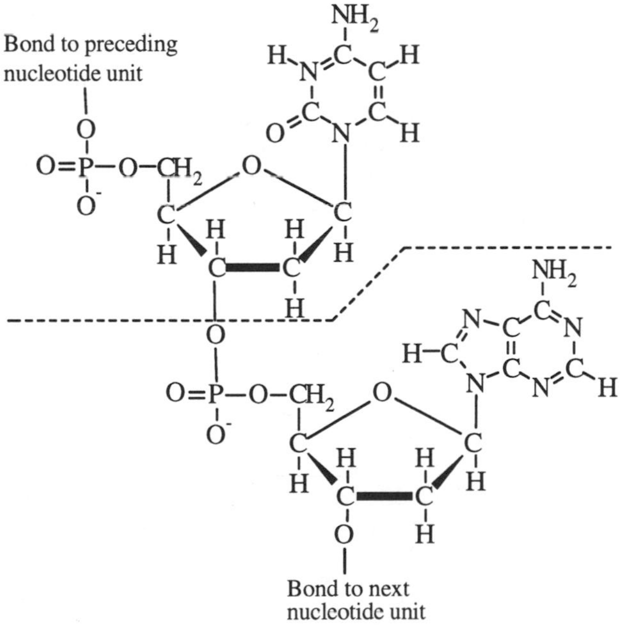

Figure 3.13. Segment of the DNA polymer showing two nucleotides (separated by a dashed line).

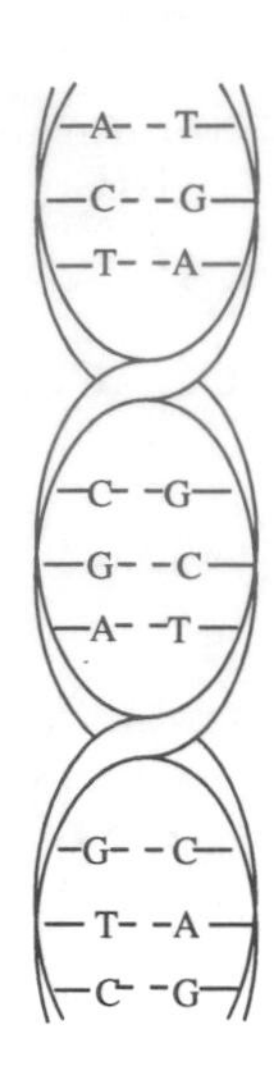

Figure 3.14. Representation of the double helix structure of DNA showing the allowed base pairs held together by hydrogen bonding between the phosphate/sugar polymer "backbones" of the two strands of DNA. The letters stand for adenine (A), cytosine (C), guanine (G), and thymine (T). The dashed lines, ---, represent hydrogen bonds.

The molecule of DNA acts as a coded message contained in and transmitted by nucleic acids. It is written by the sequence of bases from which the nucleic acids are composed, somewhat analogous to a telegraph message of dots, dashes, and spaces in between. The key aspect of DNA structure that enables storage and replication of this information is the double helix structure of DNA mentioned above.

Nucleic Acids in Protein Synthesis

Whenever a new cell is formed, the DNA in its nucleus must be accurately reproduced from the parent cell. Life processes are absolutely dependent upon accurate protein synthesis as regulated by cell DNA. The DNA in a single cell must be capable of directing the synthesis of up to 3000 or even more different proteins. The directions for the synthesis of a single protein are contained in a segment of DNA called a *gene* (Section 3.10). The process of transmitting information from DNA to a newly synthesized protein involves the following steps:

- The DNA undergoes **replication**. This process involves separation of a segment of the double helix into separate single strands, which then replicate such that guanine is opposite cytosine (and *vice versa*) and adenine is opposite thymine (and *vice versa*). This process continues until a complete copy of the DNA molecule has been produced.
- The newly replicated DNA produces **messenger RNA (m-RNA)**, a complement of the single strand of DNA, by a process called **transcription**.
- A new protein is synthesized using m-RNA as a template to determine the order of amino acids in a process called **translation**.

Modified DNA

DNA molecules may be modified by the unintentional addition or deletion of nucleotides or by substituting one nucleotide for another. The result is a *mutation* (see Section 3.10) that is transmittable to offspring. Mutations can be induced by chemical substances. This is a concern from a toxicological viewpoint because of the detrimental effects of many mutations and because substances that cause mutations often cause cancer as well. DNA malfunction may result in birth defects. The failure to control cell reproduction results in cancer. Radiation from X rays and radioactivity also disrupts DNA and may cause mutation.

CHAPTER SUMMARY

The chapter summary below is presented in a programmed format to review the main points covered in this chapter. It is used most effectively by filling in the blanks, referring back to the chapter as necessary. The correct answers are given at the end of the summary.

Five characteristics of the cell are [1] ____

____

____

Some basic constituents of cells are [2] ____________________________________

____

A crucial distinction between organisms is whether or not they can make their own food, or must rely for food on biomass, and whether or not they require molecular O_2 for metabolism. At least 4 categories of organisms based upon these distinctions are [3] ____

____.

Evidence that archaebacteria have existed from earliest times is that they exist under [4] __________________________. The most common bacterially mediated reaction in water and soil is [5] ____. Heterotrophic organisms that exist in branched strands called hyphae and digest food by excreting digestive enzymes are [6] __________________________. The animal kingdom is divided into the two major categories of [7] ______________________. The two major means of regulation and control of processes in organisms are [8] _____ ____. The five major categories of plant hormones are [9] ______________________________

____.

The ways in which organisms process and convert energy and materials are categorized as [10] ______________________. The basic chemical reaction for photosynthesis may be written as [11] ____. Changes in DNA structure that can be transferred from parent to offspring are called [12] ____________________. Biochemistry is defined as [13] ____________________

____.

Proteins are composed of smaller molecules called [14] ____________________________. The order of amino acids in the protein molecule determine its [15] ________________ structure. Structures of proteins are further divided into the categories of [16] ________ ____. Damage caused to proteins by heating, heavy metals, and other factors that disturb the protein structure is called [17] ______________________________. Biomolecules with simple formulas of CH_2O, including glucose, starch, and cellulose are called [18] ____________________

________________. Biomolecules defined more by their behavior toward solvents than by their chemical composition are called [19] ________________. The function of enzymes is [20] ________________________________. Enzymes are so specific in their action because [21] ____. Nucleic acids function to [22] ____. The three major types of small molecules that join together to produce DNA and RNA are [23] ______________________________________.

Answers

1 it (1) is composed of macromolecules containing carbon, hydrogen, oxygen, and nitrogen along with phosphorus and sulfur, (2) is capable of metabolism, (3) regulates itself, (4) interacts with its environment, and (5) reproduces

2 cell membrane, cell nucleus, cytoplasm, mitochondria, ribosomes, cell walls, vacuoles, chloroplasts

3 autotrophs, heterotrophs, aerobic, and anaerobic

4 extreme conditions

5 the oxidation of organic matter by aerobic respiration, $O_2 + \{CH_2O\} \rightarrow CO_2(g) + H_2O$

6 fungi

7 invertebrates and vertebrates

8 molecular messengers and the nervous systems of animals

9 auxins, gibberellins, abscisic acid, cytokinins, and ethylene

10 metabolism

11 $6CO_2 + 6H_2O + h\nu \rightarrow C_6H_{12}O_6 + 6O_2$

12 mutations

13 that branch of chemistry that deals with the chemical properties, composition, and biologically mediated processes of complex substances in living systems

14 amino acids

15 primary

16 secondary, tertiary, and quaternary structures

17 denaturation

18 carbohydrates

19 lipids

20 to catalyze biochemical reactions

21 they fit their substrates exactly ("lock and key" action)

22 store and pass on essential genetic information that controls reproduction and protein synthesis

23 simple sugars, cyclic nitrogenous bases, and phosphates

QUESTIONS AND PROBLEMS

1. Describe the significance of the discovery made by Watson and Crick.
2. Distinguish among aerobic, anaerobic, and facultative organisms.
3. What is the basis for the name of archaebacteria, and why do they exist in extreme environments?
4. What two factors make cyanobacteria so generally self-sufficient?
5. What is the significance of Reaction 3.3.1?
6. Write reactions for the bacterially mediated production of methane, reduction of sulfate, and denitrification.
7. How are fungi, organisms that cannot chew and ingest wood, able to use cellulose in wood as a food source?
8. What are the two major means of regulation and control in organisms?
9. In which sense are nerve impulses electrochemical phenomena
10. What mediates virtually all aspects of metabolism, that is, what kind of substance enables metabolic reactions to occur?
11. Distinguish between macronutrients and micronutrients.
12. What is the toxicological importance of lipids? How do lipids relate to hydrophobic ("water-disliking") pollutants and toxicants?
13. What is the function of a hydrolase enzyme?
14. Distinguish between eukaryotic cells and prokaryotic cells.
15. The formula of simple sugars is $C_6H_{12}O_6$. The simple formula of higher carbohydrates is $C_6H_{10}O_5$. Of course, many of these units are required to make a molecule of starch or cellulose. If higher carbohydrates are formed by joining together molecules of simple sugars, why is there a difference in the ratios of C, H, and O atoms in the higher carbohydrates as compared to the simple sugars?
16. Why does wood contain so much cellulose?
17. What would be the chemical formula of a *tri*saccharide made by the bonding together of three simple sugar molecules?
18. The general formula of cellulose may be represented as $(C_6H_{10}O_5)_x$. If the molecular weight of a molecule of cellulose is 400,000, what is the estimated value of x?
19. During one month a factory for the production of simple sugars, $C_6H_{12}O_6$, by the hydrolysis of cellulose processes one million kilograms of cellulose. The percentage of cellulose that undergoes the hydrolysis reaction is 40%. How many kg of water are consumed in the hydrolysis of cellulose each month?
20. What is the structure of the largest group of atoms common to all amino acid molecules?
21. Glycine and phenylalanine can join together to form two different dipeptides. What are the structures of these two dipeptides?

22. One of the ways in which two parallel protein chains are joined together, or cross linked, is by way of an —S—S— link. What amino acid to you think might be most likely to be involved in such a link? Explain your choice.
23. Fungi, which break down wood, straw, and other plant material, have what are called "exoenzymes." Fungi have no teeth and cannot break up plant material physically by force. Knowing this, what do you suppose an exoenzyme is? Explain how an enzyme might operate in the process by which fungi break down something as tough as wood.
24. Many fatty acids of lower molecular weight have a bad odor. Speculate as to the reasons that rancid butter has a bad odor. What chemical compound is produced that has a bad odor? What sort of chemical reaction is involved in its production?
25. The action of bile salts is a little like that of soap. What function do bile salts perform in the intestine? Look up the action of soaps, and explain how bile salts may function somewhat like soap.
26. Distinguish among the three metabolic processes of respiration, fermentation, and photosynthesis.
27. Look up the structures of ribose and deoxyribose. Explain where the "deoxy" came from in the name, deoxyribose.
28. In what respect is an enzyme and its substrate like two opposite strands of DNA?
29. Why does an enzyme no longer work if it is denatured?
30. What does the hydrophilic *vs.* hydrophobic character of toxic substances have to do with their toxicities.
31. Define metabolic processes and the two major categories into which they may be divided.
32. Distinguish among simple sugars, disaccharides and polysaccharides. List two examples of each.
33. Define lipids, triglycerides, fatty acids, waxes, and phosphoglycerides.
34. Define bile salts and suggest how they might be tied to enzyme action.

SUPPLEMENTARY REFERENCES

Geldreich, Edwin E., *Microbial Quality of Water Supply in Distribution Systems*, CRC Press/Lewis Publishers, Boca Raton, FL, 1996.

Lynn, Les, *Environmental Biology*, Kendall-Hunt, Northport, NY, 1995.

Muilenberg, Michael L. and Harriet A. Burge, *Aerobiology*, CRC Press/Lewis Publishers, Boca Raton, FL, 1996.

Subramanian, K. S. and G. V. Iyengar, Eds., *Environmental Biomonitoring: Exposure Assessment and Specimen Banking*, American Chemical Society, Washington, D.C., 1997.

Wang, Wuncheng, Joseph W. Gorsuch, and Jane S. Hughes, *Plants for Environmental Studies*, CRC Press/Lewis Publishers, Boca Raton, FL, 1997.

4 THE ANTHROSPHERE AND THE ENVIRONMENT

4.1. INTRODUCTION

This chapter addresses the technological, engineering, and industrial aspects of environmental science. It is absolutely essential to consider these areas in studying environmental science because of the enormous influence that they have on the environment. Humans will use technology to provide the food, shelter, and goods that they need for their well-being and survival. The challenge is to interweave technology with considerations of the environment and ecology such that the two are mutually advantageous.

Technology, properly applied, is an enormously positive influence for environmental protection. The most obvious such application is in air and water pollution control. Necessary as "end-of-pipe" measures are for the control of air and water pollution, it is much better to use technology in manufacturing processes to prevent the formation of pollutants. Technology is being used increasingly to develop highly efficient processes of energy conversion, renewable energy resource utilization, and conversion of raw materials to finished goods. In the transportation area, properly applied technology in areas such as high speed train transport can enormously increase the speed, energy efficiency, and safety of means for moving people and goods.

4.2. WHERE HUMANS LIVE

The dwellings of humans have an enormous influence on their well-being and on the surrounding environment. In relatively affluent societies the quality of living space has improved dramatically during the last century. Homes have become much more spacious per occupant and have become largely immune to the extremes of weather conditions. Such homes are equipped with a huge array of devices, such as indoor plumbing, climate control, communications equipment, and entertainment centers. The comfort factor for occupants has increased enormously.

The construction and use of modern homes and the other buildings in which people spend most of their time place tremendous strains on their environmental support systems and cause a great deal of environmental damage. Typically, as part of the siting and construction of new homes, shopping centers, and other buildings, the landscape is rearranged drastically at the whims of developers. Topsoil is removed, hills are cut down, and low places are filled in in an attempt to make the surrounding environment fit to a particular architectural scheme. The construction of modern buildings consumes large amounts of resources, such as concrete, steel, plastic, and glass, as well as the energy required to make synthetic building materials. The operation of a modern building requires additional large amounts of energy and of materials, such as water. It has been pointed out that all too often the design and operation of modern homes and other buildings takes place "out of the context" of the surroundings and the people who must work in and occupy the buildings.[1]

Although the above discussion of the environmental aspects of modern buildings can be construed as largely "bad news," the "good news" is that there is a large potential to design, construct, and operate homes and other buildings in a manner consistent with environmental preservation and improvement. One obvious way in which this can be done is to reevaluate the kinds of materials used in buildings. The manufacture of steel, for example, requires mining iron ore and coking coal, consumption of the energy required to make the steel, and dealing with the emissions and wastes associated with steel production. Similarly, the synthesis and fabrication of plastics used in buildings requires large amounts of petroleum and energy. Substitution of renewable materials, such as wood, and nonfabricated materials, such as quarried stone, can save large amounts of energy and minimize environmental impact. In some parts of the world sun-dried adobe blocks made from soil are practical building materials that require little energy to fabricate.

Recycling of building materials and of whole buildings can save large amounts of materials and minimize environmental damage. At a low level, stone, brick, and concrete can be used as fill material upon which new structures may be constructed. Bricks are often recyclable, and recycled used bricks often make useful and quaint materials for walls and patios. Given careful demolition practices, wood can often be recycled. Buildings can be designed with recycling in mind. This means using architectural design conducive to adding stories and annexes and to rearranging existing space. Utilities may be placed in readily accessible passageways rather than being imbedded in structural components in order to facilitate later changes and additions.

Technological advances can be used to make buildings much more environmentally friendly. Advanced window design using multiple panes and infrared-blocking glass can significantly reduce energy consumption. Modern insulation materials are highly effective. Advanced heating and air conditioning systems operate with a high degree of efficiency. Automated and computerized control of building utilities, particularly those used for cooling and heating, can significantly reduce energy consumption by regulating temperatures and lighting to the desired levels at specific locations and times in the building.

Advances in making buildings airtight and extremely well insulated can lead to problems with indoor air quality. Carpets, paints, panelling, and other manufactured components of buildings give off organic vapors, such as formaldehyde, solvents, and monomers used to make plastics and fabrics. In a poorly insulated building that is not very airtight such indoor air pollutants cause few if any problems for the building occupants. However, extremely airtight buildings can accumulate harmful levels of indoor air pollutants. Therefore, building design and operation to minimize accumulation of toxic indoor air pollutants is receiving a much higher priority.

4.3. HOW HUMANS MOVE

Few aspects of modern industrialized society have had as much influence on the environment as developments in transportation. These effects have been both direct and indirect. The direct effects are those resulting from the construction and use of transportation systems. The most obvious example of this is the tremendous effects that the widespread use of automobiles, trucks, and buses have had upon the environment. Entire landscapes have been entirely rearranged to construct highways, interchanges, and parking lots. Emissions from the internal combustion engines used in automobiles are the major source of air pollution in many urban areas.

The indirect environmental effects of widespread use of automobiles are enormous. The automobile has made possible the "urban sprawl" that is characteristic of residential and commercial patterns of development in the U.S., and in many other industrialized countries as well. The paving of vast areas of watershed and alteration of runoff patterns have contributed to flooding and water pollution. Discarded, worn-out automobiles have caused significant waste disposal problems. Vast enterprises of manufacturing, mining, and petroleum production and refining required to support the "automobile habit" have been very damaging to the environment.

On the positive side, however, applications of advanced engineering and technology to transportation can be of tremendous benefit to the environment. Modern rail and subway transportation systems, concentrated in urban areas and carefully connected to airports for longer distance travel, can enable the movement of people rapidly, conveniently, and safely, with minimum environmental damage. Although pitifully few in number in respect to the need for them, examples of such systems are emerging in progressive cities, showing the way to environmentally friendly transportation systems of the future.

The Telecommuter Society

A new development that is just now beginning to reshape the way humans move, where they live, and how they live, is the growth of a **telemcommuter society**, composed of workers who do their work at home and "commute" through their computers, modems, FAX machines, and the internet connected by way of high-speed telephone communication lines. These new technologies, along with several other developments in modern society, have made such a work pattern possible and desirable. An increasing fraction of the work force deals with information in their jobs. In principle, information can be handled just as well from a home office as it can from a centralized location, often an hour or more commuting distance from the worker's dwelling.

Actually, home and its immediate surroundings were where most work was done prior to the industrial revolution, whose assembly lines and large centralized factories demanded that workers come to a particular location for their work shifts. This work pattern and the prosperity that it brought with it resulted in the establishment of a huge suburban population that had left the cities and farms. The idyllic dream of suburbia has all too often given way to urban sprawl, traffic congestion, and long, tedious commutes. The associated environmental problems have been enormous.

Within the next approximately 10 years, it is estimated that almost 20% of the U.S. work force, a total of around 30 million people, may be working out of their homes. This tendency has been accelerated in recent years by the downsizing of many U.S. industries and "outsourcing" of work to private parties, sometimes from the companies to the workers that have been laid off. There are disadvantages, of course. People working at home may become isolated, careless in their habits, and "desocialized" from the workplace environment and the "office politics" that can play a role in advancement and training. Indeed, it is possible to visualize another type of home workplace activity in which "telecommuter counselors" advise home workers about how to manage such problems.

The changes that will result from widespread use of telecommuting are many and profound. With properly sited housing, workers can live in a rural environment with minimal disturbance to the surroundings. The flow of information has essentially no environmental costs compared to the movement of people on daily long commutes. The potential benefits to family life are obvious and may play a strong role in reshaping society positively. Whole communities may be transformed by telecommuters. Rather than being populated by people who are gone most of the day to a job far away, telecommuter communities will be occupied by people who are available on flexible schedules during the day and who are not too exhausted by long, tiring commutes to engage in civic activities at night. Telecommuters are likely to be intelligent, ambitious, and self-motivated, with all that these characteristics imply for civic activities. Cultural activities should thrive in environmentally friendly telecommuter communities combining the best of rural, suburban, and urban life.

4.4. HOW HUMANS COMMUNICATE

It has become an overworked cliché that we live in an information age. Nevertheless, the means to acquire, store, and communicate information are expanding at an incredible pace. This phenomenon is having a tremendous effect upon society and has the potential to have numerous effects upon the environment.

The major areas to consider in respect to information are its acquisition, recording, computing, storing, displaying, and communicating. Consider, for example, the detection of a pollutant in a major river. Data pertaining to the nature and concentration of the pollutant may be obtained with a combination gas chromatograph and mass spectrometer. Computation by digital computer is employed to determine the identity and concentration of the pollutant. The data can be stored on a magnetic disk, displayed on a video screen, and communicated instantaneously all over the world by satellite and fiber optic cable.

All the aspects of information and communication listed above have been tremendously augmented by recent technological advances. Perhaps the greatest such advance has been that of silicon integrated circuits. Optical memory consisting of information recorded and read by microscopic beams of laser light has enabled the storage of astounding quantities of information on a single compact disk. The use of optical fibers to transmit information digitally by light has resulted in a comparable advance in the communication of information.

The central characteristic of communication in the modern age is the combination of telecommunications with computers called **telematics**.[2] Automatic teller machines use telematics to make cash available to users at locations far from the customer's bank. Information used for banking, for business transactions, and in the media depends upon telematics.

There exists a tremendous potential for good in the applications of the "information revolution" to environmental improvement. An important advantage is the ability to acquire, analyze, and communicate information about the environment. For example, such a capability enables detection of perturbations in environmental systems, analysis of the data to determine the nature and severity of the pollution problems causing such perturbations, and rapid communication of the findings to all interested parties.

4.5. WHAT HUMANS CONSUME

Food

The most basic human need is the need for food. Without adequate supplies of food, the most pristine and beautiful environment becomes a hostile place for human life. The industry that provides food is **agriculture**, an enterprise concerned primarily with growing crops and livestock.

The environmental impact of agriculture is enormous. One of the most rapid and profound changes in the environment that has ever taken place was the conversion of vast areas of the North American continent from forests and grasslands to cropland. Throughout most of the continental United States, this conversion took place predominantly during the 1800s. The effects of it were enormous. Huge acreages of forest lands that had been stable since the last Ice Age were suddenly deprived of stabilizing tree cover and subjected to water erosion. Prairie lands put to the plow were destabilized and subjected to extremes of heat, drought, and wind that resulted in the blowing away of topsoil, culminating in the Dust Bowl of the 1930s.

In recent decades, valuable farmland has faced a new threat posed by the urbanization of rural areas. Prime agricultural land has been turned into subdivisions and paved over to create parking lots and streets. Increasing urban sprawl has led to the need for more highways. In a vicious continuing circle, the availability of new highway systems has enabled even more development. The ultimate result of this pattern of development has been the removal of once productive farmland from agricultural use.

On a positive note, agriculture has been a sector in which environmental improvement has seen some notable advances during the last 50 to 75 years. This has occurred largely under the umbrella of soil conservation. The need for soil conservation became particularly obvious during the Dust Bowl years of the 1930s, when it appeared that much of the agricultural production capacity of the U.S. would be swept away from drought-stricken soil by erosive winds. In those times and areas in which wind erosion was not a problem, water erosion took its toll. Ambitious programs of soil conservation have largely alleviated these problems. Wind erosion has been minimized by practices such as low-tillage agriculture, strip cropping in which crops are grown in strips alternating with strips of summer-fallowed crop stubble, and reconversion of marginal cultivated land to pasture. The application of low-tillage agriculture and the installation of terraces and grass waterways have greatly reduced water erosion.

Food production and consumption are closely linked with industrialization and the growth of technology. It is an interesting observation that those countries that develop high population densities prior to major industrial development experience two major changes that strongly impact food production and consumption:[3]

1. Cropland is lost as a result of industrialization; if the industrialization is rapid, increases in grain crop productivity cannot compensate fast enough for the loss of cropland to prevent a significant fall in production.
2. As industrialization raises incomes, the consumption of livestock products increases, such that demand for grain to produce more meat, milk, and eggs rises significantly.

To date, the only three countries that have experienced rapid industrialization after achieving a high population density are Japan, Taiwan, and South Korea. In each case, starting as countries that were largely self sufficient in grain supplies, these nations lost 20-30 percent of their grain production and became heavy grain importers over an approximately three-decade time period. The effects of these changes on global grain supplies and prices was relatively small because of the limited population of these countries—the largest, Japan, had a population of only about 100 million. Since approximately 1990, however, China has been experiencing economic growth at a rate of about 10% per year. With a population of 1.2 billion people, China's economic activity has an enormous effect on global markets. It may be anticipated that this economic growth, coupled with a projected population increase of more than 400 million people during the next 30 years, will result in a demand for grain and other food supplies that will cause disruptive food shortages and dramatic price increases. As an indication of what might happen, China experienced a 24 percent rate of inflation during 1994, largely due to higher prices for scarce food commodities.

In addition to the destruction of farmland to build factories, roads, housing, and other parts of the infrastructure associated with industrialization, there are other factors that tend to decrease grain production as economic activity increases. One of the major ones of these is air pollution, which decreases crop production. Water pollution can seriously curtail fish harvests. Intensive agriculture uses large quantities of water for irrigation. If groundwater is used for irrigation, aquifers may become rapidly depleted.

The discussion above points out several factors that are involved in supplying food to a growing world population. There are numerous complex interactions among the industrial, societal, and agricultural sectors. Changes in one inevitably result in changes in other sectors.

4.6. INFRASTRUCTURE

The **infrastructure** refers to the utilities, facilities, and systems used in common by members of a society and upon which the society is dependent for its normal function. The infrastructure includes both physical components—roads, bridges, and pipelines—and the instructions—laws, regulations, and operational procedures—under which the physical infrastructure operates. Parts of the infrastructure may be publicly owned, such as the U.S. Interstate Highway system and some European railroads, or privately owned, as is the case with virtually all railroads in the U.S. Some of the major components of the infrastructure of a modern society are the following:[4]

- Transportation systems, including railroads, highways, and air transport systems

- Energy generating and distribution systems
- Buildings
- Telecommunications systems
- Water supply and distribution systems
- Waste treatment and disposal systems, including those for municipal wastewater, municipal solid refuse, and industrial wastes

In general, the infrastructure refers to the facilities that large segments of a population must use in common in order for a society to function. In a sense, the infrastructure is analogous to the operating system of a computer. A computer operating system determines how individual applications operate and the manner in which they distribute and store the documents, spreadsheets, and illustrations created by the applications. Similarly, the infrastructure is used to move raw materials and power to factories and to distribute and store their output. An outdated, cumbersome computer operating system with a tendency to crash is detrimental to the efficient operation of a computer. In a similar fashion an outdated, cumbersome, broken-down infrastructure causes society to operate in a very inefficient manner and is subject to catastrophic failure.

For a society to be successful it is of the utmost importance to maintain a modern, viable infrastructure. Such an infrastructure is consistent with environmental protection. Properly designed utilities and other infrastructural elements, such as water supply systems and wastewater treatment systems, minimize pollution and environmental damage. Components of the infrastructure are subject to deterioration. To a large extent this is due to natural aging processes. Fortunately, many of these processes can be slowed or even reversed. Corrosion of steel structures, such as bridges, is a big problem for infrastructures; however, use of corrosion-resistant materials and maintenance with corrosion-resistant coatings can virtually stop this deterioration process. The infrastructure is subject to human insult, such as vandalism, misuse, and neglect. Often the problem begins with the design and basic concept of a particular component of the infrastructure. For example, many river dikes destroyed by flooding should never have been built because they attempt to thwart to an impossible extent the natural tendency of rivers to flood periodically.

Technology plays a major role in building and maintaining a successful infrastructure. Many of the most notable technological advances applied to the infrastructure were made from 150 to 100 years ago. By 1900 railroads, electric utilities, telephones, and steel building skeletons had been developed. The net effect of most of these technological innovations was to enable humankind to "conquer," or at least temporarily subdue nature. The telephone and telegraph helped to overcome isolation, high speed rail transport and later air transport conquered distance, and dams were used to control rivers and water flow.

The development of new and improved materials is having a significant influence on the infrastructure.[5] From about 1970 to 1985 the strength of steel commonly used in construction nearly doubled. During the latter 1900s significant advances were made in the properties of structural concrete. Superplasticizers enabled mixing cement with less water, resulting in a much less porous, stronger concrete product. Polymeric and metallic fibers used in concrete made it much stronger. For dams and other applications in which a material stronger than earth but not as strong as conventional concrete is required, roller-compacted concrete consisting of a mixture of cement with silt or clay has been found to be useful. The silt or clay used is obtained on site with the result that both construction costs and times are lowered.

The major challenge in designing and operating the infrastructure in the future will be to use it to work with the environment and to enhance environmental quality to the benefit of humankind. Obvious examples of environmentally friendly infrastructures are state-of-the-art sewage treatment systems, high-speed rail systems that can replace inefficient highway transport, and stack gas emission control systems in power plants. More subtle approaches with a tremendous potential for making the infrastructure more environmentally friendly include employment of workers at a computer terminal in their homes so that they do not need to commute, instantaneous electronic mail that avoids the necessity for moving letters physically, and solar electric powered installations to operate remote signals and relay stations, which avoids having to run electric power lines to them.

Whereas advances in technology and the invention of new machines and devices enabled rapid advances in the development of the infrastructure during the 1800s and early 1900s, it may be anticipated that advances in electronics and computers will have a comparable effect in the future. One of the areas in which the influence of modern electronics and computers is most visible is in telecommunications. Dial telephones and mechanical relays were perfectly satisfactory in their time, but have been made totally obsolete by innovations in electronics, computer control, and fiber optics. Air transport controlled by a truly modern, state-of-the-art computerized control system (which, unfortunately, is not yet fully installed in the U.S.) could enable present airports to handle many more airplanes safely and efficiently, thus reducing the need for airport construction. Sensors for monitoring strain, temperature, movement, and other parameters can be imbedded in the structural members of bridges and other structures. Information from these sensors can be processed by computer to warn of failure and to aid in proper maintenance. Many similar examples could be cited.

Although the payoff is relatively long term, intelligent investment in infrastructure pays very high rewards. In addition to the traditional rewards in economics and convenience, properly designed additions and modifications to the infrastructure can pay large returns in environmental improvement as well.

4.7. TECHNOLOGY AND ENGINEERING

Technology refers to the ways in which humans do and make things with materials and energy. In the modern era, technology is to a large extent the product of engineering based on scientific principles. Science deals with the discovery, explanation, and development of theories pertaining to interrelated natural phenomena of energy, matter, time, and space. Based on the fundamental knowledge of science, engineering provides the plans and means to achieve specific practical objectives. Technology uses these plans to carry out the desired objectives.

Technology has a long history, and, indeed, goes back into pre-history to times when humans used primitive tools made from stone, wood, and bone. As humans settled in cities, human and material resources became concentrated and focussed, such that technology began to develop at an accelerating pace. Technology got a tremendous boost from the discovery that metals could be worked and shaped, giving rise to the pursuit of **metallurgy**. Metallurgy probably began with processing of native elemental copper around 4000 B.C., followed by the widespread use of bronze, an alloy of tin and copper. The Bronze Age lasted until around 1200 B.C., when iron replaced bronze for tools and weapons. Other early technological innovations predating the rise of Greek and Roman civilizations included domestication of the horse, discovery of the wheel, architecture to enable construction of substantial buildings, control of water for canals and irrigation, and writing for communication.

A major advance during the Greek and Roman eras was the development of **machines**, including the windlass, pulley, inclined plane, screw, catapult for throwing missiles in warfare, and water screw for moving water. Later, the water wheel was developed for power, which was transmitted by wooden gears. Many technological innovations came from China, sometimes several hundred or even a thousand years before they appeared in Europe. As examples, China had rotating fans for ventilation by around 200 A.D., printing with wood blocks by around 740, and gunpowder about a century later. The 1800s saw an explosion in technology. Among the major advances during this century were widespread use of steam power, steam-powered railroads, the telegraph, telephone, electricity as a power source, textiles, use of iron and steel in building and bridge construction, cement, photography, and invention of the internal combustion engine, which revolutionized transportation in the following century. It may be argued that in a relative sense advances in technology during the 1800s were at least as great as those that have occurred since 1900, and certainly laid the groundwork for the vast advances that took place during the last 100 years.

Since about 1900, advancing technology has been characterized by vastly increased uses of energy; greatly increased speed in manufacturing processes, information transfer, computation, transportation, and communication; automated control; a vast new variety of chemicals; new and improved materials for new applications; and, more recently, the widespread application of computers to manufacturing, communication, and transportation. Arguably, the greatest impact during the last 100 years has been the application of electronics to technology. Electronics as it is now understood began with the discovery of the radio by Guglielmo Marconi in 1896. Electronics got an enormous boost in the early 1900s with the development of vacuum tubes that had the ability to control and amplify electronic signals and to make radio transmission possible over vast distances. A special vacuum tube, the cathode ray tube, made television and radar possible. Solid-state devices that are remarkably small and fast have supplanted vacuum tubes in virtually all applications and have made modern computers possible.

The development of electronics probably illustrates better than anything else the revolution in technology resulting from applications of basic and applied science. In modern times science and technology are inseparable and synergistic. The principles of science are applied to make technological advances possible. For example, radio communication and electronics was made possible by understanding of electromagnetic waves.

In transportation, the development of passenger-carrying airplanes has affected an astounding change in the ways in which people get around. In addition, large amounts of high-priority freight are now moved by air.

The technological advances of the present century are largely attributable to improved materials. For example, since before World War II, airliners have been made of special strong alloys of aluminum; these are being supplanted by even more advanced composites. Synthetic materials with a significant impact on modern technology include plastics, fiber reinforced materials, composites, and ceramics.

Until very recently, technological advances were made largely without heed to environmental impacts. Now, however, the greatest technological challenge is to reconcile technology with environmental consequences. The survival of humankind and of the planet that supports it now requires that the established two-way interaction between science and technology become a three-way relationship including environmental protection. That is, of course, a major theme of this book.

Engineering

Engineering uses fundamental knowledge acquired through science to provide the plans and means to achieve specific objectives in areas such as manufacturing, communication, and transportation. At one time engineering could be divided conveniently between military and civil engineering. With increasing sophistication, civil engineering evolved into even more specialized areas, such as mechanical engineering, chemical engineering, electrical engineering, and environmental engineering. Other engineering specialties include aerospace engineering, agricultural engineering, biomedical engineering, CAD/CAM (computer-aided design and computer-aided manufacturing engineering), ceramic engineering, industrial engineering, materials engineering, metallurgical engineering, mining engineering, plastics engineering, and petroleum engineering.

Mechanical engineering is the branch of engineering that deals with machines and the manner in which they handle forces, motion, and power. This discipline arose as a separate area with the development of the enormous capabilities of the steam engine in the early 1800s. The major objective of mechanical engineering is to develop and improve machines that produce goods and services. The scientific principles upon which engineering is based include consideration of the laws that govern forces and motion (dynamics); the thermodynamic laws that govern energy, power, and heat; transfer of materials and fluids; and other factors, including vibration control, materials properties, and wear minimization through proper lubrication. In addition to mechanical components, other machine components with which mechanical engineering must deal are electric, electronic, hydraulic, and fluidic components.

There are several objectives of machine design. Machines must produce whatever is needed in high quality to minimize expensive rejects. Speed of production is of utmost importance. Finally, costs must be minimized to enable adequate return on capital. An important factor involved in this endeavor is that of materials that can maintain exacting tolerances with minimum wear under sometimes severe conditions (see "Materials Science" in Section 4.8 and in Chapter 21). Control of machines is particularly important. This is especially true with greatly increased use of automation and robotics (Section 4.9 and Chapter 21). The availability of inexpensive, fast, sophisticated computers (Section 21.8) is revolutionizing control of machines and the processes that they carry out.

In the past, many of the machines and processes developed through mechanical engineering have contributed to environmental degradation and pollution. The availability of gargantuan earth-moving equipment has enabled strip-mining, destruction of wildlife habitat, and damming of natural streams. Efficient machine-equipped factories have often produced massive amounts of pollution and have provided noisy and dangerous conditions for workers. As with other branches of engineering, a major emphasis has now been placed on mechanical engineering designed to minimize environmental impact and to improve environmental quality. Examples of this include machinery designed to minimize noise, much improved energy efficiency in machines, and the uses of earth-moving equipment for environmentally beneficial purposes, such as restoration of strip-mined lands and construction of wetlands.

Electrical engineering grew from the rapidly developing electrical power industry starting in the late 1800s. The theory of electrical engineering is largely based on the mathematical formulation of the laws of electricity by the Scotsman James Clerk Maxwell in 1864, though the profession began in a primitive manner with the invention of the telegraph in 1837. Electrical engineering is concerned with the generation, transmission, and utilization of electrical energy. Of particular importance in this area has been the early utilization of alternating current, which is more complex to use, but more efficient than direct current.

Proper application of electrical engineering can be very helpful in the environmental area. Efficient generation, distribution, and utilization of electrical energy constitute one of the most promising avenues of endeavor leading to environmental improvement. The modern practice of burying electrical transmission lines in urban areas has minimized visual pollution from unsightly overhead power lines.

Electronics engineering deals with phenomena based on the behavior of electrons in vacuum tubes and other devices. As such, it is very much concerned with electromagnetic radiation ("radio waves" and microwaves) and transmission of radiofrequency signals through the air and space. Much of the early work in electronics engineering was based on fundamental studies of the behavior of electrons in vacuum tubes dating from the first decades of the 1900s. Though slow, prone to failure (because of the tendency of their hot filaments to burn out), and power consumptive by modern standards, these devices enabled development of practical radios, television, radar, and chemical instrumentation.

The fact that modern times are largely an "electronic era" is due to the invention of the solid-state electronic device known as the **transistor** in 1948. This enabled a truly remarkable development that has changed society enormously in the latter part of this century—the **silicon integrated circuit**. Integrated circuits are made by depositing and etching microscopic electronic circuits containing many transistors, capacitors, and resistors on single silicon chips as small as a pinhead and rarely larger than a dime. Although the cost of integrated circuit chips has remained about constant since 1960, the number of transistors possible on each chip has doubled regularly and may eventually reach a billion or so. This phenomenon has led to the vast array of consumer electronic devices, computers, and control systems that are a fact of modern life—and this is probably only the beginning.

The principles of electronics applied through electronic engineering for environmental improvement are enormous. Automated factories (see Section 21.6) can turn out goods with lowest possible consumption of energy and materials, while minimizing air and water pollutants and production of hazardous wastes. During the last two decades electronic control of electrical power production, distribution, and utilization has enabled greater production of light and usable energy without the construction of massive numbers of new power plants. Sophisticated electronic control and the development of electronically-based photovoltaic cells are enabling practical utilization of solar energy. Nuclear power generation, which can certainly play a major role in generating electrical energy without production of greenhouse gases (Chapter 11), can be made virtually fail-safe by electronic systems that do not doze, daydream, or have bad habits that are always potential problems with systems that rely primarily on the performance of fallible humans.

Chemical engineering uses the principles of chemical science, physics, and mathematics to design and operate processes that generate products and materials through controlled chemical reactions. Historically, a key concept of chemical engineering has been that of **unit processes**, such as mixing, distillation, filtration, and heat exchange that are combined in a variety of ways to carry out chemical processes. These in turn are based on laws of thermodynamics, chemical kinetics, mass and heat transfer, and fluid flow.

Thermodynamics, the science of heat and energy phenomena, transfers, and conversions, is of crucial importance in chemical engineering. Thermodynamics enables computation of heat required and produced in chemical production, deals with the feasibility of partition of materials between phases (see mass transfer below), and provides information regarding the effects of temperature on chemical equilibrium (degree to which a reaction goes to completion). Whereas thermodynamics deals with equilibrium situations, **chemical kinetics** deals with speeds of chemical reactions. It explains whether or not a reaction proceeds at a sufficient rate to be practical or so rapidly as to be hazardous, and how long reactants must remain together for a desired reaction to occur. A key aspect of chemical kinetics is **catalysis**, in which catalysts, which enable chemical reactions to occur, are either mixed with reactant mixtures (homogeneous catalysis) or held on solid surfaces (heterogeneous catalysis).

Mass transfer involves separation of phases of materials and transfer of materials between phases. The most straightforward example of mass transfer is distillation in which one component of a mixture is vaporized and condensed. Sorption of pollutant organic matter from aqueous solution onto activated carbon, precipitation and filtration of solids, and extraction of materials from aqueous solution into an organic solvent are all unit operations involving mass transfer. Much of chemical engineering deals with **heat transfer**, which in chemical plants is accomplished by conduction, convection, radiation, and as latent heat (condensation/evaporation, particularly of water).

Unlike other manufacturing and assembly operations that deal with the manipulation of objects and solids, virtually all of the material transferred in a chemical plant is in the form of fluids, so **fluid flow** becomes a major consideration. Consideration must be given to whether or not a fluid can be induced to flow at a sufficiently rapid rate to be practical, as well as what may be done to make a fluid flow adequately. Another major consideration is the loss of energy as fluids flow through pipes, over solid catalysts, and through chemical reactors.

Control of the processes described above is of particular importance in chemical engineering; a poorly regulated process can very rapidly get out of control and ruin the product or cause a fire or explosion. The transfers of fluids and energy that occur in chemical plants are particularly amenable to automated and computerized control.

4.8. ACQUISITION OF RAW MATERIALS

Manufacturing involves the processing of a wide variety of materials. Some materials are used with relatively little processing, whereas others are the result of highly sophisticated manufacturing processes. The acquisition and processing of materials has a number of environmental implications. There are numerous potential environmental effects of mining. These include removal and distribution of overburden in strip mining, disturbance of watersheds and aquifers, and release of pollutants to waterways. On the other hand, the development of sophisticated materials through the application of materials science has the potential to be very beneficial to the environment. As an example, underground fuel storage tanks made of polymer-reinforced fiberglass are corrosion proof and will not leak unless subjected to drastic physical insult. Their use virtually eliminates leakage, which was a significant problem with older steel tanks and makes it unnecessary to use anticorrosive coatings, which themselves pose a potential for soil and water pollution.

Raw Materials

Raw materials are discussed in greater detail in Chapter 16 and in Section 21.4. Raw materials consist of the minerals, fuel, wood, fiber, and other substances required for manufacturing processes. Such materials can be obtained from either **extractive** (nonrenewable) or from **renewable** sources.

Extractive sources of raw materials are those in which substances are dug or pumped from the Earth's crust. Such materials are irreplaceable, so that their wise use and conservation are crucial to the well-being of future generations. They include both inorganic materials and organic substances. Common examples of the former are iron ore, sulfur, and phosphate, whereas examples of organic materials extracted from Earth are coal, natural gas, and crude oil.

The prime example of a renewable material is wood. From an environmental viewpoint wood is an ideal resource. Forests conserve soil by preventing soil erosion and provide recreational areas and watersheds. The removal of carbon dioxide from air by photosynthesis carried out by trees is a significant mechanism for the removal of greenhouse-gas carbon dioxide from the atmosphere. Natural rubber and cotton are two other examples of renewable resources. The use of these substances saves irreplaceable petroleum that otherwise would be consumed in manufacturing synthetic rubber and synthetic fabrics.

Unfortunately, many resources are simply not renewable. It is impossible to grow aluminum, nickel, phosphorus, sulfur, or any of the other essential elemental resources. For these irreplaceable resources, use minimization, substitution by more abundant alternate resources, conservation, and recycling are of utmost importance.

Manufactured Materials

To an increasing degree manufactured materials with special properties are being used for structures, machines and other applications. The use of such materials has both positive and negative aspects from the environmental viewpoint. For example, most such manufactured materials require significant amounts of energy and may use irreplaceable petroleum as a raw material. On the other hand, some such materials are much more enduring and suitable than those which they replace.

The study of the synthesis, composition, properties and applications of manufactured materials is part of the discipline of **materials science**. Materials science involves a wide variety of materials, which are discussed in more detail in Section 21.5. Prominent among these are polymers, which are made of large macromolecules synthesized from smaller monomer molecules. There are natural polymeric materials, such as cotton and wood, and synthetic polymers, including polyethylene, nylon, and dacron. Ceramics include a variety of inorganic materials that usually contain silicon and oxygen and are generally formed by high-temperature processes. A fast-growing area of materials science deals with composites consisting of two or more materials, often quite dissimilar in their properties, bound together to provide a material with properties superior to those of its separate constituents. Generally composites consist of reinforcing materials, such as glass fibers or metal wires, bound in some sort of moldable matrix, such as plastic or ceramic.

4.9. MANUFACTURING

An **industry** is an enterprise that makes a kind of goods or provides a particular service needed by humans for their existence and well-being. Various industries have developed because specialization of human activities makes for the greatest efficiency in providing goods and services. The kinds of industries that a country or region has depend upon the availability of needed attributes, such as raw materials, human resouces, or availability of transport.

Classification of Industries

Industries fall into various classes. In an early stage of development, **basic-need industries** providing essentials of food, clothing, fuel, and shelter are emphasized, but become less important relative to more discretionary industries as wealth increases. The most common example of this is the high percentage of people engaged in agriculture at early stages of development, which dwindles to a very low figure (currently only about 3% in the U.S.) as the industrial and economic base becomes more developed. Somewhat arbitrarily, industries may be divided among the following categories:

- **Food production**: Agriculture and fishing
- **Extractive mineral industries** consisting of those involved with the mining of minerals, such as those used as sources of metals (energy sources are addressed in separate categories here).
- **Renewable resource industries**: Forestry, production of nonfood crops, such as cotton.
- **Renewable energy industry**, a small, but of necessity, growing industry dealing with the utilization of renewable energy resources, such as solar energy, wind power, and biomass energy.
- **Extractive energy industry** consisting of coal mining, uranium ore mining, petroleum, and natural gas.
- **Manufacturing**: Conversion of raw materials or articles to higher-value goods.
- **Construction**: Building and erection of dwellings, buildings, railroads, highways, and other components of the infrastructure.
- **Utilities**: Electricity distribution systems, natural gas.
- **Communications**: Telecommunications, media communications.
- **Transportation**: Rail, highway, air, barge, ship.
- **Wholesale and retail trade**, which provides the interface between the production of goods and their sale for consumption and use.

- **Finance**: Banks and other entities that provide the financial resources and transactions required for industries and trade.
- **Services**: Law, medicine, motels, recreation, and many others.
- **Government**: National, regional (state, provincial), city, and local entities that provide needed services and regulation.

Industrial growth in developing societies can be described by the sequence[6] of (1) traditional society devoted largely to the most fundamental economic needs of food and shelter; (2) preconditions for take-off; (3) take-off, a growth period characterized by heavy investment in industrial development; (4) drive to maturity in which the industrial base becomes mature and well diversified, and (5) age of high mass consumption, which emphasizes high consumption of consumer goods and a large service industry. It may be hoped that a sixth stage will become dominant in which societies achieve equilibrium with the environment and the resources that must sustain it. Such a stage would be characterized by limited consumption of disposable consumer goods, highly efficient utilization of energy, land use consistent with environmental harmony, zero population growth, and a high quality of life.

Manufacturing

Once a device or product is designed and developed, it must be made—synthesized or manufactured. This may consist of the synthesis of a chemical from raw materials, casting of metal or plastic parts, assembly of parts into a device or product, or any of the other things that go into producing a product that is needed in the marketplace.

Manufacturing activities have a tremendous influence on the environment. Energy, petroleum to make petrochemicals, and ores to make metals must be dug from, pumped from, or grown on the ground to provide essential raw materials. The potential for environmental pollution from mining, petroleum production, and intensive cultivation of soil is enormous. Huge land-disrupting factories and roads must be built to transport raw materials and manufactured products. The manufacture of goods carries with it the potential to cause significant air and water pollution and production of hazardous wastes. The earlier in the design and development process that environmental considerations are taken into account, the more "environmentally friendly" a manufacturing process will be.

Automation, Robotics, and Computers in Manufacturing

Three relatively new developments that have revolutionized manufacturing and that continue to do so are automation, robotics, and computers. These topics are introduced here and addressed at greater length in Chapter 21.

Automation

The use of automatic components in the performance of repetitive tasks, such as in assembly lines, defines the topic of **automation**. Automation uses mechanical and electrical devices integrated into systems to replace or extend human physical and mental activities. Primitive forms were known in ancient times; an early example consists of float devices used to control water levels in Roman plumbing systems. A key component of an automated system is the **control system**, which regulates the response of components of a system as a function of conditions, particularly those of time or location.

The simplest level of automation is **mechanization**, in which a machine is designed to increase the strength, speed, or precision of human activities. A back hoe for dirt excavation is an example of mechanization. **Open-loop**, **multifunctional** devices perform tasks according to preset instructions, but without any feedback regarding whether or how the task was done. **Closed-loop**, **multifunctional** devices use process feedback information to adjust the process on a continuous basis. The highest level of automation is **artificial intelligence** in which information is combined with simulated reasoning to arrive at a solution to a new problem or perturbation that may arise in the process.

The greatest application of automation is in manufacturing and assembly. One of two major approaches employed is **fixed automation** used for large numbers of repetitive actions over a long period of time in which the mechanical devices and electrical circuitry of the device determine what it does. Such a device has to be rebuilt to change its function in any major respect. **Programmable automation** is a more flexible approach in which instructions to the machine can be varied so that it can perform different functions.

Not all of the effects of automation on society and on the environment are necessarily good. One obvious problem is increased unemployment and attendant social unrest resulting from displaced workers. Another is the ability that automation provides to enormously increase the output of consumer goods at more affordable prices. This capability greatly increases demands for raw materials and energy, putting additional strain on the environment. To attempt to address such concerns by cutting back on automation is unrealistic, so societies must learn to live with it and to use it in beneficial ways. There are many beneficial applications of technology. Automated processes can result in much more efficient utilization of energy and materials for production, transportation, and other human needs. A prime example is the greatly increased gasoline mileage achieved during the last approximately 20 years by the application of computerized, automated control of automobile engines. Automation in manufacturing and chemical synthesis is used to produce maximum product from minimum raw material. Production of air and water pollutants and of hazardous wastes can be minimized by the application of automated processes (see Chapter 23). By replacing workers in dangerous locations, automation can contribute significantly to worker health and well-being.

Robotics

Robotics (see Section 21.7) refers to the use of machines to simulate human movements and activities. **Robots** are machines that perform such functions using computer-driven mechanical components to grip, move, reorient, and manipulate objects. Modern robots are characterized by intricately related mechanical, electronic, and computational systems. A robot can perform a variety of functions according to pre-programmed instructions that can be changed according to human direction or in response to changed circumstances.

There are a variety of mechanical mechanisms associated with robots. These are servomechanisms in which low-energy signals from electronic control devices are used to direct the actions of a relatively large and powerful mechanical system.

Robot arms may bend relative to each other through the actions of flexible joints. Specialized end effectors are attached to the ends of robot arms to accomplish specific functions. The most common such device is a gripper used like a hand to grasp objects.

Sensory devices and systems are crucial in robotics to sense position, direction, speed, and other factors required to control the functions of the robot. Sensors may be used to respond to sound, light, and temperature. One of the more sophisticated types of sensors involves a form of vision. Images captured by video camera can be processed by computer to provide information required by the robot.

Robots interact strongly with their environment. In addition to sensing their surroundings, robots must be able to respond to it in desired ways. In so doing, robots rely on sophisticated computer control. Rapid developments in computer hardware, power, and software continue to increase the ability of robots to interact with their environment. Commonly, instructions to robots are provided by computer programs programmed by humans. It is now possible in many cases to lead a robot through its desired motions and have it "learn" the sequence by computer.

Robots are now used for numerous applications. The main ones of these are for moving materials and objects, performing operations in manufacturing, assembly, and inspection. A promising use of robots is in surroundings that are hazardous to humans. For example, robots can be used to perform tasks in the presence of hazardous substances that would threaten human health and safety.

Computers

The explosive growth of digital computer hardware and software is one of the most interesting and arguably the most influential phenomena of our time. Computers have found applications throughout manufacturing. The most important of these are outlined here.

Computer-aided design (**CAD**) is employed to convert an idea to a manufactured product. Whereas innumerable sketches, engineering drawings, and physical mockups used to be required to bring this transition about, computer graphics are now used. Thus computers can be used to provide a realistic visual picture of a product, to analyze its characteristics and performance, and to redesign it based upon the results of computer analysis. The capabilities of computers in this respect are enormous. As an example, Boeing's large, extremely complex 770 passenger airliner, which entered commercial service in 1995, was designed by computer and brought to production without construction of a full-scale mockup.

Closely linked to CAD is **computer-aided manufacturing**, **CAM** which employs computers to plan and control manufacturing operations, for quality control, and to manage entire manufacturing plants. The CAD/CAM combination continues to totally change manufacturing operations to the extent that it may be called a "new industrial revolution."

The application of computers has had a profound influence on environmental concerns. One example is the improved accuracy of weather forecasting that has resulted from sophisticated and powerful computer programs and hardware. Related to this are the uses of weather satellites, which could not be placed in orbit or operated without computers. Satellites operated by computer control are used to monitor pollutants and map their patterns of dispersion. Computers are widely used in modelling to mimic complex ecosystems, climate, and other environmentally relevant systems.

Computers and their networks are susceptible to mischief and sabotage by outsiders. The exploits of "computer hackers" in breaking into government and private sector computers have been well documented. Important information has been stolen and the operation of computers has been seriously disrupted by hackers with malicious intent. Most computer operations are connected with others through the internet, enabling communication with employees at remote locations and instant contact with suppliers and customers. The problem of deliberate disruption is potentially so great that in 1997 companies throughout the world spent about $6 billion on outside experts in computer security, a figure that is expected to double around the year 2000. The U.S. Federal Bureau of Investigation (FBI) now trains its new agents in cyberspace crime, and maintains special computer crime squads in New York, Washington, and San Francisco.

Numerous kinds of protection are available for computer installations. Such protection comes in the form of both software and hardware. Special encryption software can be used to put computer messages in code that is hard to break. Hardware and software barriers to unauthorized corporate computer access, "firewalls," continue to become more sophisticated and effective.

4.10. HUMAN IMPACT AND ENVIRONMENTAL POLLUTION

The challenge facing modern, technologically based societies is to achieve and maintain a high standard of living and quality of life without ruining and indeed while enhancing the Earth support system upon which society depends for its existence. Ultimately, this has to be done on a global basis, although much can be done nationally, locally, and individually. In this respect, the education and training of people who will be directing the technology on which society operates is of utmost importance. Traditional, narrowly based education in areas such as chemistry, economics, engineering, and even ecology are not suitable to prepare people for this challenge. The development of a "booming economy" defined in a traditional sense and characterized by high rates of production, consumption, and development can be very bad for the environment and is ultimately unsustainable. Engineering solutions that do not consider environmental protection are similarly undesirable. A purely environmental solution and a "back-to-the-land" approach can result in great hardship and are unacceptable to the majority of people. What is needed, then, are systems that interweave all the aspects of a modern industrialized society with an environmentally enlightened approach.

The term **sustainable development** has been used to describe industrial development that can be sustained without environmental damage and to the benefit of all people. A "post-industrial world" built around sustainable societies has been outlined.[7] Although sustainable development has become widely used, it has been pointed out[8] that some consider the term to be "an oxymoron without substance." Clearly, if humankind is to survive with a reasonable standard of living, something like "sustainable development" must evolve in which use of nonrenewable resources is minimized insofar as possible, and the capability to produce renewable resources (for example, by promoting soil conservation to maintain the capacity to grow biomass) is enhanced.[9] This will require significant behavioral changes, particularly in limiting population growth and curbing humankind's appetite for increasing consumption of goods and energy.

CHAPTER SUMMARY

The chapter summary below is presented in a programmed format to review the main points covered in this chapter. It is used most effectively by filling in the blanks, referring back to the chapter as necessary. The correct answers are given at the end of the summary.

The development of highly efficient processes of energy conversion, renewable energy resource utilization, and conversion of raw materials to finished goods all represent [1]____. Wood, and non-fabricated materials, such as quarried stone, are [2]______________ materials that can save large amounts of energy and minimize environmental impact in building construction. Advances in making buildings airtight and extremely well insulated can lead to problems with indoor air quality because of [3]____. "Urban sprawl," paving of large areas of watershed, alteration of runoff patterns, and vast enterprises of manufacturing, mining, and petroleum production and refining are all examples of the [4]________________________ effects of the automobile. The major areas to consider in respect to information are its[5]____. The area of telematics is the combination of [6]____. Among the effects the conversion of vast areas of the North American continent from forests and grasslands to cropland that occurred in the U.S. predominantly during the 1800s are [7]____. Some notable environmental improvements in agriculture during the last several decades include [8]____. In countries that have reached a high population prior to industrialization, food production has [9]________________________________ as industrialization has occurred. The [10]__________________ refers to the utilities, facilities, and systems used in common by members of a society and upon which the society is dependent for its normal function. [11]_______________________ of steel structures, such as bridges, is a big problem for infrastructures. State-of-the-art sewage treatment systems, high-speed rail systems that can replace inefficient highway transport, and stack gas emission control systems in power plants are all examples of [12]____ infrastructures. The ways in which humans do and make things with materials and energy defines [13]___________________________. Early inventions including the windlass, pulley, inclined plane, screw, catapult for throwing missiles in warfare, and water screw for moving water are examples of [14]____. The use of fundamental knowledge acquired through science to provide the plans and means to achieve specific objectives in areas such as manufacturing, communication, and transportation defines [15]___________________________________. The branch of engineering that deals with machines and the manner in which they handle forces, motion, and power is [16]____. The modern "electronic era" began with the invention of the [17]____ in 1948. Processes such as mixing, distillation, filtration, and heat exchange that are combined in a variety of ways to carry out chemical processes on an industrial scale are [18]________________________________. The science of heat and energy phenomena, transfers, and conversions is that of [19]__________________________. Minerals, fuel, wood, fiber, and other substances required for manufacturing processes are [20]____. The study of the synthesis, composition, properties and applications of manufactured materials are covered by the discipline of [21]_______________________________________. Two or more materials, often quite unalike in their properties, bound together to provide a material with properties superior to those of its separate constituents are [22]__________________________. A(n) [23]______________________________ is an enterprise that makes a kind of goods, or provides a particular service needed by humans for their existence and well-being. Three relatively new developments that have revolutionized manufacturing and that continue to do so are [24]____. Robots are machines that simulate [25]____. The CAD/CAM stands for [26]____.

Answers

[1] positive contributions of technology to environmental improvement.

[2] renewable

[3] organic vapors used to make plastics and fabrics

[4] detrimental environmental

[5] acquisition, recording, computing, storing, displaying, and communicating

[6] telecommunications with computers

[7] forest lands were suddenly deprived of stabilizing tree cover and subjected to water erosion and plowed prairie lands were destabilized and subjected to extremes of heat, drought, and wind that resulted in the blowing away of topsoil

[8] soil conservation, low-tillage agriculture, strip cropping, and reconversion of marginal cultivated land to pasture

[9] gone down

[10] infrastructure

[11] Corrosion

[12] environmentally friendly

[13] technology

[14] machines

[15] engineering

[16] mechanical engineering

[17] transistor

[18] unit processes

19 thermodynamics
20 raw materials
21 materials science
22 composites
23 industry
24 automation, robotics, and computers
25 human movements and activities
26 computer-aided design/computer-aided manufacturing

QUESTIONS AND PROBLEMS

1. Explain "end-of-pipe" measures to control pollution. What is a better alternative?
2. Considering the material in Section 4.2, explain how improvements in the human "microenvironment" can place strains on the "macroenvironment."
3. List and explain the direct and indirect effects on the environment from a highway-based transportation system.
4. In Section 4.4 there is a discussion of "the major areas to consider in respect to information." Discuss how these areas may pertain to weather forecasting.
5. Discuss some of the adverse effects on the environment from the development of agriculture in the U.S, during the 1800s. In what sense is agriculture, itself, now being threatened?
6. What are the consequences of a country undergoing rapid industrial development *after* it has achieved a high population density.
7. What is meant by the infrastructure? List the specific parts of the infrastructure upon which you depend.
8. Compare the ways in which the development and use of computers that is now ongoing are analogous to the development and application of electricity in the late 1800s and early 1900s.
9. How is engineering related to science, and how are both related to technology?
10. In what sense may it be argued that relative advances in technology were as great during the 1800s as they have been during the 1900s.
11. Which invention in the early 1900s enabled rapid developments in electronics?
12. What device from the late 1940s has enabled the explosive growth in electronics that has occurred since then?
13. Distinguish between extractive and renewable energy resources.
14. In what sense is wood an ideal raw material environmentally?
15. Define what is meant by an industry.
16. What is automation? How do computers contribute to modern automation?
17. Distinguish between automation and robotics. How are the two related?

18. List some specific ways in which CAD/CAM is contributing to a "new industrial revolution"?

LITERATURE CITED

[1] Lenssen, Nicholas and David M. Roodman, "Making Better Buildings," Chapter 6 in *State of the World 1995*, Lester R. Brown, Ed., W. W. Norton and Company, New York, NY, 1995.

[2] Gille, Dean, "Combining Communications and Computing: Telematics Infrastructures," Chapter 10 in *Cities and their Vital Systems: Infrastructure Past, Present, and Future*, Jesse H. Ausubel, and Robert Herman, Eds., National Academy Press, Washington, D.C., 1988, pp. 233-257.

[3] Brown, Lester R., *Who Will Feed China?*, W. W. Norton and Company, New York, NY, 1995.

[4] Ausubel, Jesse H. and Robert Herman, Eds., *Cities and Their Vital Systems: Infrastructure Past, Present, and Future*, National Academy Press, Washington, D.C., 1988.

[5] Ibbs, C. William and Diego Echeverry, "New Construction Technologies for Rebuilding the Nation's Infrastructure," *Cities and Their Vital Systems: Infrastructure Past, Present, and Future*, Jesse H. Ausubel, and Robert Herman, Eds., National Academy Press, Washington, D.C., 1988, Chapter 14, pp. 294-311.

[6] Rostow, Walt W., *The Stages of Economic Growth*, 2nd ed., Cambridge University Press, Cambridge, England, 1971.

[7] Pirages, Dennis C., Ed., *Building Sustainable Societies: A Blueprint for a Post-Industrial World*, M. E. Sharpe Publishing, Armonk, NY, 1996.

[8] Mazur, Allan, "Reconsidering Sustainable Development," *Chemical and Engineering News*, January 3, 1994, pp 26-27.

[9] Meadows, Donella, Dennis Meadows, and Jørgen Randers, *Beyond the Limits: Confronting Global Collapse, Envisioning a Sustainable Future*, Chelsea Green Publishing, Post Mills, VT, 1992.

SUPPLEMENTARY REFERENCES

Gupta, Ram S., *Environmental Engineering and Science: An Introduction*, Government Institutes, Inc., Rockville, MD, 1997.

Kiely, Ger, *Environmental Engineering*, McGraw-Hill, New York, NY, 1996.

Kinni, Theodore B., *America's Best: Industry Week's Guide to World-Class Manufacturing Plants*, Wiley, New York, NY, 1996.

Miller, Richard K. and Marcia C. Rupnow, *Environmental and Waste Management Robotics*, Future Tech Surveys, Lilburn, GA, 1991.

Quattrochi, Dale A. and Michael F. Goodchild, *Scale in Remote Sensing and GIS*, CRC Press/Lewis Publishers, Boca Raton, FL, 1997.

Ray, Bill T., *Environmental Engineering*, PWS Publishers, Dracut, MA, 1995.

WATER

5 CHARACTERISTICS OF WATER AND BODIES OF WATER

5.1. INTRODUCTION

Throughout history, the quality and quantity of water available to humans have been vital factors in determining their well-being. Whole civilizations have disappeared because of water shortages resulting from changes in climate. Even in temperate climates, fluctuations in precipitation cause problems. Droughts in Africa during the 1980s caused catastrophic crop failures and starvation, and California suffered a five-year drought during the late 1980s. The summer of 1993 saw record floods along the Missouri, Mississippi, and other rivers in Minnesota, Iowa, Illinois, and Missouri. During the 1996/97 winter unprecedented flooding occurred in parts of California and in the Pacific Northwest, and near-record flooding of the Ohio River devastated thousands in March of 1997. The April, 1997, flooding of the Red River in North Dakota and Minnesota, caused by melting of a huge accumulation of winter snow and complicated by a late spring blizzard and ice jams, reached a once-in-500-year status.

Waterborne diseases, such as cholera and typhoid, killed millions of people in the past and still cause great misery in less-developed countries in which a high percentage of diseases are waterborne. Snails growing profusely in the canals distributing water from Lake Nasser created by the Aswan High Dam on the Nile River of Egypt are vectors for parasitic flatworms called blood flukes that cause debilitating shistosomiasis disease. As of 1997 it was estimated that 1 in 12 Egyptians and as many as 200 million people total in Asia, Africa, and the Caribbean region were afflicted with this malady.[1] In the complicated life cycle of the organism causing shistosomiasis, snail-borne larvae enter pores in humans exposed to warm, slow-moving water and travel to the liver, where they grow to about 1 centimenter long. Eggs from the parasite hatch in blood vessels around the intestine or bladder, from which the juvenile parasites can be released back to the water with feces or urine. Internal bleeding and cancer can result from the growth of the organism in the human body, and some victims die. Fortunately, praziquantel, a drug that became available in 1988, can cure the disease with one dose costing little more than a dime.

Ambitious programs of dam and dike construction have reduced flood damage, but they have had a number of undesirable side effects in some areas, such as inundation of farmland by reservoirs, and unsafe dams prone to failure. Problems with water-supply quantity and quality remain and in some respects are becoming more serious. These problems include increased water use due to population growth, contamination of drinking water by improperly discarded hazardous wastes (see Chapter 23), and poisoning of wildlife by water pollutants.

A report entitled "Comprehensive Assessment of the Freshwater Resources of the World" issued by the United Nations early in 1997 warned of a global shortage of water that has become a limiting factor in the economic and social development of up to 80 countries containing 40 percent of the world's population. The report noted that, whereas a 20 percent rate of utilization of water for household, agricultural, and industrial purposes is considered acceptable, more than 1 billion people live in regions where usage exceeds this rate, and a significant fraction of the population lives in countries with more than 40 percent water utilization. The report suggested that the percentage of water used in these countries for irrigation be lowered from its present level of 70 percent and that a high priority be given to conserving water, developing new supplies, and purifying more wastewater for reuse. It also suggested that water be marketed as a commodity so that free market forces would determine supply and demand.

The importance of water cannot be overemphasized. This chapter is concerned with the properties of water and bodies of water. It considers groundwater and water in rivers, lakes, estuaries, and oceans. Any consideration of aquatic environmental science requires some understanding of the sources, transport, characteristics, and composition of water. The chemical and biochemical processes that occur in water and the chemical species found in it are strongly influenced by the environment in which the water is found. The chemistry of water exposed to the atmosphere is quite different from that of water at the bottom of a lake. Microorganisms play an essential role in determining the chemical composition of water. Thus, in discussing the environmental science of water, it is necessary to consider the many general factors that influence water.

Fresh water is studied under the category of limnology. **Limnology** is the science that addresses the physics, chemistry, meteorology, and biology of fresh waters with particular emphasis on ponds and lakes. **Oceanography** deals with the special characteristics of water in oceans.

5.2. DISTRIBUTION OF WATER: THE HYDROLOGIC CYCLE

The world's water supply is found in the five parts of the **hydrologic cycle** (Figure 5.1). Most of the water is found in the oceans and a relatively large amount is also contained in the solid state as ice, snowpacks, glaciers, and the polar ice caps. Surface water is found in lakes, streams, and reservoirs. Groundwater is located in aquifers underground. Another fraction of water is present as water vapor in the atmosphere (clouds).

There is a strong connection between the *hydrosphere,* where water is found, and the *lithosphere*, or land; human activities affect both. For example, disturbance of land by conversion of grasslands or forests to agricultural land or intensification of agricultural production may reduce vegetation cover, decreasing **transpiration** (loss of water vapor by plants) and affecting the microclimate. The result is increased rain runoff, erosion, and accumulation of silt in bodies of water. The nutrient cycles may be accelerated, leading to nutrient enrichment of surface waters. This, in turn, can profoundly affect the chemical and biological characteristics of bodies of water.

The lithosphere is shaped and molded by physical and chemical processes involving water. Enormous forces are exerted by moving liquid water in rivers and solid water in glaciers as water completes the hydrologic cycle on its path back to the ocean. Rock, gravel, sand, and silt are moved in dissolved and suspended form and in the beds of rivers, and can form vast deltas where the material is deposited. Canyons are carved by rivers and moraines are left by glaciers. Groundwater dissolves mineral matter to form caverns and sinkholes.

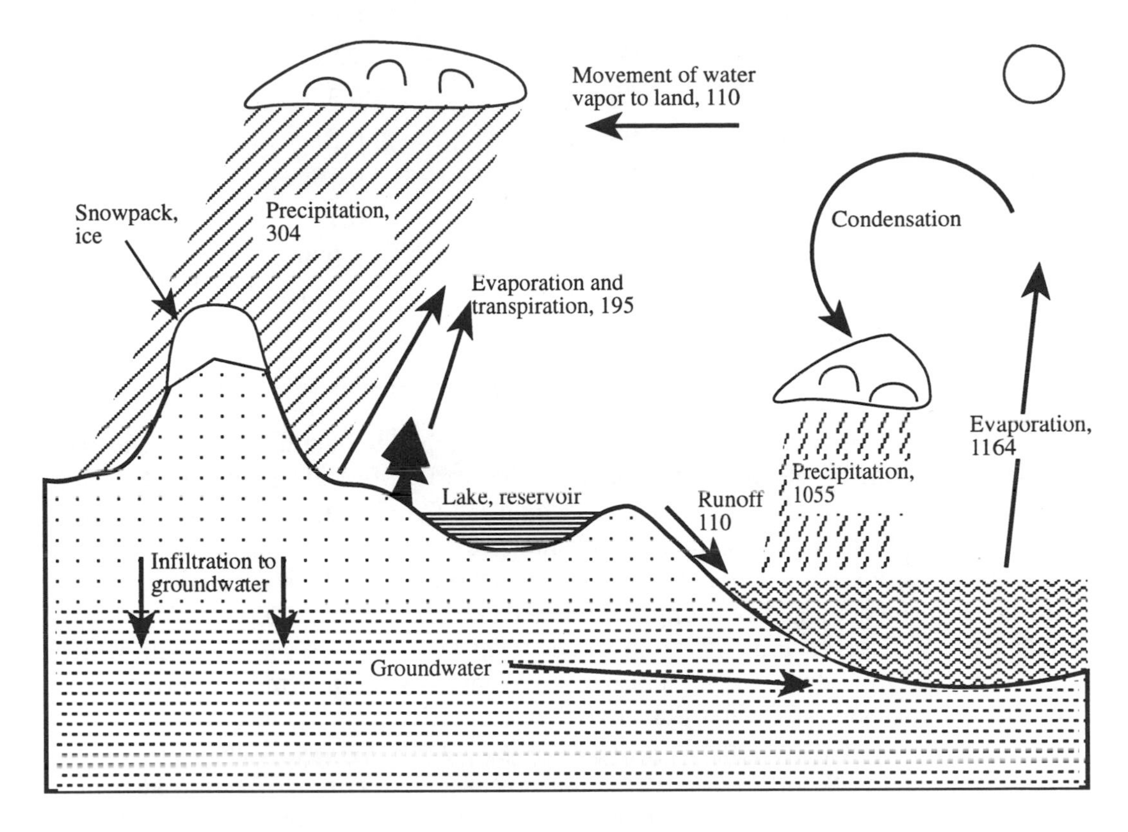

Figure 5.1. The hydrologic cycle, quantities of water in trillions of liters per day.

Distribution of Water

Excluding chemically bound water, the total amount of water on Earth is about 1.4 billion cubic kilometers (1.4 x 10^9 km^3). Of this amount, about 97.6% is present as salt water in Earth's oceans. This leaves about 33 million km^3 to be distributed elsewhere on Earth as shown in Table 5.1. Even of this amount about 87% is present in solid form, predominantly as polar snowcap and another 12% as groundwater. Therefore, just slightly over 1% of all Earth's fresh water is distributed among surface water, atmospheric water, and biospheric water. This very small fraction comprises water in lakes, including water in the Great Lakes of North America, water in all the Earth's vast rivers (Mississippi, Congo, Amazon), groundwater to a depth of 1 km, water in the atmosphere, and water in the biosphere.

Earth's water can be considered in several **compartments**. The amounts of water and the **residence times** of water in these compartments vary greatly. As noted previously, the largest of these compartments consists of the **oceans**, containing about 97% of all Earth's chemically unbound water with a residence time of about 3000 years. Oceans serve as a huge reservoir for water and as the source of most water vapor that enters the hydrologic cycle. As vast heat sinks, oceans have a tremendous moderating effect on climate.

Table 5.1. Distribution of Earth's Water Other Than Ocean Water

Location	Quantity, liters	Percent total fresh water
Snow, snowpack, ice, glaciers	2.90×10^{19}	86.9
Accessible groundwater	4.00×10^{18}	12.0
Lakes, reservoirs, ponds	1.25×10^{17}	0.37
Saline lakes	1.04×10^{17}	0.31
Soil moisture	6.50×10^{16}	0.19
Moisture in living organisms	6.50×10^{16}	0.19
Atmosphere	1.30×10^{16}	0.039
Wetlands	3.60×10^{15}	0.011
Rivers, streams, canals	1.70×10^{15}	0.0051

The majority of Earth's water not held in the oceans is bound as **snow** and **ice**. Antarctica contains about 85% of all the ice in the world. Most of the rest is contained in the permanent ice pack in the Arctic ocean and in the Greenland ice pack; a small fraction is present in mountain glaciers and snowpack.

Groundwater is water held below the surface in porous rock formations called **aquifers**. It influences, and is strongly influenced by the mineral matter with which it is in contact. It dissolves minerals from and deposits them on rock surfaces with which it is in contact. Groundwater is replenished by water flowing in from the surface, and it discharges into bodies of water that are below its level.

Bodies of fresh water include **lakes, ponds**, and **reservoirs**. Water flows from higher elevations back to the ocean through **rivers** and **streams**. The rate at which water flows in a stream is called its **discharge**. For the Mississippi River the average discharge is 50 billion liters per hour. Collectively, water in lakes, ponds, reservoirs, rivers, and streams is called **surface water**. Groundwater and surface water have appreciably different characteristics. Many substances either dissolve in surface water or become suspended in it on its way to the ocean. Surface water in a lake or reservoir that contains the mineral nutrients essential for algal growth may support a heavy growth of algae. Surface water with a high level of biodegradable organic material, used as food by bacteria, normally contains a large population of bacteria. All these factors have a profound effect upon the quality of surface water.

An environmentally important compartment of water consists of **wetlands**, in which the water table is essentially at surface level. Wetlands consist of marshes, meadows, bogs, and swamps that usually support lush plant life and a high population of animals as well. Wetlands serve as a reservoir and stabilized supply of water. They are crucial nurseries for numerous forms of wildlife.

The **atmosphere** is the smallest compartment of water and the one with the shortest residence time of about 10 days. Atmospheric water is of utmost importance in the movement of water from the oceans inland in the hydrologic cycle. The atmosphere provides the crucial precipitation that gives water, required by all land organisms, to sustain river flow, fill lakes, and replenish groundwater. Furthermore, latent heat contained in atmospheric water, and released when water vapor condenses to form rain, is a major energy transport medium and one of the main ways that solar energy is moved from the equator toward Earth's poles (see Section 9.5).

Water enters the atmosphere by **evaporation** from liquid water, **transpiration** from plants, and **sublimation** from snow and ice. Water in the atmosphere is present as water vapor and as suspended droplets of liquid and ice. Water vapor in the atmosphere is called **humidity**, and the percentage of water vapor compared to the maximum percentage that can be held at a particular temperature is called the **relative humidity**. **Condensation** occurs at a temperature called the **dew point** when water molecules leave the vapor state and form liquid or ice particles. This process is aided by the presence of **condensation nuclei** consisting of small particles of sea salt (produced by the evaporation of water from ocean spray), bacterial cells, ash, spores, and other matter upon which water vapor condenses. Condensation, alone, does not guarantee precipitation in the form of rain or snow because the condensed water vapor may remain suspended in **clouds**; precipitation occurs when the conditions are right for the cloud particles to coalesce to particles large enough to fall.

Topographical conditions can strongly influence the degree and distribution of precipitation. A striking example of this is provided by the effects of coastal mountain ranges upon precipitation. Moisture-laden air flowing in from the ocean is forced up the sides of coastal mountain ranges, cooling as it does so, and releasing rain as it becomes supersaturated. On the other side of the range, the air warms so that the level of water vapor becomes much less than the saturation concentration, the water stays in the vapor form, clouds disappear, and rain does not fall. The area of low rainfall on the leeward side of a coastal mountain range is called a **rain shadow** (see Figure 5.2).

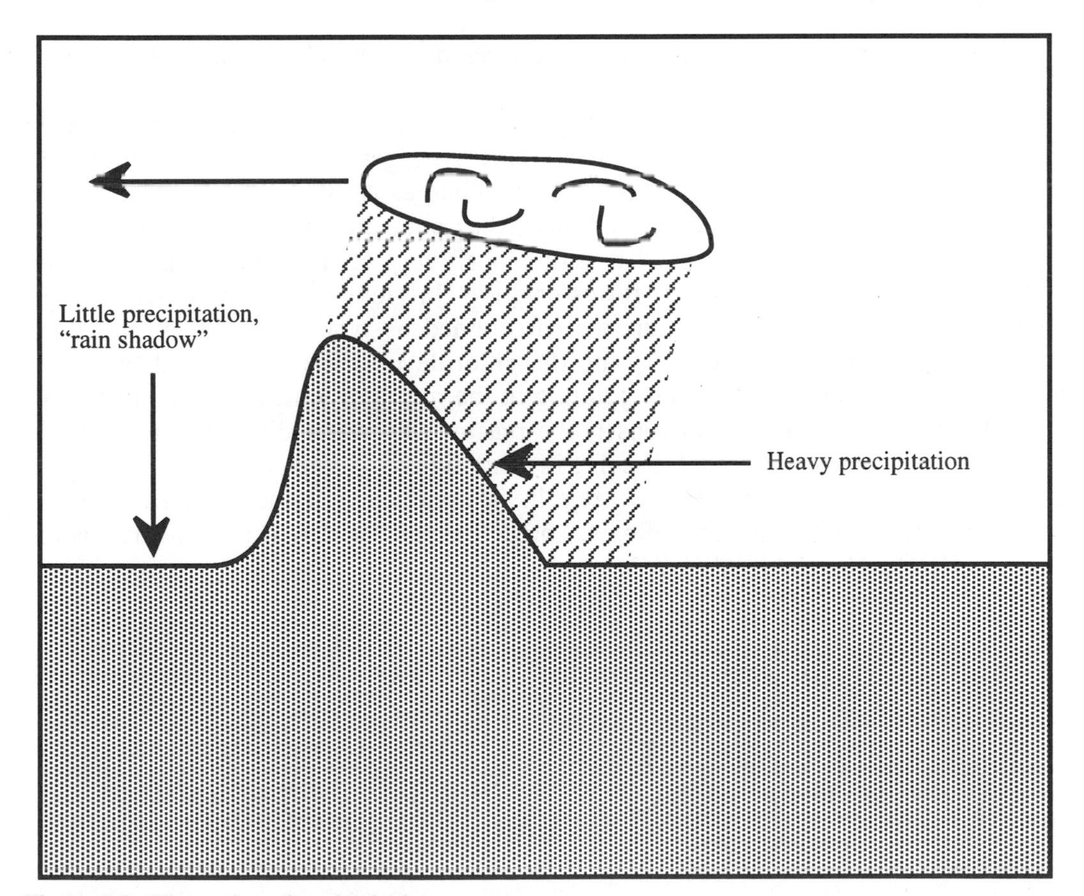

Figure 5.2. Illustration of a rain shadow.

5.3 . THE PROPERTIES OF WATER, A UNIQUE SUBSTANCE

At room temperature H_2O is a colorless, tasteless, odorless liquid. It boils at 100°C (212°F) and freezes at O°C (32°F). Water by itself is a very stable compound that is hard to decompose by heating. However, when electrically conducting ions are present in water, a current may be passed through the water, causing it to electrolyze and break up into hydrogen gas and oxygen gas.

Water is an excellent solvent for a variety of materials; these include many ionic compounds (acids, bases, salts). Some gases dissolve well in water, particularly those that react with it chemically. Sugars and many other biologically important compounds are also soluble in water. Greases and oils generally are not soluble in water but dissolve in organic solvents instead.

Some of water's solvent properties can best be understood by considering the structure and bonding of the water molecule represented below:

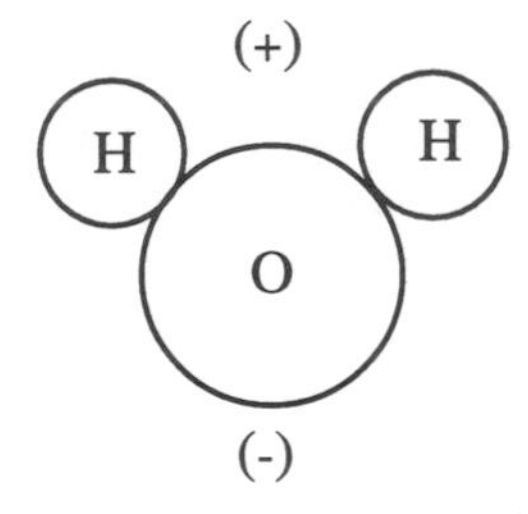

The water molecule is made up of two hydrogen atoms bonded to an oxygen atom. The three atoms are not in a straight line; instead, as shown above, they form an angle of 105°. Because of water's bent structure and the fact that the oxygen atom attracts the negative electrons more strongly than do the hydrogen atoms, the water molecule behaves like a body having opposite electrical charges at either end or pole. Such a body is called a *dipole*. Due to the fact that it has opposite charges on its two ends, the water dipole may be attracted to either positively or negatively charged ions. When NaCl dissolves in water, it forms positive Na^+ ions and negative Cl^- ions in solution. The positive sodium ions are surrounded by water molecules with their negative ends pointed at the ions, and the chloride ions are surrounded by water molecules with their positive ends pointing at the negative ions, as shown in Figure 5.3. This kind of attraction for ions is the reason why water dissolves many ionic compounds and salts that do not dissolve in other liquids. Some noteworthy examples are sodium chloride in the ocean; waste salts in urine; calcium bicarbonate, which is very important in lakes and in geological processes; and widely used industrial acids (such as HNO_3, HCl, and H_2SO_4).

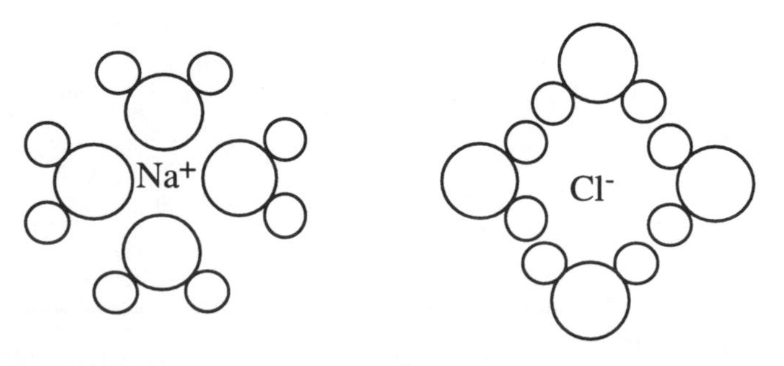

Figure 5.3. Polar water molecules surrounding Na^+ ion (left) and Cl^- ion (right).

In addition to being a polar molecule, the water molecule has another important property that gives it many of its special characteristics: the ability to form **hydrogen bonds** (see Figure 5.4). Hydrogen bonds are a special type of bond that can form between the hydrogen in one water molecule and the oxygen in another water molecule. This bonding takes place because the oxygen has a partly negative charge and the hydrogen, a partly positive charge. Hydrogen bonds, shown in Figure 5.4 as dashed lines, hold the water molecules together in large groups. Hydrogen bonds also help to hold some solute molecules or ions in solution. This happens when hydrogen bonds form between the water molecules and hydrogen or oxygen atoms on the solute molecule (see Figure 5.4).

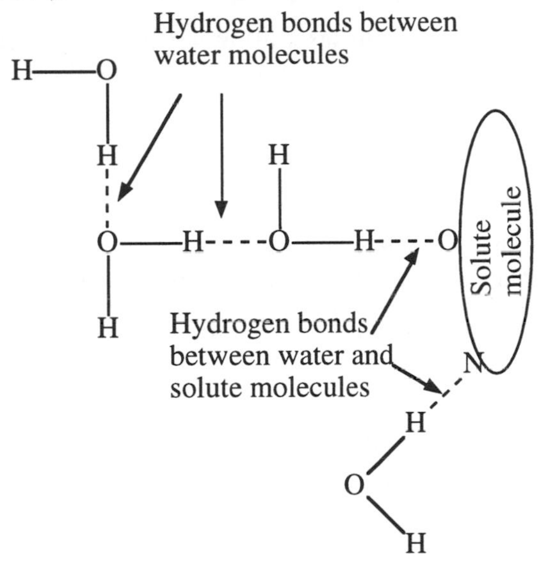

Figure 5.4. Hydrogen bonding between water molecules and between water molecules and a solute molecule in solution.

Because of its polar molecular structure and ability to form hydrogen bonds, water has a number of unique properties, including those that are essential to life. These characteristics are summarized in Table 5.2.

Water is an excellent solvent for many materials; thus it is the basic transport medium for nutrients and waste products in life processes. The extremely high dielectric constant of water relative to other liquids has a profound effect upon its solvent properties, in that most ionic materials are dissociated in water. With the exception of liquid ammonia, water has the highest heat capacity of any liquid or solid, 1 cal x g^{-1} x deg^{-1}. Because of this high heat capacity, a relatively large amount of heat is required to appreciably change the temperature of a mass of water; hence, a body of water can have a stabilizing effect upon the temperature of nearby geographic regions. In addition, this property prevents sudden large changes of temperature in large bodies of water and thereby protects aquatic organisms from the shock of abrupt temperature variations. The extremely high heat of vaporization of water, 585 cal/g at 20°C, likewise stabilizes the temperature of bodies of water and the surrounding geographic regions. It also influences the transfer of heat and water vapor between bodies of water and the atmosphere. Water has its maximum density at 4°C, a temperature above its freezing point. The fortunate consequence of this fact is that ice floats, so that few large bodies of water ever freeze solid. Furthermore, the pattern of vertical circulation of water in lakes, a determining factor in their chemistry and biology, is governed largely by the unique temperature-density relationship of water.

Table 5.2. Important Properties of Water

Property	Effects and significance
Excellent solvent	Transport of nutrients and waste products, making biological processes possible in an aqueous medium
Highest dielectric constant of any common liquid	High solubility of ionic substances and their ionization in solution
Higher surface tension than any other liquid	Controlling factor in physiology; governs drop and surface phenomena
Transparent to visible and longer-wavelength fraction of ultraviolet light	Colorless, allowing light required for photosynthesis to reach considerable depths in bodies of water
Maximum density as a liquid at 4°C	Ice floats; vertical circulation restricted in stratified bodies of water
Higher heat of evaporation than any other material	Determines transfer of heat and water molecules between the atmosphere and bodies of water
Higher latent heat of fusion than any other liquid except ammonia	Temperature stabilized at the freezing point of water
Higher heat capacity than any other liquid except ammonia	Stabilization of temperatures of organisms and geographical regions

5.4. STANDING BODIES OF WATER

The physical condition of a body of water strongly influences the chemical and biological processes that occur in water. **Surface water** occurs primarily in streams (see Section 5.5), lakes, and reservoirs. Lakes may be classified as oligotrophic, eutrophic, or dystrophic, an order that often parallels the life cycle of the lake. **Oligotrophic** lakes are deep, generally clear, deficient in nutrients, and without much biological activity. **Eutrophic** lakes have more nutrients, support more life, and are more turbid. **Dystrophic** lakes are shallow, clogged with plant life, and normally contain colored water with a low pH. **Wetlands** are flooded areas in which the water is shallow enough to enable growth of bottom-rooted plants.

Some constructed reservoirs are very similar to lakes, while others differ a great deal from them. Reservoirs with a large volume relative to their inflow and outflow are called **storage reservoirs.** Reservoirs with a large rate of flow-through compared to their volume are called **run-of-the-river reservoirs**. The physical, chemical, and biological properties of water in the two types of reservoirs may vary appreciably. Water in storage reservoirs more closely resembles lake water, whereas water in run-of-the-river reservoirs is much like river water. The effects of impounding water are addressed in Section 5.9.

Estuaries constitute another type of body of water, consisting of arms of the ocean into which streams flow. The mixing of fresh and salt water gives estuaries unique chemical and biological properties. Estuaries are the breeding grounds of much marine life, which makes their preservation very important.

Water's unique temperature-density relationship results in the formation of distinct layers within nonflowing bodies of water, as shown in Figure 5.5. During the summer a surface layer (**epilimnion**) is heated by solar radiation and, because of its lower density, floats upon the bottom layer, or **hypolimnion**. This phenomenon is called **thermal stratification.** When an appreciable temperature difference exists between the two layers, they do not mix, but behave independently and have very different chemical and biological properties. The epilimnion, which is exposed to light, may have a heavy growth of algae. As a result of exposure to the atmosphere and (during daylight hours) because of the photosynthetic activity of algae, the epilimnion contains relatively higher levels of dissolved oxygen, and is said to be *aerobic*. In the hypolimnion, bacterial action on biodegradable organic material consumes oxygen and may cause the water to become *anaerobic*, that is, essentially free of oxygen. As a consequence, chemical species in a relatively reduced form tend to predominate in the hypolimnion.

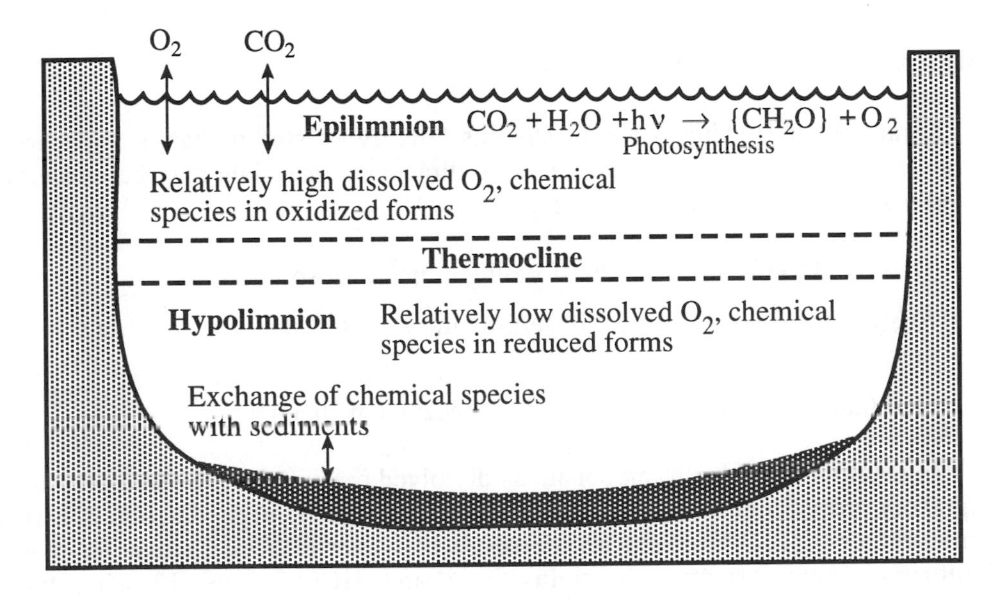

Figure 5.5. Stratification of a lake.

The shear-plane, or layer between epilimnion and hypolimnion, is called the **thermocline**. During the autumn, when the epilimnion cools, a point is reached at which the temperatures of the epilimnion and hypolimnion are equal. This disappearance of thermal stratification causes the entire body of water to behave as a hydrological unit, and the resultant mixing is known as **overturn**. An overturn also generally occurs in the spring. During the overturn, the chemical and physical characteristics of the body of water become much more uniform, and a number of chemical, physical, and biological changes may result. Biological activity may increase from the mixing of nutrients. Changes in water composition during overturn may cause disruption in water-treatment processes.

The chemistry and biology of the Earth's vast oceans are unique because of the ocean's high salt content, great depth, and other factors. Oceanographic chemistry is a discipline in its own right. The environmental problems of the oceans have increased greatly in recent years because of ocean dumping of pollutants, oil spills, and increased utilization of natural resources from the oceans.

5.5. FLOWING WATER

Surface water that flows in streams and rivers originates from precipitation that falls initially on areas of land called the **watershed**. Watershed protection has become one of the most important aspects of water conservation and management. To a large extent the quantity and quality of available water depends upon the nature of the watershed. An important characteristic of a good watershed is the ability to retain water for a significant length of time. This reduces flooding, allows for a steady flow of runoff water, and maximizes recharge of water into groundwater reservoirs (aquifer recharge, see Section 5.6). Runoff is slowed and stabilized by several means. One is to minimize cultivation and forest cutting on steeply sloping portions of the watershed, another is to use terraces and grass-planted waterways on cultivated land. The preservation of wetlands (Section 5.2) maximizes aquifer recharge, stabilizes runoff, and reduces turbidity of the runoff water. Small impoundments in the feeder streams of the watershed have similar beneficial effects.

Sedimentation by Flowing Water

The action of flowing water in streams cuts away stream banks and carries sedimentary materials for great distances. Sedimentary materials may be carried by flowing water in streams as the following:

- **Dissolved load** from sediment-forming minerals in solution
- **Suspended load** from solid sedimentary materials carried along in suspension
- **Bed load** dragged along the bottom of the stream channel

The transport of calcium carbonate as dissolved calcium bicarbonate provides a straightforward example of dissolved load. Water with a high dissolved carbon dioxide content (usually present as the result of bacterial action) in contact with calcium carbonate formations contains Ca^{2+} and HCO_3^- ions. Flowing water containing calcium in this form may become more basic by loss of CO_2 to the atmosphere, consumption of CO_2 by algal growth, or contact with dissolved base, resulting in the deposition of insoluble $CaCO_3$:

$$Ca^{2+} + 2HCO_3^- \rightarrow CaCO_3(s) + CO_2(g) + H_2O \qquad (5.5.1)$$

Most flowing water that contains dissolved load originates underground, where it dissolves minerals from the rock strata that it flows through.

Most sediments are transported by streams as suspended load, obvious from the appearance of “mud” in the flowing water of rivers draining agricultural areas or finely divided rock in Alpine streams fed by melting glaciers. Under normal conditions, finely divided silt, clay, or sand make up most of the suspended load, although larger particles are transported in rapidly flowing water. The degree and rate of movement of suspended sedimentary material in streams are functions of the velocity of water flow and the settling velocity of the particles in suspension.

Bed load is moved along the bottom of a stream by the action of water “pushing” particles along. Particles carried as bed load do not move continuously. The grinding action of such particles is an important factor in stream erosion.

Typically, about 2/3 of the sediment carried by a stream is transported in suspension, about 1/4 in solution, and the remaining relatively small fraction as bed load. The ability of a stream to carry water increases with both the overall rate of flow of the water (mass per unit time) and the velocity of the water. Both of these are higher under flood conditions, so floods are particularly important in the transport of sediments.

Streams mobilize sedimentary materials through **erosion**, **transport** materials along with stream flow, and release them in a solid form during **deposition**. Deposits of stream-borne sediments are called **alluvium**. As conditions such as lowered stream velocity begin to favor deposition, larger, more settleable particles are released first. This results in **sorting** such that particles of a similar size and type tend to occur together in alluvium deposits. Much sediment is deposited in flood plains where streams overflow their banks.

Free-Flowing Rivers

Rivers in their natural state are free-flowing. Unfortunately, the free-flowing characteristics of some of the world's most beautiful rivers have been lost to development for power generation, water supply, and other purposes. Many beautiful river valleys have been flooded by reservoirs, and other rivers have been largely spoiled by straightening channels and other measures designed to improve navigation. One of the greater losses from dam construction has consisted of highly productive farmland in river floodplains. Esthetically, an unfortunate case was the flooding early in the 1900s of the Hetch-Hetchy Valley in Yosemite National Park by a dam designed to produce hydroelectric power and water for San Francisco. More recently proposals have been made to drain the valley in an attempt to restore it to some of its original beauty.

5.6. GROUNDWATER

Most **groundwater** originates as **meteoritic** water from precipitation in the form of rain or snow and enters underground aquifers through **infiltration** (Figure 5.6). The rock and soil layer in which all pores are filled with liquid water is called the **zone of saturation**, the top of which is defined as the **water table**. Water infiltrates into aquifers in areas called **recharge zones**. Groundwater may dissolve minerals from the formations through which it passes. Most microorganisms originally present in groundwater are gradually filtered out as it seeps through mineral formations. Occasionally, the content of undesirable salts may become excessively high in groundwater, although it is generally superior to surface water as a domestic water source. Groundwater is a vital resource in its own right; it plays a crucial role in geochemical processes, such as the formation of secondary minerals. The nature, quality, and mobility of groundwater are all strongly dependent upon the rock formations in which the water is held. Physically, an important characteristic of such formations is their **porosity**, which determines the percentage of rock volume available to contain water. A second important physical characteristic is **permeability**, which describes the ease of flow of the water through the rock. High permeability is usually associated with high porosity. However, clays, which are common secondary mineral constituents of soil, tend to have low permeability even when a large percentage of the volume is filled with water.

Groundwater that is used is usually taken from **water wells**. As discussed in more detail in Section 13.11, poor design and mismanagement of water wells can result in problems of water pollution, land subsidence where the water is pumped out, and severely decreased production. As an example, when soluble iron(II) or manganese(II) are present in groundwater, exposure to air at the well wall can result in the formation of deposits of insoluble iron(III) and manganese(IV) oxides produced by bacterially catalyzed oxidation:

$$4Fe^{2+}(aq) + O_2(aq) + 10H_2O \rightarrow 4Fe(OH)_3(s) + 8H^+ \quad (5.6.1)$$

$$2Mn^{2+}(aq) + O_2(aq) + (2x + 2)H_2O \rightarrow 2MnO_2{\cdot}xH_2O(s) + 4H^+ \quad (5.6.2)$$

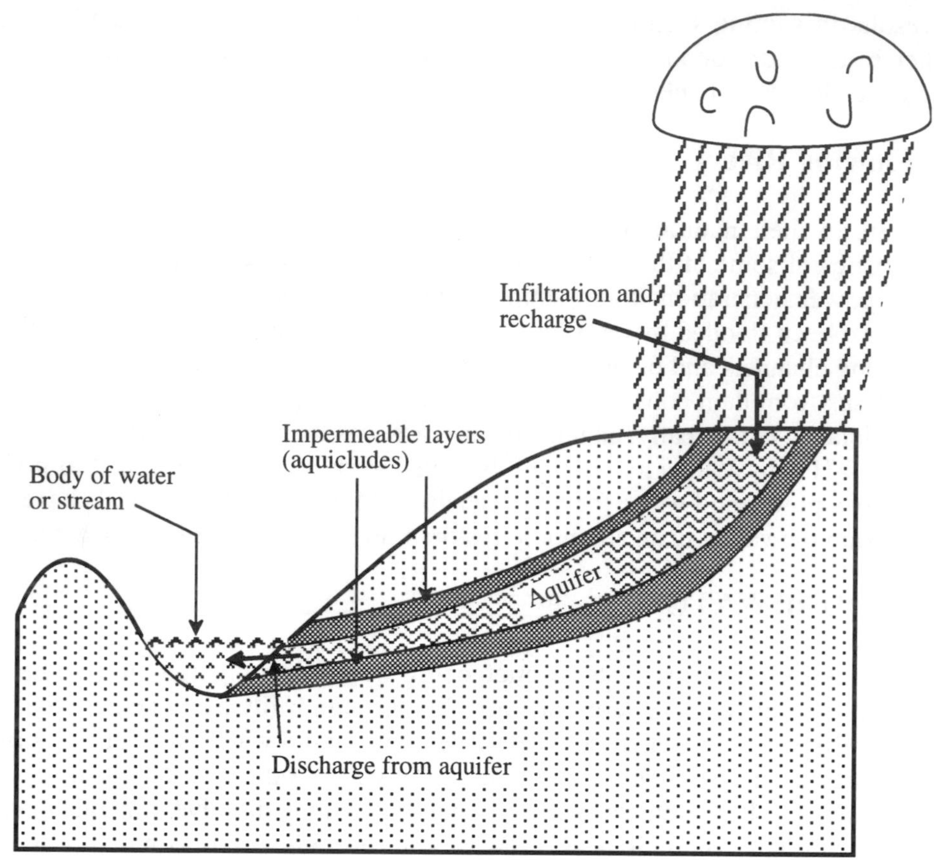

Figure 5.6. Groundwater in an aquifer.

Deposits of Fe(III) and Mn(IV) produced by these processes coat the surfaces from which water flows into the well with a scale that is relatively impermeable to water. The deposits fill the spaces that water must traverse to enter the well. As a result, they can seriously impede the flow of water into the well from the water-bearing aquifer. This creates major water source problems for municipalities using groundwater for water supply. As a result of this problem, chemical or mechanical cleaning, drilling of new wells, or even acquisition of new water sources may be required.

5.7. WATER SUPPLY AND RESOURCES

Available, renewable water supplies are largely determined by **runoff** equal to total precipitation minus that lost by evaporation/transpiration and infiltration. Average annual runoff does not give a complete picture of water availability, however. One reason is that some of the highest runoff occurs in areas where unpleasant climate and poor soil make it difficult to live and utilize the available water. Another reason is the highly seasonal nature of precipitation in many regions that receive large amounts of precipitation such that there are rainy seasons with too much water and flooding alternating with dry seasons in which drought conditions prevail. Therefore, a more meaningful measure of available water is **stable runoff**, which refers basically to available water. Periodic droughts, which in the continental United States seem to occur in approximately 30-year cycles, also greatly complicate water supply.

The water that humans use is primarily fresh surface water and groundwater. In arid regions, a small fraction of the water supply comes from the ocean, a source that is likely to become more important as the world's supply of fresh water dwindles relative to demand. Saline or brackish groundwaters may also be utilized in some areas.

In the continental U. S., on average about 1.48×10^{13} liters of water per day (76 cm per year) fall as precipitation. Of that amount, approximately 1.02×10^{13} L per day, or 53 cm per year, are lost by evaporation and transpiration. Thus, the water theoretically available for use is approximately 4.4×10^{12} liters per day, or only 23 cm per year; of this amount about 4×10^{11} L per day, or slightly more than 2 cm per year infiltrates the ground and may be later tapped as a groundwater supply; the remainder ends up as surface water and is eventually returned to the ocean.

A major problem with water supply is its non-uniform distribution with location and time. As shown in Figure 5.7, precipitation falls unevenly in the continental U.S., as it does in the rest of the world. This is a problem because people in areas with low precipitation often consume more water than people in regions with more rainfall. Rapid population growth in the more arid southwestern states of the U.S. during the last four decades has further aggravated the problem. Water shortages are becoming more acute in the southwestern U.S., which contains six of the nation's eleven largest cities (Los Angeles, Houston, Dallas, San Diego, Phoenix, and San Antonio). Other problem areas include Florida, where overdevelopment of coastal areas threatens Lake Okeechobee, and the Northeast, plagued by deteriorating water systems.

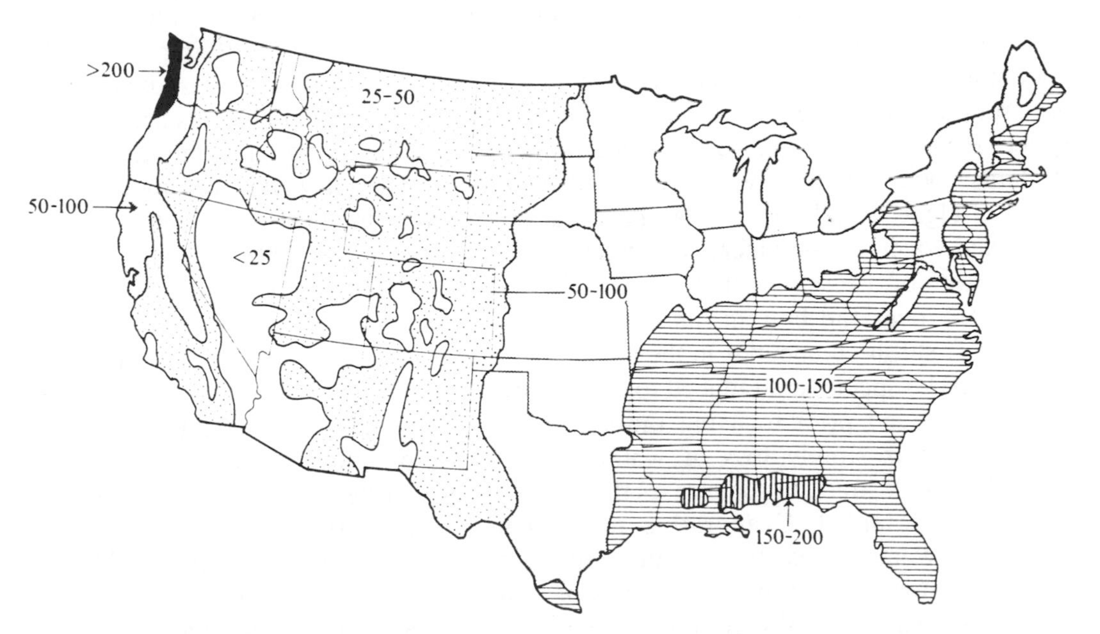

Figure 5.7. Distribution of precipitation in the continental U.S., showing average annual rainfall in centimeters.

A particularly striking example of water use—and misuse—is the extraordinarily heavy pumping of water from the Ogallala aquifer which underlies much of the U.S. High Plains, ranging from the Texas panhandle to Nebraska. The water in this aquifer is thought to be "fossil water" left from the melting of glaciers after the last Ice Age. The quantity is enormous, several times that of all Earth's surface fresh water. However, intense pumping of water from this resouce for irrigation has dropped the water table greatly—in some areas up to about 3 meters per year—with no hope of recharge. In less than a century, this vast resource, which could have served the water needs of municipalities and to irrigate high-value crops for centuries, has been depleted to grow corn, alfalfa, and other crops not well adapted to the High Plains region.

5.8. IMPOUNDMENT AND TRANSFER OF WATER

Efforts to cope with demand for water have resulted in significant alterations in the pattern of water flow and distribution. Vast irrigation systems were known to antiquity, notably in irrigation from the Nile River in ancient Egypt and by the Incas in pre-Columbian South America. In a feat of engineering the Romans built large aqueducts, some of which are still in use more than 20 centuries later. Typically water is impounded in a reservoir produced by damming a river. From the reservoir the water may be moved through pipes, tunnels, or canals. One of the most massive projects utilizing all such measures is the one providing the water supply for Los Angeles and southern California. This project started with the transfer of water from the Owens Valley in northern California, a distance of about 250 miles. It now involves transfer of water from the Feather and Sacramento Rivers, as well as from the Colorado River. The huge Aswan High Dam forming Lake Nasser has stabilized the flow of Egypt's Nile River. China has undertaken a huge project to move water over 600 miles from the Yangtze River to the area around Bejing. Another huge project designed largely to provide hydroelectric power is under construction in Quebec.

Impounding water in reservoirs may have some profound effects upon water quality resulting from factors such as different velocities, changed detention time, and altered surface-to-volume ratios relative to the streams that were impounded. Beneficial changes due to impoundment include decreases in organic matter, turbidity, and hardness (calcium and magnesium content). Some detrimental changes are lower oxygen levels due to decreased reaeration, decreased mixing, accumulation of pollutants, lack of bottom scour produced by flowing stream water, and increased growth of algae. Algal growth may be enhanced when suspended solids settle from impounded water, causing increased exposure of the algae to sunlight. Stagnant water in the bottom of a reservoir may be of low quality. Oxygen levels frequently go to almost zero near the bottom, and odorous hydrogen sulfide is produced by the reduction of sulfur compounds in the low oxygen environment. Insoluble iron(III) and manganese(IV) species are reduced to soluble iron(II) and manganese(II) ions, which must be removed prior to using the water. A major detrimental effect of water impoundment is simply loss of water by evaporation and by infiltration into surrounding rock strata. In some cases this can amount to 10% or more of the flow of the river impounded. Evaporative losses can cause increased water salinity. Impounded water drops its load of suspended silt, which will eventually fill a reservoir. Unforeseen detrimental effects of impoundment are illustrated vividly by Lake Nasser. The silt deposited in the lake is no longer available to provide nutrients for the fields irrigated by the river water downstream. Esthetically, irrigation canals have been found to be totally inadequate substitutes for wild, free-flowing rivers.

One of the greater environmental disasters resulting from water diversion has been the deterioration of the Aral Sea, a landlocked body of saline water in the former Soviet Union Republics of Uzbekistan and Kazakhstan in the south central part of Asia. Beginning with diversion of water flowing into the sea from the Amu Dar'ya and Syr Dar'ya Rivers to irrigate cotton in 1918, the water flow into the Aral sea was severely curtailed. By 1990 the Sea had lost 2/3 of its volume, the salinity of the water had increased proportionally, and the once thriving body of water had essentially "died." Salty dust blown from its shore area has contaminated surrounding farmland. This dead body of water serves as a stark reminder of the consequences of thoughtless technology that does not adequately consider environmental consequences.

5.9. WATER USE AND CONSUMPTION

At present, the continental U.S. uses 1.6 x 10^{12} liters per day, or 8 centimeters of its 76 cm average annual precipitation. This amounts to an almost 10-fold increase from a usage of 1.66 x 10^{11} liters per day at the turn of the century. Even more striking is the per capita increase from about 40 liters per day in 1900 to as much as several thousand liters per day now. (Per capita use in less industrialized nations is still about the same as in the U.S. a century ago.) Much of the increase in use in industrialized nations is accounted for by high agricultural and industrial use, each of which accounts for slightly more than 40% of total consumption. Municipal use consumes the remaining 10-12%.

The greatest amount of water withdrawn for industrial applications is that used to cool power plants and other energy-related facilities. This almost always amounts to over half the water withdrawn by industry and may approach 100 percent. Such use is relatively benign because usually no more than 5% of the water withdrawn is lost to evaporation, and the remainder—assuming it does not become contaminated—is available for other applications. Some manufacturing operations do require very large amounts of water—as examples, about 8,000 L to produce a kilogram of aluminum and an astounding 350,000–400,000 L to manufacture an automobile.

Most agricultural use of water is for irrigation. Sometimes as little as 10% of the water withdrawn for irrigation actually reaching the crops, with the rest lost to evaporation and infiltration. Furthermore, irrigation water may become seriously contaminated, such as by salts leached from the soil and by fertilizer. However, more efficient irrigation and cropping practices have a high potential for water conservation, while still serving irrigation needs effectively. (One such technique is "trickle irrigation," in which just enough water drips on the plant roots to keep them moist. Buildup of salts in soil can be a problem with low-consumption irrigation techniques.)

Water is not destroyed, but it can be lost for practical use. The three ways in which this may occur are the following:

- **Evaporative losses**, such as occur during spray irrigation and when water is used for evaporative cooling.
- **Infiltration** of water into the ground, often in places and ways that preclude its later uses as groundwater.
- **Degradation** from pollutants, such as salts picked up by water used for irrigation.

The total of the three factors above is called water **consumption**. In many cases it is only a fraction of **withdrawal** consisting of the amount of water that is taken and run through a water system for some purpose. An example in which consumption is only a small fraction of widthdrawal is when water is used for cooling, in which all but a small fraction lost to evaporation is returned to a stream or body of water slightly warmer, but undamaged.

Some very marked changes in the pattern of water use are inevitable. The impact may be particularly severe upon agriculture, as illustrated, for example, by drastic curtailment of irrigation water to California agriculture in 1991. In the southwestern U.S., for example, agriculture accounts for the bulk of total water usage — 85 percent in California, 90 percent in New Mexico, 89 percent in Arizona, and 68 percent in Texas. In some areas industries and municipalities are willing to buy their water at prices up to ten times that paid for irrigation water. The increased cost of water could have marked effects on food prices and availability in the U.S.

Water continues to be the subject of heated disputes among land owners and governmental agencies. The state of South Dakota has protested the release of water from reservoirs in the state to maintain barge traffic on the lower Missouri River. Suggestions to transfer water from Washington, Oregon, or northern California to meet growing demand in southern California generate heated discussion. Numerous international disputes have arisen over water supplies, and these can be expected to become more common as water becomes more scarce relative to demand.

As with other scarce resources, **conservation** provides much of the answer for water shortages. Conservation can begin at the point where precipitation falls by proper management of the watershed (see Section 5.5). Proper agricultural practices that minimize and slow runoff, retain soil moisture, and maximize aquifer recharge are very helpful. Use of low tillage agriculture that retains water in plant residues on the surface is very helpful. Construction of terraces, small ponds, and similar water-retaining structures are helpful conservation practices. Other water conservation measures may be applied to minimize domestic consumption. These include use of shower heads that consume minimum amounts of water, low-consumption toilets, appliances (washing machines, dishwashers) designed for low water consumption, and lawns that are planted with native grass species that require minimal water.

A limiting factor in water conservation and reuse is buildup of dissolved salt content. Furthermore, much of the water available for use is saltwater from the ocean or from brackish groundwater. Therefore, salt removal, **desalination**, can be an effective way of augmenting, recycling, or extending water supplies. This topic is addressed further in Chapter 8, "Water Treatment." Distillation can be used to vaporize and condense pure water from a saltwater source. Despite highly efficient methods of distillation that have been developed, it does consume significant amounts of energy. The most widely available, cost-effective means of desalination is provided by reverse osmosis. Described in more detail in Section 8.9 and illustrated by Figure 8.4, reverse osmosis consists essentially of squeezing water through a membrane that allows water, but not salts, to pass through. Reverse osmosis yields a largely salt-free permeate and a brine that is concentrated in salt.

5.10. WATER TECHNOLOGY AND INDUSTRIAL USE

Next to air, water is the cheapest and most universally available raw material. It has found a vast number of uses as a consequence. Some of these are mentioned here, and water technology is briefly addressed.

Water is used for some of its unique physical properties. The most common of these is the high heat of vaporization/condensation of water. Because of this characteristic, enormous amounts of energy can be put into water by evaporating it to steam, enabling it to be transferred over long distances by pipe to remote locations where the heat can be released by allowing the water to condense. Therefore, water in the form of steam has become a favorite means of transferring heat. The high heat of fusion for ice enables the use of solid water to absorb heat.

The heat energy imparted to steam may be converted to mechanical energy very efficiently in a turbine or piston engine (the latter largely of historical interest). In modern times steam is largely generated in **water-tube** boilers in which liquid water is introduced in small (2-8 cm inside diameter) tubes and converted to steam by hot combustion gases circulating outside the tubes. Larger components of the system, particularly the water drums and steam drums, are isolated from the fire, so that if they fail, their contents will not blast into the firebox and cause an explosion. The failure of a water tube is a relatively less serious event.

A modern water-tube boiler does more than just convert liquid water to steam. Furnace walls are cooled with water, which is converted to steam. Steam is heated after it is generated by passing hot flue gas over a **superheater** consisting of tubes containing steam to further increase efficiency. To further increase efficiency, combustion air is pre-heated, and flue gas is cooled, with an **air heater**. The aspects of water and steam used to generate steam power are outlined in Figure 5.8.

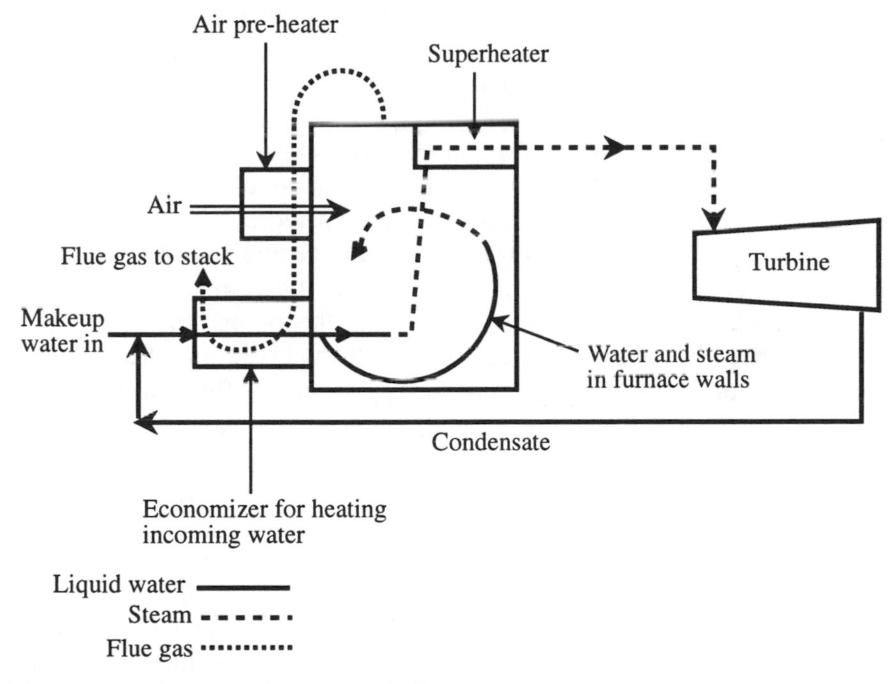

Figure 5.8. Aspects of water and steam in a boiler.

Boiler feedwater used to raise steam must meet some stringent requirements to prevent scaling, fouling, and corrosion. Often this is accomplished to a large extent by steam condensate that is recirculated through the system so that the only additonal water required is **makeup water**. Makeup water is treated to remove insoluble contaminants and those that become insoluble when the water is heated. Hardness in the form of dissolved calcium and magnesium is removed by treatment with phosphate salts:

$$5Ca^{2+} + 3PO_4^{3-} + OH^- \rightarrow Ca_5OH(PO_4)_3(s) \quad (5.10.1)$$

Silicon must be kept to low levels because it can carry over with steam and cause damaging deposits of SiO_2 on turbine blades. Corrosive, dissolved oxygen and carbon dioxide can be removed by heating the feedwater in an open boiler system. Traces of residual oxygen can be removed by reaction with added hydrazine:

$$N_2H_4 + O_2 \rightarrow 2H_2O + N_2 \quad (5.10.2)$$

Anticorrosive agents, such as cyclohexylamine are added to boiler makeup water.

Water is an effective mechanical agent. It can be used to mine and process some kinds of minerals. Examples are sand and gravel harvested from river beds by pumped water. Water under ultra-high pressure can be used to cut materials very effectively and precisely.

One of the greatest uses of water is for its solvent properties. Especially with its surface tension reduced by adding a surface-active agent, water becomes a powerful solvent that is used in a large number of cleaning operations. Water has even begun to replace organic solvents, which pose pollution and disposal problems, in washing small parts, such as electronic constituents. Mixed with suspended lubricants, water is very useful as a lubricant and cooling agent, such as in metal-stamping operations. As a solvent for chemical reactants, water serves as the solvent medium for a number of important chemical synthesis and processing applications. Substances can be purified by dissolving them in water, then evaporating some of the water off to leave a quantity of the substance in a purified form.

Water is a chemical ingredient for a number of industrial chemical reactions. It is required for the hardening of Portland Cement to make concrete. It can be used as the reagent in treating some kinds of hazardous wastes by hydrolysis (see Section 24.5).

Waterpower is the use of moving water for the generation of energy. The water so used may be water descending by the force of gravity from a dammed stream, or it may be rising and falling water from tides. It is the oldest source of non-animal energy, and has been in use for many centuries. Prior to invention of the steam engine, waterwheels used falling water to provide energy, often for flour mills, and for many decades after steam power was developed, waterpower was competitive in some applications.

At the current time, waterpower is most widely used to power electricity generators, an application called **hydroelectric power**. By 1993, total hydroelectric capacity in the U.S. stood at about 100,000 megawatts representing 13% of the total power-generating capacity (a large coal-fired or nuclear power plant typically has a capacity of 1000 MW). If all available sites in the U.S., including Alaska, were utilized for hydroelectric power, the capacity could be doubled. However, for environmental and economic reasons this will not happen. A useful adaptation of hydroelectric power is **pumped storage** in which water is used to run turbines attached to generators to generate power at times of high demand; the process is reversed under low-demand conditions such that the generator acts as a motor and the turbine as a pump to pump water to an elevated storage reservoir.

The economic and environmental advantages of waterpower are obvious. Not the least of these is that the "fuel" is free. Environmentally, there are no emissions or ash with which to deal. There are numerous disadvantages. Under conditions of extreme drought, water may become unavailable, so a system dependent upon waterpower is vulnerable, though much less so than wind-powered, or solar-powered systems. Unfortunately, the vast reservoirs required for most waterpower developments destroy free-flowing rivers and are detrimental to fish migration, such as that required for salmon reproduction. Most remaining available sites are in remote regions from which the transfer of electricity requires massive power lines, which also present environmental problems.

Much of what may be regarded as water technology deals with water purification and treatment. These topics are addressed in Chapter 8.

CHAPTER SUMMARY

The chapter summary below is presented in a programmed format to review the main points covered in this chapter. It is used most effectively by filling in the blanks, referring back to the chapter as necessary. The correct answers are given at the end of the summary.

The science that addresses the physics, chemistry, meteorology, and biology of fresh waters with particular emphasis on ponds and lakes is [1]________________. The five parts of the hydrologic cycle in which the world's water supply is found are [2]____. Transpiration is an example of the strong interaction between the hydrosphere and the [3]__________________. Just slightly over [4]____________ percent of all Earth's fresh water is distributed among surface water, atmospheric water, and biospheric water. Two molecular and bonding features that strongly influence the behavior of water are [5]____. Eight "important properties of water" are [6]____. A body of water consisting of arms of the ocean into which streams flow is [7]________________________. Three ways in which sedimentary materials may be carried by flowing water in streams are [8]____. The rock and soil layer in which all the pores are filled with liquid water is called the [9]____________________, the top of which is the [10]__________________. A reaction of soluble iron that can form deposits that seriously impede the flow of water into a well is [11]____. A huge nonreplenishable fresh water resource that has been notably misused in the U.S. is [12]________________________. A water quality problem with stagnant water in the bottom of a reservoir arises from low levels of [13]__________________. Three ways in which water can be lost for practical use are [14]____. Water is used in industry for some of its unique physical properties, the most common of which, other than its solvent properties, is its [15]____. The reaction $5Ca^{2+} + 3PO_4^{3-} + OH^- \rightarrow Ca_5OH(PO_4)_3(s)$ is used to treat makeup water to remove [16]__________________________________. The reaction $N_2H_4 + O_2 \rightarrow 2H_2O + N_2$ is used to treat water to remove [17]____. Examples of water as an effective mechanical agent are [18]____. To make water an especially effective solvent, it is useful to add a [19]_______________________________. When waterpower is used to power electricity generators, the application is called [20]______________________________.

Answers

1 limnology

2 oceans, ice, surface fresh water, groundwater, water vapor in the atmosphere

3 lithosphere or biosphere

4 1

5 its polar molecular structure and tendency to form hydrogen bonds

6 excellent solvent, high dielectric constant, high surface tension, transparent, maximum density as a liquid, high heat of evaporation, high latent heat of fusion, and high heat capacity

7 an estuary

8 dissolved load, suspended load, and bed load

9 zone of saturation

10 water table

11 $4Fe^{2+}(aq) + O_2(aq) + 10H_2O \rightarrow 4Fe(OH)_3(s) + 8H^+$

12 the Ogallala aquifer

13 dissolved oxygen

14 evaporative losses, infiltration, and degradation

15 high heat of vaporization/condensation

16 hardness

17 dissolved oxygen

18 mining, such as for sand and gravel, and use of ultra-high pressure water for cutting

19 surface-active agent (surfactant)

20 hydroelectric power

QUESTIONS AND PROBLEMS

1. Compare the physical conditions of water in a deep lake in summer with those in a free-flowing stream, and explain how these affect the environmental chemistry and biology of water.

2. Some evidence suggests that rainfall is increased downwind from sources of pollution, such as large industrial cities. Offer a possible explanation in terms of what this chapter says about the process by which precipitation is formed in the atmosphere.

3. Modify Figure 5.1, "The Hydrologic Cycle," to show involvement of the anthrosphere.
4. Western areas of the U.S. states of Oregon and Washington are quite wet, whereas eastern regions of these states beyond the Cascade Mountains are generally very dry. Give a possible explanation. What phenomenon does this observation illustrate?
5. Try to relate some of the "Important Properties of Water" in Table 5.2 to the nature of the water molecule and to hydrogen bonds.
6. Distinguish between dissolved load, suspended load, and bed load in streams.
7. What are deposits formed from suspended solids in streams called? Why do particles of a similar size and type tend to occur together in such deposits?
8. Under what circumstances does water in a stream or lake tend to enter an aquifer? What are the circumstances under which the opposite occurs?
9. Water infiltrating through soil may pick up high concentrations of dissolved CO_2 from bacterial decay before flowing into rock. If the bedrock is limestone, $CaCO_3$, a sinkhole may form. Explain.
10. Above a reservoir constructed by damming a free-flowing mountain stream, the stream water contains a large concentration of O_2, some NO_3^-, and some suspended solid $Fe(OH)_3$. Water from the bottom of the reservoir is fed through turbines, and the water exiting from the turbines contains little O_2, some NH_4^+, and some dissolved Fe^{2+}. Explain.
11. What are the advantages of "trickle irrigation"?
12. How is use made of the high heat of evaporation/condensation of water used industrially.
13. List as many industrial uses of water as you can based on such things as its heat, solvent, and mechanical properties.
14. Explain how pumped storage of water can lead to more efficient electricity production and utilization.

LITERATURE CITED

[1] Jehl, Douglas, "Yes, This Is the Nile, but Don't Go Near the Water," *New York Times*, March 25, 1997, p. A4.

SUPPLEMENTARY REFERENCES

Beer, Tom, *Environmental Oceanography*, 2nd ed., CRC Press/Lewis Publishers, Boca Raton, FL, 1997.

Egna, Hillary and Claude Boyd, *Dynamics of Pond Aquaculture*, CRC Press/Lewis Publishers, Boca Raton, FL, 1997.

Horne, Alexander J. and Charles R. Goldman, *Limnology*, 2nd ed., McGraw-Hill, Inc., New York, NY, 1994.

Kresic, Neven, *Quantitative Solutions in Hydrogeology and Groundwater Modeling*, CRC Press/Lewis Publishers, Boca Raton, FL, 1997.

Laenen, Antonius and David A. Dunnette, *River Quality: Dynamics and Restoration*, CRC Press/Lewis Publishers, Boca Raton, FL, 1997.

LaMoreaux, P. E., *Environmental Hydrogeology*, CRC Press/Lewis Publishers, Boca Raton, FL, 1997.

Miller, G. Tyler, Jr., *Living in the Environment*, 10th ed., Wadsworth Publishing Co., Belmont, CA, 1997.

Palmer, Christopher M., *Principles of Contaminant Hydrogeology*, 2nd ed., CRC Press/Lewis Publishers, Boca Raton, FL, 1996.

Singh, Vijay P., *Environmental Hydrology*, Kluwer Academic Publishers, Norwell, Ma, 1995.

Ward, Andrew and William Elliot, Eds., *Environmental Hydrology*, CRC Press/Lewis Publishers, Boca Raton, FL, 1995.

6 AQUATIC BIOLOGY, MICROBIOLOGY, AND CHEMISTRY

6.1. AQUATIC LIFE

Before discussing the living organisms (**biota**) in water, it is useful to define some terms that apply to them. **Plankton** are small plants, animals, and single-celled organisms that float, drift, or move weakly under their own power near the surface of water. At the bottom of the food chain are photosynthetic plankton or **phytoplankton**. Small animals that float freely or propel themselves weakly are called **zooplankton**. **Invertebrate planktivores** (minute crustaceans, insect larvae) and **vertebrate planktivores** (small fish, minnows) feed on zooplankton. **Water plants** anchored to the bottom with roots inhabit shallow regions of bodies of water. Among these kinds of plants are spike rushes that grow in very shallow water, reeds that grow in somewhat deeper water, pond lilies that float in water, and, at still greater depths, plants such as pondweed that are actually submerged. **Periphyton**, more picturesquely named **aufwuchs**, are small organisms, such as diatoms, algae, and water moss, that grow attached to twigs, rocks, and debris in water. Water supports a large variety of animal life, including fish, sponges, snails, hydras, and insects. **Arthropods**, animals with hard, segmented exoskeletons and jointed legs, are normally abundant in water. Especially prominent among arthropods in water are **crustaceans** ("water insects" in a sense), such as crayfish, lobsters, crabs, shrimps, and water fleas. Animals that can move about entirely on their own power, such as fish, are called **nekton**.

Organisms that live in water may be classified as either autotrophic or heterotrophic. **Autotrophic** biota utilize solar or chemical energy to fix elements from simple, nonliving inorganic material into complex life molecules that compose living organisms. Algae are typical autotrophic aquatic organisms. Generally, CO_2, NO_3^-, and $H_2PO_4^-/HPO_4^{2-}$ are sources of C, N, and P, respectively, for autotrophic organisms. Organisms that utilize solar energy to synthesize organic matter from inorganic materials are called **producers**.

Heterotrophic organisms utilize the organic substances produced by autotrophic organisms as energy sources and as the raw materials for the synthesis of their own biomass. **Grazers** feed on living organic matter. **Detritovores** are organisms that feed on nonliving plant and animal matter called **detritus**. **Decom-**

posers (or **reducers**) are a subclass of detritovores consisting chiefly of bacteria and fungi, which ultimately break down material of biological origin to the simple compounds originally fixed by the autotrophic organisms.

The ability of a body of water to produce living material is known as its **productivity.** Productivity results from a combination of physical and chemical factors (see below). Water of low productivity generally is desirable for water supply or for swimming. Relatively high productivity is required for the support of fish. Excessive productivity can result in choking by weeds and can cause odor problems. The growth of algae may become quite high in very productive waters, with the result that the concurrent decomposition of dead algae reduces oxygen levels in the water to very low values. This set of conditions is commonly called **eutrophication**.

Physical and Chemical Factors in Aquatic Life

Aquatic organisms are strongly influenced by the physical and chemical properties of the body of water in which they live. *Temperature, transparency,* and *turbulence* are the three main physical properties affecting aquatic life. Very low water temperatures result in very slow biological processes, whereas very high temperatures are fatal to most organisms. A difference of only a few degrees can produce large differences in the kinds of organisms present. Thermal discharges of hot, spent, cooling water from power plants frequently kill heat-sensitive fish while increasing the growth of fish and other species that are adapted to higher temperatures. The transparency of water is particularly important in determining the growth of algae. Thus, turbid water may not be very productive of biomass, even though it has the nutrients, optimum temperature, and other conditions needed. Turbulence is an important factor in mixing and transport processes in water. Some small organisms (**plankton**) depend upon water currents for their own mobility. Water turbulence is largely responsible for the transport of nutrients to living organisms and of waste products away from them. It plays a role in the transport of oxygen, carbon dioxide, and other gases through a body of water and in the exchange of these gases at the water-atmosphere interface. Moderate turbulence is generally beneficial to aquatic life.

Dissolved oxygen (DO) frequently determines the extent and kinds of life in a body of water. Oxygen deficiency is fatal to many aquatic animals such as fish. The presence of oxygen can be equally fatal to many kinds of anaerobic bacteria.

Biochemical oxygen demand, BOD, the amount of oxygen utilized when the organic matter in a given volume of water is degraded biologically, is another important water-quality parameter. A body of water with a high biochemical oxygen demand, and no means of rapidly replenishing the oxygen, obviously cannot sustain organisms that require oxygen.

Carbon dioxide is produced by respiratory processes in waters and sediments and can also enter water from the atmosphere. Carbon dioxide is required for the photosynthetic production of biomass by algae and in some cases is a limiting factor. High levels of carbon dioxide produced by the degradation of organic matter in water can cause excessive algal growth and productivity.

The levels of nutrients in water frequently determine its productivity. Aquatic plant life requires an adequate supply of carbon (CO_2), nitrogen (nitrate), phosphorus ($H_2PO_4^-$, HPO_4^{2-}), and trace elements such as iron. In many cases, phosphorus is the limiting nutrient and is generally controlled in attempts to limit excess productivity.

The salinity of water also determines the kinds of life forms present. Irrigation waters may pick up harmful levels of salt. Marine life obviously requires or tolerates salt water, whereas many freshwater organisms are intolerant of salt.

6.2. LIFE IN THE OCEAN

As a medium for life, the marine environment differs substantially from freshwater environments. The most obvious such difference is the salt content of the ocean, consisting mostly of sodium chloride, NaCl. This affects both the organisms in the ocean and the oceanic environment. Because of osmotic phenomena, freshwater organisms cannot live in the sea, and *vice versa.* Due to its high dissolved salt content, the temperature/density relationships of ocean water are unlike those of fresh water. Other factors that influence life in the ocean include tides, waves, its huge volume, and its great depth in some places.

The most shallow of the three life zones in oceans is the **photic zone** in which there is sufficient light to enable a high level of photosynthesis. Organic matter from the photic zone falls to lower zones as **detrital matter**, which serves as a food source for the organisms below the photic zone. The next layer down is the dimly illuminated **mesopelagic zone** inhabited by various sea creatures, such as sharks. Lower still is the **bathypelagic zone** characterized by enormously high pressures and a total absence of sunlight. This zone is inhabited by unique organisms, some of which generate light by bioluminescence. The **benthic region** on the sea bottom is populated by a variety of worms, crustaceans, crabs, and clams. Life can be especially abundant in the vicinity of **hydrothermal vents** ("black smokers"), which warm the surrounding water and emit hot water rich in metal sulfides and H_2S. Unique bacteria in the vicinity of the vents synthesize biomass chemosynthetically, gaining energy by mediating the oxidation of sulfide to sulfate, SO_4^{2-}, and providing the food needed by other organisms.

Compared to land and fresh water, the biomass productivity of the ocean is relatively low, largely because of a shortage of nutrients. Not surprisingly, productivity is highest in areas where upwelling currents provide nutrients, and in coastal areas.

With the exception of the unique life forms around hydrothermal vents noted above, primary productivity in the oceans is from *phytoplankton*, such as dinoflagellates and diatoms, consisting of small floating photosynthetic organisms. The next kind of organism up the food chain consists of *herbivorous zooplankton*, largely planktonic arthropods that feed directly on phytoplankton. These in turn are fed upon by *carnivorous zooplankton*, a diverse group consisting largely of larvae of larger organisms. The herbivorous and carnivorous zooplankton are fed upon by larger nekton—fish, sharks, whales. In general, larger creatures of this type feed on smaller ones, although there are some exceptions—the huge baleen whale feeds predominantly on smaller organisms, which it filters from water.

6.3. LIFE AT THE INTERFACE OF SEAWATER WITH FRESH WATER AND WITH LAND

Areas in which seawater meets the shore or where fresh water flows into the sea are especially active locations for life. These two areas are addressed briefly here. These regions are particularly vulnerable to environmental disruption, such as from oil spills from tankers, or by human modifications of the physical nature of the interface.

The ocean meets with land on sandy or rocky coastlines. A particularly prominent phenomenon that occurs at this interface is alternate exposure of shoreline rock, sand, and mud to seawater and to air, the consequence of wave and tidal action. This results in a striking zonation of life forms.

On rocky shores, the zone between the high-tide and low-tide marks, alternately covered with seawater and exposed to the atmosphere by tidal action, is called the **littoral zone**, which can be further subdivided into subzones. The most common types of plants that grows in the littoral zone are seaweeds of the genus *Fucus*, and related types characterized by their olive-brown color, branched fronds, and air bladders, which enable them to float. Prominent among animal life in this region are periwinkles (single-shelled organisms that move by means of a wide, muscular foot), mussels (bivalve mollusks), and barnacles (marine crustaceans that cling strongly to rocks and other objects).

Sandy and muddy (mudflat) coastal areas offer the opportunity for the growth of a large number of burrowing animals. Among these in sand are various kinds of crabs, shrimp, clams, and sand dollars. Many of the animals that live in sand and mud are very small, ranging from about 60 μm to slightly less than 1 mm. These include nematodes (unsegmented roundworms), copepods (a type of crustaceans), ostracods (seed shrimp), and gastrotrichs (small bottle-shaped organisms with cilia on their undersides).

A specialized type of seashore structure supporting unique marine ecosystems consists of **coral reefs** built up by calcium carbonate deposits produced by coral and other marine organisms. Coral reefs form in tropical regions where a firm geological formation is available at shallow depths, conditions that often exist around volcanic islands. These structures provide habitats for the coral itself, associated algae, crustaceans, echinoderms, mollusks, sponges, and fish.

Estuaries are the locations where fresh water from rivers mixes with seawater from the ocean. These regions show gradations of salinity, which varies with the ebb and flow of tides. The food chain in estuaries is based on both detrital food sources and phytoplankton. They are especially important as nurseries for marine fish and shellfish, in part because potential predators from the ocean are intolerant of the lower salinity of estuarine waters.

The alluvial plains of estuary regions that are alternatively covered with seawater, then drained as the tides rise and fall, may support a heavy growth of grass and other salt-tolerant plants and are called **salt marshes**. These marshes are the sites of tidal creeks and tidal pools. Among the kinds of plants that they support are salt marsh cordgrass, saltgrass, marsh hay cordgrass, black grass, and marsh elder. Among the animals that grow in salt marshes are mussels, fiddler crabs (highly adaptable because they can breathe with either lungs or gills), marsh perwinkles, and sandhoppers. Exposed by low tide, these tasty animals attract predators, such as egrets, gulls, and herons.

6.4. FRESHWATER LIFE

A variety of life inhabits bodies of fresh water, such as ponds, lakes, and reservoirs. The types of organisms are strongly influenced by physical characteristics of the body of water. One of the most important of these is the stratification into the upper epilimnion and hypolimnion layers (see Section 5.4 and Figure 5.5). The mixing of these layers that occurs in the spring and fall—the **overturn**—is very important for aquatic life because it brings nutrients to the surface, thereby increasing fertility. Several zones of a body of water are particularly important to its biota. Rooted plants grow in the **littoral zone** in which sunlight penetrates to the bottom. Further from shore, the region in which light penetrates sufficiently to support significant photosynthesis, but not to the bottom, is called the **limnetic zone**, populated by phytoplankton, zooplankton, and fish. Open water below the limnetic zone is called the **profundal zone**. Biological activity in the profundal zone is variable, but often relatively low because it is too deep for photosynthesis, and detrital matter raining down from higher levels does not stay long enough to support much life. However, biological activity is usually high in the **benthic zone** at the bottom of a body of water because of the large amount of detrital biomass that it receives. Organisms that predominate in this region are decomposers, detritovores, and, in benthic waters below the profundal zone, anaerobic bacteria.

The nature and abundance of life in a standing body of water depend significantly on whether the body of water is oligotrophic, eutrophic, or dystrophic (see Section 5.4). Because they are nutrient-poor, oligotrophic lakes lack the inorganic phosphate and nitrate nutrients required to support a high level of photosynthesis and the input of organic matter to support large populations of detritovores. Nutrient-rich eutrophic lakes support a high level of photosynthetic activity, and the biomass produced also supports a large population of organisms that feed on it and its decomposition products. Dystrophic lakes are so rich in organic matter washed in from surrounding areas that they support a high level of biological activity in their littoral zones. They have relatively low planktonic activity, however.

Life in flowing-water streams and rivers is dependent upon the input of detrital food from the land, although relatively clear streams may generate significant amounts of biomass photosynthetically. Another major factor affecting such life is the influence of often strong currents in flowing water. Organisms have to adapt to currents so that they are not swept downstream, which they do by having streamlined or flat shapes, or by being attached to rocks or to the stream bed. Insect larvae frequently stay under stones where the current is weak and where they can cling to the stone surface. Aquatic water moss and algae may have strong "holdfasts" with which they are attached to stones.

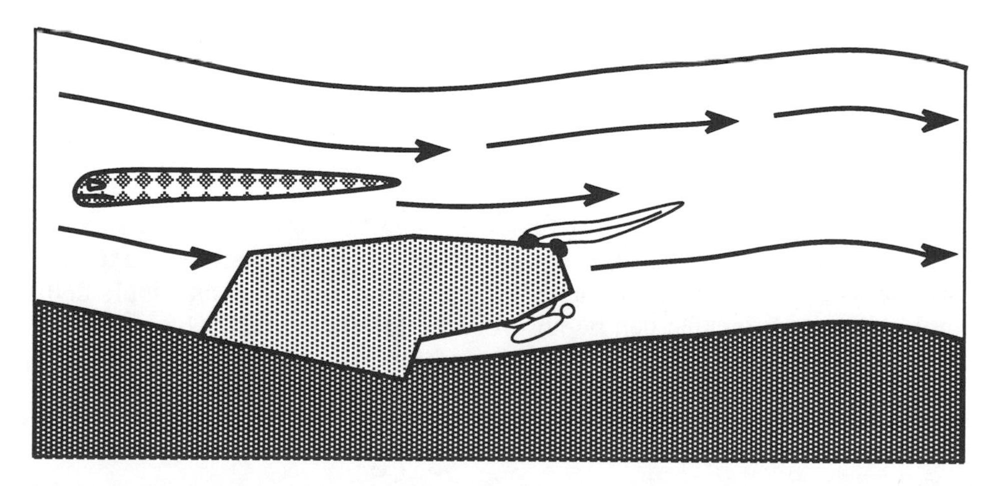

Figure 6.1. Water-dwelling organisms adapt to strong currents in various ways, such as streamlining, holding tight to surfaces, or staying under rocks.

Organic food material in streams may be in the form of large or small particles (detritus) or as dissolved organic matter. Various organisms have developed mechanisms to deal with these food sources. **Shredders** tear apart larger particles of organic matter; their feces and the shredded particles that they don't ingest become part of the supply of fine-particle organic matter. The finely divided organic matter, in turn, is ingested by **collectors** ("fine particulate detritovores") that collect the food by filtering and gathering mechanisms. Aquatic **grazers** feed on algae that grow on solids in the streams, and **gougers** burrow into larger particles of wood. **Piercers** suck the juices from bottom-rooted plants. Aquatic bacteria and fungi grow on particulate organic matter, such as dead leaves. At the top of the food chain, **predators** feed on other organisms.

6.5. MICROORGANISMS IN WATER

Microorganisms in water may consist of bacteria, fungi, protozoans, and algae, all types of organisms discussed in Chapter 3. Microorganisms have a strong influence on water. The algae are the primary producers that generate biomass, $\{CH_2O\}$, photosynthetically. Fungi and bacteria are responsible for breaking down and mineralizing organic matter and are essential in nutrient recycling. Bacteria are responsible for the major elemental oxidation-reduction conversions that occur in water. Various protozoans exhibit characteristics of bacteria, fungi, and algae.

Algae in Water

Algae in water may be considered as generally microscopic organisms that subsist on inorganic nutrients and produce organic matter from carbon dioxide by photosynthesis. The general nutrient requirements of algae are carbon (from CO_2 or HCO_3^-), nitrogen (generally as NO_3^-), phosphorus (as some form of orthophosphate), sulfur (as SO_4^{2-}), and some trace elements. The crucial role played by algae in water is their production of organic matter by algal photosynthesis as described by the reaction

$$CO_2 + H_2O \xrightarrow{h\nu} \{CH_2O\} + O_2(g) \tag{6.5.1}$$

where $\{CH_2O\}$ represents a unit of carbohydrate and $h\nu$ stands for the energy of a quantum of light.

Fungi in Water

The most important function of fungi in the environment is the breakdown of insoluble cellulose in wood and other plant materials by the action of extracellular cellulase enzyme. Excreted from the cell, this enzyme hydroyzes insoluble cellulose to soluble carbohydrates that can be absorbed by the fungal cell. The carbohydrates are oxidized biochemically by **aerobic respiration** represented by the reaction,

$$\{CH_2O\} + O_2(g) \rightarrow CO_2 + H_2O \tag{6.5.2}$$

which is in a sense the opposite of photosynthesis. Although the role of fungi growing directly in water is limited, they largely determine the composition of natural waters and wastewaters because of the large amount of their decomposition products that enter water. An example of such a product is humic material, which interacts with hydrogen ions and metals (see Section 6.12).

Bacteria in Water

Bacteria obtain the energy and raw materials needed for their metabolic processes and reproduction by mediating chemical reactions. Bacterial species have evolved that utilize many of these reactions that are possible in water. As a consequence of their participation in such reactions, bacteria are involved in many biogeochemical processes in water and soil. Bacteria are essential participants in the important elemental cycles in nature, including those of nitrogen, carbon, and sulfur. They are responsible for the formation of many mineral deposits, including some of iron and manganese. On a smaller scale, some of these deposits form through bacterial action in natural water systems and even in pipes used to transport water.

Bacterial Growth

Because of their large surface/volume ratios that enable very rapid exchange of nutrients and waste products with their surroundings, and because of their short reproduction times (less than 1 hour in many cases), bacterial populations may increase very rapidly. This is illustrated by the **population curve** for bacterial growth shown in Figure 6.2, illustrating numbers of bacteria resulting from an initially small number of bacteria present at time zero. Following a **lag phase** of little or no growth, is a **log phase**, or exponential phase, during which the population doubles over a regular time interval called the **generation time**. It is during the log phase that very large amounts of chemical species may be transformed in water during a short time period, such as occurs when dissolved oxygen in water is rapidly used up by aerobic respiration (Reaction 6.5.2). The log phase terminates and the **stationary phase** begins when a limiting factor, such as exhaustion of nutrients occurs. The stationary phase is followed by the **death phase**.

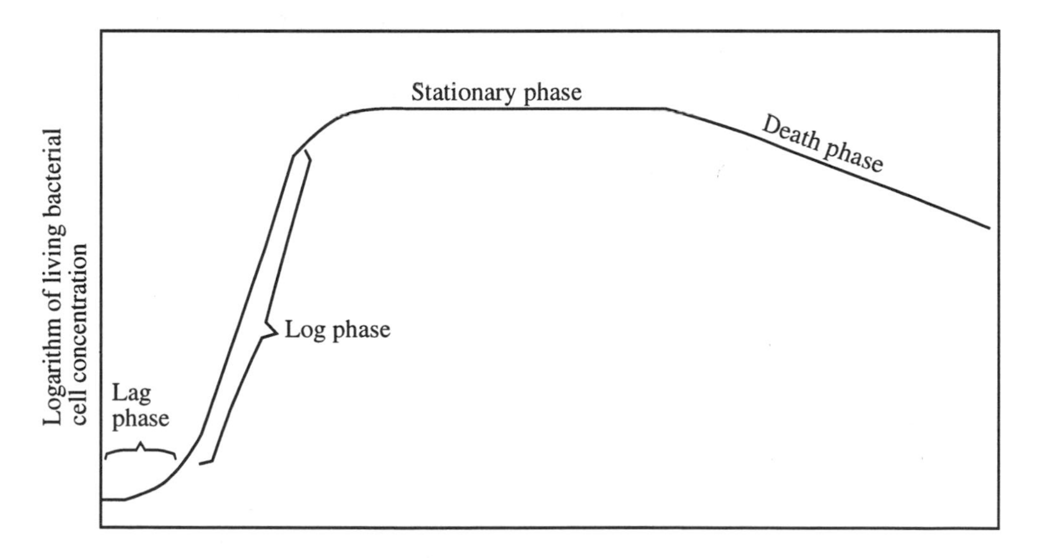

Figure 6.2. Population curve for a bacterial culture.

Factors Affecting Bacterial Growth

Figure 6.3 illustrates the effect of **substrate concentration** on the growth of bacteria, such as those that live in water. It is seen that growth increases in a linear fashion up to a saturation value, beyond which increasing substrate levels do not cause increased growth. This curve would be for a constant population of bacteria, superimposed on which would be increasing bacterial population.

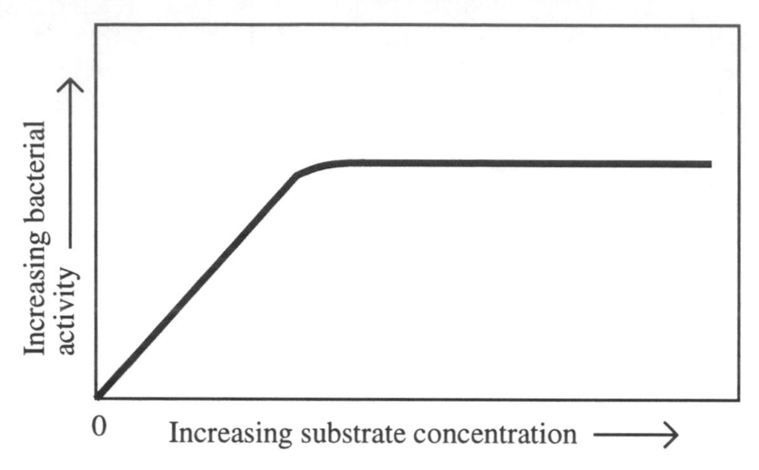

Figure 6.3. Effect of increasing substrate concentration on bacterial growth.

Temperature has a strong effect upon bacterial activity, growth, and metabolism. As shown in Figure 6.4, a curve of growth rate as a function of temperature is skewed toward the high temperature end of the curve and exhibits an abrupt dropoff beyond the temperature maximum. This occurs because enzymes are destroyed by being denatured at temperatures not far above the optimum. The temperature/growth curves of bacteria vary with the species. The temperature range for optimum growth of bacteria is remarkably wide, with some bacteria being able to grow at 0°C, and others existing at temperatures as high as 80°C. **Psychrophilic bacteria** are bacteria having temperature optima below approximately 20°C. The temperature optima of **mesophilic bacteria** lie between 20°C and 45°C. Bacteria having temperature optima above 45°C are called **thermophilic bacteria**.

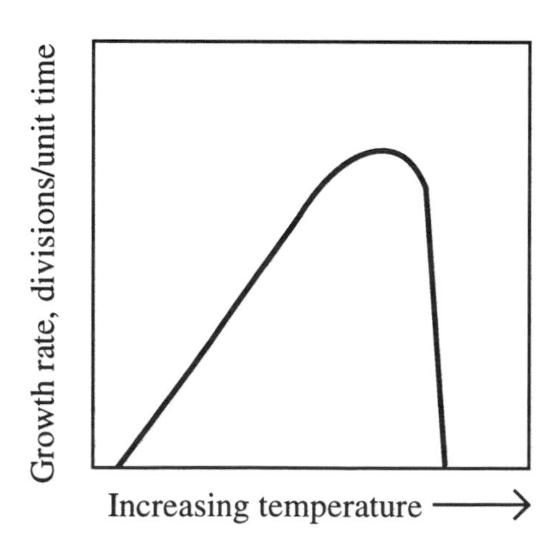

Figure 6.4. Bacterial growth as a function of temperature.

Figure 6.5 is a plot of pH *vs.* bacterial growth rate. Although the optimum pH will vary somewhat, enzymes that govern bacterial activity typically have a pH optimum around neutrality. Enzymes tend to become denatured at pH extremes. This behavior likewise is reflected in plots of bacterial metabolism as a function of pH. For some bacteria, such as those that generate sulfuric acid by the oxidation of sulfide (production of pollutant acid mine water), the pH optimum may be quite acidic.

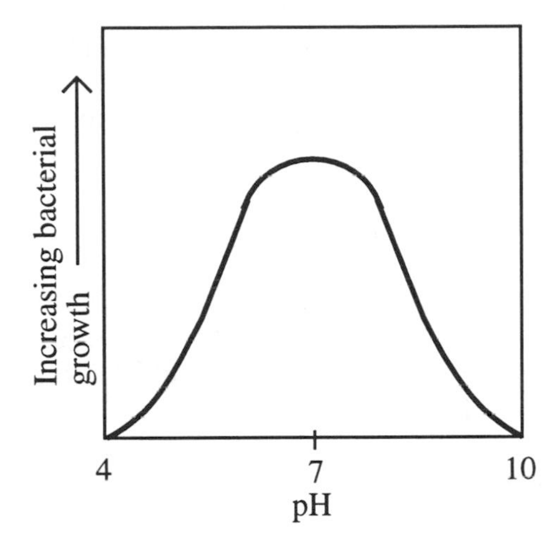

Figure 6.5. Bacterial growth rate as a function of pH.

6.6. MICROORGANISMS AND ELEMENTAL TRANSITIONS

As a consequence of their participation in oxidation/reduction reactions, microorganisms, particularly bacteria, are involved in many biogeochemical processes in water. There is not room in this section to cover these in detail. However, some of the more important oxidation/reduction reactions mediated by microorganisms in water are briefly discussed here.

Bacterially mediated reactions of carbon include photosynthesis by algae and photosynthetic bacteria (Reaction 6.5.1) and aerobic respiration (Reaction 6.5.2). Anaerobic respiration of organic matter produces methane:

$$2\{CH_2O\} \rightarrow CH_4 + CO_2 \quad (6.6.1)$$

Biodegradation of carbonaceous organic matter by bacteria is addressed in Section 6.7.

The most important bacterially mediated transformations of nitrogen species in water are nitrogen fixation, whereby molecular nitrogen is fixed as organic nitrogen; nitrification, the process of oxidizing ammonia to nitrate; nitrate reduction, the process by which nitrogen in nitrate ion is reduced to form compounds having nitrogen in a lower oxidation state; and denitrification, the reduction of nitrate and nitrite to N_2, with a resultant net loss of nitrogen gas to the atmosphere. The overall microbial process for **nitrogen fixation**, the binding of atmospheric nitrogen in a chemically combined form, is the following:

$$3\{CH_2O\} + 2N_2 + 3H_2O + 4H^+ \rightarrow 3CO_2 + 4NH_4^+ \quad (6.6.2)$$

Among the few aquatic bacteria that can fix atmospheric nitrogen are *Azotobacter*, several species of *Clostridium*, and *Cyanobacteria*. **Nitrification**, the conversion of NH_4^+ to NO_3^-, occurs by the following reaction:

$$2O_2 + NH_4^+ \rightarrow NO_3^- + 2H^+ + H_2O \quad (6.6.3)$$

This reaction is important in nature because it provides nitrogen in the nitrate form that most plants utilize. **Denitrification** is the process by which nitrate is reduced by bacteria to gaseous nitrogen and returned to the atmosphere:

$$4NO_3^- + 5\{CH_2O\} + 4H^+ \rightarrow 2N_2 + 5CO_2 + 7H_2O \quad (6.6.4)$$

Some bacteria metabolize inorganic sulfur species in water. Acting in conjunction with other bacteria, *Desulfovibrio* can reduce sulfate ion to H_2S:

$$SO_4^{2-} + 2\{CH_2O\} + 2H^+ \rightarrow H_2S + 2CO_2 + 2H_2O \quad (6.6.5)$$

Because of the high concentration of sulfate ion in seawater, bacterially mediated formation of H_2S causes pollution problems in some coastal areas and is a major source of atmospheric sulfur. In waters where sulfide formation occurs, the sediment is often black in color due to the formation of FeS.

Some bacteria can oxidize hydrogen sulfide to higher oxidation states, particularly to sulfate ion:

$$H_2S + 2O_2 \rightarrow 2H^+ + SO_4^{2-} \quad (6.6.6)$$

This process produces strong sulfuric acid, and one of the colorless sulfur bacteria, *Thiobacillus thiooxidans*, is tolerant of solutions containing up to 1 mole of H^+ per liter, a remarkable acid tolerance.

Of all the metals, iron is the one most commonly acted upon by bacteria in water. A variety of bacteria, including *Ferrobacillus*, *Gallionella*, and some forms of *Sphaerotilus*, utilize iron compounds in obtaining energy for their metabolic needs. These bacteria catalyze the oxidation of iron(II) to iron(III) by molecular oxygen:

$$4FeCO_3(s) + O_2 + 6H_2O \rightarrow 4Fe(OH)_3(s) + 4CO_2 \quad (6.6.7)$$

One consequence of bacterial action on iron compounds is acid mine drainage from coal mines. Acid mine water results from the presence of sulfuric acid produced from pyrite, FeS_2, a sulfur-containing mineral associated with coal. The first of these bacterially mediated reactions is the oxidation of pyrite:

$$2FeS_2(s) + 2H_2O + 7O_2 \rightarrow 2Fe^{2+} + 4H^+ + 4SO_4^{2-} \quad (6.6.8)$$

The next step is the bacterially catalyzed oxidation of iron(II) ion to iron(III) ion,

$$4Fe^{2+} + O_2 + 4H^+ \rightarrow 4Fe^{3+} + 2H_2O \quad (6.6.9)$$

The Fe^{3+} ion reacts chemically with pyrite to dissolve it,

$$FeS_2(s) + 14Fe^{3+} + 8H_2O \rightarrow 15Fe^{2+} + 2SO_4^{2-} + 16H^+ \quad (6.6.10)$$

and this reaction in conjunction with Reaction 6.6.8 constitutes a cycle for the dissolution of pyrite. $Fe(H_2O)_6{}^{3+}$ is an acidic ion, and at pH values much above 3, the iron(III) precipitates as the hydrated iron(III) oxide:

$$Fe^{3+} + 3H_2O \leftrightarrow Fe(OH)_3(s) + 3H^+ \qquad (6.6.11)$$

The beds of streams afflicted with acid mine drainage often are covered with "yellowboy," an unsightly deposit of amorphous, semigelatinous $Fe(OH)_3$. The most damaging component of acid mine water, however, is sulfuric acid, H_2SO_4. It is directly toxic because of its strong acidity and has other undesirable effects.

6.7. MICROBIAL DEGRADATION OF ORGANIC MATTER

The biodegradation of organic matter in the aquatic and terrestrial environments is a crucial environmental process. The biodegradation of organic matter by microorganisms occurs by way of a number of stepwise, microbially catalyzed reactions. These reactions will be discussed here individually with examples.

Oxidation

The degradation of hydrocarbons by microbial oxidation is an important environmental process because it is the primary means by which petroleum wastes are eliminated from water and soil. Bacteria capable of degrading hydrocarbons include *Micrococcus*, *Pseudomonas*, *Mycobacterium*, and *Nocardia*. Oxidation of hydrocarbons is discussed here, but the same principles apply to oxidation of hydrocarbon groups on other organic compounds. **Oxidation** occurs by the action of oxygenase enzymes (see Section 3.15 for a discussion of enzymes). **Epoxidation** consists of adding an oxygen atom between two C atoms in an unsaturated system as shown below:

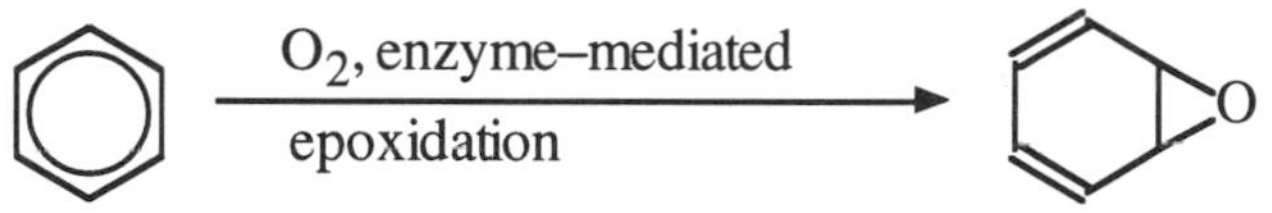

O_2, enzyme–mediated
epoxidation
O
(6.7.1)

This is a particularly important means of metabolic attack upon aromatic rings that abound in many synthetic compounds encountered as water pollutants.

The most common initial step in the microbial oxidation of alkanes involves conversion of a terminal $–CH_3$ group to a $–CO_2$ group. After formation of a carboxylic acid from the alkane, further oxidation normally occurs along the chain, two carbon atoms at a time, by a process called β-oxidation:

$$CH_3CH_2CH_2CH_2CO_2H + 3O_2 \rightarrow CH_3CH_2CO_2H + 2CO_2 + 2H_2O \qquad (6.7.2)$$

Alkane degradability varies with the nature of the alkanes, and microorganisms show a strong preference for straight-chain hydrocarbons because branching inhibits β-oxidation at the site of the branch.

Despite their chemical stability, aromatic rings are susceptible to microbial oxidation. The overall process leading to ring cleavage is the following in which cleavage is preceded by addition of –OH to adjacent carbon atoms:

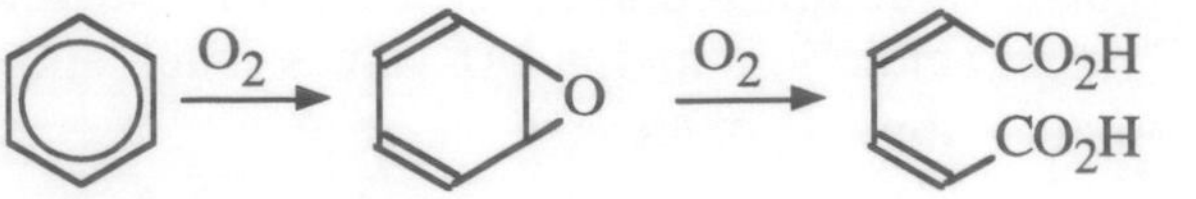

$$\text{C}_6\text{H}_6 \xrightarrow{O_2} \text{benzene epoxide} \xrightarrow{O_2} \text{HO}_2\text{C-CH=CH-CH=CH-CO}_2\text{H} \quad (6.7.3)$$

The biodegradation of petroleum is essential to the elimination of oil spills (about 5 x 10^6 metric tons per year). This oil is degraded by both marine bacteria and filamentous fungi. In some cases, the rate of degradation is limited by available nitrate and phosphate. The physical form of crude oil makes a large difference in its degradability. Degradation in water occurs at the water-oil interface. Therefore, thick layers of crude oil prevent contact with bacterial enzymes and O_2. Apparently, bacteria synthesize an emulsifier that keeps the oil dispersed in the water as a fine colloid and therefore accessible to the bacterial cells.

Hydroxylation often accompanies microbial oxidation. It is the attachment of –OH groups to hydrocarbon chains or rings. It can follow epoxidation as shown by the following rearrangement reaction for benzene epoxide:

$$\text{benzene epoxide (H, O, H)} \xrightarrow{\text{Rearrangement}} \text{C}_6\text{H}_5\text{OH}$$

Hydroxylation can consist of the addition of more than one -OH group. An example of epoxidation and hydroxylation is the metabolic production of the 7,8-diol-9,10-epoxide of benzo(a)pyrene as illustrated below:

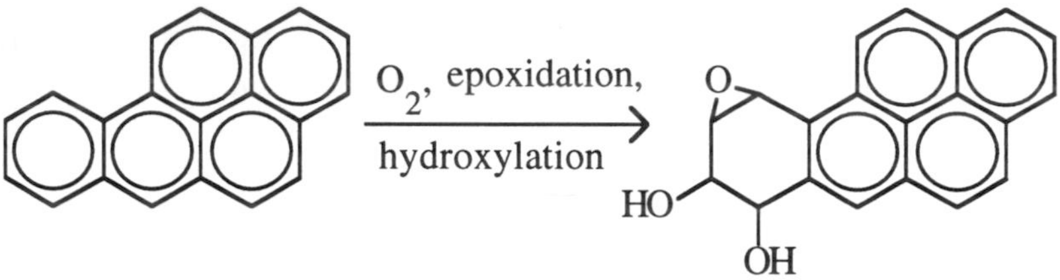

$$\text{benzo(a)pyrene} \xrightarrow[\text{hydroxylation}]{O_2\text{, epoxidation,}} \text{7,8-diol-9,10-epoxide (O, HO, OH)}$$

The 7,8-diol-9,10-epoxide product is a cancer-causing agent.

Other Biochemical Degradation Processes

Hydrolysis, the addition of H_2O accompanied by cleavage of a molecule into two species, is a major step in microbial degradation of many pollutant compounds, especially pesticidal esters, amides, and organophosphate esters. **Hydrolase enzymes** bring about hydrolysis. Typically, one species of *Pseudomonas* hydrolyzes malathion as follows:

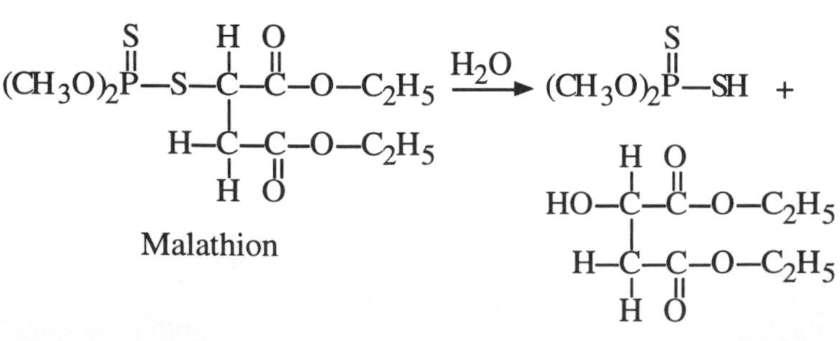

$$\underset{\text{Malathion}}{(CH_3O)_2P(=S)\text{—S—CH(C(=O)—O—C}_2H_5)\text{—CH}_2\text{—C(=O)—O—C}_2H_5} \xrightarrow{H_2O} (CH_3O)_2P(=S)\text{—SH} + \text{HO—CH(C(=O)—O—C}_2H_5)\text{—CH}_2\text{—C(=O)—O—C}_2H_5 \quad (6.7.4)$$

Reductions are carried out by **reductase enzymes** in bacteria; for example, nitroreductase enzyme catalyzes the reduction of the nitro group, $-NO_2$, to amine groups, NH_2. Microbially mediated **dehalogenation** reactions involve the replacement of a halogen atom (F, Cl, Br, I) with –OH.

Many environmentally significant organic compounds contain alkyl groups, such as the methyl ($-CH_3$) group, attached to atoms of O, N, and S. An important step in the microbial metabolism of many of these compounds is **dealkylation**, replacement of alkyl groups by H as shown in Figure 6.6. Examples of these kinds of reactions with environmentally significant organic compounds include O-dealkylation of methoxychlor insecticides, N-dealkylation of carbaryl insecticides, and S-dealkylation of dimethyl mercaptan.

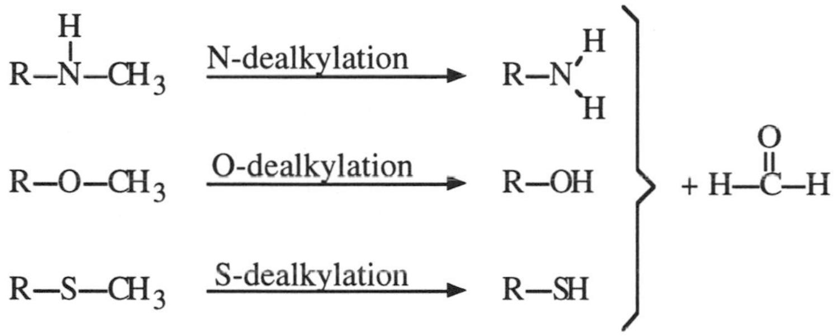

$$R-NH-CH_3 \xrightarrow{\text{N-dealkylation}} R-NH_2$$

$$R-O-CH_3 \xrightarrow{\text{O-dealkylation}} R-OH \quad + \; H-\overset{\displaystyle O}{\overset{\|}{C}}-H$$

$$R-S-CH_3 \xrightarrow{\text{S-dealkylation}} R-SH$$

Figure 6.6. Metabolic dealkylation reactions shown for the removal of CH_3 from N, O, and S atoms in organic compounds.

6.8. ACID-BASE PHENOMENA IN AQUATIC CHEMISTRY

Alkalinity

The capacity of water to accept H^+ ions (protons) is called **alkalinity**. Alkalinity is important in water treatment and in the chemistry and biology of natural waters. Frequently, the alkalinity of water must be known to calculate the quantities of chemicals to be added in treating the water. Highly alkaline water often has a high pH and generally contains elevated levels of dissolved solids. These characteristics may be detrimental for water to be used in boilers, food processing, and municipal water systems. Alkalinity serves as a pH buffer and reservoir for inorganic carbon, thus helping to determine the ability of a water to support algal growth and other aquatic life. It is used by limnologists, who study freshwater systems, as a measure of water fertility. Generally, the basic species responsible for alkalinity in water are bicarbonate ion, carbonate ion, and hydroxide ion:

$$HCO_3^- + H^+ \rightarrow CO_2 + H_2O \tag{6.8.1}$$

$$CO_3^{2-} + H^+ \rightarrow HCO_3^- \tag{6.8.2}$$

$$OH^- + H^+ \rightarrow H_2O \tag{6.8.3}$$

As an example of a water-treatment process in which water alkalinity is important, consider the use of *filter alum*, $Al_2(SO_4)_3{\bullet}18H_2O$, as a coagulant. The hydrated aluminum ion is acidic, and when added to water it reacts with base to form gelatinous aluminum hydroxide,

$$Al(H_2O)_6^{3+} + 3OH^- \rightarrow Al(OH)_3(s) + 6H_2O \quad (6.8.4)$$

which settles and carries suspended matter with it. This reaction removes alkalinity from the water. Sometimes the addition of more alkalinity is required to prevent the water from becoming too acidic.

Acidity

Acidity as applied to natural water systems is the capacity of the water to neutralize OH^-. Most natural waters are alkaline, and acidic water usually indicates severe pollution. Acidity generally results from the presence of weak acids such as $H_2PO_4^-$, CO_2 (see below), H_2S, proteins, fatty acids, and acidic metal ions, particularly Fe^{3+}. *Free mineral acid* is applied to strong acids such as H_2SO_4 and HCl in water. Pollutant acid mine water contains an appreciable concentration of free mineral acid.

The acidic character of some hydrated metal ions may contribute to acidity, for example:

$$Al(H_2O)_6^{3+} \leftrightarrow Al(H_2O)_5OH^{2+} + H^+ \quad (6.8.5)$$

Some industrial wastes, for example, pickling liquor used to remove corrosion from steel, contain acidic metal ions and often some excess strong acid. For such wastes the determination of acidity is important in calculating the amount of lime, or other chemicals, that must be added to neutralize the acid.

Carbon Dioxide in Water

Carbon dioxide, CO_2, is a weak acid in water. Because of the presence of carbon dioxide in air and its production from microbial decay of organic matter, dissolved CO_2 is present in virtually all natural waters and wastewaters. Carbon dioxide is a weak acid so that rainfall from even an absolutely unpolluted atmosphere is slightly acidic due to the presence of dissolved CO_2.

The concentration of gaseous CO_2 in the atmosphere varies with location and season; it is increasing by about 1 part per million (ppm) by volume per year. For purposes of calculation here, the concentration of atmospheric CO_2 will be taken as 350 ppm (0.0350%) in dry air. At 25°C water in equilibrium with unpolluted air containing 350 ppm carbon dioxide has a $CO_2(aq)$ concentration of 1.146×10^{-5} M (see Henry's law calculation of gas solubility in Section 5.3), and this value will be used for subsequent calculations.

Although CO_2 in water is often represented as H_2CO_3, most carbon dioxide dissolved in water is present simply as molecular $CO_2(aq)$. The CO_2–HCO_3^-–CO_3^{2-} system in water may be described by the equations,

$$CO_2 + H_2O \leftrightarrow HCO_3^- + H^+ \quad (6.8.6)$$

$$K_{a1} = \frac{[H^+][HCO_3^-]}{[CO_2]} = 4.45 \times 10^{-7} \quad pK_{a1} = 6.35 \quad (6.8.7)$$

$$HCO_3^- \leftrightarrow CO_3^{2-} + H^+ \tag{6.8.8}$$

$$K_{a2} = \frac{[H^+][CO_3^{2-}]}{[HCO_3^-]} = 4.69 \times 10^{-11} \quad pK_{a2} = 10.33 \tag{6.8.9}$$

where $pK_a = -\log K_a$. The predominant species formed by CO_2 dissolved in water depends upon pH. This is best shown by a **distribution of species diagram** with pH as a master variable as illustrated in Figure 6.7. Such a diagram shows the major

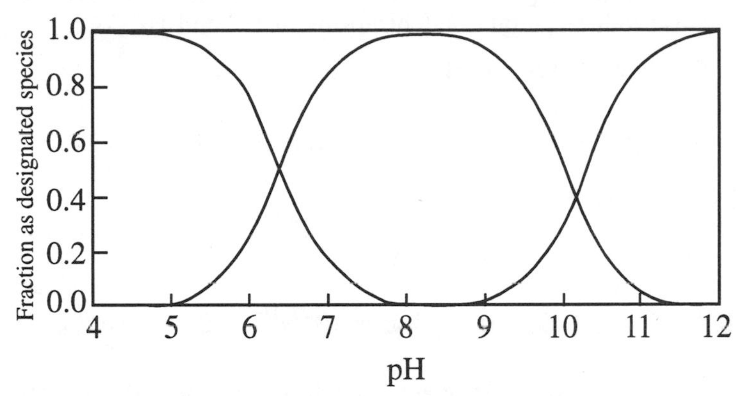

Figure 6.7. Distribution of species diagram for the CO_2–HCO_3^-–CO_3^{2-} system in water.

species present in solution as a function of pH. For CO_2 in aqueous solution, the diagram is a series of plots of the fractions present as CO_2, HCO_3^-, and CO_3^{2-} as a function of pH. These fractions, designated as α_x, are given by the following expressions:

$$\alpha_{CO_2} = \frac{[CO_2]}{[CO_2] + [HCO_3^-] + [CO_3^{2-}]} \tag{6.8.10}$$

$$\alpha_{HCO_3^-} = \frac{[HCO_3^-]}{[CO_2] + [HCO_3^-] + [CO_3^{2-}]} \tag{6.8.11}$$

$$\alpha_{CO_3^{2-}} = \frac{[CO_3^{2-}]}{[CO_2] + [HCO_3^-] + [CO_3^{2-}]} \tag{6.8.12}$$

Substitution of the expressions for K_{a1} and K_{a2} into the α expressions gives the fractions of species as a function of acid dissociation constants and hydrogen ion concentration:

$$\alpha_{CO_2} = \frac{[H^+]^2}{[H^+]^2 + K_{a1}[H^+] + K_{a1}K_{a2}} \tag{6.8.13}$$

$$\alpha_{HCO_3^-} = \frac{K_{a1}[H^+]}{[H^+]^2 + K_{a1}[H^+] + K_{a1}K_{a2}} \tag{6.8.14}$$

$$\alpha_{CO_3^{2-}} = \frac{K_{a1}K_{a2}}{[H^+]^2 + K_{a1}[H^+] + K_{a1}K_{a2}} \tag{6.8.15}$$

Calculations from these expressions show that for pH significantly below pK_{a1}, α_{CO_2} is essentially 1; when pH = pK_{a1}, $\alpha_{CO_2} = \alpha_{HCO_3^-}$; when pH = $1/2(pK_{a1} + pK_{a2})$, $\alpha_{HCO_3^-}$ is at its maximum value of 0.98; when pH = pK_{a2}, $\alpha_{HCO_3^-} = \alpha_{CO_3^{2-}}$; and for pH significantly above pK_{a2}, $\alpha_{CO_3^{2-}}$ is essentially 1. The distribution of species diagram in Figure 3.1 shows that hydrogen carbonate (bicarbonate) ion (HCO_3^-) is the predominant species in the pH range found in most waters, with CO_2 predominating in more acidic waters.

As mentioned above, the value of $[CO_2(aq)]$ in water at 25°C in equilibrium with air that is 350 ppm CO_2 is 1.146×10^{-5} M. The carbon dioxide dissociates partially in water to produce equal concentrations of H^+ and HCO_3^-:

$$CO_2 + H_2O \leftrightarrow HCO_3^- + H^+ \tag{6.8.16}$$

The concentrations of H^+ and HCO_3^- are calculated from K_{a1}:

$$K_{a1} = \frac{[H^+][HCO_3^-]}{[CO_2]} = \frac{[H^+]^2}{1.146 \times 10^{-5}} = 4.45 \times 10^{-7} \tag{6.8.17}$$

$$[H^+] = [HCO_3^-] = (1.146 \times 10^{-5} \times 4.45 \times 10^{-7})^{1/2} = 2.25 \times 10^{-6} \quad pH = 5.65$$

This calculation explains why pure water that has equilibrated with the unpolluted atmosphere is slightly acidic with a pH somewhat less than 7.

6.9. PHASE INTERACTIONS AND SOLUBILITY

Homogeneous chemical reactions occurring entirely in aqueous solution are rather rare in natural waters and wastewaters. Instead, most significant chemical and biochemical phenomena in water involve interactions between species in water and another phase, as shown by the examples illustrated in Figure 6.8. A typical such proc-

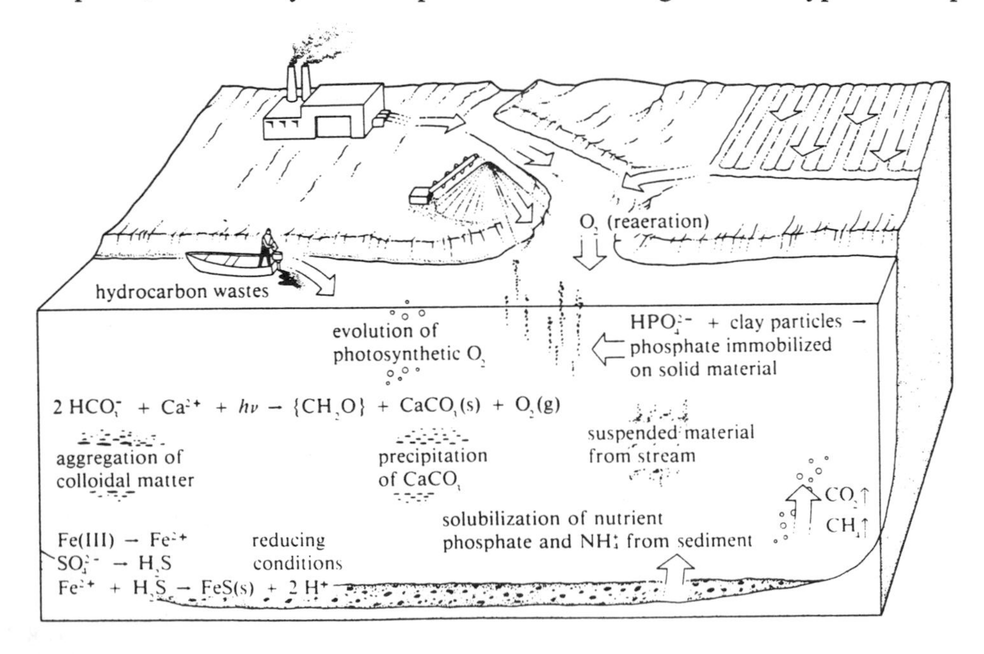

Figure 6.8. Most important environmental chemical processes in water involve interactions between water and another phase.

ess is production of solid biomass through the photosynthetic activity of algae occurring within a suspended algal cell and involving exchange of dissolved solids and gases between the surrounding water and the cell. Similar exchanges occur when bacteria degrade organic matter (often in the form of small particles) in water. Chemical reactions occur that produce solids or gases in water. Iron and many important trace-level elements are transported through aquatic systems as colloidal chemical compounds or are sorbed to solid particles. Pollutant hydrocarbons and some pesticides may be present on the water surface as an immiscible liquid film. Sediment can be washed physically into a body of water.

In addition to water, the phases in a body of water may be divided between *sediments* (bulk solids) and *suspended colloidal material*. An important aspect of phase interactions involves *solubilities* of gases and solids in water. *Colloidal material*, consisting of very fine particles of solids, gases, or immiscible liquids suspended in water, is involved with many significant aquatic chemical phenomena. It is very reactive because of its high surface area to volume ratio.

Gases

The solubilities of gases in water are calculated with **Henry's Law**, which states that *the solubility of a gas in a liquid is proportional to the partial pressure of that gas in contact with the liquid.* Dissolved gases—O_2 for fish and CO_2 for photosynthetic algae—are crucial to the welfare of living species in water. Some gases in water can also cause problems; for example, fish may die from bubbles of nitrogen formed in their blood after they have been exposed to water supersaturated with N_2. Volcanic carbon dioxide evolved from Lake Nyos in the African country of Cameroon asphyxiated 1,700 people in 1986.

Without enough dissolved oxygen, many kinds of aquatic organisms cannot exist in water. Dissolved oxygen is consumed by the degradation of organic matter in water. Many fish kills are caused not from the direct toxicity of pollutants but by a deficiency of oxygen because of its consumption in the biodegradation of pollutants.

Most elemental O_2 comes from the atmosphere, which is 20.95% oxygen by volume of dry air. The solubility of O_2 decreases with increasing water temperature. The concentration of O_2 in water at 25°C in equilibrium with air at atmospheric pressure is only 8.32 mg/L. Thus, water in equilibrium with air cannot contain a high level of dissolved oxygen compared to many other solute species. Oxygen-consuming processes in the water may cause the dissolved oxygen level to rapidly approach zero unless some efficient mechanism for the reaeration of water is operative, such as turbulent flow in a shallow stream or air pumped into the aeration tank of an activated sludge, secondary-sewage treatment facility. At higher temperatures, the decreased solubility of O_2, combined with the increased respiration rate of aquatic organisms, frequently causes a condition in which a higher demand for oxygen accompanied by its lower solubility in water results in severe oxygen depletion.

Carbon Dioxide and Carbonate Species in Water

An important aquatic system involving interchanges among gaseous, aquatic, and solid mineral phases is the one that relates gaseous carbon dioxide, carbon dioxide and related species dissolved in water, and solid mineral carbonates, particularly limestone ($CaCO_3$) and dolomite ($CaCO_3 \cdot MgCO_3$). Carbon dioxide (CO_2), bicarbonate ion (HCO_3^-), and carbonate ion (CO_3^{2-}) have an extremely important influence upon the chemistry of water. Many minerals are deposited as salts of the carbonate ion, CO_3^{2-}. Algae in water utilize dissolved CO_2 and HCO_3^- in the synthesis of biomass. The equilibrium of dissolved CO_2 with the atmosphere,

$$CO_2(\text{water}) \longleftrightarrow CO_2(\text{atmosphere}) \tag{6.9.1}$$

and equilibrium of CO_3^{2-} ion between aquatic solution and solid carbonate minerals,

$$MCO_3(\text{slightly soluble carbonate salt}) \longleftrightarrow M^{2+} + CO_3^{2-} \tag{6.9.2}$$

have a strong buffering (stabilizing) effect upon the pH of water.

Carbon dioxide is only about 0.035% by volume of normal dry air. As a consequence of the low level of atmospheric CO_2, water totally lacking in alkalinity in equilibrium with the atmosphere contains only a very low level of carbon dioxide. A large share of the carbon dioxide found in water is a product of the breakdown of organic matter by bacteria (see Reaction 6.5.2). The formation of HCO_3^- and CO_3^{2-} greatly increases the solubility of carbon dioxide. High concentrations of free carbon dioxide in water may adversely affect respiration and gas exchange of aquatic animals and may even be fatal.

As water seeps through layers of decaying organic matter while infiltrating the ground, it may dissolve a great deal of CO_2 produced by the respiration of organisms in the soil. Later, as water goes through limestone formations, it dissolves calcium carbonate because of the presence of the dissolved CO_2, the process by which limestone caves are formed:

$$CaCO_3(s) + CO_2(aq) + H_2O \longleftrightarrow Ca^{2+} + 2HCO_3^- \tag{6.9.3}$$

Sediments

Sediments, which typically consist of mixtures of clay, silt, sand, organic matter, and various minerals, may vary in composition from pure mineral matter to predominantly organic matter. Physical, chemical, and biological processes may all result in the deposition of sediments in the bottom regions of bodies of water. Sedimentary material may be simply carried into a body of water by erosion or through sloughing (caving in) of the shore. Thus, clay, sand, organic matter, and other materials may be washed into a lake and settle out as layers of sediment.

Sediments may be formed by simple precipitation reactions. A common example is the formation of calcium carbonate sediment when water rich in carbon dioxide and containing a high level of calcium along with HCO_3^- anions loses carbon dioxide to the atmosphere,

$$Ca^{2+} + 2HCO_3^- \rightarrow CaCO_3(s) + CO_2(g) + H_2O \tag{6.9.4}$$

or when the pH is raised by a photosynthetic reaction:

$$Ca^{2+} + 2HCO_3^- + h\nu \rightarrow \{CH_2O\} + CaCO_3(s) + O_2(g) \tag{6.9.5}$$

The latter reaction is an example of the influence of aquatic life on sediment formation. Another example is the bacterially mediated oxidation of iron(II) to iron(III) to produce a precipitate of insoluble iron(III) hydroxide:

$$4Fe^{2+} + 10H_2O + O_2 \rightarrow 4Fe(OH)_3(s) + 8H^+ \quad (6.9.6)$$

Colloids in Water

Many minerals, some organic pollutants, proteinaceous materials, some algae, and some bacteria are suspended in water as very small **colloidal particles**. Such particles have some characteristics of both species in solution and larger particles in suspension, range in diameter from about 0.001 micrometer (μm) to about 1 μm, and scatter white light as a light blue hue observed at right angles to the incident light. The characteristic light-scattering phenomenon of colloids results from their being the same order of size as the wavelength of light and is called the **Tyndall effect**. The unique properties and behavior of colloidal particles are strongly influenced by their physical-chemical characteristics, including high surface area relative to their volume, high interfacial energy, and high surface/charge density ratio.

There are three classes of colloidal particles as illustrated in Figure 6.9. **Hydrophilic colloids** generally consist of macromolcculcs, such as proteins and synthetic polymers, that are characterized by strong interaction with water. **Hydrophobic colloids** interact to a lesser extent with watcr and are stable because of their positive or negative electrical charges. Examples of hydrophobic colloids are clay particles, petroleum droplets, and very small gold particles. **Association colloids** consist of

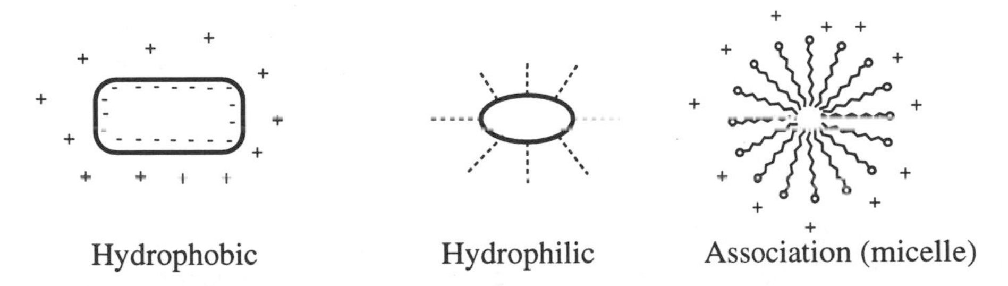

Figure 6.9. Representations of hydrophobic, hydrophilic, and association colloidal particles. The negatively charged hydrophobic and association colloidal particles are shown surrounded by positively charged counter ions. The dashed lines in the hydrophilic colloidal partices show bonds to water. In the micelle of the association colloid, the jagged lines represent organophilic (usually hydrocarbon) "tails" and the circles represent anionic "heads" attracted to water.

special aggregates of ions and molecules called **micelles**. To understand how micelle formation occurs, consider sodium stearate, a typical soap with the following formula:

$$CH_3CH_2CH_2CH_2CH_2CH_2CH_2CH_2CH_2CH_2CH_2CH_2CH_2CH_2CH_2CH_2CH_2CO_2^-Na^+$$

This molecule (ion) has a long organic "tail" consisting of the chain of CH_2 groups and a charged ionic "head," $CO_2^-\ Na^+$. As a result of this structure, stearate anions in water tend to form clusters consisting of as many as 100 anions grouped together with their hydrocarbon "tails" on the inside of a spherical colloidal particle and their ionic "heads" on the surface in contact with water and with Na^+ counterions. This results in the formation of colloidal particles called **micelles** (Figure 6.9).

The stability of colloids is a prime consideration in determining their behavior. It is involved in important aquatic chemical phenomena, including the formation of sediments, dispersion and agglomeration of bacterial cells, and dispersion and removal of pollutants (such as crude oil from an oil spill).

6.10. OXIDATION-REDUCTION

Oxidation-reduction (redox) reactions are those involving changes of oxidation states of reactants. Such reactions are easiest to visualize as the transfer of electrons from one species to another. For example, soluble cadmium ion, Cd^{2+}, is removed from wastewater by reaction with metallic iron. The overall reaction is

$$Cd^{2+} + Fe \rightarrow Cd + Fe^{2+} \quad (6.10.1)$$

This reaction is the sum of two **half-reactions**, a reduction half-reaction in which cadmium ion accepts two electrons and is reduced to cadmium metal,

$$Cd^{2+} + 2e^- \rightarrow Cd \quad (6.10.2)$$

and an oxidation half-reaction in which elemental iron is oxidized:

$$Fe \rightarrow Fe^{2+} + 2e^- \quad (6.10.3)$$

When these two half-reactions are added algebraically, the electrons cancel on both sides and the result is the overall reaction given in Equation 6.10.1.

Oxidation-reduction phenomena are highly significant in the environmental chemistry and microbiology of natural waters and wastewaters. The reduction of oxygen by organic matter in a lake,

$$\{CH_2O\}\text{(becomes oxidized)} + O_2\text{(becomes reduced)} \rightarrow CO_2 + H_2O \quad (6.10.4)$$

results in oxygen depletion which can be fatal to fish. The rate at which sewage is oxidized is crucial to the operation of a waste-treatment plant. Reduction of insoluble iron(III) to soluble iron(II),

$$Fe(OH)_3(s) + 3H^+ + e^- \rightarrow Fe^{2+} + 3H_2O \quad (6.10.5)$$

in a reservoir contaminates the water with iron, which is hard to remove in the water-treatment plant. Oxidation of NH_4^+ to NO_3^- in water,

$$NH_4^+ + 2O_2 \rightarrow NO_3^- + 2H^+ + H_2O \quad (6.10.6)$$

is essential for getting the ammonium nitrogen into a form assimilable by algae in the water. Many other examples can be cited of the ways in which the types, rates, and equilibria of redox reactions largely determine the nature of important solute species in water.

The relative tendency for a water solution to be oxidizing or reducing is due to the activity of electrons. A high electron activity denotes reducing, and a low electron activity indicates oxidizing conditions. Because electron activity in water varies over many orders of magnitude, environmental chemists find it convenient to discuss oxidizing and reducing tendencies in terms of pE, a parameter analogous to pH and defined conceptually as the negative log of the electron activity:

$$pE = -\log(a_{e^-}) = \frac{E}{0.0591} \text{ (at 25°C)} \tag{6.10.7}$$

In the above equation E is an electrode potential, which in principle at least can be measured with electrodes.

The nature of chemical species in water is usually a function of both pE and pH. A good example of this may be illustrated by a simplified pE-pH diagram for iron in water, assuming that iron is in one of the four forms of Fe^{2+} ion, Fe^{3+} ion, solid $Fe(OH)_3$,or solid $Fe(OH)_2$ as shown in Figure 6.10. Water in which the pE is higher than that shown by the upper dashed line is thermodynamically unstable toward oxidation, and below the lower dashed line water is thermodynamically unstable toward reduction. It is seen that Fe^{3+} ion is stable only in a very oxidizing, acidic medium such as that encountered in acid mine water, whereas Fe^{2+} ion is stable over a relatively large region as reflected by the common occurrence of soluble iron(II) in oxygen-deficient groundwaters. Highly insoluble $Fe(OH)_3$ is the predominant iron species over a very wide pE-pH range.

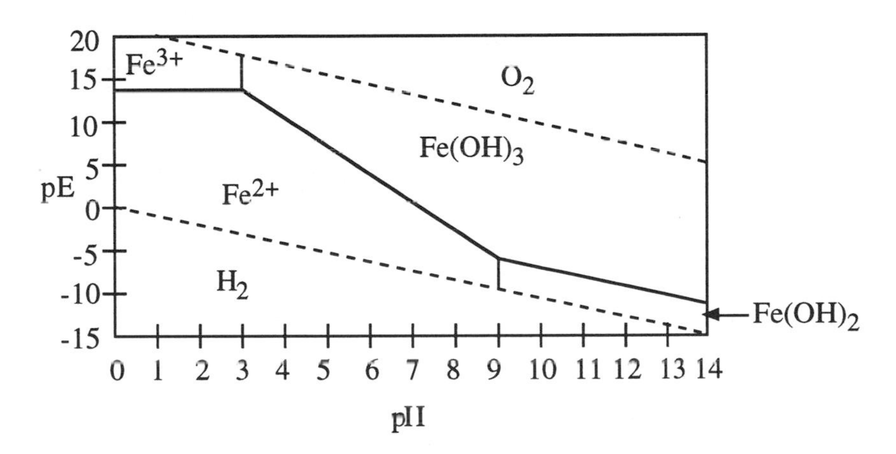

Figure 6.10. A simplified pE-pH diagram for iron in water (maximum total soluble iron concentration 1.0×10^{-5} M).

6.11. METAL IONS AND CALCIUM IN WATER

The formula of a metal ion in aqueous solution usually is written M^{n+}, which signifies the simple hydrated metal cation $M(H_2O)_x{}^{n+}$. A bare metal ion, such as Mg^{2+}, cannot exist as a separate entity in water. In order to secure the highest stability of their outer electron shells, metal ions in water are bonded, or *coordinated*, to water molecules or other stronger bases (electron-donor partners) that might be present.

Metal ions in aqueous solution seek to reach a state of maximum stability through chemical reactions, including acid-base,

$$Fe(H_2O)_6{}^{3+} \leftrightarrow FeOH(H_2O)_5{}^{2+} + H^+ \tag{6.11.1}$$

precipitation,

$$Fe(H_2O)_6{}^{3+} \leftrightarrow Fe(OH)_3(s) + 3H_2O + 3H^+ \tag{6.11.2}$$

and oxidation-reduction reactions:

$$Fe(H_2O)_6^{2+} \leftarrow\rightarrow Fe(OH)_3(s) + 3H_2O + e^- + 3H^+ \quad (6.11.3)$$

Calcium and Hardness in Water

Of the cations found in most freshwater systems, calcium generally has the highest concentration. The chemistry of calcium, although complicated enough, is simpler than that of the transition metal ions found in water. Calcium is a key element in many geochemical processes, and minerals constitute the primary sources of calcium ion in waters. Among the primary contributing minerals are gypsum, $CaSO_4 \bullet 2H_2O$; anhydrite, $CaSO_4$; dolomite, $CaCO_3 \bullet MgCO_3$; and calcite and aragonite, which are different mineral forms of $CaCO_3$.

Water containing a high level of carbon dioxide readily dissolves calcium from its carbonate minerals:

$$CaCO_3(s) + CO_2(aq) + H_2O \leftarrow\rightarrow Ca^{2+} + 2HCO_3^- \quad (6.11.4)$$

When the above equation is reversed and CO_2 is lost from the water, calcium carbonate deposits are formed. The concentration of CO_2 in water determines the extent of dissolution of calcium carbonate.

Calcium ion, along with magnesium and sometimes iron(II) ion, accounts for **water hardness**. The most common manifestation of water hardness is the curdy precipitate formed by soap in hard water. *Temporary hardness* is due to the presence of calcium and bicarbonate ions in water and may be eliminated by boiling the water, thus causing the reversal of Reaction 6.11.4:

$$Ca^{2+} + 2HCO_3^- \leftarrow\rightarrow CaCO_3(s) + CO_2(g) + H_2O \quad (6.11.5)$$

Increased temperature in water having temporary hardness may force the above reaction to the right by evolving CO_2 gas. As a result, a white precipitate of calcium carbonate may form in boiling water having temporary hardness. The equilibrium between dissolved carbon dioxide and calcium carbonate minerals is important in determining several natural water chemistry parameters such as alkalinity, pH, and dissolved calcium concentration (Figure 6.11).

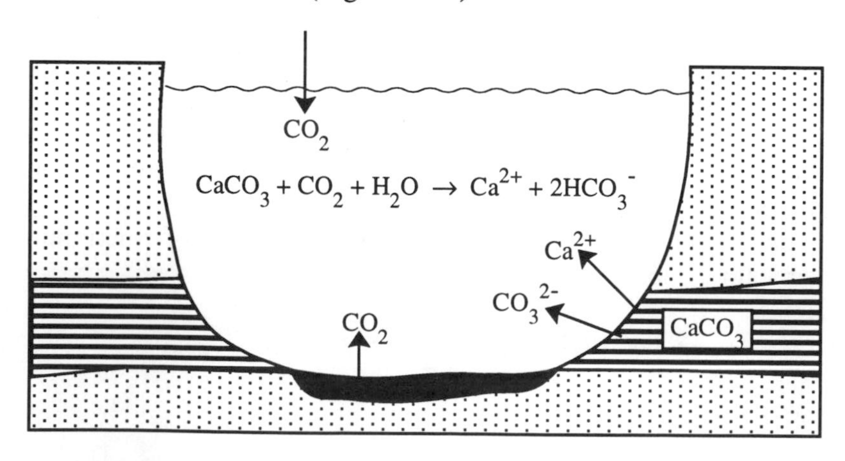

Figure 6.11. Carbon dioxide-calcium carbonate equilibria.

6.12. COMPLEXATION AND SPECIATION OF METALS

The properties of metals dissolved in water depend largely upon the nature of metal species dissolved in the water. Therefore, **speciation** of metals plays a crucial role in their environmental chemistry in natural waters and wastewaters. In addition to the hydrated metal ions, for example, $Fe(H_2O)_6{}^{3+}$ and hydroxo species such as $FeOH(H_2O)_5{}^{2+}$ (see Reaction 6.11.1), metals may exist in water reversibly bound to inorganic anions or to organic compounds as **metal complexes**, or they may be present as **organometallic** compounds containing carbon-to-metal bonds. The solubilities, transport properties, and biological effects of such species are often vastly different from those of the metal ions themselves.

A metal ion in water may combine with an electron donor (Lewis base) to form a **complex** or **coordination compound** (or ion). Thus, cadmium ion in water combines with a ligand, cyanide ion, to form a complex ion as shown below:

$$Cd^{2+} + CN^- \leftrightarrow CdCN^+ \tag{6.12.1}$$

Additional cyanide ligands may be added to form the progressively weaker (more easily dissociated) complexes $Cd(CN)_2$, $Cd(CN)_3{}^-$, and $Cd(CN)_4{}^{2-}$.

In this example, the cyanide ion is a **unidentate ligand**, meaning that it has only one site that bonds to the cadmium metal ion. Of considerably more importance than complexes of unidentate ligands in natural waters are complexes with **chelating agents.** A chelating agent has more than one atom that may be bonded to a central metal ion at one time to form a ring structure. Chelates of metals with chelating groups from humic substances are shown in Figure 6.12. In general, since a chelating agent may bond to a metal ion in more than one place simultaneously, chelates are more stable than complexes involving unidentate ligands. Metal chelate stability tends to increase with the number of chelating sites available on the ligand.

Humic substances are the most important class of naturally occurring complexing agents. These are degradation-resistant materials formed during the decomposition of vegetation that occur as deposits in soil, marsh sediments, peat, coal, lignite, or other locations where large quantities of vegetation have decayed. The types of humic substances are classified on the basis of their solubilities: **Humin** is insoluble, **humic acid** is soluble only in base, and **fulvic acid** is soluble in both acid and base. Because of their acid-base, sorptive, and complexing properties, both the soluble and insoluble humic substances have a strong effect upon the properties of water. In general, fulvic acid dissolves in water and exerts its effects as the soluble species. Humin and humic acid remain insoluble and affect water quality through exchange of species, such as cations or organic materials, with water.

Some feeling for the nature of humic substances may be obtained by considering the structure of a hypothetical molecule of fulvic acid below:

The binding of metal ions by humic substances is one of the most important environmental qualities of humic substances. Iron and aluminum are very strongly bound to humic substances, whereas magnesium is rather weakly bound. This binding can occur as chelation between a carboxyl group and a phenolic hydroxyl group, as chelation between two carboxyl groups, or as complexation with a carboxyl group (see Figure 6.12).

Fulvic-acid complexes of metals in natural waters have a number of effects. Among these is that they keep some of the biologically important transition-metal ions in solution and are particularly involved in iron solubilization and transport. Fulvic acid-type compounds are associated with color in water. These yellow materials, called **Gelbstoffe**, frequently are encountered along with soluble iron.

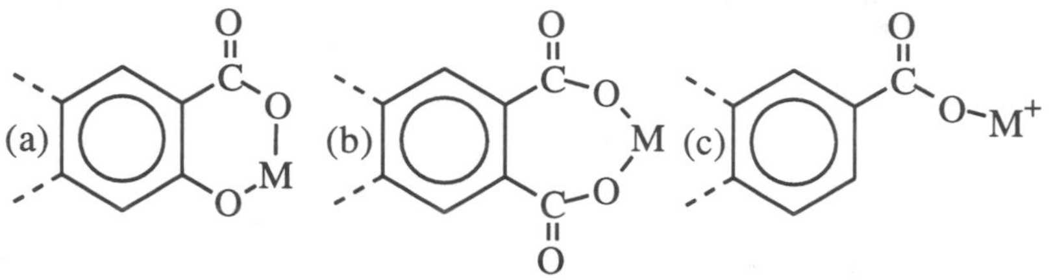

Figure 6.12. Binding of a metal ion, M^{2+}, by humic substances: (a) by chelation between carboxyl and phenolic hydroxyl, (b) by chelation between two carboxyl groups, and (c) by complexation with a carboxyl group.

Organometallic Compounds

Another type of environmentally important metal species consists of **organometallic compounds**, differing from complexes and chelates in that the organic portion is bonded to the metal by a carbon-metal bond and the organic ligand is frequently not capable of existing as a stable separate species. Example organometallic compound species are monomethylmercury ion and dimethylmercury:

Hg^{2+}	$HgCH_3^+$	$Hg(CH_3)_2$
Mercury(II) ion	Monomethylmercury ion	Dimethylmercury

Organometallic compounds may enter the environment directly as pollutant industrial chemicals and some, including organometallic mercury, tin, selenium, and arsenic compounds, are synthesized biologically by bacteria. Some of these compounds are particularly toxic because of their mobilities in living systems and abilities to cross cell membranes.

CHAPTER SUMMARY

The chapter summary below is presented in a programmed format to review the main points covered in this chapter. It is used most effectively by filling in the blanks, referring back to the chapter as necessary. The correct answers are given at the end of the summary.

Algae are photosynthetic, autotrophic, aquatic organisms meaning that they [1]______

____

____.

The three main physical properties affecting aquatic life are [2]________________

________________________. Plankton depend upon [3]_________________ for their mobility. The most shallow of the three life zones in oceans is the [4]____________________ characterized by water in which there is sufficient light to enable a high level of photosynthesis. Two areas of the ocean that are especially active locations for life are where [5]____.
The littoral zone is located between [6]_________________________ and is characterized by alternate exposure to [7]____________________________________. One of the most important features of many freshwater bodies of water, such as lakes, is their [8]__________________________________. Life in flowing-water streams and rivers that are not very transparent is dependent upon [9]____. The most important function of fungi in the environment is [10]____. Bacteria obtain the energy and raw materials needed for their metabolic processes and reproduction by [11]____.
If the number of a specific kind of bacteria in a water sample were 1.5×10^6, 3.0×10^6, and 6.0×10^6 initially, at 1/2 hour, and at 1 hour respectively, the bacteria are in the [12]________________ phase of growth. A plot of bacterial growth as a function of temperature exhibits [13]__________________________ beyond the temperature maximum. As applied to bacterial processes, the reaction $2\{CH_2O\} \rightarrow CH_4 + CO_2$ illustrates [14]____________________________________. As applied to the degradation of petroleum hydrocarbons, the microbially mediated reaction $CH_3CH_2CH_2CH_2CO_2H + 3O_2 \rightarrow CH_3CH_2CO_2H + 2CO_2 + 2H_2O$ is known as [15]_______________________________. The reaction $HCO_3^- + H^+ \rightarrow CO_2 + H_2O$ illustrates the HCO_3^- ion contributing to an important water quality parameter called [16]____________________________. The ionization of $Al(H_2O)_6^{3+}$ ion in water would most likely make the water [17]_______________________. CO_2 reacts with H_2O to give [18]____. Rather than occurring in solution, most significant chemical and biochemical phenomena in water involve [19]________________________. Henry's Law states that the solubility of a gas in a liquid is proportional to [20]____. The formation of sediments in the bottom regions of bodies of water may result from [21]____ processes.
Colloidal particles in water go up to about [22]________________ in size and have characteristics of both [23]____. A species, such as $CH_3(CH)_{16}CO_2^-$, having a long hydrocarbon chain attached to an ionic group would be expected to form [24]________________________ colloids in water consisting of [25]_______________________________. The relative tendency for a water solution to be oxidizing or reducing is due to the activity of [26]____________________________________, which is higher when the solution is [27]_________________________________. Rather than existing in water solution as simple metal ions, M^{n+}, metals dissolved in water are [28]________________________.
Of the cations found in most fresh-water systems, [29]_____________________________ generally has the highest concentration. Calcium ion, along with magnesium and sometimes iron(II) ion, accounts for water [30]____. Other than

hydrated metal ions, metals may be present as [31]____________________. Degradation-resistant materials with chelating properties formed during the decomposition of vegetation that occur as deposits where large quantities of vegetation have decayed are [32]____. Organo-metallic compounds differ from complexes and chelates in that they have metal bonded to [33]____________________. Some organometallic compounds are synthesized biologically by [34]____________________________.

Answers

1 utilize solar or chemical energy to fix elements from simple, nonliving inorganic material into complex life molecules

2 temperature, transparency, and turbulence

3 water currents

4 photic zone

5 seawater meets the shore or where fresh water flows into the sea

6 low-tide and high-tide marks

7 seawater and the atmosphere

8 stratification into the upper epilimnion and hypolimnion layers

9 the input of detrital food from the land

10 the breakdown of insoluble cellulose in wood and other plant materials by the action of extracellular cellulase enzyme

11 mediating chemical reactions

12 log

13 an abrupt dropoff

14 anaerobic respiration

15 β-oxidation

16 alkalinity

17 acidic

18 HCO_3^- and H^+ ions

19 interactions between species in water and another phase

20 the partial pressure of that gas in contact with the liquid

21 physical, chemical, and biological

22 1 μm

23 solutions and suspensions

24 association

25 micelles

26 electrons

27 reducing

28 hydrated

29 calcium

30 hardness

31 metal complexes or organometallic compounds

32 humic substances

33 carbon

34 bacteria

QUESTIONS AND PROBLEMS

1. What are plankton? In which sense are phytoplankton at the bottom of the food chain?
2. Define productivity of water. Name and describe the undesirable pollution condition resulting from excess productivity.
3. Distinguish between DO and BOD in water.
4. Explain why the plot of bacterial growth as a function of temperature is skewed?
5. How does overturn in water affect aquatic life?
6. With regard to animal life in streams, distinguish among shredders, collectors, grazers, and gougers.
7. What is the most important function of fungi insofar as water is concerned?
8. What are humic substances and why are they significant in water?
9. How do organometallic compounds differ from complexes?
10. In what sense is the interface of seawater and fresh water of particular importance for aquatic life?
11. What are oxidation/reduction reactions? Give some examples of such reactions in water.
12. The numbers of a specific kind of bacteria in water as a function of time were found to be the following:

Elapsed time, hours	Millions of bacteria per milliliter	Elapsed time, hours	Millions of bacteria per milliter
0.0	0.103	2.0	1.59
0.5	0.199	2.5	2.13
1.0	0.400	3.0	2.49
1.5	0.815	3.5	2.52

What do these numbers indicate about the phase or phases of the population curve of the bacteria?

13. What is a metal complex? Give an example of one.
14. Explain using chemical reactions how bacteria are involved with the nitrogen cycle.
15. What are epoxidation and β-oxidation, and how are they involved with biodegradation?
16. Write the reaction by which an epoxide group and two -OH groups are biochemically attached to an organic compound to produce a cancer-causing agent.
17. What is water alkalinity? What substance dissolved in solution is most responsible for water alkalinity? What are the benefits of alkalinity?
18. What is the significance of other phases, such as mineral matter or suspended bacteria, in water?
19. How is Henry's Law used?
20. Water hardness is due to calcium ions. Show a chemical reaction by which hardness is added to water that also explains the formation of limestone caves.
21. What are colloidal particles? What are their major effects? What are the types of colloidal particles?
22. Association colloidal particles, such as those formed by soap ions, will hold miniscule droplets of oil within the particles (micelles). Offer an explanation for this phenomenon.
23. Consider water containing iron at a low pE of -6 and a pH of 7. What would be the stable iron species? What iron species would form if the water were exposed to atmospheric oxygen?

SUPPLEMENTARY REFERENCES

Mathias, Jack A., *Integrated Fish Farming* CRC Press/Lewis Publishers, Boca Raton, FL, 1997.

Mitchell, Ralph, *Environmental Microbiology*, Wiley, New York, NY, 1992.

Ostrander, Gary, *Techniques in Aquatic Toxicology*, CRC Press/Lewis Publishers, Boca Raton, FL, 1996.

7 WATER POLLUTION

7.1. NATURE AND TYPES OF WATER POLLUTANTS

Water pollution was introduced briefly in Section 1.10. Water pollutants can be divided among some general classifications, as summarized in Table 7.1. Most of these categories of pollutants, and several subcategories, are discussed in this chapter. An enormous amount of material is published on this subject each year, and it is impossible to cover it all in one chapter. In order to be up to date on this subject the reader may want to survey journals and books dealing with water pollution, such as those listed in the Supplementary References section at the end of this chapter.

Table 7.1. General Types of Water Pollutants

Class of pollutant	Significance
Trace elements	Health, aquatic biota
Metal-organic combinations	Metal transport
Inorganic pollutants	Toxicity, aquatic biota
Asbestos	Human health
Algal nutrients	Eutrophication
Radionuclides	Toxicity
Acidity, alkalinity, salinity (in excess)	Water quality, aquatic life
Sewage	Water quality, oxygen levels
Biochemical oxygen demand	Water quality, oxygen levels
Trace organic pollutants	Toxicity
Pesticides	Toxicity, aquatic biota, wildlife
Polychlorinated biphenyls	Possible biological effects
Chemical carcinogens	Incidence of cancer
Petroleum wastes	Effect on wildlife, esthetics
Pathogens	Health effects
Detergents	Eutrophication, wildlife, esthetics
Sediments	Water quality, aquatic biota, wildlife
Taste, odor, and color	Esthetics

7.2. ELEMENTAL POLLUTANTS

A **trace element** is one that occurs in a medium such as water at a level of perhaps less than a percent, typically at a level of a few parts per million or less. The term **trace substance** is a more general one applied to both elements and chemical compounds. Table 7.2 summarizes the more important trace elements encountered in natural waters. Some of these are recognized as nutrients required for animal and plant life. Of these, many are essential at low levels but toxic at higher levels. This is typical behavior for many substances in the aquatic environment, a point that must be kept in mind in judging whether a particular element is beneficial or detrimental. Some of these elements, such as lead or mercury, have such toxicological and environmental significance that they are discussed in detail in separate sections.

Some of the **heavy metals** (Section 7.3) are among the most harmful of the elemental pollutants. These elements, which are in general the metals in the lower right-hand corner of the periodic table, include essential elements like iron as well as toxic metals like lead, cadmium, and mercury. Most of them have a tremendous affinity for sulfur and attack sulfur bonds in enzymes, thus immobilizing the enzymes so that they do not function. Protein carboxylic acid ($-CO_2H$) and amino ($-NH_2$) groups are also chemically bound by heavy metals. Cadmium, copper, lead, and mercury ions bind to cell membranes, hindering transport processes through the cell wall. Heavy metals may also precipitate phosphate biocompounds or catalyze their decomposition. The biochemical effects of metals are discussed in Chapter 20.

Some of the **metalloids**, elements on the borderline between metals and non-metals, are significant water pollutants. Arsenic, selenium, and antimony are of particular interest.

Inorganic chemicals manufacture has the potential to contaminate water with trace elements. Among the industries regulated for potential trace element pollution of water are those producing chlor-alkali, hydrofluoric acid, sodium dichromate (sulfate process and chloride ilmenite process), aluminum fluoride, chrome pigments, copper sulfate, nickel sulfate, sodium bisulfate, sodium hydrosulfate, sodium bisulfite, titanium dioxide, and hydrogen cyanide.

7.3. HEAVY METALS

Cadmium

Pollutant **cadmium** in water may arise from industrial discharges and mining wastes. Cadmium is widely used in metal plating. Chemically, cadmium is very similar to zinc, and these two metals frequently undergo geochemical processes together. Both metals are found in water in the +2 oxidation state. Cadmium and zinc are common water and sediment pollutants in harbors surrounded with industrial installations.

The effects of acute cadmium poisoning in humans are very serious. Among them are high blood pressure, kidney damage, destruction of testicular tissue, and destruction of red blood cells. It is believed that much of the physiological action of cadmium arises from its chemical similarity to zinc. Specifically, cadmium may replace zinc in some enzymes, thereby altering the stereostructure of the enzyme and impairing its catalytic activity. Disease symptoms ultimately result.

Table 7.2. Occurrence and Significance of Trace Elements in Natural Waters

Element	Sources	Effects and significance
Arsenic	Mining by-product, pesticides, chemical waste	Toxic, possibly carcinogenic
Beryllium	Coal, nuclear power and space industries	Acute and chronic toxicity, possibly carcinogenic
Boron	Coal, detergent formulations, industrial wastes	Toxic to some plants
Cadmium	Industrial discharge, mining waste, metal plating, water pipes	Replaces zinc biochemically, causes high blood pressure and kidney damage, destroys testicular tissue and red blood cells, toxic to aquatic biota.
Chromium	Metal plating, cooling-tower water additive (chromate), normally found as Cr(VI) in polluted water	Essential trace element (glucose tolerance factor), possibly carcinogenic as Cr(VI)
Copper	Metal plating, industrial and domestic wastes, mining, mineral leaching	Essential trace element, not very toxic to animals, toxic to plants and algae at moderate levels
Fluorine (fluoride ion)	Natural geological sources, industrial waste, water additive	Prevents tooth decay at about 1 mg/L, causes mottled teeth and bone damage at around 5 mg/L in water
Iodine (iodide)	Industrial waste, natural brines, seawater intrusion	Prevents goiter
Iron	Corroded metal, industrial wastes, in contact with iron minerals	Essential nutrient (component of hemoglobin), not very toxic, damages materials (bathroom fixtures and clothing)
Lead	Industrial sources, mining, plumbing, fuels (coal)	Toxicity (anemia, kidney disease, nervous system), wildlife destruction
Manganese	Mining, industrial waste, acid mine drainage, microbial action on manganese minerals at low pE	Relatively nontoxic to animals, toxic to plants at higher levels, stains materials (bathroom fixtures and clothing)
Mercury	Industrial waste, mining, coal	Acute and chronic toxicity
Molybdenum	Industrial waste, natural sources, cooling-tower water additive	Toxic to animals, essential for plants

Selenium	Natural geological sources, sulfur, coal	Essential at low levels, toxic at higher levels, causes "alkali disease" and "blind staggers" in cattle, possibly carcinogenic
Silver	Geological sources, mining, electroplating, film-processing wastes	Causes blue-grey discoloration of skin, mucous membranes, eyes
Zinc	Industrial waste, metal	Essential element in many metallo-enzymes, aids wound healing, toxic to plants at higher levels, major component of sewage sludge, limiting land disposal of sludge

Lead

Lead occurs in water in the +II oxidation state and arises from a number of industrial and mining sources. Lead from leaded gasoline used to be a major source of atmospheric and terrestrial lead, much of which eventually enters natural water systems. In addition to pollutant sources, lead-bearing limestone and galena (PbS) contribute lead to natural waters in some locations.

Despite greatly increased total use of lead by industry, evidence from hair samples and other sources indicates that body burdens of this toxic metal have decreased during recent decades. This may be the result of less lead used in plumbing, paint, gasoline, and other products that come in contact with food or drink.

Acute lead poisoning in humans causes severe dysfunction in the kidneys, reproductive system, liver, and the brain and central nervous system. Sickness or death results. Lead poisoning from environmental exposure is thought to have caused mental retardation in many children. Mild lead poisoning causes anemia. The victim may have headaches and sore muscles and may feel generally fatigued and irritable.

The potential exists for lead to pose problems in drinking water in cases where old lead pipe is still in use. Lead used to be used in solder and some pipe-joint formulations, enabling contact of household water with lead. Water that has stood in household plumbing for some time may accumulate spectacular levels of lead (along with zinc, cadmium, and copper) and should be let run for a while before use.

Mercury

Mercury generates the most concern of any of the heavy-metal pollutants. Mercury is found as a trace component of many minerals. Fossil fuel coal and lignite contain mercury, often at levels of 100 parts per billion or even higher, a matter of some concern with increased use of these fuels for energy resources. The primary use of mercury metal has been as an electrode in the electrolytic generation of chlorine gas. Large quantities of inorganic mercury(I) and mercury(II) compounds are used annually. Organic mercury compounds used to be widely applied as pesticides, particularly fungicides. These mercury compounds include aryl mercurials such as phenyl mercuric dimethyldithiocarbamate

$$C_6H_5\text{-}Hg\text{-}S\text{-}C(=S)\text{-}N(CH_3)_2$$

(used in paper mills as a slimicide and as a mold retardant for paper), and alkylmercurials such as ethylmercuric chloride, C_2H_5HgCl, used as a seed fungicide. The alkyl mercury compounds tend to resist degradation and are generally considered to be more of an environmental threat than either the aryl or inorganic compounds.

The toxicity of mercury was tragically illustrated in the Minamata Bay area of Japan during the period 1953-1960. A total of 111 cases of mercury poisoning and 43 deaths were reported among people who had consumed seafood from the bay that had been contaminated with mercury waste from a chemical plant that drained into Minamata Bay. Congenital defects were observed in 19 babies whose mothers had consumed seafood contaminated with mercury. The level of metal in the contaminated seafood was 5-20 parts per million.

Because there are few major natural sources of mercury, and since most inorganic compounds of this element are relatively insoluble, it used to be assumed that mercury was not a serious water pollutant in water. However, in 1970 alarming mercury levels were discovered in fish in Lake Saint Clair between Michigan and Ontario, Canada. A subsequent survey by the U.S. Federal Water Quality Administration revealed a number of other waters contaminated with mercury. It was found that several chemical plants, particularly caustic-chemical plants, were each releasing up to 14 or more kilograms of mercury in wastewater each day.

The unexpectedly high concentrations of mercury found in water and in fish tissues result from the formation of soluble monomethylmercury ion, CH_3Hg^+, and volatile dimethylmercury, $(CH_3)_2Hg$, by anaerobic bacteria in sediments. Mercury from these compounds becomes concentrated in fish lipid (fat) tissue by as much as a thousand-fold compared to the surrounding water. The methylating agent by which inorganic mercury is converted to methylmercury compounds is methylcobalamin, a vitamin B_{12} analog:

$$HgCl_2 \xrightarrow{\text{Methylcobalamin}} CH_3HgCl + Cl^- \qquad (7.3.1)$$

It is believed that the bacteria that synthesize methane produce methylcobalamin as an intermediate in the synthesis. Thus, waters and sediments in which anaerobic decay is occurring provide the conditions under which methylmercury production occurs. In neutral or alkaline waters, the formation of dimethyl mercury, $(CH_3)_2Hg$, is favored. This volatile compound can escape to the atmosphere.

7.4. METALLOIDS

The most significant water pollutant metalloid element is arsenic, a toxic element that has been the chemical villain of more than a few murder plots. Acute arsenic poisoning can result from the ingestion of more than about 100 mg of the element. Chronic poisoning occurs with the ingestion of small amounts of arsenic over a long period of time. There is some evidence that this element is also carcinogenic.

Arsenic occurs in the Earth's crust at an average level of 2-5 ppm. The combustion of fossil fuels, particularly coal, introduces large quantities of arsenic into the environment, much of it reaching natural waters. Arsenic occurs with phosphate minerals and enters into the environment along with some phosphorus compounds. Some formerly used pesticides, particularly those from before World War II, contain highly toxic arsenic compounds. The most common of these are lead arsenate, $Pb_3(AsO_4)_2$; sodium arsenite, Na_3AsO_3; and Paris Green, $Cu_3(AsO_3)_2$. Another major source of arsenic is mine tailings. Arsenic produced as a by-product of copper, gold, and lead refining greatly exceeds the commercial demand for arsenic, and it accumulates as waste material. Arsenic occurs in shellfish and is ingested in acceptable amounts by people who eat shellfish as part of their diet.

Like mercury, arsenic may be converted to more mobile and toxic methyl derivatives—methylarsinic acid, dimethylarsinic acid, and dimethylarsine by bacteria:

$CH_3AsO(OH)_2$	$(CH_3)_2AsO(OH)$	$(CH_3)_2AsH$
Methylarsinic acid	Dimethylarsinic acid	Dimethylarsine

7.5. ORGANICALLY BOUND METALS AND METALLOIDS

Complexation and chelation (Section 6.12) have a strong influence on heavy metals' behavior in natural waters and wastewaters. The formation of organometallic methylmercury and organometalloid methylarsenic compounds were discussed in Sections 7.3 and 7.4, respectively. Both chelation and organometallic compound formation involve the combination of metals and organic entities in water. It must be stressed that the interaction of metals with organic compounds is of utmost importance in determining the role played by the metal in an aquatic system.

There are two major types of metal-organic interactions to be considered in an aquatic system. The first of these is complexation, usually chelation when organic ligands are involved. A reasonable definition of complexation by organics applicable to natural water and wastewater systems is a system in which a species is present that reversibly dissociates to a metal ion and an organic complexing species as a function of hydrogen ion concentration:

$$ML + 2H^+ \rightarrow M^{2+} + H_2L \qquad (7.5.1)$$

In this equation, M^{2+} is a metal ion and H_2L is the acidic form of a complexing—frequently chelating—ligand, L^{2-}.

Organometallic compounds, on the other hand, contain metals bound to organic entities by way of a carbon atom and do not dissociate reversibly at lower pH or greater dilution. Furthermore, the organic component, and sometimes the particular oxidation state of the metal involved, may not be stable apart from the organometallic compound. In general, as illustrated in Figure 7.1, organometallic compounds may be classified as those in which the organic group is an alkyl group, such as ethyl, $-C_2H_5$; a π electron donor, such as ethylene; or carbon monoxide, CO. Various combinations of these three types are common, the most prominent of which are arene carbonyl species in which a metal atom is bonded to both an aryl entity, such as benzene, and to several carbon monoxide molecules.

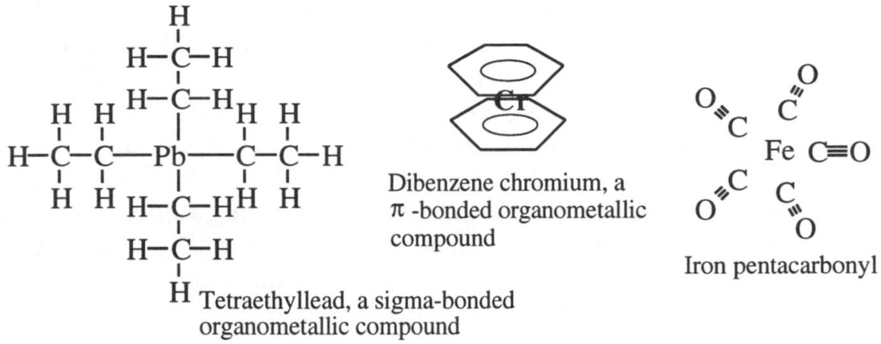

Tetraethyllead, a sigma-bonded organometallic compound

Dibenzene chromium, a π-bonded organometallic compound

Iron pentacarbonyl

Figure 7.1. Types of organometallic compounds.

Of all the metals, tin has the greatest number of organometallic compounds in commercial use, with global production on the order of 40,000 metric tons per year. Major industrial uses of organotin compounds include applications of tin compounds in fungicides, acaricides, disinfectants, antifouling paints, stabilizers to lessen the effects of heat and light in PVC plastics, catalysts, and precursors for the formation of films of SnO_2 on glass. Tributyl tin chloride and related tributyl tin (TBT) compounds have bactericidal, fungicidal, and insecticidal properties and are of particular environmental significance because of growing use as industrial biocides. In addition to synthetic organotin compounds, methylated tin species can be produced biologically in the environment. As a result of their widespread production, distribution, and use, organotin compounds, such as tetra-*n*-butyltin, $Sn(C_4H_9)_4$, are common water pollutants.

The interaction of trace metals with organic compounds in natural waters is too vast an area to cover in detail in this chapter; however, it may be noted that metal-organic interactions may involve organic species of both pollutant (such as EDTA) and natural (such as fulvic acids) origin. These interactions are influenced by, and sometimes play a role in, redox equilibria; formation and dissolution of precipitates; colloid formation and stability; acid-base reactions; and microorganism-mediated reactions in water. Metal-organic interactions may increase or decrease the toxicity of metals in aquatic ecosystems, and they have a strong influence on the growth of algae in water.

7.6. INORGANIC SPECIES

Some important inorganic water pollutants not discussed in other parts of this chapter are considered here. These include cyanide, ammonia, carbon dioxide, hydrogen sulfide, nitrite, and sulfite.

Cyanide, a deadly poisonous substance, exists in water as HCN, a weak acid. The cyanide ion has a strong affinity for many metal ions, forming relatively less toxic ferrocyanide, $Fe(CN)_6^{4-}$, with iron(II), for example. Volatile HCN is very toxic and has been used in gas chamber executions in the U.S. Cyanide is widely used in industry, especially for metal cleaning and electroplating. It is also one of the main gas and coke scrubber effluent pollutants from gas works and coke ovens. Cyanide is widely used in certain mineral-processing operations. With all of these uses the potential exists for water contamination by cyanide.

Excessive levels of ammoniacal nitrogen can cause water-quality problems. **Ammonia** is the initial product of the decay of nitrogenous organic wastes, and its presence frequently indicates the presence of such wastes. It is a constituent of many groundwaters and is sometimes added to drinking water, where it reacts with chlorine to provide disinfectant residual chlorine (see Section 8.11). Most ammonia in water is present as ammonium ion, NH_4^+, rather than as NH_3.

Hydrogen sulfide, H_2S, is a product of the anaerobic decay of organic matter containing sulfur. It is also produced in the anaerobic reduction of sulfate by microorganisms (see Reaction 6.6.5) and is evolved as a gaseous pollutant from geothermal waters. Wastes from chemical plants, paper mills, textile mills, and tanneries may also contain H_2S. Its presence is easily detected by its characteristic rotten-egg odor. The sulfide ion, S^{2-}, produced from H_2S has tremendous affinity for many heavy metals, and precipitation of metallic sulfides often accompanies production of H_2S.

Free **carbon dioxide**, CO_2, is frequently present in water at high levels due to decay of organic matter. It is also added to softened water during water treatment as part of a recarbonation process (see Reaction 8.7.8). Too much carbon dioxide lowers the pH of water and may make it more corrosive and harmful to aquatic life.

Nitrite ion, NO_2^-, occurs in water as an intermediate oxidation state of nitrogen. Its pE range of stability (Section 6.10) is relatively narrow. Nitrite is added to some industrial process water to inhibit corrosion; it is rarely found in drinking water at levels over 0.1 mg/L.

Sulfite ion, SO_3^{2-}, which may exist as HSO_3^- ion, depending on pH, is found in some industrial wastewaters. Sodium sulfite is commonly added to boiler feedwaters as an oxygen scavenger.

Asbestos in Water

The toxicity of inhaled asbestos is well established. The fibers scar lung tissue and cancer eventually develops, often 20 or 30 years after exposure. It is not known for sure whether asbestos is toxic in drinking water. This has been a matter of considerable concern because of the dumping of taconite (iron ore tailings) containing asbestos-like fibers into Lake Superior. The fibers have been found in drinking waters of cities around the lake. After having dumped the tailings into Lake Superior since 1952, the Reserve Mining Company at Silver Bay on Lake Superior solved the problem in 1980 by constructing a 6-square-mile containment basin inland from the lake. This $370-million facility keeps the taconite tailings covered with a 3-meter layer of water to prevent escape of fiber dust.

7.7. ALGAL NUTRIENTS AND EUTROPHICATION

The term **eutrophication**, derived from the Greek word meaning "well-nourished," describes a condition of lakes or reservoirs involving excess algal growth, which may eventually lead to severe deterioration of the body of water. The first step in eutrophication of a body of water is an input of plant nutrients (Table 7.3) from watershed runoff or sewage. The nutrient-rich body of water then produces a great deal of plant biomass by photosynthesis, along with a smaller amount of animal biomass. Dead biomass accumulates in the bottom of the lake, where it partially decays, recycling nutrient carbon dioxide, phosphorus, nitrogen, and potassium. If the lake is not too deep, bottom-rooted plants begin to grow, accelerating the accumulation of solid material in the basin. Eventually a marsh is formed, which finally fills in to produce a meadow or forest.

Table 7.3. Essential Plant Nutrients: Sources and Functions

Nutrient	Source	Function
Macronutrients		
Carbon (CO_2)	Atmosphere, decay	Biomass constituent
Hydrogen	Water	Biomass constituent
Oxygen	Water	Biomass constituent
Nitrogen (NO_3^-)	Decay, atmosphere (from nitrogen-fixing organisms), pollutants	Protein constituent
Phosphorus (phosphate)	Decay, minerals, pollutants	DNA/RNA constituent
Potassium	Minerals, pollutants	Metabolic function
Sulfur (sulfate)	Minerals	Proteins, enzymes
Magnesium	Minerals	Metabolic function
Calcium	Minerals	Metabolic function
Micronutrients		
B, Cl, Co, Cu, Fe, Mo, Mn, Na, Si V, Zn	Minerals, pollutants	Metabolic function and/or constituent of enzymes

Eutrophication occurs naturally and is responsible for phenomena such as the formation of huge deposits of coal and peat. However, human activity can greatly accelerate the process with often detrimental results. To understand why this is so, refer to Table 7.3, which shows the chemical elements needed for plant growth. Most of these are present at a level more than sufficient to support plant life in the average lake or reservoir. Hydrogen and oxygen come from the water itself. Carbon is provided by CO_2 from the atmosphere or from decaying vegetation. Sulfate, magnesium, and calcium are normally present in abundance from mineral strata in contact with the water. The micronutrients are required at only very low levels (for example, approximately 40 ppb for copper). Therefore, the nutrients most likely to be limiting are the "fertilizer" elements: nitrogen, phosphorus, and potassium. These are all present in sewage and are, of course, found in runoff from heavily fertilized fields. They are also constituents of various kinds of industrial wastes. Each of these elements can also come from natural sources—phosphorus and potassium from mineral formations, and nitrogen fixed by bacteria, or discharge of lightning in the atmosphere.

Generally, the single plant nutrient most likely to be limiting is phosphorus, and it is generally named as the culprit in excessive eutrophication. Household detergents used to be a common source of phosphate in wastewater, and eutrophication control has concentrated upon eliminating phosphates from detergents, removing phosphate at the sewage-treatment plant, and preventing phosphate-laden sewage effluents (treated or untreated) from entering bodies of water. In some cases, nitrogen or even carbon may be limiting nutrients. This is particularly true of nitrogen in seawater.

The whole eutrophication picture is a complex one. Ironically, in a food-poor world, nutrient-rich wastes from over-fertilized fields or from sewage are causing excessive plant growth in many lakes and reservoirs. This is an example of how often pollutants are simply resources (in this case, plant nutrients) gone to waste.

7.8. ACIDITY, ALKALINITY, AND SALINITY

Aquatic biota are sensitive to extremes of pH. Largely because of osmotic effects, they cannot live in a medium having a salinity to which they are not adapted. Thus, a fresh-water fish soon succumbs in the ocean, and sea fish normally cannot live in fresh water. Excess salinity soon kills plants not adapted to it. There are, of course, ranges in salinity and pH in which organisms live. These ranges frequently may be represented by a reasonably symmetrical curve, along the fringes of which an organism may exist without really thriving.

The most common source of **pollutant acid** in water is acid mine drainage. The sulfuric acid in such drainage arises from the microbial oxidation of pyrite (FeS_2) or other sulfide minerals. The values of pH encountered in acid-polluted water may fall below 3, a condition deadly to most forms of aquatic life except the culprit bacteria mediating the pyrite and iron(II) oxidation. Industrial wastes frequently contribute strong acid to water. Sulfuric acid produced by the air oxidation of pollutant sulfur dioxide enters natural waters as acidic rainfall. In cases where the water does not have contact with a basic mineral, such as limestone, the water pH may become dangerously low. This condition occurs in some Canadian lakes, for example.

Acid Mine Waters

Bacterial action on metal compounds generates acid mine drainage, a common and damaging water pollutant. Acid mine water results from the presence of sulfuric acid produced by the oxidation of pyrite, FeS_2. Microorganisms are closely involved in the overall process, which consists of several reactions. These were discussed along with elemental transitions mediated by microorganisms in Section 6.6. The major pollutant species produced with acid mine water are strong sulfuric acid, H_2SO_4; acidic hydrated iron(III) ion, $Fe(H_2O)_6^{3+}$; and damaging, unsightly precipitates of amorphous, semigelatinous hydrated iron(III) oxide, $Fe(OH)_3$.

Alkalinity

In many geographic areas, the soil and mineral strata are alkaline and impart a high **alkalinity**, frequently accompanied by a high pH, to water. Human activity can cause release of soil alkalinity into water, such as by exposure of alkaline overburden from strip mining to surface water or groundwater. Excess alkalinity exhibits a characteristic fringe of white salts at the edges of a body of water.

Water **salinity** may be increased by a number of human activities. Water passing through a municipal water system picks up salt from a number of processes; for example, recharging water softeners with sodium chloride. Salts can leach from mining spoil piles. Irrigation adds a great deal of salt to water, a phenomenon responsible for the Salton Sea in California, and is a source of conflict between the U. S. and Mexico over saline contamination of the Rio Grande and Colorado rivers. Saline seeps occur in some of the Western U .S. states when water seeps into a slight depression in tilled, sometimes irrigated, fertilized land, carrying salts (particularly sodium, magnesium, and calcium sulfates) along with it. The water evaporates in the dry summer, leaving a salt-laden area that no longer supports much plant growth. With time, these areas spread, removing once fertile crop land from production.

7.9. OXYGEN, OXIDANTS, AND REDUCTANTS

Oxygen is a vitally important species in water. In water, oxygen is consumed rapidly by the oxidation of organic matter, $\{CH_2O\}$:

$$\{CH_2O\} + O_2 \xrightarrow{\text{Microorganisms}} CO_2 + H_2O \tag{7.9.1}$$

Unless the water is reaerated efficiently, such as by turbulent flow in a shallow stream, it rapidly becomes depleted in oxygen and will not support higher forms of aquatic life.

In addition to the microorganism-mediated oxidation of organic matter, oxygen in water may be consumed by the biooxidation of nitrogenous material,

$$NH_4^+ + 2O_2 \rightarrow 2H^+ + NO_3^- + H_2O \tag{7.9.2}$$

and by the chemical or biochemical oxidation of chemical reducing agents:

$$4Fe^{2+} + O_2 + 10H_2O \rightarrow 4Fe(OH)_3(s) + 8H^+ \tag{7.9.3}$$

$$2SO_3^{2-} + O_2 \rightarrow 2SO_4^{2-} \tag{7.9.4}$$

The degree of oxygen consumption by microbially mediated oxidation of contaminants in water is called the **biochemical oxygen demand** (or biological oxygen demand), **BOD**. This parameter is commonly measured by determining the quantity of oxygen utilized by suitable aquatic microorganisms during a five-day period and is used to measure short-term oxygen demand exerted by a pollutant.

The addition of oxidizable pollutants to streams produces a typical **oxygen sag curve** as shown in Figure 7.2. Initially, a well-aerated, unpolluted stream is relatively free of oxidizable material; the oxygen level is high; and the bacterial population is relatively low. With the addition of oxidizable pollutant, the oxygen level drops because reaeration cannot keep up with oxygen consumption. In the decomposition zone, the bacterial population rises. The septic zone is characterized by a high bacterial population and very low oxygen levels. It terminates when the oxidizable pollutant is exhausted, then the recovery zone begins in which bacteria decrease and dissolved oxygen increases until the water regains its original condition.

7.10. ORGANIC POLLUTANTS

Sewage

As shown in Table 7.4, sewage from domestic, commercial, food-processing, and industrial sources contains a wide variety of pollutants, including organic pollutants. Some of these, particularly oxygen-demanding substances (see Section 7.9), oil, grease, and solids, are removed by primary and secondary sewage-treatment processes. Others, such as salts, heavy metals, and degradation-resistant (refractory) organics, are not efficiently removed.

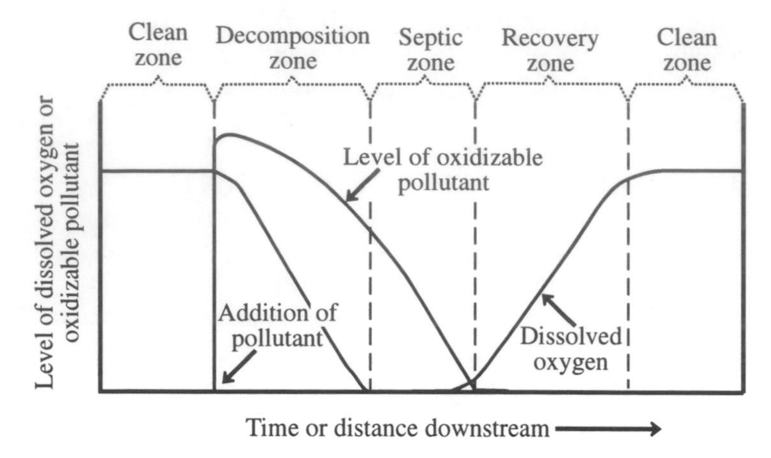

Figure 7.2. Oxygen sag curve resulting from the addition of oxidizable pollutant material to a stream.

Table 7.4. Some of the Primary Constituents of Municipal Sewage

Constituent	Potential sources	Effects in water
Oxygen-demanding substances	Mostly organic materials, particularly human feces	Consume dissolved oxygen
Refractory organics	Industrial wastes, household products	Toxic to aquatic life
Viruses	Human wastes	Cause disease (possibly cancer); major deterrent to sewage recycle through water systems
Detergents	Household detergents	Esthetics, prevent grease and oil removal, toxic to aquatic life
Phosphates	Detergents	Algal nutrients
Grease and oil	Cooking, food processing, industrial wastes	Esthetics, harmful to some aquatic life
Salts	Human wastes, water softeners, industrial wastes	Increase water salinity
Heavy metals	Industrial wastes, chemical laboratories	Toxicity
Chelating agents	Some detergents, indusdustrial wastes	Heavy metal ion solubilization and transport
Solids	All sources	Esthetics, harmful to aquatic life

Disposal of inadequately treated sewage can cause severe problems. For example, offshore disposal of sewage produces beds of sewage residues. Now that most municipal sewage is treated in sewage treatment plants, a significant problem can be the sludge produced as a product of the sewage treatment processes employed. This sludge contains organic material that continues to degrade slowly; refractory organics; and heavy metals. The amounts of sludge produced are truly staggering, ranging up to a million tons of sludge or more each year for a populous metropolitan area. A major consideration in the safe disposal of such amounts of sludge is the presence of potentially dangerous components such as heavy metals.

Careful control of sewage sources is needed to minimize sewage pollution problems. Particularly, heavy metals and refractory organic compounds need to be controlled at the source to enable use of sewage, or treated sewage effluents, for irrigation, recycle to the water system, or groundwater recharge.

Soaps, Detergents, and Detergent Builders

Soaps, detergents, and associated chemicals are potential sources of organic pollutants. These pollutants are discussed briefly here.

Soaps are salts of higher fatty acids, such as sodium stearate, $C_{17}H_{35}COO^{-}Na^{+}$. The cleaning action of soap results largely from its emulsifying power caused by the dual nature of the soap anion. A soap ion consists of an ionic carboxyl "head" and a long hydrocarbon "tail":

O
||
C–O⁻Na⁺

In the presence of oils, fats, and other water-insoluble organic materials, the "tail" of the anion tends to dissolve in the organic matter, whereas the "head" remains in aquatic solution. Thus, the soap emulsifies, or suspends, organic material in water. In the process, the anions form colloidal soap micelles, as shown in Figure 6.9. Soap lowers the surface tension of water, thus making the water "wetter."

The primary disadvantage of soap as a cleaning agent comes from its reaction with divalent cations to form insoluble salts of fatty acids:

$$2C_{17}H_{35}COO^{-}Na^{+} + Ca^{2+} \rightarrow Ca(C_{17}H_{35}CO_2)_2(s) + 2Na^{+} \quad (7.10.1)$$

These insoluble products, usually salts of magnesium or calcium, are not at all effective as cleaning agents. In addition, the insoluble "curds" form unsightly deposits on clothing and in washing machines. If sufficient soap is used, all of the divalent cations may be removed by their reaction with soap, and the water containing excess soap will have good cleaning qualities. This is the approach commonly used when soap is employed in the bathtub or wash basin, where the insoluble calcium and magnesium salts can be tolerated. However, in applications such as washing clothing, the water must be softened by the removal of calcium and magnesium or their complexation by substances such as polyphosphates.

Although the formation of insoluble calcium and magnesium salts has resulted in the virtual elimination of soap as a cleaning agent for clothing, dishes, and most other materials, it has distinct advantages from the environmental standpoint. As soon as soap gets into sewage or an aquatic system, it generally precipitates as calcium and magnesium salts. Hence, any effects that soap might have in solution are eliminated. As it eventually biodegrades, the soap is completely eliminated from the environment. Therefore, aside from the occasional formation of unsightly scum, soap does not cause any substantial pollution problems.

Synthetic **detergents** have good cleaning properties and do not form insoluble salts with "hardness ions" such as calcium and magnesium. Such synthetic detergents have the additional advantage of being the salts of relatively strong acids, and, therefore, they do not precipitate out of acidic waters as insoluble acids, an undesirable characteristic of soaps.

Synthetic detergents usually have a surface-active agent, or surfactant, added to them that lowers the surface tension of water to which the detergent is added, making the water "wetter." Until the early 1960s, the most common surfactant used was an alkyl benzene sulfonate, ABS, a sulfonation product of an alkyl derivative of benzene:

$$Na^{+}\,{}^{-}O-S(=O)_2-C_6H_4-CH(CH_3)-CH_2-CH(CH_3)-CH_2-CH(CH_3)-CH_2-CH(CH_3)-CH_2-CH(CH_3)-CH_3$$

ABS was only very slowly biodegradable because of its branched-chain structure. The persistence of this nondegraded surfactant caused "heads" of foam to appear in glasses of drinking water in areas where sewage was recycled through the domestic water supply. Spectacular beds of foam appeared near sewage outflows, and sometimes the entire aeration tank of an activated sludge sewage treatment plant would be smothered by a blanket of foam. Other undesirable effects of persistent detergents upon waste-treatment processes were lowered surface tension of water; deflocculation of colloids; flotation of solids; emulsification of grease and oil; and destruction of useful bacteria. Consequently, ABS was replaced by biodegradable α-benzenesulfonate (LAS) surfactant having the general structure:

$$H-(CH_2)_8-CH(C_6H_4-S(=O)_2-O^{-}\,{}^{+}Na)-(CH_2)_3-H$$

where the benzene ring may be attached at any point on the alkyl chain except at the ends. LAS is more biodegradable than ABS because the alkyl portion of LAS is not branched, nor does it contain in its hydrocarbon chain any carbon atoms bonded to four other carbon atoms, a structural feature that is extremely detrimental to biodegradability. Since LAS has replaced ABS in detergents, problems arising from the surface-active agent in the detergents (such as toxicity to fish fingerlings) have greatly diminished and the levels of surface-active agents found in water have decreased markedly.

Most of the environmental problems later attributed to detergents arose from builders rather than surface-active agents. Builders added to detergent formulations bind to hardness ions, making the detergent solution alkaline and greatly improving the action of the detergent surfactant. Polyphosphates were once widely used in builders. Other ingredients include anticorrosive sodium silicates, amide foam stabilizers, soil-suspending carboxymethylcellulose, diluent sodium sulfate, and water absorbed by other components. Of these materials, the polyphosphates have caused the most concern as environmental pollutants, although these problems have largely been resolved.

Biorefractory Organic Pollutants

Millions of tons of organic compounds are manufactured globally each year. Significant quantities of several thousand such compounds appear as water pollutants. Most of these compounds, particularly the less biodegradable ones, are substances to which living organisms have not been exposed until recent years. Frequently, their effects upon organisms are not known, particularly for long-term exposures at very low levels. The potential of synthetic organics for causing genetic damage, cancer, or other ill effects is uncomfortably high. On the positive side, organic pesticides enable a level of agricultural productivity without which millions would starve.

Biorefractory organics are the organic compounds of most concern in wastewater, particularly when they are found in sources of drinking water. These are nonbiodegradable, low-molecular-mass compounds of low volatility, such as organochlorine compounds and some aryl hydrocarbons. Many of these compounds have been found in drinking water, and some are known to cause taste and odor problems in water. Biorefractory compounds are not completely removed by biological treatment, and water contaminated with these compounds may require treatment by physical and chemical means, including air stripping, solvent extraction, ozonation, and carbon absorption.

Pesticides in Water

The introduction of DDT during World War II marked the beginning of a period of very rapid growth in pesticide use. Pesticides are employed for many different purposes.[1] Chemicals used in the control of invertebrates include insecticides, molluscicides for the control of snails and slugs, and nematicides for the control of microscopic roundworms. Vertebrates are controlled by rodenticides, which kill rodents; avicides used to repel birds; and piscicides used in fish control. Herbicides are used to kill plants. Plant growth regulators, defoliants, and desiccants are used for various purposes in the cultivation of plants. Fungicides are used against fungi, bactericides against bacteria, and algicides against algae. Annual U.S. pesticide production is several hundred million kilograms of active ingredients. Although insecticide production has remained about level during the last two or three decades, herbicide production has increased greatly as chemicals have increasingly replaced cultivation of land in the control of weeds. Large quantities of pesticides enter water either directly, in applications such as mosquito control, or indirectly, primarily from drainage of agricultural lands. Prominent among these pesticides are chlorinated hydrocarbons, organic phosphates, and carbamates, which are derived from carbamic acid,

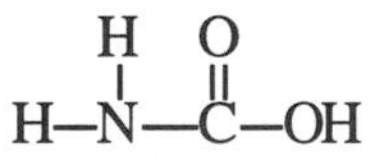

$$H{-}N(H){-}C({=}O){-}OH$$

The toxicities of pesticides vary widely. Parathion,

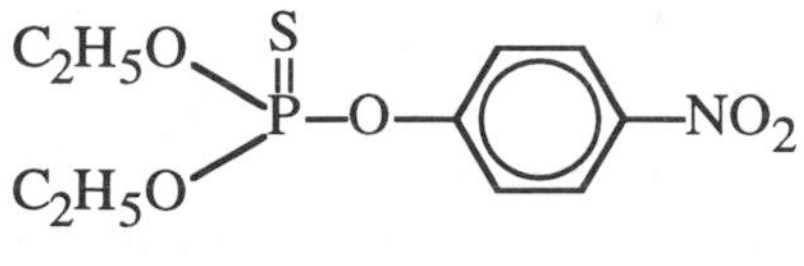

$$(C_2H_5O)_2P({=}S){-}O{-}C_6H_4{-}NO_2$$

banned for most uses in 1991, is so toxic that as little as 120 mg of it has been known to kill an adult human and a dose of 2 mg has killed a child. Most accidental poisonings have occurred by absorption through the skin. Since its use began, several hundred people have been killed by parathion.

Much less toxic **malathion** shows how differences in structural formula can cause pronounced differences in the properties of organophosphate pesticides. Malathion has two carboxyester linkages, which are hydrolyzable by carboxylase enzymes to relatively nontoxic products as shown by the following reaction:

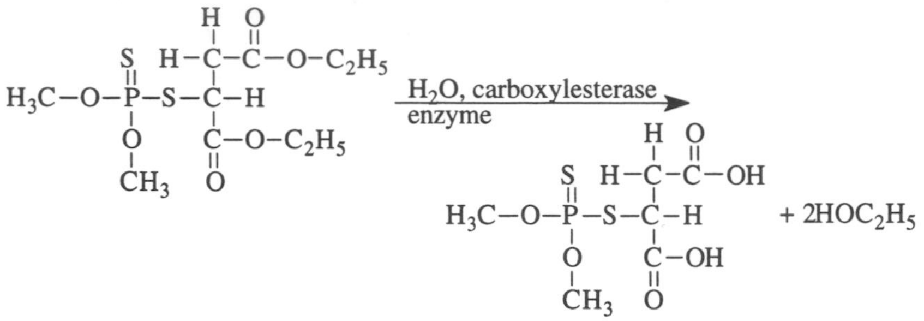

$$H_3C{-}O{-}\underset{\underset{CH_3}{|}}{\underset{|}{O}}\overset{S}{\overset{\|}{P}}{-}S{-}\underset{\underset{O}{\|}}{\underset{C{-}O{-}C_2H_5}{|}}\overset{H{-}\overset{H}{\overset{|}{C}}{-}\overset{O}{\overset{\|}{C}}{-}O{-}C_2H_5}{\overset{|}{C}}{-}H \xrightarrow[\text{enzyme}]{H_2O,\ \text{carboxylesterase}} H_3C{-}O{-}\underset{\underset{CH_3}{|}}{\underset{|}{O}}\overset{S}{\overset{\|}{P}}{-}S{-}\underset{\underset{O}{\|}}{\underset{C{-}OH}{|}}\overset{H{-}\overset{H}{\overset{|}{C}}{-}\overset{O}{\overset{\|}{C}}{-}OH}{\overset{|}{C}}{-}H + 2HOC_2H_5 \quad (7.10.2)$$

The enzymes that accomplish malathion hydrolysis are possessed by mammals, but not by insects, so that mammals can detoxify malathion, whereas insects cannot; therefore, malathion is selectively toxic to insects. Malthion's LD_{50} (dose required to kill 50% of test subjects) for adult male rats is about 100 times that of parathion, one of the more toxic organophosphate insecticides once commonly used.

The chlorinated hydrocarbon DDT (dichlorodiphenyltrichloroethane or 1,1,1-trichloro-2,2-di-(4-chlorophenyl)-ethane),

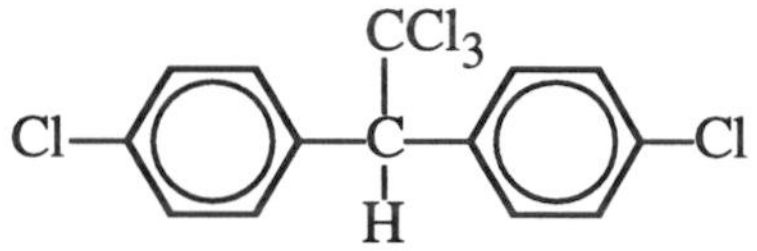

$$Cl{-}C_6H_4{-}\underset{H}{\underset{|}{\overset{CCl_3}{\overset{|}{C}}}}{-}C_6H_4{-}Cl \quad \text{DDT}$$

was used in massive quantities following World War II. Although its toxicity is generally low, its persistence and accumulation in food chains have led to a ban on DDT use in the United States; it is still employed in some countries.

A number of water pollution and health problems have been associated with the manufacture of pesticides. For example, degradation-resistant hexachlorobenzene (structural formula shown below) once used to make other pesticides has often been found in water:

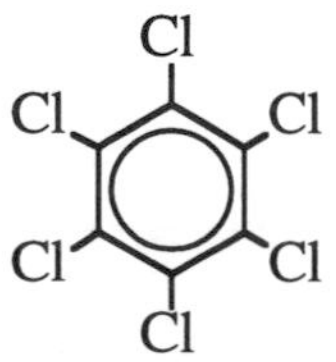

$$C_6Cl_6$$

The most notorious byproducts of pesticide manufacture are **polychlorinated dibenzodioxins**, which have the same basic structure as that of TCDD (2,3,7,8-tetrachlorodibenzo-*p*-dioxin shown in Figure 7.3), but different numbers and arrangements of chlorine atoms on the ring structure. Commonly referred to as "dioxins," these species have a high environmental and toxicological significance.[2]

From 1 to 8 Cl atoms may be substituted for H atoms on dibenzo-*p*-dioxin, giving a total of 75 possible chlorinated derivatives. Of these, the most notable pollutant and hazardous waste compound is **2,3,7,8-tetrachlorodibenzo-*p*-dioxin (TCDD)**, shown in Figure 7.3, and often referred to simply as "**dioxin**." This compound, which is one of the most toxic of all synthetic substances to some animals, was produced as a low-level contaminant in the manufacture of some aryl, oxygen-containing organohalide compounds such as chlorophenoxy herbicides and hexachlorophene (Figure 7.4) manufactured by processes used until the 1960s.

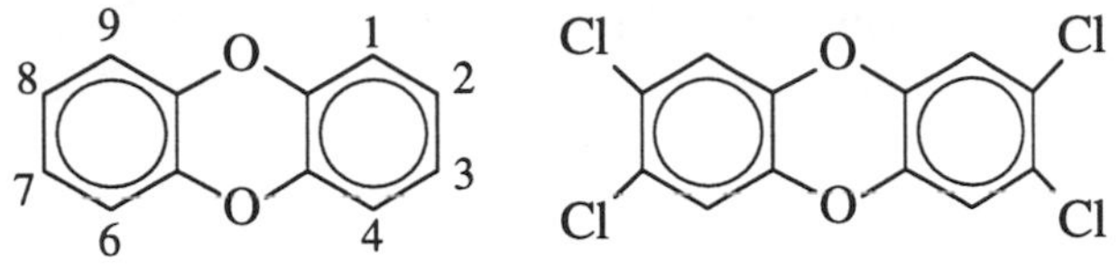

Dibenzo-*p*-dioxin 2,3,7,8-Tetrachlorodibenzo-*p*-dioxin

Figure 7.3 Dibenzo-*p*-dioxin and 2,3,7,8-tetrachlorodibenzo-*p*-dioxin (TCDD), often called simply "dioxin." In the structure of dibenzo-*p*-dioxin each number refers to a numbered carbon atom to which an H atom is bound and the names of derivatives are based upon the carbon atoms where another group has been substituted for the H atoms, as is seen by the structure and name of 2,3,7,8-tetrachlorodibenzo-*p*-dioxin.

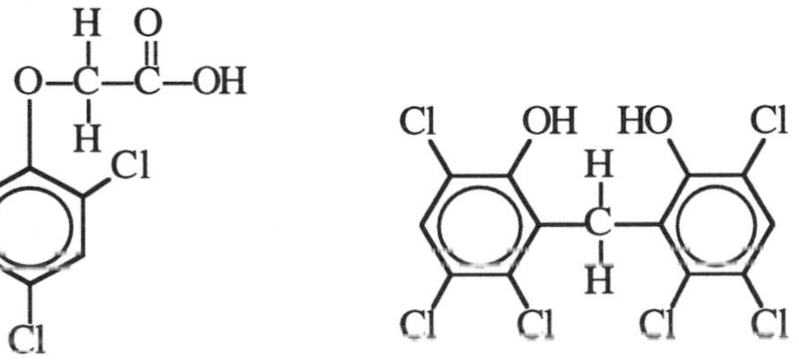

2,4–Dichlorophenoxy-acetic acid (and esters) Hexachlorophene

Figure 7.4. Two chemicals whose manufacture resulted in the production of byproduct TCDD.

The chlorophenoxy herbicides, including 2,4,5-trichlorophenoxyacetic acid (2,4,5-T), were manufactured on a large scale for weed and brush control and as military defoliants. Fungicidal and bactericidal hexachlorophene was once widely applied to crops in the production of vegetables and cotton and was used as an antibacterial agent in personal care products, an application that has been discontinued because of toxic effects and possible TCDD contamination.

TCDD has a very low vapor pressure of only 1.7×10^{-6} mm Hg at 25°C, a high melting point of 305°C, and a water solubility of only 0.2 μg/L. It is stable thermally up to about 700°C, has a high degree of chemical stability, and is poorly biodegradable. It is very toxic to some animals, with an LD_{50} of only about 0.6 μg/kg body mass in male guinea pigs. (The type and degree of its toxicity to humans is largely unknown; it is known to cause a severe skin condition called chloracne.) Because of its properties, TCDD is a stable, persistent environmental pollutant and hazardous waste constituent of considerable concern. It has been identified in some municipal incineration emissions and has been a widespread environmental pollutant from improper waste disposal.

The most notable case of TCDD contamination resulted from efforts to control dust by spraying waste oil mixed with TCDD on roads and horse arenas in Missouri in the early 1970s. As a result of this contamination, the U. S. EPA bought out the entire TCDD-contaminated town of Times Beach, Missouri, in March, 1983, at a cost of $33 million and set up a soil incinerator that operated on the town site until early 1997. TCDD has been released in a number of industrial accidents, the most massive of which exposed several tens of thousands of people to a cloud of chemical emissions spread over an approximately 3-square-mile area at the Givaudan-La Roche Icmesa manufacturing plant near Seveso, Italy, in 1976.

One of the greater environmental disasters ever to result from pesticide manufacture involved the production of Kepone, structural formula

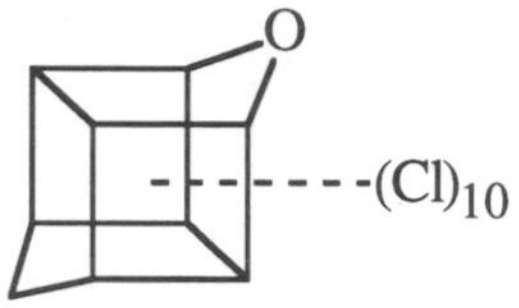

O

$(Cl)_{10}$

Used for the control of banana-root borer, tobacco wireworm, ants, and cockroaches, kepone exhibits acute, delayed, and cumulative toxicity in birds, rodents, and humans, and it causes cancer in rodents. Kepone was manufactured in Hopewell, Virginia, during the mid-1970s, where workers were exposed to it and are alleged to have suffered health problems as a result. As much as 53,000 kg of Kepone may have been dumped from the manufacturing operation into the Hopewell sewage system during the years that the plant was operated, causing the sewage treatment plant to become inoperative at times. The sewage effluent was discharged to the James River, resulting in extensive environmental dispersion and toxicity to aquatic organisms. Decontamination of the river would have required dredging and detoxification of 135 million cubic meters of river sediment at a prohibitively high cost of several billion dollars.

Polychlorinated Biphenyls

First discovered as environmental pollutants in 1966, polychlorinated byphenyls (PCB compounds) have been found throughout the world in water, sediments, bird tissue, and fish tissue. These compounds constitute an important class of special wastes. They are made by substituting from 1 to 10 Cl atoms onto the biphenyl aromatic structure as shown on the left in Figure 7.5. This substitution can produce 209 different compounds (congeners), of which one example is shown in Figure 7.5.

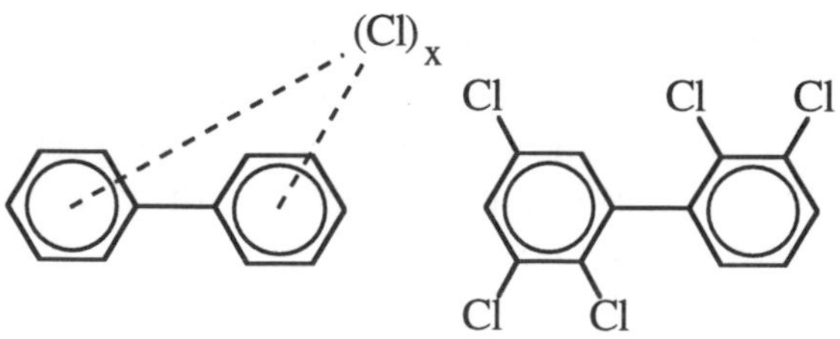

$(Cl)_x$

Cl Cl Cl

Cl Cl

Figure 7.5. General formula of polychlorinated biphenyls (left), where X may range from 1 to 10, and a specific 5-chlorine congener (right).

Polychlorinated biphenyls have very high chemical, thermal, and biological stability; low vapor pressure; and high dielectric constants. These properties have led to the use of PCBs as coolant-insulation fluids in transformers and capacitors, for the impregnation of cotton and asbestos, as plasticizers, and as additives to some epoxy paints. The same properties that made extraordinarily stable PCBs so useful also contributed to their widespread dispersion and accumulation in the environment. Under regulations issued in the U.S. under the authority of the Toxic Substances Control Act passed in 1976, the manufacture of PCBs was discontinued in the U.S. and their uses and disposal were strictly controlled.

Several chemical formulations have been developed to substitute for PCBs in electrical applications. Disposal of PCBs from discarded electrical equipment and other sources remains a problem, particularly since PCBs can survive ordinary incineration by escaping as vapors through the smokestack. However, they can be destroyed by special incineration processes.

PCBs are especially prominent pollutants in the sediments of the Hudson River as a result of waste discharges from two capacitor manufacturing plants, which operated about 60 km upstream from the southernmost dam on the river from 1950 to 1976. The river sediments downstream from the plants exhibit PCB levels of about 10 ppm, 1-2 orders of magnitude higher than levels commonly encountered in river and estuary sediments. Recent evidence suggests that significant amounts of the PCBs have undergone biodegradation by the action of anaerobic bacteria in the sediments, and there is some hope that the process may be augmented by the addition of oxygen and essential nutrients, thus resulting in additional aerobic biodegradation.

7.11. RADIONUCLIDES IN THE AQUATIC ENVIRONMENT

The massive production of **radionuclides** (radioactive isotopes) by weapons and nuclear reactors since World War II has been accompanied by increasing concern about the effects of radioactivity upon health and the environment. Radionuclides are produced as fission products of heavy nuclei of such elements as uranium or plutonium. They also result from the reaction of neutrons with stable nuclei. These phenomena are illustrated in Figure 7.6. Radionuclides are formed in large quantities as waste products in nuclear power generation, and their ultimate disposal has caused much controversy. Artificially produced radionuclides are also

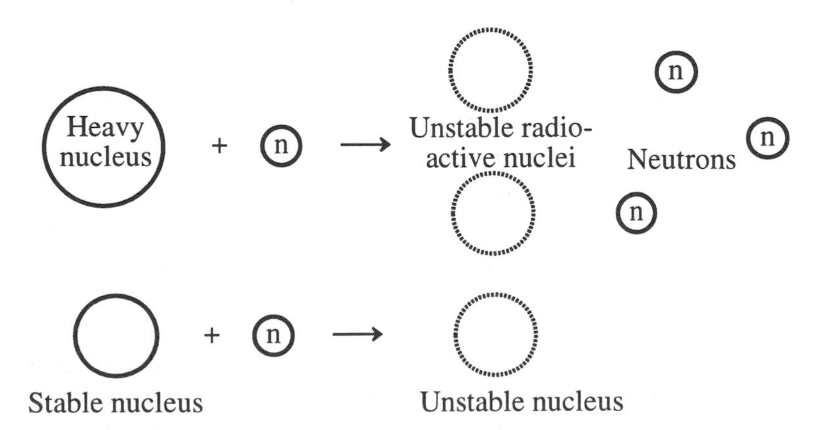

Figure 7.6. A heavy nucleus, such as that of ^{235}U, may absorb a neutron and break up (undergo fission), yielding lighter radioactive nuclei. A stable nucleus may absorb a neutron to produce a radioactive nucleus.

widely used in industrial and medical applications, particularly as "tracers" to follow the fates of specific compounds. With so many possible pollutant sources of radionuclides, as well as some significant natural sources, the transport, reactions, and biological concentration of radionuclides in aquatic ecosystems pose some environmental concerns.

Radionuclides differ from other nuclei in that they emit **ionizing radiation**—alpha particles, beta particles, and gamma rays. The most massive of these emissions is the **alpha particle**, a helium nucleus of atomic mass 4, consisting of two neutrons and two protons and denoted ${}^{4}_{2}\alpha$. An example of alpha production is found in the radioactive decay of uranium-238, atomic number 92 and atomic mass 238, which loses an alpha particle, atomic number 2 and atomic mass 4, to yield a thorium nucleus, atomic number 90 and atomic mass 234:

$$ {}^{238}_{92}U \rightarrow {}^{234}_{90}Th + {}^{4}_{2}\alpha + \text{energy} \quad (7.11.1) $$

Beta radiation consists of either highly energetic, negative electrons, which are designated ${}^{0}_{-1}\beta$, or positive electrons, called positrons and designated ${}^{0}_{-1}\beta$. Radioactive chlorine-38 loses a negative beta particle to become argon-38:

$$ {}^{38}_{17}Cl \rightarrow {}^{38}_{18}Ar + {}^{0}_{-1}\beta \quad (7.11.2) $$

Since the negative beta particle has essentially no mass and a -1 charge, the stable argon-38 isotope product has the same mass and a charge 1 greater than chlorine-38.

Gamma rays are electromagnetic radiation similar to X-rays, though more energetic. Since the energy of gamma radiation is often a well-defined property of the emitting nucleus, it may be used in some cases for the qualitative and quantitative analysis of radionuclides.

Alpha particles, beta particles, and gamma rays are called **ionizing radiation** because they produce ions in materials. Because the large alpha particles do not penetrate matter deeply, but cause an enormous amount of ionization along their short path of penetration, they present little hazard outside the body, but are very dangerous to ingest. Beta particles are more penetrating than alpha particles, but produce much less ionization per unit path length. Gamma rays are much more penetrating than particulate radiation. Their degree of penetration is proportional to their energy.

The number of nuclei disintegrating in a short time interval—the **decay rate**—is directly proportional to the number of radioactive nuclei present. Therefore, each radionuclide has a characteristic **half-life**, the period of time during which half of a given number of atoms of a specific kind of radionuclide decay. Ten half-lives are required for the loss of 99.9% of the activity of a radionuclide.

Radiation damages living organisms by initiating harmful chemical reactions in tissues. For example, bonds are broken in the macromolecules that carry out life processes. In cases of acute radiation poisoning, bone marrow, which produces red blood cells, is destroyed and the concentration of red blood cells is diminished. Radiation-induced genetic damage is of great concern, and may not become apparent until many years after exposure. For the majority of the population, unavoidable exposure to natural radiation exceeds that from artificial sources.

Table 7.5 summarizes the major natural and artificial radionuclides likely to be encountered in water. The study of the ecological and health effects of radionuclides

Table 7.5. Radionuclides in Water

Radionuclide	Half-life	Nuclear reaction, description, source
Naturally occurring and from cosmic reactions		
Carbon-14	5730 years	$^{14}N(n,p)^{14}C$,[a] thermal neutrons from cosmic or nuclear-weapon sources reacting with N_2
Silicon-32	~ 300 years	$^{40}Ar(p,x)^{32}Si$, nuclear spallation (splitting of the nucleus) of atmospheric argon by cosmic-ray protons
Potassium-40	~ 1.4×10^9 years	0.0119% of natural potassium
Naturally occurring from ^{238}U series		
Radium-226	1620 years	Diffusion from sediments, atmosphere
Lead-210	21 years	$^{226}Ra \rightarrow$ 6 steps $\rightarrow$ ^{210}Pb
Thorium-230	75,200 years	$^{238}U \rightarrow$ 3 steps $\rightarrow$ ^{230}Th produced *in situ*
Thorium-234	24 days	$^{238}U \rightarrow$ ^{234}Th produced *in situ*
From reactor and weapons fission		
Strontium-90	28 years	^{90}Sr, ^{131}I, and ^{137}Cs are the fission-product radioisotopes of greatest significance because of their high yields and biological activity.
Iodine-131	8 days	
Cesium-137	30 years	
Barium- 140	13 days	The isotopes from ^{140}Ba through ^{85}Kr are listed in generally decreasing order of fission yield.
Zirconium-95	65 days	
Cerium-141	33 days	
Strontium-89	51 days	
Ruthenium-103	40 days	
Krypton-85	10.3 years	
Cobalt-60	5.25 years	From nonfission neutron reactions in reactors
Manganese-54	310 days	From nonfission neutron reactions in reactors
Iron-55	2.7 years	$^{56}Fe(n,2n)^{55}Fe$, from high-energy neutrons acting on iron in weapon hardware
Plutonium-239	24,300 years	$^{235}U(n,\gamma)^{239}Pu$, neutron capture by uranium

[a] This notation denotes the isotope nitrogen-14 reacting with a neutron, n, giving off a proton, p, and forming the isotope carbon-14; other nuclear reactions may be similarly deduced from the notation shown. (Note that x represents nuclear fragments from the spallation reaction.)

involves consideration of many factors. Among these are the type and energy of radiation emitter and the half-life of the source. In addition, the degree to which the particular element is absorbed by living species and the chemical interactions and transport of the element in aquatic ecosystems are important factors. In general, radionuclides with intermediate half-lives are the most dangerous because they persist long enough to enter living systems while still retaining a high activity. Because they may be incorporated within living tissue, radionuclides of "life elements" are particularly dangerous. Much concern has been expressed over radioactive strontium-90, a common waste product of nuclear testing. This element is interchangeable with calcium in bone. When aboveground nuclear weapons testing was a common practice during the height of the "Cold War," strontium-90 fallout dropped onto pasture and crop land and was ingested by cattle. Eventually, it entered the bodies of infants and children by way of cow's milk.

Some radionuclides found in water, primarily radium and potassium-40, originate from natural sources, particularly leaching from minerals. Others come from pollutant sources, primarily nuclear power plants and testing of nuclear weapons. The levels of radionuclides found in water typically are measured in units of picoCuries/liter, where a Curie is 3.7×10^{10} disintegrations per second, and a picoCurie is 1×10^{-12} as great, or 3.7×10^{-2} disintegrations per second (2.2 disintegrations per minute). The radionuclide of most concern in drinking water is **radium**, Ra. Areas in the United States where significant radium contamination of water has been observed include the uranium-producing regions of the western U.S., Missouri, Iowa, Illinois, Wisconsin, Minnesota, Florida, North Carolina, Virginia, and the New England states.

CHAPTER SUMMARY

The chapter summary below is presented in a programmed format to review the main points covered in this chapter. It is used most effectively by filling in the blanks, referring back to the chapter as necessary. The correct answers are given at the end of the summary.

The general types of water pollutants most likely to be involved with metal transport are [1]________________________________, and those most likely to be involved with eutrophication are [2]________________________________. Lead, cadmium, and mercury are all examples of water pollutants called [3]____________ ____________________, whereas arsenic is an example of [4]________________. One of these pollutants that is chemically very similar to zinc is [5]_____________ ____________, one that potentially can get into water from plumbing made of it is [6]__________, and one that is converted to methylated organometallic forms by bacteria is [7]________________. A metalloid pollutant that occurs with phosphate minerals and enters into the environment along with some phosphorus compounds is [8]________________. The two major types of metal-organic interactions to be considered in an aquatic system are [9]______________________________________ ______________________. Of all the metals, [10]_____________ has the greatest number of organometallic compounds in commercial use. Bactericidal, fungicidal, and insecticidal tributyl tin chloride and related tributyl tin (TBT) compounds are of particular environmental significance because of their growing use as [11]_________ ______________________. Some classes and specific examples of inorganic water pollutants are [12]____ ____ ____. A highly toxic water pollutant that is widely used in certain mineral-processing operations is [13]________ _____. A water pollutant that is the initial product of the decay of nitrogenous organic wastes is [14]_____________ and one produced by bacteria acting on sulfate is [15]______________________. A substance that causes lung cancer when inhaled, but which may not be particularly dangerous in drinking water is [16]____________ __________. Eutrophication describes a condition of lakes or reservoirs involving [17]______________________. Generally, the single plant nutrient that is generally named as the culprit in excessive eutrophication is [18]_________________. The most common source of pollutant acid in water is [19]___________________ ____________ formed by the microbial oxidation of [20]_________________ __________________. Another water quality parameter discussed along with salinity that is, however, generally not introduced directly into water by human activity is [21]________________. The recharging of water softeners can introduce excessive amounts of [22]_______________ into water. The overall biologically mediated reaction by which oxygen is consumed in water is [23]______________ _________________________. Biochemical oxygen demand is a measurement of [24]____ ____. Primary and secondary sewage treatment processes remove [25]____________________________ _________ from water. Substances that are especially troublesome in improperly disposed sewage sludge are [26]__________________________________ ____.

A soap anion exhibits a dual nature because it consists of [27]______________ ___________________ that likes water and a [28]_______________________ ________________. Soap's primary disadvantage as a cleaning agent is [29]______ ____. The surface-active agent, or surfactant, in detergents acts by [30]_______________________ ____.

The advantage of LAS surface active agent over ABS is that the former is [31]______ ________________. Many of the environmental problems that currently are associated with the use of detergents arise from the [32]_________________ in detergents. Nonbiodegradable, low-molecular-weight organic compounds of low volatility are commonly called [33]_____________________________. Some general classes of pesticides are [34]_____________________________________ ____ ____. Chemically, both parathion and malathion are classified as [35]____________________________, but the latter is safer because [36]______________________________________. Although the toxicity of DDT is low, it is environmentally damaging because of its [37]____________________________________. The most environmentally damaging by-products of pesticide manufacture are [38]____________________ ________________ of which the most notorious example is [39]____________ ____________________. A general and specific example of PCB structures are [40]____. PCBs are especially common in sediments of the [41]______________________ as the result of the operation of capacitor plants. Two ways of producing radionuclides are [42]______________________________________. Radionuclides produce [43]______

________________ radiation in the three major forms of [44]____. A time interval characteristic of radionuclides is called the [45]____________________ defined as [46]____. Radiation damages living organisms by [47]____. In general, radionuclides with intermediate half-lives are the most dangerous because [48]____.
The radionuclide of most concern in drinking water is [49]______________________.

Answers

1 metal-binding organic species

2 algal nutrients

3 heavy metals

4 metalloids

5 cadmium

6 lead

7 mercury

8 arsenic

9 complexation and organometallic compound formation

10 tin

11 industrial biocides

12 those that contribute acidity, alkalinity, or salinity to water; algal nutrients; cyanide ion; ammonia; carbon dioxide; hydrogen sulfide; nitrite; and sulfite

13 cyanide

14 ammonia (ammonium ion)

15 hydrogen sulfide

16 asbestos

17 excess algal growth

18 phosphate

19 acid mine drainage

20 pyrite

21 alkalinity

22 salinity

23 $\{CH_2O\} + O_2 \xrightarrow{\text{Microorganisms}} CO_2 + H_2O$

24 the degree of oxygen consumption by microbially mediated oxidation of contaminants in water

25 oxygen-demanding substances, oil, grease, and solids

26 organic material which continues to degrade slowly, refractory organics, and heavy metals

27 an ionic carboxyl "head"

28 long hydrocarbon "tail"

29 its reaction with divalent cations to form insoluble salts of fatty acids

30 lowering the surface tension of water to which the detergent is added, making the water "wetter"

31 biodegradable

32 builders

33 biorefractory compounds

34 insecticides, molluscicides, nematicides, rodenticides, herbicides, plant growth regulators, defoliants, plant desiccants, fungicides, bactericides, and algicides.

35 organophosphates

36 mammals can detoxify it

37 persistence and accumulation in food chains

38 polychlorinated dibenzodioxins

39 TCDD

40 structures shown in Figure 7.5

41 Hudson River

42 as fission products of heavy nuclei and by the reaction of neutrons with stable nuclei

43 ionizing

44 alpha particles, beta particles, and gamma rays

45 half-life

46 time during which half of a given number of atoms of a specific kind of radionuclide decay

47 initiating harmful chemical reactions in tissues

48 they persist long enough to enter living systems while still retaining a high activity

49 radium

QUESTIONS AND PROBLEMS

1. What do mercury and arsenic have in common in regard to their interactions with bacteria in sediments?

2. What are some characteristics of radionuclides that make them especially hazardous to humans?

3. To what class do pesticides containing the group below belong?

$$-\overset{H}{\overset{|}{N}}-\overset{O}{\overset{\|}{C}}-$$

4. What is the primary detrimental effect upon organisms of salinity in water arising from dissolved NaCl and Na_2SO_4?

5. Consider the following compound:

$$Na^{+}\,{}^{-}O-\overset{O}{\overset{\|}{\underset{\underset{O}{\|}}{S}}}-C_6H_4-\overset{H}{\underset{H}{C}}-\overset{H}{\underset{H}{C}}-\overset{H}{\underset{H}{C}}-\overset{H}{\underset{H}{C}}-\overset{H}{\underset{H}{C}}-\overset{H}{\underset{H}{C}}-\overset{H}{\underset{H}{C}}-\overset{H}{\underset{H}{C}}-\overset{H}{\underset{H}{C}}-\overset{H}{\underset{H}{C}}-H$$

Which of the following characteristics is not possessed by the compound: (a) one end of the molecule is hydrophilic and the other end is hydrophobic, (b) surface-active qualities, (c) the ability to lower surface tension of water, (d) good biodegradability, (e) tendency to persist in water.

6. Give a specific example of each of the following general classes of water pollutants: (a) trace elements, (b) organically bound metal, (c) pesticides

7. A sample of water contaminated by the accidental discharge of a radionuclide used for medicinal purposes showed an activity of 23,956 counts per second at the time of sampling and 2,993 cps exactly 46 days later. Estimate the half-life of the radionuclide.

8. What are the two reasons that soap is environmentally less harmful than ABS surfactant used in detergents?

9. Why can one not give an exact chemical formula of the specific compound designated as PCB?

10. Match each compound in the left column with the description corresponding to it in the right column.

(a) CdS

(1) Pollutant released to a U.S. stream by a poorly controlled manufacturing process.

(b) $(CH_3)_2AsH$

(2) Insoluble form of a toxic trace element likely to be found in anaerobic sediments.

(c) 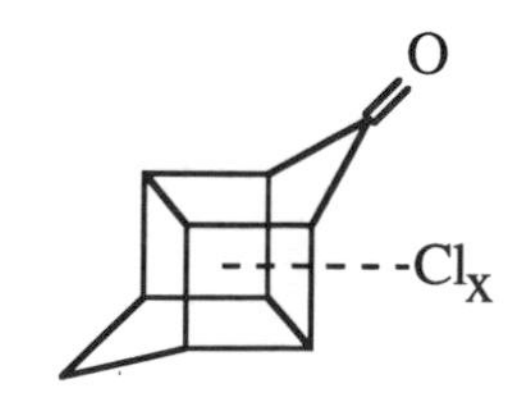

(3) Common environmental pollutant formerly used as a transformer coolant.

(d) 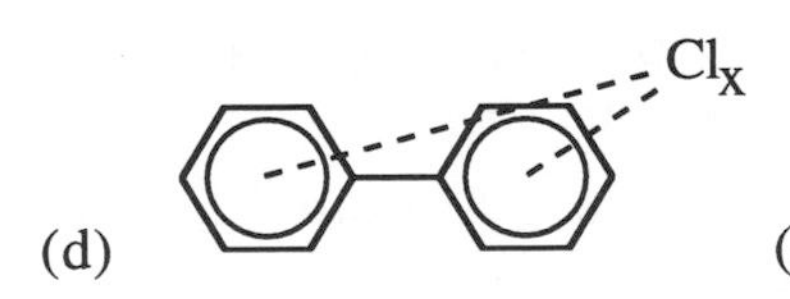

(4) Chemical species thought to be produced by bacterial action.

11. A radioisotope has a nuclear half-life of 24 hours and a biological half-life of 16 hours (half of the element is eliminated from the body in 16 hours). A person accidentally swallowed sufficient quantities of this isotope to give an initial "whole body" count rate of 1000 counts per minute. What was the count rate after 16 hours?

12. A certain pesticide is fatal to fish fingerlings at a level of 0.50 parts per million in water. A leaking metal can containing 5.00 kg of the pesticide was dumped into a stream with a flow of 10.0 liters per second moving at 1 kilometer per hour. The container leaked pesticide at a constant rate of 5 mg/sec. For what distance (in km) downstream was the water contaminated by fatal levels of the pesticide by the time the container is empty?

13. A pesticide sprayer got stuck while trying to ford a stream flowing at a rate of 136 liters per second. Pesticide leaked into the stream for exactly 1 hour and at a rate that contaminated the stream at a uniform 0.25 ppm of methoxychlor. How much pesticide was lost from the sprayer during this time?

LITERATURE CITED

[1] Milne, G. W. A., *CRC Handbook of Pesticides*, CRC Press/Lewis Publishers, Boca Raton, FL, 1995.

[2] Esposito, M. Pat, "Dioxin Wastes," Section 4.3 in *Standard Handbook of Hazardous Waste Treatment and Disposal*, Harry M. Freeman, Ed., McGraw Hill, New York, 1989, pp. 4.25-4.34.

SUPPLEMENTARY REFERENCES

Abel, P. D., *Water Pollution Biology*, 2nd Edition, Taylor & Francis, London, UK, 1997.

Dennison, Mark S., *Storm Water Discharges*, CRC Press/Lewis Publishers, Boca Raton, FL, 1996.

Environmental Science and Technology, journal published monthly by the American Chemical Society, Washington, D.C.

Gee, Shirley J., Bruce D. Hammock, and Jeanette M. Van Emon, *Environmental Immunochemical Analysis for Detection of Pesticides and Other Chemicals: A Users Guide*, Noyes Publications, Park Ridge, NJ, 1997.

Kamrin, Michael A., *Pesticide Profiles: Toxicity, Environmental Impact, and Fate*, CRC Press/Lewis Publishers, Boca Raton, FL, 1997.

Nathanson, Jerry A., *Basic Environmental Technology: Water Supply, Waste Management, and Pollution Control*, 2nd Edition, Prentice Hall, Upper Saddle River, NJ, 1996.

Olem, H., Ed., *Diffuse Pollution (Water Science and Technology Series, Vol 28)*, Elsevier Science Ltd, Amsterdam, Netherlands, 1996.

Seiber, James N., Ed., *Fumigants: Environmental Fate, Exposure, and Analysis*, ACS Symposium Series 652, American Chemical Society, Washington, D.C., 1997.

Vigil, Kenneth M., *Clean Water: The Citizen's Complete Guide to Water Quality and Water Pollution Control*, Columbia Cascade Publishing Co., Reston, VA, 1996.

Water Environment Federation Task Force on Developing Source Control, *Developing Source Control Programs for Commercial and Industrial Wastewater*, Water Environment Federation, Chicago, IL, 1996

8 WATER TREATMENT

8.1. WATER TREATMENT AND WATER USE

The treatment of water may be divided into three major categories:

- Purification for domestic use
- Treatment for specialized industrial applications
- Treatment of wastewater to make it acceptable for release or reuse.

The type and degree of treatment are strongly dependent upon the source and intended use of the water. Water for domestic consumption must be thoroughly disinfected to eliminate disease-causing microorganisms, but may contain appreciable levels of dissolved calcium and magnesium (hardness). Water to be used in boilers may contain bacteria, but must be quite soft to prevent scale formation. Wastewater being discharged into a large river may require less rigorous treatment than water to be reused in an arid region. As world demand for limited water resources grows, more sophisticated and extensive means will have to be employed to treat water.

Most physical and chemical processes used to treat water involve similar phenomena, regardless of their application to the three main categories of water treatment listed above. Therefore, after introductions to water treatment for municipal use, industrial use, and disposal, each major kind of treatment process is discussed as it applies to all of these applications.

8.2. MUNICIPAL WATER TREATMENT

The function of a modern water treatment plant is to produce clear, safe, even tasteful drinking water from raw water that may consist of a murky liquid pumped from a polluted river laden with mud and swarming with bacteria; well water, much too hard for domestic use and containing high levels of stain-producing dissolved iron and manganese; or other questionable sources of raw water. A schematic diagram of a typical municipal water treatment plant is shown in Figure 8.1. This particular facility treats water containing excessive hardness and a high level of iron. The raw water taken from wells first goes to an aerator. Contact of the water with air removes volatile solutes such as hydrogen sulfide, carbon dioxide, methane, and volatile odorous substances such as thiomethane (CH_3SH) and bacterial metabolites.

Contact with oxygen also aids iron removal by oxidizing soluble iron(II) to insoluble iron(III). After aeration, lime is added as CaO or $Ca(OH)_2$ to raise the pH, causing the formation of precipitates containing Ca^{2+} and Mg^{2+} hardness ions. These precipitates settle from the water in a primary basin. Much of the solid material remains in suspension and may be caused to settle by the addition of coagulants, such as iron(III) and aluminum sulfates, which form gelatinous metal hydroxides, activated silica, or synthetic organic polyelectrolytes. The settling of the coagulated solids occurs in a secondary basin after the addition of carbon dioxide to lower the pH. Sludge from both the primary and secondary basins is pumped to a sludge lagoon. The water is finally chlorinated, filtered, and pumped to the city water mains.

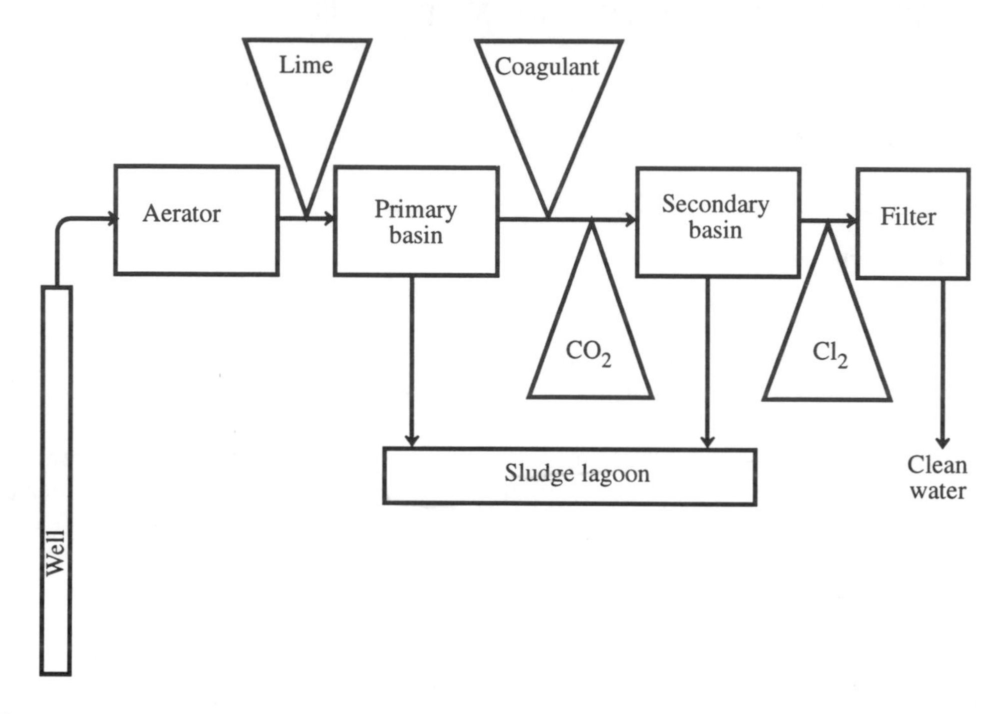

Figure 8.1. Schematic of a municipal water treatment plant.

8.3. TREATMENT OF WATER FOR INDUSTRIAL USE

Water is widely used in various process applications in industry. Other major industrial applications are boiler feedwater and cooling water. The kind and degree of treatment of water in these applications depends upon the end use. As examples, cooling water may require only minimal treatment, removal of corrosive substances and scale-forming solutes is essential for boiler feedwater, and water used in food processing must be free of pathogens and toxic substances. Improper treatment of industrial water can cause problems, such as corrosion, scale formation, reduced heat transfer in heat exchangers, reduced water flow, and product contamination. These effects may cause reduced equipment performance or equipment failure, increased energy costs due to inefficient heat utilization or cooling, increased costs for pumping water, and product deterioration. Obviously, the effective treatment of water at minimum cost for industrial applications is a very important area of water treatment.

Numerous factors must be taken into consideration in designing and operating an industrial water treatment facility. These include the following:

- Water requirement
- Quantity and quality of available water sources
- Sequential use of water (successive uses for applications requiring progressively lower water quality)
- Water recycle
- Discharge standards

The various specific processes employed to treat water for industrial use are discussed in later sections of this chapter. **External treatment**, usually applied to the plant's entire water supply, involves processes such as aeration, filtration, and clarification to remove material from water that may cause problems. Such substances include suspended or dissolved solids, hardness, and dissolved gases. Following this basic treatment, the water may be divided into different streams, some to be used without further treatment and the rest to be treated for specific applications.

Internal treatment is designed to modify the properties of water for specific applications. Examples of internal treatment include several processes, the most common of which are the following:

- Removal of dissolved oxygen by reaction with hydrazine (N_2H_4) or sulfite (SO_3^{2-}):

$$2SO_3^{2-} + O_2 \rightarrow 2SO_4^{2-} \qquad (8.3.1)$$

$$N_2H_4 + O_2 \rightarrow 2H_2O + N_2(g) \qquad (8.3.2)$$

- Addition of chelating agents to react with dissolved Ca^{2+} and prevent formation of calcium deposits
- Addition of precipitants, such as phosphate used for calcium removal
- Treatment with dispersants to inhibit scale
- Addition of corrosion inhibitors
- Adjustment of pH
- Disinfection for food processing uses or to prevent bacterial growth in cooling water

8.4. SEWAGE TREATMENT

Typical municipal sewage contains oxygen-demanding materials, sediments, grease, oil, scum, pathogenic bacteria, viruses, salts, algal nutrients, pesticides, refractory organic compounds, heavy metals, and an astonishing variety of flotsam ranging from children's socks to sponges. It is the job of the waste treatment plant to remove as much of this material as possible.

Characteristics used to describe sewage include turbidity (international turbidity units), suspended solids (ppm), total dissolved solids (ppm), acidity (H^+ ion concentration or pH), and dissolved oxygen (in ppm O_2). Biochemical oxygen demand (Section 7.8) is used as a measure of oxygen-demanding substances.

Current processes for the treatment of wastewater may be divided into three main categories: primary treatment, secondary treatment, and tertiary treatment, each of which is discussed separately. Also discussed are total wastewater treatment systems, based largely upon physical and chemical processes.

Waste from a municipal water system is normally treated in a **publicly owned treatment works, POTW**. In the United States these systems are allowed to discharge only effluents that have attained a certain level of treatment, as mandated by federal law. As discussed in Chapter 24, one of the objectives of hazardous waste treatment is to remove water from the waste and purify it such that it can be safely discharged to a POTW.

Primary Waste Treatment

Primary treatment of wastewater consists of the removal of insoluble matter such as grit, grease, and scum from water. The first step in primary treatment is normally screening. Screening removes or reduces the size of trash and large solids that get into the sewage system. These solids are collected on screens and scraped off for subsequent disposal. Most screens are cleaned with power rakes. Comminuting devices shred and grind solids in the sewage. Particle size may be reduced to the extent that the particles can be returned to the sewage flow.

Grit in wastewater consists of such materials as sand and coffee grounds, which do not biodegrade well and generally have a high settling velocity. **Grit removal** is practiced to prevent its accumulation in other parts of the treatment system, to reduce clogging of pipes and other parts, and to protect moving parts from abrasion and wear. Grit normally is allowed to settle in a tank under conditions of low flow velocity, and it is then scraped mechanically from the bottom of the tank.

Primary sedimentation removes both settleable and floatable solids. During primary sedimentation it is important for flocculent particles to aggregate for better settling, a process that may be aided by the addition of chemicals. The material that floats in the primary settling basin, known collectively as grease, consists of oils, waxes, free fatty acids, and insoluble soaps containing calcium and magnesium. Normally, some of the grease settles with the sludge and some floats to the surface, where it may be removed by a skimming device.

Secondary Waste Treatment by Biological Processes

The most obvious harmful effect of biodegradable organic matter in wastewater is BOD, consisting of a biochemical oxygen demand for dissolved oxygen by microorganism-mediated degradation of the organic matter (see Section 7.8). **Secondary wastewater treatment** is designed to remove BOD, usually by taking advantage of the same kind of biological processes that would otherwise consume oxygen in water receiving the wastewater. Secondary treatment by biological processes takes many forms but consists basically of the following: Microorganisms provided with added oxygen are allowed to degrade organic material in solution or in suspension until the BOD of the waste has been reduced to acceptable levels. The waste is oxidized biologically under conditions controlled for optimum bacterial growth and at a site where this growth does not influence the environment.

A common biological process for treating wastewater consists of **fixed-film biological (FFB)** reactors. The oldest and simplest of these is the **trickling filter** (Figure 8.2) in which wastewater is sprayed over rocks or other solid support material covered with microorganisms, such that contact of the wastewater with air is allowed and degradation of organic matter occurs by the action of the microorganisms. A more modern approach is the **rotating biological reactor**, consisting of groups of large plastic disks mounted close together on a rotating shaft such that growths of microorganisms on the disks are alternately exposed to air and wastewater. The greatest advantage of these processes is their low energy consumption because air does not have to be pumped into the system.

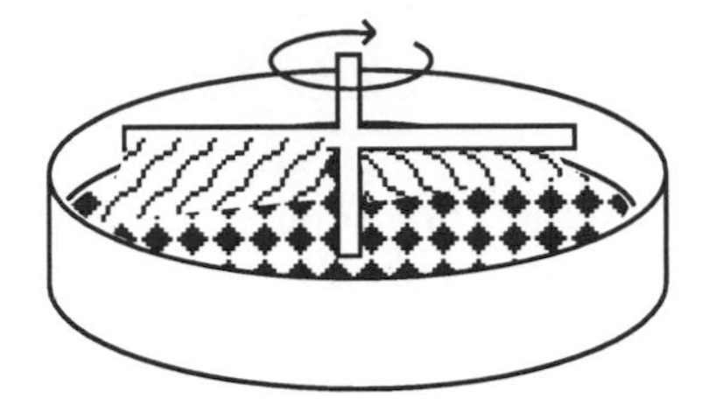

Figure 8.2. Trickling filter for secondary waste treatment.

The **activated sludge process**, Figure 8.3, is probably the most versatile and effective of all waste treatment processes. Microorganisms in the aeration tank convert organic material in wastewater to microbial biomass and CO_2. Organic nitrogen is converted to ammonium ion or nitrate. Organic phosphorus is converted to orthophosphate. The microbial cell matter formed as part of the waste degradation processes is normally kept in the aeration tank until the microorganisms are past the log phase of growth (Section 6.5), at which point the cells flocculate relatively well to form settleable solids. These solids collect in the bottom part of a settler and a fraction of them is discarded. Part of the solids, the return sludge, is recycled to the head of the aeration tank and comes into contact with fresh sewage. The combination of a high concentration of "hungry" cells in the return sludge and a rich food source in the influent sewage provides optimum conditions for the rapid degradation of organic matter. The activated sludge process removes organic carbon from water by conversion to CO_2 and by incorporation into biomass.

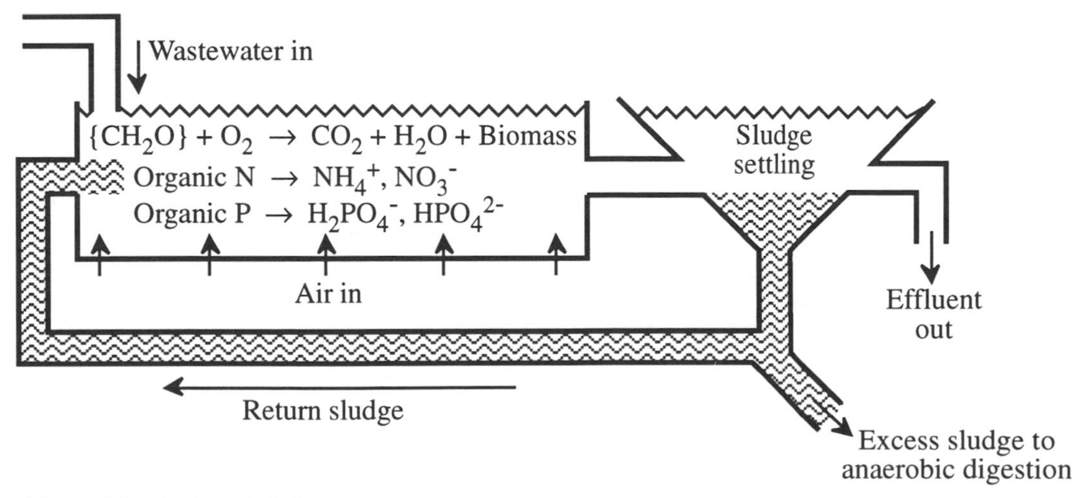

Figure 8.3. Activated sludge process.

The disposal of waste sludge from an activated sludge plant can be a problem, primarily because it is only about 1% solids and contains many undesirable components. Normally, partial water removal is accomplished by drying on sand filters, vacuum filtration, or centrifugation. The dewatered sludge may be incinerated or used as landfill, both of which are facing increased regulatory and public opposition. To a certain extent, sewage sludge may be digested in the absence of oxygen by methane-producing anaerobic bacteria to produce methane and carbon dioxide,

$$2\{CH_2O\} \rightarrow CH_4 + CO_2 \quad (8.4.1)$$

a process that reduces both the volatile-matter content and the volume of the sludge by about 60%. A carefully designed plant may produce enough methane to provide for all of its power needs, which can be a very important factor in reducing costs.

One of the most desirable means of sludge disposal is to use it to fertilize and condition soil. However, care has to be taken that excessive levels of heavy metals are not applied to the soil as sludge contaminants. Problems with various kinds of sludges resulting from water treatment are discussed further in Section 8.10.

Tertiary Waste Treatment

Tertiary waste treatment (sometimes called **advanced waste treatment**) is a term used to describe a variety of processes performed on the effluent from secondary waste treatment so that the water can be reused. The contaminants removed by tertiary waste treatment fall into the general categories of (1) suspended solids; (2) dissolved organic compounds; and (3) dissolved inorganic materials, including the important class of algal nutrients. Each of these categories presents its own problems with regard to water quality. Suspended solids are primarily responsible for residual biological oxygen demand in secondary sewage effluent waters. The dissolved organics are the most hazardous from the standpoint of potential toxicity. The major problem with dissolved inorganic materials is that presented by algal nutrients, primarily nitrates and phosphates. In addition, potentially hazardous toxic metals may be found among the dissolved inorganics.

In addition to these chemical contaminants, secondary sewage effluent often contains a number of disease-causing microorganisms, requiring disinfection in cases where humans may later come into contract with the water. Among the bacteria that may be found in secondary sewage effluent are organisms causing tuberculosis, dysenteric bacteria (*Bacillus dysenteriae*, *Shigella dysenteriae*, *Shigella paradysenteriae*, *Proteus vulgaris*), cholera bacteria (*Vibrio cholerae*), bacteria causing mud fever (*Leptospira icterohemorrhagiae*), and bacteria causing typhoid fever (*Salmonella typhosa*, *Salmonella paratyphi*). In addition, viruses causing diarrhea, eye infections, infectious hepatitis, and polio may be encountered. Ingestion of sewage still causes disease, even in more developed nations.

Physical-Chemical Treatment of Municipal Wastewater

Complete physical-chemical wastewater treatment systems, which rely upon chemicals and energy-intensive processes for water treatment, can be used instead of biological treatment systems. Around 1970 it appeared that such systems might become seriously competitive with biological waste treatment systems. However, with the first "energy crisis" of 1973 the costs for energy and chemicals, many of which are based upon petroleum or require energy-intensive processes for their manufacture, made physical-chemical treatment of municipal wastewater much less attractive. Basically, a physical-chemical treatment process involves:

- Removal of scum and solid objects.
- Clarification, generally with the addition of a coagulant, and frequently with the addition of other chemicals (such as lime for phosphorus removal).
- Filtration to remove filterable solids.
- Activated carbon adsorption.
- Disinfection.

8.5. INDUSTRIAL WASTEWATER TREATMENT

Wastewater to be treated must be characterized fully, particularly with a thorough chemical analysis of possible waste constituents and their chemical and metabolic products. The biodegradability of wastewater constituents should also be determined. The options available for the treatment of wastewater are summarized briefly in this section and discussed in greater detail in later sections.

One of two major ways of removing organic wastes is biological treatment by an activated sludge, or related process (see Section 8.3 and Figure 8.3). It may be necessary to acclimate microorganisms to the degradation of constituents that are not normally biodegradable. Consideration needs to be given to possible hazards of biotreatment sludges, such as those containing excessive levels of heavy metal ions. The other major process for the removal of organics from wastewater is sorption by activated carbon (see Section 8.8), usually in columns of granular activated carbon. Activated carbon and biological treatment can be combined with the use of powdered activated carbon in the activated sludge process. The powdered activated carbon sorbs some constituents that may be toxic to microorganisms and is collected with the sludge. An important consideration in using activated carbon to treate wastewater is the hazard that spent activated carbon may present from the wastes it retains. These hazards may include those of toxicity or reactivity; for example, spent activated carbon used to sorb explosives manufacture wastes may be very dangerously reactive, especially after the carbon has dried. Regeneration of the carbon is expensive and can be hazardous in some cases.

Wastewater can be treated by a variety of chemical processes, including acid/base neutralization, precipitation, and oxidation/reduction. In some cases these treatment steps must precede biological treatment; for example, wastewater exhibiting extremes of pH must be neutralized in order for microorganisms to thrive in it. Cyanide in the wastewater may be oxidized with chlorine and organics with ozone, hydrogen peroxide promoted with ultraviolet radiation, or dissolved oxygen at high temperatures and pressures. Heavy metals may be precipitated with base, carbonate, or sulfide.

Wastewater can be treated by several physical processes. In some cases, simple density separation and sedimentation can be used to remove water-immiscible liquids and solids. Filtration is frequently required, and flotation by gas bubbles generated on particle surfaces may be useful. Wastewater solutes can be concentrated by evaporation, distillation, and membrane processes, including reverse osmosis, hyperfiltration, and ultrafiltration. Organic constituents can be removed by solvent extraction, air stripping, or steam stripping. Synthetic resins that attract organic constituents are useful for removing some pollutant solutes from wastewater. Cation exchange resins are effective for the removal of heavy metals.

8.6. REMOVAL OF SOLIDS

Relatively large solid particles are removed from water by simple **settling** and **filtration**. A special type of filtration procedure known as **microstraining** is especially effective in the removal of the very small particles. These filters are woven from stainless steel wire so fine that it is barely visible. This enables preparation of filters with openings only 60-70 μm across. These openings may be reduced to 5-15 μm by partial clogging with small particles, such as bacterial cells. The cost of this treatment is likely to be substantially lower than the costs of competing processes. High flow rates at low back pressures are normally achieved.

The settling and filtration of colloidal solids from water usually require **coagulation**. Salts of aluminum and iron are the coagulants most often used in water treatment. Of these, alum or filter alum is most commonly used. This substance is a hydrated aluminum sulfate, $Al_2(SO_4)_3{\cdot}18H_2O$. When added to water, the aluminum ion in this salt hydrolyzes by reactions that consume alkalinity in the water, such as:

$$Al(H_2O)_6^{3+} + 3HCO_3^- \rightarrow Al(OH)_3(s) + 3CO_2 + 6H_2O \tag{8.6.1}$$

The gelatinous hydroxide thus formed carries suspended material with it as it settles. In addition, however, it is likely that positively charged hydroxyl-bridged dimers such as,

$$(H_2O)_4Al\begin{matrix} \overset{H}{O} \\ \underset{H}{O} \end{matrix}Al(H_2O)_4^{4+}$$

and higher polymers are formed that interact specifically with colloidal particles, bringing about coagulation. Metal ions in coagulants also react with virus proteins and destroy up to 99% of the virus in water. Iron salts that precipitate gelatinous $Fe(OH)_3(s)$ may be used as coagulants in a manner analogous to aluminum. Sodium silicate partially neutralized by acid aids coagulation, particularly when used with alum.

Natural and synthetic polyelectrolytes act to flocculate colloidal-size particles. Among the natural compounds so used are starch and cellulose derivatives, proteinaceous materials, and gums composed of polysaccharides. Synthetic polymers that are effective flocculants are now widely used. Neutral polymers and both anionic and cationic polyelectrolytes have been employed successfully as flocculants in various applications.

An important class of solids that must be removed from wastewater consists of suspended solids in secondary sewage effluent that arise primarily from sludge that was not removed in the settling process. These solids account for a large part of the BOD in the effluent and may interfere with other aspects of tertiary waste treatment. For example, these solids may clog membranes in reverse osmosis water treatment processes. The quantity of material involved may be rather high. Processes designed to remove suspended solids often will remove 10-20 mg/L of organic material from secondary sewage effluent. In addition, a small amount of the inorganic material is removed as well.

8.7. REMOVAL OF CALCIUM AND OTHER METALS

Calcium and magnesium salts, which generally are present in water as bicarbonates or sulfates, cause water hardness manifested by the formation of an insoluble "curd" from the reaction of soap with Ca^{2+} or Mg^{2+} (see Section 6.11). Another problem caused by hard water is the formation of mineral deposits. For example, when water containing calcium and bicarbonate ions is heated, insoluble calcium carbonate is formed:

$$Ca^{2+} + 2HCO_3^- \rightarrow CaCO_3(s) + CO_2(g) + H_2O \tag{8.7.1}$$

This product coats the surfaces of hot-water systems, clogging pipes and reducing heating efficiency. Dissolved salts such as calcium and magnesium bicarbonates and sulfates can be especially damaging in boiler feedwater. Clearly, the removal of water hardness—called **water softening**—is essential for many uses of water.

Several processes are used for softening water. On a large scale in a centralized water treatment plant the lime-soda process utilizing lime, $Ca(OH)_2$, and soda ash, Na_2CO_3, is employed. Calcium is precipitated as $CaCO_3$ and magnesium as $Mg(OH)_2$. When the calcium is present primarily as "bicarbonate hardness," it can be removed by the addition of $Ca(OH)_2$ alone:

$$Ca^{2+} + 2HCO_3^- + Ca(OH)_2 \rightarrow 2CaCO_3(s) + 2H_2O \tag{8.7.2}$$

When bicarbonate ion is not present at substantial levels, a source of CO_3^{2-} must be provided at a relatively high pH by the addition of Na_2CO_3. For example, calcium present as the chloride can be removed from water by the addition of soda ash:

$$Ca^{2+} + 2Cl^- + 2Na^+ + CO_3^{2-} \rightarrow CaCO_3(s) + 2Cl^- + 2Na^+ \tag{8.7.3}$$

The precipitation of magnesium as the hydroxide requires a higher pH than the precipitation of calcium as the carbonate:

$$Mg^{2+} + 2OH^- \rightarrow Mg(OH)_2(s) \tag{8.7.4}$$

The high pH required may be provided by the basic carbonate ion from soda ash:

$$CO_3^{2-} + H_2O \rightarrow HCO_3^- + OH^- \tag{8.7.5}$$

The water softened by lime-soda softening plants usually suffers from two defects. First, because of super-saturation effects, some $CaCO_3$ and $Mg(OH)_2$ usually remain in solution. If not removed, these compounds will precipitate at a later time and cause harmful deposits or undesirable cloudiness in water. The second problem results from the use of highly basic sodium carbonate, which gives the product water an excessively high pH, up to pH 11. To overcome these problems, the water is **recarbonated** by bubbling CO_2 into it. The carbon dioxide neutralizes excess OH^- to bring the pH within the range 7.5-8.5 and converts the slightly soluble calcium carbonate and magnesium hydroxide to their soluble bicarbonate forms:

$$CaCO_3(s) + CO_2 + H_2O \rightarrow Ca^{2+} + 2HCO_3^- \quad (8.7.6)$$

$$Mg(OH)_2(s) + 2CO_2 \rightarrow Mg^{2+} + 2HCO_3^- \quad (8.7.7)$$

Water adjusted to a pH, alkalinity, and Ca^{2+} concentration very close to $CaCO_3$ saturation is *chemically stabilized.* It neither deposits $CaCO_3$ in water pipes, which can clog them, nor dissolves protective $CaCO_3$ coatings from the pipe surfaces. With a Ca^{2+} concentration much below $CaCO_3$ saturation, water is called *aggressive*.

Calcium may be removed from water very efficiently by the addition of orthophosphate:

$$5Ca^{2+} + 3PO_4^{3-} + OH^- \rightarrow Ca_5OH(PO_4)_3(s) \quad (8.7.8)$$

It should be pointed out that the chemical formation of a slightly soluble product for the removal of undesired solutes such as hardness ions, phosphate, iron, and manganese must be followed by sedimentation in a suitable apparatus. Frequently, coagulants must be added, and filtration employed for complete removal of these sediments.

Water may be softened by ion exchange, the reversible transfer of ions between aquatic solution and a solid material capable of bonding ions. The water is passed over a solid cation exchanger in the sodium ion form, represented by $Na^{+-}\{Cat(s)\}$, which trades the Na^+ ions bound with the solid for Ca^{2+} ions in water:

$$2Na^{+-}\{Cat(s)\} + Ca^{2+} \rightarrow Ca^{2+-}\{Cat(s)\}_2 + 2Na^+ \quad (8.7.9)$$

The Na^+ ions released cause no problems with the end use of the water, such as for laundry.

Water softening by cation exchange is widely used, effective, and economical, although it does cause some deterioration of water quality arising from the contamination of wastewater by sodium chloride. Such contamination results from the periodic need to regenerate a water softener with sodium chloride, in order to displace calcium and magnesium ions from the resin and replace these hardness ions with sodium ions:

$$Ca^{2+-}\{Cat(s)\}_2 + 2Na^+ + 2Cl^- \rightarrow 2Na^{+-}\{Cat(s)\} + Ca^{2+} + 2Cl^- \quad (8.7.10)$$

During the regeneration process, a large excess of sodium chloride must be used—several kilograms for a home water softener. Appreciable amounts of dissolved sodium chloride can be introduced into sewage by this route.

Chelation or, as it is sometimes known, *sequestration* (see Section 6.12), is an effective method of softening water without actually having to remove calcium and magnesium from solution. A complexing agent is added that binds with and greatly reduces the concentrations of free hydrated cations, as shown by the chelation of calcium ion with excess EDTA anion (Y^{4-}):

$$Ca^{2+} + Y^{4-} \rightarrow CaY^{2-} \quad (8.7.11)$$

This reduces the concentration of hydrated calcium ion, preventing the precipitation of calcium carbonate:

$$Ca^{2+} + CO_3^{2-} \rightarrow CaCO_3(s) \quad (8.7.12)$$

Removal of Iron, Manganese, and Heavy Metals

Soluble iron and, to a lesser extent, manganese are found in many groundwaters because of reducing conditions that favor the soluble +2 oxidation state of these metals. Both metals are detrimental to water quality because of their staining tendencies. The basic method for removing both of these metals depends upon oxidation to higher insoluble oxidation states. The oxidation is generally accomplished by aeration and is favored by a high pH.

Heavy metals such as copper, cadmium, mercury, and lead are found in wastewaters from a number of industrial processes. Because of the toxicity of many heavy metals, their concentrations must be reduced to very low levels prior to release of the wastewater. A number of approaches are used in heavy metals removal.

Lime treatment for water softening discussed above removes heavy metals as insoluble hydroxides, basic salts, or coprecipitates along with calcium carbonate or ferric hydroxide. This process does not completely remove mercury, cadmium, or lead, so their removal is aided by the addition of sulfide, taking advantage of the tendency of heavy metals to act as sulfide-seekers:

$$Cd^{2+} + S^{2-} \rightarrow CdS(s) \tag{8.7.13}$$

Lime precipitation does not normally permit recovery of metals, produces large quantities of potentially hazardous sludge, and is sometimes undesirable from the economic viewpoint.

Electrodeposition (reduction of metal ions to metal by electrons at an electrode), *reverse osmosis* (see Section 8.9), and *ion exchange* are frequently used for metal removal. Solvent extraction using organic-soluble chelating substances is also effective in removing many metals. **Cementation**, a process by which a metal deposits by reaction of its ion with a more readily oxidized metal, may be employed:

$$Cu^{2+} + Fe \text{ (iron scrap)} \rightarrow Fe^{2+} + Cu \tag{8.7.14}$$

Activated carbon adsorption effectively removes some metals from water at the part per million level. Sometimes a chelating agent is sorbed to the charcoal to increase metal removal.

Even when not specifically designed for the removal of heavy metals, most waste treatment processes remove appreciable quantities of the more troublesome heavy metals encountered in wastewater. These metals accumulate in the sludge from biological treatment, so sludge disposal must be given careful consideration.

8.8. REMOVAL OF DISSOLVED ORGANICS

Very low levels of exotic organic compounds in drinking water are suspected of contributing to cancer and other maladies. Some of these are chlorinated organic compounds produced by chlorination of organics in water, especially humic substances. Removal of organics to very low levels prior to chlorination has been found to be effective in preventing formation of compounds of the chloroform ($CHCl_3$) type called trihalomethanes. In addition, many organic compounds survive, or are produced by, secondary wastewater treatment.

The standard method for the removal of dissolved organic material is adsorption on granular or powdered activated carbon, a material that is made from a variety of carbonaceous materials, including wood, pulp-mill char, peat, and lignite. The carbon is produced by charring the raw material in the absence of air below 600°C, followed by an activation step consisting of partial oxidation. Carbon dioxide may be employed as an oxidizing agent at 600-700°C (Reaction 8.8.1) or the carbon may be oxidized by water at 800-900°C (Reaction 8.8.2):

$$CO_2 + C \rightarrow 2CO \qquad (8.8.1)$$

$$H_2O + C \rightarrow H_2 + CO \qquad (8.8.2)$$

These processes develop porosity, increase the surface area, and leave the C atoms in arrangements that have affinities for organic compounds.

A major reason for the effectiveness of activated carbon as an adsorbent is its tremendous surface area. A solid cubic foot of carbon particles may have a combined pore and surface area of approximately 10 square miles!

Removal of organics may also be accomplished by adsorbent synthetic polymers. Such polymers as Amberlite XAD-4 have hydrophobic surfaces and strongly attract relatively insoluble organic compounds, such as chlorinated pesticides. Oxidation of dissolved organics holds some promise for their removal. Ozone, hydrogen peroxide, molecular oxygen (with or without catalysts), chlorine and its derivatives, permanganate, or ferrate can be used.

8.9. REMOVAL OF DISSOLVED INORGANICS

In order for complete water recycling to be feasible, inorganic solute removal is essential. The effluent from secondary waste treatment generally contains 300-400 mg/L more dissolved inorganic material than does the municipal water supply. It is obvious, therefore, that 100% water recycle without removal of inorganics would cause the accumulation of an intolerable level of dissolved material. Even when water is not destined for immediate reuse, the removal of the inorganic nutrients phosphorus and nitrogen is highly desirable to reduce eutrophication downstream. In some cases, the removal of toxic trace metals is needed.

One means for removing inorganics from water is distillation, although the energy required is generally too high for the process to be economically feasible, and volatile substances, such as ammonia and odorous compounds, may be carried into the product. Freezing produces a very pure water, but is considered uneconomical with present technology. Membrane processes considered most promising for bulk removal of inorganics from water are electrodialysis and reverse osmosis. Ion exchange is also effective for inorganics removal.

Ion Exchange

The ion exchange method for softening water is described in Section 8.7. The ion exchange process used for removal of inorganic salts, represented below as dissolved ions of M^+ and X^-, consists of passing the water successively over a solid cation exchanger and a solid anion exchanger so that the ions are replaced by water:

$$H^{+-}\{Cat(s)\} + M^+ + X^- \rightarrow M^{+-}\{Cat(s)\} + H^+ + X^- \qquad (8.9.1)$$

$$OH^{-+}\{An(s)\} + H^+ + X^- \rightarrow X^{-+}\{An(s)\} + H_2O \qquad (8.9.2)$$

In these reactions $^{-}\{Cat(s)\}$ represents the solid cation exchanger and the notation $OH^{-+}\{An(s)\}$ represents the solid anion exchanger. The cation exchanger is regenerated with strong acid and the anion exchanger with strong base.

Demineralization by ion exchange generally produces water of a very high quality. Unfortunately, some organic compounds in wastewater foul ion exchangers, and microbial growth on the exchangers can diminish their efficiency. In addition, regeneration of the resins is expensive, and the concentrated wastes from regeneration require disposal in a manner that will not damage the environment.

Reverse Osmosis

Reverse osmosis is a very useful technique for the removal of dissolved inorganics from water.[1] Basically, reverse osmosis consists of forcing pure water through a semipermeable membrane that allows the passage of water but not of other material. This process depends on the preferential sorption of water on the surface of the membrane, which is composed of porous cellulose acetate or polyamide. Pure water from the sorbed layer is forced through pores in the membrane under pressure. The principle of reverse osmosis is illustrated in Figure 8.4.

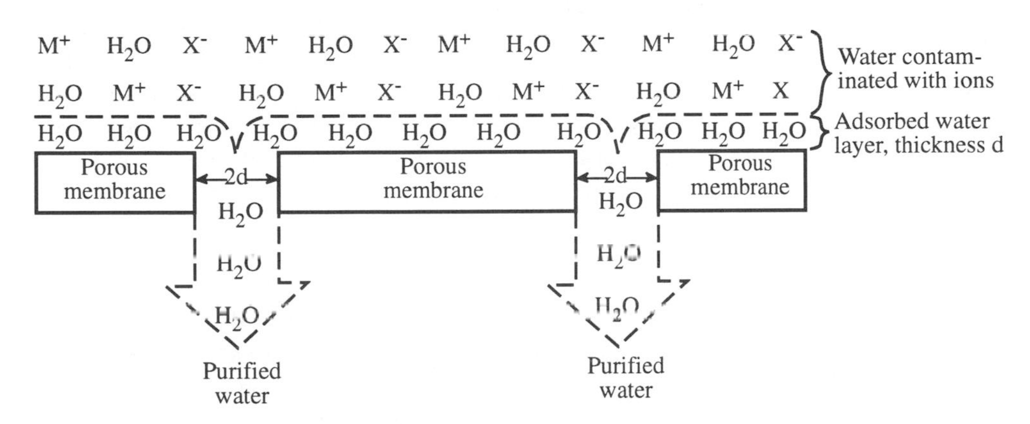

Figure 8.4. Solute removal from water by reverse osmosis. A layer of water of thickness d is adsorbed onto the membrane surface and is squeezed through the very fine pores as pure water, leaving impurities behind.

Phosphorus Removal

Advanced waste treatment normally requires removal of phosphorus to reduce algal growth. Municipal wastes typically contain approximately 25 mg/L of phosphate (as orthophosphates, polyphosphates, and insoluble phosphates), and the efficiency of phosphate removal must be quite high to prevent algal growth. Normally, the activated sludge process removes about 20% of the phosphorus from sewage.

Chemically, phosphate is most commonly removed by precipitation. Lime, $Ca(OH)_2$, is the chemical usually used to precipitate phosphorus:

$$5Ca(OH)_2 + 3HPO_4^{2-} \rightarrow Ca_5OH(PO_4)_3(s) + 3H_2O + 6OH^- \quad (8.9.3)$$

Nitrogen Removal

Next to phosphorus, nitrogen is the algal nutrient most commonly removed as part of advanced wastewater treatment. Nitrogen in municipal wastewater generally is present as organic nitrogen or ammonia. Ammonia is the primary nitrogen product produced by most biological waste treatment processes. It may be stripped in the form of NH_3 gas from the water by air after the pH has been raised to approximately 11.5 by the addition of lime (which also serves to remove phosphate).

Nitrification followed by denitrification is a promising technique for the removal of nitrogen from wastewater. The first step is an essentially complete conversion of ammonia and organic nitrogen to nitrate under strongly aerobic conditions, achieved by more extensive than normal aeration of the sewage:

$$NH_4^+ + 2O_2 \text{ (Nitrifying bacteria)} \rightarrow NO_3^- + 2H^+ + H_2O \qquad (8.9.4)$$

The second step is the reduction of nitrate to nitrogen gas. This reaction is also bacterially catalyzed and requires a carbon source and a reducing agent such as methanol, CH_3OH.

$$6NO_3^- + 5CH_3OH + 6H^+ \text{ (denitrifying bacteria)} \rightarrow 3N_2(g) + 5CO_2 + 13H_2O \qquad (8.9.5)$$

The denitrification process may be carried out either in a tank or on a carbon column. In pilot plant operations, conversions of 95% of the ammonia to nitrate and 86% of the nitrate to nitrogen have been achieved.

8.10. SLUDGE

Perhaps the most pressing water treatment problem at this time has to do with sludge collected or produced during water treatment. Finding a safe place to put the sludge or a use for it has proven troublesome, and the problem is aggravated by the growing numbers of water treatment systems. Improper disposal of wastes continues to be a subject of public and governmental concern.

Some sludge is present in wastewater and may be collected from it. Such sludge includes human wastes, garbage grindings, organic wastes and inorganic silt and grit from storm water runoff, and organic and inorganic wastes from commercial and industrial sources. There are two major kinds of sludge generated in a waste treatment plant. The first of these is organic sludge from activated sludge, trickling filter, or rotating biological reactors. The second is inorganic sludge from the addition of chemicals, such as in phosphorus removal described in the preceding section.

Most commonly, sewage sludge is subjected to anaerobic digestion in a digester designed to allow bacterial action to occur in the absence of air. This reduces the mass and volume of sludge and ideally results in the formation of a stabilized humus. Disease agents are also destroyed in the process.

Following digestion, sludge is generally conditioned and thickened to concentrate and stabilize it and make it more dewaterable. Relatively inexpensive processes, such as gravity thickening, may be employed to get the moisture content down to about 95%. Sludge may be further conditioned chemically by the addition of iron or aluminum salts, lime, or polymers.

Sludge dewatering is employed to convert the sludge from an essentially liquid material to a damp solid containing not more than about 85% water. This may be accomplished on sludge drying beds consisting of layers of sand and gravel. Mechanical devices may also be employed, including vacuum filtration, centrifugation, and filter presses. Heat may be used to aid the drying process.

Some of the alternatives for the ultimate disposal of sludge include land spreading, ocean dumping,[2] and incineration. Each of these choices has disadvantages, such as the presence of toxic substances in sludge spread on land or the high fuel cost of incineration.

Rich in nutrients, waste sewage sludge contains around 5% N, 3% P, and 0.5% K on a dry-weight basis and can be used to fertilize and condition soil. The humic material in the sludge improves the physical properties and cation-exchange capacity of the soil. Among the factors limiting this application of sludge are excess nitrogen pollution of runoff water and groundwater, survival of pathogens, and the presence of heavy metals in the sludge.

A variety of chemical sludges are produced by various water treatment and industrial processes. Among the most abundant of such sludges is alum sludge produced by the hydrolysis of Al(III) salts used in the treatment of water, which creates gelatinous aluminum hydroxide:

$$Al^{3+} + 3OH^-(s) \rightarrow Al(OH)_3(aq) \qquad (8.10.1)$$

Alum sludges normally are 98% or more water and are very difficult to dewater.

Both iron(II) and iron(III) compounds are used for the precipitation of impurities from wastewater via the precipitation of $Fe(OH)_3$. The sludge contains $Fe(OH)_3$ in the form of soft, fluffy precipitates that are difficult to dewater beyond 10 or 12% solids.

The addition of lime, $Ca(OH)_2$, or quicklime, CaO, to water is used to raise the pH to about 11.5 and causes the precipitation of $CaCO_3$, along with metal hydroxides and phosphates. Calcium carbonate is readily recovered from lime sludges and can be recalcined to produce CaO, which can be recycled through the system.

Metal hydroxide sludges are produced in the removal of metals such as lead, chromium, nickel, and zinc from wastewater by raising the pH to such a level that the corresponding hydroxides or hydrated metal oxides are precipitated. The disposal of these sludges is a substantial problem because of their toxic heavy metal content. Reclamation of the metals is an attractive alternative for these sludges.

Pathogenic (disease-causing) microorganisms may persist in the sludge left from the treatment of sewage. Many of these organisms present potential health hazards, and there is risk of public exposure when the sludge is applied to soil. The most significant organisms in municipal sewage sludge include the following: (1) indicators of pollution, including fecal and total coliform; (2) pathogenic bacteria, including *Salmonellae* and *Shigellae*; (3) enteric (intestinal) viruses, including enterovirus and poliovirus; and (4) parasites, such as *Entamoeba histolytica* and *Ascaris lumbricoides*. Therefore, it is necessary both to be aware of pathogenic microorganisms in municipal wastewater treatment sludge and to find a means of reducing the hazards caused by their presence.

Several ways are recommended to significantly reduce levels of pathogens in sewage sludge. These include prolonged aerobic digestion, air drying, composting, or lime stabilization in which sufficient lime is added to raise the pH of the sludge to 12 or higher.

8.11. WATER DISINFECTION

Chlorine is the most commonly used disinfectant employed for killing bacteria in water. Added to water, chlorine rapidly hydrolyzes according to the reaction

$$Cl_2 + H_2O \rightarrow H^+ + Cl^- + HOCl \quad (8.11.1)$$

where HOCl is hypochlorous acid. Sometimes, hypochlorite salts are substituted for chlorine gas as a disinfectant. Calcium hypochlorite, $Ca(OCl)_2$, is commonly used. The hypochlorites are safer to handle than gaseous chlorine.

The two chemical species formed by chlorine in water, HOCl and OCl^-, are known as **free available chlorine**. Free available chlorine is very effective in killing bacteria. In the presence of ammonia, HOCl reacts with NH_4^+ ion to form monochloramine (NH_2Cl), dichloramine ($NHCl_2$), and trichloramine (NCl_3). The chloramines are called **combined available chlorine**. Chlorination practice frequently provides for formation of combined available chlorine which, although a weaker disinfectant than free available chlorine, is more readily retained as a disinfectant throughout the water distribution system.

Chlorine Dioxide

Chlorine dioxide, ClO_2, is an effective water disinfectant that is of particular interest because, in the absence of impurity Cl_2, it does not produce impurity trihalomethanes in water treatment. In acidic and neutral water, respectively, the two half-reactions for ClO_2 acting as an oxidant are the following:

$$ClO_2 + 4H^+ + 5e^- \leftrightarrow Cl^- + 2H_2O \quad (8.11.2)$$

$$ClO_2 + e^- \leftrightarrow ClO_2^- \quad (8.11.3)$$

In the neutral pH range, chlorine dioxide in water remains largely as molecular ClO_2 until it contacts a reducing agent with which to react. Chlorine dioxide is a gas that is violently reactive with organic matter and explosive when exposed to light. For these reasons, ClO_2 is not shipped, but is generated on-site by processes such as the reaction of chlorine gas with solid sodium hypochlorite:

$$2NaClO_2(s) + Cl_2(g) \leftrightarrow 2ClO_2(g) + 2NaCl(s) \quad (8.11.4)$$

A high content of elemental chlorine in the product may require its purification to prevent unwanted side-reactions from Cl_2.

As a water disinfectant, chlorine dioxide does not chlorinate or oxidize ammonia or other nitrogen-containing compounds. Some concern has been raised over possible health effects of its main degradation byproducts, ClO_2^- and ClO_3^-.

Ozone

Ozone is sometimes used as a disinfectant in place of chlorine, particularly in Europe. Figure 8.5 shows the basic components of an ozone water treatment system. Basically, air is filtered, cooled, dried, and pressurized, then subjected to an electrical discharge of approximately 20,000 volts. The ozone produced is then pumped into a contact chamber where water contacts the ozone for 10-15 minutes. Concern over possible production of toxic organochlorine compounds by water chlorination processes has increased interest in ozonation. Furthermore, ozone is more destructive to viruses than is chlorine. Unfortunately, the solubility of ozone in water is relatively low, which limits its disinfective power.

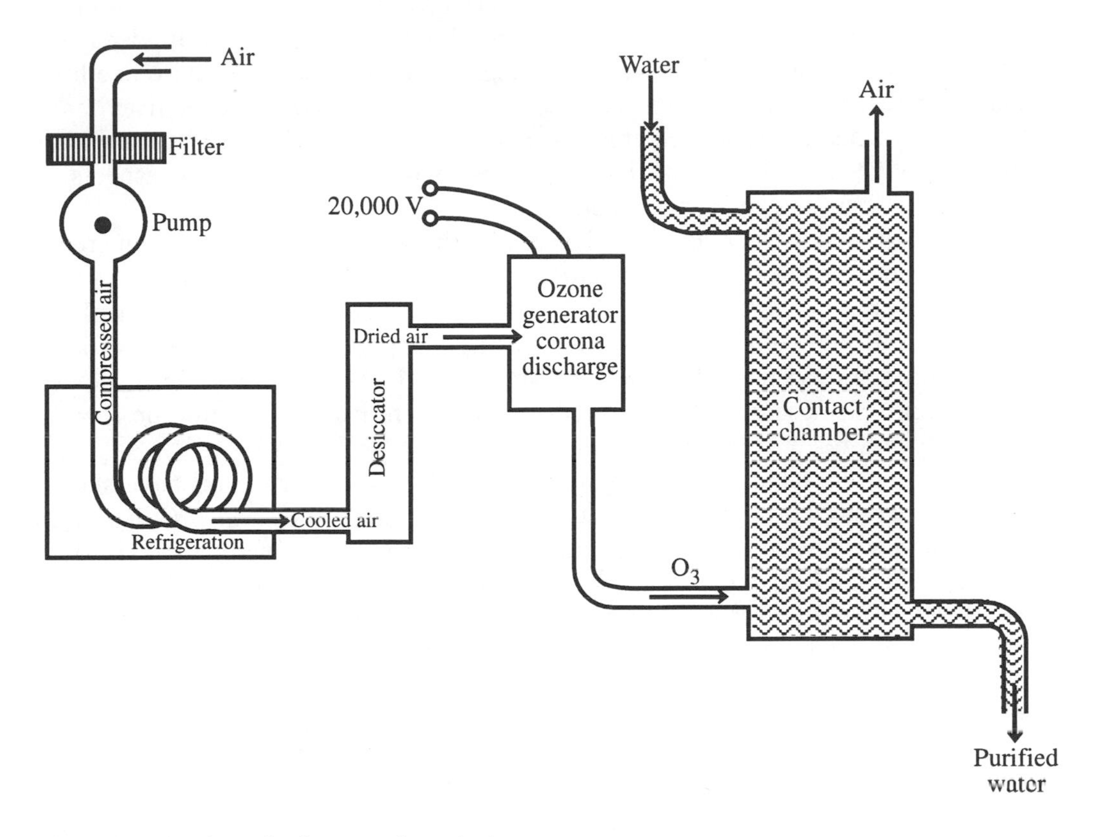

Figure 8.5. A schematic diagram of a typical ozone water treatment system.

8.12. NATURAL WATER PURIFICATION PROCESSES

Virtually all the materials that waste treatment processes are designed to eliminate may be absorbed by soil or degraded in soil; in fact, most are essential for soil fertility. The most basic thing that wastewater provides to plants is the water that is essential to plant growth. In addition to water, wastewater is normally very rich in essential plant nutrients, which are usually provided by fertilizers. These nutrients include phosphorus, largely as inorganic phosphate ions ($H_2PO_4^-$ and HPO_4^{2-} ions), nitrogen (primarily as ammonium nitrogen, NH_4^+), and potassium (K^+ion). Wastewater may also contain organically bound phosphorous and nitrogen that undergo biologically-mediated mineralization processes to provide the inorganic forms of these elements required by plants. Wastewater also contains essential trace elements and vitamins. Stretching the point a bit, the degradation of organic wastes provides the CO_2 essential for photosynthetic production of plant biomass.

Soil may be viewed as a natural filter for wastes. Most organic matter is readily degraded in soil and, in principle, soil constitutes an excellent treatment system for water. Soil can function to provide primary, secondary, and tertiary treatment of wastewater and has the additional advantage of not requiring expensive disposal of sludge after treatment. Soil has physical, chemical, and biological characteristics that can enable wastewater detoxification, biodegradation, chemical decomposition, and physical and chemical fixation.[3] Soil is a natural medium for a number of living organisms that may have an effect upon biodegradation of wastewaters, including those that contain industrial wastes. Of these, the most important are bacteria, including some kinds of *Agrobacterium*, *Arthrobacteri*, *Bacillus*, *Flavobacterium*, and *Pseudomonas*. Actinomycetes and fungi are important in decay of vegetable matter and may be involved in biodegradation of wastes. Other unicellular organisms that may be present in or on soil are protozoa and algae. Soil animals, such as earthworms, affect soil parameters such as soil texture. The growth of plants in soil may have an influence on its waste treatment potential in such aspects as uptake of soluble wastes and erosion control.

Early civilizations, such as the Chinese, used human organic wastes to increase soil fertility, and the practice continues today. The ability of soil to purify water was noted well over a century ago. In 1850 and 1852, J. Thomas Way, a consulting chemist to the Royal Agricultural Society in England, presented two papers to the Society entitled "Power of Soils to Absorb Manure." Mr. Way's experiments showed that soil is an ion exchanger. Much practical and theoretical information on the ion exchange process resulted from his work.

Experiments involving the direct application of wastewater to soil have shown appreciable increases in soil productivity. A process called *overland flow* has been used in which wastewater containing pulverized solids is allowed to trickle over sloping soil. Suspended solids, BOD, and nutrients are largely removed. This method is most applicable to rural communities in relatively warm climates. Approximately one acre of land is required to handle the sewage from 200 persons.

If such systems are not properly designed and operated, odor can become an overpowering problem. The author of this book is reminded of driving into a small town, recalled from some years before as a very pleasant place, and being assaulted with a virtually intolerable odor. The disgruntled residents pointed to a large spray irrigation system on a field in the distance — unfortunately upwind — spraying liquified pig manure as part of an experimental feedlot waste treatment operation. The experiment was not deemed a success and was discontinued by the investigators, presumably before they met with violence from the local residents.

Industrial Wastewater Treatment by Soil

Wastes that are amenable to land treatment are biodegradable organic substances, particularly those contained in municipal sewage and in wastewater from some industrial operations, such as food processing. However, through acclimation over a long period of time, soil bacterial cultures may develop that are effective in degrading normally recalcitrant compounds that occur in industrial wastewater. Acclimated microorganisms are found particularly at contaminated sites, such as those where soil has been exposed to crude oil for many years.

Land treatment is most used for petroleum refining wastes and is applicable to the treatment of fuels and wastes from leaking underground storage tanks. It can also be applied to biodegradable organic chemical wastes, including some organohalide compounds. Land treatment is not suitable for the treatment of wastes containing acids, bases, toxic inorganic compounds, salts, heavy metals, and organic compounds that are excessively soluble, volatile, or flammable.

CHAPTER SUMMARY

The chapter summary below is presented in a programmed format to review the main points covered in this chapter. It is used most effectively by filling in the blanks, referring back to the chapter as necessary. The correct answers are given at the end of the summary.

The major categories for water treatment are [1]______________________________ ______________________________. External treatment of water is usually applied to [2]______________________________ whereas internal treatment is designed to [3]______________________________ ______________________________. Current processes for the treatment of wastewater may be divided into the three main categories of [4]______________________________ ______________________________. Primary treatment of water is designed to remove [5]______________________________. The major constituent removed from water by secondary wastewater treatment is [6]______________________________. The reaction by which sewage sludge may be digested in the absence of oxygen by methane-producing anaerobic bacteria is [7]______________________________ and it reduces [8]______________________________. Tertiary waste treatment normally refers to [9]______________________________ ______________________________, and the three general kinds of contaminants that it removes are [10]______________________________ ______________________________. Five major operations in a physical-chemical wastewater treatment process are [11]______________________________ ______________________________. Two factors that have prevented development of physical-chemical wastewater treatment are [12]______________________________. Some physical processes used in industrial wastewater treatment are [13]______________________________ ______________________________. Before colloidal solids can be removed by filtration, they usually must be subjected to [14]______________________________. A reaction by which water becomes softened when heated is [15]______________________________. The reaction by which water containing "bicarbonate hardness," can be treated by addition of lime is [16]______________________________. When bicarbonate ion is not present at substantial levels, softening water by $CaCO_3$ removal requires [17]______________________________. Two reactions involving a solid ion exchanger by which salts may be removed from water are [18]______________________________ ______________________________. The basic method for removing both soluble iron and manganese from water is [19]______________________________. The removal of mercury, cadmium, or lead by lime treatment is aided by addition of [20]______________________________. The standard method of removing organic compounds from water is [21]______________________________. Some methods for removing dissolved inorganic material from water are [22]____.

A water purification process that consists of forcing pure water through a semipermeable membrane that allows the passage of water but not of other material is [23]__________________________. Phosphorus is removed in advanced wastewater treatment to [24]______________________________________. Phosphorus removal is usually accomplished by the addition of [25]____________ for which the reaction is [26]____. At a very high aeration rate in an activated sludge treatment process, phosphate is commonly removed because [27]____.

The two overall biological reactions by which nitrogen originally present as NH_4^+ can be removed from water are [28]____ and [29]____. Anaerobic digestion of sewage sludge in a digester serves to [30]____. The main plant nutrients contained in sewage sludge are [31]____. Sludge created by the water treatment reaction $Al^{3+} + 3OH^-(s) \rightarrow Al(OH)_3(aq)$ is known as alum sludge and causes problems because it is [32]____. The most commonly used water disinfectant is [33]_______ which reacts with water according to the reaction [34]____. Free available chlorine consists of [35]_______________________ in water, and combined available chlorine consists of [36]____. Chlorine dioxide is of particular interest for water disinfection because it does not produce [37]____. An oxidizing disinfectant that does not contain chlorine is [38]____________________ produced by [39]____. Soil has physical, chemical, and biological characteristics that can enable [40]____ of impurities in wastewater. The soil characteristics that are important in determining its use for land treatment of wastes are [41]____. Acclimated microorganisms adapted to degradation of organic compounds are found most commonly at [42]____.

Answers

[1] purification for domestic use, treatment for specialized industrial applications, and treatment of wastewater to make it acceptable for release or reuse

[2] the plant's entire water supply

[3] modify the properties of water for specific applications

[4] primary treatment, secondary treatment, and tertiary treatment

[5] insoluble matter such as grit, grease, and scum from water

6 biochemical oxygen demand

7 $2\{CH_2O\} \rightarrow CH_4 + CO_2$

8 both the volatile-matter content and the volume of the sludge by about 60%

9 a variety of processes performed on the effluent from secondary waste treatment

10 suspended solids, dissolved organic compounds, and dissolved inorganic materials

11 removal of scum and solid objects; clarification, generally with addition of a coagulant, and frequently with the addition of other chemicals (such as lime for phosphorus removal); filtration to remove filterable solids; activated carbon adsorption; and disinfection

12 high costs of chemicals and energy

13 density separation, filtration, flotation, evaporation, distillation, reverse osmosis, hyperfiltration, ultrafiltration, solvent extraction, air stripping, or steam stripping

14 coagulation

15 $Ca^{2+} + 2HCO_3^- \rightarrow CaCO_3(s) + CO_2(g) + H_2O$

16 $Ca^{2+} + 2HCO_3^- + Ca(OH)_2 \rightarrow 2CaCO_3(s) + 2H_2O$

17 a source of CO_3^{2-} at a relatively high enough pH

18 $H^{+-}\{Cat(s)\} + Na^+ + Cl^- \rightarrow Na^{+-}\{Cat(s)\} + H^+ + Cl^-$ and $OH^{-+}\{An(s)\} + H^+ + Cl^- \rightarrow Cl^{-+}\{An(s)\} + H_2O$

19 oxidation

20 sulfide

21 activated carbon sorption

22 distillation, electrodialysis, ion exchange, reverse osmosis

23 reverse osmosis

24 reduce algal growth

25 lime

26 $5Ca(OH)_2 + 3HPO_4^{2-} \rightarrow Ca_5OH(PO_4)_3(s) + 3H_2O + 6OH^-$

27 the CO_2 is swept out, the pH rises, and reactions occur, such as $5Ca^{2+} + 3HPO_4^{2-} + H_2O \rightarrow Ca_5OH(PO_4)_3(s) + 4H^+$

28 $NH_4^+ + 2O_2$ (Nitrifying bacteria) $\rightarrow NO_3^- + 2H^+ + H_2O$

29 $6NO_3^- + 5CH_3OH + 6H^+$ (denitrifying bacteria) $\rightarrow 3N_2(g) + 5CO_2 + 13H_2O$

30 reduce the mass and volume of sludge and destroy disease agents

31 5% N, 3% P, and 0.5% K on a dry-weight basis

32 very difficult to dewater

33 Cl_2

34 $Cl_2 + H_2O \rightarrow H^+ + Cl^- + HOCl$

35 HOCl and OCl^-

36 the chloramines

37 trihalomethanes

38 ozone

39 an electrical discharge through dry air

40 detoxification, biodegradation, chemical decomposition, and physical and chemical fixation

41 physical form, ability to retain water, aeration, organic content, acid-base characteristics, and oxidation-reduction behavior

42 sites contaminated with the kinds of wastes degraded

QUESTIONS AND PROBLEMS

1. What is the purpose of the return sludge step in the activated sludge process?
2. What are the two processes by which the activated sludge process removes soluble carbonaceous material from sewage?
3. Why might hard water be desirable as a medium if phosphorus is to be removed by an activated sludge plant operated under conditions of high aeration?
4. How does reverse osmosis differ from simple sieve separation or ultrafiltration?
5. How many liters of methanol would be required daily to remove the nitrogen from a 200,000-L/day sewage treatment plant producing an effluent containing 50 mg/L of nitrogen? Assume that the nitrogen has been converted to NO_3^- in the plant. The denitrifying reaction is Reaction 8.9.5.
6. Discuss some of the advantages of physical-chemical treatment of sewage as opposed to biological wastewater treatment. What are some disadvantages?
7. Why is recarbonation necessary when water is softened by the lime-soda process?
8. Assume that a waste contains 300 mg/L of biodegradable $\{CH_2O\}$ and is processed through a 200,000-L/day sewage-treatment plant that converts 40% of the waste to CO_2 and H_2O. Calculate the volume of air (at 25°C, 1 atm) required for this conversion. Assume that the O_2 is transferred to the water with 20% efficiency.
9. If all of the $\{CH_2O\}$ in the plant described in Question 8 could be converted to methane by anaerobic digestion, how many liters of methane (STP) could be produced daily?
10. Assuming that aeration of water does not result in the precipitation of calcium carbonate, which of the following would not be removed by aeration: hydrogen sulfide, carbon dioxide, volatile odorous bacterial metabolites, alkalinity, iron.
11. Suggest a source of microorganisms to use in a waste treatment process. Where should an investigator look for microorganisms to use in such an application? What are some kinds of wastes for which soil is particularly unsuitable as a treatment medium?

12. In which of the following water supplies would moderately high water hardness be most detrimental: municipal water, irrigation water, boiler feedwater, drinking water (in regard to potential toxicity)?

13. Discuss how soil may be viewed as a natural filter for wastes. How does soil aid waste treatment? How may waste treatment be of benefit to soil in some cases?

14. Match each water contaminant in the left column with its preferred method of removal in the right column.

(a) Mn^{2+}	(1) Activated carbon
(b) Ca^{2+} and HCO_3^-	(2) Raise pH by addition of Na_2CO_3
(c) Trihalomethane compounds	(3) Addition of lime
(d) Mg^{2+}	(4) Oxidation

15. A cementation reaction employs iron to remove Cd^{2+} present at a level of 350 mg/L from a wastewater stream. Given that the atomic mass of Cd is 112.4 and that of Fe is 55.8, how many kg of Fe are consumed in removing all the Cd from 4.50×10^6 liters of water?

16. Consider municipal drinking water from two different kinds of sources, one a flowing, well-aerated stream with a heavy load of particulate matter and the other an anaerobic groundwater. Describe possible differences in the water treatment strategies for these two sources of water.

17. In treating water for industrial use, consideration is often given to "sequential use of the water." What is meant by this term? Give some plausible examples of sequential use of water.

18. Active biomass is used in the secondary treatment of municipal wastewater. Describe three ways of supporting a growth of the biomass, contacting it with wastewater, and exposing it to air.

19. Using appropriate chemical reactions for illustration, show how calcium present as the dissolved HCO_3^- salt in water is easier to remove than other forms of hardness, such as dissolved $CaCl_2$.

20. Label each of the following as external treatment (ex) or internal treatment (in): ( ) aeration, ( ) addition of inhibitors to prevent corrosion, ( ) adjustment of pH, ( ) filtration, ( ) clarification, ( ) removal of dissolved oxygen by reaction with hydrazine or sulfite, ( ) disinfection for food processing.

21. Label each of the following as primary treatment (pr), secondary treatment (sec), or tertiary treatment (tert): ( ) screening, ( ) comminuting, ( ) grit removal, ( ) removal of dissolved inorganic materials, ( ) BOD removal, ( ) activated carbon filtration removal of dissolved organic compounds.

22. Both activated sludge waste treatment and natural processes in streams and bodies of water remove degradable material by biodegradation. Explain why activated sludge treatment is so much more effective.

23. Of the following, the one that does not belong with the rest is ( ) removal of scum and solid objects, ( ) clarification, ( ) filtration, ( ) degradation with activated sludge, ( ) activated carbon adsorption, ( ) disinfection.

24. Explain why complete physical-chemical wastewater treatment systems are better than biological systems in dealing with toxic substances and overloads.

25. What are the two major ways in which dissolved carbon (organic compounds) is removed from water in industrial wastewater treatment. How do these two approaches differ fundamentally?

26. What is the reaction for the hydrolysis of aluminum ion in water? How is this reaction used for water treatment?

27. Explain why coagulation is used with filtration.

28. What are two major problems that arise from the use of excessively hard water?

29. Show with chemical reactions how the removal of bicarbonate hardness with lime results in a net removal of ions from solution, whereas removal of non-bicarbonate hardness does not.

30. What two purposes are served by adding CO_2 to water that has been subjected to lime-soda softening?

31. Why is cation exchange normally used without anion exchange for softening water?

32. Show with chemical reactions how oxidation is used to remove soluble iron and manganese from water.

33. Show with chemical reactions how lime treatment, sulfide treatment, and cementation are used to remove heavy metals from water.

34. How is activated carbon prepared? What are the chemical reactions involved? What is remarkable about the surface area of activated carbon?

35. How is the surface of the membrane employed involved in the process of reverse osmosis?

36. Describe with a chemical reaction how lime is used to remove phosphate from water. Although other chemicals may be used for phosphate removal, suggest why lime may be favored.

37. Denitrification involving the microbial reduction of nitrate ion is one of the best ways to remove excess nutrient nitrogen from water. Why is nitrification required as a preliminary step in removal of nitrogen from water by biological denitrification?

38. What are some possible beneficial uses for sewage sludge? What are some of its characteristics that may make such uses feasible? Suggest some of the potential problems that may arise from using sewage sludge for these purposes.

39. Distinguish between free available chlorine and combined available chlorine in water disinfection.

40. What are one major advantage and one major disadvantage of using chlorine dioxide for water disinfection?

41. What are one major advantage and one major disadvantage of using ozone for water disinfection?

LITERATURE CITED

1. Blanton, T. Clay, Dale Rohe, Joseph G. Jacangelo, and Benito J. Mariñas, "Emerging Membrane Processes for Drinking Water Treatment," *Waterworld News*, January/February, 1991, pp. 10-13.

2. "Ocean Dumping," *SFI Bulletin*, No. 403, Sport Fishing Institute, Washington, D.C., April, 1989, pp. 1-3.

3. D'Itri, F. M., J. A. Martinez, and M. A. Lámbari, *Municipal Wastewater in Agriculture*, Ann Arbor Science Publishers, Inc., Ann Arbor, MI., 1981.

SUPPLEMENTARY REFERENCES

American Society of Civil Engineers, *Management of Water Treatment Plant Residuals*, American Society of Civil Engineers, New York, NY, 1996.

Berne, Francois and Jean Cordonnier, *Industrial Water Treatment: Refining, Petrochemicals and Gas Processing Techniques*, Gulf Publishing Co, Summit, NJ , 1995.

Doerr, William and Rajeev Krishnan, Eds., *How to Implement Industrial Water Reuse: A Systematic Approach*, American Institute of Chemical Engineers, New York, NY, 1995.

Eikkum, A. S. and R. W. Seabloom, Eds., *Alternative Wastewater Treatment*, D. Reidel Publishing Co., Hingham, MA, 1982.

Gillies, M. T., Ed., *Potable Water from Wastewater*, Noyes Data Corp., Park Ridge, NJ, 1981.

Hammer, Donald A., Ed., *Constructed Wetlands for Wastewater Treatment*, Lewis Publishers, Inc., Chelsea, MI, 1989.

HDR Engineering, *Handbook of Public Water Systems*, Van Nostrand Reinhold, New York, NY, 1997.

Perkins, Richard J., *Onsite Wastewater Disposal*, Lewis Publishers, Chelsea, MI, 1989.

Roques, Henri, Ed., *Chemical Water Treatment: Principles and Practice*, VCH Publishers, New York, NY, 1996.

Sussman, S., "Industrial Water Treatment," in *Kirk-Othmer Concise Encyclopedia of Chemical Technology*, David Eckroth, Ed., John Wiley and Sons, New York, NY, 1985.

Winkler, M. A., *Biological Treatment of Wastewater*, John Wiley and Sons, Inc., New York, NY, 1981.

THE ATMOSPHERE

9 METEOROLOGY AND CHARACTERISTICS OF THE ATMOSPHERE

9.1. IMPORTANCE OF THE ATMOSPHERE

The atmosphere is a protective blanket that nurtures life on the Earth and protects it from the hostile environment of outer space. The atmosphere is the source of carbon dioxide for plant photosynthesis and of oxygen for respiration. It provides the nitrogen that nitrogen-fixing bacteria and ammonia-manufacturing chemical plants use to produce chemically bound nitrogen, an essential component of life molecules. The atmosphere is an ideal medium for life and was largely produced in its present form by life processes. As a basic part of the hydrologic cycle (Figure 5.1) the atmosphere transports water from the oceans to land, thus acting as the condenser in a vast solar-powered still. Circulation of air, water vapor, and heat in the atmosphere redistributes solar energy from the equator toward the polar regions, and water from the oceans to land masses. Unfortunately, the atmosphere also has been used as a dumping ground for many pollutant materials—ranging from sulfur dioxide to refrigerant Freon—a practice that causes damage to vegetation and materials, shortens human life, and alters the characteristics of the atmosphere itself.

The atmosphere serves a vital protective function. It absorbs most of the cosmic rays from outer space and protects organisms from their effects. It also absorbs most of the electromagnetic radiation from the sun, allowing transmission of significant amounts of radiation only in the regions of 300-2500 namometers (near-ultraviolet, visible, and near-infrared radiation) and 0.01-40 meters (radio waves). By absorbing electromagnetic radiation below 300 nm the atmosphere filters out damaging ultraviolet radiation that would otherwise be very harmful to living organisms. Furthermore, because it reabsorbs much of the infrared radiation by which absorbed solar energy is re-emitted to space, the atmosphere stabilizes the Earth's temperature, preventing the tremendous temperature extremes that occur on planets and moons lacking substantial atmospheres.

9.2. COMPOSITION OF THE ATMOSPHERE

Atmospheric science deals with the movement of air masses in the atmosphere, atmospheric heat balance, and atmospheric chemical composition and reactions. In order to understand atmospheric chemistry and air pollution, it is important to have an overall appreciation of the atmosphere, its composition, and physical characteristics as discussed in the first parts of this chapter.

Atmospheric Composition

In percentages by volume, dry air within several kilometers of ground level consists of two **major components**,

- Nitrogen, 78.08 %
- Oxygen, 20.95 %

minor components,

- Argon, 0.934 %
- Carbon dioxide, 0.035 %

noble gases,

- Neon, 1.818×10^{-3} %
- Helium, 5.24×10^{-4} %
- Krypton, 1.14×10^{-4} %
- Xenon, 8.7×10^{-6} %

and **trace gases** as given in Table 9.1. Atmospheric air contains varying percentages of water as discussed below.

Table 9.1. Atmospheric Trace Gases in Dry Air Near Ground Level

Gas or species	Volume percent[a]	Major sources	Process for removal from the atmosphere
CH_4	1.6×10^{-4}	Biogenic[b]	Photochemical[c]
CO	$\sim 1.2 \times 10^{-5}$	Photochemical, anthropogenic[d]	Photochemical
N_2O	3×10^{-5}	Biogenic	Photochemical
NO_x[e]	10^{-10}-10^{-6}	Photochemical, lightning, anthropogenic	Photochemical
HNO_3	10^{-9}-10^{-7}	Photochemical	Washed out by precipitation
NH_3	10^{-8}-10^{-7}	Biogenic	Photochemical, washed out by precipitation
H_2	5×10^{-5}	Biogenic, photochemical	Photochemical
H_2O_2	10^{-8}-10^{-6}	Photochemical	Washed out by precipitation
$HO\cdot$[f]	10^{-13}-10^{-10}	Photochemical	Photochemical
$HO_2\cdot$[f]	10^{-11}-10^{-9}	Photochemical	Photochemical
H_2CO	10^{-8}-10^{-7}	Photochemical	Photochemical
CS_2	10^{-9}-10^{-8}	Anthropogenic, biogenic	Photochemical
OCS	10^{-8}	Anthropogenic, biogenic, photochemical	Photochemical
SO_2	$\sim 2 \times 10^{-8}$	Anthropogenic, photochemical, volcanic	Photochemical
I_2	0-trace	—	—
CCl_2F_2[g]	2.8×10^{-5}	Anthropogenic	Photochemical
H_3CCCl_3[h]	$\sim 1 \times 10^{-8}$	Anthropogenic	Photochemical

[a] Levels in the absence of gross pollution.
[b] From biological sources.
[c] Reactions induced by the absorption of light energy (see Section 9.9 and Chapter 10).
[d] Sources arising from human activities.
[e] Sum of NO and NO_2.
[f] Reactive free radical species with one unpaired electron, described later in Chapter 10; these are transient species whose concentrations become much lower at night.
[g] A chlorofluorocarbon, Freon F-12.
[h] Methyl chloroform.

Atmospheric Water

The water vapor content of the troposphere is normally within a range of 1–3% by volume with a global average of about 1%. However, air can contain as little as 0.1% or as much as 5% water. The percentage of water in the atmosphere decreases rapidly with increasing altitude. Water circulates through the atmosphere in the hydrologic cycle as shown in Figure 5.1.

Water vapor absorbs infrared radiation even more strongly than does carbon dioxide, thus greatly influencing the Earth's heat balance. Clouds formed from water vapor reflect light from the sun and have a temperature-lowering effect. On the other hand, water vapor in the atmosphere acts as a kind of "blanket" at night, retaining heat.

When ice particles in the atmosphere change to liquid droplets or when these droplets evaporate, heat is absorbed from the surrounding air. Reversal of these processes results in heat release to the air (as latent heat). This may occur many miles from the place where heat was absorbed and is a major mode of energy transport in the atmosphere. It is the predominant type of energy transition involved in thunderstorms, hurricanes, and tornadoes (see Section 9.5).

On a global basis, rivers drain only about one-third of the precipitation that falls on the Earth's continents. This means that two-thirds of the precipitation is lost as combined evaporation and transpiration. During the summer, this evapotranspiration may exceed precipitation because of the large quantities of water stored in the root zone of the soil. To a degree, evapotranspiration furnishes atmospheric water vapor necessary for cloud formation and precipitation. Therefore, large-scale deforestation, soil damage (such as by plowing up grasslands in semi-arid areas), and irrigation affect regional climate and rainfall. As noted in Section 9.3, the cold tropopause serves as a barrier to the movement of water into the stratosphere.

9.3. REGIONS AND STRATIFICATION OF THE ATMOSPHERE

As anyone who has exercised at high altitudes well knows, the density of the atmosphere decreases sharply with increasing altitude as a consequence of the gas laws and gravity. More than 99% of the total mass of the atmosphere is found within approximately 30 km (about 20 miles) of the Earth's surface. Such an altitude is miniscule compared to the Earth's diameter, so it is not an exaggeration to characterize the atmosphere as a "tissue-thin" protective layer. Although the total mass of the global atmosphere is enormous, approximately 5.14 x 10^{15} metric tons, it is still only approximately one millionth of the Earth's total mass.

The fact that atmospheric pressure decreases as an approximately exponential function of altitude largely determines the characteristics of the atmosphere. Plots of pressure at a specified altitude, P_h, and temperature *versus* altitude are shown in Figure 9.1. The plot of P_h is nonlinear because of variations arising from temperature differences and the mixing of air masses. The plot reflects nonlinear variations in temperature with altitude that are discussed later in this section.

The characteristics of the atmosphere vary widely with altitude, time (season), location (latitude), and even solar activity. Extremes of pressure and temperature are illustrated in Figure 9.1. At very high altitudes the atmosphere is so sparsely populated by gas molecules that normally reactive species, such as atomic oxygen, O, persist for long periods of time. That occurs because at the extremely low pressures at these altitudes the **mean free path** travelled by a reactive species before it collides with a potential reactant is quite high. A particle with a mean free path of 1 x 10^{-6} cm at sea level has a mean free path greater than 1 x 10^{6} cm at an altitude of 500 km, where the pressure is lower by many orders of magnitude.

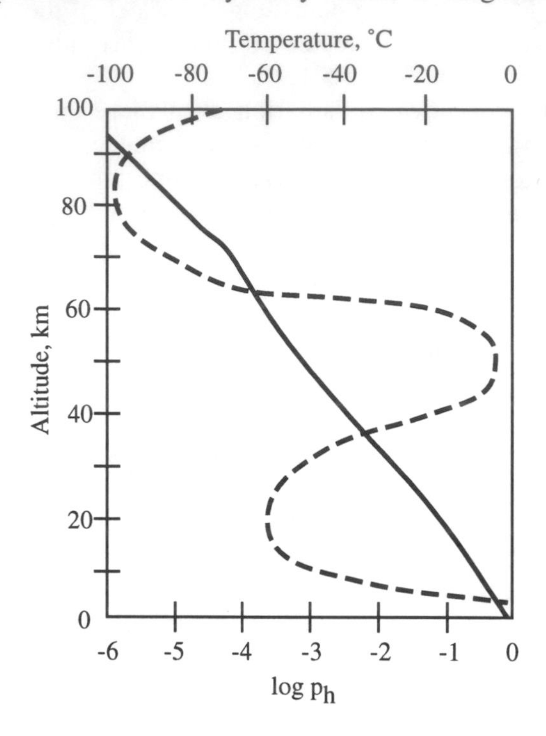

Figure 9.1. Variation of pressure (solid line, expressed as log of ratio of pressure to atmospheric pressure at sea level) and temperature (dashed line) with altitude.

Stratification of the Atmosphere

As shown in Figure 9.2, the atmosphere is stratified on the basis of temperature/density relationships resulting from interrelationships between physical and photochemical (light-induced chemical phenomena) processes in air.

The **troposphere** extends from sea level to an altitude of 8-16 km (lower over the poles, higher over the equator). It contains about 3/4 of the atmosphere's mass, has a generally homogeneous composition of major gases other than water, and exhibits decreasing temperature with increasing altitude from the heat-radiating surface of the earth. The upper limit of the troposphere, which has a temperature minimum of about -56°C, varies in altitude by a kilometer or more with atmospheric temperature, underlying terrestrial surface, and time. The homogeneous composition of the troposphere results from constant mixing by circulating air masses. However, the water vapor content of the troposphere is extremely variable because of cloud formation, precipitation, and evaporation of water from terrestrial water bodies.

In the very cold **tropopause** at the top of the troposphere, water vapor condenses to ice and cannot reach altitudes at which it would photodissociate through the action of intense high-energy ultraviolet radiation. If this happened, the hydrogen produced would escape the Earth's atmosphere and be lost. (Much of the hydrogen and helium originally present in the Earth's atmosphere was lost by this process.)

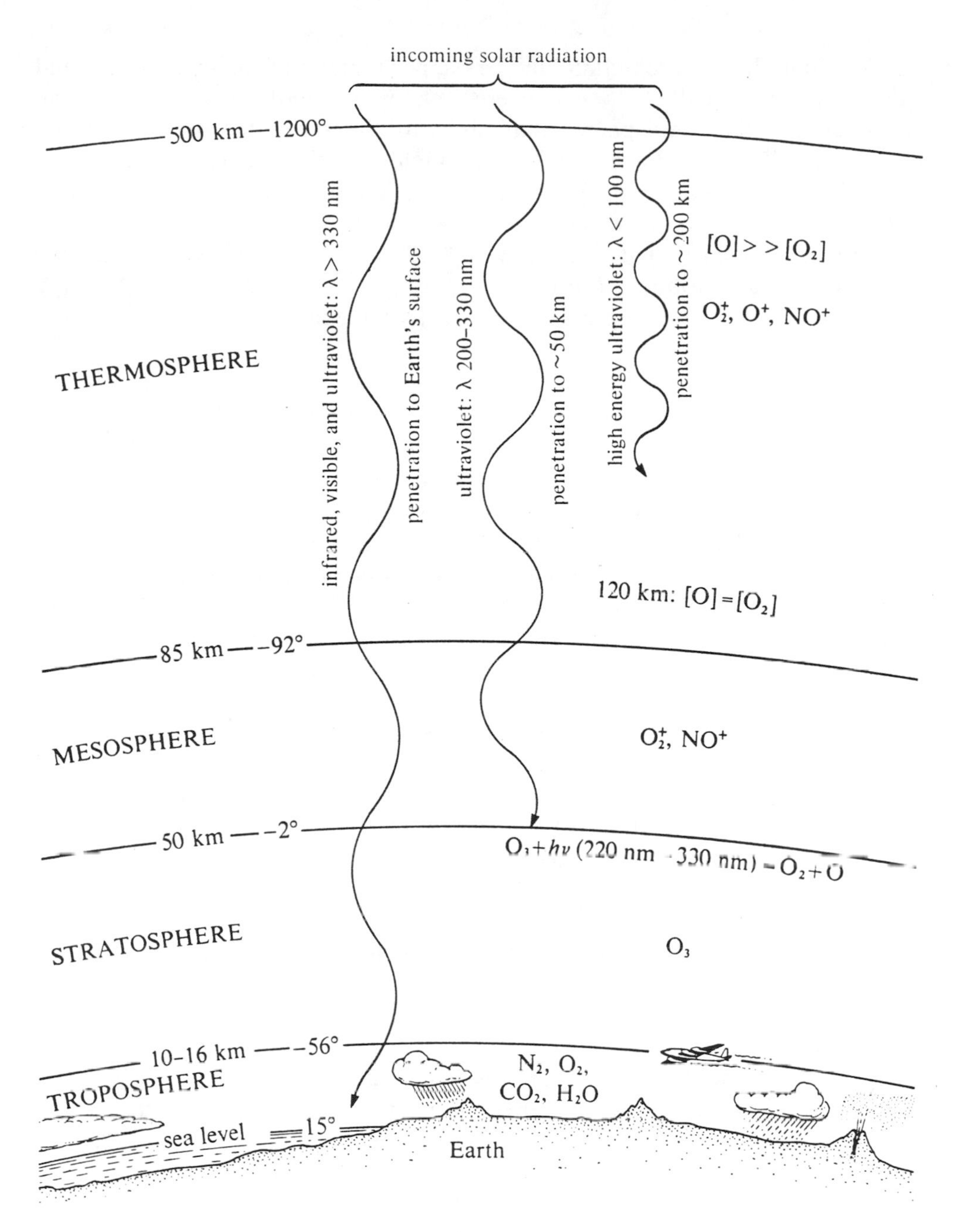

Figure 9.2. Major regions of the atmosphere (not to scale).

The atmospheric layer directly above the troposphere is the **stratosphere**, in which the temperature rises to a maximum of about -2°C with increasing altitude. This phenomenon is due to the presence of ozone, O_3, which may reach a level of around 10 ppm by volume in the mid-range of the stratosphere. The heating effect is caused by the absorption of ultraviolet radiation energy from the sun by ozone, a phenomenon discussed later in this chapter. The fraction of ozone is about 1000 times as great, and the fraction of water vapor only about 1/1000 as much in the stratosphere as compared to the troposphere.

The absence of high levels of radiation-absorbing species in the **mesosphere** immediately above the stratosphere results in a further temperature decrease to about –92°C at an altitude around 85 km. The upper regions of the mesosphere and higher define a region called the exosphere from which molecules and ions can completely escape the atmosphere. Extending to the far outer reaches of the atmosphere is the **thermosphere**, in which the highly rarified gas reaches temperatures up to 1200°C by the absorption of very energetic radiation of wavelengths less than approximately 200 nm by gas species in this region.

At altitudes of approximately 50 km and up, ions are so prevalent that the region is called the **ionosphere**. These ions are formed by the absorption of highly energetic ultraviolet radiation, represented by *hν*, by molecules:

$$N_2 + h\nu \rightarrow N_2^+ + e^- \qquad (9.3.1)$$

The presence of the ionosphere has been known since about 1901, when it was discovered that radio waves could be transmitted over long distances, where the curvature of the Earth makes line-of-sight transmission impossible. These radio waves bounce off the ionosphere. At night, there is less intense solar radiation to form ions, which recombine, raising the lower limit of the ionosphere and enabling longer-distance radio transmission. The emission of light from highly energized (excited) molecules and ions in the ionosphere is responsible for the spectacular displays of northern lights, the aurora borealis.

9.4. ENERGY TRANSFER IN THE ATMOSPHERE

The physical and chemical characteristics of the atmosphere and the critical heat balance of the Earth are determined by energy and mass transfer processes in the atmosphere. These phenomena are addressed in this section.

Incoming solar energy is largely in the visible region of the spectrum. The shorter wavelength blue solar light is scattered relatively more strongly by molecules and particles in the upper atmosphere, which is why the sky is blue as it is viewed by scattered light. Similarly, light that has been transmitted through scattering atmospheres appears red, particularly around sunset and sunrise, and under circumstances in which the atmosphere contains a high level of particles. The solar energy flux reaching the atmosphere is huge, amounting to 1.34×10^3 watts per square meter (19.2 kcal per minute per square meter) perpendicular to the line of solar flux at the top of the atmosphere, as illustrated in Chapter 1, Figure 1.3. This value is the **solar constant** and may be termed **insolation**, which stands for "incoming solar radiation." If all this energy reached the Earth's surface and were retained, the planet would have vaporized long ago. As it is, the complex factors involved in maintaining the Earth's heat balance within very narrow limits are crucial to retaining conditions of climate that will support present levels of life on Earth. The great changes of climate that resulted in ice ages during some periods, or tropical conditions during others, were caused by variations of only a few degrees in average temperature. Marked climate changes within recorded history have been caused by much smaller average temperature changes. The mechanisms by which the Earth's average temperature is retained within its present narrow range are complex and not completely understood, but the main features are explained here.

About half of the solar radiation entering the atmosphere reaches Earth's surface either directly or after scattering by clouds, atmospheric gases, or particles. The other half of this radiation is either reflected directly back or absorbed in the atmosphere and its energy radiated back into space at a later time as infrared radiation. Most of the solar energy reaching the surface is absorbed and it must be returned to space in order to maintain heat balance. In addition, a very small amount of energy (less than 1% of that received from the sun) reaches the Earth's surface by convection and conduction processes from Earth's hot mantle, and this, too, must be lost.

Energy transport, which is crucial to eventual reradiation of energy from the Earth is accomplished by three major mechanisms—conduction, convection, and radiation. **Conduction** of energy occurs through the interaction of adjacent atoms or molecules without the bulk movement of matter. **Convection** involves the movement of whole masses of air, which may be either relatively warm or cold. It is the mechanism by which abrupt temperature variations occur when large masses of air move across an area. As well as carrying **sensible heat** due to the kinetic energy of molecules, convection carries **latent heat** in the form of water vapor, which releases heat as it condenses. An appreciable fraction of the Earth's surface heat is transported to clouds in the atmosphere by conduction and convection before being lost ultimately by radiation.

Radiation of energy occurs through electromagnetic radiation in the infrared region of the spectrum. As the only way in which energy is transmitted through a vacuum, radiation is the means by which all energy lost from the planet to maintain its heat balance is ultimately returned to space. The electromagnetic radiation that carries energy away from the Earth is of a much longer wavelength than the sunlight that brings energy to the Earth. This is a crucial factor in maintaining the Earth's heat balance and one susceptible to upset by human activities. The maximum intensity of incoming radiation occurs at 0.5 micrometers (500 nanometers) in the visible region, with essentially none outside the range of 0.2 μm to 3 μm. This range encompasses the whole visible region and small parts of the ultraviolet and infrared adjacent to it. Outgoing radiation is in the infrared region, primarily between 2 μm and 40 μm, and with maximum intensity at about 10 μm. Thus the Earth loses energy by electromagnetic radiation of a much higher wavelength (lower energy per photon) than that of the radiation by which it receives energy.

Earth's Radiation Budget

The Earth's radiation budget is illustrated in Figure 9.3. The average surface temperature is maintained at a relatively comfortable 15°C because of an atmospheric "greenhouse effect" in which water vapor and, to a lesser extent carbon dioxide, reabsorb much of the outgoing radiation and reradiate about half of it back to the surface. Were this not the case, the surface temperature would average around -18°C. Most of the absorption of infrared radiation is done by water molecules in the atmosphere. Absorption is weak in the regions 7-8.5 μm and 11-14 μm and nonexistent between 8.5 μm and 11 μm, leaving a "hole" in the infrared absorption spectrum through which radiation may escape. Carbon dioxide, though present at a much lower concentration than water vapor, absorbs strongly between 12 μm and 16.3 μm, and plays a key role in maintaining the heat balance. There is concern that an increase in the carbon dioxide level in the atmosphere could prevent sufficient energy loss to cause a perceptible and damaging increase in the Earth's temperature. This phenomenon, mentioned in respect to its effect on global climate in Section 9.7 and discussed in more detail in Chapter 12, is popularly known as the **greenhouse effect** and may occur from elevated CO_2 levels caused by increased use of fossil fuels and the destruction of massive quantities of forests.

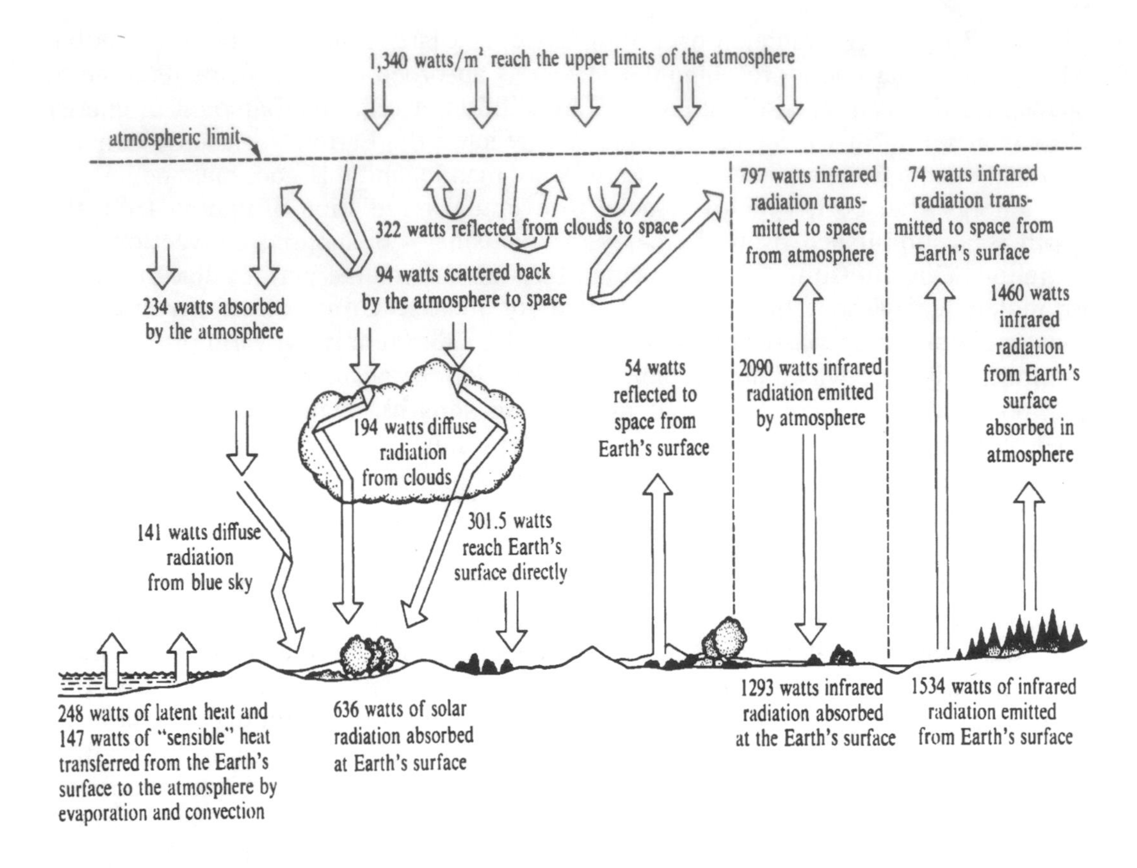

Figure 9.3. Earth's radiation budget expressed on the basis of portions of the 1,340 watts/m^2 composing the solar flux.

9.5. ATMOSPHERIC MASS TRANSFER, METEOROLOGY, AND WEATHER

Meteorology is the science of atmospheric phenomena, encompassing the study of the movement of air masses as well as physical forces in the atmosphere—heat, wind, and transitions of water, primarily liquid to vapor, or *vice versa*. Meteorological phenomena affect, and in turn are affected by, the chemical properties of the atmosphere. For example, meteorological phenomena determine whether or not power plant stack gas containing pollutant sulfur dioxide is dispersed high in the atmosphere, with little direct effect upon the immediate surroundings, or settles as an unpleasant mass of contaminated air in the vicinity of the power plant. Los Angeles largely owes its susceptibility to smog to the meteorology of the Los Angeles basin, which holds hydrocarbons and nitrogen oxides long enough to cook up an unpleasant brew of damaging chemicals under the intense rays of the sun (see the discussion of photochemical smog in Chapter 12).

Short-term variations in the state of the atmosphere are described as **weather**. The weather is defined in terms of seven major factors: temperature, clouds, winds, humidity, horizontal visibility (as affected by fog, etc.), type and quantity of precipitation, and atmospheric pressure. All of these factors are closely interrelated. Longer-term variations and trends within a particular geographical region in those factors that compose weather are described as *climate*, a term defined and discussed in Section 9.7.

The driving force behind weather and climate is the distribution and ultimate reradiation to space of solar energy. A large fraction of solar energy is converted to latent heat by evaporation of water into the atmosphere. Because of its unique heat properties (Section 5.3 and Table 5.2), water is capable of absorbing (and later releasing) enormous amounts of heat per unit mass. Thus, the latent heat of evaporation of water is one of the main ways in which energy is redistributed in the atmosphere.

Condensation of water vapor, which forms clouds, must occur prior to the formation of precipitation in the form of rain or snow. For this condensation to happen, air must be cooled below a temperature called the **dew point**, and **condensation nuclei** must be present. These nuclei are hydroscopic substances such as salts, sulfuric acid droplets, and some organic materials, including bacterial cells. Air pollution in some forms is a significant source of condensation nuclei. As water condenses from atmospheric air, large quantities of heat are released. This is a particularly significant means for transferring energy from the ocean to land. Solar energy falling on the ocean is converted to latent heat by the evaporation of water, then the water vapor moves inland where it condenses. The latent heat released when the water condenses warms the surrounding land mass.

Solar energy received by Earth is largely redistributed by the movement of huge masses of air with different pressures, temperatures, and moisture contents. Horizontally moving air is called **wind**, whereas vertically moving air is referred to as an **air current**. Atmospheric air moves constantly, with behavior and effects that reflect the laws governing the behavior of gases. First of all, gases will move horizontally and/or vertically from regions of *high atmopheric pressure* to those of *low atmospheric pressure*. Furthermore, expansion of gases causes cooling, whereas compression causes warming. A mass of warm air tends to move from Earth's surface to higher altitudes, where the pressure is lower; in so doing, it expands *adiabatically* (that is, without exchanging energy with its surroundings) and becomes cooler. If there is no condensation of moisture from the air, the cooling effect is about 10°C per 1000 meters of altitude, a figure known as the **dry adiabatic lapse rate**. A cold mass of air at a higher altitude does the opposite; it sinks and becomes warmer at about 10°C/1000 m. Often, however, when there is sufficient moisture in rising air, water condenses from it, releasing latent heat. This partially counteracts the cooling effect of the expanding air, giving a **moist adiabatic lapse rate** of about 6°C/1000 m. Parcels of air do not rise and fall, or even move horizontally, in a completely uniform way, but exhibit eddies, currents, and various degrees of turbulence.

As noted above, *wind* is air moving horizontally, whereas *air currents* are created by air moving up or down. Wind occurs because of differences in air pressure from high pressure regions to low pressure areas. Air currents are largely **convection currents** formed by differential heating of air masses. Air that is over a solar heated land mass is warmed, becomes less dense, therefore rises, and is replaced by cooler and more dense air. Wind and air currents are strongly involved with air pollution phenomena. Wind carries and disperses air pollutants. Prevailing wind direction is an important factor in determining the areas most affected by an air pollution source. Wind is an important renewable energy resource (see the discussion of renewable energy resources in Chapter 22). Furthermore, wind plays an important role in the propagation of life by dispersing spores, seeds, and organisms, such as spiders.

Topographical Effects

Topography, the surface configuration and relief features of the Earth's surface may strongly affect winds and air currents. Differential heating and cooling of land surfaces and bodies of water can result in **local convective winds**, including land breezes and sea breezes at different times of the day along the seashore, as well as breezes associated with large bodies of water inland. Mountain topography causes complex and variable localized winds. The masses of air in mountain valleys heat up during the day causing upslope winds and cool off at night causing downslope winds. Upslope winds flow over ridge tops in mountainous regions. The blocking of wind and of masses of air by mountain formations some distance inland from seashores can trap bodies of air, particularly when temperature inversion conditions occur (see Section 9.6).

Influence of Ocean Currents

The circulation of water in ocean currents has a significant effect on weather. The most prominent such effect is that of the **Gulf Stream**, which carries warm water from the Gulf of Mexico to northern Europe. Heat transferred from the Gulf Stream water to air in Europe serves to make much of Europe far more hospitable to human habitation than it would otherwise be. The cooled water sinks and moves westward in a current moving in the direction of Laborador before turning south to return to the Gulf of Mexico. Other massive oceanic water circulation patterns occur in the Pacific Ocean and affect weather along the Pacific coast. One manifestation of this pattern in the pacific is a warm mass of surface water in the eastern tropical Pacific Ocean known as El Niño.

Oscillations in the circulation of oceanic water lasting two or three decades are associated with weather cycles on land. In Europe a cool period in the 1950s and 1960s was followed by two decades of comparatively warm weather. In 1996 a distinct cooling of the water in the North Atlantic current resulted in a very cold winter in Europe, a pattern that may persist for some years. On a longer time span large fluctuations in the climate of land bordering the North Atlantic have probably been associated with temperature fluctuations in the circulating ocean water. Several centuries of unusually warm weather in Medieval times enabled crops to thrive in England and allowed the Vikings to establish colonies in Greenland. These times were followed by about 300 years of unusually cold weather, the "Little Ice Age" that caused great hardship in Europe.

Movement of Air Masses

Basically, weather is the result of the interactive effects of (1) redistribution of solar energy, (2) horizontal and vertical movement of air masses with varying moisture contents, and (3) evaporation and condensation of water, accompanied by uptake and release of heat. To see how these factors determine weather—and ultimately climate—on a global scale, first consider the cycle illustrated in Figure 9.4. This figure shows solar energy being absorbed by a body of water, and causing some water to evaporate. The warm, moist mass of air thus produced moves from a region of high pressure to one of low pressure and cools by expansion as it rises in

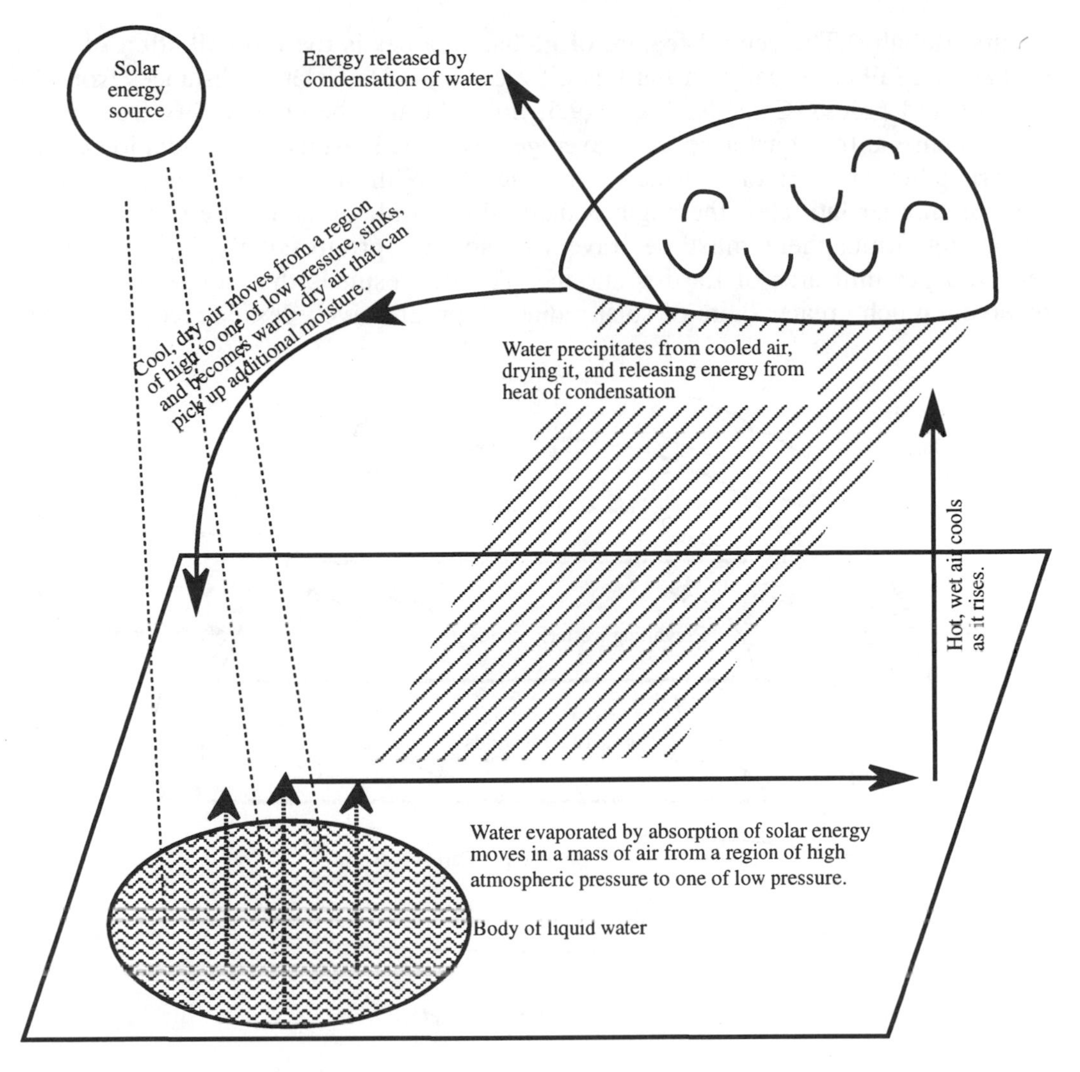

Figure 9.4. Circulation patterns involved with movement of air masses and water; uptake and release of solar energy as latent heat in water vapor.

what is called a **convection column**. As the air cools, water condenses from it and energy is released; this is a major pathway by which energy is transferred from the Earth's surface to high in the atmosphere. This process converts air from warm, moist air to cool, dry air. Its movement to high altitudes results in a degree of "crowding" of air molecules in the air parcel and creates a zone of relatively high pressure high in the troposphere at the top of the convection column. This air mass, in turn, moves from the upper-level region of high pressure to one of low pressure; in so doing, it subsides, thus creating an upper-level low-pressure zone, and becomes warm, dry air in the process. The pileup of this air at the surface creates a surface high-pressure zone, where the cycle described above began. The warm, dry air in this surface high-pressure zone again picks up moisture, and the cycle repeats.

Global Weather

The factors discussed above that determine and describe the movement of air masses are involved in the massive movement of air, moisture, and energy that occurs globally. The central feature of global weather is the redistribution of solar energy that falls unequally on Earth at different latitudes (relative distances from the equator and poles). Consider Figure 9.5. Sunlight and the energy flux from it are most intense at the equator because, averaged over the seasons, solar radiation comes in perpendicular to Earth's surface at the equator. With increasing distance from the equator (higher latitudes) the angle is increasingly oblique and more of the energy-absorbing atmosphere must be traversed, so that progressively less energy is received per unit area of Earth's surface. The net result is that equatorial regions receive a much greater share of solar radiation, progressively less is received farther

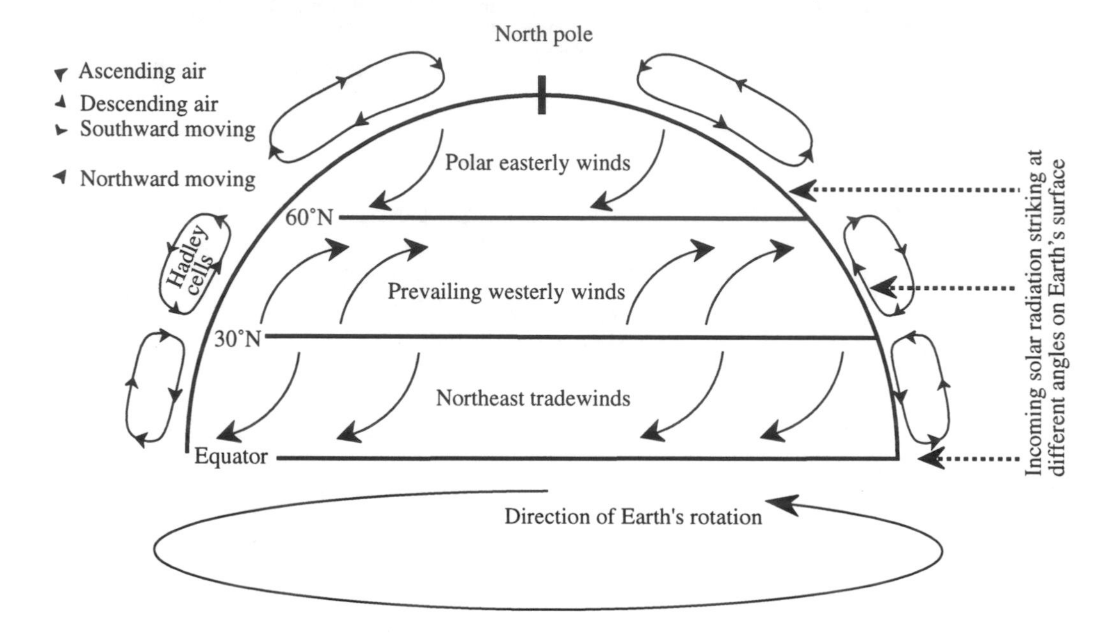

Figure 9.5. Global circulation of air in the northern hemisphere.

from the equator, and the poles receive a comparatively miniscule amount. The excess heat energy in the equatorial regions causes the air to rise. The air ceases to rise when it reaches the stratosphere because in the stratosphere the air becomes warmer with higher elevation. As the hot equatorial air rises in the troposphere, it cools by expansion and loss of water, then sinks again. The air circulation patterns in which this occurs are called **Hadley cells**. As shown in Figure 9.5, there are three major groupings of these cells, which result in very distinct climatic regions on Earth's surface. The air in the Hadley cells does not move straight north and south, but is deflected by Earth's rotation and by contact with the rotating earth; this is the **Coriolis effect**, which results in spiral-shaped air circulation patterns, called cyclonic or anticyclonic, depending upon the direction of rotation. These give rise to different directions of prevailing winds, depending on latitude. The boundaries between the massive bodies of circulating air shift markedly over time and season, resulting in significant weather instability.

The movement of air in Hadley cells combined with other atmospheric phenomena results in the development of massive **jet streams** that are in a sense, shifting rivers of air that may be several kilometers deep and several tens of km wide. Jet streams move through discontinuities in the tropopause (see Section 9.3) generally from west to east at velocities around 200 km/hr (well over 100 mph); in so doing, they redistribute huge amounts of air and have a strong influence on weather patterns.

The air and wind circulation patterns described above redistribute massive amounts of energy. If it were not for this effect, the equatorial regions would be unbearably hot, and the regions closer to the poles intolerably cold. About half of the heat that is redistributed is carried as sensible heat by air circulation, almost 1/3 is carried by water vapor as latent heat, and the remaining approximately 20 percent by ocean currents.

Weather Fronts and Storms

The interface between two masses of air that differ in temperature, density, and water content is called a **front**. A mass of cold air moving such that it displaces one of warm air is a **cold front**, and a mass of warm air displacing one of cold air is a **warm front**. Since cold air is more dense than warm air, the air in a cold mass of air along a cold front pushes under warmer air. This causes the warm, moist air to rise, such that water condenses from it. The condensation of water releases energy, so that the air rises further. The net effect can be formation of massive cloud formations (thunderheads) that may reach stratospheric levels. These spectacular thunderheads may produce heavy rainfall and even hail, and sometimes violent storms with strong winds, including tornadoes. Warm fronts cause somewhat similar effects as warm, moist air pushes over colder air. However, the front is usually much broader, and the weather effects milder, typically resulting in widespread drizzle, rather than intense rainstorms.

Swirling **cyclonic storms**, such as typhoons, hurricanes, and tornadoes, are created in low-pressure areas by rising masses of warm, moist air. As such air cools, water vapor condenses, and the latent heat released warms the air more, sustaining and intensifying its movement upward in the atmosphere. Air rising from surface level creates a low-pressure zone into which surrounding air moves. The movement of the incoming air assumes a spiral pattern, thus causing a cyclonic storm.

9.6. INVERSIONS AND AIR POLLUTION

The complicated movement of air across the Earth's surface is a crucial factor in the creation and dispersal of air pollution phenomena. When air movement ceases, air stagnation can occur with a resultant buildup of air pollutants in localized regions. Although the temperature of air relatively near the Earth's surface normally decreases with increasing altitude, certain atmospheric conditions can result in the opposite condition—increasing temperature with increasing altitude. Such conditions are characterized by high atmospheric stability and are known as **temperature inversions**. Because they limit the vertical circulation of air, temperature inversions result in air stagnation and the trapping of air pollutants in localized areas.

Inversions can occur in several ways. In a sense, the whole atmosphere is inverted by the warm stratosphere, which floats atop the troposphere, with relatively little mixing. An inversion can form when a warm air mass (warm front) overrides a cold air mass (cold front). **Radiation inversions** are likely to form in still air at night when the Earth is no longer receiving solar radiation. The air closest to the Earth cools faster than the air higher in the atmosphere, which remains warm, thus less dense. Furthermore, cooler surface air tends to flow into valleys at night, where it is overlain by warmer, less dense air. **Subsidence inversions**, often accompanied by radiation inversions, can become very widespread. These inversions can form in the vicinity of a surface high-pressure area when high-level air subsides to take the place of surface air blowing out of the high-pressure zone. The subsiding air is warmed as it compresses and can remain as a warm layer several hundred meters above ground level. A **marine inversion** is produced during the summer months when cool air laden with moisture from the ocean blows onshore and under warm, dry inland air.

As noted above, inversions contribute significantly to the effects of air pollution. This is because, as shown in Figure 9.6, inversions prevent mixing of air pollutants, thus keeping the pollutants in one area. This not only prevents the pollutants from escaping, but also acts like a container in which additional pollutants accumulate. Furthermore, in the case of secondary pollutants formed by atmospheric chemical processes, such as photochemical smog (see Chapter 12), the pollutants may be kept together such that they react with each other and with sunlight to produce even more noxious products.

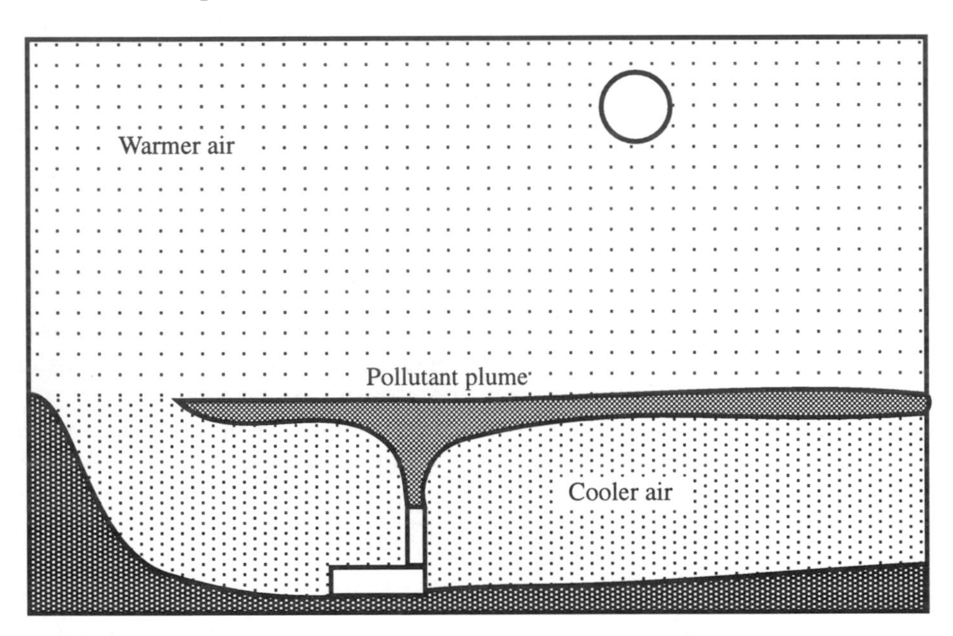

Figure 9.6. Illustration of pollutants trapped in a temperature inversion.

9.7. GLOBAL ASPECTS OF WEATHER AND CLIMATE

Perhaps the single most important influence on Earth's environment is **climate**, consisting of long-term weather patterns over large geographical areas. Although Earth's atmosphere is huge and has an enormous ability to resist and correct for detrimental change, it is possible that human activities are reaching a point at which they may be adversely affecting climate. Human effects on climate are addressed in Chapter 12, which discusses "The Endangered Global Atmosphere."

As a general rule, climatic conditions are characteristic of a particular region. This does not mean that climate remains the same throughout the year, of course, because it varies with season. One important example of such variation is the **monsoon**, seasonal variations in wind patterns between oceans and continents. The climates of Africa and the Indian subcontinent are particularly influenced by monsoons. In the latter, for example, summer heating of the Indian land mass causes air to rise, thereby creating a low-pressure area that attracts warm, moist air from the ocean. This air rises on the slopes of the Himalayan mountains, which also block the flow of colder air from the north, moisture from the air condenses, and monsoon rains carrying enormous amounts of precipitation fall. Thus, from May until into August, summer monsoon rains fall in India, Bangladesh, and Nepal. Reversal of the pattern of winds during the winter months causes these regions to have a dry season, but produces winter monsoon rains in the Philippine islands, Indonesia, New Guinea, and Australia.

Summer monsoon rains are responsible for tropical rain forests in central Africa. The interface between this region and the Sahara Desert varies over time. When the boundary is relatively far north, rain falls on the Sahel desert region at the interface, crops grow, and the people do relatively well. When the boundary is more to the south, a condition that may last for several years, devastating droughts and even starvation may occur.

It is known that there are fluctuations, cycles, and cycles imposed on cycles in climate. The causes of these variations are not completely understood, but they are known to be substantial, and even devastating to civilization. The last **ice age**, which ended only about 10,000 years ago and which was preceded by several similar ice ages, produced conditions under which much of the present land mass of the Northern Hemisphere was buried under thick layers of ice and uninhabitable. A "mini-ice age" occurred during the 1300s, causing crop failures and severe hardship in northern Europe. In modern times the El Niño Southern Oscillation occurs during a period of several years when a large, semipermanent tropical low-pressure area shifts into the central Pacific region from its more common location in the vicinity of Indonesia. This shift modifies prevailing winds, changes the pattern of ocean currents, and affects upwelling of ocean nutrients with profound effects on weather, rainfall, and fish and bird life over a vast area of the Pacific from Australia to the west coasts of South and North America.

9.8. MICROCLIMATE

The preceding section described climate on a large scale, ranging up to global dimensions. The climate that organisms and objects on the surface are exposed to close to the ground, under rocks, and surrounded by vegetation, is often quite different from the surrounding macroclimate. Such highly localized climatic conditions are termed the **microclimate**. Microclimate effects are largely determined by the uptake and loss of solar energy very close to Earth's surface and by the fact that air circulation due to wind is much lower at the surface. During the day, solar energy absorbed by relatively bare soil heats the surface, but is lost only slowly because of very limited air circulation at the surface. This provides a warm blanket of surface air several centimeters thick, and a thinner layer of warm soil. At night, radiative loss of heat from the surface of soil and vegetation can result in surface temperatures several degrees colder than the air about 2 meters above ground level. These lower temperatures result in condensation of **dew** on vegetation and the soil surface, thus providing a relatively more moist microclimate near ground level. Heat absorbed during early morning evaporation of the dew tends to prolong the period of cold experienced right at the surface.

Vegetation substantially affects microclimate. In relatively dense growths, circulation may be virtually zero at the surface because vegetation severely limits convection and diffusion. The crown surface of the vegetation intercepts most of the solar energy, so that maximum solar heating may be a significant distance up from Earth's surface. The region below the crown surface of vegetation thus becomes one of relatively stable temperature. In addition, in a dense growth of vegetation, most of the moisture loss is not from evaporation from the soil surface, but rather from transpiration from plant leaves. The net result is the creation of temperature and humidity conditions that provide a favorable living environment for a number of organisms, such as insects and rodents.

Another factor influencing microclimate is the degree to which the slope of land faces north or south. South-facing slopes of land in the northern hemisphere receive greater solar energy. Advantage has been taken of this phenomenon in restoring land strip-mined for brown coal in Germany by terracing the land such that the terraces have broad south slopes and very narrow north slopes. On the south-sloping portions of the terrace, the net effect has been to extend the short summer growing season by several days, thereby significantly increasing crop productivity. In areas where the growing season is longer, better growing conditions may exist on a north slope because it is less subject to temperature extremes and to loss of water by evaporation and transpiration.

A particularly marked effect on microclimate is that induced by urbanization. In a rural setting, vegetation and bodies of water have a moderating effect, absorbing modest amounts of solar energy and releasing it slowly. The stone, concrete, and asphalt pavement of cities have an opposite effect, strongly absorbing solar energy, and reradiating heat back to the urban microclimate. Rainfall is not allowed to accumulate in ponds, but is drained away as rapidly and efficiently as possible. Human activities generate significant amounts of heat and produce large quantities of CO_2 and other greenhouse gases that retain heat. The net result of these effects is that a city is capped by a **heat dome** in which the temperature is as much as 5°C warmer than in the surrounding rural areas, such that large cities have been described as "heat islands." The rising warmer air over a city brings in a breeze from the surrounding area and causes a local greenhouse effect that probably is largely counterbalanced by reflection of incoming solar energy by particulate matter above cities. Overall, compared to climatic conditions in nearby rural surroundings, the city microclimate is warmer and foggier, overlain with more cloud cover a greater percentage of the time, subject to more precipitation, and generally less humid.

9.9. ATMOSPHERIC CHEMISTRY

Atmospheric chemistry addresses chemical processes that occur in the atmosphere. This topic is introduced briefly here and covered in more detail in Chapter 10, "Atmospheric Chemistry." The study of atmospheric chemical reactions is difficult. One of the primary obstacles encountered in studying atmospheric chemistry is that the chemist generally must deal with incredibly low concentrations, so that the detection and analysis of reaction products is quite difficult. Simulating high-altitude conditions in the laboratory can be extremely hard because of interferences, such as those from species given off from container walls under conditions of very low pressure. Many chemical reactions that require a third body (usually a molecule of N_2 or O_2) to absorb excess energy occur very slowly in the upper atmosphere, where there is a sparse concentration of third bodies, but occur readily in a container whose walls effectively absorb energy. Container walls may serve as catalysts for some important reactions, or they may absorb important species and react chemically with the more reactive ones.

Photochemical Processes

The absorption of light by chemical species can bring about reactions, called **photochemical reactions**, that do not otherwise occur under the conditions (particularly the temperature) of the medium in the absence of light. Thus, photochemical reactions, even in the absence of a chemical catalyst, occur at temperatures much lower than those that otherwise would be required. Photochemical reactions, which are induced by intense solar radiation, play a very important role in determining the nature and ultimate fate of a chemical species in the atmosphere.

Nitrogen dioxide, NO_2, is one of the most photochemically active species found in a polluted atmosphere and is an essential participant in the smog-formation process. A species such as NO_2 may absorb light of energy $h\nu$, producing an **electronically excited molecule**,

$$NO_2 + h\nu \rightarrow NO_2^* \quad (9.9.1)$$

designated in the reaction above by an asterisk, *. The photochemistry of nitrogen dioxide is discussed in greater detail in Sections 11.5 and 12.6.

Electronically excited molecules are one of the three relatively reactive and unstable species that are encountered in the atmosphere and are strongly involved with atmospheric chemical processes. The other two species are atoms or molecular fragments with unshared electrons, called **free radicals**, and **ions** consisting of charged atoms or molecular fragments.

9.10. ATMOSPHERIC PARTICLES

Particles are common significant components of the atmosphere, particularly the troposphere. Colloidal-sized particles in the atmosphere are called **aerosols**. Most aerosols from natural sources have a diameter of less than 0.1 μm. These particles originate in nature from sea sprays, smokes, dusts, and the evaporation of organic materials from vegetation. Other typical particles of natural origin in the atmosphere are bacteria, fog, pollen grains, and volcanic ash.

As shown in Figure 9.7, atmospheric particles undergo a number of processes in the atmosphere. Small colloidal particles are subject to *diffusion processes*. Smaller particles *coagulate* together to form larger particles. *Sedimentation* and *scavenging* by raindrops and other forms of precipitation are the major mechanisms by which particles are removed from the atmosphere. Particles also react with atmospheric gases.

Many important atmospheric phenomena involve aerosol particles, including electrification phenomena, cloud formation, and fog formation. Particles help determine the heat balance of the Earth's atmosphere by reflecting light. Probably the most important function of particles in the atmosphere is their action as nuclei for the formation of ice crystals and water droplets. Current efforts at rain-making arc centered around the addition of condensing particles to atmospheres supersaturated with water vapor. Dry ice was used in early attempts; later, silver iodide, which forms huge numbers of very small particles, has been used.

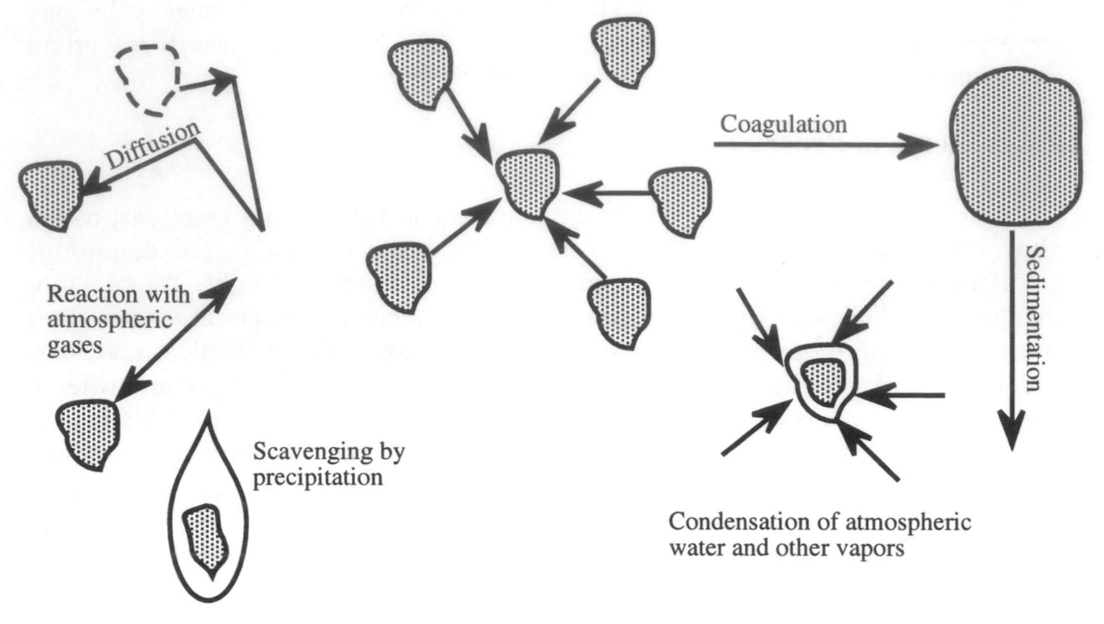

Figure 9.7. Processes that particles undergo in the atmosphere.

Particles are involved in many chemical reactions in the atmosphere. Neutralization reactions, which occur most readily in solution, may take place in water droplets suspended in the atmosphere. Small particles of metal oxides and carbon have a catalytic effect on oxidation reactions. Particles may also participate in oxidation reactions induced by light.

CHAPTER SUMMARY

The chapter summary below is presented in a programmed format to review the main points covered in this chapter. It is used most effectively by filling in the blanks, referring back to the chapter as necessary. The correct answers are given at the end of the summary.

Dry air consists of [1]__________ percent nitrogen, [2] __________ percent oxygen, [3]__________ percent argon, and [4]__________ percent carbon dioxide. The three major sources of trace gases in the atmosphere are [5]__________________________

____.

More than [6]_______ percent of the total mass of the atmosphere is found within approximately 30 km of the Earth's surface. This mass is approximately [7] _______ _______ metric tons. With higher altitude the mean free path of a molecule or atom in the atmosphere becomes [8]____________________. The atmosphere is stratified on the basis of [9]______________________. The lowest layer of the atmosphere is the [10]_____________, characterized by [11]_________________

____

____,

and above it is the [12]______________ containing the essential [13]___________ layer. The value of the solar constant is [14]_____________________. Energy transport in the atmosphere occurs by way of [15]_________________________

______________. As well as carrying sensible heat due to the kinetic energy of molecules, convection carries [16]__________ heat in the form of [17]______ ______________________________.
Radiation of energy occurs through [18]__________________, which can travel through a vacuum in space. The average surface temperature of Earth is maintained at a relatively comfortable 15°C because of [19]______________ ______________. Much of the outgoing radiation is reabsorbed by [20]____ ____________________ and reradiated. The science of atmospheric phenomena, encompassing the study of the movement of air masses as well as physical forces in the atmosphere such as heat, wind, and transitions of water, is called [21]____________. The two things required for condensation of water vapor from air are [22]______________________________ ______________________________.
Long-term weather patterns over large geographical areas are called [23] _______. The uptake and loss of solar energy very close to Earth's surface combined with the fact that air circulation from wind very close to the surface is relatively low determine the [24] ____________. Cities are often capped by a [25] _________ in which the temperature is as much as [26] ________ than in surrounding rural areas. Stagnant air masses leading to air pollution incidents, such as formation of photochemical smog, are produced by [27]__________________. In order for photchemical reactions to occur, [28]______________________________, which results in production of [29]______________________________. In addition to this highly reactive species, two other kinds of active species formed when light is absorbed are [30]______________________________. The water vapor content of the troposphere is normally within a range of [31]________ percent by volume with a global average of about [32]________ percent. Water vapor [33]____________ infrared radiation, but clouds formed from it [34]____ ______ light from the sun. Energy taken up and released when water evaporates and condenses in the atmosphere is particulary important in [35]____________ ____________________. Colloidal-sized particles in the atmosphere are called [36]____________. Atmospheric particles undergo a number of processes including [37]______________________________ ______________________________.
Among the important atmospheric phenomena involving atmospheric aerosol particles are [38]______________________________ ____________. Among the chemical reactions involving atmospheric particles are [39]______________________________ ______________________________ ______________________________.

Answers

[1] 78.08

[2] 20.95

[3] 0.934

4 0.035

5 biogenic, photochemical, and anthropogenic

6 99

7 5.14×10^{15}

8 longer

9 temperature/density relationships

10 troposphere

11 a generally homogeneous composition of major gases other than water and decreasing temperature with increasing altitude

12 stratosphere

13 ozone

14 1.34×10^3 watts per square meter

15 conduction, convection, and radiation

16 latent

17 water vapor which releases heat as it condenses

18 electromagnetic radiation

19 an atmospheric "greenhouse effect"

20 carbon dioxide and water vapor

21 meteorology

22 air must be cooled below the dew point, and nuclei of condensation must be present

23 climate

24 microclimate

25 heat dome

26 5°C warmer

27 temperature inversions

28 light must be absorbed

29 electronically excited molecules

30 free radicals and ions

31 1–3

32 1

33 absorbs

34 reflect

35 weather phenomena involving energy

36 aerosols

37 diffusion, coagulation, sedimentation, scavenging, and reaction with atmospheric gases

38 electrification phenomena, cloud formation, and fog formation

39 neutralization reactions in water droplets and catalysis of oxidation reactions by particles of metal oxides and carbon

QUESTIONS AND PROBLEMS

1. Why is there a temperature minimum at the boundary of the troposphere and the stratosphere? What phenomenon is responsible for the temperature maximum at the boundary of the stratosphere and the mesosphere?

2. What function does a third body serve in an atmospheric chemical reaction? What are the most common third bodies in the atmosphere?

3. Why does the lower boundary of the ionosphere lift at night?

4. Measured in μm, what are the lower wavelength limits of solar radiation reaching the Earth; the wavelength at which maximum solar radiation reaches the earth; and the wavelength at which maximum energy is radiated back into space?

5. Of the gases neon, sulfur dioxide, helium, oxygen, and nitrogen, which shows the most variation in its atmospheric concentration?

6. The sunlight incident upon a 1-square-meter area perpendicular to the line of transmission of the solar flux just above the Earth's atmosphere provides energy at a rate most closely equivalent to: (a) that required to power a pocket calculator, (b) that required to provide a moderate level of lighting for a 40-person capacity classroom illuminated with fluorescent lights, (c) that required to propel a 2500 pound automobile at 55 mph, (d) that required to power a 100-watt incandescent light bulb, (e) that required to heat a 40-person classroom to 70°F when the outside temperature is -10°F.

7. What is the distinction between the symbols * and • in discussing chemically active species in the atmosphere?

8. Given the total mass of Earth's atmosphere and the percentage by volume that is carbon dioxide, attempt to calculate the mass of CO_2 in the atmosphere. This exercise will require a review of the gas laws from Chapter 2. It may also be useful to calculate the average molecular mass of air assuming that it is composed only of oxygen and nitrogen.

9. After completing the calculation above, calculate the mass of pure carbon that would have to be burned to double the carbon dioxide content of the atmosphere.

10. Discuss in which sense the troposphere is homogenous. In regard to which important species is it **not** homogeneous? Why are regions of the atmosphere above the troposphere relatively less homogeneous?

11. What vital protective function is served by the stratosphere? Illustrate the answer with examples of phenomena such as chemical and photochemical reactions.

12. What vital protective function is served by the tropopause?

13. What is the area in square kilometers that receives solar energy undiminished by atmospheric absorption equivalent to the output of a 1,000 megawatt power plant?

14. Distinguish among the atmospheric energy transport processes of conduction, convection, and radiation.

15. Distinguish between sensible and latent heat. Why is the latter particularly important in atmospheric energy exchange processes?

16. What areas are covered by meteorology? How are meteorologic phenomena involved in air pollution phenomena? In this respect, why are temperature inversions particularly important?

17. Explain why the study of atmospheric chemical reactions is particularly difficult.

18. Explain what causes photochemical reactions and how they can lead to chain reactions.

19. Distinguish between the two following hypothetical chemical species: X* and X·.

20. What does $h\nu$ represent in photochemical terms? How may it lead to the occurrence of chemical reactions?

SUPPLEMENTAL REFERENCES

Anthes, Richard A., *Meteorology*, 7th Edition, Prentice Hall, Upper Saddle River, NJ, 1996

Berner, Elizabeth Kay and Robert A. Berner, *Global Environment: Water, Air, and Geochemical Cycles*, Prentice Hall, Upper Saddle River, NJ, 1994

Budyko, M. I., *The Earth's Climate*, Academic Press, New York, NY, 1982.

Goody, Richard, *Principles of Atmospheric Physics and Chemistry*, Oxford University Press, New York, NY, 1995.

Moran, Joseph M. and Michael D. Morgan, *Essentials of Weather*, Prentice Hall, Upper Saddle River, NJ, 1994.

Seinfeld, John H., *Atmospheric Chemistry and Physics of Air Pollution*, John Wiley and Sons, Inc., New York, NY, 1986.

Warneck, Peter, *Chemistry of the Natural Atmosphere*, Academic Press, San Diego, CA, 1988.

10 ATMOSPHERIC CHEMISTRY

10.1. INTRODUCTION TO ATMOSPHERIC CHEMISTRY

The quality of the air that living organisms breathe, the nature and level of air pollutants, visibility and atmospheric esthetics, and even climate are dependent upon chemical phenomena that occur in the atmosphere. These, in turn are strongly tied to absorption of solar energy, interactions between the gas phase of the atmosphere and small solid particles suspended in it, and interchange of chemical species with the geosphere (see Figure 10.1). Chemical reactions in the atmosphere and the chemical nature of atmospheric chemical species are the topic of **atmospheric chemistry**, which is introduced in this chapter.

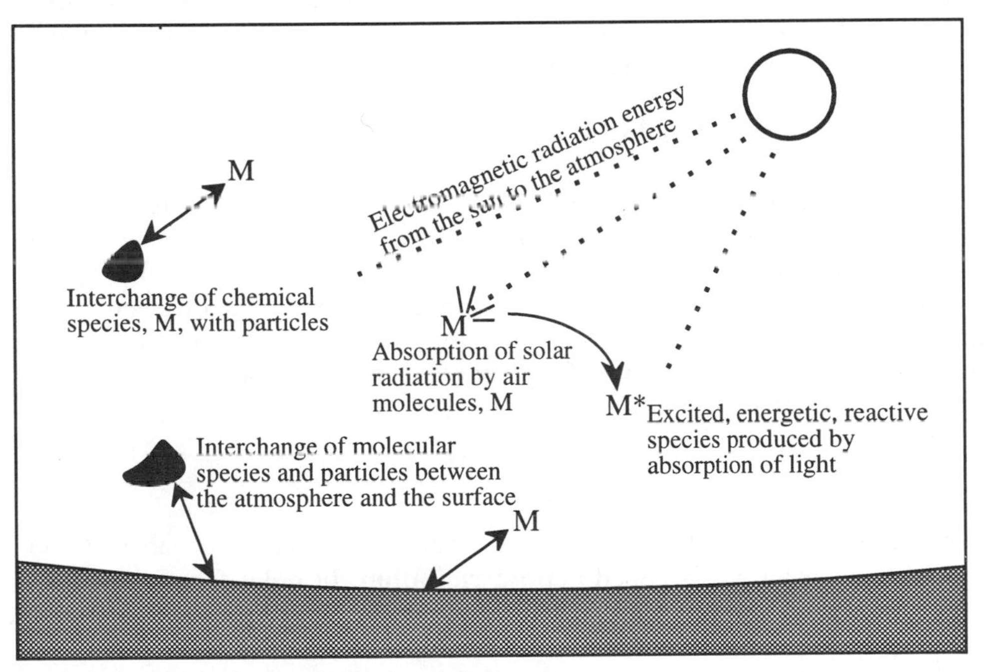

Figure 10.1 Illustration of atmospheric chemical processes.

Atmospheric chemistry involves the unpolluted atmosphere, highly polluted atmospheres, and a wide range of gradations in between. The same general phenomena govern all and produce one huge atmospheric cycle, in which there are numerous subcycles. Gaseous atmospheric chemical species fall into the following somewhat arbitrary and overlapping classifications: inorganic oxides (CO, CO_2, NO_2, SO_2); oxidants (O_3); reductants (CO, SO_2, H_2S); organics (in the unpolluted atmosphere, CH_4 is the predominant organic species, whereas alkanes, alkenes, and aryl compounds are common around sources of organic pollution); photochemically active species (NO_2, formaldehyde); acids (H_2SO_4); bases (NH_3); salts (NH_4HSO_4,); and unstable reactive species (electronically excited NO_2, HO• radical). In addition, both solid and liquid particles play a strong role in atmospheric chemistry as sources and sinks for gas-phase species, as sites for surface reactions (solid particles), and as bodies for aqueous-phase reactions (liquid droplets). Two constituents of utmost importance in atmospheric chemistry are radiant energy from the sun, predominantly in the ultraviolet region of the spectrum, and the hydroxyl radical, HO•. The former provides a way to pump a high level of energy into a single gas molecule to start a series of atmospheric chemical reactions, and the latter is the most important reactive intermediate and "currency" of daytime atmospheric chemical phenomena. NO_3 radicals are important intermediates in nighttime atmospheric chemistry. These are addressed in more detail in this chapter and Chapters 11 and 12.

As discussed in Section 9.9, it is difficult to study atmospheric chemical reactions in the laboratory. In addition to the very low concentrations of species involved, walls required to contain gases in the laboratory evolve contaminant species, serve as third bodies to absorb energy from atmospheric chemical reactions, and act as catalysts. These effects are particularly pronounced when attempts are made to replicate the low-pressure conditions of the upper atmosphere.

10.2. PHOTOCHEMICAL PROCESSES

The absorption of electromagnetic solar radiation ("light," discussed in Section 1.9) by chemical species causes **photochemical reactions**. Photochemical reactions give atmospheric chemistry a unique quality and largely determine the nature and ultimate fate of atmospheric chemical species. The ability of electromagnetic radiation to cause photochemical reactions to occur is a function of its energy, E, which increases with increasing frequency (ν) and decreasing wavelength (λ) according to the relationship,

$$E = h\nu \tag{10.2.1}$$

where h is Planck's constant, 6.62×10^{-27} erg sec. Radiation capable of causing photochemical reactions is called **actinic radiation**. In order for a photochemical reaction to occur, a single unit of photochemical energy from actinic radiation, called a **quantum** and having an energy of $h\nu$, must be absorbed by the reacting species. If the absorbed light is in the visible region of the sun's spectrum, the absorbing species is colored. Colored NO_2 is a common example of such a species in the atmosphere.

An important characteristic of photochemical reactions is the degree to which a particular end result occurs from the absorption of photons. This is expressed by the **quantum yield**. For the hypothetical case in which product "B" is produced as the result of absorption of photons by reactant "A," the quantum yield is the following:

$$\text{Quantum yield} = \frac{\text{Number of molecules of B}}{\text{Number of photons absorbed by A}} \tag{10.2.2}$$

Nitrogen dioxide, NO_2, is one of the most photochemically active species found in a polluted atmosphere. When a molecule such as NO_2 absorbs actinic radiation of energy $h\nu$,

$$NO_2 + h\nu \rightarrow NO_2^* \tag{10.2.3}$$

an **electronically excited molecule** designated in the reaction above by an asterisk, *, may be produced. The photochemistry of nitrogen dioxide is discussed in greater detail in Sections 10.4 and 10.9 of this chapter. Chapter 12 discusses how the photodissociation of NO_2 produces reactive O atoms that initiate the formation of pollutant photochemical smog.

Electronically excited molecules and atoms are reactive and unstable species in the atmosphere that participate in a wide range of atmospheric chemical processes. Two other generally reactive and unstable species in the atmosphere are **free radicals** composed of atoms or molecular fragments with unshared electrons, and **ions** consisting of charged atoms or molecular fragments. The participation of these three kinds of species in atmospheric chemical processes are discussed in later sections of this chapter and in Chapters 11 and 12.

Electronically excited molecules produced when unexcited ground-state molecules absorb energetic electromagnetic radiation in the ultraviolet or visible regions of the spectrum may possess several possible excited states. Generally, however, ultraviolet or visible radiation is energetic enough to excite molecules only to several of the lowest energy levels. The nature of the excited state may be understood by considering the disposition of electrons in a molecule. Most molecules have an even number of electrons. As mentioned in Section 2.4, electrons in molecules and atoms occupy orbitals, with a maximum of two electrons with opposite spin in the same orbital. The absorption of light may promote one of these electrons to a vacant higher-energy orbital. In some cases the promoted electron retains a spin opposite to that of its former partner, giving rise to an **excited singlet state**. In other cases the spin of the promoted electron is reversed, such that it has the same spin as its former partner; this gives rise to an **excited triplet state**.

Ground state	Singlet state	Triplet state
	↓	↑
↑ ↓	↑	↑
↑ ↓	↑ ↓	↑ ↓

These excited states are relatively energized compared to the ground state; therefore, excited chemical species are reactive. Their participation in atmospheric chemical reactions, such as those involved in smog formation, are discussed later in detail.

Electromagnetic radiation absorbed in the infrared region is not sufficiently energetic to break chemical bonds, but does cause the receptor molecules to gain vibrational and rotational energy. The energy absorbed as infrared radiation ultimately is dissipated as heat and raises the temperature of the whole atmosphere. As discussed in Chapter 12, the absorption of infrared radiation is very important in the retention of energy radiated from the Earth's surface.

The reactions that occur following absorption of a photon of light sufficiently energetic to produce an electronically excited species are largely determined by the way in which the excited species loses its excess energy. This may occur by one of several processes divided into two general classes. Of these, **photophysical processes** are those that do not involve chemical bond breakage or loss of electrons and include loss of energy from the excited molecule by electromagnetic radiation as it returns to the ground state (fluorescence or phosphorescence), transfer of energy to other molecules, or transfer of energy within the absorbing molecule; the last two processes result in dissipation of the excess energy as heat. Photochemical reactions occur as a result of de-excitation processes that involve chemical bond breakage or ion formation, particularly the following:

- **Photodissociation** of the excited molecule (the process responsible for the predominance of atomic oxygen in the upper atmosphere)

$$O_2^* \rightarrow O + O \qquad (10.2.4)$$

- **Direct reaction** with another species

$$O_2^* + O_3 \rightarrow 2O_2 + O \qquad (10.2.5)$$

- **Photoionization** through loss of an electron

$$N_2^* \rightarrow N_2^+ + e^- \qquad (10.2.6)$$

Photochemically excited molecules may undergo processes other than those listed above. One of these is **intramolecular rearrangement**, such as occurs when 2-nitrobenzaldehyde absorbs light:

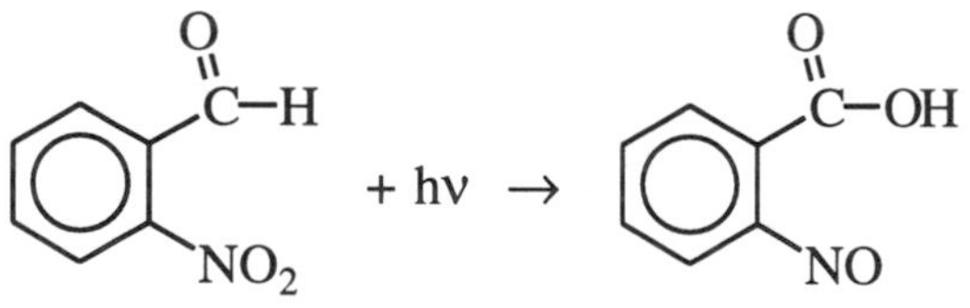

$$C_6H_4(NO_2)C(=O)H + h\nu \rightarrow C_6H_4(NO)C(=O)OH \qquad (10.2.7)$$

This reaction occurs so readily that it is used to measure light intensity. **Photo-isomerization** can also occur, such as shown by the example below in which the H atom and CH_3 group switch positions when the compound absorbs a photon:

Absorption of *hν* causes isomerization as H_3C and H switch positions.

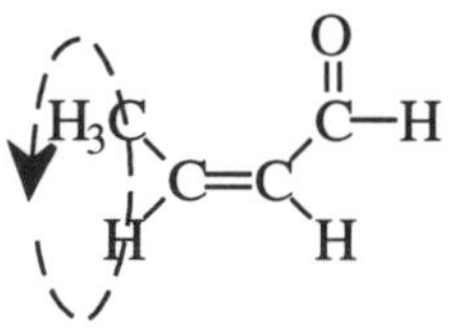

$$(H_3C)(H)C{=}C(H)C(=O)H$$

Photochemically excited molecules may also undergo **dimerization**, in which two molecules join together.

Insofar as gas-phase reactions in the troposphere are concerned, the most important of the processes listed above is photodissociation. This is because photodissociation converts relatively stable and unreactive molecular species to reactive atoms and free radicals that participate in additional reactions, including chain reactions (see Section 10.3).

10.3. CHEMICAL PROCESSES AND CHAIN REACTIONS IN THE ATMOSPHERE

As noted in the preceding section, energetic electromagnetic radiation in the atmosphere may produce atoms or groups of atoms with unpaired electrons called free radicals:

$$H_3C{-}\overset{\overset{\displaystyle O}{\|}}{C}{-}H + h\nu \rightarrow H_3C\bullet + H\dot{C}O \qquad (10.3.1)$$

In the formula above, the single dot, •, represents the unpaired electron that makes free radicals so chemically reactive. Free radicals are involved with most significant atmospheric chemical phenomena and are of the utmost importance in the atmosphere. Because of their unpaired electrons and the strong pairing tendencies of electrons under most circumstances, free radicals are highly reactive; therefore, they generally have short lifetimes. It is important to distinguish between high reactivity and instability. A totally isolated free radical or a single atom, such as an O atom, would be quite stable; it wants to react, but there is nothing around for it to react with. Therefore, free radicals and single atoms from diatomic gases (such as O from O_2) tend to persist under the rarefied conditions of very high altitudes because they can travel long distances before colliding with another reactive species. However, unlike free radicals, electronically excited species have a finite, generally very short, lifetime because they can lose energy through radiation without having to react with another species.

Chain Reactions

A key aspect of chemical processes in the atmosphere is that of **chain reactions**. These occur when a series of reactions involving particular reactive intermediates, usually free radicals, goes through a number of cycles. Most commonly, an atmospheric chain reaction begins photochemically, such as by the photodissociation of NO_2:

$$NO_2 + h\nu \rightarrow NO + O \qquad (10.3.2)$$

Radicals can take part in chain reactions in which one of the products of each reaction is a radical. Eventually, through processes such as reaction with another radical, one of the radicals in a chain is destroyed and the chain ends:

$$H_3C\bullet + H_3C\bullet \rightarrow C_2H_6 \qquad (10.3.3)$$

This process is a **chain-terminating reaction**. Reactions involving free radicals are responsible for smog formation, discussed in Chapter 12.

Hydroxyl and Hydroperoxyl Radicals in the Atmosphere

The hydroxyl radical, HO•, which is the single most important reactive intermediate species in atmospheric chemical processes, is formed by several mechanisms. At higher altitudes it is produced by photolysis of water:

$$H_2O + h\nu \rightarrow HO\bullet + H \tag{10.3.4}$$

In the presence of organic matter, hydroxyl radical is produced in abundant quantities as an intermediate in the formation of photochemical smog. To a certain extent in the atmosphere, and for laboratory experimentation, HO• is made by the photolysis of nitrous acid vapor:

$$HONO + h\nu \rightarrow HO\bullet + NO$$

In the relatively unpolluted troposphere, hydroxyl radical is produced as the result of the photolysis of ozone, followed by the reaction of a fraction of the excited oxygen atoms with water molecules:

$$O_3 + h\nu(\lambda < 315\text{ nm}) \rightarrow O^* + O_2 \tag{10.3.5}$$

$$O^* + H_2O \rightarrow 2HO\bullet \tag{10.3.6}$$

Among the important atmospheric trace species that react with hydroxyl radical and are thus removed from the atmosphere are carbon monoxide, sulfur dioxide, hydrogen sulfide, methane, and nitric oxide.

Hydroxyl radical is most frequently removed from the troposphere by reaction with methane or carbon monoxide:

$$CH_4 + HO\bullet \rightarrow H_3C\bullet + H_2O \tag{10.3.7}$$

$$CO + HO\bullet \rightarrow CO_2 + H \tag{10.3.8}$$

The reactive methyl radical, $H_3C\bullet$, and the hydrogen atom produced in the preceding reactions undergo additional reactions in the atmosphere. The global concentration of hydroxyl radical, averaged diurnally and seasonally, is estimated to range from 2×10^5 to 1×10^6 radicals per cm^3 in the troposphere. Because of the greater humidity and higher incident sunlight, which result in elevated O* levels, the concentration of HO• is relatively higher in tropical regions. The Southern Hemisphere probably has about a 20% higher level of HO• than does the Northern Hemisphere because there is more HO•–consuming CO produced by human activities in the Northern Hemisphere.

10.4. OXIDATION PROCESSES IN THE ATMOSPHERE

The 21 percent (dry basis) by volume content of molecular O_2 makes the atmosphere thermodynamically oxidizing. One prominent manifestation of this condition is the tendency for oxidizable materials to corrode when exposed to the atmosphere. Iron, for example, exposed to moist air tends to rust:

$$4Fe + 3O_2 + xH_2O \rightarrow 2Fe_2O_3 \cdot xH_2O \quad (10.4.1)$$

From the standpoint of atmospheric chemistry, however, the oxidizing tendency of the atmosphere is shown by the conversion of reduced molecular species to oxidized forms. It is this feature of the atmosphere exposed to sunlight that results in the formation of photochemical smog. Among the simple molecular species that enter the atmosphere in relatively reduced forms and that are oxidized are the following:

$$2CO + O_2 \rightarrow 2CO_2 \quad (10.4.2)$$

$$CH_4 + 2O_2 \rightarrow CO_2 + 2H_2O \quad (10.4.3)$$

$$4NO + 3O_2 + 2H_2O \rightarrow 4HNO_3 \quad (10.4.4)$$

$$2SO_2 + O_2 + 2H_2O \rightarrow 2H_2SO_4 \quad (10.4.5)$$

$$H_2S + 2O_2 \rightarrow H_2SO_4 \quad (10.4.6)$$

Although shown here in a very simple form, these reactions actually represent processes that may involve many steps, photochemistry, and reactive intermediates, particularly hydroxyl radical. Oxidation reactions may also occur on particle surfaces and in solution in aqueous aerosol droplets, which are strongly exposed to atmospheric oxygen. Another aspect of these reactions is that the products are often acidic—mildly acidic CO_2 from carbon-containing species, and strongly acidic nitric and sulfuric acid from nitrogen oxides and gaseous sulfur species, respectively.

Reducing agents, such as those shown above, may be quite stable in dry air that is not exposed to sunlight. However, the absorption of photons from solar radiation starts processes that result in oxidation. As a simple example, the photochemical dissociation of nitrogen dioxide,

$$NO_2 + h\nu \rightarrow NO + O \quad (10.4.7)$$

can produce reactive O atoms that can react with oxidizable molecules,

$$CH_4 + O \rightarrow H_3C\cdot + HO\cdot \quad (10.4.8)$$

to begin the series of reactions that forms the final oxidized products (in this case CO_2 and H_2O). Intermediate hydroxyl radical, $HO\cdot$, can abstract H atoms from hydrocarbons,

$$CH_4 + HO\cdot \rightarrow H_3C\cdot + H_2O \quad (10.4.9)$$

or add to molecules such as NO_2,

$$HO\cdot + NO_2 \rightarrow HNO_3 \quad (10.4.10)$$

to bring about oxidations. Chain reactions can be involved, such as the following sequence that regenerates NO_2 from NO:

$$H_3C\cdot(\text{from Reaction 10.4.8}) + O_2 \rightarrow H_3COO\cdot \quad (10.4.11)$$

$$H_3COO\cdot + NO \rightarrow NO_2(\text{back to Reaction 10.4.7}) + H_3CO\cdot \quad (10.4.12)$$

The NO_2 product may undergo photochemical dissociation to produce O atoms and again initiate processes that result in oxidation.

A feature of the photochemical atmosphere, particularly when it is polluted by nitrogen oxides and hydrocarbons, is the generation of strong oxidant molecules. The most common example of a strong organic oxidant species is peroxyacetyl nitrate, PAN, formed from the reaction of $H_3CC(O)OO^{\bullet}$ radical with NO_2:

$$H_3CC(O)OO^{\bullet} + NO_2 + M(\text{energy-absorbing third molecule}) \rightarrow CH_3C(O)OONO_2 + M \quad (10.4.13)$$

The most prominent inorganic oxidant is ozone, generated by reactions such as,

$$O + O_2 + M(\text{energy-absorbing third molecule}) \rightarrow O_3 + M \quad (10.4.14)$$

One of the ways in which ozone acts as an oxidant is to add to unsaturated compounds to form reactive ozonides:

$$H_3C{-}CH{=}CH_2 + O_3 \rightarrow H_3C{-}CH(O{-}O)(O)CH_2 \text{ (ozonide ring: C–O–O–C bridged by O)} \quad (10.4.15)$$

10.5. ACID-BASE REACTIONS IN THE ATMOSPHERE

Acid-base reactions occur between acidic and basic species in the atmosphere. The atmosphere is normally at least slightly acidic because of the presence of a low level of carbon dioxide, which dissolves in atmospheric water droplets and dissociates slightly:

$$CO_2(g) \xrightarrow{\text{water}} CO_2(aq) \quad (10.5.1)$$

$$CO_2(aq) + H_2O \rightarrow H^+ + HCO_3^- \quad (10.5.2)$$

Atmospheric sulfur dioxide forms a somewhat stronger acid when it dissolves in water:

$$SO_2(g) + H_2O \rightarrow H^+ + HSO_3^- \quad (10.5.3)$$

In terms of pollution, however, strongly acidic HNO_3 and H_2SO_4 formed by the atmospheric oxidation of N oxides, SO_2, and H_2S (see Reactions 10.4.4–10.4.6) are much more important because they lead to the formation of damaging acid rain (see Section 12.4).

As reflected by the generally acidic pH of rainwater, basic species are relatively less common in the atmosphere. Particulate calcium oxide, hydroxide, and carbonate can get into the atmosphere from ash and ground rock and can react with acids such as in the following reaction:

$$Ca(OH)_2(s) + H_2SO_4(aq) \rightarrow CaSO_4(s) + H_2O \tag{10.5.4}$$

The most important basic species in the atmosphere is gas-phase ammonia, NH_3. The greatest source of atmospheric ammonia is from biodegradation of nitrogen-containing biological matter and from bacterial reduction of nitrate:

$$NO_3^-(aq) + 2\{CH_2O\}(biomass) + H^+ \rightarrow NH_3(g) + 2CO_2 + H_2O \tag{10.5.5}$$

Ammonia is particularly important as a base in the atmosphere because it is the only water-soluble base present at significant levels in the atmosphere. Dissolved in atmospheric water droplets, it plays a strong role in neutralizing atmospheric acids:

$$NH_3(aq) + HNO_3(aq) \rightarrow NH_4NO_3(aq) \tag{10.5.6}$$

$$NH_3(aq) + H_2SO_4(aq) \rightarrow NH_4HSO_4(aq) \tag{10.5.7}$$

These reactions have three effects: (1) They result in the presence of NH_4^+ ion in the atmosphere as dissolved or solid salts, (2) they serve in part to neutralize acidic constituents of the atmosphere, and (3) they produce relatively corrosive ammonium salts.

10.6. IONS IN THE ATMOSPHERE

One of the characteristics of the upper atmosphere above the troposphere which is difficult to duplicate under laboratory conditions is the presence of significant levels of electrons and positive ions. Because of the rarefied conditions, these ions may exist in the upper atmosphere for long periods before recombining to form neutral species.

At altitudes of approximately 50 km and up, ions are so prevalent that the region is called the **ionosphere**. The presence of the ionosphere has been known since about 1901, when it was discovered that radio waves could be transmitted over long distances, where the curvature of the Earth makes line-of-sight transmission impossible. These radio waves bounce off the ionosphere.

Ultraviolet radiation is the primary producer of ions in the ionosphere:

$$N_2 + h\nu \rightarrow N_2^+ + e^- \tag{10.6.1}$$

Other ions are produced by secondary reactions, such as the following:

$$N_2^+ + O \rightarrow NO^+ + N \tag{10.6.2}$$

In darkness, the positive ions produced by photochemical processes slowly recombine with free electrons. The process is especially rapid in the lower regions of the ionosphere, where the concentration of species is relatively high. Thus, the lower boundary of the ionosphere lifts at night and makes possible the transmission of radio waves over much greater distances.

The Earth's magnetic field has a strong influence upon the ions in the upper atmosphere. Probably the best-known manifestation of this phenomenon is found in the Van Allen belts, discovered in 1958. These regions consist of two belts of ionized particles circling the Earth as shown in Figure 10.2. If they are visualized as two doughnuts, then the axis of the Earth's magnetic field extends through the holes in the doughnuts. In the inner belt, the highly energetic ionizing radiation consists of protons. In the outer belt, it consists of electrons.

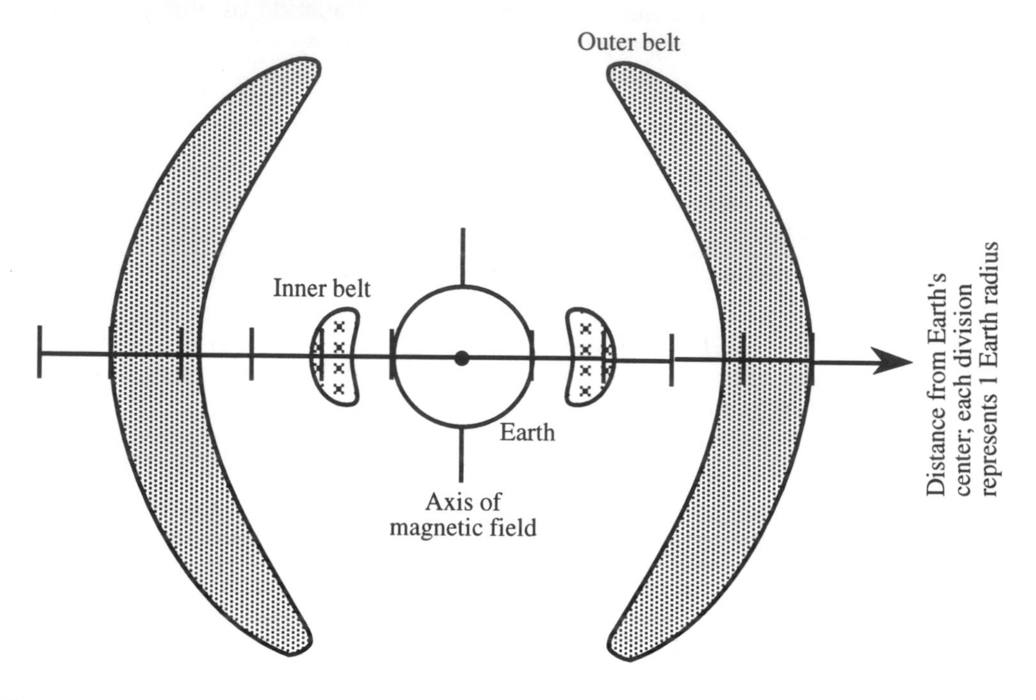

Figure 10.2. Cross section of the Van Allen belts encircling the Earth.

Ions, produced in the upper atmosphere primarily by the action of energetic electromagnetic radiation, may also be produced in the troposphere by the shearing of water droplets during precipitation. The shearing may be caused by the compression of descending masses of cold air or by strong winds over hot, dry land masses. The last phenomenon is known as the foehn, sharav (in the Near East), or Santa Ana (in southern California). These hot, dry winds cause severe discomfort. The ions produced by them consist of electrons and positively charged molecular species. With sufficient humidity, these ions quickly become surrounded by clusters of one to eight water molecules forming what are called small air ions that may last for up to several minutes. Small air ions undergo further reactions, including incorporation of trace gases from the atmosphere. Eventually, they are neutralized by combining with ions of the opposite charge or with uncharged condensation particles.

10.7. EVOLUTION OF THE ATMOSPHERE

Earth's atmosphere originally was very different from its present state and has changed largely through biological activity and accompanying chemical changes. According to theories about the origin of life that have been widely accepted, at least until recently, approximately 3.5 billion years ago, when the first primitive life molecules were formed, the atmosphere was chemically reducing, consisting primarily of methane, ammonia, water vapor, and hydrogen. Bombardment by intense, bond-breaking ultraviolet light, along with lightning and radiation from radionuclides, provided the energy to bring about chemical reactions that resulted in the production of relatively complicated molecules, including even amino acids and sugars. Thus a rich chemical mixture was formed in the sea from which life molecules evolved. Initially, these very primitive life forms derived their energy from fermentation of organic matter formed by chemical and photochemical processes, then gained the ability to produce organic matter, "$\{CH_2O\}$," by photosynthesis,

$$CO_2 + H_2O + h\nu \rightarrow \{CH_2O\} + O_2(g) \quad (10.7.1)$$

and the stage was set for the massive biochemical transformation that resulted in the production of almost all the atmosphere's oxygen.

The oxygen initially produced by photosynthesis was probably quite toxic to primitive life forms. However, much of this oxygen was converted to iron oxides by reaction with soluble iron(II):

$$4Fe^{2+} + O_2 + 4H_2O \rightarrow 2Fe_2O_3 + 8H^+ \quad (10.7.2)$$

This process formed enormous deposits of iron oxides, the existence of which provides convincing evidence for the liberation of free O_2 in the primitive atmosphere.

Eventually, enzyme systems developed that enabled organisms to mediate the reaction of waste-product oxygen with oxidizable organic matter in the sea. Later, this mode of waste-product disposal was utilized by organisms to produce energy by respiration, which is now the mechanism by which nonphotosynthetic organisms obtain energy. In time, O_2 accumulated in the atmosphere. In addition to providing an abundant source of oxygen for respiration, the accumulated atmospheric oxygen formed an ozone shield (see Section 10.8). The ozone shield absorbs bond-rupturing ultraviolet radiation. With the ozone shield protecting tissue from destruction by high-energy ultraviolet radiation, the Earth became a much more hospitable environment for life, and life forms were enabled to move from the sea to land.

In the late 1990s an alternate to the above theory for the origin of life gained credence. This occurred because of discoveries of abundant life forms at very great depths around hydrothermal vents on the sea floor. Some authorities contend that these conditions, hostile though they are to more familiar life forms, were those under which the self-replicating molecules required for life began.

10.8. REACTIONS OF ATMOSPHERIC OXYGEN

Some of the primary features of the exchange of oxygen among the atmosphere, lithosphere, hydrosphere, and biosphere are summarized in Figure 10.3. The oxygen cycle is critically important in atmospheric chemistry, geochemical transformations, and life processes.

Oxygen in the troposphere plays a strong role in processes that occur on the Earth's surface. Atmospheric oxygen takes part in energy-producing reactions, such as the burning of fossil fuels:

$$CH_4(\text{in natural gas}) + 2O_2 \rightarrow CO_2 + 2H_2O \quad (10.8.1)$$

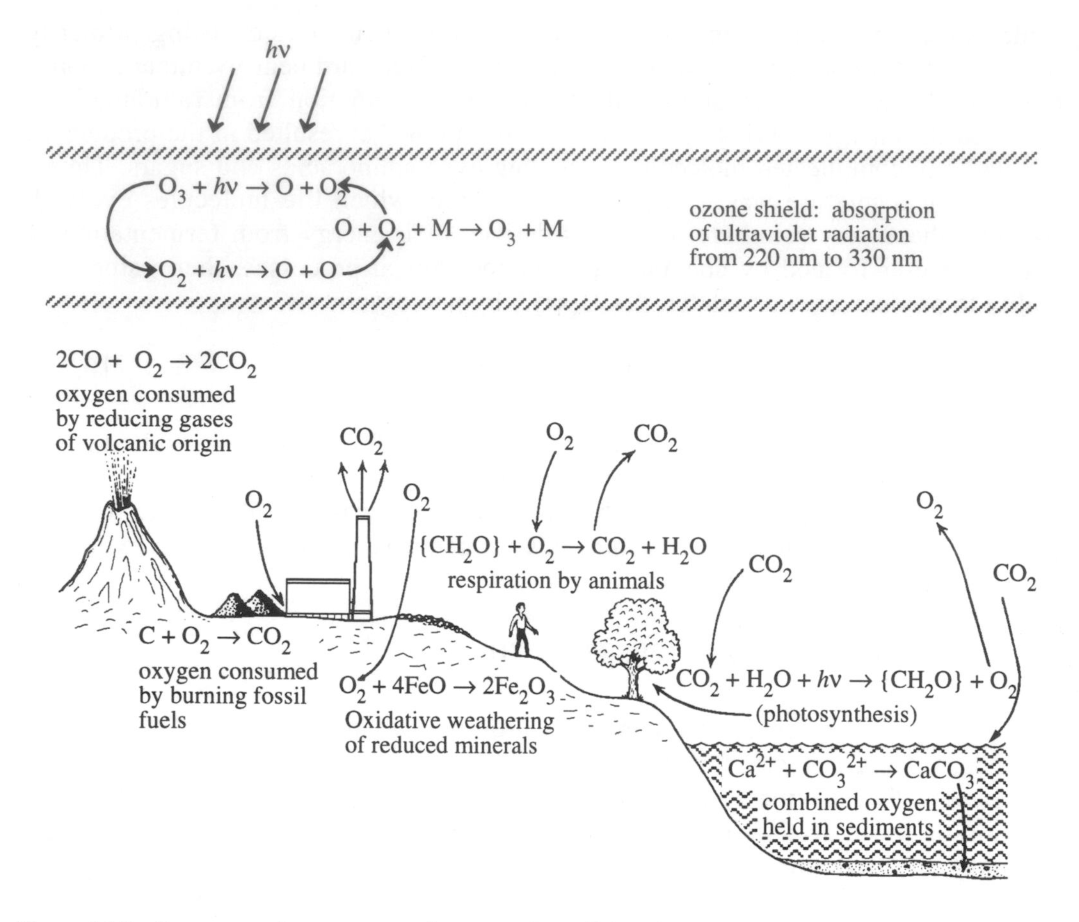

Figure 10.3. Oxygen exchange among the atmosphere, lithosphere, hydrosphere, and biosphere.

Atmospheric oxygen is utilized by aerobic organisms in the degradation of organic material. Some oxidative weathering processes consume oxygen. An important example is the oxidative weathering of mineral iron(II) to iron(III):

$$4FeO + O_2 \rightarrow 2Fe_2O_3 \quad (10.8.2)$$

Oxygen is returned to the atmosphere through plant photosynthesis:

$$CO_2 + H_2O + h\nu \rightarrow \{CH_2O\} + O_2 \quad (10.8.3)$$

All molecular oxygen now in the atmosphere is thought to have originated through the action of photosynthetic organisms, which shows the importance of photosynthesis in the oxygen balance of the atmosphere. It can be shown that most of the carbon fixed by these photosynthetic processes is dispersed in mineral formations as humic material (Sections 6.12 and 14.2); only a very small fraction is deposited in fossil fuel beds. Therefore, although fossil fuel combustion consumes large amounts of O_2, there is no danger of running out of atmospheric oxygen.

Because of the extremely rarefied atmosphere and the effects of ionizing radiation, elemental oxygen in the upper atmosphere exists to a large extent in forms other than diatomic O_2. In addition to O_2, the upper atmosphere contains oxygen atoms, O; excited oxygen molecules, O_2*; and ozone, O_3.

Atomic oxygen, O, is stable primarily in the thermosphere, where the atmosphere is so rarefied that the three-body collisions necessary for the chemical reaction of atomic oxygen seldom occur (the third body in this kind of three-body reaction absorbs energy to stabilize the products). Atomic oxygen is produced by a photochemical reaction:

$$O_2 + h\nu \rightarrow O + O \tag{10.8.4}$$

The oxygen-oxygen bond is strong and ultraviolet radiation in the wavelength regions 135-176 nm and 240-260 nm is most effective in causing dissociation of molecular oxygen. Because of photochemical dissociation, O_2 is virtually nonexistent at very high altitudes and less than 10% of the oxygen in the atmosphere at altitudes exceeding approximately 400 km is present in the molecular form.

Oxygen atoms in the atmosphere can exist in the ground state (O) and in excited states (O*). Excited oxygen atoms are produced by the photolysis of ozone at wavelengths below 308 nm (Reaction 10.8.5) or by highly energetic chemical reactions such as the one illustrated in Reaction 10.8.6:

$$O_3 + h\nu(\lambda < 308\text{ nm}) \rightarrow O^* + O_2 \tag{10.8.5}$$

$$O + O + O \rightarrow O_2 + O^* \tag{10.8.6}$$

Excited atomic oxygen emits visible light at wavelengths of 636 nm, 630 nm, and 558 nm. This emitted light is partially responsible for **airglow**, a very faint electromagnetic radiation continuously emitted by the Earth's atmosphere. Although its visible component is extremely weak, airglow is quite intense in the infrared region of the spectrum.

Oxygen ion, O^+, which may be produced by ultraviolet radiation acting upon oxygen atoms,

$$O + h\nu \rightarrow O^+ + e^- \tag{10.8.7}$$

is the predominant positive ion in some regions of the ionosphere.

Ozone and the Ozone Layer

Ozone, O_3, has an essential protective function because it absorbs harmful ultraviolet radiation in the stratosphere and serves as a radiation shield, protecting living beings on the Earth from the effects of excessive amounts of such radiation. It is produced by a photochemical reaction,

$$O_2 + h\nu(\text{energetic ultraviolet radiation}) \rightarrow O + O \tag{10.8.8}$$

followed by a three-body reaction,

$$O + O_2 + M \rightarrow O_3 + M(\text{increased energy}) \tag{10.8.9}$$

in which M is another molecule, usually N_2 or O_2, which absorbs the excess energy given off by the reaction and enables the ozone molecule to stay together. The region of maximum ozone concentration occurs in the stratosphere at an altitude of 25-30 km where it may reach levels of 10 ppm.

Ozone absorbs ultraviolet light very strongly in the region 220-330 nm. If this light were not absorbed by ozone, severe damage would result to exposed forms of life on the Earth. Absorption of electromagnetic radiation by ozone converts the radiation's energy to heat and is responsible for the temperature maximum encountered at the boundary between the stratosphere and the mesosphere at an altitude of approximately 50 km. The reason that the temperature maximum occurs at a higher altitude than that of the maximum ozone concentration arises from the fact that ozone is such an effective absorber of ultraviolet light. Therefore, most of this radiation is absorbed in the upper stratosphere, where it generates heat, and only a small fraction reaches the lower altitudes, which remain relatively cool.

Thermodynamically, the overall reaction,

$$2O_3 \rightarrow 3O_2 \qquad (10.8.10)$$

is favored so that ozone is inherently unstable. Its decomposition in the stratosphere is catalyzed by a number of natural and pollutant trace constituents, including NO, NO_2, H, HO•, HOO•, ClO, Cl, Br, and BrO. Ozone decomposition also occurs on solid surfaces, such as metal oxides and salts produced by rocket exhausts.

Ozone is an undesirable pollutant in the troposphere. It is toxic, and a mild overdose causes labored breathing, a feeling of chest pressure, cough, and irritated eyes. In addition to its toxicological effects, which are discussed in Chapter 20, ozone damages materials, such as rubber.

10.9. REACTIONS OF ATMOSPHERIC NITROGEN

The 78% by volume of nitrogen contained in the atmosphere constitutes an inexhaustible reservoir of that essential element. The nitrogen cycle and nitrogen fixation by microorganisms were discussed in Chapter 6. A small amount of nitrogen is thought to be fixed (chemically bound to other elements) in the atmosphere by lightning, and some is also fixed by combustion processes, as in the internal combustion engine.

Before the use of synthetic fertilizers reached its current high levels, chemists were concerned that denitrification processes in the soil would lead to nitrogen depletion on the Earth. Now, with millions of tons of synthetically fixed nitrogen being added to the soil each year, concern has shifted to possible excess accumulation of nitrogen in soil, fresh water, and the oceans.

Unlike oxygen, which is almost completely dissociated to the monatomic form in higher regions of the thermosphere, molecular nitrogen is not readily dissociated by ultraviolet radiation. However, at altitudes exceeding approximately 100 km, atomic nitrogen is produced by photochemical reactions:

$$N_2 + h\nu \rightarrow N + N \qquad (10.9.1)$$

Most stratospheric ozone is probably removed by the action of nitric oxide, which reacts with ozone as follows:

$$O_3 + NO \rightarrow NO_2 + O_2 \qquad (10.9.2)$$

$$NO_2 + O \rightarrow NO + O_2 \quad \text{(regeneration of NO from } NO_2\text{)} \qquad (10.9.3)$$

Pollutant oxides of nitrogen, particularly NO_2, are key species involved in air pollution and the formation of photochemical smog. For example, NO_2 is readily dissociated photochemically to NO and reactive atomic oxygen:

$$NO_2 + h\nu \rightarrow NO + O \tag{10.9.4}$$

This reaction is the most important primary photochemical process involved in smog formation. The roles played by nitrogen oxides in smog formation and other forms of air pollution are discussed in Chapters 11 and 12.

10.10. ATMOSPHERIC CARBON DIOXIDE

Although only about 0.035% (350 ppm) of air consists of carbon dioxide, it is the atmospheric "nonpollutant" species of most concern. As mentioned in Section 9.4, carbon dioxide, along with water vapor, is primarily responsible for the absorption of infrared energy re-emitted by the Earth such that some of this energy is reradiated back to the Earth's surface. Current evidence suggests that changes in the atmospheric carbon dioxide level will substantially alter the Earth's climate through the greenhouse effect (Chapter 12).

Valid measurements of overall atmospheric CO_2 can only be taken in areas remote from industrial activity. Such areas include Antarctica and the top of Mauna Loa Mountain in Hawaii. Measurements of carbon dioxide levels in these locations over the last 40 years suggest an annual increase in CO_2 of about 1 ppm per year.

The most obvious factor contributing to increased atmospheric carbon dioxide is consumption of carbon-containing fossil fuels. In addition, release of CO_2 from the biodegradation of biomass and uptake by photosynthesis are important factors determining overall CO_2 levels in the atmosphere. The role of photosynthesis is illustrated by the seasonal cycle in carbon dioxide levels in the northern hemisphere. Maximum values occur in April and minimum values in late September or October. These oscillations are due to the "photosynthetic pulse," influenced most strongly by forests in middle latitudes. Forests have a much greater influence than other vegetation because in general forest trees carry out more photosynthesis than other kinds of plants, such as prairie grasses. Furthermore, forests store enough fixed, but readily oxidizable carbon in the form of wood and humus to have a marked influence on atmospheric CO_2 content. Thus, during the summer months, forests carry out enough photosynthesis to reduce the atmospheric carbon dioxide content markedly. During the winter, metabolism of biota, such as bacterial decay of humus, releases a significant amount of CO_2. Therefore, the current worldwide trend toward destruction of forests and conversion of forest lands to agricultural uses will contribute substantially to a greater overall increase in atmospheric CO_2 levels.

With current trends, it is likely that global CO_2 levels will double by the middle of the next century, which may well raise the Earth's mean surface temperature by 1.5–4.5°C. Such a change might have more potential to cause massive irreversible environmental changes than any other disaster short of global nuclear war.

Chemically and photochemically, CO_2 is a comparatively insignificant species because of its relatively low concentrations and low photochemical reactivity. The infrared radiation absorbed by carbon dioxide is not energetic enough to cause photochemical reactions to occur. The photodissociation of CO_2 by energetic solar ultraviolet radiation is probably a source of CO in the upper atmosphere:

$$CO_2 + h\nu \rightarrow CO + O \tag{10.10.1}$$

10.11. ATMOSPHERIC WATER

Gaseous water in the upper atmosphere is involved in the formation of hydroxyl and hydroperoxyl radicals as mentioned in Section 10.3. Condensed water vapor in the form of very small droplets is important in atmospheric chemistry. The harmful effects of some air pollutants—for instance, the corrosion of metals by acid-forming gases—require the presence of water, which may come from the atmosphere. Atmospheric water vapor has an important influence upon pollution-induced fog formation under some circumstances. Water vapor interacting with pollutant particulate matter in the atmosphere may reduce visibility to undesirable levels through the formation of aerosol particles (see Sections 10.12 and 11.2).

Most stratospheric water comes from the photochemical oxidation of methane:

$$CH_4 + 2O_2 + h\nu \xrightarrow[\text{(several steps)}]{} CO_2 + 2H_2O \qquad (10.11.1)$$

The water thus produced serves as a source of stratospheric hydroxyl radical as shown by the following reaction:

$$H_2O + h\nu \rightarrow HO\cdot + H \qquad (10.11.2)$$

10.12. ATMOSPHERIC PARTICLES AND ATMOSPHERIC CHEMISTRY

Particles as normal atmospheric constituents are discussed in Section 9.9, and particulate pollutants are covered in Section 11.2. Figure 10.4 summarizes the main atmospheric chemical processes that particles undergo in the atmosphere. As indicated by Figure 10.4, particles are involved in many chemical reactions in the atmosphere. Neutralization reactions, which occur most readily in solution, may take place in water droplets suspended in the atmosphere. Small particles of metal oxides and carbon have a catalytic effect on oxidation reactions. Particles may also participate in oxidation reactions induced by light.

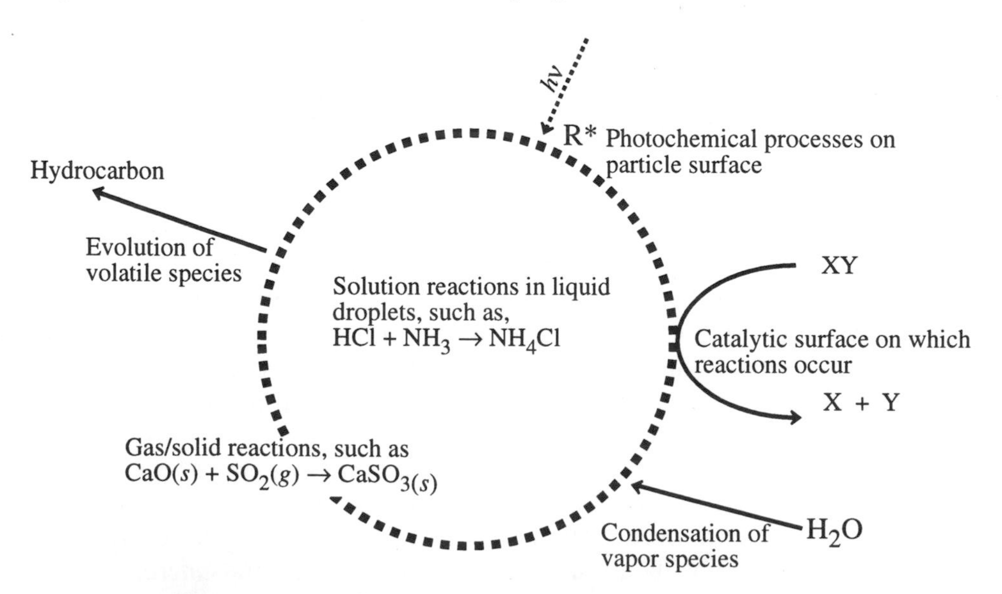

Figure 10.4. Particles in the atmosphere participate in a number of atmospheric chemical processes.

CHAPTER SUMMARY

The chapter summary below is presented in a programmed format to review the main points covered in this chapter. It is used most effectively by filling in the blanks, referring back to the chapter as necessary. The correct answers are given at the end of the summary.

The four major categories of atmospheric chemical species are [1]______________
____
____.
The ability of electromagnetic radiation to cause photochemical reactions to occur is a function of its energy, E, which increases with increasing [2]____________ and decreasing [3]____________ according to the relationship [4]______________.
An expression of the degree to which the absorption of quanta of electromagnetic radiation causes photochemical reactions to occur is the [5]______________. Three relatively reactive and unstable species that are encountered in the atmosphere that are strongly involved with atmospheric chemical processes are [6]______________
____. The three main kinds of photochemical reactions that occur as a result of de-excitation processes involving chemical bond breakage or ion formation are [7]________________
____.
Free radicals are highly reactive because of the strong [8]________________ of single electrons. Atmospheric [9]____________________________ occur when reactions involving particular reactive intermediates, usually free radicals, go through a series of cycles. The single most important reactive intermediate species in atmospheric chemical processes is [10]________________. From the standpoint of atmospheric chemistry, the [11]______________ tendency of the atmosphere is shown by the conversion of reduced molecular species to [12]________________ forms. The photochemical dissociation of nitrogen dioxide, NO_2, can produce reactive [13]______________ that can react with oxidizable molecules. A photochemical atmosphere polluted by nitrogen oxides and hydrocarbons generates strong [14]____________ molecules. The most prominent inorganic oxidant in the atmosphere is [15]__________________. In terms of its acid-base character, the unpolluted atmosphere is normally slightly [16]______________ because of the presence of a low level of [17]______________________. [18]______________ is the only water-soluble base present at significant levels in the atmosphere. A portion of the atmosphere at altitudes of approximately 50 km and up, the bottom boundary of which raises at night is the [19]________________. The Earth's early atmosphere was originally believed to be chemically [20]________________ and became oxidizing by the action of [21]______________. In addition to molecular O_2 the upper atmosphere contains the following oxygen species: [22]______________
____. The main process by which elemental oxygen is returned to Earth's atmosphere in the oxygen cycle is [23]______
______________. The two reactions by which stratospheric ozone is produced are [24].____
____.
The annual increase in CO_2 in the atmosphere is about [25]______________ per

year. The main source of water in the stratosphere is [26]______________________ ______________________. Major chemical processes or effects of particles in the atmosphere are [27]____ ____.

Answers

[1] inorganic oxides, oxidants, reductants, organics, photochemically active species, acids, bases, salts, and unstable reactive species

[2] frequency (ν)

[3] wavelength (λ)

[4] $E = h\nu$

[5] quantum yield

[6] electronically excited molecules, free radicals, and ions

[7] photodissociation, direct reaction, photoionization

[8] pairing tendencies

[9] chain reactions

[10] the hydroxyl radical, HO•

[11] oxidizing

[12] oxidized

[13] O atoms

[14] oxidant

[15] ozone

[16] acidic

[17] carbon dioxide

[18] Ammonia, NH_3

[19] ionosphere

[20] reducing

[21] photosynthesis

[22] oxygen atoms, O; excited oxygen molecules, O_2^*; and ozone, O_3

[23] photosynthesis

[24] $O_2 + h\nu \rightarrow O + O$ and $O + O_2 + M \rightarrow O_3 + M$

[25] 1 ppm

[26] the photochemical oxidation of methane

[27] surface catalytic reactions, surface photochemical reactions, solution reactions, condensation of vapors, evolution of volatile species, gas/solid reactions

QUESTIONS AND PROBLEMS

1. What does an asterisk, *, denote after the formula of a species, such as NO_2*, in the atmosphere?
2. In what sense are *h*ν and HO• "of utmost importance" in the atmosphere?
3. What does the dot denote after a species such as HO•? Why are such species often highly reactive?
4. What are the three classes of "relatively reactive and unstable species in the atmosphere?" Describe their characteristics and give examples of each.
5. What are the three main reactions or processes by which photochemically excited species lose their excess energy?
6. What is a chain reaction? What is a common photochemical reaction that initiates chain reactions in the atmosphere?
7. With which two common molecular species does atmospheric hydroxyl radical react leading to its removal from the atmosphere?
8. What kinds of pollutant species are likely to form in an atmosphere contaminated with nitrogen oxides and hydrocarbons and subjected to sunlight?
9. What are three effects of the reaction of NH_3 with acids in the atmosphere?
10. What is the ionosphere? How is it formed? Why was it known to exist long before rockets were developed to reach its altitude?
11. Cite the evidence showing the importance of life on the nature of Earth's atmosphere.
12. How does elemental oxygen react in the stratosphere? What is a very significant product of these reactions? What protective function does it serve?
13. What are three oxygen species other than unexcited O_2 that exist in the upper atmosphere?
14. What is an atmospheric substance that is essential to our well-being in the stratosphere, but toxic in the troposphere? Explain.
15. Explain why there is an annual oscillation in carbon dioxide levels in the atmosphere. These oscillations are more pronounced in the Northern than in the Southern Hemisphere. Offer a possible explanation for that observation.
16. What phenomenon is responsible for the temperature maximum at the boundary of the stratosphere and the mesosphere?
17. Why might it be expected that the reaction of a free radical with NO_2 is a chain-terminating reaction (consider the total number of electrons in NO_2).
18. Suppose that 22.4 liters of air at STP is used to burn 1.50 g of carbon to form CO_2, and that the gaseous product is adjusted to STP. What are the volume and the average molecular mass of the resulting mixture?
19. Of the species O, HO*•, NO_2*, H_3C•, and N^+, which could most readily revert to a nonreactive, "normal" species in total isolation?

20. A 12.0-liter sample of air at 25°C and 1.00 atm pressure was collected and dried. After drying, the volume of the sample was exactly 11.50 L. What was the percentage *by weight* of water in the original air sample?

21. At an altitude of 50 km, the average atmospheric temperature is essentially 0°C. What is the average number of air molecules per cubic centimeter of air at this altitude?

22. Give possible examples of neutralization reactions and oxidation reactions catalyzed by particles in the atmosphere. What kinds of particles would be required for each of these types of reactions?

23. Distinguish between excited singlet states and excited triplet states. In what sense are they both "excited"?

24. Define the photochemical phenomena represented by each of the following:

    (a) $O_2^* \rightarrow O + O$

    (b) $O_2^* + O_3 \rightarrow 2O_2 + O$

    (c) $N_2^* \rightarrow N_2^+ + e^-$

25. Why are free radicals so highly reactive? Despite their high reactivity, why do free radicals tend to persist for significant lengths of time at high altitudes?

26. What is the kind of reaction below called? Explain.

    $H_3C^\bullet + H_3C^\bullet \rightarrow C_2H_6$

SUPPLEMENTAL REFERENCES

Barker, John R., Ed., *Progress and Problems in Atmospheric Chemistry*, World Scientific Publishing Co., River Edge, NJ, 1995.

Finlayson-Pitts, Barbara J. and James N. Pitts, *Atmospheric Chemistry: Fundamentals and Experimental Techniques*, John Wiley & Sons, New York, NY, 1986.

Goody, Richard, *Principles of Atmospheric Physics and Chemistry*, Oxford University Press, Marblehead, MA, 1995.

Hobbs, Peter V., *Basic Physical Chemistry for the Atmospheric Sciences*, Cambridge University Press, New York, NY, 1995.

Prinn, Ronald G., Ed., *Global Atmospheric-Biospheric Chemistry*, Plenum Publishing Corporation, Edison, NJ, 1994.

Seinfeld, John H., *Atmospheric Chemistry and Physics of Air Pollution*, John Wiley and Sons, Inc., New York, NY, 1986.

Seinfeld, John H. and Spyros N. Pandis, *Atmospheric Chemistry and Physics: Air Pollution to Climate*, John Wiley and Sons, New York, NY, 1997.

Sloane, Christine S. and Thomas W. Tesche, Eds., *Atmospheric Chemistry: Models and Predictions for Climate and Air Quality*, CRC Press/Lewis Publishers, Boca Raton, FL, 1991.

11 AIR POLLUTION AND ITS CONTROL

11.1. INTRODUCTION

This chapter addresses inorganic and organic air pollutants of various kinds. It discusses their origins, the processes that they undergo in the atmosphere, their effects, and their treatment. Some air pollutants are potentially so damaging that they may affect the global atmosphere, and even life on Earth. These are discussed in Chapter 12, "The Endangered Global Atmosphere."

One of the biggest concerns regarding air pollution is its effect upon plants.[1] In many areas of the world exposure to air pollutants has significantly reduced yields of crops grown for food and other purposes. Some forests have been seriously damaged by air pollution. An example is the killing of trees in Germany's Black Forest by exposure to acidic precipitation.

The first of several kinds of inorganic air pollutants addressed here are particles, often formed from reactions of gaseous pollutants that enter the atmosphere as the result of human activities. Gaseous air pollutants include CO, SO_2, NO, and NO_2, as well as less abundant NH_3, N_2O, N_2O_5, H_2S, Cl_2, HCl, and HF. (Their quantities are relatively small compared to the amount of CO_2 in the atmosphere, the possible environmental effects of which are discussed in Chapter 12.)

There is a strong connection between inorganic and organic substances in the atmosphere. For example, inorganic NO_2 photodissociates to start the processes by which organic vapors form aldehydes, oxidants, and other substances in photochemical smog. As another example, smog-generated oxidants convert inorganic SO_2 to much more acidic sulfuric acid, the major contributor to acid precipitation.

Organic pollutants may strongly affect atmospheric quality. Such pollutants may come from both natural and artificial sources. In some cases contaminants from both kinds of sources interact to produce a pollution effect, such as occurs when terpene hydrocarbons from trees interact with NO_x from autos to make photochemical smog.

The effects of atmospheric pollutants may be divided between **direct effects**, from **primary pollutants**, such as cancer caused by exposure to vinyl chloride, and effects resulting from **secondary pollutants** produced by atmospheric reactions of primary pollutants, such as photochemical smog or acid rain. Generally, secondary pollutants are more important. In some localized situations, particularly the workplace, direct effects of organic air pollutants may be equally important.

11.2. PARTICLES IN THE ATMOSPHERE

Atmospheric particles, commonly called **particulates**, range in size from about one-half millimeter down to molecular dimensions, and consist of a large variety of solid or liquid materials and discrete objects. Particles are the most visible and obvious form of air pollution. Atmospheric **aerosols** are suspensions in air of solid or liquid particles below 100 μm in diameter. Pollutant particles of 0.001-100 μm size are commonly suspended in the air near sources of pollution, such as the urban atmosphere, industrial plants, highways, and power plants. Very small, solid particles include carbon black, silver iodide, combustion nuclei, and sea-salt nuclei formed by the loss of water from droplets of seawater. Larger particles include cement dust, wind-blown soil dust, foundry dust, and pulverized coal. Some of the terms commonly used to describe atmospheric particles are summarized in Table 11.1.

Table 11.1. Important Terms Describing Atmospheric Particles

Term	Meaning
Aerosol	Colloidal-sized atmospheric particles
Condensation aerosol	Formed by condensation of vapors or reactions of gases
Dispersion aerosol	Formed by grinding of solids, atomization of liquids, or dispersion of dusts
Fog	Term denoting high level of water droplets
Haze	Denotes decreased visibility due to the presence of particles
Mists	Liquid particles (raindrops, fog, sulfuric acid droplets)
Smoke	Particles formed by incomplete combustion of fuel

Chemical Processes for Inorganic Particle Formation

Metal oxides make up a large class of inorganic particles in the atmosphere. They are formed when fuels are burned that contain metals, such as organic vanadium in residual fuel oil or pyrite (FeS_2) in coal:

$$3FeS_2 + 8O_2 \rightarrow Fe_3O_4 + 6SO_2 \qquad (11.2.1)$$

A common process for the formation of aerosol mists involves the oxidation of atmospheric sulfur dioxide to H_2SO_4. The sulfuric acid product is a hygroscopic substance that accumulates atmospheric water to form small liquid droplets in which it may react with basic air pollutants to form salts:

$$2SO_2 + O_2 + 2H_2O \rightarrow 2H_2SO_4 \qquad (11.2.2)$$

$$H_2SO_4(\text{droplet}) + 2NH_3(g) \rightarrow (NH_4)_2SO_4\ (\text{droplet}) \qquad (11.2.3)$$

$$H_2SO_4(\text{droplet}) + CaO(s) \rightarrow CaSO_4(\text{ droplet}) \qquad (11.2.4)$$

Under low humidity conditions water is lost from droplets of these salt solutions and a solid aerosol is formed.

The Composition of Inorganic Particles

Figure 11.1 illustrates the basic factors responsible for the composition of inorganic particulate matter. In general, the proportions of elements in atmospheric particulate matter reflect relative abundances of elements in the parent material. The composition of particulate matter reflects both the elemental compositon of its source and chemical reactions that may change the composition.

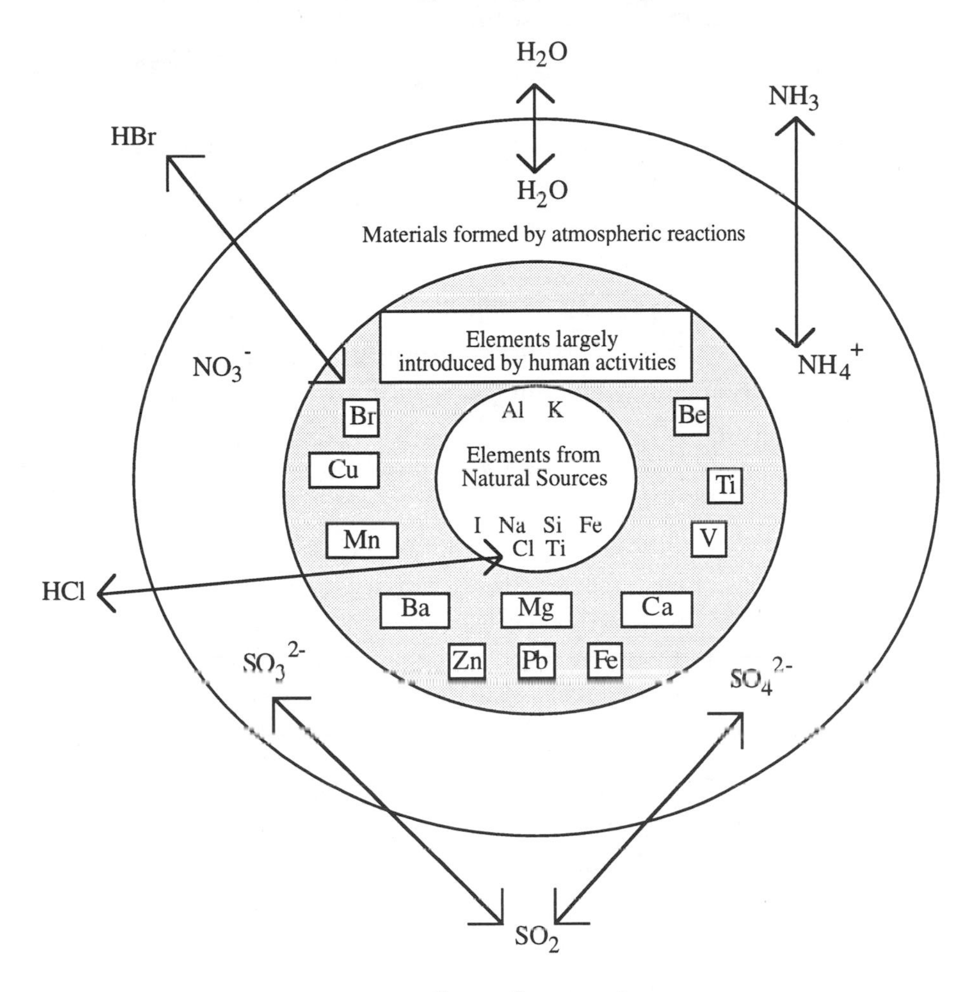

Figure 11.1. Some of the components of inorganic particulate matter and their origins.

The chemical composition of atmospheric particulate matter is quite diverse. Among the constituents of inorganic particulate matter found in polluted atmospheres are salts, oxides, nitrogen compounds, sulfur compounds, various metals, and radionuclides. In coastal areas sodium and chlorine get into atmospheric particles as sodium chloride from sea spray. The major trace elements that typically occur at levels above 1 $\mu g/m^3$ in particulate matter are those largely from terrestrial sources—aluminum, calcium, carbon, iron, potassium, sodium, and silicon. Lesser quantities of copper, lead, titanium, and zinc and even lower levels of antimony, beryllium, bismuth, cadmium, cobalt, chromium, cesium, lithium, manganese, nickel, rubidium, selenium, strontium, and vanadium are commonly observed. The likely sources of some of these elements are given below:

- **Al, Fe, Ca, Si**: Soil erosion, rock dust, coal combustion
- **C**: Particles of carbon consist of soot, carbon black, coke, and graphite produced primarily by incomplete combustion of carbonaceous fuels. Because of its good adsorbent properties, carbon can be a carrier of gaseous and other particulate pollutants, and particulate carbon surfaces may catalyze some heterogeneous atmospheric reactions, including the important conversion of SO_2 to sulfate.
- **Na, Cl**: Marine aerosols, chloride from incineration of organohalide polymer wastes
- **Sb, Se** (antimony and selenium): Very volatile elements, possibly from the combustion of oil, coal, or refuse
- **V**: Combustion of residual petroleum (present at very high levels in residues from Venezuelan crude oil)
- **Zn**: Tends to occur in small particles, probably from combustion
- **Hg** (mercury): Toxic, volatile, mobile heavy metal from coal and refuse combustion, commonly in the vapor form
- **Pb**: Combustion of leaded fuels and wastes containing lead, pollutant heavy metal of greatest concern in the atmosphere

Fly Ash

Much of the mineral particulate matter in a polluted atmosphere is in the form of **fly ash** consisting of inorganic material and elemental carbon produced during the combustion of high-ash fossil fuel. Fly ash enters furnace flues and is efficiently collected in a properly equipped stack system. However, some escapes through the stack and enters the atmosphere. Unfortunately, the fly ash thus released tends to consist of smaller particles that do the most damage to human health, plants, and visibility. The constituents of fly ash are oxides of aluminum, calcium, iron, and silicon; elemental carbon (soot, carbon black); and usually minor constituents, including magnesium, sulfur, titanium, phosphorus, potassium, and sodium.

Radioactive Particles

Radionuclides, which were discussed in some detail as water pollutants in Section 7.10, are significant air pollutants as well. A major natural source of radionuclides in the atmosphere is **radon**, a noble gas product of radium decay. Radon may enter the atmosphere as either of two isotopes, ^{222}Rn (half-life 3.8 days) and ^{220}Rn (half-life 54.5 seconds). Both emit alpha particles (energetic, positively charged helium nuclei) in decay chains that terminate with stable isotopes of lead. The initial decay products, ^{218}Po and ^{216}Po, are nongaseous and adhere readily to atmospheric particulate matter, so some of the radioactivity in these particles is of natural origin. Furthermore, cosmic rays act on nuclei in the atmosphere to produce other radionuclides, including ^{7}Be, ^{10}Be, ^{14}C, ^{39}Cl, ^{3}H, ^{22}Na, ^{32}P, and ^{33}P.

The combustion of fossil fuels introduces radioactivity into the atmosphere in the form of radionuclides contained in fly ash. Large coal-fired power plants lacking ash-control equipment may introduce up to several hundred milliCuries (a measure of radioactivity) of radionuclides into the atmosphere each year, far more than either an equivalent nuclear or oil-fired power plant.

Before the practice was discontinued, the above-ground detonation of nuclear weapons added large amounts of radioactive particulate matter to the atmosphere. Because of food contamination and biouptake, the most serious fission contaminant products from this source were ^{90}Sr, ^{131}I, and ^{137}Cs.

Organic Particles in the Atmosphere

A significant portion of organic particulate matter is produced by internal combustion engines in complicated processes that involve pyrosynthesis and nitrogenous compounds. These products may include nitrogen-containing compounds and oxidized hydrocarbon polymers. Lubricating oil and its additives may also contribute to organic particulate matter. The organic particles of greatest concern are **polycyclic aromatic hydrocarbons** (PAH), which consist of condensed-ring aryl molecules produced by pyrolysis or partial combustion of organic compounds. The most often cited example of a PAH compound is benzo(a)pyrene, a compound that the body can metabolize to a carcinogenic form:

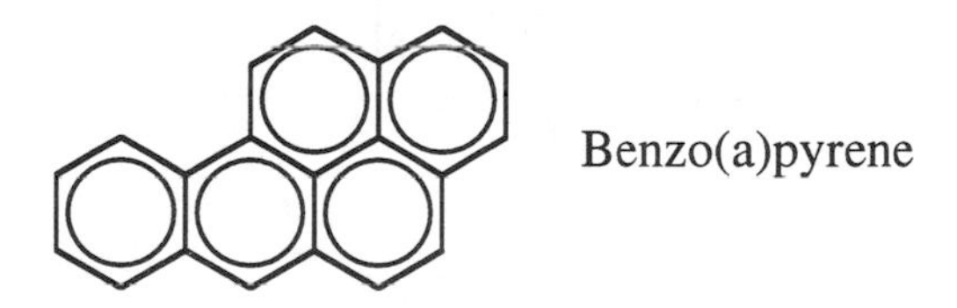

Effects of Particles

Atmospheric particles have numerous effects. The most obvious of these are reduction and distortion of visibility. Particles of 0.1 µm–1 µm size cause interference phenomena because they are about the same dimensions as the wavelengths of visible light, so their light-scattering properties are especially significant. Particles provide active surfaces upon which heterogeneous atmospheric chemical reactions can occur and nucleation bodies for the condensation of atmospheric water vapor, thereby causing significant weather and pollution effects.

Atmospheric particles inhaled through the respiratory tract may damage health. The most dangereous are very small particles, which are likely to reach the lungs and be retained by them. A strong correlation has been found between increases in the daily mortality rate and acute episodes of air pollution, including those in which particulate pollutants are present in high concentrations. The respiratory system may be damaged directly by particulate matter. Substances in inhaled particles may enter the blood system or lymph system in the alveoli of the lungs and be carried throughout the body to act as systemic poisons on other organs.

Control of Particulate Emissions

The removal of particulate matter from gas streams is the most widely practiced means of air pollution control. A number of devices have been developed for this purpose. The simplest means of particle control is **sedimentation** using chambers in which particles fall out under the influence of gravity. **Inertial mechanisms** for particle removal cause a gas stream to spin so that the particles forced outward by centrifugal force collect on walls of the device. **Fabric filters** used in **baghouses** consist of fabrics that allow the passage of gas but retain particulate matter. Periodically the fabric composing the filter is shaken to remove the particles and to reduce back-pressure to acceptable levels. A **venturi scrubber** passes gas through a converging section, throat, and diverging section as shown in Figure 11.2. Injection of scrubbing liquid at right angles to incoming gas breaks the liquid into very small droplets, which scavenge particles from the gas stream. **Electrostatic precipitators** (Figure 11.3) place an electrical charge on particles and collect them on a grounded surface.

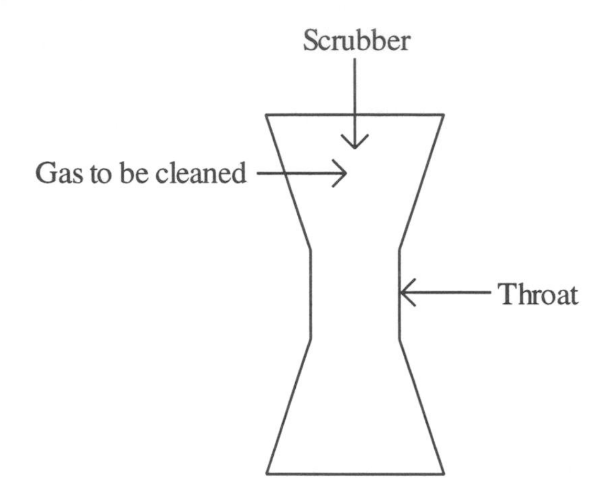

Figure 11.2. Venturi scrubber.

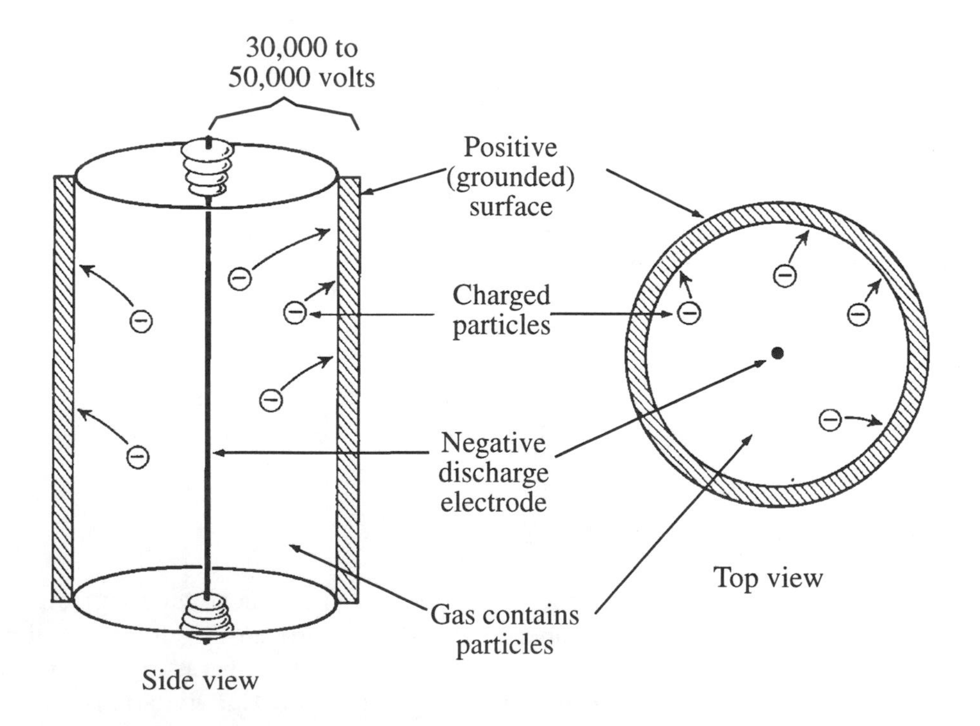

Figure 11.3. Schematic diagram of an electrostatic precipitator.

11.3. CARBON MONOXIDE

Carbon monoxide, CO, the toxicity of which is covered in Chapter 20, causes problems in cases of locally high concentrations. Because of carbon monoxide emissions from internal combustion engines, the highest levels of this toxic gas tend to occur in congested urban areas at times when the greatest number of people are exposed, such as during rush hours.

Control of Carbon Monoxide Emissions

The internal combustion engine in motor vehicles is the main source of localized pollutant carbon monoxide emissions. Modern automobiles use computerized engine control and catalytic exhaust reactors to cut down on carbon monoxide emissions. Excess air is pumped into the exhaust gas, and the mixture is passed through a catalytic converter in the exhaust system, resulting in oxidation of CO to CO_2.

Fate of Atmospheric CO

The residence time of carbon monoxide in the atmosphere is of the order of 4 months. It is generally agreed that carbon monoxide is removed from the atmosphere by reaction with hydroxyl radical, HO• :

$$CO + HO^\bullet \rightarrow CO_2 + H \qquad (11.3.1)$$

The H atom goes through chain reactions that regenerate HO•, which can oxidize additional CO.

11.4. SULFUR DIOXIDE AND GASEOUS SULFUR COMPOUNDS

Figure 11.4 shows the main aspects of the global sulfur cycle. This cycle involves primarily H_2S, SO_2, SO_3, and sulfates. Of these species, SO_2 is the most important because of its high abundance and facile oxidation to highly acidic H_2SO_4.

Sulfur Dioxide Reactions in the Atmosphere

Many factors, including temperature, humidity, light intensity, atmospheric transport, and surface characteristics of particulate matter, may influence the atmospheric chemical reactions of sulfur dioxide. Like many other gaseous pollutants, sulfur dioxide undergoes chemical reactions resulting in the formation of particulate matter. Whatever the processes involved, much of the sulfur dioxide in the atmosphere ultimately is oxidized to sulfuric acid and sulfate salts, particularly ammonium sulfate and ammonium hydrogen sulfate.

Effects of Atmospheric Sulfur Dioxide

Though not terribly toxic to most people, low levels of sulfur dioxide in air do have some health effects. Sulfur dioxide's primary effect is upon the respiratory tract, producing irritation and increasing airway resistance, especially to people with respiratory weaknesses and to sensitized asthmatics. Therefore, exposure to the gas may increase the effort required to breathe. Mucus secretion is also stimulated by exposure to air contaminated by sulfur dioxide.

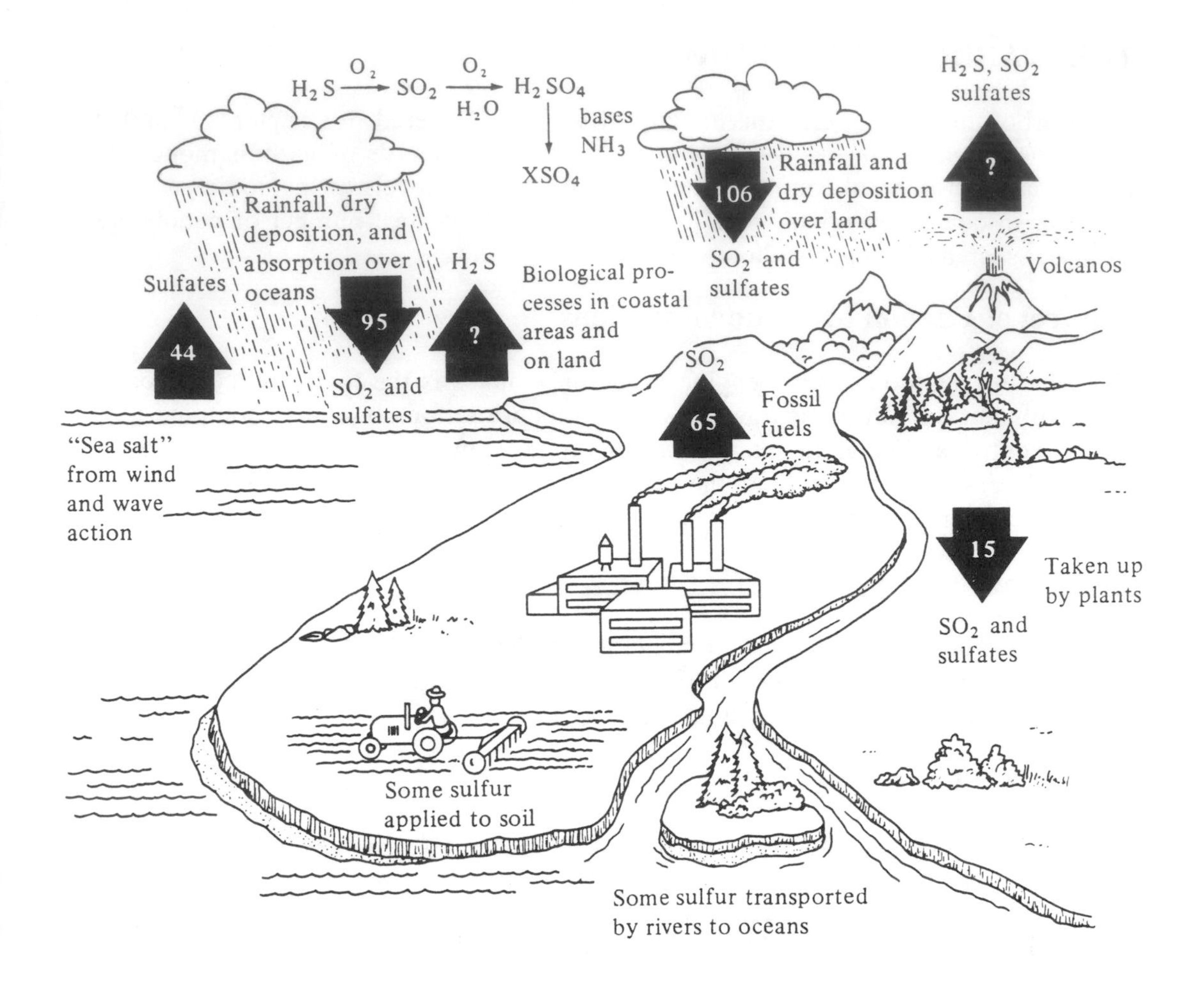

Figure 11.4. The atmospheric sulfur cycle.

Atmospheric sulfur dioxide is harmful to plants. Acute exposure to high levels of the gas kills leaf tissue (leaf necrosis). Chronic exposure of plants to sulfur dioxide causes chlorosis, a bleaching or yellowing of the normally green portions of the leaf. Sulfur dioxide in the atmosphere is converted to sulfuric acid, so that in areas with high levels of sulfur dioxide pollution, plants may be damaged by sulfuric acid aerosols. Such damage appears as small spots where sulfuric acid droplets have impinged leaves.

Sulfur Dioxide Removal

A number of processes are being used to remove sulfur and sulfur oxides from fuel before combustion and from stack gas after combustion. Most of these efforts concentrate on coal, since it is the major source of sulfur oxides pollution. Discrete particles of pyritic sulfur may be physically separated from coal, and chemical methods may also be employed to remove sulfur. **Fluidized bed combustion** of granular coal in a bed of finely divided limestone or dolomite maintained in a fluid-like condition by air injection eliminates SO_2 emissions at the point of combustion. Heat calcines the limestone,

$$CaCO_3 \rightarrow CaO + CO_2 \tag{11.4.1}$$

and the lime produced absorbs SO_2:

$$CaO + SO_2 \rightarrow CaSO_3 \quad (11.4.2)$$

Several wet or dry processes can be used for the removal of sulfur dioxide from stack gas. These include throwaway systems in which the SO_2-absorbing reagents are discarded, and recovery systems in which the reagents are regenerated and recycled.

A common throwaway system is one in which a lime slurry is injected into the flue gas beyond the boilers. Sulfur dioxide reacts with the lime,

$$Ca(OH)_2 + SO_2 \rightarrow CaSO_3 + H_2O \quad (11.4.3)$$

and the calcium sulfite product and unreacted lime are removed before the stack gas is discharged to the atmosphere. Although these scrubbers effectively remove both SO_2 and fly ash when operating properly, they are subject to corrosion and scaling problems, and disposal of the large amounts of lime sludge produced poses formidable obstacles.

Recovery systems in which sulfur dioxide or elemental sulfur are removed from the spent sorbing material, which is recycled, are much more desirable from an environmental viewpoint than are throwaway systems. One such system uses sodium sulfite to absorb SO_2 followed by regeneration of the reagent with heat:

$$Na_2SO_3 + H_2O + SO_2 \rightarrow 2NaHSO_3 \quad (11.4.4)$$

$$2NaHSO_3 + \text{heat} \rightarrow Na_2SO_3 + H_2O + SO_2 \quad (11.4.5)$$

Sulfur dioxide recovered from a stack-gas-scrubbing process can be converted to hydrogen sulfide (H_2S) by reaction with synthesis gas (H_2, CO, CH_4) followed by the Claus reaction to produce marketable elemental sulfur:

$$SO_2 + (H_2, CO, CH_4) \rightarrow H_2S + CO_2 \quad (11.4.6)$$

$$2H_2S + SO_2 \rightarrow 2H_2O + 3S \quad (11.4.7)$$

Hydrogen Sulfide, Carbonyl Sulfide, and Carbon Disulfide

Hydrogen sulfide is produced by microbial decay of sulfur compounds and microbial reduction of sulfate (see Section 6.6), from geothermal steam, as a by-product of wood pulping, and from a number of miscellaneous natural and anthropogenic sources. Most atmospheric hydrogen sulfide is rapidly converted to SO_2 and to sulfates. The organic homologs of hydrogen sulfide, the mercaptans, enter the atmosphere from decaying organic matter and have particularly objectionable odors.

Hydrogen sulfide pollution from artificial sources is not as much of an overall air pollution problem as sulfur dioxide pollution. However, there have been several acute incidents of hydrogen sulfide emissions resulting in damage to human health and even fatalities. The most notorious such incident occurred in Poza Rica, Mexico, in 1950. Accidental release of hydrogen sulfide from a plant used for the recovery of sulfur from natural gas caused the deaths of 22 people and the hospitalization of over 300.

Hydrogen sulfide at levels well above ambient concentrations destroys immature plant tissue. This type of plant injury is readily distinguished from that due to other phytotoxins. More sensitive species are killed by continuous exposure to around 3000 parts per billion H_2S, whereas other species exhibit reduced growth, leaf lesions, and defoliation.

Damage to certain kinds of materials is a very expensive effect of hydrogen sulfide pollution. Paints containing basic lead carbonate pigment, $2PbCO_3{\cdot}Pb(OH)_2$ (no longer used), were particularly susceptible to darkening by H_2S. A black layer of copper sulfide forms on copper metal exposed to H_2S. Eventually, this layer is replaced by a green coating of basic copper sulfate such as $CuSO_4{\cdot}3Cu(OH)_2$. The green "patina," as it is called, is very resistant to further corrosion. Such layers of corrosion can seriously impair the function of copper contacts on electrical equipment. Hydrogen sulfide also forms a black sulfide coating on silver.

Carbonyl sulfide, COS, is now recognized as a component of the atmosphere at a tropospheric concentration of approximately 500 parts per trillion by volume. It is, therefore, a significant sulfur species in the atmosphere. Both COS and carbon disulfide, CS_2, are oxidized in the atmosphere by reactions initiated by the hydroxyl radical.

11.5. GASEOUS NITROGEN COMPOUNDS IN THE ATMOSPHERE

The three oxides of nitrogen normally encountered in the atmosphere are nitrous oxide (N_2O), nitric oxide (NO), and nitrogen dioxide (NO_2). Microbially generated nitrous oxide is relatively unreactive and probably does not significantly influence important chemical reactions in the lower atmosphere. However, colorless, odorless nitric oxide (NO) and pungent red-brown nitrogen dioxide (NO_2), collectively designated NO_x, are very important in polluted air. Regionally high pollutant NO_2 concentrations can result in severe air quality deterioration. Practically all anthropogenic NO_x enters the atmosphere as NO from the combustion of fossil fuels in both stationary and mobile sources. The contribution of automobiles to nitric oxide production in the U.S. and some other countries has become somewhat lower in the last decade as newer automobiles with nitrogen oxide pollution controls have become more common.[2]

Atmospheric Reactions of NO_x

Nitrogen dioxide is a very reactive and significant species in the atmosphere. It absorbs light throughout the ultraviolet and visible spectrum penetrating the troposphere. At wavelengths below 398 nm, photodissociation to oxygen atoms occurs,

$$NO_2 + h\nu \rightarrow NO + O \tag{11.5.1}$$

and the highly reactive O atom product sets off several significant inorganic reactions, and many atmospheric reactions involving organic species. Atmospheric chemical reactions convert NO_x to nitric acid, inorganic nitrate salts, organic nitrates, and oxidant peroxyacetyl nitrate (see Chapter 12, Reaction 12.6.22). These species cycle among each other, as shown in Figure 11.5. Although NO is the primary form in which NO_x is released to the atmosphere, the conversion of NO to NO_2 is relatively rapid in the troposphere.

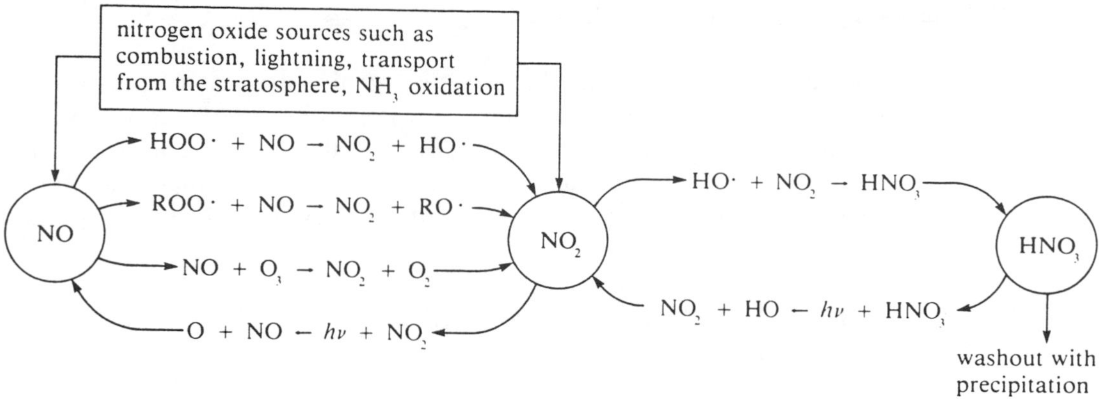

nitrogen oxide sources such as combustion, lightning, transport from the stratosphere, NH_3 oxidation

$HOO\cdot + NO \rightarrow NO_2 + HO\cdot$

$ROO\cdot + NO \rightarrow NO_2 + RO\cdot$

$NO + O_3 \rightarrow NO_2 + O_2$

$O + NO \leftarrow h\nu + NO_2$

NO

NO_2

$HO\cdot + NO_2 \rightarrow HNO_3$

$NO_2 + HO \leftarrow h\nu + HNO_3$

HNO_3

washout with precipitation

Figure 11.5. Principal reactions among NO, NO_2, and HNO_3 in the atmosphere. ROO• represents an organic peroxyl radical, such as the methylperoxyl radical, CH_3OO•.

Harmful Effects of Nitrogen Oxides

Nitric oxide, NO, is less toxic than NO_2. (In recent years the role of NO as an important intermediate in biochemical processes has gained recognition.) Acute exposure to NO_2 can be quite harmful to human health and sufficiently high exposures to this gas can be fatal. For exposures ranging from several minutes to one hour, a level of 50-100 ppm of NO_2 causes inflammation of lung tissue for a period of 6-8 weeks, after which time the subject normally recovers. Exposure of the subject to 150-200 ppm of NO_2 causes *bronchiolitis fibrosa obliterans*, a condition fatal within 3-5 weeks after exposure. Death generally results within 2-10 days after inhalation of air containing 500 ppm or more of NO_2. Although extensive damage to plants is observed in areas receiving heavy exposure to NO_2, most of this damage probably comes from secondary products of nitrogen oxides, such as PAN formed in smog (see Chapter 12).

Control of Nitrogen Oxides

NO production in combustion is favored by high temperatures and by high excess oxygen concentrations. Therefore, measures taken to reduce these conditions are used to lower NO production. Reduction of flame temperature to prevent NO formation is accomplished by adding recirculated exhaust gas, cool air, or inert gases. Low excess-air firing used to reduce NO_x emissions during the combustion of fossil fuels employs the minimum amount of excess air required for oxidation of the fuel, so that less oxygen is available for the reaction

$$N_2 + O_2 \rightarrow 2NO \qquad (11.5.2)$$

in the high temperature region of the flame. To minimize production of NO, a two-stage combustion process may be used. The first stage is fired at a relatively high temperature with a substoichiometric amount of air, and NO formation is limited by the absence of excess oxygen. In the second stage, burnout of hydrocarbons, soot, and CO is completed at a relatively low temperature in excess air, the low temperature preventing formation of NO.

Ammonia as an Atmospheric Pollutant

Ammonia is the only non-oxygenated gaseous inorganic nitrogen compound that is likely to be a significant atmospheric pollutant. It was mentioned in Section 10.5 as the only water-soluble base present at significant levels in the atmosphere. It is toxic and can be a significant localized pollutant in specific cases. Its most important effect in the atmosphere is its formation of corrosive pollutants NH_4NO_3, NH_4HSO_4, and $(NH_4)_2SO_4$.

11.6. FLUORINE, CHLORINE, AND THEIR GASEOUS COMPOUNDS

Fluorine, hydrogen fluoride, and other volatile fluorides are produced in the manufacture of aluminum, and hydrogen fluoride is a by-product in the conversion of fluorapatite (rock phosphate) to phosphoric acid, superphosphate fertilizers, and other phosphorus products. Hydrogen fluoride gas is a dangerously toxic substance that is so corrosive that it even reacts with glass. It is irritating to body tissues, and the respiratory tract is very sensitive to it. Brief exposure to HF vapors at the part-per-thousand level may be fatal. The acute toxicity of F_2 is even higher than that of HF. Chronic exposure to high levels of fluorides causes fluorosis, the symptoms of which include mottled teeth and pathological bone conditions.

Plants are particularly susceptible to the effects of gaseous fluorides. Fluorides from the atmosphere appear to enter the leaf tissue through the stomata. Fluoride is a cumulative poison in plants, and exposure of sensitive plants to even very low levels of fluorides for prolonged periods results in damage. Characteristic symptoms of fluoride poisoning are chlorosis (fading of green color due to conditions other than the absence of light), edge burn, and tip burn. Conifers (such as pine trees) afflicted with fluoride poisoning may develop reddish-brown necrotic needle tips at distances of several miles from the pollutant sources.

Chlorine and Hydrogen Chloride

Chlorine gas, Cl_2, can be quite damaging on a local scale. Chlorine was the first poisonous gas deployed in World War I. It is widely used as a manufacturing chemical, in the plastics industry, for example, as well as for water treatment and as a bleach. Therefore, possibilities for its release exist in a number of locations. Highly toxic chlorine is a mucous-membrane irritant, spills of which have caused fatalities among exposed persons. It is very reactive and a powerful oxidizing agent. Chlorine dissolves in atmospheric water droplets, yielding hydrochloric acid and hypochlorous acid, an oxidizing agent:

$$H_2O + Cl_2 \rightarrow H^+ + Cl^- + HOCl \qquad (11.6.1)$$

Hydrogen chloride, HCl, is emitted from a number of sources. Incineration of chlorinated plastics, such as polyvinylchloride, releases HCl as a combustion product.

```
       Cl H  H  H  Cl H  H  H  Cl H
       |  |  |  |  |  |  |  |  |  |
 ......C--C--C--C--C--C--C--C--C--C...... Polyvinylchloride
       |  |  |  |  |  |  |  |  |  |
       H  H  Cl H  H  H  Cl H  H  H
```

11.7. ORGANIC COMPOUNDS FROM NATURAL SOURCES

Most organics in the atmosphere come from natural sources, with only about 1/7 of the total atmospheric hydrocarbons originating from human activities. This ratio is primarily the result of the huge quantities of methane produced by anaerobic bacteria in the decomposition of organic matter in water, sediments, and soil:

$$2\{CH_2O\} \text{ (bacterial action)} \rightarrow CO_2(g) + CH_4(g) \qquad (11.7.1)$$

Flatulent emissions from domesticated animals, arising from bacterial decomposition of food in their digestive tracts, add about 85 million metric tons of methane to the atmosphere each year. Methane is a natural constituent of the atmosphere and is present at a level of about 1.4 parts per million (ppm) in the troposphere. Although its atmospheric chemical effects are minimal, methane is a major and growing contributor to greenhouse gases, and each molecule of methane added to the atmosphere contributes much more to greenhouse warming than does a molecule of carbon dioxide (see Section 12.2).

Vegetation is the most important natural source of atmospheric hydrocarbons other than methane. Ethene (ethylene), C_2H_4, is released to the atmosphere by a variety of plants, which use ethene as a molecular messenger (see Section 3.8). Most of the hydrocarbons emitted by plants (predominantly trees, such as citrus and pine trees) are **terpenes**. As exemplified by the structures of α-pinene, isoprene, and limonene,

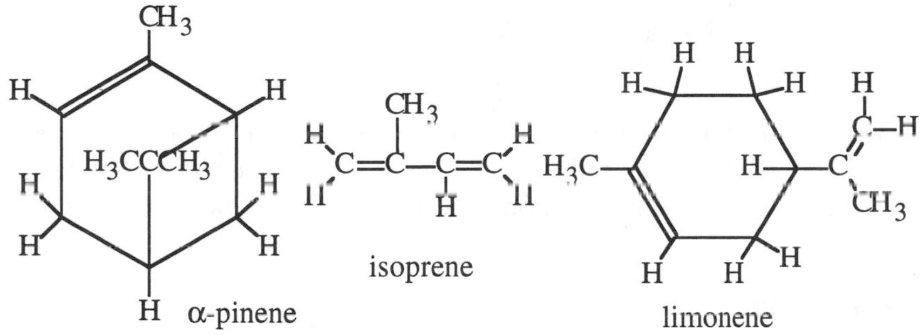

α-pinene　　isoprene　　limonene

terpenes contain alkenyl (double) bonds, usually two or more per molecule. Therefore, terpenes are among the most reactive compounds in the atmosphere. Terpenes react very rapidly with hydroxyl radical, HO•, and with other oxidizing agents in the atmosphere, particularly ozone, O_3. Such reactions form aerosols, which probably cause the blue haze in the atmosphere above some heavy growths of vegetation.

Perhaps the greatest variety of compounds emitted by plants consist of **esters**, such as coniferyl benzoate below:

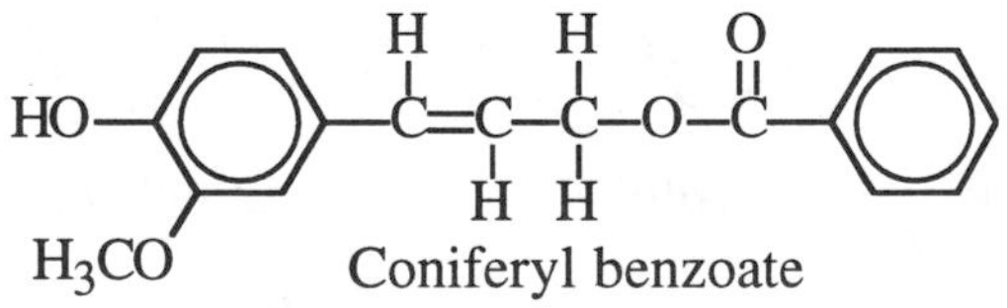

Coniferyl benzoate

However, they are released in such small quantities that they have little influence upon atmospheric chemistry. Esters are primarily responsible for the fragrances associated with much vegetation.

11.8. POLLUTANT HYDROCARBONS

Ethene and terpenes, which were discussed in the preceding section, are **hydrocarbons**, organic compounds containing only hydrogen and carbon. As defined in Chapter 2, hydrocarbons may be **alkanes**; **alkenes**, such as ethene; **alkynes** (compounds with triple bonds); and **aryl (aromatic) compounds**, such as naphthalene, which have characteristic benzene rings.

Because of their widespread use in fuels, hydrocarbons predominate among organic atmospheric pollutants. Petroleum products, primarily gasoline, are the source of most of the anthropogenic pollutant hydrocarbons found in the atmosphere. Hydrocarbons from fuel may enter the atmosphere either directly or as byproducts of the partial combustion of other hydrocarbons. The latter are particularly important because they tend to be unsaturated and relatively reactive.

Atmospheric alkenes come from a variety of processes, including emissions from internal combustion engines and turbines, foundry operations, and petroleum refining. Several alkenes, including the ones shown below, are among the top 50 chemicals produced each year, with worldwide production of several billion kg/year:

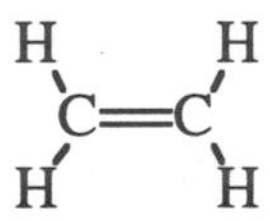

Ethylene 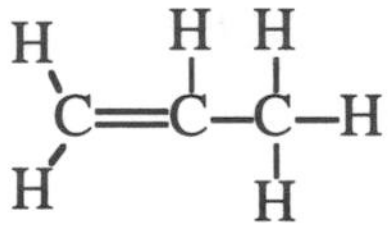 Propylene 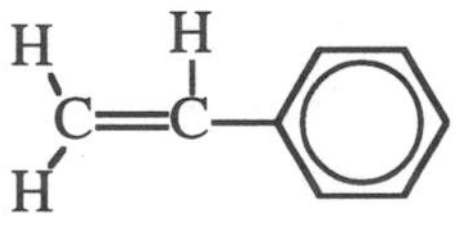 Styrene

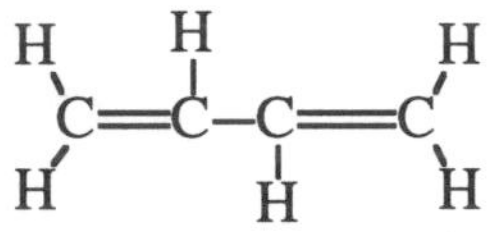

Butadiene

In addition to the direct release of alkenes in manufacturing processes, now generally well controlled, these hydrocarbons are commonly produced by the partial combustion and "cracking" at high temperatures of alkanes, particularly in the internal combustion engine.

Aryl Hydrocarbons

Aryl hydrocarbons may be divided into the two major classes: those that have only one benzene ring and those with multiple rings. The latter are **polycyclic aromatic (aryl) hydrocarbons, PAH**. The aryl hydrocarbons shown at the end of this paragraph are among the top 50 chemicals manufactured. With many industrial applications, plus production of aryl combustion by-products, aryl compounds are common atmospheric pollutants.

Benzene 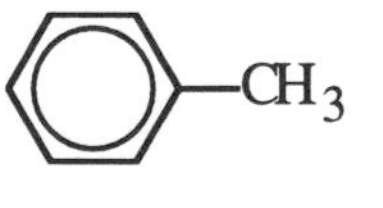 Toluene 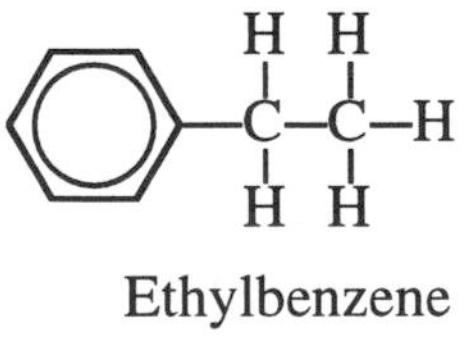 Ethylbenzene

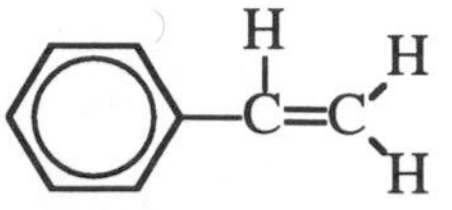

Styrene 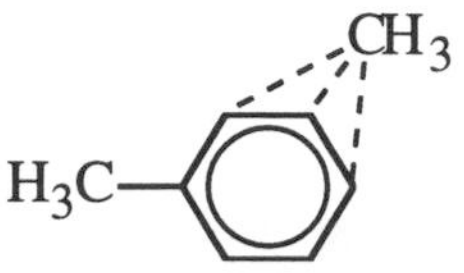 Xylene (3 isomers) 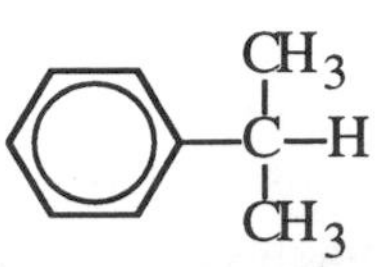 Cumene

11.9. OXYGEN-CONTAINING ORGANIC COMPOUNDS

Aldehydes and Ketones

Carbonyl compounds, consisting of aldehydes and ketones, are often the first species formed, other than unstable reaction intermediates, in the photochemical oxidation of atmospheric hydrocarbons (see photochemical smog in Chapter 12). The simplest and most widely produced of the carbonyl compounds is **formaldehyde**:

```
      O
      ||
      C         Formaldehyde
    H   H
```

The structures of some other industrially important aldehydes and ketones are shown below:

```
  H O       H   H O        H O H          H O H H
  | ||       \  | ||       | || |         | || | |
H-C-C-H       C=C-C-H    H-C-C-C-H      H-C-C-C-C-H
  |          /             |    |         |    | |
  H         H              H    H         H    H H

Acetaldehyde  Acrolein      Acetone     Methylethyl ketone
```

Aldehydes are second only to NO_2 as atmospheric sources of free radicals produced by the absorption of light. This is because the carbonyl group is a **chromophore**, a molecular group that readily absorbs light. It absorbs well in the near-ultraviolet region of the spectrum, and the excited species resulting therefrom may initiate chain reactions involved in photochemical smog formation.

Having both double bonds and carbonyl groups, alkenyl aldehydes are especially reactive in the atmosphere. The most common of these found in the atmosphere is acrolein (see above), a powerful lachrymator (tear producer) that is used as an industrial chemical and produced as a combustion by-product.

Alcohols and Aryl Alcohols

Of the alcohols, methanol, ethanol, 2-propanol, and ethylene glycol (Figure 11.6) rank among the top 50 chemicals, with annual worldwide production of the about a billion kg or more. A number of aliphatic alcohols have been reported in the atmosphere. Because of their volatility, the lower alcohols, especially methanol and ethanol, predominate as atmospheric pollutants. Alcohols can undergo photochemical reactions, beginning with abstraction of hydrogen by hydroxyl radical. Mechanisms for scavenging alcohols from the atmosphere are relatively efficient because the lower alcohols are quite water soluble and the higher ones have low vapor pressures.

```
   H          H H            H H           H H H
   |          | |            | |           | | |
 H-C-OH     H-C-C-OH      HO-C-C-OH      H-C-C-C-H      (benzene ring)
   |          | |            | |           | | |             |
   H          H H            H H           H OHH             OH

Methanol     Ethanol     Ethylene glycol  2-Propanol       Phenol
```

Figure 11.6. The more common lower molecular-mass alcohols, and phenol, an aryl alcohol.

Alkenyl alcohols in the atmosphere are by-products of combustion. Typical of these is 2-buten-1-ol, below, which has been detected in automobile exhausts:

```
   H  H     H
   |  |     |
H–C–C=C–C–OH
   |     |  |
   H     H  H
```

Phenols, which are aryl alcohols, are more noted as water pollutants than as air pollutants. Phenol (Figure 11.6) is among the top 50 chemicals produced. Its greatest use is for the manufacture of resins and polymers, such as Bakelite copolymer of phenol and formaldehyde. Phenol and other phenolic compounds are produced by the pyrolysis of coal and are major by-products of coking. In local situations involving coal coking and similar operations phenols can be troublesome air pollutants.

Oxides

Ethylene oxide and propylene oxide rank among the 50 most widely produced industrial chemicals and have a limited potential to enter the atmosphere as pollutants. Ethylene oxide is a moderately to highly toxic sweet-smelling, colorless, flammable, explosive gas used as a chemical intermediate, sterilant, and fumigant. It is a mutagen and a carcinogen to experimental animals.

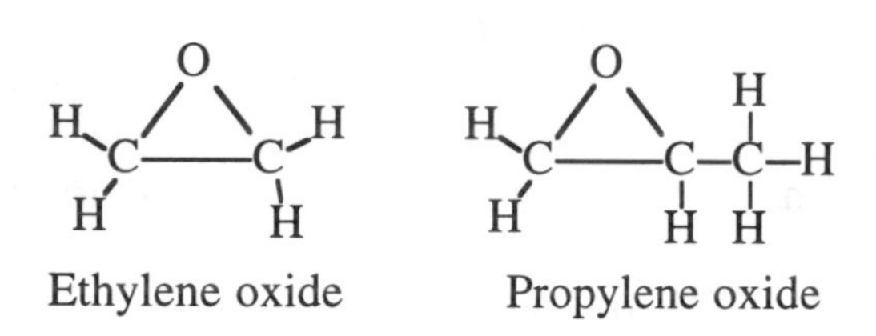

Ethylene oxide Propylene oxide

Ethers

Ethers are relatively uncommon atmospheric pollutants; however, the flammability hazard of diethyl ether vapor in an enclosed work space is well known. Some air pollutant ethers are produced by internal combustion engines. The cyclic ether and industrial solvent, tetrahydrofuran, occurs as an air contaminant. The most widely used gasoline octane booster, methyltertiarybutyl ether, MTBE, is a potential air pollutant, but its hazard is limited by its low vapor pressure.

```
   H     H          H  H     H  H       H     H     H  H
   |     |          |  |     |  |        \    |     |  |
H–C–O–C–H      H–C–C–O–C–C–H         C=C–O–C–C–H
   |     |          |  |     |  |        /          |  |
   H     H          H  H     H  H       H           H  H

Dimethyl ether        Diethyl ether            Vinylethyl ether
```

```
    H     H                       H
    |     |                       |
H–C–––C–H               H  H–C–H H
H–C     C–H           H–C–O–C–––C–H
 H   O   H                H  H–C–H H
                               |
Tetrahydrofuran                H
                    Methyltertiarybutyl ether (MTBE)
```

Carboxylic Acids

Carboxylic acids have at least one of the functional groups,

$$-\overset{\displaystyle O}{\overset{\|}{C}}-OH$$

attached to an alkane, alkene, or aryl hydrocarbon moiety. A carboxylic acid, pinonic acid (structure below), is produced by the photochemical oxidation of natu-

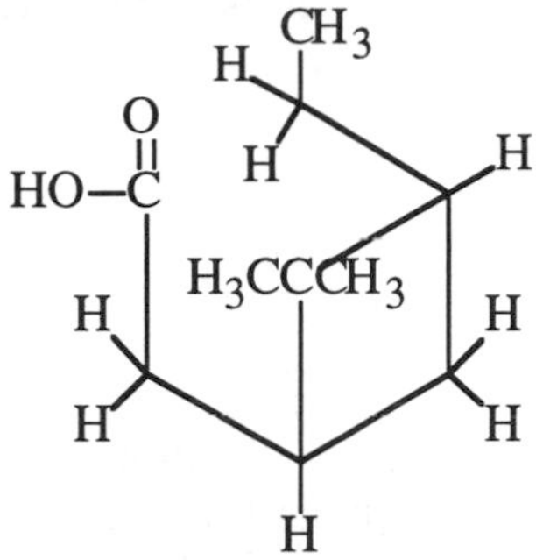

Pinonic acid

rally produced α-pinene (see structure in Section 11.7). Most of the many carboxylic acids found in the atmosphere are probably the result of the photochemical oxidation of other organic compounds through gas-phase reactions or by reactions of other organic compounds dissolved in aqueous aerosols. These acids are often the end products of photochemical oxidation because their low vapor pressures and their significant water-solubilities make them susceptible to scavenging from the atmosphere.

11.10. ORGANOHALIDE COMPOUNDS

Organohalides consist of halogen-substituted hydrocarbon molecules, each of which contains at least one atom of F, Cl, Br, or I. They may be saturated (**alkyl halides**), unsaturated (**alkenyl halides**), or aryl (**aryl halides**). Organohalides exhibit a wide range of physical and chemical properties. Structural formulas of several organohalides widely used industrially and commonly encountered in the atmosphere are given in Figure 11.7.

Volatile **chloromethane** (methyl chloride) is consumed in the manufacture of silicones. **Dichloromethane** is a volatile liquid with excellent solvent properties for nonpolar organic solutes. It has been used as a solvent for the decaffeination of coffee, in paint strippers, as a blowing agent in urethane polymer manufacture, and to depress vapor pressure in aerosol formulations. **Dichlorodifluoromethane** is one of the chlorofluorocarbon compounds used as a refrigerant and involved in stratospheric ozone depletion. One of the more common industrial chlorinated solvents is **1,1,1-trichloroethane**. **Vinyl chloride** is consumed in large quantities as a raw material to manufacture pipe, hose, wrapping, and other products fabricated from polyvinylchloride plastic. This highly flammable, volatile, sweet-smelling gas is one of the few known human carcinogens, and it has been proven to cause angiosarcoma, a rare form of liver cancer.

Aryl halide compounds have many uses that have resulted in substantial human exposure and environmental contamination. The most environmentally significant aryl halides are polychlorinated biphenyls, PCBs, a group of compounds formed by the chlorination of biphenyl,

$$+ \text{x } Cl_2 \rightarrow \quad Cl_x \qquad (11.10.1)$$

which have extremely high physical and chemical stabilities and other qualities that have led to their being used in many applications, including heat transfer fluids, hydraulic fluids, and dielectrics. Although not very volatile, PCBs can get into the atmosphere from high temperature sources, such as incinerators, and be transported with atmospheric particles.

Alkyl halides

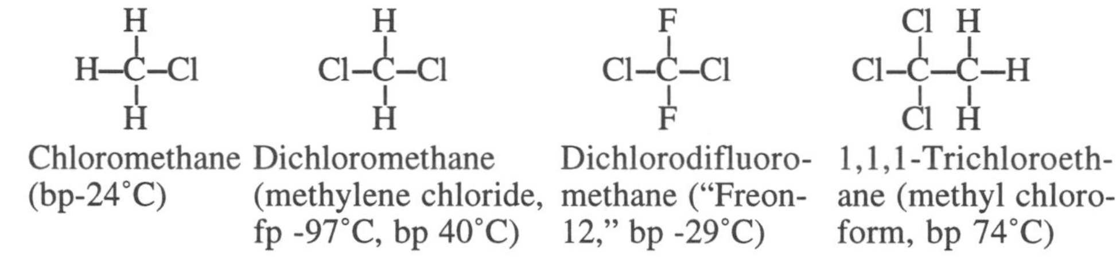

Alkenyl halides

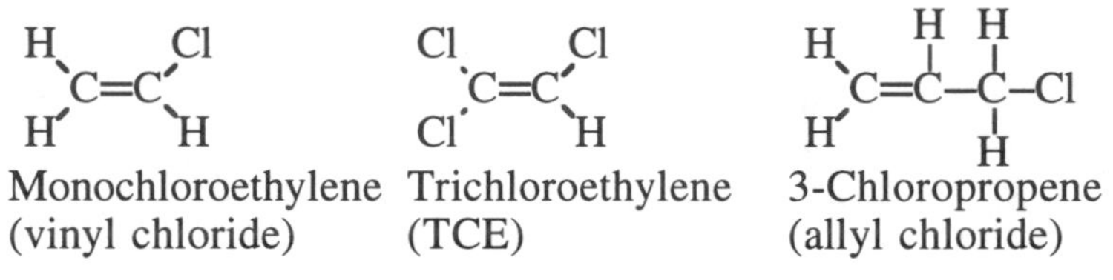

Aryl halides

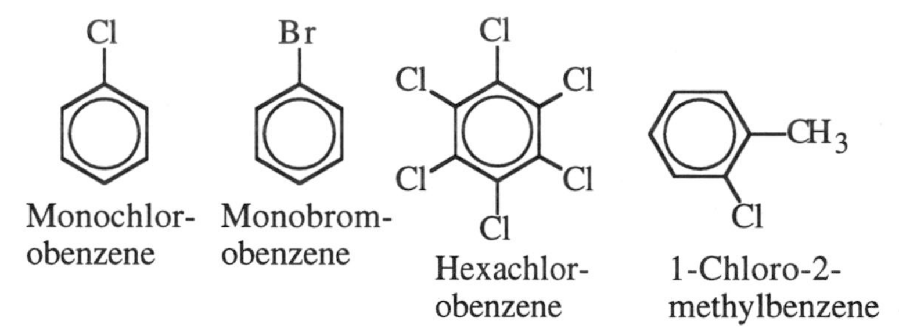

Figure 11.7. Alkyl halides and alkenyl halides commonly encountered in the atmosphere.

The fluorine-containing air pollutants with the greatest potential for damage to the atmosphere are the **chlorofluorocarbons (CFC)**, commonly called Freons, which have been used as fluids in refrigeration mechanisms, as blowing agents in the fabrication of flexible and rigid foams, and many other products. Chlorofluorocarbons are volatile 1- and 2-carbon compounds, such as $CCl_3FCCl_2F_2$, that contain Cl and F bonded to carbon. These compounds are notably stable and nontoxic. **Halons** are related compounds that contain bromine, the most common ones of which are $CBrClF_2$ (Halon-1211), $CBrF_3$ (Halon-1301), and $C_2Br_2F_4$ (Halon-2402). Halons are excellent fire extinguishing agents for which there are no good substitutes. CFCs and halons have the potential to seriously deplete Earth's protective stratospheric ozone layer, and they are discussed in this context in Chapter 12.

11.11. ORGANOSULFUR COMPOUNDS

Substitution of alkyl or aryl hydrocarbon groups such as phenyl and methyl for H on hydrogen sulfide, H_2S, leads to a number of different organosulfur compounds. Substitution for one H yields thiols, or mercaptans, R–SH; substitution for two Hs yields sulfides, also called thioethers, R–S–R. Structural formulas of examples of these compounds are shown below:

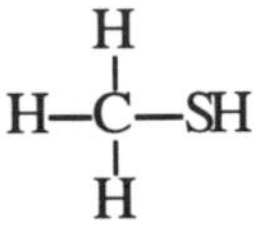

Methanethiol 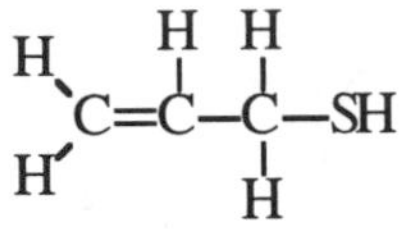   2-Propene-1-thiol 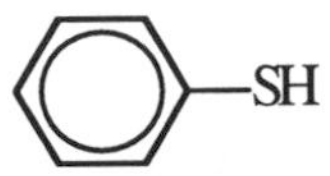   Benzenethiol

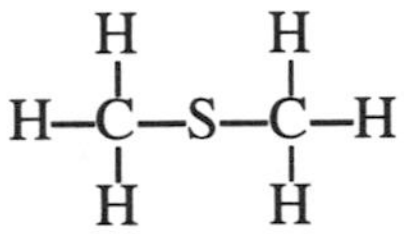

Dimethylsulfide    Thiophene 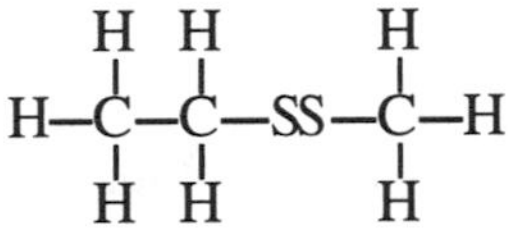   Ethylmethyldisulfide

Although not highly significant as atmospheric contaminants on a large scale—organosulfur compounds' effects on atmospheric chemistry are minimal in areas such as aerosol formation or production of acid precipitation components—these compounds are the worst of all in producing odor. Therefore, organosulfur compounds can cause local air pollution problems because of their bad odors. Major sources of organosulfur compounds in the atmosphere include microbial degradation of sulfur-containing substances, wood pulping, volatile matter evolved from growing plants, animal wastes, packing house and rendering plant wastes, starch manufacture, sewage treatment, and petroleum refining.

As with all H-containing organic species in the atmosphere, reaction of organosulfur compounds with hydroxyl radical is a first step in their atmospheric photochemical reactions. The sulfur from both mercaptans and sulfides ends up as SO_2 and ultimately as sulfuric acid or sulfate salts.

11.12. ORGANONITROGEN COMPOUNDS

Organic nitrogen compounds that may be found as atmospheric contaminants include **amines**, **amides**, **nitriles**, **nitro compounds**, or **heterocyclic nitrogen compounds**. Structures of common examples of each of these five classes of compounds reported as atmospheric contaminants are shown in Figure 11.8.

Lower molecular-mass amines are volatile. These amines are prominent among the compounds giving rotten fish their characteristic odor—an obvious reason why air contamination by amines is undesirable. The simplest and most important aryl amine is aniline, used in the manufacture of dyes, amides, photographic chemicals, and drugs. A number of amines are widely used industrial chemicals and solvents, so that industrial sources have the potential to contaminate the atmosphere with these chemicals. Decaying organic matter, especially protein wastes, produces amines, so that rendering plants, packing houses, and sewage treatment plants are important sources of these substances.

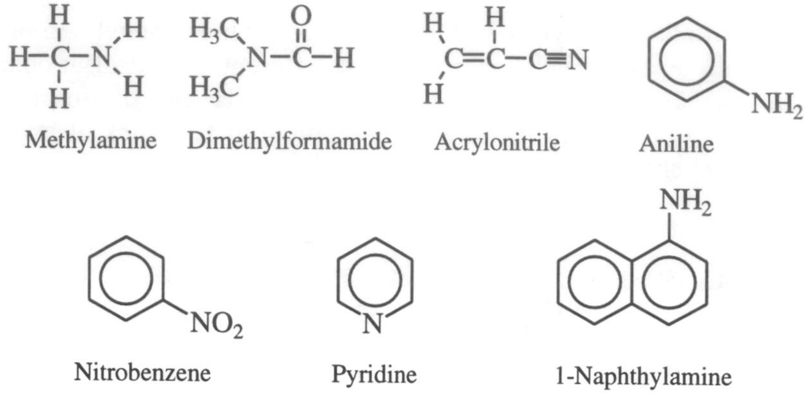

Methylamine Dimethylformamide Acrylonitrile Aniline

Nitrobenzene Pyridine 1-Naphthylamine

Figure 11.8. Some common organonitrogen compounds.

Aryl amines are of particular concern as atmospheric pollutants, particularly in the workplace, because some, such as 1-naphtylamine shown above, are known to cause urethral tract cancer (particularly of the bladder) in exposed individuals. Some of the aryl amines, including 1-naphtylamine, are among the few compounds that are known to be human carcinogens based on observations of cancer in humans. This occurred as the result of exposure to workers to the compounds from coal tar used to make dyes in Germany around 1900. Aryl amines are widely used as chemical intermediates, antioxidants, and curing agents in the manufacture of polymers (rubber and plastics), drugs, pesticides, dyes, pigments, and inks.

In the atmosphere, amines can be attacked by hydroxyl radical and undergo further reactions. As bases, their acid-base chemistry in the atmosphere may be important, particularly in the presence of acids in acidic precipitation.

The amide most likely to be encountered as an atmospheric pollutant is dimethylformamide. It is widely used commercially as a solvent for the synthetic polymer, polyacrylonitrile (Orlon, Dacron). Most amides have relatively low vapor pressures, which limit their entry into the atmosphere.

Nitriles, which are characterized by the —C≡N group, have been reported as air contaminants, particularly from industrial sources. Both acrylonitrile and acetonitrile, CH_3CN, have been reported in the atmosphere as a result of synthetic rubber manufacture. As expected from their volatilities and high levels of industrial production, most of the nitriles reported as atmospheric contaminants are low molecular-mass aliphatic or alkenyl nitriles, or aryl nitriles with only one benzene ring. Acrylonitrile, used to make polyacrylonitrile polymer, is the only nitrogen-containing organic chemical among the top 50 chemicals, with annual worldwide production exceeding 1 billion kg.

Among the nitro compounds (RNO_2) reported as air contaminants are nitromethane, nitroethane, and nitrobenzene. These are common nitro compounds that are produced from industrial sources. Highly oxygenated compounds containing the NO_2 group, particularly peroxyacetyl nitrate (PAN, the formation of which is shown in Reaction 10.4.13 and discussed as an oxidizing constituent of photochemical smog in Chapter 12), are end-products of the photochemical oxidation of hydrocarbons in urban atmospheres.

CHAPTER SUMMARY

The chapter summary below is presented in a programmed format to review the main points covered in this chapter. It is used most effectively by filling in the blanks, referring back to the chapter as necessary. The correct answers are given at the end of the summary.

Gaseous pollutants that enter the atmosphere in the greatest quantities are [1]________ ____. The effects of pollutants in the atmosphere may be divided between [2]______________________________ and the formation of [3] ______________________________, such as photochemical smog. Colloidal-sized atmospheric particles constitute atmospheric [4]________________ of which liquid particles make up [5]__________________, and particles formed by incomplete combustion of fuel constitute [6]__________________. When gaseous atmospheric sulfur dioxide is oxidized, particles form because of the tendency of the product to accumulate [7]________________. In atmospheric particles the elements that tend to come from soil erosion, rock dust, and coal combustion are [8]________ ____________________, incomplete combustion of fuels tends to produce particles containing [9]_____________, and marine aerosols contain [10]________________. Inorganic material and elemental carbon produced during the combustion of high-ash fossil fuel and collected from furnace flue systems are [11]________________. The pollutant metal of greatest concern in the urban atmosphere is [12]____________. A major natural source of atmospheric radioactivity is the noble gas [13]_________. Organic hydrocarbon particles consisting of condensed rings are called [14]________ ______________________. Generally the most readily apparent influence that aerosol particles have upon air quality results from their [15]________________ __________, which are most apparent for particles of a size comparable to [16] ________________________________. Devices that remove aerosol particles from stack gas with a high-voltage electrical charge are called [17]________________ _____________. Much of the CO present in the atmosphere is produced as [18]_____ ____. Modern automobiles use [19]__________________________________ reactors to cut down on carbon monoxide emissions. The atmospheric species in the global sulfur cycle are primarily [20]__________________________, of which [21]________ ____________ is the most important because of its high abundance and facile oxidation to highly acidic [22]________________. In a recovery system for sulfur dioxide removal from stack gas, some of the sulfur dioxide can be converted [23]____ ______, which in turn can undergo the reaction [24]________________________ __________________ to produce commercially marketable [25]______________ _________. A pollutant inorganic sulfur gas that can cause darkening of some kinds of paints and coatings on some metal surfaces is [26]______________________. __________. The three oxides of nitrogen normally encountered in the atmosphere are [27]__________________________. The formula NO_x stands for [28]__________ __________________________ in the atmosphere. When exposed to ultraviolet radiation of a wavelength less than 398 nm, nitrogen dioxide undergoes the reaction [29]____. Low excess-air firing of furnaces is used to reduce [30]________________ emissions. The only non-oxygenated gaseous inorganic nitrogen compound that is likely to be a significant atmospheric pollutant is [31]____________________. Incineration of polyvinylchloride can release pollutant [32]__________ as a combustion product. The fact that most organics (hydrocarbons) in the atmosphere come from natural sources is primarily the result of the release of huge quantities of [33]____________ produced largely by the overall biochemical process represented as [34]__________________________________. Most of the hydrocarbons emitted by plants, such as citrus and pine trees, are [35]________________. Toluene, styrene, and cumene are all examples of potential air pollutant [36]____________________________. Often the first species formed, other than unstable reaction intermediates, in the photochemical oxidation of atmospheric hydrocarbons are [37]____. Halogen-substituted hydrocarbon molecules, some of which may become air pollutants, are called [38]____________________. The fluorine-containing air pollutants with the greatest potential for damage to the atmosphere are the [39]____________________. Amines, amides, nitriles, and nitro compounds are all examples of [40]___________________________.

Answers

1 CO, SO_2, NO, and NO_2
2 direct effects
3 secondary pollutants
4 aerosols
5 mists
6 smoke
7 water vapor
8 Al, Fe, Ca, Si
9 carbon, C
10 Na and Cl
11 fly ash
12 lead
13 radon
14 polycyclic aromatic hydrocarbons, PAH
15 optical effects
16 the wavelengths of visible light
17 electrostatic precipitators
18 an intermediate in the oxidation of methane by hydroxyl radical
19 catalytic exhaust
20 H_2S, SO_2, SO_3, and sulfates
21 SO_2

22 H_2SO_4

23 H_2S

24 $2H_2S + SO_2 \rightarrow 2H_2O + 3S$

25 elemental sulfur

26 H_2S

27 N_2O, NO, and NO_2

28 the sum of NO and NO_2

29 $NO_2 + h\nu \rightarrow NO + O$

30 NO_x

31 ammonia, NH_3

32 HCl

33 methane

34 $2\{CH_2O\}$ (bacterial action) $\rightarrow CO_2(g) + CH_4(g)$

35 terpenes

36 aryl hydrocarbons

37 carbonyl compounds, consisting of aldehydes and ketones

38 organohalides

39 chlorofluorocarbons (CFCs or Freons)

40 organic nitrogen compounds

QUESTIONS AND PROBLEMS

1. How is NO_2 involved in the processes that result in conversion of organic vapors to aldehydes, oxidants, and other substances characteristic of photochemical smog?
2. What is the major source of pollutant metal oxide particles in the atmosphere?
3. What is asbestos and why is it of particular concern as an air pollutant?
4. What is the origin of radioactive radon in the atmosphere?
5. How are venturi scrubbers used to protect air quality?
6. Why do the highest levels of carbon monoxide tend to occur in congested urban areas?
7. What is the main species that scavenges carbon monoxide from the atmosphere?
8. What is sulfur dioxide's primary effect on humans?
9. In what sort of system is the reaction $Ca(OH)_2 + SO_2 \rightarrow CaSO_3 + H_2O$ used to control a common atmospheric pollutant. What is at least one big disadvantage of such a system?

10. How does nitrous oxide, N_2O, get into the atmosphere? What is its significance as an atmospheric pollutant?
11. What are the major inorganic reaction processes involving NO_x in the atmosphere?
12. Explain and justify the statement that "Nitrogen dioxide is a very reactive and significant species in the atmosphere."
13. What is the most important effect of ammonia in the atmosphere?
14. How does pollutant chlorine in the atmosphere react with atmospheric water?
15. Explain how methane released into the troposphere results in the formation of water vapor in the stratosphere.
16. What are terpenes, and how are they significant in the atmosphere?
17. Why do aryl hydrocarbons have a relatively high potential to pollute the atmosphere?
18. What are the two major classes of carbonyl compounds? What is their particular significance in the atmosphere?
19. Why are the lower alcohols, such as methanol or ethanol, so readily removed from the atmosphere?
20 What is an organohalide compound that is known to be a human carcinogen? Which class of organohalide compounds has a particularly high potential to destroy stratospheric ozone?
21. What characteristic of organosulfur compounds makes them particularly undesirable air pollutants?
22. Which functional group is characteristic of nitriles?

LITERATURE CITED

[1] Yunus, Mohammad and Muhammad Iqbal, Eds., *Plant Response to Air Pollution*, John Wiley & Sons, New York, NY, 1996.

[2] Faiz, Asif, Christopher S. Weaver, and Michael P. Walsh, *Air Pollution from Motor Vehicles: Standards and Technologies for Controlling Emissions*, World Bank, New York, NY, 1996

SUPPLEMENTARY REFERENCES

Boubel, Richard W., Donald L. Fox, D. Bruce Turner, and Arthur C. Stern, *Fundamentals of Air Pollution*, 3rd ed, Academic Press Textbooks, San Diego, CA, 1994.

Colls, J., *Air Pollution*, Chapman & Hall, London, UK, 1997.

Gammage, Richard B. and Barry A. Berven, *Indoor Air and Human Health*, CRC Press/Lewis Publishers, Boca Raton, FL, 1996.

Hesketh, Howard E., *Air Pollution Control: Traditional and Hazardous Pollutants*, Technomic Publishing Company, New York, NY, 1996.

Heumann, William, J., Ed., *Industrial Air Pollution Control Systems*, McGraw Hill Publishing Co., New York, NY, 1997.

Leonard, R. Leon, *Air Quality Permitting*, CRC Press/Lewis Publishers, Boca Raton, FL, 1997.

Seinfeld, John H. and Spyros N. Pandis, *Atmospheric Chemistry and Physics: Air Pollution to Climate*, John Wiley & Sons, New York, NY, 1997.

Turco, Richard P., *Earth Under Siege: Air Pollution and Global Change*, Oxford University Press, New York, NY, 1996.

12 THE ENDANGERED GLOBAL ATMOSPHERE

12.1. ANTHROPOGENIC CHANGE IN THE ATMOSPHERE

There is a very strong connection between life forms on Earth and the nature of Earth's climate, which determines its suitability for life. As proposed by James Lovelock, a British chemist, this forms the basis of the **Gaia hypothesis**, which contends that the atmospheric O_2/CO_2 balance established and maintained by organisms determines and maintains Earth's climate and other environmental conditions.

In Section 10.7 there is a discussion of the evolution of the atmosphere as determined by life on Earth. Although details of how this process has come about are being debated, especially with recent theories regarding the origin of life under the harsh conditions that are now observed on ocean floors, there is no doubt that living organisms have had a profound influence on the atmosphere in the past. The most massive of the changes caused were those arising from photosynthesis, which produces biomass, "$\{CH_2O\}$," and molecular oxygen, O_2:

$$CO_2 + H_2O + h\nu \rightarrow \{CH_2O\} + O_2(g) \qquad (12.1.1)$$

This process converted Earth's atmosphere from a chemically reducing to a chemically oxidizing state and precipitated enormous deposits of insoluble oxidized iron:

$$4Fe^{2+} + O_2 + 4H_2O \rightarrow 2Fe_2O_3 + 8H^+ \qquad (12.1.2)$$

In addition to providing O_2 that most non-photosynthetic organisms use for respiration, the photosynthetically released oxygen formed stratospheric ozone, O_3, which absorbs damaging ultraviolet radiation from the sun, enabling living organisms to move from water onto land where they are directly exposed to sunlight.

Other instances of climatic change and regulation induced by organisms can be cited. An example is the maintenance of atmospheric carbon dioxide at low levels through the action of photosynthetic organisms (note from Reaction 12.1.1 that photosynthesis removes CO_2 from the atmosphere). But, at an ever accelerating pace during the last 200 years, another organism, humankind, has engaged in a number of activities that are altering the atmosphere profoundly.[1] These are summarized as follows:

- Industrial activities, which emit a variety of atmospheric pollutants including SO_2, particulate matter, photochemically reactive hydrocarbons, chlorofluorocarbons, and inorganic substances (such as toxic heavy metals)
- Burning of large quantities of fossil fuel, which can introduce CO_2, CO, SO_2, NO_X, hydrocarbons (including CH_4), and particulate soot, polycyclic aromatic hydrocarbons, and fly ash into the atmosphere
- Transportation practices, which emit CO_2, CO, NO_X, photochemically reactive (smog-forming) hydrocarbons, and polycyclic aromatic hydrocarbons
- Alteration of land surfaces, including deforestation
- Burning of biomass and vegetation, including tropical and subtropical forests and savanna grasses, which produces atmospheric CO_2, CO, NO_X, and particulate soot and polycyclic aromatic hydrocarbons
- Agricultural practices, which produce methane (from the digestive tracts of domestic animals and from the cultivation of rice in waterlogged anaerobic soils) and N_2O from bacterial denitrification of nitrate-fertilized soils

These kinds of human activities have significantly altered the atmosphere, particularly in regard to its composition of minor constituents and trace gases. Major effects have been the following:

- Increased acidity in the atmosphere
- Production of pollutant oxidants in localized areas of the lower troposphere (see Photochemical Smog, Section 12.5)
- Elevated levels of infrared-absorbing gases (greenhouse gases)
- Threats to the ultraviolet-filtering ozone layer in the stratosphere
- Increased corrosion of materials induced by atmospheric pollutants

In 1957 photochemical smog was only beginning to be recognized as a serious problem, acid rain and the greenhouse effect were scientific curiosities, and the ozone-destroying potential of chlorofluorocarbons had not even been imagined. In that year, Revelle and Suess[2] prophetically referred to human perturbations of the Earth and its climate as a massive "geophysical experiment." The effects that this experiment may have on the global atmosphere are discussed in this chapter. For additional information the reader is referred to a series of excellent articles from *Science* dealing with urban air pollution (smog),[3] acid precipitation,[4] threats to the stratospheric ozone layer,[5] and global warming.[6]

12.2. GREENHOUSE GASES AND GLOBAL WARMING

This section deals with infrared-absorbing trace gases (other than water vapor) in the atmosphere that contribute to global warming — the "greenhouse effect" — by allowing incoming solar radiant energy to penetrate to the Earth's surface while reabsorbing infrared radiation emanating from it. Levels of these "greenhouse gases" have increased at a rapid rate during recent decades and are continuing to do so. Concern over this phenomenon has intensified since about 1980. This is because ever since accurate temperature records have been kept, the 1980s have been the warmest 10-year period recorded. On an annual basis, 1988 was the warmest year ever recorded, 1987 was second and 1981 third.[7] However, some authorities argue that recent observations of warming trends are largely artifacts of the location of temperature-measuring facilities in urban areas subject to localized anthropogenic heating. They contend that satellite measurements of temperature are much more meaningful. Although such measurements taken since the 1970s have shown annual fluctuations, they have not revealed any overall trend in global temperature. In addition to being a scientific issue, greenhouse warming of the atmosphere is also becoming a major policy and political issue.

There are many uncertainties surrounding the issue of greenhouse warming. However, several things about the phenomenon are certain. It is known that CO_2 and other greenhouse gases, such as CH_4, absorb infrared radiation by which Earth loses heat. The levels of these gases have increased markedly since about 1850 as nations have become industrialized and as forest lands and grasslands have been converted to agriculture. Chlorofluorocarbons, which also are greenhouse gases, were not even introduced into the atmosphere until the 1930s. Although trends in levels of these gases are well known, their effects on global temperature and climate are much less certain. The phenomenon has been the subject of much computer modelling. Most models predict global warming of 1.5-5°C, about as much again as has occurred since the last ice age. Such warming would have profound effects on rainfall, plant growth, and sea levels, which might rise as much as 0.5-1.5 meters.

Carbon dioxide is the gas most commonly thought of as a greenhouse gas; it is responsible for about half of the atmospheric heat retained by trace gases. It is produced primarily by burning of fossil fuels and deforestation accompanied by burning and biodegradation of biomass. On a molecule-for-molecule basis, methane, CH_4, is 20–30 times more effective in trapping heat than is CO_2. Other trace gases that contribute are chlorofluorocarbons and N_2O.

Analyses of gases trapped in polar ice samples indicate that preindustrial levels of CO_2 and CH_4 in the atmosphere were approximately 260 parts per million and 0.70 ppm, respectively. Over the last 300 years these levels have increased to current values of around 350 ppm and 1.7 ppm, respectively; most of the increase by far has taken place at an accelerating pace over the last 100 years. (A note of interest is the observation, based upon analyses of gases trapped in ice cores, that the atmospheric level of CO_2 at the peak of the last ice age about 18,000 years past was 25 percent <u>below</u> preindustrial levels.) About half of the increase in carbon dioxide in the last 300 years can be attributed to deforestation, which still accounts for approximately 20 percent of the annual increase in this gas. Carbon dioxide is increasing by about 1 ppm per year. Methane is going up at a rate of almost 0.02 ppm/year.[8,9] The comparatively very rapid increase in methane levels is attributed to a number of human activities, among which are direct leakage of natural gas, by-product emissions from coal mining and petroleum recovery, and release from the burning of savannas and tropical forests. Biogenic sources resulting from human activities produce large amounts of atmospheric methane. These include methane from bacteria degrading organic matter, such as municipal refuse in landfills; methane evolved from anaerobic biodegradation of organic matter in rice paddies; and methane emitted as the result of bacterial action in the digestive tracts of ruminant animals.

Both positive and negative feedback mechanisms may be involved in determining the rates at which carbon dioxide and methane build up in the atmosphere. Laboratory studies indicate that increased CO_2 levels in the atmosphere cause accelerated uptake of this gas by plants undergoing photosynthesis, which tends to slow buildup of atmospheric CO_2. Given adequate rainfall, plants living in a warmer climate that would result from the greenhouse effect would grow faster and take up more CO_2. This could be an especially significant effect on forests, which have a high CO_2-fixing ability. However, the projected rate of increase in carbon dioxide levels is so rapid that forests would lag behind in their ability to fix additional CO_2. Similarly, higher atmospheric CO_2 concentrations will result in accelerated sorption of the gas by oceans. The amount of dissolved CO_2 in the oceans is about 60 times the amount of CO_2 gas in the atmosphere. However, the times for transfer of carbon dioxide from the atmosphere to the ocean are of the order of years. Because of low mixing rates, the times for transfer of oxygen from the upper approximately 100-meter layer of the oceans to ocean depths are much longer, of the order of decades. Therefore, like the uptake of CO_2 by forests, increased absorption by oceans will lag behind the emissions of CO_2. Severe drought conditions resulting from climatic warming could cut down substantially on CO_2 uptake by plants. Warmer conditions would accelerate release of both CO_2 and CH_4 by microbial degradation of organic matter. (It is important to realize that about twice as much carbon is held in soil in dead organic matter—necrocarbon—potentially degradable to CO_2 and CH_4 as is present in the atmosphere.) Global warming might speed up the rates at which biodegradation adds these gases to the atmosphere.

It is certain that atmospheric CO_2 levels will continue to increase significantly. The degree to which this occurs depends upon future levels of CO_2 production and the fraction of that production that remains in the atmosphere. Given plausible projections of CO_2 production and a reasonable estimate that half of that amount will remain in the atmosphere, projections can be made that indicate that sometime during the middle part of the next century the concentration of this gas will reach 600 ppm in the atmosphere. This is well over twice the levels estimated for pre-industrial times. Much less certain are the effects that this change will have on climate. It is virtually impossible for the elaborate computer models used to estimate these effects to accurately take account of all variables, such as the degree and nature of cloud cover. Clouds both reflect incoming light radiation and absorb outgoing infrared radiation, with the former effect tending to predominate. The magnitude of these effects depends upon the degree of cloud cover, brightness, altitude, and thickness. In the case of clouds, too, feedback phenomena occur; for example, warming induces formation of more clouds, which reflect more incoming energy. Most computer models predict global warming of at least 3.0°C and as much as 5.5°C occurring over a period of just a few decades. These estimates are sobering because they correspond to the approximate temperature increase since the last ice age 18,000 years past, which took place at a much slower pace of only about 1 or 2°C per 1,000 years.

Drought is one of the most serious problems that could arise from major climatic change resulting from greenhouse warming. Typically, a 3-degree warming would be accompanied by a 10 percent decrease in precipitation. Water shortages would be aggravated, not just from decreased rainfall, but from increased evaporation, as well. Increased evaporation results in decreased runoff, thereby reducing water available for agricultural, municipal, and industrial use. Water shortages, in turn, lead to increased demand for irrigation and to the production of lower quality, higher salinity runoff water and wastewater. In the U.S., such a problem would be especially intense in the Colorado River basin, which supplies much of the water used in the rapidly growing U.S. southwest. The magnitude of this problem is emphasized by long-term droughts in the western U.S., such as one that caused California's governor to propose drastic mandatory water conservation measures in 1991.

A variety of other problems, some of them unforeseen as of now, could result from global warming. An example is the effect of warming on plant and animal pests—insects, weeds, diseases, and rodents. Many of these would certainly thrive much better under warmer conditions.

Interestingly, another air pollutant, acid-rain-forming sulfur dioxide (see Section 12.4), may have a counteracting effect on greenhouse gases. This is because sulfur dioxide is oxidized in the atmosphere to sulfuric acid, forming a light-reflecting haze. Furthermore, the sulfuric acid and resulting sulfates act as condensation nuclei (Section 9.5) that increase the extent, density, and brightness of light-reflecting cloud cover.

12.3. DROUGHT AND DESERTIFICATION

Drought is a condition of water shortage resulting from a long-term deficiency in rainfall. In many parts of the world cycles of drought are natural and expected, and their adverse effects are minimized with proper planning. In the continental U.S., for example, periods of drought lasting several years occurred in the 1870s, 1900s, 1930s (the most severe on record), 1950s, and 1970s. The "dust bowl" years of the 1930s were characterized by devastating dust storms that ripped topsoil from the land and virtually ruined huge areas of once productive prairie land. This disaster gave rise to one of the earliest environmental movements and resulted in huge federal programs designed to protect irreplaceable farmland from drought and wind erosion. Shelter belts of trees were planted to slow the sweep of wind across the land. Farming practices were altered such that strips of land were left uncultivated in alternate years with crop stubble exposed to catch and hold sparse rain and winter snow. Use of the plow, which severely disturbs soil to considerable depth and buries crop residues under a layer of dirt, gave way to tillage methods that leave a surface residue of crop biomass, which captures precipitation and anchors soil against wind erosion.

Whereas periodic droughts are normal and largely manageable events, a much more serious condition can occur through human mismanagement. This phenomenon is **desertification** in which vegetation is removed from once fertile land, streams and groundwater sources dry up, and the atmospheric, terrestrial, and living environments assume characteristics of desert conditions. There are several causes of desertification. The most troubling potential cause is greenhouse warming discussed in the preceding section. "Slash-and-burn" agriculture by which trees and other vegetation are stripped from rain forests for short-term production of pasture and agricultural land is currently the largest contributor to desertification. Excessive grazing is ruinous to soil and pushes land along the path to desert formation. Desertification has a strong tendency toward positive feedback, meaning that it feeds on itself. Decreased plant cover leads to erosion and rapid loss of water from soil, which in turn further decreases the capacity of land to support plant life. Obviously, desertification is a major environmental problem that must be dealt with firmly and vigorously, if Earth is to sustain its present populations.

12.4. ACID RAIN

Precipitation made acidic by the presence of acids stronger than $CO_2(aq)$ is commonly called **acid rain** or **acid precipitation**, a term that applies to all kinds of acidic aqueous precipitation, including fog, dew, snow, and sleet. In a more general sense, **acid deposition** refers to the deposition on the Earth's surface of aqueous acids, acid gases (such as SO_2), and acidic salts (such as NH_4HSO_4).[4] Therefore, deposition in solution form is *acid precipitation*, and deposition of dry gases and compounds is *dry deposition*. Sulfur dioxide, SO_2, contributes more to the acidity of precipitation than does CO_2 present at higher levels in the atmosphere for two reasons. The first of these is that sulfur dioxide is significantly more soluble in water, as indicated by its Henry's Law constant (a quantitative measure of gas solubility, see Section 6.9) of 1.2 mol x L^{-1} x atm^{-1} compared to 3.38 x 10^{-2} mol x L^{-1} x atm^{-1} for CO_2. Secondly, the value of K_{a1} for $SO_2(aq)$,

$$SO_2(aq) + H_2O \leftrightarrow H^+ + HSO_3^- \tag{12.4.1}$$

$$K_{a1} = \frac{[H^+][HSO_3^-]}{[SO_2]} = 1.7 \times 10^{-2} \tag{12.4.2}$$

is more than four orders of magnitude higher than the value of 4.45 x 10^{-7} for K_{a1} of CO_2. This simply means that SO_2 is a stronger acid in water than CO_2.

Although acid rain can originate from the direct emission of strong acids, such as HCl gas or sulfuric acid mist, most of it is a secondary air pollutant produced by the atmospheric oxidation of acid-forming gases such as the following:

$$SO_2 + {}^1/_2O_2 + H_2O \xrightarrow[\text{ing of several steps}]{\text{Overall reaction consist-}} \{2H^+ + SO_4^{2-}\}(aq) \tag{12.4.3}$$

$$2NO_2 + {}^1/_2O_2 + H_2O \xrightarrow[\text{ing of several steps}]{\text{Overall reaction consist-}} 2\{H^+ + NO_3^-\}(aq) \tag{12.4.4}$$

Chemical reactions such as these play a dominant role in determining the nature, transport, and fate of acid precipitation. As the result of such reactions the chemical properties (acidity, ability to react with other substances) and physical properties (volatility, solubility) of acidic atmospheric pollutants are altered drastically. For example, even the small fraction of NO that does dissolve in water does not react significantly. However, its ultimate oxidation product, HNO_3, though volatile, is highly water-soluble, strongly acidic, and very reactive with other materials. Therefore, it tends to be removed readily from the atmosphere and to do a great deal of harm to plants, corrodable materials, and other things that it contacts.

Although emissions from industrial operations and fossil fuel combustion are the major sources of acid-forming gases, acid rain has also been encountered in areas far from such sources. This is due in part to the fact that acid-forming gases are oxidized to acidic constituents and deposited over several days, during which time the air mass containing the gas may have moved a thousand kilometers or more. It is likely that the burning of biomass, such as is employed in "slash-and-burn" agriculture evolves the gases that lead to acid formation in more remote areas. In arid regions, dry acid gases or acids sorbed to particles may be deposited, with effects similar to those of acid rain deposition.

Acid rain spreads out over areas of several hundred to several thousand kilometers. This classifies it as a *regional* air pollution problem compared to a largely *local* air pollution problem for smog and a *global* one for ozone-destroying chlorofluorocarbons and greenhouse gases. Regional air pollution problems include those caused by soot, smoke, and fly ash from combustion sources and fires (forest fires). Nuclear fallout from weapons testing or from reactor fires (of which, fortunately, their has been only one major one to date—the one at Chernobyl in the Soviet Union) may also be regarded as a regional phenomenon.

Acid precipitation shows a strong geographic dependence, as illustrated in Figure 12.1, representing the pH of precipitation in the continental U.S. The preponderance of acidic rainfall in the northeastern U.S. is obvious. Analyses of the movements of air masses have shown a correlation between acid precipitation and prior movement of an air mass over major sources of anthropogenic sulfur and nitrogen oxides emissions. This is particularly obvious in southern Scandinavia, which receives a heavy burden of air pollution from densely populated, heavily industrialized areas in Europe.

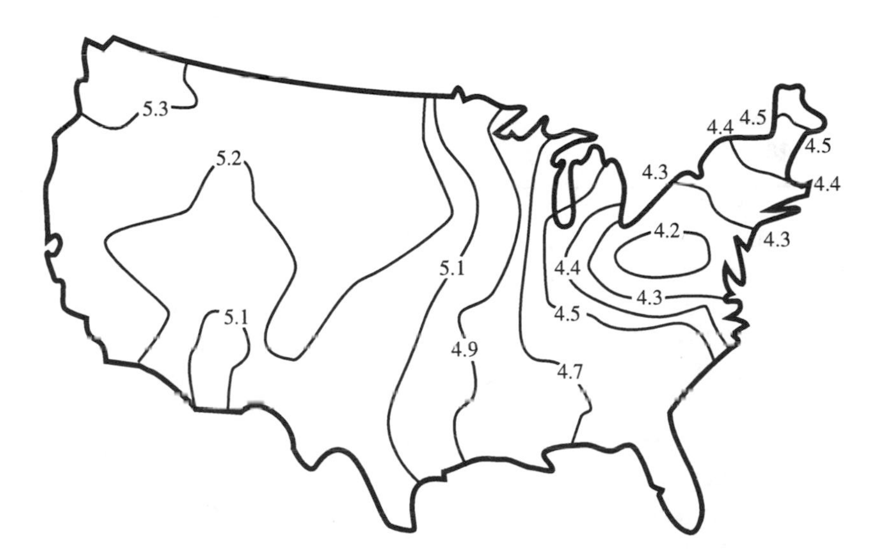

Figure 12.1. Isopleths of pH illustrating a hypothetical precipitation-pH pattern in the lower 48 continental United States. Actual values found vary with the time of year and climatic conditions.

Acid rain is not a new phenomenon; it has been observed for well over a century, with many of the older observations from Great Britain. The first manifestations of this phenomenon were elevated levels of SO_4^{2-} in precipitation collected in industrialized areas. More modern evidence was obtained from analyses of precipitation in Sweden in the 1950s and of U.S. precipitation a decade or so later. A vast research effort on acid rain was conducted in North America by the National Acid Precipitation Assessment Program, which resulted from the U.S. Acid Precipitation Act of 1980.

The longest study of acid precipitation in the U.S. has been conducted at the U.S. Forest Service Hubbard Brook Experimental Forest in New Hampshire's White Mountains. It is downwind from major U.S. urban and industrial centers so it is a prime candidate to receive acid precipitation. This is reflected by mean annual pH values ranging from 4.0 to 4.2 during the 1964-74 period, during which time the annual hydrogen ion input ($[H^+]$ x volume) increased by 36%.

Table 12.1 shows typical major cations and anions in pH-4.25 precipitation. Although actual values encountered vary greatly with time and location of collection, this table does show some major features of ionic solutes in precipitation. From the predominance of sulfate anion, it is apparent that sulfuric acid is the major contributor to acid precipitation. Nitric acid makes up a smaller but growing contribution to the acid present. Hydrochloric acid ranks third.

Table 12.1. Typical Values of Ion Concentrations in Acidic Precipitation

Cations		Anions	
Ion	Concentration equivalents/L x 10^6	Ion	Concentration equivalents/L x 10^6
H^+	56	SO_4^{2-}	51
NH_4^+	10	NO_3^-	20
Ca^{2+}	7	Cl^-	12
Na^+	5		Total 83
Mg^{2+}	3		
K^+	2		
	Total 83		

An important factor in the study of acid rain and sulfur pollution involves the comparison of primary sulfate species (those emitted directly by point sources) and secondary sulfate species (those formed from gaseous sulfur compounds, primarily by the atmospheric oxidation of SO_2). A low primary-sulfate content indicates transport of the pollutant from some distance, whereas a high primary-sulfate content indicates local emissions. This information can be useful in determining the effectiveness of SO_2 control in reducing atmospheric sulfate, including sulfuric acid. Primary and secondary sulfates can be measured using the oxygen-18 content of the sulfates. This content is higher in sulfate emitted directly from a power plant than it is in sulfate formed by the oxidation of SO_2. This technique can yield valuable information on the origins and control of acid rain.

Ample evidence exists of the damaging effects of acid rain. The major such effects are the following:

- Direct phytotoxicity to plants from excessive acid concentrations. (Evidence of direct or indirect phytoxicity of acid rain is provided by the declining health of eastern U. S. and Scandinavian forests and especially by damage to Germany's Black Forest.)
- Phytotoxicity from acid-forming gases, particularly SO_2 and NO_2, that accompany acid rain
- Indirect phytotoxicity, such as from Al^{3+} liberated from soil
- Destruction of sensitive forests
- Respiratory effects on humans and other animals
- Acidification of lake water with toxic effects to lake flora and fauna, especially fish fingerlings

- Corrosion to exposed structures, electrical relays, equipment, and ornamental materials. Because of the effect of hydrogen ion, limestone, $CaCO_3$, is especially susceptible to damage from acid rain:

$$2H^+ + CaCO_3(s) \rightarrow Ca^{2+} + CO_2(g) + H_2O$$

- Associated effects, such as reduction of visibility (increased haziness) by acidic sulfate aerosols and the influence of sulfate aerosols on physical and optical properties of clouds. (As mentioned in Section 12.2, intensification of cloud cover and changes in the optical properties of cloud droplets—specifically, increased reflectance of light—resulting from acid sulfate in the atmosphere may even have a mitigating effect on greenhouse warming of the atmosphere.)

Soil sensitivity to acid precipitation can be estimated from cation exchange capacity (CEC, Chapter 14). Soil with a high CEC can exchange ions such as Ca^{2+} ion on soil for H^+ ions in acidic water, thus reducing the acidity of the water. Soil is generally insensitive if free carbonates are present or if it is flooded frequently.

Forms of precipitation other than rainfall, such as snow, may be acidic. Acidic fog can be especially damaging because it is very penetrating. In early December, 1982, Los Angeles experienced a severe, two-day episode of acid fog. This fog consisted of a heavy concentration of acidic mist particles at ground level, which reduced visibility and were very irritating to breathe. The pH of the water in these particles was 1.7, much lower than ever before recorded for acid precipitation.

12.5. OZONE LAYER DESTRUCTION

Recall from Section 9.5 that stratospheric ozone, O_3, serves as a shield to absorb harmful ultraviolet radiation in the stratosphere, protecting living beings on the Earth from the effects of excessive amounts of such radiation. The two reactions by which stratospheric ozone are produced are

$$O_2 + h\nu \rightarrow O + O \qquad (\lambda < 242.4\text{ nm}) \qquad (12.5.1)$$

$$O + O_2 + M \rightarrow O_3 + M\text{ (energy-absorbing } N_2 \text{ or } O_2) \qquad (12.5.2)$$

and it is destroyed by photodissociation

$$O_3 + h\nu \rightarrow O_2 + O \qquad (\lambda < 325\text{ nm}) \qquad (12.5.3)$$

and a series of reactions from which the net result is the following:

$$O + O_3 \rightarrow 2O_2 \qquad (12.5.4)$$

Ozone in the stratosphere is present at a steady-state concentration resulting from the balance of ozone production and destruction by the above processes. The quantities of ozone involved are interesting. A total of about 350,000 metric tons of ozone are formed and destroyed daily. Ozone never makes up more than a small fraction of the gases in the ozone layer. In fact, if all the atmosphere's ozone were in a single layer at surface temperature and pressure conditions of approximately 273 K and 1 atm, it would be only 3 mm thick!

Ozone absorbs ultraviolet radiation very strongly in the region 220-330 nm. Therefore, it is effective in filtering out dangerous UV-B radiation, 290 nm $< \lambda <$ 320 nm. (UV-A radiation, 320 nm-400 nm, is relatively less harmful and UV-C radiation, $<$ 290 nm, does not penetrate to the troposphere.) If UV-B were not absorbed by ozone, severe damage would result to exposed forms of life on the Earth. Absorption of electromagnetic radiation by ozone converts the radiation's energy to heat and is responsible for the temperature maximum encountered at the boundary between the stratosphere and the mesosphere at an altitude of approximately 50 km. The reason that the temperature maximum occurs at a higher altitude than that of the highest ozone concentration is that ozone is so effective in absorbing ultraviolet radiation that most of this radiation is absorbed in the upper stratosphere, where it generates heat, and only a small fraction reaches the lower altitudes, which remain relatively cool.

Increased intensities of ground-level ultraviolet radiation caused by stratospheric ozone destruction would have some significant adverse consequences. One major effect would be on plants, including crops used for food. The destruction of microscopic plants that are the basis of the ocean's food chain (phytoplankton) could severely reduce the productivity of the world's seas. Human exposure would result in an increased incidence of cataracts. The effect of most concern to humans is the elevated occurrence of skin cancer in individuals exposed to ultraviolet radiation. This is because UV-B radiation is absorbed by cellular DNA (see Chapter 20) resulting in photochemical reactions that alter the function of DNA so that the genetic code is improperly translated during cell division. This can result in uncontrolled cell division leading to skin cancer. People with light complexions lack protective melanin, which absorbs UV-B radiation, and are especially susceptible to its effects. The most common type of skin cancer resulting from ultraviolet exposure is squamous cell carcinoma, which forms lesions that are readily removed and has little tendency to spread (metastasize). Readily metastasized malignant melanoma caused by absorption of UV-B radiation is often fatal. Fortunately, this form of skin cancer is relatively uncommon.

The major culprit in ozone depletion consists of chlorofluorocarbon (CFC) compounds, commonly known as "Freons." These volatile compounds have been used and released to a very large extent in recent decades. The major use associated with CFCs is as refrigerant fluids. Other applications have included solvents, aerosol propellants, and blowing agents in the fabrication of foam plastics. The same extraordinarily high chemical stability that makes CFCs nontoxic enables them to persist for years in the atmosphere and to enter the stratosphere. In the stratosphere the photochemical dissociation of CFCs by intense ultraviolet radiation,

$$CF_2Cl_2 + h\nu \rightarrow Cl\bullet + CClF_2\bullet \tag{12.5.5}$$

yields chlorine atoms, each of which can go through chain reactions, particularly the following:

$$Cl\bullet + O_3 \rightarrow ClO\bullet + O_2 \tag{12.5.6}$$

$$\underline{ClO\bullet + O \rightarrow Cl\bullet + O_2} \tag{12.5.7}$$

$$O_3 + O \rightarrow 2O_2 \text{ (net reaction for ozone destruction)}$$

The net effect of these reactions is catalysis of the destruction of several thousand molecules of O_3 for each Cl atom produced. Because of their widespread use and persistence, the two CFCs of most concern in ozone destruction are CFC-11 and CFC-12, $CFCl_3$ and CF_2Cl_2, respectively. Even in the intense ultraviolet radiation of the stratosphere the most persistent chlorofluorocarbons have lifetimes of the order of 100 years.

The most prominent instance of ozone layer destruction is the so-called "Antarctic ozone hole" that has been documented in recent years.[10] This phenomenon is manifested by the appearance during the Antarctic's late winter and early spring of severely depleted stratospheric ozone (up to 50%) over the polar region. The reasons why this occurs are related to the normal effect of NO_2 in limiting Cl-atom-catalyzed destruction of ozone by combining with ClO:

$$ClO + NO_2 \rightarrow ClONO_2 \quad (12.5.8)$$

In the polar regions, particularly Antarctica, NO_x gases are removed along with water by freezing in polar stratospheric clouds at temperatures below -70°C as compounds such as $HNO_3{\cdot}3H_2O$. Furthermore, chlorine species can be liberated from $ClONO_2$ and other chlorine compounds (HCl) by reactions in the cloud ice, followed by photodissociation to yield ozone-destroying atomic chlorine:

$$ClONO_2 + H_2O \rightarrow HOCl + HNO_3 \quad (12.5.9)$$

$$ClONO_2 + HCl \rightarrow Cl_2 + HNO_3 \quad (12.5.10)$$

$$HOCl + h\nu \rightarrow HO{\cdot} + Cl \quad (12.5.11)$$

$$Cl_2 + h\nu \rightarrow Cl + Cl \quad (12.5.12)$$

12.6. PHOTOCHEMICAL SMOG

Photochemical smog is a major local or regionalized air pollution phenomenon characterized by oxidants, irritating vapors, and visibility-obscuring particles that occurs in urban areas where the combination of pollution-forming emissions and appropriate atmospheric conditions are right for its formation. Though not a threat to the global atmosphere as such, in some urban areas photochemical smog is highly detrimental to health and to the quality of life. *Smog* originally was used to describe the unpleasant combination of smoke and fog laced with sulfur dioxide, a chemically reducing atmosphere, which was formerly prevalent in London when high-sulfur coal was the primary fuel used in that city. However, the photochemical smog discussed in this section is chemically quite different because of its oxidizing qualities and the formation of oxidants in the air, particularly ozone, is indicative of smog formation.

Photochemical smog has a long history. Exploring what is now southern California in 1542, Juan Rodriguez Cabrillo named San Pedro Bay "The Bay of Smokes" because of the heavy haze that covered the area. Complaints of eye irritation from anthropogenically polluted air in Los Angeles were recorded as far back as 1868. Characterized by reduced visibility, eye irritation, cracking of rubber, and deterioration of materials, smog became a serious nuisance in the Los Angeles area during the 1940s. It is now recognized as a major air pollution problem in many areas of the world.

The species in the atmosphere that give smog its noxious character do not enter the atmosphere directly, but are produced by photochemical processes acting on precursor atmospheric pollutants. Because of this, the constituents of smog are **secondary pollutants**, in contrast to a material such as sulfur dioxide, which is a primary air pollutant. The three ingredients required to generate photochemical smog are ultraviolet light, reactive hydrocarbons, and nitrogen oxides, the latter two of which are produced as emissions from internal combustion engines. Although the automobile is the major source of these pollutants, hydrocarbons may come from biogenic sources, of which α-pinene and isoprene (structural formulas shown in Section 11.7) from trees are the most abundant. In order for high levels of smog to form, relatively stagnant air must be subjected to sunlight under low humidity conditions in the presence of pollutant nitrogen oxides and hydrocarbons.

The driving force behind smog formation is the tendency for hydrocarbons to be eliminated from the atmosphere by a number of chemical and photochemical reactions. Starting from relatively innocuous hydrocarbon precursors, these reactions are responsible for the formation of many noxious secondary pollutant products and intermediates that make up photochemical smog. The processes by which this occurs are driven by the natural tendency for the oxygen-rich atmosphere to be oxidizing, particularly through photochemical processes. The oxidation process terminates with formation of CO_2, solid organic particulate matter that settles from the atmosphere, or water-soluble products (for example, acids, aldehydes), which are removed by rain. Inorganic species such as ozone or nitric acid are by-products of these reactions.

Stated succinctly, "The urban atmosphere is a giant chemical reactor in which pollutant gases such as hydrocarbons and oxides of nitrogen and sulfur react under the influence of sunlight to create a variety of products."[3] Although not as great a threat to the global atmosphere as some of the other air pollutants discussed in this chapter, smog does pose significant hazards to living things and materials in local urban areas in which millions of people are exposed.

Overview of Smog Formation

As shown in Figure 12.2, smoggy atmospheres show characteristic variations with time of day in levels of NO, NO_2, hydrocarbons, aldehydes, and oxidants. Examination of the figure shows that, shortly after sunrise, the level of NO in the atmosphere decreases markedly, a decrease that is accompanied by a peak in the concentration of NO_2. During midday (significantly, after the concentration of NO has fallen to a very low level), the levels of aldehydes and oxidants become relatively high. The concentration of total hydrocarbons in the atmosphere peaks sharply in the morning, then decreases during the remaining daylight hours.

Mechanisms of Smog Formation

Figure 12.3 shows the overall reaction scheme for smog formation, which is based upon the photochemically initiated reactions that occur in an atmosphere containing nitrogen oxides, reactive hydrocarbons, and oxygen. The time variations in levels of hydrocarbons, ozone, NO, and NO_2 are explained by the following overall reactions:

1. Primary photochemical reaction producing oxygen atoms:

$$NO_2 + h\nu\ (\lambda < 420\ nm) \rightarrow NO + O \qquad (12.6.1)$$

2. Reactions involving oxygen species (M is an energy-absorbing third body, usually a molecule of N_2 or O_2):

$$O_2 + O + M \rightarrow O_3 + M \qquad (12.6.2)$$

(This reaction produces ozone, O, the single species most characteristic of smog.)

$$O_3 + NO \rightarrow NO_2 + O_2 \qquad (12.6.3)$$

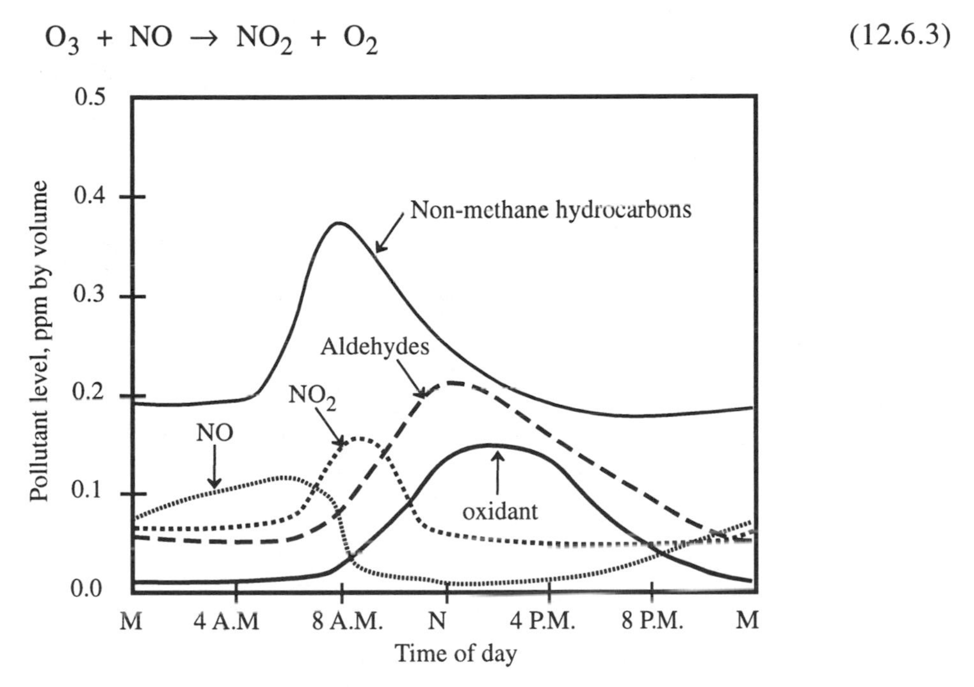

Figure 12.2. Generalized plot of atmospheric concentrations of species involved in smog formation as a function of time of day.

3. Production of organic free radicals from hydrocarbons, RH:

$$O + RH \rightarrow R^{\bullet} + \text{other products} \qquad (12.6.4)$$

$$O_3 + RH \rightarrow R^{\bullet} + \text{and/or other products} \qquad (12.6.5)$$

($R^{\bullet}$ is a free radical that may or may not contain oxygen; the simplest such free radical is the methyl radical, $H_3C^{\bullet}$)

4. Chain propagation, branching, and termination by a variety of reactions such as the following:

$$NO + ROO^{\bullet} \rightarrow NO_2 + \text{and/or other products} \qquad (12.6.6)$$

$$NO_2 + R^{\bullet} \rightarrow \text{products (for example, PAN)} \qquad (12.6.7)$$

The latter kind of reaction is the most common chain-terminating process in smog because NO_2 is a stable free radical present at high concentrations.

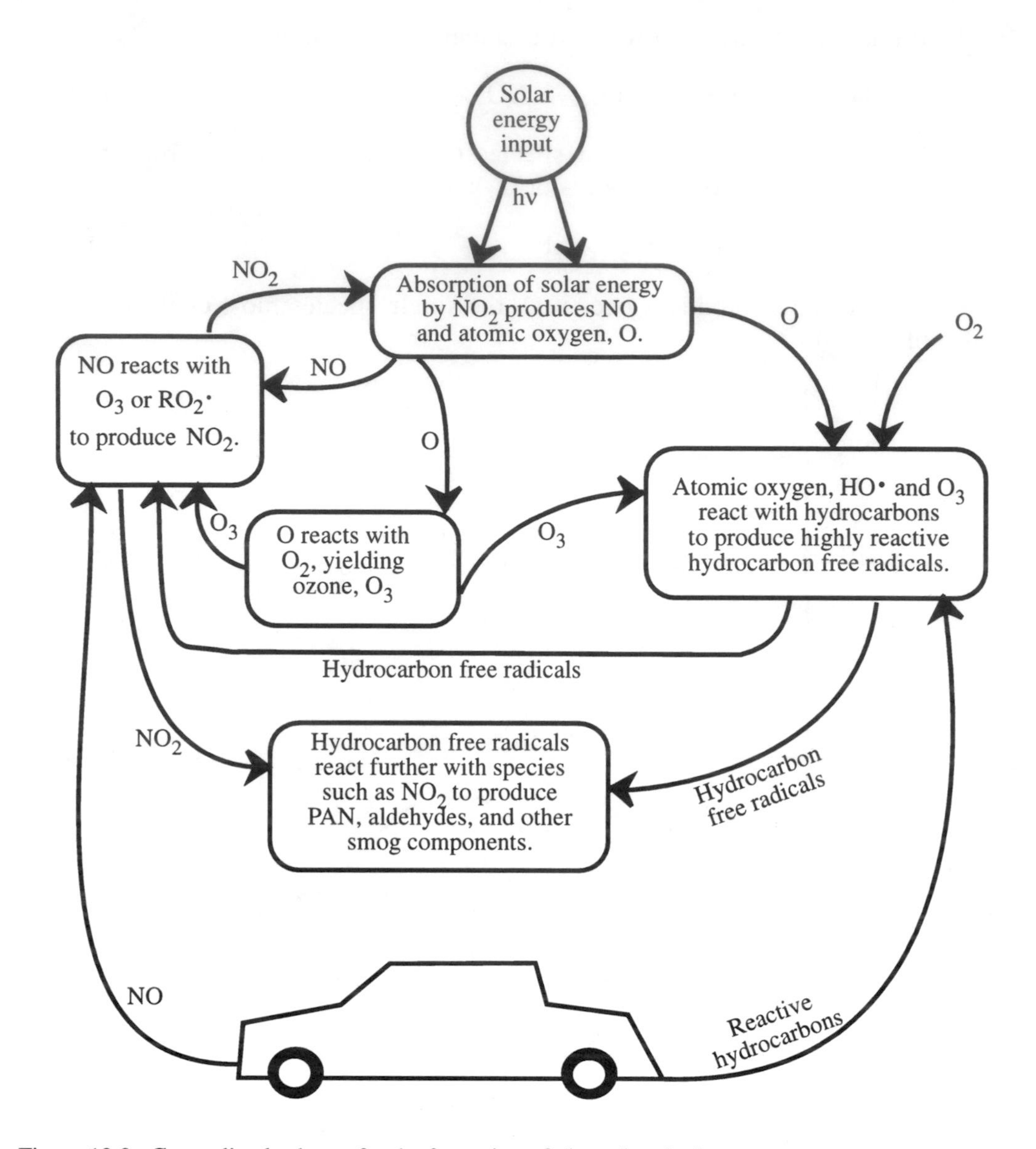

Figure 12.3. Generalized scheme for the formation of photochemical smog.

The most important reactive intermediate species in smog is hydroxyl radical, HO•, generated, for example, by the reaction of atomic oxygen with a hydrocarbon,

$$RH + O \rightarrow R\bullet + HO\bullet \tag{12.6.8}$$

Among the inorganic species with which the hydroxyl radical reacts are oxides of nitrogen,

$$HO\bullet + NO_2 \rightarrow HNO_3 \tag{12.6.9}$$

$$HO\bullet + NO + M \rightarrow HNO_2 + M \tag{12.6.10}$$

(reactions that produce nitric and nitrous acid) and carbon monoxide:

$$CO + HO\bullet + O_2 \rightarrow CO_2 + HOO\bullet \qquad (12.6.11)$$

The last reaction is significant in that it is responsible for the disappearance of most atmospheric CO (see Section 11.3) and because it produces the hydroperoxyl radical $HOO\bullet$, a significant reactive species in the smog-forming process.

The two most significant classes of inorganic products from smog are sulfates and nitrates. Inorganic sulfates and nitrates, along with sulfur and nitrogen oxides can contribute to acidic precipitation, corrosion, reduced visibility, and adverse health effects. Nitric acid formed by chemical processes in a smoggy atmosphere reacts with ammonia in the atmosphere to form ammonium nitrate:

$$NH_3 + HNO_3 \rightarrow NH_4NO_3 \qquad (12.6.12)$$

Other nitrate salts may also be formed.

Nitric acid and nitrates are among the more damaging end products of smog. In addition to possible adverse effects on plants and animals, they cause severe corrosion problems. Electrical relay contacts and small springs associated with electrical switches are especially susceptible to damage from nitrate-induced corrosion.

Organic Species in Smog Formation

Reactions of organic species are particularly important in smog formation. When aliphatic hydrocarbons, RH, react with active oxidants, particularly $HO\bullet$ radical,

$$RH + HO\bullet + O_2 \rightarrow ROO\bullet + H_2O \qquad (12.6.13)$$

reactive oxygenated organic radicals, $ROO\bullet$, are produced., For example, if the hydrocarbon, RH, were methane, the reaction would be the following:

$$CH_4 + HO\bullet + O_2 \rightarrow CH_3OO\bullet + H_2O \qquad (12.6.14)$$

Reactions such as

$$RH + HO\bullet \rightarrow R\bullet + H_2O \qquad (12.6.15)$$

are **abstraction reactions** involving the removal of an atom, usually hydrogen, by reaction with an active species. Abstraction reactions occur mostly on alkanes. Alkenes, which are generally much more reactive than alkanes, undergo **addition reactions**. Typically, the hydroxyl radical reacts with an alkene such as propylene to form another reactive free radical:

$$HO\bullet + H\text{–}\underset{H}{\overset{H}{C}}\text{–}\overset{H}{C}{=}C\langle^{H}_{H} \rightarrow H\text{–}\underset{H}{\overset{H}{C}}\text{–}\underset{H}{\overset{\bullet}{C}}\text{–}\underset{H}{\overset{H}{C}}\text{–}OH \qquad (12.6.16)$$

Ozone adds to unsaturated compounds to form reactive ozonides:

$$O_3 + H\text{–}\underset{H}{\overset{H}{C}}\text{–}\overset{H}{C}{=}C\langle^{H}_{H} \rightarrow H\text{–}\underset{H}{\overset{H}{C}}\text{–}\underset{H}{C}\langle^{OO}_{O}\rangle C\langle^{H}_{H} \qquad (12.6.17)$$

One of the most important reaction sequences in the smog-formation process begins with the abstraction by HO• of a hydrogen atom from a hydrocarbon and leads to the oxidation of NO to NO_2 as follows:

$$RH + HO^\bullet \rightarrow R^\bullet + H_2O \tag{12.6.18}$$

The alkyl radical, R•, reacts with O_2 to produce a peroxyl radical, ROO•:

$$R^\bullet + O_2 \rightarrow ROO^\bullet \tag{12.6.19}$$

This strongly oxidizing species very effectively oxidizes NO to NO_2,

$$ROO^\bullet + NO \rightarrow RO^\bullet + NO_2 \tag{12.6.20}$$

thus explaining the once-puzzling rapid conversion of NO to NO_2 in an atmosphere in which the latter is undergoing photodissociation. The alkoxyl radical product, RO•, is not so stable as ROO•. In cases where the oxygen atom is attached to a carbon atom that is also bonded to H, a carbonyl compound is likely to be formed by the following type of reaction:

$$H_3CO^\bullet + O_2 \rightarrow H{-}\overset{\overset{\displaystyle O}{\|}}{C}{-}H + HOO^\bullet \tag{12.6.21}$$

Aldehydes produced by reactions such as the above are responsible for much of the noxious character of smog, particularly its tendencies to irritate nasal and eye tissue. In addition, the rapid production of photosensitive carbonyl compounds from alkoxyl radicals is an important stimulant for further atmospheric photochemical reactions.

Peroxyacyl nitrates (PAN) are highly significant air pollutants formed by an addition reaction with NO_2:

$$R{-}\overset{\overset{\displaystyle O}{\|}}{C}{-}OO^\bullet + NO_2 \rightarrow R{-}\overset{\overset{\displaystyle O}{\|}}{C}{-}OO{-}NO_2 \tag{12.6.22}$$

When R is the methyl group, the product is peroxyacetyl nitrate, an organic oxidant particularly characteristic of smog. Alkyl nitrates and alkyl nitrites may be formed by the reaction of alkoxyl radicals (RO•) with nitrogen dioxide and nitric oxide, respectively:

$$RO^\bullet + NO_2 \rightarrow RONO_2 \tag{12.6.23}$$

$$RO^\bullet + NO \rightarrow RONO \tag{12.6.24}$$

Addition reactions with NO_2 such as these are important in terminating the reaction chains involved in smog formation.

Effects of Smog

The harmful effects of smog occur mainly in the areas of (1) human health and comfort, (2) damage to materials, (3) effects on the atmosphere, and (4) toxicity to plants. The exact degree to which exposure to smog affects human health is not known, although substantial adverse effects are suspected. Pungent-smelling, smog-produced ozone is known to be toxic. Ozone at 0.15 ppm causes coughing, wheezing, bronchial constriction, and irritation to the respiratory mucous system in healthy, exercising individuals. Other potential health effects of ozone from smog are discussed in Chapter 20. Peroxyacyl nitrates and aldehydes found in smog are eye irritants. Materials are adversely affected by some smog components. Rubber has a high affinity for ozone and is cracked and aged by it. Indeed, the cracking of rubber used to be employed as a test for the presence of ozone.

Even lightly populated nonindustrial areas are subject to the effects of smog brought about by human activities. Particularly, the practice of burning savanna grasses for agricultural purposes causes smog. This burning produces NO_x and reactive hydrocarbons that are required for smog formation. Furthermore, these grasses grow in tropical regions, which have the intense sunlight required for smog formation. The net result is rapid development of smoggy conditions as manifested by ozone levels several times normal background values.

The Urban Aerosol and Acid Fog

The most apparent manifestation of smog is visibility-obscuring **urban aerosol**. Many of the particles composing this aerosol are made from gases by chemical processes and are therefore quite small, usually less than 2 µm.. Particles of such a size are especially harmful because they scatter light most efficiently and are the most respirable. Aerosol particles formed from smog often contain toxic constituents, such as respiratory tract irritants and mutagens. The urban aerosol also contains particle constituents that originate from processes other than smog formation, among which are sulfuric acid droplets, salts, metals, and polycyclic aromatic hydrocarbons. Highly corrosive ammonium salts, such as NH_4HSO_4, are common constituents of urban aerosol particles. Water is always present, even in low humidity atmospheres, and is usually a constituent of urban aerosol particles. Carbon and polycyclic aromatic hydrocarbons from partial combustion and diesel engine emissions are generally abundant constituents, and particulate elemental carbon is usually most responsible for absorbing light in the urban aerosol.

A kind of urban aerosol particulate matter formed under smoggy conditions that is of particular concern is **acid fog**, which may have pH values below 2 due to the presence of H_2SO_4 or HNO_3. This material is part of the acid rain phenomenon discussed in Section 12.4 Acid fog formation occurs because the gas-phase oxidation of SO_2 and NO_x produces strong acids, which form very small aerosol particles. These, in turn, act as condensation nuclei for water vapor. Acid-base phenomena occur in the droplets, and they act as scavengers to remove ionic species from air. Because fog aerosol particles form in areas of intense acid gas pollution near the surface, the concentrations of acids and ionic species in fog aerosol droplets tend to be much higher than in cloud aerosol droplets at higher altitudes.

Effects of Smog on Plants and Crops

In view of worldwide shortages of food, the known harmful effects of smog on plants is of particular concern. These effects are largely due to oxidants in the smoggy atmosphere. The three major oxidants invovled are ozone, PAN, and nitrogen oxides. Of these, PAN has the highest toxicity to plants, attacking younger leaves and causing "bronzing" and "glazing" of their surfaces. Exposure for several hours to an atmosphere containing PAN at a level of only 0.02-0.05 ppm will damage vegetation. The sulfhydryl group of proteins in organisms is susceptible to damage by PAN, which reacts with such groups as both an oxidizing agent and an acetylating agent. Fortunately, PAN is usually present at only low levels. Nitrogen oxides occur at relatively high concentrations during smoggy conditions, but their toxicity to plants is relatively low. The low toxicity of nitrogen oxides and the usually low levels of PAN leave ozone as the greatest smog-produced threat to plant life.

In addition to health effects and damage to materials, one of the greater problems caused by smog is destruction of crops and reduction of crop yields. The annual cost of these effects in California, alone, is about $15 billion. Typical of the phytotoxicity of O_3, ozone damage to a lemon leaf is typified by chlorotic stippling (characteristic yellow spots on a green leaf), as represented in Figure 12.4. Reduction in plant growth may occur without visible lesions on the plant. Brief exposure to approximately 0.06 ppm of ozone may temporarily cut photosynthesis rates in some plants in half. Crop damage from ozone and other photochemical air pollutants in California alone is estimated to cost millions of dollars each year. The geographic distribution of damage to plants in California is illustrated in Figure 12.5.

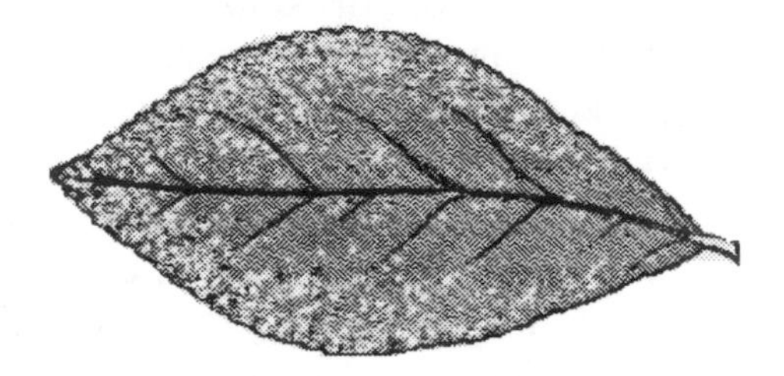

Figure 12.4. Representation of ozone damage to a lemon leaf. In color, the spots appear as yellow chlorotic stippling on the green upper surface caused by ozone exposure.

Figure 12.5. Geographic distribution of plant damage from smog in California.

12.7. NUCLEAR WINTER

Nuclear winter is a term used to describe a catastrophic atmospheric effect that might occur after a massive exchange of nuclear firepower between major powers. The heat from the nuclear blasts and from resulting fires would result in powerful updrafts carrying combustion products to stratospheric regions.[11] The reflection and scattering of sunlight by particles carried into the stratosphere would result in several years of much lower temperatures and freezing temperatures even during summertime,[12] such as occurred in 1816, "the year without a summer," following the astoundingly massive Tambora, Indonesia, volcanic explosion of 1815. Brutally cold years around 210 B.C. that followed a similar volcanic incident in Iceland were recorded in ancient China. In addition to the direct suffering caused, massive starvation would result from crop failures accompanying years of nuclear winter. The incidents cited above clearly illustrate the climatic effects of huge quantities of particulate matter ejected high into the atmosphere.

Evidence exists to suggest that military explosives can result in the introduction of large quantities of particulate matter into the atmosphere. For example, carpet bombings of cities, such as the tragic, militarily pointless fire bombing of Dresden, Germany, near the end of World War II, have produced huge firestorms that created their own wind causing a particle-laden updraft into the atmosphere. Of course, the effect of a full-scale nuclear exchange would be many-fold higher.

An idea of the potential climatic effect resulting from a full-scale nuclear exchange may be obtained by considering the magnitude of the blasts that might be involved. Only two nuclear bombs have been used in warfare, both dropped on cities in Japan in 1945. The Hiroshima fission bomb had the explosive force of 12 kilotons of TNT explosive. Its blast, fireball, and instantaneous emissions of neutrons and gamma radiation, followed by fires and exposure to radioactive fission products, killed about 100,000 people and destroyed the city on which it was dropped. By comparison with this 12 kiloton bomb, modern fusion bombs are typically rated at 500 kilotons, and 10 megaton weapons are common. A full-scale nuclear exchange might involve a total of the order of 5,000 megatons of nuclear explosives. As a result, unimaginable quantities of soot from the partial combustion of wood, plastics, paving asphalt, petroleum, forests, and other combustibles would be carried to the stratosphere. At such high altitudes tropospheric removal mechanisms (see Figure 9.8) are not effective because there is not enough water in the stratosphere to produce rainfall to wash particles from the air and convection processes are very limited. Much of the particulate matter would be in the μm size range in which light is reflected, scattered, and absorbed most effectively and settling is very slow. Therefore, vast areas of the Earth would be overlain by a stable cloud of particles and the fraction of sunlight reaching the Earth's surface would be drastically reduced, resulting in a dramatic cooling effect. There would be other effects as well. The extreme heat and pressure in the fireball would result in the fixation of nitrogen as ozone-destroying nitrogen oxides:

$$O_2 + N_2 \rightarrow 2NO \tag{12.7.1}$$

The timing and location of nuclear blasts are very important in determining their climatic effects. Atmospheric testing of nuclear weapons, including a 58-megaton monster detonated by the Soviet Union, has had little atmospheric effect. Such tests were carried out at widely spaced intervals on deserts, small tropical islands, and other places with little combustible matter. In contrast, military use of nuclear weapons would involve a high concentration of firepower, both in time and in space, on industrial and military targets consisting largely of combustibles. Furthermore, destruction of hardened military sites requires blasts that disrupt large quantities of soil, rock, and concrete, which are pulverized, vaporized, and blown into the atmosphere.

On a hopeful note, the East-West conflict that has dominated world politics and threatened nuclear war since the mid-1900s has now abated and the probability of nuclear warfare seems to have diminished. However, the outbreak of war in the Middle East in 1991, vicious ethnic conflicts in the former Yugoslavia in 1993, nuclear proliferation, disintegration and fragmentation of authority in great powers with vast nuclear arsenals, racial hatred, and a "trigger-happy" state of mind among even educated people who should know better should still give us cause for concern in respect to the prospect of "nuclear winter."

12.8. VISITORS FROM SPACE

Of all the possible atmospheric catastrophes that can occur, arguably the most threatening would be one caused by collision of a large asteroid with Earth. Convincing evidence now exists that mass extinctions of species in the past have resulted from Earth being hit by asteroids several kilometers in diameter. Such an event would cause much the same effects as those from "nuclear winter" described in the preceding section, though with a large asteroid the effects would be much more pronounced.

The possible effects of space objects hitting Earth are the subject of much current investigation. In the mid-1980s, Louis A. Frank, a space physicist from the University of Iowa, postulated a continuous rain of house-sized "snowballs" entering Earth's atmosphere from space to explain the observation of spots on images of the atmosphere taken from the *Dynamic Explorer 1* spacecraft. In 1997 the U.S. National Aeronautics and Space Administration announced observation of hydroxyl-radical-rich paths high in the atmosphere observed by the far-ultraviolet Earth Camera aboard NASA's *Polar* spacecraft clearly indicating paths in which mini-comets composed of ice have hit the atmosphere and vaporized several thousands of kilometers above Earth's surface.[13] Characterized by Frank as a "gentle cosmic rain," much different from "killer asteroids," these visitors from space have caused much excitement among atmospheric scientists. The frequency of these collisions is such that it is possible that very large cosmic "snowballs" (or very small comets) may have been the source of much of Earth's water and that organic matter carried by them may have been involved in the origin of life.

12.9. WHAT IS TO BE DONE?

Of all environmental hazards, there is little doubt that major disruptions in the atmosphere and climate have the greatest potential for catastrophic and irreversible environmental damage. If levels of greenhouse gases and reactive trace gases continue to increase at present rates, significant environmental effects are virtually certain. On a hopeful note, the bulk of these emissions arise from industrialized nations, which—in principle—can apply the resources needed to reduce them substantially. The best example to date has been the 1987 "Montreal Protocol on Substances that Deplete the Ozone Layer," an international treaty through which a large number of nations agreed to cut chlorofluorocarbon emissions by 50% by the year 2000, a goal that may even be exceeded. This agreement may pave the way for more encompassing agreements covering carbon dioxide and other trace gases.

More ominous, however, is the combination of population pressure and desire for better living standards on a global basis. Consider, for example, the demand that these two factors place on energy resources, and the environmental disruption that may result. In many densely populated developing nations, high-sulfur coal is the most accessible, cheapest source of energy. It is understandably difficult to persuade populations faced with real hunger to forego short-term economic gain for the sake of long-term environmental quality. Destruction of rain forests by "slash-and-burn" agricultural methods does make economic sense to those engaged in subsistence farming to obtain badly needed hard currency, which can be earned by converting forest to pasture land and exporting fast-food-hamburger beef to wealthier nations, but in the overall context of global environmental economics, such policies are tragically shortsighted.

An interesting phenomenon in recent years has been the increase in anti-environmental rhetoric by elements suggesting that environmental problems, especially those that might pose threats to the global atmosphere, have been grossly exaggerated. Such claims and the rebuttals to them have been summarized in an article in *Chemical and Engineering News.*[14] This article points out that a ranking U.S. congressman has called warnings of global warming "liberal claptrap," other parties have attributed the Montreal protocol to a conspiracy involving the U.S. National Aeronautics and Space Administration in league with large chemical companies, and at least one conservative economist has concluded that because money must be spend to eliminate CFCs, the science used to justify their curtailment must be wrong.

Some of the most vehement criticism of environmentalism has centered on the threat of CFCs to stratospheric ozone. Invoking "junk science," some critics have contended that the relatively heavy CFC molecules cannot rise into the stratosphere, that the loss of ozone has been greatly exaggerated and arises entirely from natural causes, and that, if chlorine is even involved in ozone loss, it comes from sea spray and volcanic eruptions.

Present day anti-environmentalism has been attributed to backlash against "green" policies and politics—some of which admittedly have been extreme and unreasoned—and has been given the name of "brownlash" in a book by Ehrlich and Ehrlich.[15] This book considers and rebuts a wide range of "brownlash" rhetoric in environmental areas including mineral, food, and energy resources; human population; biological issues, including endangered species and biological diversity; water pollution; air pollution; global climate change; ozone depletion; exposure to toxic substances; and hazardous wastes. It also addresses political issues pertaining to the environment, the role of the media, and environmental economics. Economics and ideology appear to be the driving forces behind much of the ongoing "brownlash." The search for true answers to environmental issues has been aggravated by the tendency of organizations on all sides to hire their own environmental experts and to use this expertise to advocate particular viewpoints, rather than seeking and disseminating accurate knowledge of environmental issues.

Serious Concern Over Changes in Climate

As outlined in an article entitled "Storm Warnings Rattle Insurers,"[16] insurance companies have become quite concerned about the possibility of significant changes in global climate, especially because of potential effects on the frequency and severity of damaging storms. In 1996-97 there were at least six weather disasters that cost over a billion dollars each. These included (1) a catastrophic drought in the Southern Plains that began in the fall of 1995 and lasted through the summer of 1996; (2) a blizzard followed by flooding that occurred in the northeastern U.S., the mid-Atlantic states, and the Appalachian mountain areas in January of 1996; (3) flooding in the Pacific Northwest in February, 1996; (4) Hurricane Fran, which caused 36 deaths and over $5 billion in damage during September, 1996; (5) severe flooding in the northern west coast region in December, 1996, and January, 1997; and an unprecedented 500-year flood complicated by freezing weather and ice jams that hit the Dakotas and Minnesota in April, 1997, virtually wiping out the city of Grand Forks, North Dakota.

Although, as discussed in Sections 12.2 and 12.3 of this chapter, drought is the most frequently mentioned possible effect of greenhouse warming, the frequency and severity of storms, often accompanied by high levels of precipitation, have the insurance companies particularly concerned. During the 1990s "100-year" weather events, those that are expected to occur statistically only once each century, have become so common in the U.S. that the term has begun to lose its meaning. As shown in Figure 12.6, the last century has seen a significant increase in precipitation in the lower 48 continental United States. These observations are consistent with currently accepted models of the weather effects of greenhouse warming, which predict that more precipitation will come in the form of brief, heavy precipitation events, such as thunderstorms (heavy convective storms) rather than through gentle rainfall that comes over a longer time period. The debate continues over whether the apparent weather anomalies observed during recent years denote a marked change in climate or are simply normal fluctuations in weather. However, the insurance companies, whose prosperity and even survival depend upon accurate statistical analysis of often subtle risk factors, seem to be concerned that human effects on climate are real and that the resulting losses may be substantial.

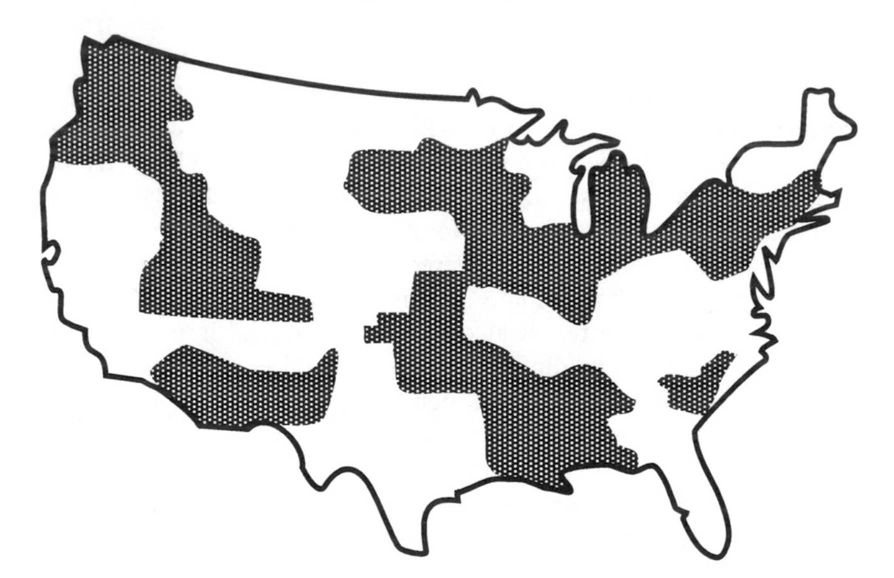

Figure 12.6. Sections the lower 48 United States in which precipitation levels have increased by 10-20% since about 1900 (shown as shaded regions). Some areas, particularly North Dakota, eastern Montana, Wyoming, and California have experienced decreases in precipitation of a similar magnitude. This map is based on data gathered by the National Oceanic and Atmospheric Administration's National Climatic Data Center.

Prudent Measures

What is to be done? First of all, it is important to keep in mind that the atmosphere has a strong ability to cleanse itself of pollutant species. Fine particulate matter and water-soluble gases, including greenhouse-gas CO_2, and acid-gas SO_2, are removed with precipitation. For most gaseous contaminants, oxidation precedes or accompanies removal processes. To a degree, oxidation is carried out by O_3. To a larger extent, the most active atmospheric oxidant is hydroxyl radical, HO•. This atmospheric scavenger species reacts with all important trace gasses, except for CO_2 and chlorofluorocarbons. It is now generally recognized that HO• is an almost universal atmospheric cleansing agent. Given this crucial role of HO• radical, any pollutants that substantially reduce its concentration in the atmosphere are potentially troublesome. A concern over CO emissions to the atmosphere is the reactivity of HO• with CO which could result in removal of HO• from the atmosphere:

$$CO + HO^{\bullet} \rightarrow CO_2 + H \qquad (12.8.1)$$

Of all the big threats to the global climate, it is virtually certain that humankind will have to try to cope with greenhouse warming and the climatic effects thereof, such as drought and increased frequency and severity of storms. Measures to be taken in dealing with this problem fall into the three following categories:

- **Minimization** by reducing emissions of greenhouse gases, switching to alternate energy sources, increasing energy conservation, and reversing deforestation. It is especially sensible to use measures that have major benefits in addition to reduction of greenhouse warming. Such measures include, as examples, reforestation, restoration of grasslands, increased energy conservation, and a massive shift to solar energy sources.
- **Counteracting measures**, such as injecting light-reflecting particles into the upper atmosphere.
- **Adaptation**, particularly through increased efficiency and flexibility of the distribution and use of water, which might be in very short supply in many parts of the world as a consequence of greenhouse warming. Important examples are implementation of more efficient irrigation practices and changes in agriculture to grow crops that require less irrigation. Emphasis on adaptation is favored by those who contend that not enough is known about the types and severity of global warming to justify massive expenditures on minimization and counteractive measures. In any case, adaptation will certainly have to be employed as a means of coping with global warming.

A common measure taken against the effects of another atmospheric hazard, ultraviolet radiation, provides an example of adaptation. This measure is the use of sunscreens placed on the skin as lotions to filter out UV-B radiation. The active ingredient of sunscreen must absorb ultraviolet light effectively. But this is not enough because it is the absorption of photons of ultraviolet radiation by skin that makes it so dangerous in the first place. Therefore, active compounds in sunscreen formulations must also dissipate the absorbed energy in a harmless way. This is accomplished by an organic compound called *o*-hydroxybenzophenone contained in sunscreens, which absorbs ultraviolet energy, rearranges to accomodate the extra energy introduced into it by absorbing the ultraviolet photon, then dissipates the energy to its surroundings, reverting back to its original molecular form.

The "**tie-in strategy**" has been proposed as a sensible approach to dealing with the kinds of global environmental problems discussed in this chapter. Dating back to the early 1980s, this approach advocates taking measures consisting of "high-leverage actions" which are designed to prevent problems from occurring and which have substantial merit even if the major problems that they are designed to avoid do not materialize. An example is implementation of environmentally sound substitutes for fossil fuels to lower atmospheric CO_2 output and prevent greenhouse warming. Even if it turns out that the greenhouse effect is exaggerated, such substitutes would save the Earth from other kinds of environmental damage, such as disruption of land by strip mining coal or preventing oil spills from petroleum transport. Definite economic and political benefits would also accrue from lessened dependence on uncertain, volatile petroleum supplies. Increased energy efficiency would diminish both greenhouse gas and acid rain production, while lowering costs of production and reducing the need for expensive and environmentally disruptive new power plants. The implementation of these kinds of tie-in strategies requires some degree of incentive beyond normal market forces, and, therefore, is opposed by some on ideological grounds. A good example is opposition to mandatory fuel mileage standards for automobiles, which many view as unjustified interference with their personal freedom to have as large and wasteful a vehicle as their finances can stand. However, to quote Schneider,[6] "a market that does not include the costs of environmental disruptions can hardly be called a free market."

International Efforts in Environmental Protection

In a speech given during a visit to Costa Rica's Braulio Carillo National Park on May 9, 1997, U.S. President Clinton praised Cost Rica's environmental stewardship, which has been exemplary for a small, less industrialized nation. Costa Rica has protected approximately one-fourth of its area for nature preserves, much of it in the form of irreplaceable rain forest. The preservation of rain forests along with their photosynthetic carbon dioxide fixing capacity is one of the most effective measures that can be taken to reduce the potential for greenhouse warming. Among its other environmental accomplishments, Costa Rica gets about 85% of its electricity from renewable resources. The rich plant life in Costa Rica's preserved areas provides a significant source of natural pharmaceutical agents, and the country sells licensing rights to pharmaceutical companies to do "biological prospecting" for new drugs. The country's efforts at environmental protection have been rewarded economically in that tourism largely drawn by nature preserves has now become Costa Rica's largest source of income.

CHAPTER SUMMARY

The chapter summary below is presented in a programmed format to review the main points covered in this chapter. It is used most effectively by filling in the blanks, referring back to the chapter as necessary. The correct answers are given at the end of the summary.

The idea that the atmospheric O_2/CO_2 balance established and maintained by organisms determines and maintains Earth's climate and other environmental conditions is called the [1] ______________________________. Activities of humans during the last 200 years that have altered the atmosphere significantly include [2]__________

____

____.

Major effects that have resulted from these activities include [3]__________________

____

____. The greenhouse effect results from reabsorption of Earth's [4]________________________________

____.

The gas most commonly regarded as a greenhouse gas is [5]__________________. A potential greenhouse gas emitted by the anaerobic biodegradation of organic matter in rice paddies is [6]____________________. Most computer models predict global warming of about [7]____________________ over the next several decades. The condition from greenhouse warming likely to do most harm is [8]__________ __________, which in turn could lead to the almost permanent loss of productive land through the process of [9]__________________________. Precipitation made acidic by the presence of acids stronger than $CO_2(aq)$ is commonly called [10]__________ __________, which is part of the more general phenomenon of [11]__________ ____________________. Rather than being produced directly most acid rain is a [12]____________________. The two most common precursors to acid rain are [13]____. Among the most damaging effects of acid rain are [14]______________________________________

____.

The overall reactions by which stratospheric ozone is destroyed are [15]__________

____.

Ozone absorbs ultraviolet radiation very strongly in the region [16]______________. Destruction of stratospheric ozone would result in increased intensities of [17]______ ______________________________. The greatest threat to the stratospheric ozone layer has been from [18]____, which begin the process of ozone destruction starting with the photochemical reaction [19]____. The three ingredients required to generate photochemical smog are [20]______________

____.

"A giant chemical reactor in which pollutant gases such as hydrocarbons and oxides of nitrogen and sulfur react under the influence of sunlight to create a variety of products" describes the [21]______________________________. During midday under smog-forming conditions the levels of [22]__________________________ in the atmosphere become relatively high. The photochemical reaction that initiates smog formation is [23]____. The single species most characteristic of smog is [24]______________________. Two types of reactions that hydroxyl radical undergoes with hydrocarbons are [25]______

____.

The four areas in which photochemical smog has its most harmful effects are [26]____

____

____. The catastrophic atmospheric effect that might occur after a massive exchange of nuclear firepower between major powers has been termed [27]______________________________________. The most convincing evidence that this phenomenon might actually occur is provided by

historical records of [28]____.
The measures to be taken in dealing with greenhouse warming and the effects of it fall into the categories of [29]____. A strategy that advocates taking measures consisting of "high-leverage actions" which are designed to prevent problems, such as those from greenhouse warming from occurring and which have substantial merit even if the major problems that they are designed to avoid do not materialize is called [30]____.

Answers

[1] Gaia hypothesis

[2] industrial activities, burning of large quantities of fossil fuel, transportation practices and their atmosperic emissions, alteration of land surfaces, burning of biomass and vegetation, and agricultural practices which produce methane

[3] increased acidity in the atmosphere, production of pollutant oxidants, elevated levels of greenhouse gases, threats to the ozone layer in the stratosphere, and increased corrosion of materials

[4] outgoing energy in the form of infrared radiation

[5] carbon dioxide

[6] methane

[7] 3.0°C-5.5°C

[8] drought

[9] desertification

[10] acid rain

[11] acid deposition

[12] secondary pollutant

[13] SO_2 and NO_2

[14] phytotoxicity, destruction of sensitive forests, respiratory effects, acidification of lake water, and corrosion

[15] $O_3 + h\nu \rightarrow O_2 + O$ and $O + O_3 \rightarrow 2O_2$

[16] 220-330 nm

[17] ground-level ultraviolet radiation

[18] chlorofluorocarbon (CFC) compounds

[19] $CF_2Cl_2 + h\nu \rightarrow Cl\cdot + CClF_2\cdot$ (photodissociation of atomic Cl)

[20] ultraviolet light, reactive hydrocarbons, and nitrogen oxides

[21] urban atmosphere

[22] aldehydes and oxidants

23 $NO_2 + h\nu \ (\lambda < 420 \text{ nm}) \rightarrow NO + O$

24 ozone, O_3

25 abstraction and addition

26 human health and comfort, damage to materials, effects on the atmosphere, and toxicity to plants

27 nuclear winter

28 massive volcanic eruptions

29 minimization, counteracting measures, and adaptation

30 a tie-in strategy

QUESTIONS AND PROBLEMS

1. How do modern transportation problems contribute to the kinds of atmospheric problems discussed in this chapter?
2. What is the rationale for classifying most acid rain as a secondary pollutant?
3. Distinguish among UV-A, UV-B, and UV-C radiation. Why does UV-B pose the greatest danger in the troposphere?
4. How does the extreme cold of stratospheric clouds in Antarctic regions contribute to the Antarctic ozone hole?
5. How does the oxidizing nature of ozone from smog contribute to the damage that it does to cell membranes?
6. What may be said about the time and place of the occurrence of maximum ozone levels from smog with respect to the origin of the primary pollutants that result in smog formation?
7. What is the basis for "nuclear winter"?
8. What is meant by a "tie-in strategy"?
9. List two ways in which modern agricultural practices contribute to the production of atmospheric methane.
10. Describe how humans have been conducting a "massive geophysical experiment" with Earth.
11. Describe how cloud formation may exercise a degree of self-correction on the greenhouse effect.
12. Explain how acid rain is a regional air pollution problem compared to a local or global problem.
13. What is phytotoxicity? Give an example of indirect phytotoxicity from acid rain.
14. What reactive species is produced from chlorofluorocarbons that reacts with stratospheric ozone? What is the reaction? How does it lead to ozone layer destruction?

15. What are the conditions that lead to the formation of photochemical smog? How is photochemical smog manifested?

16. Other than ozone, what are two major inorganic products from smog? What are their effects?

LITERATURE CITED

[1] Graedel, Thomas E. and Paul J. Crutzen, "The Changing Atmosphere," in *Managing Planet Earth*, special issue of *Scientific American*, September, 1989, pp. 58-68.

[2] Revelle, Roger and Hans E. Suess, "Carbon Dioxide Exchange Between Atmosphere and Ocean and the Question of an Increase of Atmospheric CO during the Past Decades," *Tellus*, **9**, 18 (1957).

[3] Seinfeld, John H., "Urban Air Pollution: State of the Science," *Science*, **243**, February 10, 1989, pp. 745-752.

[4] Schwartz, Stephen E., "Acid Deposition: Unraveling a Regional Phenomenon," *Science*, **243**, February 10, 1989, pp. 753-763.

[5] McElroy, Michael B. and Ross J. Salawitch, "Changing Composition of the Global Stratosphere," *Science*, **243**, February 10, 1989, pp. 763-770.

[6] Schneider, Stephen H., "The Greenhouse Effect: Science and Policy," *Science*, **243**, February 10, 1989, pp. 751-781.

[7] Schneider, Stephen, H., "The Changing Climate," in *Managing Planet Earth*, special issue of *Scientific American*, September, 1989, pp. 70-79.

[8] Pearce, F., "Methane, the Hidden Greenhouse Gas," *New Scientist*, May 6, 1989, pp. 37-41.

[9] Khalil, M. A. K. and R. A. Rasmussen, "Atmospheric Methane: Recent Global Trends," *Environmental Science and Technology*, **24**, 549-553 (1991).

[10] Stolarski, Richard S. and Paul J. Crutzen, "The Antarctic Ozone Hole," *Scientific American*, January, 1988.

[11] Crutzen, P. J. and J. W. Birks, "The Atmosphere after a Nuclear War: Twilight at Noon," *Ambio*, **11**, 114-125 (1982).

[12] Sagan, Carl and Richard Turco, *A Path Where No Man Thought: Nuclear Winter and the End of the Arms Race*, Random House, New York, NY, 1990.

[13] Wilson, Elizabeth, "Minicomets Produce 'Gentle Cosmic Rain,'" *Chemical and Engineering News*, June 2, 1997, p. 7.

[14] Cicerone, Ralph J., "Antienvironmental Backlash," *Chemical and Engineering News*, April 7, 1997, pp. 53-5.

[15] Ehrlich, Paul R. and Anne H. Ehrlich, *Betrayal of Science and Reason: How Anti-Environmental Rhetoric Threatens Our Future*, Island Press, Washington, DC, 1996.

[16] Hileman, Bette, "Storm Warnings Rattle Insurers," *Chemical and Engineering News*, April 14, 1997, pp. 28-31.

SUPPLEMENTARY REFERENCES

Armentrout, Patricia, *The Ozone Layer*, Rourke Publishing Group, New York, NY, 1997.

Bazzaz, Fakhri and William Sombrook, *Global Climate Change and Agricultural Production*, John Wiley & Sons, Somerset, NJ, 1996.

Brower, Michael, *Cool Energy: The Renewable Solution to Global Warming*, Union of Concerned Scientists, Cambridge, MA, 1989.

Brown, Paul, *Global Warming: Can Civilization Survive?*, Blandford Press, New York, NY, 1997.

Cook, Elizabeth, *Ozone Protection in the United States: Elements of Success*, World Resources Institute, Washington, D.C., 1996.

Edmonds, Alex, *Closer Look at Acid Rain*, Copper Beech Books, New York, NY, 1997.

Edmonds, Alex, Selina Wood, Atessa Barwidck, and Edgar-Hyde, *The Ozone Hole (Closer Look At)*, Copper Beech Books, New York, NY, 1997.

Kemp, David D., *Global Environmental Issues: A Climatological Approach*, Routledge, Chapman and Hall, Inc., New York, NY, 1990.

Klaassen, Ger, *Acid Rain and Environmental Degradation: The Economics of Emission Trading*, Edward Elgar Publishing Co., New York, NY, 1996.

Le Bras, Georges, Ed., *Chemical Processes in Atmospheric Oxidation: Laboratory Studies of Chemistry Related to Tropospheric Ozone*, Springer Verlag, New York, NY, 1997.

Lyman, Francesca, *The Greenhouse Trap*, Beacon Press, Boston, MA, 1990.

Mabey, Nick, Stephen Hall, Clare Smith, and Sujata Gupta, *Argument in the Greenhouse: The International Economics of Controlling Global Warming*, Routledge, London and New York, 1997.

Moomaw, William R. and Irfint M. Mintzer, *Strategies for Limiting Global Climate Change*, World Resources Institute Publications, Washington, D.C., 1989.

Nilsson, Annika, *Ultraviolet Reflections: Life Under a Thinning Ozone Layer*, John Wiley & Sons, Somerset, NJ, 1996

Oppenheimer, Michael and Robert H. Boyle, *Dead Heat: The Race Against the Greenhouse Effect*, Basic Books, New York, NY, 1990.

Park, Chris C., *Chernobyl: The Long Shadow*, Routledge, London and New York, 1989.

Paterson, Matthew, *Global Warming and Global Politics (Environmental Politics)*, Routledge, London and New York, 1996.

Ramakrishna, Kilaparti, *Global Warming and International Law: A Review of Strategies*, Island Press, New York, NY, 1997.

Ray, Dixy Lee, Lou Guzzo, and Jeff Riggenbach, *Trashing the Planet: How Science Can Help Us Deal With Acid Rain, Depletion of the Ozone and Nuclear Waste*, (audio tape), Blackstone Audio Books, New York, NY, 1997.

Roleff, Tamara L., Scott Barbour, and Karin Swisher, Eds., *Global Warming: Opposing Viewpoints*, Greenhaven Press, New York, NY, 1997.

Samuel, R., *Global Warming and the Built Environment*, Chapman & Hall, London, UK, 1996.

Seinfeld, John H., *Atmospheric Chemistry and Physics of Air Pollution*, John Wiley and Sons, Inc., Somerset, NJ, 1986.

Watts, Robert G., *Engineering Response to Global Climate Change*, CRC Press/Lewis Publishers, Boca Raton, FL, 1997.

White, James C., Ed., *Acid Rain: The Relationship Between Sources and Receptors*, Elsevier, New York, NY, 1988.

EARTH

13 THE GEOSPHERE

13.1. INTRODUCTION

The **geosphere**, or solid Earth, is that part of the Earth upon which humans live and from which they extract most of their food, minerals, and fuels.[1] Once thought to have an almost unlimited buffering capacity against the perturbations of humankind, the geosphere is now known to be rather fragile and subject to harm by human activities. For example, some billions of tons of Earth material are mined or otherwise disturbed each year in the extraction of minerals and coal. Two atmospheric pollutant phenomena—excess carbon dioxide and acid rain (see Chapter 12)—have the potential to cause major changes in the geosphere. Too much carbon dioxide in the atmosphere may cause global heating ("greenhouse effect"), which could significantly alter rainfall patterns and turn currently productive areas of the Earth into desert regions. The low-pH characteristic of acid rain can bring about drastic changes in the solubilities and oxidation-reduction rates of minerals. Erosion caused by intensive cultivation of land is washing away vast quantities of topsoil from fertile farmlands each year. In some areas of industrialized countries, the geosphere has been the dumping ground for toxic chemicals (see the discussion of hazardous wastes in Chapters 23 and 24). Ultimately, the geosphere must provide disposal sites for the nuclear wastes of the more than 400 nuclear reactors now operating worldwide. It may be readily seen that the preservation of the geosphere in a form suitable for human habitation is one of the greatest challenges facing humankind.

Environmental geology deals with the relationship of the geosphere to the other environmental spheres that it influences and is influenced by, including humankind and its technology.[2] Included in environmental geology are the following:

- Ways in which human activities and technology impact the geosphere, and the manner in which such impacts may be minimized or be made beneficial.
- Utilization of resources from the geosphere, such as minerals, fossil fuels, groundwater, and rock.
- Evaluation, prediction, and minimization of natural hazards, including earthquakes, landslides, floods, and volcanoes.

A particularly important, pertinent, and controversial aspect of environmental geology is that of **land use** consisting of the ways in which land is employed for the purposes of humankind. The importance of proper land use can be seen from the numerous examples of land misuse. These include in modern times such abuses as shopping centers and residential developments constructed on prime agricultural land, poorly restored strip mines, and loss of topsoil from improperly cultivated farmland. Incidents of land misuse go back to ancient times. Biblical accounts of lands that abounded with crops and vinyards described areas in present-day Syria, Lebanon, and Palestine that were once thriving with vegetation, livestock, and renewable resources. Many of these areas have lost their productive capacity because of misuse of the land, erosion, overgrazing, and poor agricultural practices. Agricultural and other technologies are now being used with some success, though at great cost, to restore some of these areas to productivity. A major challenge of land use planning is to utilize the principles of environmental geology to minimize such abuses in the future.

Earth Science

The geosphere interfaces intimately with the hydrosphere and the atmosphere (Figure 13.1). **Earth science** considers these three closely related spheres and how they influence each other. The study of water as it interacts with the solid Earth may be divided among hydrology, pertaining to nonoceanic liquid water above and below ground; glaciology, dealing with ice and snowpack on Earth's surface; and oceanography. Meteorology addesses phenomena in the atmosphere, and the science of climate is climatology. Geology deals with the solid earth as a whole. As covered later in this chapter, geology is itself divided into several major categories.

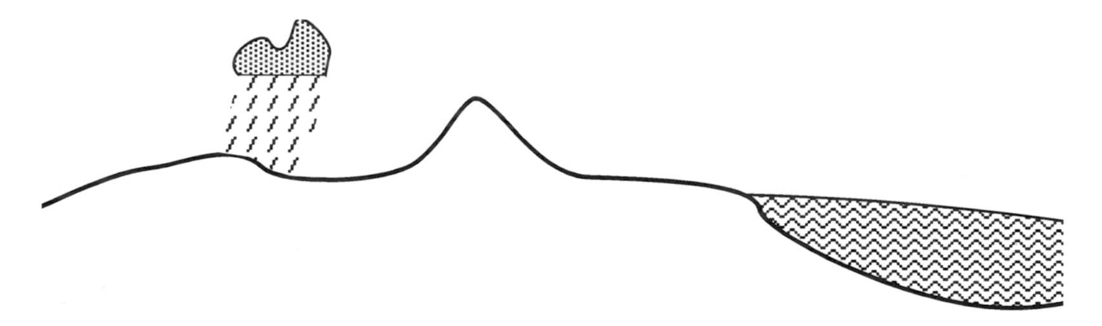

Figure 13.1. The geosphere has a very close relationship with both the hydrosphere and the atmosphere.

The interface between the geosphere and the atmosphere at Earth's surface is very important to the environment. Human activities on the Earth's surface may affect climate, most directly through the change of surface albedo, defined as the percentage of incident solar radiation reflected by a land or water surface. Whereas the albedo of fresh snow is 85-90%, that of a surface paved with asphalt is only 8%. In some heavily developed areas, anthropogenic (human-produced) heat release is comparable to the solar input. The anthropogenic energy release over the 60 square kilometers of Manhattan Island averages about 4 times the solar energy falling on the area; over the 3,500 km^2 of Los Angeles the anthropogenic energy release is about 13% of the solar flux.

One of the greatest impacts of humans upon the geosphere is the creation of desert areas through abuse of land that receives only marginal amounts of rainfall. This process, called **desertification** (Section 12.3) is manifested by declining groundwater tables, salinization of topsoil and water, reduction of surface waters, unnaturally high soil erosion, and desolation of native vegetation. The problem is severe in some parts of the world, particularly Africa's Sahel (southern rim of the Sahara), where the Sahara advanced southward at a particularly rapid rate during the period 1968-73, contributing to widespread starvation in Africa during the 1980s. Large, arid areas of the western U.S. are experiencing at least some desertification as the result of human activities combined with some recent severe droughts. As the populations of the Western states increase, one of the greatest challenges facing the residents is to prevent additional conversion of land to desert.

The most important part of the geosphere for life on earth is soil. It is the medium upon which plants grow, and virtually all terrestrial organisms depend upon it for their existence. The productivity of soil is strongly affected by environmental conditions and pollutants. Because of the importance of soil, all of Chapter 14 is devoted to it.

With increasing population and industrialization, one of the more important aspects of human use of the geosphere has to do with the protection of water sources. Mining, agricultural, chemical, and radioactive wastes all have the potential for contaminating both surface water and groundwater. Sewage sludge spread on land may contaminate water by release of nitrate and heavy metals. Landfills may likewise be sources of contamination. Leachates from unlined pits and lagoons containing hazardous liquids or sludges may pollute drinking water.

It should be noted that many soils have the ability to assimilate and neutralize pollutants. Various chemical and biochemical phenomena in soils, including acid-base reactions, oxidation reduction processes, hydrolysis, precipitation, sorption, and biochemical degradation, operate to reduce the harmful nature of pollutants. Some hazardous organic chemicals may be degraded to harmless products on soil, and heavy metals may be sorbed by it. In general, however, extreme care should be exercised in disposing of chemicals, sludges, and other potentially hazardous materials on soil, particularly where the possibility of water contamination exists.

13.2. THE NATURE OF SOLIDS IN THE GEOSPHERE

The earth is divided into layers, including the solid iron-rich inner core, molten outer core, mantle, and crust. Environmental science is most concerned with the **lithosphere**, which consists of the outer mantle and the **crust**. The latter is the earth's outer skin that is accessible to humans. It is extremely thin compared to the diameter of the earth, ranging from 5 to 40 km thick.

Most of the solid earth crust consists of rocks. Rocks are composed of minerals, where a **mineral** is a naturally occurring inorganic solid with a definite internal crystal structure and chemical composition. A **rock** is a solid, cohesive mass of pure mineral or an aggregate of two or more minerals.

Igneous, Sedimentary, and Metamorphic Rock

At elevated temperatures deep beneath Earth's surface, rocks and mineral matter melt to produce a molten substance called **magma**. Cooling and solidification of magma produces **igneous rock**. Common igneous rocks are granite, basalt, quartz (SiO_2), feldspar ($(Ca,Na,K)AlSi_3O_8$), and magnetite (Fe_3O_4). Exposure of igneous rocks formed under water-deficient, chemically reducing conditions of high temperature and high pressure to wet, oxidizing, low-temperature and low-pressure conditions at the surface causes the rocks to disintegrate by a process called **weathering**. Weathering of igneous rocks is slow because they tend to be hard with a low porosity and low reactivity.

Erosion from wind, water, or glaciers picks up materials from weathering rocks and deposits it as **sediments** or **soil**. A process called **lithification** describes the conversion of sediments to **sedimentary rocks**. In contrast to the parent igneous rocks, sediments and sedimentary rocks are porous, soft, and chemically reactive. Sedimentary rocks may be **detrital rocks** consisting of solid particles eroded from igneous rocks as a consequence of weathering; quartz is the most likely to survive weathering and transport from its original location chemically intact. A second kind of sedimentary rocks consists of **chemical sedimentary rocks** produced by the precipitation or coagulation of dissolved or colloidal weathering products. **Organic sedimentary rocks** contain residues of plant and animal remains. Carbonate minerals of calcium and magnesium—**limestone** or **dolomite** —are especially abundant in sedimentary rocks. **Metamorphic rock** is formed by the action of heat and pressure on sedimentary, igneous, or other kinds of metamorphic rock that are not in a molten state.

Rock Cycle

The interchanges and conversions among igneous, sedimentary, and metamorphic rocks, as well as the processes involved therein, are described by the **rock cycle**. A rock of any of these three types may be changed to a rock of any other type. Or a rock of any of these three kinds may be changed to a different rock of the same general type in the rock cycle. The rock cycle is illustrated in Figure 13.2.

Petrology

The study of the origin, evolution, identification, and classification of rocks is known as **petrology**. Petrology deals with the three major kinds of rocks classified according to their origins—igneous, sedimentary, and metamorphic rock.

Numerous factors are considered under igneous petrology. One of two fundamental kinds of igneous rocks consists of **plutonic** rock that has solidified from liquid magma over long time periods of hundreds of thousands or even millions of years deep within the Earth. The most common such rock is granite, which has large crystalline structures that are visible without magnification. The other major kind of igneous rock is **lava**, which has flowed onto Earth's surface in a molten state and solidified rapidly so that it consists of very fine crystals or is even a natural glass.

The two main kinds of rocks considered under sedimentary petrology are carbonate rocks, consisting mostly of calcium carbonate, $CaCO_3$, or dolomite, $CaCO_3{\bullet}MgCO_3$, and clastic rocks composed of a variety of weathered and eroded materials, such as sandstones, claystones, siltstones, glacial till, and various conglomerates. In addition to description and classification, sedimentary petrology deals with the conditions under which the parent materials of the sedimentary rocks were formed and transported, the history and conditions of rock formation, the cementation of the rock to form a consolidated mass, physical influences, such as compaction, after the rock was formed, and chemical modification of the rock.

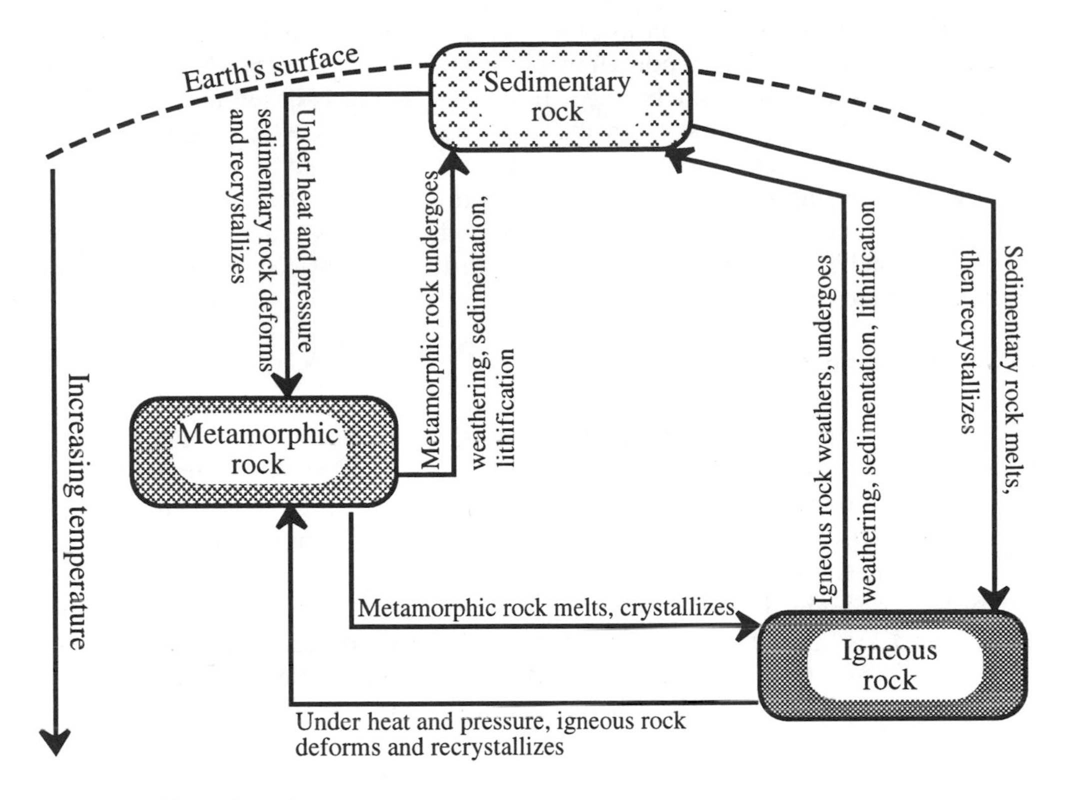

Figure 13.2. The rock cycle.

Metamorphic petrology addresses the ways in which rocks have been changed in form without going through a molten state. Metamorphosis commonly occurs as the result of elevated temperature and pressure. One type occurs when the temperature is increased by intrusions of igneous rock, and another type in which temperature and pressure are increased in large regions usually associated with mountain belts. An important influence on the formation of metamorphic rocks is exercised by the intrusion and loss of mineral solutions during the process of metamorphosis.

Structure and Properties of Minerals

The combination of two characteristics is unique to a particular mineral. These characteristics are a defined chemical composition, as expressed by the mineral's chemical formula, and a specific crystal structure. The **crystal structure** of a mineral refers to the way in which the atoms are arranged relative to each other. It cannot be determined from the appearance of visible crystals of the mineral, but requires structural methods such as X-ray structure determination. Different minerals may have the same chemical composition, or they may have the same crystal structure, but both may not be identical for truly different minerals.

Physical properties of minerals can be used to classify them. The characteristic external appearance of a pure crystalline mineral is its **crystal form**. Because of space constrictions on the ways that minerals grow, the pure crystal form of a mineral is often not expressed. **Color** is an obvious characteristic of minerals, but can vary widely due to the presence of impurities. The appearance of a mineral surface in reflected light describes its **luster**. Minerals may have a metallic luster or appear partially metallic (or submetallic), vitreous (like glass), dull or earthy, resinous, or pearly. The color of a mineral in its powdered form as observed when the mineral is rubbed across an unglazed porcelain plate is known as **streak**. **Hardness** is expressed on Mohs scale, which ranges from 1 to 10 and is based upon 10 minerals that vary from talc, hardness 1, to diamond, hardness 10. **Cleavage** denotes the manner in which minerals break along planes and the angles in which these planes intersect. For example, mica cleaves to form thin sheets. Most minerals **fracture** irregularly, although some fracture along smooth curved surfaces or into fibers or splinters. **Specific gravity**, density relative to that of water, is another important physical characteristic of minerals.

Kinds of Minerals

Although over two thousand minerals are known, only about 25 **rock-forming minerals** make up most of the earth's crust. The nature of these minerals may be better understood with a knowledge of the elemental composition of the crust. Oxygen and silicon make up 49.5% and 25.7% by mass of the earth's crust, respectively. Therefore, most minerals are **silicates** such as quartz, SiO_2, or potassium feldspar, $KAlSi_3O_8$. In descending order of abundance the other elements in the earth's crust are aluminum (7.4%), iron (4.7%), calcium (3.6%), sodium (2.8%), potassium (2.6%), magnesium (2.1%), and other (1.6%). Table 13.1 summarizes the major kinds of minerals in the earth's crust. The most abundant of these are the feldspars, which are aluminosilicates of potassium, sodium, and calcium. Feldspars are modified by weathering processes to form secondary clay minerals (see below).

Table 13.1. Major Mineral Groups in the Earth's Crust

Mineral group	Examples	Formula
Silicates	Quartz	SiO_2
	Olivine	$(Mg,Fe)_2SiO_4$
	Potassium feldspar	$KAlSi_3O_8$
Oxides	Corundum	Al_2O_3
	Magnetite	Fe_3O_4
Carbonates	Calcite	$CaCO_3$
	Dolomite	$CaCO_3 \cdot MgCO_3$
Sulfides	Pyrite	FeS_2
	Galena	PbS
Sulfates	Gypsum	$CaSO_4 \cdot 2H_2O$
Halides	Halite	NaCl
	Fluorite	CaF_2
Native elements	Copper	Cu
	Sulfur	S

Secondary minerals are formed by alteration of parent mineral matter. **Clays** are silicate minerals, usually containing aluminum, that constitute one of the most significant classes of secondary minerals. Olivine, augite, hornblende, and feldspars all form clays. **Evaporites** are soluble salts that precipitate from solution under special arid conditions, commonly as the result of the evaporation of seawater.

The most common evaporite is **halite**, NaCl. Many evaporites are hydrates, such as gypsum ($CaSO_4 \cdot 2H_2O$).

A number of mineral substances are gaseous at the magmatic temperatures of volcanoes and are mobilized with volcanic gases. These kinds of substances condense near the mouths of volcanic fumaroles and are called **sublimates**. Elemental sulfur is a common sublimate.

Numerous techniques are used to characterize rocks and minerals. Crystallography at both the atomic level and on a larger scale to determine the visible or microscopic crystalline form of minerals is an important aspect of mineralogy. The electron microprobe is an important instrument employed to characterize minerals. A grain of the mineral as small as 1 micrometer in size observed under an optical microscope may be subjected to an electron beam and the characteristic X-rays emitted used to determine the elemental composition of the mineral grain. The distribution of elements present at the boundaries of two different kinds of minerals can be used with thermodynamics to infer the conditions under which the minerals were formed.

13.3. PHYSICAL FORM OF THE GEOSPHERE

The most fundamental aspect of the physical form of the geosphere has to do with Earth's shape and dimensions. The earth is shaped as a **geoid** defined by a surface corresponding to the average sea level of the oceans and continuing as hypothetical sea levels under the continents. This shape is not a perfect sphere because of variations in the attraction of gravity at various places on Earth's surface. This slight irregularity in shape is important in surveying to precisely determine the locations of points on Earth's surface according to longitude, latitude, and elevation above sea level. Of more direct concern to humans is the nature of landforms and the processes that occur on them. This area of study is classified as **geomorphology**.

Plate Tectonics and Continental Drift

The geosphere has a highly varied, constantly changing physical form. Most of the Earth's land mass is contained in several massive continents separated by vast oceans. Towering mountain ranges spread across the continents, and in some places the ocean bottom is at extreme depths. Earthquakes, which often cause great destruction and loss of life, and volcanic eruptions, which sometimes throw enough material into the atmosphere to cause temporary changes in climate, serve as reminders that the Earth is a dynamic, living body that continues to change. There is convincing evidence, such as the close fit between the western coast of Africa and the eastern coast of South America, that widely separated continents were once joined and have moved relative to each other. This ongoing phenomenon is known as **continental drift**. It is now believed that 200 million years ago much of Earth's land mass was all part of a supercontinent, now called Gowandaland. This continent split apart to form the present-day continents of Antarctica, Australia, Africa, and South America, as well as Madagascar, the Seychelle Islands, and India.

The observations described above are explained by the theory of **plate tectonics**.[3] This theory views Earth's solid surface as consisting of several rigid plates that move relative to each other. These plates drift at an average rate of several centimeters per year atop a relatively weak, partially molten layer that is part of Earth's upper mantle called the **asthenosphere**. The science of plate tectonics explains the large-scale phenomena that affect the geosphere, including the creation and enlargement of oceans as the ocean floors open up and spread, the collision and breaking apart of continents, the formation of mountain chains, volcanic activities, the creation of islands of volcanic origin, and earthquakes.

The boundaries between these plates are where most geological activity, such as earthquakes and volcanic activity occur. These boundaries are of the three following types:

- **Divergent boundaries** where the plates are moving away from each other. Occurring on ocean floors, these are regions in which hot magma flows upward and cools to produce new solid lithosphere. This new solid material creates **ocean ridges**.
- **Convergent boundaries** where plates move toward each other. One plate may be pushed beneath the other in a **subduction zone** in which matter is buried in the asthenosphere and eventually remelted to form new magma. When this does not occur, the lithosphere is pushed up to form mountain ranges along a collision boundary.
- **Transform fault boundaries** in which two plates slide past each other. These boundaries create faults that result in earthquakes.

The phenomena described above are parts of the **tectonic cycle**, a geological cycle which describes how tectonic plates move relative to each other, magma rises to form new solid rocks, and solid lithospheric rocks sink to become melted thus forming new magma. The tectonic cycle is illustrated in Figure 13.3.

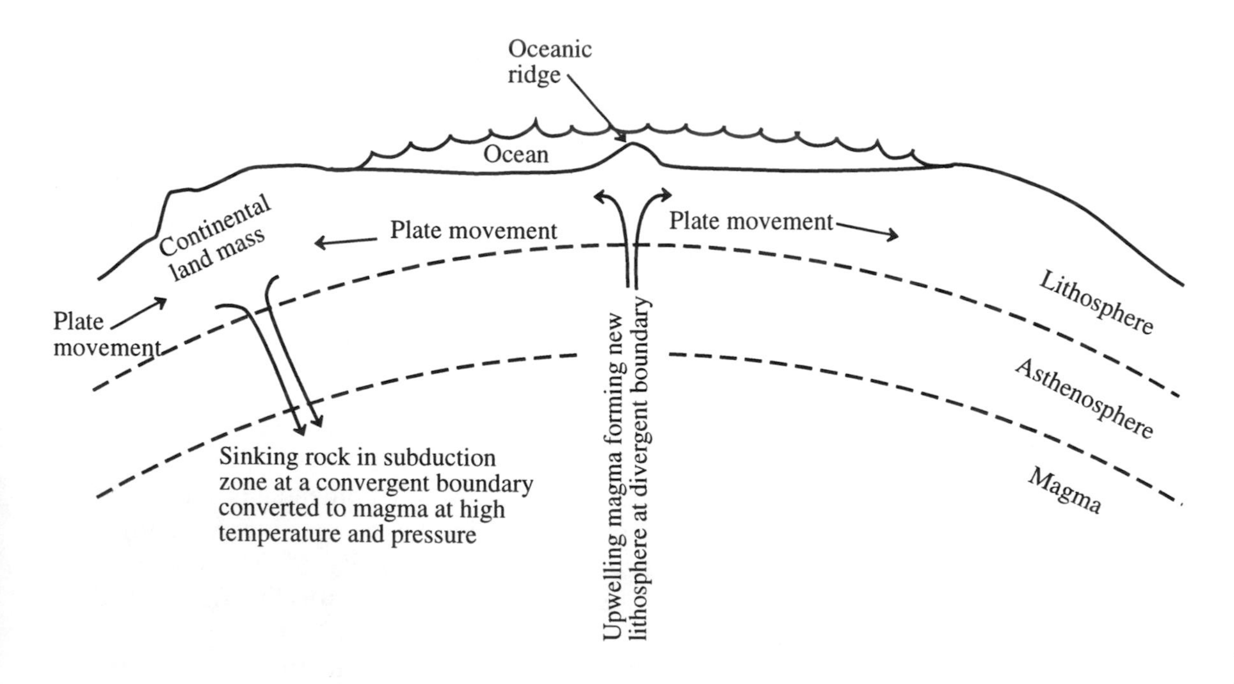

Figure 13.3. Illustration of the techtonic cycle in which upwelling magma along a boundary where two plates diverge creates new lithosphere on the ocean floor, and sinking rock in a subduction zone is melted to form magma.

Structural Geology

Earth's surface is constantly being reshaped by geological processes. The movement of rock masses during processes such as the formation of mountains results in substantial deformation of rock. At the opposite extreme of the size scale are defects in crystals at a microscopic level. **Structural geology** addresses the geometric forms of geologic structures over a wide range of size, the nature of structures formed by geological processes, and the formation of folds, faults, and other geological structures.

Primary structures are those that have resulted from the formation of a rock mass from its parent materials. Primary structures are modified and deformed to produce **secondary structures.** A basic premise of structural geology is that most layered rock formations were deposited in a horizontal configuration. Cracking of such a formation without displacement of the separate parts of the formation relative to each other produces a **joint**, whereas displacement produces a **fault** (see Figure 13.4).

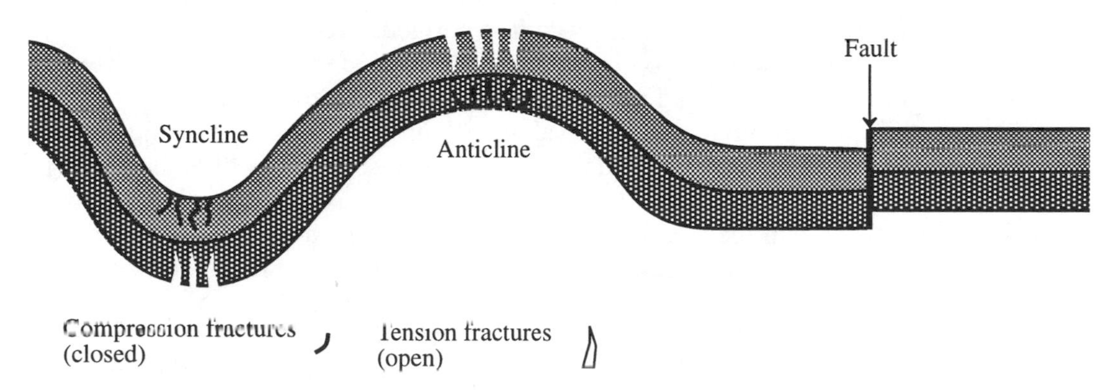

Figure 13.4. Folds (syncline and anticline) are formed by the bending of rock formations. Faults are produced by rock formations moving vertically or laterally in respect to each other.

An important relationship in structural geology is that between the force or **stress** placed upon a geological formation or object and the deformation resulting therefrom, called the **strain**. An important aspect of structural geology, therefore, is **rheology**, which deals with the deformation and flow of solids and semisolids. Whereas rocks tend to be strong, rigid, and brittle under the conditions at Earth's surface, their rheology changes such that they may become weak and pliable under the extreme conditions of temperature and pressure at significant depths below Earth's surface.

13.4. INTERNAL PROCESSES

The preceding section addressed the physical form of the geosphere. Related to the physical configuration of the geosphere are several major kinds of processes that occur that change this configuration and that have the potential to cause damage and even catastrophic effects. These can be divided into the two main categories of **internal processes** that arise from phenomena located significantly below the Earth's surface and **surface processes** that occur on the surface. Internal processes are addressed in this section and surface processes in Section 13.5.

Earthquakes

Earthquakes, occur as motion of ground resulting from the release of energy that accompanies an abrupt slippage of rock formations subjected to stress along a fault. In addition to shaking of ground, which can be quite violent, earthquakes can cause the ground to rupture, subside, or rise.[4] **Liquefaction** is an important phenomenon that occurs during earthquakes with ground that is poorly consolidated and in which the water table may be high. Liquefaction results from separation of soil particles accompanied by water infiltration such that the ground behaves like a fluid. Another devastating phenomenon consists of **tsunamis**, large ocean waves resulting from earthquake-induced movement of ocean floor. Tsunamis sweeping onshore have destroyed many homes and taken many lives, often large distances from the epicenter of the earthquake itself.

Earthquakes usually arise from plate tectonic processes and originate along plate boundaries. Basically, two huge masses of rock tend to move relative to each other, but are locked together along a fault line. This causes deformation of the rock formations, which increases with increasing stress. Eventually, the friction between the two moving bodies is insufficient to keep them locked in place, and movement occurs along an existing fault, or a new fault is formed. Freed from constraints on their movement, the rocks undergo elastic rebound, in which they partially revert to their original form. The associated movement constitutes an earthquake.

The location of the initial movement along a fault that causes an earthquake to occur is called the **focus** of the earthquake, and the surface location directly above it is the **epicenter**. Energy is transmitted from the focus by **seismic waves**. Seismic waves that travel through the interior of the Earth are called **body waves** and those that traverse the surface are **surface waves**. **Compressional waves** act much like a coiled spring by alternately compressing and expanding soil. **Shear waves** involve lateral motion, somewhat like the movement of a string on a guitar.

The loss of life and destruction of property by earthquakes makes them one of nature's more damaging natural phenomena. Literally millions of lives have been lost in past earthquakes, and damage from an earthquake in a developed urban area can easily run into billions of dollars. Therefore, the economic geology (see Section 13.12) associated with earthquakes is very important. Significant progress has been made in designing structures that are earthquake-resistant. As evidence of that, during a 1964 earthquake in Nigata, Japan, some buildings tipped over on their sides due to liquefaction of the underlying soil, but remained structurally intact! Other measures that can lessen the impact of earthquakes is the identification of areas susceptible to earthquakes, discouraging development in such areas, and educating the public about earthquake hazards. Accurate prediction would be a tremendous help in lessening the effects of earthquakes, but so far has been generally unsuccessful. Most challenging of all is the possibility of preventing major earthquakes. One unlikely possibility would be to detonate nuclear explosives deep underground along a fault line to release stress before it builds up to an excessive level. Fluid injection to facilitate slippage along a fault has also been considered.

Volcanoes

In addition to earthquakes, the other major subsurface process that has the potential to massively affect the environment consists of emissions of molten rock (lava), gases, steam, ash, and particles due to the presence of magma near the earth's surface. This phenomenon is called a **volcano** (Figure 13.5).[5] Volcanoes can be very destructive and damaging to the environment. Aspects of the potential harm from volcanoes are discussed in Section 15.3.

Volcanoes take on a variety of forms which are beyond the scope of this chapter to cover in detail. Basically, they are formed when magma rises to the surface. This frequently occurs in subduction zones created where one plate is pushed beneath another. The downward movement of solid lithospheric material subjects it to high temperatures and pressures that causes the rock in it to melt and rise to the surface as magma. Molten magma issuing from a volcano at temperatures usually in excess of 500°C and often as high as 1,400°C, is called **lava**, and is one of the more common manifestations of volcanic activity.

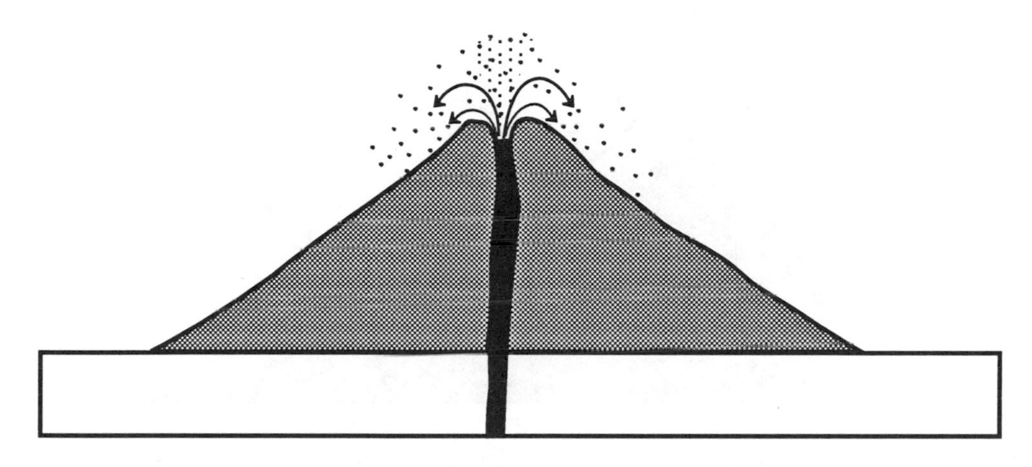

Figure 13.5. Volcanoes come in many shapes and forms. A classically shaped volcano may be a cinder cone formed by ejection of rock and lava, called pyroclastics, from the volcano to produce a relatively uniform cone.

13.5. SURFACE PROCESSES

Surface geological features are formed by upward movement of materials from Earth's crust. With exposure to water, oxygen, freeze-thaw cycles, organisms, and other influences on the surface, surface features are subject to two processes that largely determine the landscape—weathering and erosion. As noted earlier in this chapter, weathering consists of the physical and chemical breakdown of rock and erosion is the removal and movement of weathered products by the action of wind, liquid water, and ice. Weathering and erosion work together in that one augments the other in breaking down rock and moving the products. Weathered products removed by erosion are eventually deposited as sediments and may undergo diagenesis and lithification to form sedimentary rocks.

One of the most common surface processes that can adversely affect humans consists of **landslides** that occur when soil or other unconsolidated materials slide down a slope.[6] Related phenomena include rockfalls, mudflows, and snow avalanches. As shown in Figure 13.6, a landslide typically consists of an upper slump that is prevented from sliding farther by a mass of material accumulated in a lower flow. Figure 13.6 illustrates what commonly happens in a landslide when a mass of earth moves along a slip plane under the influence of gravity. The stability of earthen material on a slope depends upon a balance between the mass of slope material and the resisting force of the shear strength of the slope material. There is a tendency for the earth to move along slip planes. In addition to the earthen material itself, water, vegetation, and structures constructed by humans may increase the driving force leading to a landslide. The shear strength is, of course, a function of the geological material along the slip plane and may be affected by other factors as well, such as the presence of various levels of water and the degree and kinds of vegetation growing on the surface.

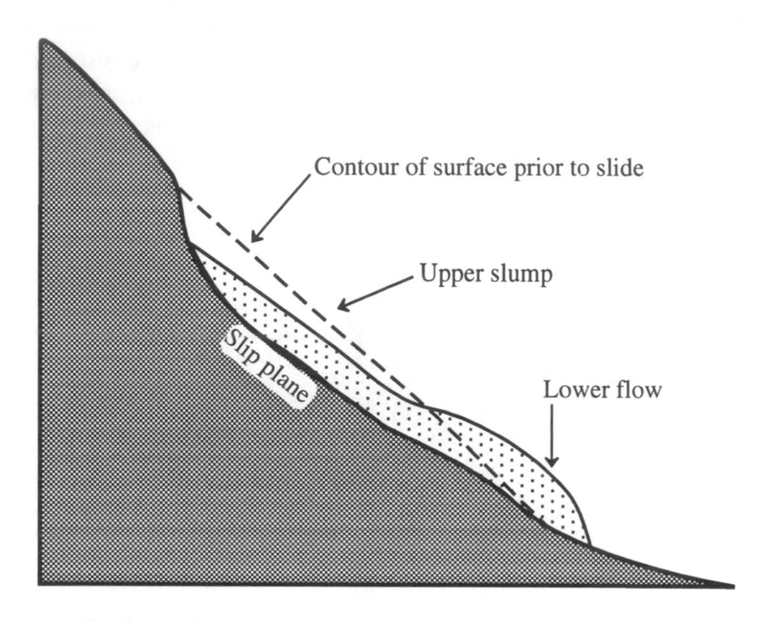

Figure 13.6. A landslide occurs when earth moves along a slip plane. Typically a landslide consists of an upper slump and lower flow. The latter serves to stabilize the slide, and when it is disturbed, such as by cutting through it to construct a road, the earth may slide farther.

The tendency of landslides to form is influenced by a number of outside factors. Climate is important because it influences the accumulation of water that often precedes a landslide as well as the presence of plants that can also influence soil stability. Although it would seem that plant roots should stabilize soil, the ability of some plants to add significant mass to the slope by accumulating water and to destabilize soil by aiding water infiltration may have an opposite effect. Disturbance of earth by road or other construction may cause landslides to occur. Earth may be shaken loose by earthquakes, causing landslides to occur.

As discussed in Section 15.4, landslides can be very dangerous to human life and their costs in property damage can be enormous. In addition to destroying structures located on the surface of sliding land or covering structures or people with earth, landslides can have catastrophic indirect effects. For example, landslides that dump huge quantities of earth into reservoirs can raise water levels almost instantaneously and cause devastating waves and floods.

Subsidence occurs when the surface level of earth sinks over a significant area. The most spectacular evidence of subsidence is manifested as large sinkholes that may form rather suddenly, sometimes swallowing trees, automobiles, and even whole buildings in the process. Overall, much more damage is caused by gradual and less extreme subsidence, which may damage structures as it occurs or result in inundation of areas near water level. Such subsidence is frequently caused by the removal of fluids, such as petroleum, from below ground.

13.6. SEDIMENTS

Vast areas of land, as well as lake and stream sediments, are formed from sedimentary rocks. The properties of these masses of material depend strongly upon their origins and transport. Water is the main vehicle of sediment transport, although wind can also be significant. Hundreds of millions of tons of sediment are carried by major rivers each year. The study of sediments and the environments in which they are formed is the science of **sedimentology**.

The action of flowing water in streams cuts away stream banks and carries sedimentary materials for great distances. Sedimentary materials may be carried by flowing water in streams as dissolved load, suspended load, or bed load. **Dissolved load** consists of solutions of minerals, such as calcium bicarbonate, which can precipitate to form a sediment of solid calcium carbonate:

$$Ca^{2+} + 2HCO_3^- \rightarrow CaCO_3(s) + CO_2(g) + H_2O \qquad (13.6.1)$$

Most flowing water containing dissolved load originates underground, where it dissolves minerals from the rock strata that it flows through. **Suspended load** consists primarily of finely divided silt, clay, or sand originating from sources such as soil eroded from land or finely divided rock released by melting glaciers. **Bed load** is made up of larger particles dragged along the bottom of the stream channel.

Typically, about 2/3 of the sediment carried by a stream is transported in suspension, about 1/4 in solution, and the remaining relatively small fraction as bed load. The ability of a stream to carry sediment increases with both the overall rate of flow of the water (mass per unit time) and the velocity of the water. Both of these are higher under flood conditions, so floods are particularly important in the transport of sediments. Streams mobilize sedimentary materials through **erosion**, **transport** materials along with stream flow, and release them in a solid form during **deposition**. Deposits of stream-borne sediments are called **alluvium**.

Clays

Clays, a group of microcrystalline secondary minerals consisting of hydrous aluminum silicates that have sheet-like structures and often are present as very small colloidal particles, are particularly important sedimentary minerals in soil and in the sediments of bodies of water. Clays predominate in the inorganic components of most soils and are very important in holding water and in plant nutrient cation exchange. Clays may hold pollutant compounds within their layered structures. A typical clay mineral is kaolinite ($Al_2Si_2O_5(OH)_4$) formed by the chemical weathering of potassium feldspar rock ($KAlSi_3O_8$):

$$2KAlSi_3O_8(s) + 2H^+ + 9H_2O \rightarrow Al_2Si_2O_5(OH)_4(s) + 2K^+(aq) + 4H_4SiO_4(aq) \qquad (13.6.2)$$

13.7. INTERACTION WITH THE ATMOSPHERE AND HYDROSPHERE

The geosphere interacts strongly with the other spheres of the environment. It is strongly influenced by them and in turn has a strong influence on each. These interactions are addressed briefly here.

The hydrosphere has more influence than any other on the geosphere. Ocean water covers more of the geosphere than does land. Fresh water from precipitation falls on land and produces streams, vast rivers, and lakes. Liquid water as groundwater occurs in huge quantities beneath the surface of the land. Water in the solid state as glacial ice was a major force in determining much of the geomorphology of Earth's surface during the ice ages and continues to do so in colder climates today. The nature of solids in the geosphere largely determines the distribution and fate of surface water and groundwater. Porous, permeable rock formations may be conducive to the inflow and storage of groundwater. The nature and elevation of surrounding geological strata determine whether a stream is an **influent stream** in which water is lost to the ground to recharge groundwater supplies in underground aquifers or an **effluent stream** in which groundwater enters the stream from surrounding aquifers to maintain stream flow.

One of the largest influences of water on the geosphere arises from erosion of land surfaces by water (Figure 13.7). Flowing water dislodges weathered rock, carries it some distance from its source, and deposits it as sediments. Water then plays a major role in the chemical processes that convert these sediments to sedimentary rock. Relatively young mountain formations tend to have sharp features. Through the erosive action of liquid water and ice, these features become rounded, and eventually the mountain formations are worn down.

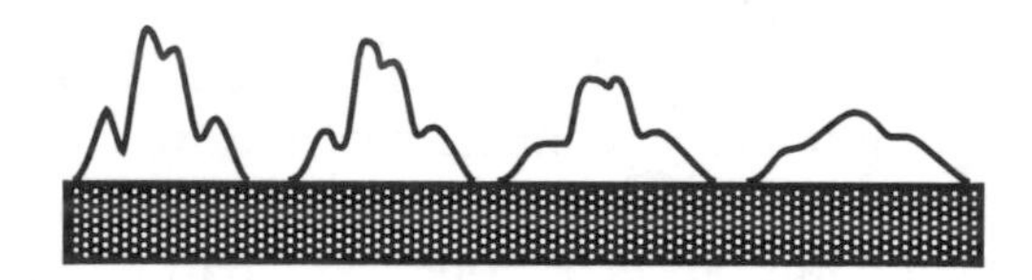

Figure 13.7. Continued erosion of mountains by water cause wears them down and rounds their features.

The most important role of the atmosphere in shaping the geosphere arises from water carried from oceans onto land in the hydrologic cycle. Climate, resulting largely from atmospheric conditions, has a strong influence on the geosphere, particularly in determining the amount of precipitation falling on land. One of the most direct atmospheric forces that influences the geosphere is from wind. Wind can be a major factor in erosion, particularly of disturbed soil under dry conditions. In addition, wind can transport and deposit solids that make up parts of the geosphere. One such solid material consists of **sand dunes**, composed of sand carried just above the surface of the ground by wind. Active sand dunes that are still moving can cause severe problems with roads and other structures. Stabilized sand dunes should be treated with care to prevent their becoming active and mobile. The other major kind of solid material carried by wind is **loess** consisting of silt, defined as finely divided sediment in a size range of 1/256 to 1/16 mm in diameter, carried and deposited by wind. Loess deposits in the United States are near major rivers. These deposits were formed from rock ground by glaciers during the Pleistocene Ice Ages and deposited in large areas along the river flood plains. As the river flows subsided when the glaciers retreated, large areas of sediment were left dry without much vegetative cover. Winds carried this material away and deposited it as loess.

The interactions of the geosphere and the anthrosphere are many. Mountainous terrain may make it impossible to grow enough food to support a significant human population. Weak geological strata can make the construction and maintenance of buildings very difficult. Earth is the source of the minerals — metal ores, fossil fuels, stone—required to sustain an economy. Human activities in turn have a tremendous influence on the geosphere. This is seen when one observes Earth's surface from above while travelling by airplane. Highways cut across terrain, dams interrupt the flow of rivers, and vast expanses of land have been converted from forests and prairies to cultivated land. Increasingly, human endeavors are being used to enhance the geosphere that has been ruthlessly exploited, especially during the last two centuries. Some of the greatest success has been in terracing farmland to minimize water erosion and topsoil loss. In some cases streams that were channelized into straight ditches are being restored with constructed bends and meanders.

13.8. LIFE SUPPORT BY THE GEOSPHERE

The connection between the biosphere and the geosphere is obvious. Most plants exist on soil, as do other forms of life, such as earthworms, fungi, and bacteria. Some kinds of rock have a biological origin. Deposits of limestone and silicaceous deposits from the shells of aquatic organisms originated through life processes. The production of oxygen by photosynthesis resulted in the oxidation of soluble iron to insoluble iron oxides, thus producing iron ore deposits. Coal, kerogen (the organic matter in oil shale), and petroleum are of biological origin.

Much of what is known about geological history and about the evolution of life on Earth is based on **paleontology**, the study of fossils in rock. Similar life forms have existed at different times, so it is possible to use observations of fossils to determine the relative ages of rock strata. Fossil records extend back approximately 3 billion years, so that very long time spans can be addressed.

Trace-level elements in the geosphere play an important role in the health of living organisms. Fluorine as fluoride ion prevents tooth decay and strengthens bone at relatively low levels, whereas at somewhat higher levels it has detrimental effects on tooth and bone. Goiter, a condition caused by enlargement of the thyroid gland, is caused by a deficiency of iodine. Selenium at very low levels is required in the diet of animals, but at levels of only a few parts per million selenium is toxic. Sickness in animals has been shown to result from either too little or too much selenium in the soil on which animal feed is grown and on the selenium content of the water that they drink. Zinc is an essential nutrient for both plants and animals. It is required by animals, for example, in order for wounds to heal properly. Too much zinc in soil, such as soil treated with excessive amounts of zinc-laden sewage sludge, can be phytotoxic (toxic to plants).

13.9. SOIL

Insofar as the human environment and life on Earth are concerned, the most important part of the Earth's crust is soil, discussed in detail in Chapter 14. **Soil** consists of particles that make up a variable mixture of minerals, organic matter, and water, capable of supporting plant life on Earth's surface. It is the final product of the weathering action of physical, chemical, and biological processes on rocks, which largely produces clay minerals (see above). The organic portion of soil consists of plant biomass in various stages of decay. High populations of bacteria, fungi, and animals such as earthworms may be found in soil. Soil contains air spaces and generally has a loose texture. Engineers who work with earthen materials view soil as divided earthen materials that can be moved without blasting.

Soils usually exhibit distinctive layers with increasing depth (Figure 13.8). These layers, called **horizons**, form as the result of complex interactions among processes that occur during weathering. Rainwater percolating through soil carries dissolved and colloidal solids to lower horizons where they are deposited. Biological processes, such as bacterial decay of residual plant biomass, produce slightly acidic CO_2, organic acids, and complexing compounds that are carried by rainwater to lower horizons where they interact with clays and other minerals, altering the properties of the minerals. The top layer of soil, typically several inches in thickness, is known as the A horizon, or **topsoil**. This is the layer of maximum biological activity in the soil and contains most of the soil's organic matter. Metal ions and clay particles in the A horizon are subject to considerable leaching, such that it is sometimes called the zone of leaching. The next layer is the B horizon, or **subsoil**. It receives material such as organic matter, salts, and clay particles leached from the topsoil, so it is called the zone of accumulation. The C horizon is composed of fractured and weathered parent rocks from which the soil originated. Reflecting its composition, the C horizon is called the zone of partially altered parent material.

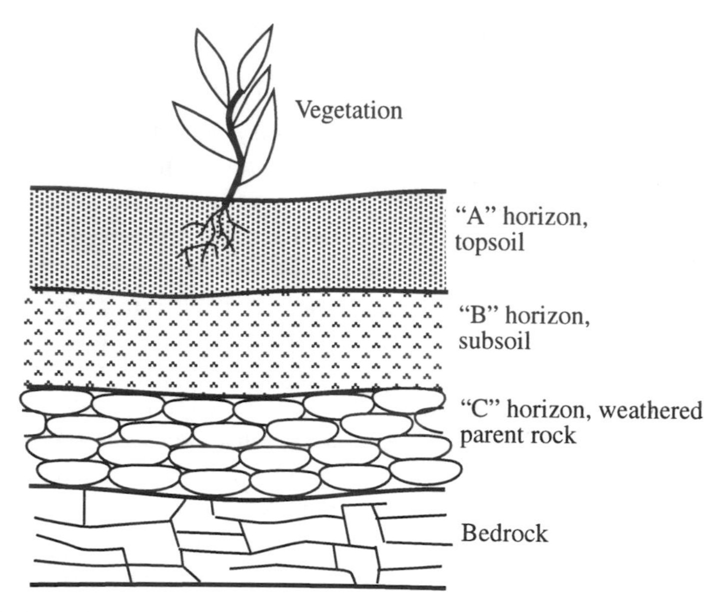

Figure 13.8. Soil profile showing soil horizons.

Soils exhibit a large variety of characteristics that are used to classify them for various purposes, including crop production, road construction, and waste disposal. The parent rocks from which soils are formed obviously play a strong role in determining the composition of soils. Other soil characteristics include strength, workability, soil particle size, permeability, and degree of maturity.

13.10. GEOCHEMISTRY

Geochemistry deals with chemical species, reactions, and processes in the lithosphere and their interactions with the atmosphere and hydrosphere.[7] The branch of geochemistry that explores the complex interactions among the rock/water/air/life (and human) systems that determine the chemical characteristics of the surface environment is **environmental geochemistry**. Obviously, geochemistry and its environmental subdiscipline are very important in environmental science.

Geochemistry addresses a large number of chemical and related physical phenomena. Some of the major areas of geochemistry are the following:

- The chemical composition of major components of the geosphere, including magma and various kinds of solid rocks.
- Processes by which elements are mobilized, moved, and deposited in the geosphere through a cycle known as the **geochemical cycle**.
- Chemical processes that occur during the formation of igneous rocks from magma.
- Chemical processes that occur during the formation of sedimentary rocks.
- Chemistry of rock weathering.
- Chemistry of volcanic phenomena.
- Role of water and solutions in geological phenomena, such as deposition of minerals from hot brine solutions.
- The behavior of dissolved substances in concentrated brines.

An important consideration in geochemistry is that of the interaction of life forms with geochemical processes addressed as biogeochemistry or organic geochemistry. The deposition of biomass and the subsequent changes that it undergoes have led to the formation of huge deposits of petroleum, coal, and oil shale. Chemical changes induced by photosynthesis have resulted in massive deposits of calcium carbonate (limestone). Deposition of the biochemically synthesized shells of microscopic animals have led to the formation of large formations of calcium carbonate and silica. Biogeochemistry is closely involved with elemental cycles, such as those of carbon.

Physical and Chemical Aspects of Weathering

Defined in Section 13.2, *weathering* is discussed here as a geochemical phenomenon. Rocks tend to weather more rapidly when there are pronounced differences in physical conditions—alternate freezing and thawing and wet periods alternating with severe drying. Other mechanical aspects are swelling and shrinking of minerals with hydration and dehydration as well as growth of roots through cracks in rocks. The rates of chemical reactions involved in weathering increase with increasing temperature.

As a chemical phenomenon, weathering can be viewed as the result of the tendency of the rock/water/mineral system to attain equilibrium. This occurs through the usual chemical mechanisms of dissolution/precipitation, acid-base reactions, complexation, hydrolysis, and oxidation-reduction.

Weathering is very slow in dry air. Water increases the rate of weathering by many orders of magnitude for several reasons. Water, itself, is a chemically active substance in the weathering process. Furthermore, water holds weathering agents in solution such that they are transported to chemically active sites on rock minerals and contact the mineral surfaces at the molecular and ionic level. Prominent among such weathering agents are CO_2, O_2, organic acids, sulfur acids ($SO_2(aq)$, H_2SO_4), and nitrogen acids (HNO_3, HNO_2). Water provides the source of H^+ ion needed for acid-forming gases to act as acids as shown by the following:

$$CO_2 + H_2O \rightarrow H^+ + HCO_3^- \quad (13.10.1)$$

$$SO_2 + H_2O \rightarrow H^+ + HSO_3^- \quad (13.10.2)$$

Rainwater is essentially free of mineral solutes. It is usually slightly acidic due to the presence of dissolved carbon dioxide or more highly acidic because of acid-rain forming constitutents. As a result of its slight acidity and lack of alkalinity and dissolved calcium salts, rainwater is *chemically aggressive* toward some kinds of mineral matter, which it breaks down by a process called **chemical weathering**. Because of this process, river water has a higher concentration of dissolved inorganic solids than does rainwater.

A typical chemical reaction involved in weathering is the dissolution of calcium carbonate (limestone) by water containing dissolved carbon dioxide:

$$CaCO_3(s) + H_2O + CO_2(aq) \rightarrow Ca^{2+}(aq) + 2HCO_3^-(aq) \quad (13.10.3)$$

Weathering may also involve oxidation reactions, such as occurs when pyrite, FeS_2, dissolves:

$$4FeS_2(s) + 15O_2(g) + (8 + 2x)H_2O \rightarrow$$
$$2Fe_2O_3 \cdot xH_2O + 8SO_4^{2-}(aq) + 16H^+(aq) \quad (13.10.4)$$

Isotopic Geochemistry

The measurement of subtle differences in the ratios of naturally occurring elemental isotopes found in rocks and minerals can yield significant information about geochemical phenomena. The activity of radioactive carbon-14, which occurs in atmospheric carbon dioxide and is incorporated into biomass by photosynthesis, has been widely used to determine the age of biological materials. The half-life of carbon-14 is 5,570 years, so this method can be used to determine the ages of organic materials less than about 30,000 years old. Small differences in the ratios of oxygen-16 to oxygen-18 in calcium carbonate deposited by the metabolic activities of marine organisms can be used to estimate the temperatures at which these deposits formed. Such information can be used to estimate past climatic conditions and in predicting future climates. Radiometric age dating is based upon the decay of radioisotopes and the formation of daughter products. The most abundant isotope of uranium, uranium-238, decays eventually to lead-206, so that determination of the uranium-238/lead-206 ratio in a rock can be used to estimate the time from which the original uranium was deposited.

13.11. WATER ON AND IN THE GEOSPHERE

Groundwater (Figure 13.9) is a vital resource in its own right that plays a crucial role in geochemical processes, such as the formation of secondary minerals. The nature, quality, and mobility of groundwater are all strongly dependent upon the rock formations in which the water is held. Physically, an important characteristic of such formations is their **porosity**, which determines the percentage of rock volume available to contain water. A second important physical characteristic is **permeability**, which describes the ease of flow of the water through the rock. High permeability is usually associated with high porosity. However, clays tend to have low permeability even when a large percentage of the volume is filled with water.

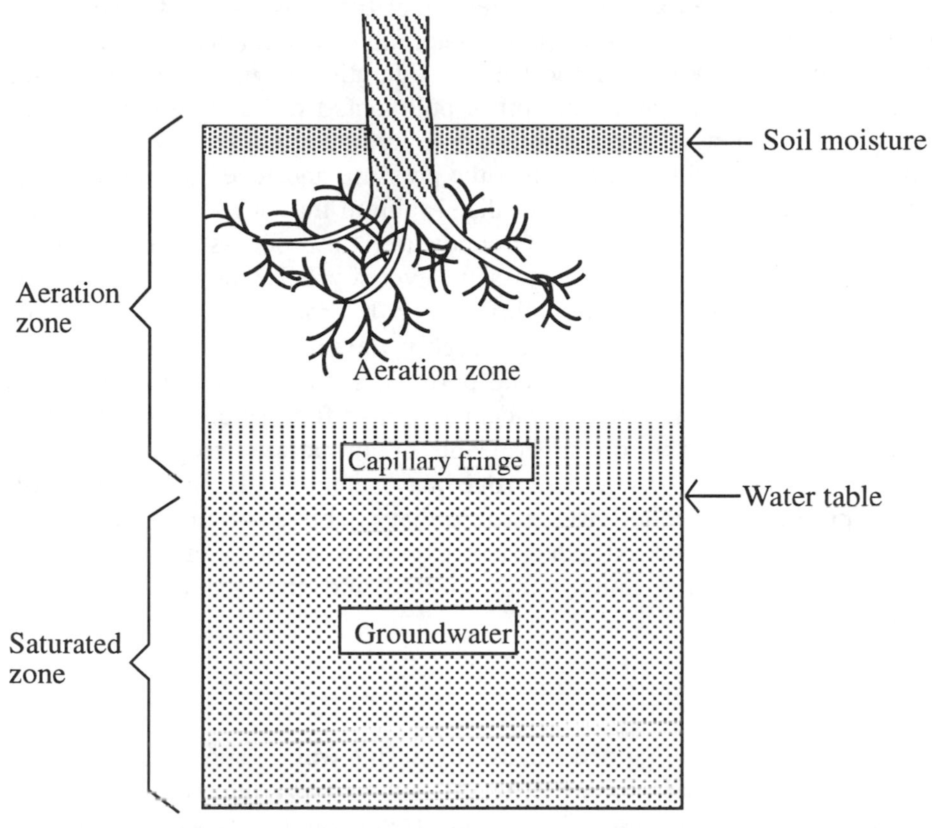

Figure 13.9. Some major features of the distribution of water underground.

Most groundwater originates as **meteoric** water from precipitation in the form of rain or snow. If water from this source is not lost by evaporation, transpiration, or to stream runoff, it may infiltrate into the ground. Initial amounts of water from precipitation onto dry soil are held very tightly as a film on the surfaces and in the micropores of soil particles in a **belt of soil moisture**. At intermediate levels, the soil particles are covered with films of water, but air is still present in larger voids in the soil. The region in which such water is held is called the **unsaturated zone** or **zone of aeration** and the water present in it is **vadose water**. At lower depths in the presence of adequate amounts of water, all voids are filled to produce a **zone of saturation**, the upper boundary of which is the **water table**. Water present in a zone of saturation is called **groundwater**. Because of its surface tension, water is drawn somewhat above the water table by capillary-sized passages in soil in a region called the **capillary fringe**.

The water table is crucial in explaining and predicting the flow of wells and springs and the levels of streams and lakes. It is also an important factor in determining the extent to which pollutant and hazardous chemicals underground are likely to be transported by water. The water table can be mapped by observing the equilibrium level of water in wells, which is essentially the same as the top of the saturated zone. The water table is usually not level, but tends to follow the general contours of the surface topography. It also varies with differences in permeability and water infiltration. The water table is at surface level in the vicinity of swamps and frequently above the surface where lakes and streams are encountered. The water level in such bodies may be maintained by the water table. **Influent** streams or reservoirs are located above the water table; they lose water to the underlying aquifer and cause an upward bulge in the water table beneath the surface water.

Groundwater **flow** is an important consideration in determining the accessibility of the water for use and transport of pollutants from underground waste sites. Various parts of a body of groundwater are in hydraulic contact so that a change in pressure at one point will tend to affect the pressure and level at another point. For example, infiltration from a heavy, localized rainfall may affect the water table at a point remote from the infiltration. Groundwater flow occurs as the result of the natural tendency of the water table to assume even levels by the action of gravity.

Groundwater flow is strongly influenced by rock permeability. Porous or extensively fractured rock is relatively highly **pervious**, meaning that water can migrate through the holes, fissures, and pores in such rock. Because water can be extracted from such a formation, it is called an **aquifer**. By contrast, an **aquiclude** is a rock formation that is too impermeable or unfractured to yield groundwater. Impervious rock in the unsaturated zone may retain water infiltrating from the surface to produce a **perched water table** that is above the main water table and from which water may be extracted. However, the amounts of water that can be extracted from such a formation are limited and the water is vulnerable to contamination.

Water Wells

Most groundwater is tapped for use by water wells drilled into the saturated zone. The use and mis-use of water from this source has a number of environmental implications. In the U.S. about 2/3 of the groundwater pumped is consumed for irrigation; lesser amounts of groundwater are used for industrial and municipal applications.

As water is withdrawn, the water table in the vicinity of the well is lowered. This **drawdown** of water creates a **zone of depression**. In extreme cases groundwater is severely depleted and surface land levels can even subside (which is one reason that Venice, Italy, is now very vulnerable to flooding). Heavy drawdown can result in infiltration of pollutants from sources such as septic tanks, municipal refuse sites, and hazardous waste dumps. Mineral deposits, such as oxides of iron and manganese, can form from well water exposed to air as it drains from the aquifer into the well. Such deposits may seriously limit the flow of water into the well.

13.12. ECONOMIC GEOLOGY

Technological applications of geology have several important economic benefits.[8] Geology is used to find and develop essential resources of raw materials extracted from Earth's crust. These include minerals, uranium used to fuel nuclear reactors, natural gas, petroleum, and coal. Knowledge of subsurface strata is essential to the siting and construction of buildings and other structures. Geological principles are used to prevent and warn of hazards, such as earth slides.

Geological science, now equipped with a wide array of high technology tools, is used to locate and develop deposits of metal ores; nonmetal minerals, such as aggregate, gypsum, clay and salt; and fossil fuels. Most economic geologists work in the petroleum industry in which one of their major functions is to maximize the probability of striking economically viable deposits of crude oil or natural gas when exploratory wells are drilled. Other important aspects of petroleum geology include structural geology, stratigraphy (the science of stratified rock formations), and sedimentary petrology.

The provision of mineral resources for human use is one of the most important aspects of economic geology and is one with a multitude of environmental implications. Chapter 16 entitled "Geospheric Resources" is devoted to the topic of the extraction of mineral resources from the geosphere.

13.13. ENGINEERING GEOLOGY

Engineering geology addresses the ways in which geological materials and structures are used and dealt with technologically in respect to structures, materials science, and other engineering aspects.[9] Practically anything that is built on or near the surface of the Earth requires consideration of geological engineering aspects. For example, if a structure is to be built on a slope, it is important to consider the geological properties of the surface on which it is to be built. The degree of the slope and the nature of the material on which it is located can be used to predict the likelihood of earth slides and to design structures to avoid these natural disasters. Engineering geology is used to design rock quarries and to determine the most efficient means of removing rocks from the ground.

Large public works projects are the human endeavors most likely to affect the geosphere because they entail earth moving, digging, boring, and other operations performed on the geosphere. In turn, the nature, costs, and safety of structures constructed as part of huge public works projects are all highly dependent upon the characteristics of the geological formations upon which they are built. Therefore, public works projects require a high degree of sophisticated geological engineering throughout their planning and construction stages, and their maintenance and operation requires consideration of geological engineering as well. Large public works projects include dams, roads, railroads, airports, pipelines, canals, tunnels, and large structures. Much of the engineering that goes into such projects involves evaluation of the geologic strata upon which the structures are located and development of measures to prevent problems. For example, dams should be located on and anchored to strong formations of rock, preferably igneous or metamorphic rock. Fractures in the rock formations along which leaks may develop should be detected and filled with a wet mixture of cement and sediment called grout. Numerous factors must be considered in highway construction. To a greater extent than with most structures, highways must make use of surrounding geological materials, which must be evaluated for their suitability for fill and roadbed. Topography (surface configuration) and slope are crucial for grading and drainage. Geologic engineering is a crucial consideration in the siting and construction of large structures, such as buildings or nuclear or fossil-fueled power plants. The underlying strata must be carefully evaluated for its load-bearing capacity and to discover unexpected features, such as faults or fractures that might shift or caverns that might cause subsidence.

CHAPTER SUMMARY

The chapter summary below is presented in a programmed format to review the main points covered in this chapter. It is used most effectively by filling in the blanks, referring back to the chapter as necessary. The correct answers are given at the end of the summary.

The relationship of the geosphere to the other environmental spheres that it influences and in turn is influenced by, including humankind and its technology, defines [1] ____________________. The percentage of incident solar radiation reflected by a land or water surface is known as [2] ____________________. A condition manifested by declining groundwater tables, salinization of topsoil and water, reduction of surface waters, unnaturally high soil erosion, and desolation of native vegetation is called [3] ____________________. Earth's outer skin that is accessible to humans is called the [4] ____________________. It is extremely thin compared to the diameter of the earth, ranging from 5 to 40 km thick. Most of the solid earth crust consists of [5] ____________________, which in turn are composed of [6] ____________________. Molten rocks are called [7] ____________________ which turn to [8] ____________________ rock when they reach the surface and eventually disintegrate by [9] ____________________ processes. Sediments undergo [10] ____________________ to produce [11] ____________________ rocks. Metamorphic rock is formed by [12] ____ ____.
Eight readily measured or observed physical properties of minerals that can be used to classify them are [13] ____ ____.
Chemically most minerals are [14] ____________________. Clays are examples of [15] ____________________ minerals formed by alteration of parent mineral matter. Continental drift is explained by the theory of [16] ____________________ ____________________. The [17] ____________________ cycle describes how tectonic plates move relative to each other, magma rises to form new solid rocks, and solid lithospheric rocks sink to become melted and reform magma. Cracking of a rock formation without displacement of the separate parts of the formation relative to each other produces a [18] ____________________ whereas displacement produces a [19] ____________________. The location of the initial movement along a fault that causes an earthquake to occur is called the [20] ____________________. of the earthquake, and the surface location directly above it is the [21] ____________________ ____________________. The molten rock that can flow from a volcano is called [22] ____________________, whereas fragments of rock and lava given off by a volcano constitute [23] ____________________ ____________________. Deposits of stream-borne sediments are called [24] ____________________.
One of the largest influences of water on the geosphere arises from [25] ____________________ of land surfaces by water. Soil is the final product of the [26] ____________________ action of physical, chemical, and biological processes on rocks.The science that deals with chemical species, reactions, and processes in the lithosphere and their interactions with the atmosphere and hydrosphere is [27] ____________________. As a chemical phenomenon, weathering can be viewed as the result of the tendency of the rock/water/mineral system to [28] ____________________.

The characteristic of the rock formations in which water is held which determines the percentage of rock volume available to contain water is called [29]____________, whereas the ease of flow of the water through the rock is called [30]_____________. Water present in a zone of saturation is called [31]_________________________. An underground formation from which water can be extracted is called an [32]________ ____________.The use of geology to find and develop essential resources of raw materials extracted from Earth's crust is addressed by [33]______________________, whereas [34]______________________ pertains to the ways in which geological materials and structures are used and dealt with technologically in respect to structures, materials science, and other similar aspects.

Answers

1 environmental geology

2 surface albedo

3 desertification

4 crust

5 rocks

6 minerals

7 magma

8 igneous

9 weathering

10 lithification

11 sedimentary

12 the action of heat and pressure on sedimentary, igneous, or other kinds of metamorphic rock that are not in a molten state

13 crystal form, color, luster, streak, hardness, cleavage, fracture, and specific gravity

14 silicates

15 secondary

16 plate tectonics

17 tectonic cycle

18 joint

19 fault

20 focus

21 epicenter

22 lava

23 pyroclastics

24 alluvium

25 erosion

26 weathering

27 geochemistry

28 attain equilibrium

29 porosity

30 permeability

31 groundwater

32 aquifer

33 economic geology

34 engineering geology

QUESTIONS AND PROBLEMS

1. Of the following, the one that is **not** a manifestation of desertification is (a) declining groundwater tables, (b) salinization of topsoil and water, (c) production of deposits of MnO_2 and $Fe_2O_3 \cdot H_2O$ from anaerobic processes, (d) reduction of surface waters, (e) unnaturally high soil erosion.

2. Why do silicates and oxides predominate among Earth's minerals?

3. Explain how the following are related: weathering, igneous rock, sedimentary rock, soil.

4. Match the following:

   1. Metamorphic rock
   2. Chemical sedimentary rocks
   3. Detrital rock
   4. Organic sedimentary rocks

   (a) Produced by the precipitation or coagulation of dissolved or colloidal weathering products
   (b) Contain residues of plant and animal remains
   (c) Formed from action of heat and pressure on sedimentary rock
   (d) Formed from solid particles eroded from igneous rocks by weathering

5. Where does most flowing water that contains dissolved load originate? Why does it tend to come from this source?

6. What is engineering geology? How can it be related to environmental improvement?

7. What is surface albedo? How may human activities affect it?

8. In what sense do humans have more direct knowledge of outer space than they do of Earth's lithosphere? Why is this so?

9. What are the three major classes of rocks? How are they involved in the rock cycle?

10. What are the observable characteristics by which minerals are classified? What is the distinction between mineral crystal structure and mineral form?

11. What is the distinction between a mineral evaporite and a sublimate?

12. What is meant by plate tectonics, and how does this concept relate to continental drift?

13. What is the distinction between a joint and a fault in rock formations?

14. What does lava manifest? What is the origin of lava?

LITERATURE CITED

1 Chernicoff, Stanley and Chip Fox, *Essentials of Geology*, Worth Publishing Co., New York, NY, 1996.

2 Montgomery, Carla W., *Environmental Geology*, 5th ed, Wm. C Brown Publishers (McGraw Hill), Boston, MA, 1997.

3 Condie, Kent C., *Plate Tectonics and Crustal Evolution*, 4th ed., Butterworth-Heinemann, Newton, MA, 1997.

4 Keller, Edward A., *Active Tectonics: Earthquakes, Uplift, and Landscape*, Prentice Hall, Upper Saddle River, NJ, 1996.

5 Llamas-Ruiz, Andres and Ali Garousi, *Volcanos and Earthquakes*, Sterling Publishing Co., New York, NY, 1997.

6 Goodwin, Peter, *Landslides, Slumps, and Creep*, Franklin Watts Publishing Co., New York, NY, 1997.

7 Brownlow, Arthur H., *Geochemistry*, Prentice Hall, Upper Saddle River, NJ, 1996.

8 Evans, Anthony, M., *An Introduction to Economic Geology and Its Environmental Impact*, Blackwell Science Inc., Cambridge, MA, 1997.

9 Rahn, Perry H., *Engineering Geology: An Environmental Approach*, Prentice Hall, Upper Saddle River, NJ, 1996.

SUPPLEMENTARY REFERENCES

Bates, Robert L. and Julia A. Jackson, Eds., *Glossary of Geology*, American Geological Institute, Boulder, CO, 1997.

Berthelin, J., Ed., *Diversity of Environmental Biogeochemistry*, Elsevier Science Publishing, New York, NY, 1991.

Colley, H., *Introduction to Environmental Geology*, Chapman & Hall, New York, NY, 1997.

Gill, Robin, *Chemical Fundamentals of Geology*, Chapman & Hall, New York, NY, 1996.

Keller, Edward A., *Environmental Geology*, Prentice Hall, Upper Saddle River, NJ, 1996.

Keys, W. Scott, *A Practical Guide to Borehole Geophysics in Environmental Investigations*, CRC Press/Lewis Publishers, Boca Raton, FL, 1997.

Klous, W. Jacquelyne, *This Dynamic Earth: The Story of Plate Tectonics*, Diane Publishing Co, Washington, D.C., 1996.

McKinney, Michael and Robert L. Tolliver, *Current Issues in Geology*, West Publishing Co., Westbury, NY, 1996.

Oliver, J. E., *Shocks and Rocks: Seismology in the Plate Tectonics Revolution: The Story of Earthquakes and the Great Earth Science Revolution of the 1960s*, American Geophysical Union, New York, NY, 1996.

Pipkin, D. D. and Bernard W. Trent, *Geology and the Environment*, West Publishing Co., Westbury, NY, 1996.

Rydin, Yvonne, *Environmental Impact of Land and Property Management*, John Wiley and Sons, New York, NY, 1996.

Seibold, E. and W. H. Berger, *The Sea Floor: An Introduction to Marine Geology*, 3rd ed., Springer Verlag, New York, NY, 1996.

Stephens, Daniel B., *Vadose Zone Hydrology*, CRC Press/Lewis Publishers, Boca Raton, FL, 1996.

Sutherland, Lin, *The Volcanic Earth: Volcanoes and Plate Tectonics: Past, Present & Future*, New South Wales University Press Ltd., New South Wales, Australia, 1996.

14 SOIL

14.1. SOIL AND AGRICULTURE

Soil and agricultural practices are strongly tied with the environment. Some of these considerations are addressed later in this chapter along with a discussion of soil erosion and conservation. Cultivation of land and agricultural practices can influence both the atmosphere and the hydrosphere. Although this chapter deals primarily with soil, the topic of agriculture in general is introduced for perspective.

Agriculture

Agriculture, the production of food by growing crops and livestock, provides for the most basic of human needs. This topic is discussed in more detail in Section 21.3. The two basic categories of agriculture are **crop farming** to produce edible substances from photosynthetically generated biomass, and **livestock farming** for the production of meat, milk, wool, hide, and other animal products. Both crops and livestock were developed from wild ancestors by early farmers. Particularly in the case of crops, output has been increased markedly during the last several decades by developing hybrids from crossing two or more true-breeding strains. Now recombinant DNA technology and genetic engineering are being used to revolutionize agriculture through the production of higher yielding crops and animals, hormones to increase milk production, development of crops resistant to herbicides applied to kill competing weeds, and similar developments.

Agriculture has a tremendous influence on the environment and has a significant potential for environmental harm. In addition to direct effects from the cultivation of land, there are indirect effects from irrigation and other measures used to increase agricultural yield. The rearing of domestic animals may have environmental effects. For example, The Netherlands' pork industry has been so productive that hog manure and its by-products have caused serious problems. Goats and sheep have destroyed pasture land in the Near East, Northern Africa, Portugal, and Spain. Of particular concern are the environmental effects of raising cattle. Significant amounts of forest land have been converted to marginal pasture land to raise beef. Production of one pound of beef requires about 4 times as much water and 4 times as much feed as does production of 1 pound of chicken. An interesting aspect of the problem is emission of greenhouse-gas methane by anaerobic bacteria in the digestive systems of cattle and other ruminant animals; cattle rank right behind wetlands and rice paddies as producers of atmospheric methane. However, cattle and other ruminant livestock do have a positive environmental/resource impact because they can use cellulose from plants as a food source to produce meat and milk, the result of the action of specialized bacteria in the stomachs of ruminant animals.

Pesticides and Agriculture

Pesticides, particularly insecticides and herbicides, are an integral part of modern agricultural production. In the U.S., agricultural pesticides are regulated under the Federal Insecticide, Fungicide and Rodenticide (FIFRA) Act. Starting in 1989 under FIFRA, pre-1972 pesticides were required to be re-registered. Since this process began, manufacturers have withdrawn from the market about 20,000 of those 40,000 products because of the expense of the safety review process. The problem has been especially severe for **minor-use pesticides** for which the market is not very large. In contrast to pesticides used on approximately 220 million acres of major crops in the U.S.—corn, soybeans, wheat, and cotton—minor-use pesticides are used on only about 8 million acres of orchards, trees, ornamental plants, turf grass, fruits, nuts, and vegetables. Despite their limited use, about 40% of the monetary value of agricultural pesticides resides with minor-use pesticides.

14.2. THE NATURE OF SOIL

Soil is the most fundamental requirement for agriculture. The composition and properties of soil were briefly introduced in Chapter 13. In this chapter the nature of soil is addressed in greater detail. The study of soil is called **pedology** or, more simply, soil science. To humans and most terrestrial organisms, soil is the most important part of the geosphere. Though only a tissue-thin layer compared to the Earth's total diameter, soil is the medium that produces most of the food required by most living things. Good soil—and a climate conducive to its productivity—is the most valuable asset a nation can have.

In addition to being the site of most food production, soil is the receptor of large quantities of pollutants, such as particulate matter from power plant smokestacks. Fertilizers, pesticides, and some other materials applied to soil often contribute to water and air pollution. Therefore, soil is a key component of environmental chemical cycles.

Soil is a variable mixture of minerals, organic matter, and water that supports plant life on the Earth's surface. As mentioned in Section 13.9, soil is the final product of the weathering action of physical, chemical, and biological processes on rocks, which largely produce clay minerals. The organic portion of soil consists of plant biomass in various stages of decay. High populations of microorganisms and animals such as earthworms may be found in soil. Soil contains air spaces and generally has a loose texture (Figure 14.1). Soils are open systems that undergo continual exchange of matter and energy with the atmosphere, hydrosphere, and biosphere.

Soil Solids

The solid fraction of typical productive soil is approximately 5% organic matter and 95% inorganic matter. Some soils, such as peat soils, may contain as much as 95% organic material. Other soils contain as little as 1% organic matter.

As shown in Chapter 13, Figure 13.8, typical soils exhibit distinctive layers called **horizons**, which form as the result of complex interactions among processes that occur during weathering. Rainwater percolating through soil carries dissolved and colloidal solids to lower horizons where they are deposited. Biological processes, such as bacterial decay of residual plant biomass, produce slightly acidic CO_2, organic acids, and complexing compounds that are carried by rainwater to lower horizons where they interact with clays and other minerals, altering the properties of the minerals. The top layer of soil, typically 10-30 cm in thickness, is the A horizon, or **topsoil**. This is the layer of maximum biological activity in the soil and contains most of the soil organic matter. Metal ions and clay particles in the A horizon are subject to considerable leaching. The next layer is the B horizon, or **subsoil**. It receives material such as organic matter, salts, and clay particles leached from the topsoil. The C horizon is composed of weathered parent rocks from which the soil originated.

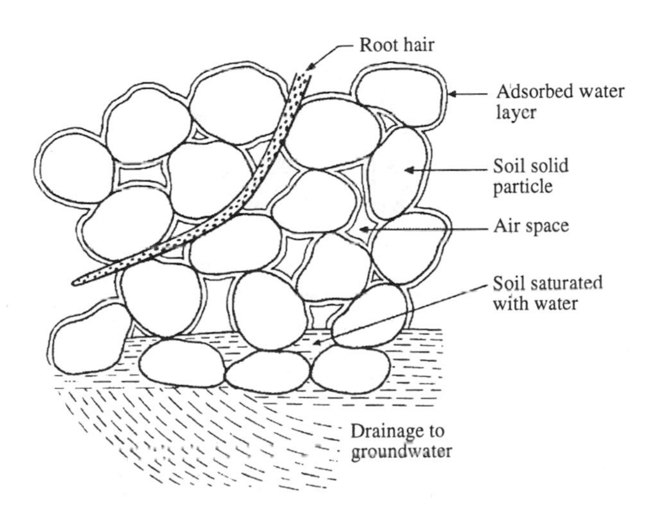

Figure 14.1. Fine structure of soil, showing solid, water, and air phases.

Soils exhibit a large variety of characteristics that are used for their classification for various purposes, including crop production, road construction, and waste disposal. The parent rocks from which soils are formed obviously play a strong role in determining the composition of soils. Other soil characteristics include strength, workability, soil particle size, permeability, and degree of maturity.

Water and Air in Soil

Water is part of the three-phase, solid-liquid-gas system making up soil. It is the basic transport medium for carrying essential plant nutrients from solid soil particles into plant roots and to the farthest reaches of the plant's leaf structure (Figure 14.2). The water enters the atmosphere from the plant's leaves, a process called **transpiration**.

Normally, because of the small size of soil particles and the presence of small capillaries and pores in the soil, the water phase is not totally independent of soil solid matter. Water present in larger spaces in soil is relatively more available to plants and readily drains away. Water held in smaller pores, or between the unit layers of clay particles is held much more strongly. Water in soil interacts strongly with organic matter and with clay minerals.

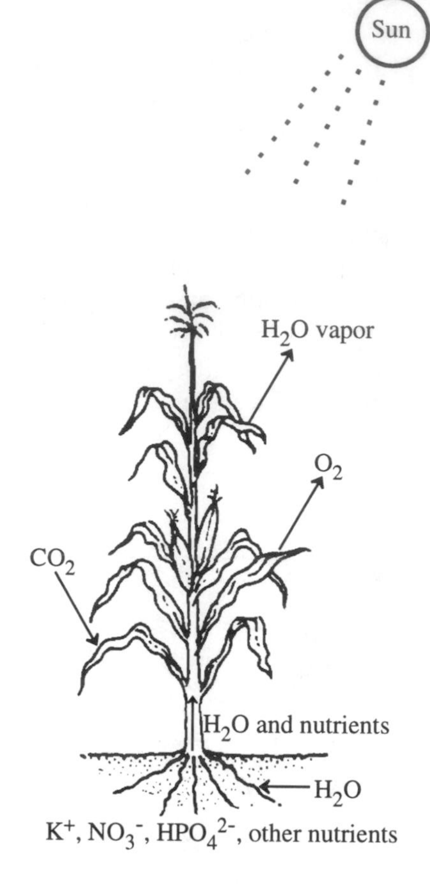

Figure 14.2. Plants transport water from the soil to the atmosphere by transpiration. Nutrients are also carried from the soil to the plant extremities by this process. Plants remove CO_2 from the atmosphere and add O_2 by photosynthesis. The reverse occurs during plant respiration.

Too much water is detrimental to the quality of soil. As soil becomes water-logged (water-saturated); it undergoes drastic changes in physical, chemical, and biological properties. Oxygen in such soil is rapidly used up by the respiration of microorganisms that degrade soil organic matter. In such soils, the bonds holding soil colloidal particles together are broken, which causes disruption of soil structure. Thus, the excess water in such soils is detrimental to plant growth, and the soil does not contain the air required by most plant roots. Most useful crops, with the notable exception of rice, cannot grow on waterlogged soils.

The exclusion of air from waterlogged soil results in the establishment of chemically reducing conditions that cause reduction of insoluble manganese and iron oxides:

$$MnO_2(s) + 4H^+ + 2e^- \rightarrow Mn^{2+} + 2H_2O \tag{14.2.1}$$

$$Fe_2O_3(s) + 6H^+ + 2e^- \rightarrow 2Fe^{2+} + 3H_2O \tag{14.2.2}$$

The soluble Fe^{2+} and Mn^{2+} ions released are toxic to plants at high levels. Their later oxidation to insoluble oxides may cause formation of deposits of Fe_2O_3 and MnO_2, which clog tile drains in fields.

The Soil Solution

The **soil solution** is the aqueous portion of soil that contains dissolved matter from soil chemical and biochemical processes in soil and from exchange with the hydrosphere and biosphere. This medium transports chemical species to and from soil particles and provides intimate contact between the solutes and the soil particles. In addition to providing water for plant growth, it is an essential pathway for the exchange of plant nutrients between roots and solid soil.

Dissolved mineral matter in soil is largely present as ions. Prominent among the cations are H^+, Ca^{2+}, Mg^{2+}, K^+, Na^+, and usually very low levels of Fe^{2+}, Mn^{2+}, and Al^{3+}. The last three cations may be present in partially hydrolized form, such as $FeOH^+$, or complexed by organic humic substance ligands. Anions that may be present are HCO_3^-, CO_3^{2-}, HSO_4^-, SO_4^{2-}, Cl^-, and F^-. In addition to being bound to H^+ in species such as bicarbonate, anions may be complexed with metal ions, such as in AlF^{2+}. Multivalent cations and anions form ion pairs with each other in soil solutions. Examples of these are $CaSO_4$ and $FeSO_4$.

Roughly 35% of the volume of typical soil is composed of air-filled pores. Whereas the normal dry atmosphere at sea level contains 21% O_2 and 0.03% CO_2 by volume, these percentages may be quite different in soil air because of the decay of organic matter:

$$\{CH_2O\} + O_2 \rightarrow CO_2 + H_2O \qquad (14.2.3)$$

This process consumes oxygen and produces CO_2. As a result, the oxygen content of air in soil may be as low as 15%, and the carbon dioxide content may be several percent. Thus, the decay of organic matter in soil increases the equilibrium level of dissolved CO_2 in groundwater. This lowers the pH and contributes to weathering of carbonate minerals, particularly calcium carbonate (see Chapter 6, Reaction 6.9.3). As discussed in Section 14.2, CO_2 also shifts the equilibrium of the process by which roots absorb metal ions from soil.

The Inorganic Components of Soil

As noted in Section 13.2, the most abundant elements in the Earth's crust are oxygen, silicon, aluminum, iron, calcium, sodium, potassium, and magnesium, so minerals composed of these elements—particularly silicon and oxygen—constitute most of the mineral fraction of the soil. Common soil mineral constituents are finely divided quartz (SiO_2), epidote ($4CaO{\cdot}3(AlFe)_2O_3{\cdot}6SiO_2{\cdot}H_2O$), albite ($NaAlSi_3O_8$), orthoclase ($KAlSi_3O_8$), geothite ($FeO(OH)$), magnetite ($Fe_3O_4$), calcium and magnesium carbonates ($CaCO_3$, $CaCO_3{\cdot}MgCO_3$), and oxides of manganese and titanium.

The weathering of parent rocks and minerals to form the inorganic soil components results ultimately in the formation of inorganic colloids. These colloids are repositories of water and plant nutrients, which may be made available to plants as needed. Inorganic soil colloids often absorb toxic substances in soil, thus playing a role in detoxification of substances that otherwise would harm plants. The abundance and nature of inorganic colloidal material in soil are obviously important factors in determining soil productivity.

The uptake of plant nutrients by roots may involve complex interactions with the water and inorganic phases. For example, a nutrient held by inorganic colloidal material has to traverse the mineral/water, and then the water/root interfaces. This process is often strongly influenced by the ionic structure of soil inorganic matter.

Organic Matter in Soil

Though typically comprising less than 5% of a productive soil, organic matter largely determines soil productivity. It serves as a source of food for micro-organisms; undergoes chemical reactions such as ion exchange; and influences the physical properties of soil. Some organic compounds even contribute to the weathering of mineral matter, the process by which soil is formed. For example, $C_2O_4^{2-}$, oxalate ion, produced as a soil fungi metabolite, present in the soil solution dissolves minerals, thus speeding the weathering process and increasing the availability of nutrient ion species. This weathering process involves oxalate complexation of iron or aluminum in minerals, represented by the reaction

$$3H^+ + M(OH)_3(s) + 2CaC_2O_4(s) \rightarrow M(C_2O_4)_2^-(aq) + 2Ca^{2+}(aq) + 3H_2O \quad (14.2.4)$$

in which M is Al or Fe. Some soil fungi produce citric acid, and other chelating organic acids, which react with silicate minerals and release potassium and other nutrient metal ions held by these minerals. The strong chelating agent 2-ketogluconic acid is produced by some soil bacteria. By solubilizing metal ions, it may contribute to the weathering of minerals. It may also be involved in the release of phosphate from insoluble phosphate compounds.

Biologically active components of the organic soil fraction include polysaccharides, amino sugars, nucleotides, and organic sulfur and phosphorus compounds. Humus, a plant degradation product that biodegrades very slowly, makes up the bulk of soil organic matter.

The accumulation of organic matter in soil is strongly influenced by temperature and by the availability of oxygen. Since the rate of biodegradation decreases with decreasing temperature, organic matter does not degrade rapidly in colder climates and tends to build up in soil. In water and in waterlogged soils, decaying vegetation does not have easy access to oxygen, and organic matter accumulates. The organic content may reach 90% in areas where plants grow and decay in soil saturated with water.

Soil Humus

Soil humus is by far the most significant organic constituent of soil. Humus, composed of a base-soluble fraction of humic and fulvic acids (described in Section 6.12) and an insoluble fraction called humin, is the residue left from plant biodegradation. Humus is largely the partial biodegradation product of lignin which, along with readily degraded cellulose, makes up the bulk of plant biomass. The process by which humus is formed is called **humification**. Part of each molecule of humic substance is nonpolar and hydrophobic, and part is polar and hydrophilic.

Humic substances influence soil properties to a degree out of proportion to their small percentage in soil. They strongly bind metals, and serve to hold micronutrient metal ions in soil. Because of their acid-base character, humic substances serve as buffers in soil. The water-holding capacity of soil is significantly increased by humic substances. These materials also stabilize aggregates of soil particles and increase the sorption of organic compounds by soil.

14.3. ACID-BASE AND ION EXCHANGE REACTIONS IN SOILS

One of the more important chemical functions of soils is cation exchange, such as

$$\text{Soil}\}NH_4^+ + H^+(aq) \longleftrightarrow \text{Soil}\}H^+ + NH_4^+(aq) \quad (14.3.1)$$

Both the mineral and organic portions of soils exchange cations. Clay minerals exchange cations because of the presence of negatively charged sites on the mineral, resulting from the substitution of an atom of lower oxidation number for one of higher number; for example, magnesium for aluminum. Organic materials exchange cations because of the presence of the carboxylate group and other basic functional groups. Humus typically has a very high capacity to exchange cations.

Cation exchange in soil is the mechanism by which potassium, calcium, magnesium, and essential trace-level metals are made available to plants. When nutrient metal ions are taken up by plant roots, hydrogen ion is exchanged for the metal ions. This process, plus the leaching of calcium, magnesium, and other metal ions from the soil by water containing carbonic acid, tends to make the soil acidic:

$$\text{Soil}\}Ca^{2+} + 2CO_2 + 2H_2O \rightarrow \text{Soil}\}(H^+)_2 + Ca^{2+}(\text{root}) + 2HCO_3^- \quad (14.3.2)$$

Soil acts as a buffer and resists changes in pH. The buffering capacity depends upon the type of soil.

Adjustment of Soil Acidity

Most common plants grow best in soil with a pH near neutrality. If the soil becomes too acidic for optimum plant growth, it may be restored to productivity by liming, ordinarily through the addition of calcium carbonate:

$$\text{Soil}\}(H^+)_2 + CaCO_3 \rightarrow \text{Soil}\}Ca^{2+} + CO_2 + H_2O \quad (14.3.3)$$

In areas of low rainfall, soils may become too basic (alkaline) due to the presence of basic salts such as Na_2CO_3. Alkaline soils may be treated with aluminum or iron sulfate, which release acid on hydrolysis:

$$2Fe^{3+} + 3SO_4^{2-} + 6H_2O \rightarrow 2Fe(OH)_3(s) + 6H^+ + 3SO_4^{2-} \quad (14.3.4)$$

Sulfur added to soils is oxidized by bacterially mediated reactions to sulfuric acid:

$$S + {}^3/_2O_2 + H_2O \rightarrow 2H^+ + SO_4^{2-} \quad (14.3.5)$$

and sulfur is used, therefore, to acidify alkaline soils. The huge quantities of sulfur now being removed from fossil fuels to prevent air pollution by sulfur dioxide may make the treatment of alkaline soils by sulfur much more attractive economically.

14.4. MACRONUTRIENTS IN SOIL

One of the most important functions of soil in supporting plant growth is to provide essential plant nutrients—macronutrients and micronutrients. Macronutrients are those elements that occur in substantial levels in plant materials or in fluids in the plant. Micronutrients (Section 14.6) are elements that are essential only at very low levels and generally are required for the functioning of essential enzymes.

The elements generally recognized as essential macronutrients for plants are carbon, hydrogen, oxygen, nitrogen, phosphorus, potassium, calcium, magnesium, and sulfur. Carbon, hydrogen, and oxygen are obtained from the atmosphere. The other essential macronutrients must be obtained from soil. Of these, nitrogen, phosphorus, and potassium are the most likely to be lacking and are commonly added to soil as fertilizers. Because of their importance, these elements are discussed separately in Section 14.5.

Calcium-deficient soils are relatively uncommon. Liming, a process used to treat acid soils (see Reaction 14.3.3), provides a more than adequate calcium supply for plants. However, calcium uptake by plants and leaching by carbonic acid (Reaction 14.3.2) may produce a calcium deficiency in soil. Acid soils may still contain an appreciable level of calcium which, because of competition by hydrogen ion, is not available to plants. Treatment of acid soil to restore the pH to near-neutrality generally remedies the calcium deficiency. In alkaline soils, the presence of high levels of sodium, magnesium, and potassium sometimes produces calcium deficiency because these ions compete with calcium for availability to plants.

Although magnesium makes up 2.1% of the Earth's crust, most of it is rather strongly bound in minerals. Generally, exchangeable magnesium is considered available to plants and is held by ion-exchanging organic matter or clays. The availability of magnesium to plants depends upon the calcium/magnesium ratio. If this ratio is too high, magnesium may not be available to plants and magnesium deficiency results.

Soils deficient in sulfur do not support plant growth well, largely because sulfur is a component of some essential amino acids and of thiamin and biotin. Sulfate ion is generally present in the soil as immobilized insoluble sulfate minerals or as soluble salts, which are readily leached from the soil and lost as soil water runoff.

14.5. NITROGEN, PHOSPHORUS, AND POTASSIUM IN SOIL

Nitrogen, phosphorus, and potassium are plant nutrients that are obtained from soil. They are so important for crop productivity that they are commonly added to soil as fertilizers.

Nitrogen

Figure 14.3 summarizes the primary sinks and pathways of nitrogen in soil. In most soils, over 90% of the nitrogen content is organic. This organic nitrogen is primarily the product of the biodegradation of dead plants and animals. It is eventually hydrolyzed to NH_4^+, which can be oxidized to NO_3^- by the action of bacteria in the soil.

Nitrogen bound to soil humus (see Section 14.2) is especially important in maintaining soil fertility. Unlike potassium or phosphate, nitrogen is not a significant product of mineral weathering (see Section 13.10). Nitrogen-fixing organisms ordinarily cannot supply sufficient nitrogen to meet peak demand. Inorganic nitrogen from fertilizers and rainwater is often largely lost by leaching. Soil humus, however, serves as a reservoir of nitrogen required by plants. It has the additional advantage that its rate of decay, hence its rate of nitrogen release to plants, roughly parallels plant growth — rapid during the warm growing season, slow during the winter months.

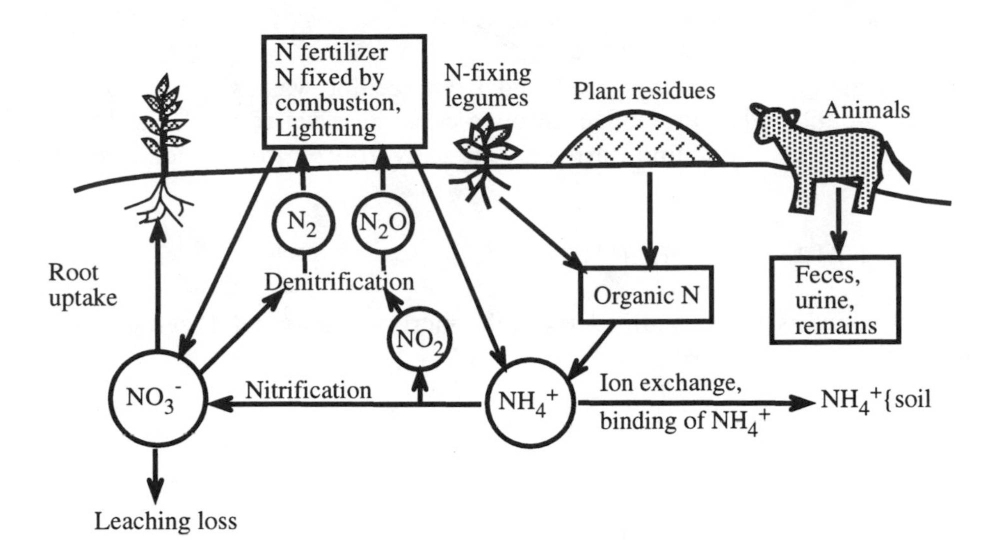

Figure 14.3. Nitrogen sinks and pathways in soil.

Nitrogen is an essential component of proteins and other constituents of living matter. Plants and cereals grown on nitrogen-rich soils not only provide higher yields, but are often substantially richer in protein and, therefore, more nutritious. Nitrogen is most generally available to plants as nitrate ion, NO_3^-. Some plants such as rice may utilize ammonium nitrogen; however, other plants are poisoned by this form of nitrogen. When nitrogen is applied to soils in the ammonium form, nitrifying bacteria perform an essential function in converting it to available nitrate ion.

Nitrogen fixation is the process by which atmospheric N_2 is converted to nitrogen compounds available to plants. Prior to the widespread introduction of nitrogen fertilizers, soil nitrogen was provided primarily by legumes. These are plants such as soybeans, alfalfa, and clover, which contain on their root structures bacteria capable of fixing atmospheric nitrogen. Leguminous plants have a symbiotic (mutually advantageous) relationship with the bacteria that provide their nitrogen. The nitrogen-fixing bacteria in legumes exist in special structures on the roots called root nodules (see Fig. 14.4). The rod-shaped bacteria that fix nitrogen are members of a special genus called *Rhizobium*. These bacteria fix nitrogen in symbiotic combination with plants.

Nitrate pollution of some surface waters and groundwater is a significant problem in some agricultural areas (see Chapter 7). Although fertilizers have been implicated in such pollution, there is evidence that feedlots are a major source of nitrate pollution. The growth of livestock populations and the concentration of livestock in feedlots have aggravated the problem. Such concentrations of cattle, coupled with the fact that a steer produces approximately 18 times as much waste material as a human, have resulted in high levels of water pollution in rural areas with small human populations. Streams and reservoirs in such areas frequently are just as polluted as those in densely populated and highly industrialized areas.

Nitrate in farm wells is a common and especially damaging manifestation of nitrogen pollution from feedlots because of the susceptibility of ruminant animals to nitrate poisoning. The stomach contents of ruminant animals such as cattle and sheep constitute a reducing medium (low pE) and contain bacteria capable of reducing nitrate ion to toxic nitrite ion:

$$NO_3^- + 2H^+ + 2e^- \rightarrow NO_2^- + H_2O \qquad (14.5.1)$$

The origin of most nitrate produced from feedlot wastes is amino nitrogen, organically bound $-NH_2$, present in nitrogen-containing waste products. During the degradation process the amino nitrogen is first hydrolyzed to ammonia, or ammonium ion:

$$RNH_2 + H_2O \rightarrow R\text{-}OH + NH_3\ (NH_4^+) \qquad (14.5.2)$$

and the product is then oxidized through microorganism-catalyzed reactions to nitrate ion:

$$NH_3 + 2O_2 \rightarrow H^+ + NO_3^- + H_2O \qquad (14.5.3)$$

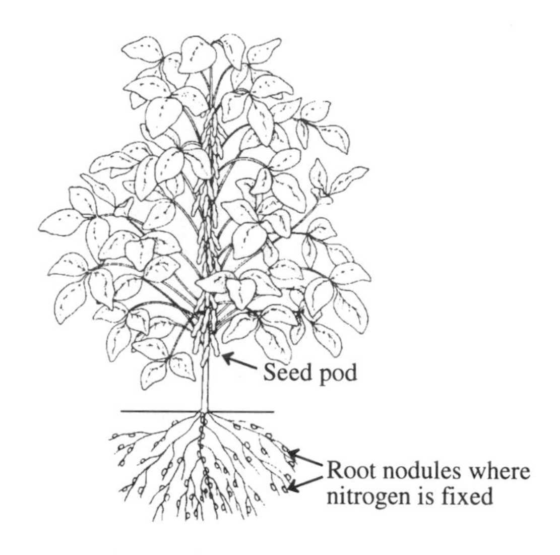

Figure 14.4. A soybean plant, showing root nodules where nitrogen is fixed.

Phosphorus

Although the percentage of phosphorus in plant material is relatively low, it is an essential component of plants. Phosphorus, like nitrogen, must be present in a simple inorganic form before it can be taken up by plants. In the case of phosphorus, the utilizable species is some form of orthophosphate ion. In the pH range that is present in most soils, $H_2PO_4^-$ and HPO_4^{2-} are the predominant orthophosphate species. Because of precipitation reactions, such as the formation of relatively insoluble hydroxyapatite,

$$3HPO_4^{2-} + 5CaCO_3(s) + 2H_2O \rightarrow Ca_5(PO_4)_3(OH)(s) + 5HCO_3^- + OH^- \qquad (14.5.4)$$

little phosphorus applied as fertilizer leaches from the soil. This is important from the standpoint of both water pollution and utilization of phosphate fertilizers.

Potassium

Relatively high levels of potassium are utilized by growing plants. Potassium activates some enzymes and plays a role in the water balance in plants. It is also essential for some carbohydrate transformations. Crop yields are generally greatly reduced in potassium-deficient soils. The higher the productivity of the crop, the more potassium is removed from soil. When nitrogen fertilizers are added to soils to increase productivity, removal of potassium is enhanced. Therefore, potassium may become a limiting nutrient in soils heavily fertilized with other nutrients.

Potassium is one of the most abundant elements in the Earth's crust, of which it makes up 2.6%; however, much of this potassium is not easily available to plants. For example, some silicate minerals such as leucite, $K_2O{\bullet}Al_2O_3{\bullet}4SiO_2$, contain strongly bound potassium. Exchangeable potassium held by clay minerals is relatively more available to plants.

Nitrogen, Phosphorus, and Potassium Fertilizers

Crop fertilizers contain nitrogen, phosphorus, and potassium as major components. Magnesium, sulfate, and micronutrients may also be added. Fertilizers are designated by numbers, such as 6-12-8, showing the respective percentages of nitrogen expressed as N (in this case 6%), phosphorus as P_2O_5 (12%), and potassium as K_2O (8%). Farm manure corresponds to an approximately 0.5-0.24-0.5 fertilizer, so it is not a very effective fertilizer. Such organic fertilizers must biodegrade to release the simple inorganic species (NO_3^-, $H_xPO_4^{x-3}$, K^+) assimilable by plants.

Most modern nitrogen fertilizers are made by the Haber process, in which N_2 and H_2 are combined over a catalyst at temperatures of approximately 500°C and pressures up to 1000 atm:

$$N_2 + 3H_2 \rightarrow 2NH_3 \quad (14.5.5)$$

The anhydrous ammonia product has a very high nitrogen content of 82%. It may be added directly to the soil, for which it has a strong affinity because of its water solubility and formation of ammonium ion:

$$NH_3(g)\ (\text{water}) \rightarrow NH_3(aq) \quad (14.5.6)$$

$$NH_3(aq) + H_2O \rightarrow NH_4^+ + OH^- \quad (14.5.7)$$

Special equipment is required to apply anhydrous NH_3 because ammonia gas is toxic. Aqua ammonia, a 30% solution of NH_3 in water, may be used with much greater safety. It is sometimes added directly to irrigation water. It should be pointed out that ammonia vapor is toxic and NH_3 is reactive with some substances. Improperly discarded or stored ammonia can be a hazardous waste.

Ammonium nitrate, NH_4NO_3, is a common solid nitrogen fertilizer. It is made by oxidizing ammonia over a platinum catalyst, converting the nitric oxide product to nitric acid, and reacting the nitric acid with ammonia. Although convenient to apply to soil, ammonium nitrate requires considerable care during manufacture and storage because it is explosive. Ammonium nitrate also poses some hazards. It is mixed with fuel oil to form an explosive that serves as a substitute for dynamite in quarry blasting and construction. This mixture was used as the explosive agent in the tragic 1995 bombing of the Federal Building in Oklahoma City.

Other compounds used as nitrogen fertilizers include sodium nitrate (obtained largely from deposits found in arid regions of Chile), calcium nitrate, potassium nitrate, and ammonium phosphates. Ammonium sulfate, a by-product of coke ovens, used to be widely applied as fertilizer. The alkali metal nitrates tend to make soil alkaline, whereas ammonium sulfate leaves an acidic residue.

Phosphate minerals are found in several states, including Idaho, Montana, Utah, Wyoming, North Carolina, South Carolina, Tennessee, and Florida. The principal mineral is fluorapatite, $Ca_5(PO_4)_3F$. The phosphate from fluorapatite is relatively unavailable to plants and fluorapatite is frequently treated with phosphoric or sulfuric acids to produce superphosphates:

$$2Ca_5(PO_4)_3F(s) + 14H_3PO_4 + 10H_2O \rightarrow 2HF(g) + 10Ca(H_2PO_4)_2{\cdot}H_2O \quad (14.5.8)$$

$$2Ca_5(PO_4)_3F(s) + 7H_2SO_4 + 3H_2O \rightarrow 2HF(g) + 3Ca(H_2PO_4)_2{\cdot}H_2O + 7CaSO_4 \quad (14.5.9)$$

The superphosphate products are much more soluble than the parent phosphate minerals. The HF produced as a by-product of superphosphate production must be contained to prevent air pollution problems.

Potassium fertilizer components consist of potassium salts, generally KCl. Such salts are found as deposits in the ground or may be obtained from some brines. Very large deposits are found in Saskatchewan, Canada. These salts are all quite soluble in water. One problem encountered with potassium fertilizers is the luxury uptake of potassium by some crops, which absorb more potassium than is really needed for their maximum growth. In a crop where only the grain is harvested, leaving the rest of the plant in the field, luxury uptake does not create much of a problem because most of the potassium is returned to the soil with the dead plant. However, when hay or forage is harvested, potassium contained in the plant as a consequence of luxury uptake is lost from the soil.

14.6. MICRONUTRIENTS IN SOIL

Boron, chlorine, copper, iron, manganese, molybdenum (for N-fixation), and zinc are considered essential plant **micronutrients**. These elements are needed by plants only at very low levels and frequently are toxic at higher levels. Most of these elements function as components of essential enzymes. Manganese, iron, chlorine, and zinc may be involved in photosynthesis.Though not established for all plants, it is possible that sodium, silicon, and cobalt may also be essential plant nutrients.

Iron and manganese occur in a number of soil minerals. Sodium and chlorine (as chloride) occur naturally in soil and are transported as atmospheric particulate matter from marine sprays. Some of the other micronutrients and trace elements are found in primary (unweathered) minerals that occur in soil.

Soil trace elements may be coprecipitated with secondary minerals that are involved in soil formation. Such secondary minerals include oxides of aluminum, iron, and manganese (precipitation of hydrated oxides of iron and manganese very efficiently removes many trace metal ions from solution); calcium and magnesium carbonates; smectites; vermiculites; and illites.

Some plants accumulate extremely high levels of specific trace metals. Those accumulating more than 1.00 mg/g of dry weight are called **hyperaccumulators**. *Aeolanthus biformifolius DeWild* growing in copper-rich regions of Shaba Province, Zaire, contains up to 1.3% copper (dry weight) and is known as a "copper flower".

14.7. WASTES AND POLLUTANTS IN SOIL

Soil receives large quantities of waste products. Much of the sulfur dioxide emitted in the burning of sulfur-containing fuels ends up on soil as sulfates. Atmospheric nitrogen oxides are converted to nitrates in the atmosphere, and the nitrates eventually are deposited on soil. Soil sorbs NO and NO_2 readily, and these gases are oxidized to nitrate in the soil. Carbon monoxide is converted to CO_2 and possibly to biomass by soil bacteria and fungi. Particulate lead from automobile exhausts is found at elevated levels in soil along heavily traveled highways. Elevated levels of lead from lead mines and smelters are found on soil near such facilities.

Soil is the receptor of many hazardous wastes from landfill leachate, lagoons, and other sources (see Section 23.10). In some cases, land farming of degradable hazardous organic wastes is practiced as a means of disposal and degradation. The degradable material is worked into the soil, and soil microbial processes bring about its degradation. As discussed in Chapter 8, sewage and fertilizer-rich sewage sludge may be applied to soil.

Volatile organic compounds (VOC), such as benzene, toluene, xylenes, dichloromethane, trichloroethane, and trichloroethylene, may contaminate soil in industrialized and commercialized areas. One of the more common sources of these contaminants is leaking underground storage tanks. Landfills built before current stringent regulations were enforced and improperly discarded solvents are also significant sources of soil VOCs.

Measurements of levels of polychlorinated biphenyls (PCBs) in soils that have been archived for several decades provide interesting insight into the contamination of soil by pollutant chemicals and subsequent loss of these substances from soil.[1] Analyses of soils from the United Kingdom dating from the early 1940s to 1992 showed that the PCB levels increased sharply from the 1940s, reaching peak levels around 1970. Subsequently, levels fell sharply, and by 1993 were back to the early 1940s concentrations. This fall was accompanied by a shift in distribution to the more highly chlorinated PCBs, which was attributed by those doing the study to evaporation and long-range transport of the lighter PCBs away from the soil. These trends parallel levels of PCB manufacture and use in the United Kingdom from the early 1940s to the present. This is consistent with the observation that relatively high concentrations of PCBs have been observed in remote Arctic and sub-Arctic regions attributed to condensation in colder climates of PCBs evaporated in warmer regions.

Soil receives enormous quantities of pesticides as an inevitable result of their application to crops. It has been estimated[2] that globally about $20 billion are spent each year on 2.5 million tons of agricultural pesticides, whereas in the U.S. the corresponding figures are $4.1 billion and 500,000 tons. The degradation and eventual fate of these enormous quantities of pesticides on soil largely determine their ultimate environmental effects. Detailed knowledge of these effects are now required for licensing of a new pesticide (in the U.S. under the Federal Insecticide, Fungicide, and Rodenticide Act, FIFRA). Among the factors to be considered are the sorption of the pesticide by soil; leaching of the pesticide into water, as related to its potential for water pollution; effects of the pesticide on microorganisms and animal life in the soil; and possible production of relatively more toxic degradation products.

Adsorption by soil is a key step in the degradation of a pesticide. The degree of adsorption and the speed and extent of ultimate degradation are influenced by a number of factors, including solubility, volatility, charge, polarity, and molecular structure and size. Adsorption of a pesticide by soil components may have several effects. Under some circumstances, it retards degradation by separating the pesticide from the microbial enzymes that degrade it, whereas under other circumstances the reverse is true. Purely chemical degradation reactions may be catalyzed by adsorption. Loss of the pesticide by volatilization or leaching is diminished. The toxicity of a herbicide to plants may be strongly affected by soil sorption.

Degradation of Pesticides on Soil

The three primary ways in which pesticides are degraded in or on soil are *chemical degradation*, *photochemical reactions*, and, most important, *biodegradation*. Various combinations of these processes may operate in the degradation of a pesticide.

Chemical degradation of pesticides has been observed experimentally in soils and clays sterilized to remove all microbial activity. Of the chemical degradation reactions, probably the most common are hydrolytic reactions of pesticides in which the molecules split apart with the addition of molecules of H_2O.

Many pesticides have been shown to undergo **photochemical reactions**, that is, chemical reactions brought about by the absorption of light (see Chapter 9). Frequently, isomers of the pesticides are formed as products. Many of the studies reported apply to pesticides in water or on thin films, and the photochemical reactions of pesticides on soil and plant surfaces remain largely a matter of speculation.

Biodegradation and the Rhizosphere

Although insects, earthworms, and plants may be involved to a minor extent in the **biodegradation** of pesticides and other pollutant organic chemicals, microorganisms have the most important role. Several examples of microorganism-mediated degradation of organic chemical species are given in Chapter 6.

The **rhizosphere**, the layer of soil in which plant roots are most active, is a particularly important part of soil in respect to biodegradation of wastes. It is a zone of increased biomass and is strongly influenced by the plant root system and the microorganisms associated with plant roots. The rhizosphere may have more than 10 times the microbial biomass per unit volume compared to nonrhizospheric zones of soil. This population varies with soil characteristics, plant and root characteristics, moisture content, and exposure to oxygen. If this zone is exposed to pollutant compounds, microorganisms adapted to their biodegradation may also be present.

Plants and microorganisms exhibit a strong synergistic relationship in the rhizosphere, which benefits the plant and enables highly elevated populations of rhizospheric microorganisms to exist. Epidermal cells sloughed from the root as it grows and carbohydrates, amino acids, and root-growth-lubricant mucigel secreted from the roots all provide nutrients for microorganism growth. Root hairs provide a hospitable biological surface for colonization by microorganisms.

The biodegradation of a number of synthetic organic compounds has been demonstrated in the rhizosphere. Understandably, studies in this area have focused on herbicides and insecticides that are widely used on crops. Among the organic species for which enhanced biodegradation in the rhizosphere has been shown are the following (associated plant or crop shown in parentheses): 2,4-D herbicide (wheat, African clover, sugarcane, flax), parathion (rice, bush bean), carbofuran (rice), atrazine (corn), diazinon (wheat, corn, peas), volatile aromatic alkyl and aryl hydrocarbons and chlorocarbons (reeds), and surfactants (corn, soybean, cattails). It is interesting to note that enhanced biodegradation of polycyclic aromatic hydrocarbons (PAH) was observed in the rhizospheric zones of prairie grasses. This observation is consistent with the fact that in nature such grasses burn regularly and significant quantities of PAH compounds are deposited on soil as a result.

14.8. SOIL LOSS AND DETERIORATION

Soil is a fragile resource that can be lost by erosion or become so degraded that it is no longer useful to support crops. **Desertification** refers to the process associated with drought and loss of fertility by which soil becomes unable to grow significant amounts of plant life. Desertification caused by human activities is a common problem globally, occurring in diverse locations, such as Argentina, the Sahara, Uzbekistan, the U.S. Southwest, Syria, and Mali. It is a very old problem dating back many centuries to the introduction of domesticated grazing animals to areas where rainfall and groundcover were marginal. The most notable example is desertification aggravated by domesticated goats in the Sahara region. Desertification involves a number of interrelated factors, including erosion, climate variations, water availability, loss of fertility, loss of soil humus, and deterioration of soil chemical properties.

A related problem is **deforestation** consisting of loss of forests. The problem is particularly acute in tropical regions, where the forests contain most of the existing plant and animal species. In addition to extinction of these species, deforestation can cause devastating deterioration of soil through erosion and loss of nutrients.

Soil erosion can occur by the action of both water and wind, although water is the primary source of erosion. Millions of tons of topsoil are carried by the Mississippi River and swept from its mouth each year. About one-third of U.S. topsoil has been lost since cultivation began on the continent. At the present time approximately one-third of U.S. cultivated land is eroding at a rate sufficient to reduce soil productivity. It is estimated that 48 million acres of land, somewhat more than 10 percent of that under cultivation, is eroding at unacceptable levels, taken to mean a loss of more than 14 tons of topsoil per acre each year. Specific areas in which the greatest erosion is occurring include northern Missouri, southern Iowa, west Texas, western Tennessee, and the Mississippi Basin. Figure 14.5 shows the pattern of soil erosion in the lower 48 continental United States.

Problems involving soil erosion were aggravated in the 1970s and early 1980s when high prices for farmland caused intensive cultivation of high-income crops, particularly corn and soybeans. These crops grow in rows with bare soil in between, which tends to wash away with each rainfall. Furthermore, the practice became widespread of planting corn and soybeans year after year, without intervening plantings of soil-restoring clover or grass. The problem of decreased productivity due to soil erosion has been masked somewhat by increased use of chemical fertilizers. As living standards and industrialization increase in highly populated developing countries, pressures on the world's limited supplies of good farmland will increase.

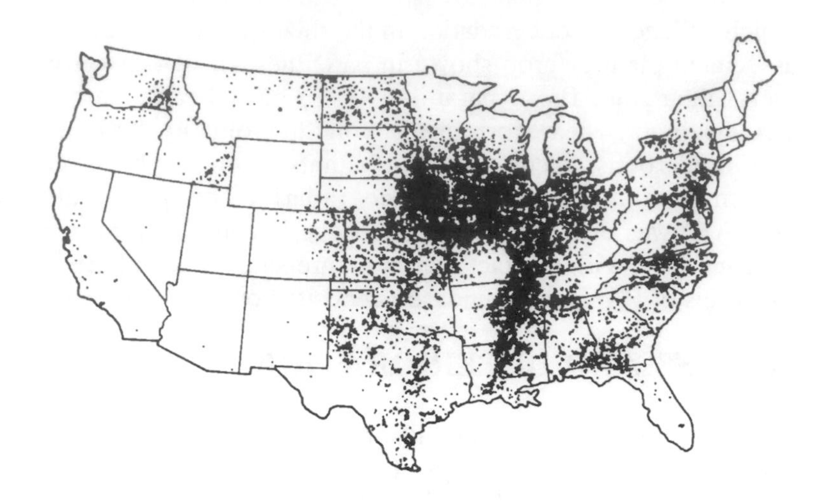

Figure 14.5. Pattern of soil erosion in the continental U.S. as of 1977. The dark areas indicate locations where the greatest erosion is occurring.

Wind erosion, such as occurs on the generally dry, high plains soils of eastern Colorado, poses another threat. After the Dust Bowl days of the 1930s, much of this land was allowed to revert to grassland, and the topsoil was held in place by the strong root systems of the grass cover. However, in an effort to grow more wheat and improve the sale value of the land, much of it has been cultivated in recent years. For example, from 1979 through 1982, more than 450,000 acres of Colorado grasslands were plowed. Much of this was done by speculators who purchased grassland at a low price of $100-$200 per acre, broke it up, and sold it as cultivated land at more than double the original purchase price. Although freshly cultivated grassland may yield well for one or two years, the nutrients and soil moisture are rapidly exhausted, and the land becomes very susceptible to wind erosion.

The preservation of soil from erosion is commonly termed **soil conservation**. There are a number of solutions to the soil erosion problem. Some are old, well-known agricultural practices, such as terracing, contour plowing, and periodically planting fields with cover crops, such as clover. For some crops **no-till agriculture** greatly reduces erosion. This practice consists of planting a crop among the residue of the previous year's crop, without plowing. Weeds are killed in the newly planted crop row by application of a herbicide prior to planting. The surface residue of plant material left on top of the soil prevents erosion.

Another, more experimental, solution to the soil erosion problem is the cultivation of perennial plants, which develop a large root system and come up each spring after being harvested the previous fall. For example, a perennial corn plant has been developed by crossing corn with a distant, wild relative, teosinte, which grows in Central America. Unfortunately, the resulting plant does not give outstanding grain yields. It should be noted that an annual plant's ability to propagate depends upon producing large quantities of seeds, which is why plants harvested for their grain (seeds) are annual plants. In contrast, a perennial plant must develop a strong root system with bulbous growths called rhizomes, which store food for the coming year. However, it is possible that the application of genetic engineering (see Section 14.9) may result in the development of perennial crops with good seed yields. The cultivation of such a crop would cut down on a great deal of soil erosion.

The best known perennial plants are trees, which are very effective in stopping soil erosion. Wood from trees can be used as biomass fuel, as a source of raw materials, and as food (see below). There is a tremendous unrealized potential for an increase in the production of biomass from trees. For example, the production of biomass from natural forests of loblolly pine trees in South Carolina used to be about 3 dry tons per hectare per year. This has now been increased at least four-fold through selection of superior trees, and 30 tons may eventually be possible. In Brazil, experiments have been conducted with a species of Eucalyptus, which has a 7-year growth cycle. With improved selection of trees, the annual yields for three successive cycles of these trees in dry tons per hectare per year were 23, 33, and 40.

The most important use for wood is, of course, as lumber for construction. This use will remain important as higher energy costs increase the costs of other construction materials, such as steel, aluminum, and cement. Wood is about 50 percent cellulose, which can be hydrolyzed by rapidly improving enzyme processes to yield glucose sugar. The glucose can be used directly as food, fermented to ethyl alcohol for fuel (gasohol), or employed as a carbon and energy source for protein-producing yeasts. Given these and other potential uses, the future of trees as an environmentally desirable and profitable crop is very bright.

Soil and Water Resources

The conservation of soil and the protection of water resources are strongly interrelated. Most fresh water falls initially on soil, and the condition of the soil largely determines the fate of the water and how much is retained in a usable condition. The land area upon which rainwater falls is called a **watershed**. In addition to collecting the water, the watershed determines the direction and rate of flow and the degree of water infiltration into groundwater aquifers (see the hydrologic cycle in Figure 5.1). Excessive rates of water flow prevent infiltration, lead to flash floods, and cause soil erosion. Measures taken to enhance the utility of land as a watershed also fortunately help prevent erosion. Some of these measures involve modification of the contour of the soil, particularly terracing, construction of waterways, and construction of water-retaining ponds. Waterways are planted with grass to prevent erosion, and water-retaining crops and bands of trees can be planted on the contour to achieve much the same goal. Reforestation and control of damaging grazing practices conserve both soil and water.

14.9. GENETIC ENGINEERING AND AGRICULTURE

The nuclei of living cells contain the genetic instruction for cell reproduction. These instructions are in the form of a special material called deoxyribonucleic acid, DNA. In combination with proteins, DNA makes up the cell chromosomes. During the 1970s the ability to manipulate DNA through genetic engineering became a reality and during the 1980s became the basis of a major industry. Such manipulation falls into the category of recombinant DNA technology. Recombinant DNA gets its name from the fact that it contains DNA from two different organisms, recombined together. This technology promises some exciting developments in agriculture.

The "green revolution" of the mid-1960s used conventional plant-breeding techniques of selective breeding, hybridization, cross-pollination, and back-crossing to develop new strains of rice, wheat, and corn, which, when combined with chemical fertilizers, yielded spectacularly increased crop yields. For example, India's output of grain increased 50 percent. By working at the cell nucleus level, however, it is now possible to greatly accelerate the process of plant breeding. Thus, plants may be developed that resist particular diseases, produce much larger quantitities of desired products, grow in seawater, or have much higher productivity. The possibility exists for developing entirely new kinds of plants.

One exciting possibility with genetic engineering is the development of plants other than legumes which fix their own nitrogen. For example, if nitrogen-fixing corn could be developed, the savings in fertilizer would be enormous. Furthermore, since the nitrogen is fixed in an organic form in plant root structures, there would be no pollutant runoff of chemical fertilizers.

Another promising possibility with genetic engineering is increased efficiency of photosynthesis. Plants utilize only about 1 percent of the sunlight striking their leaves, so there is appreciable room for improvement in that area.

Cell-culture techniques can be applied in which billions of cells are allowed to grow in a medium and develop mutants that, for example, might be resistant to particular viruses or herbicides or have other desirable qualities. If the cells with the desired qualities can be regenerated into whole plants, results can be obtained that might have taken decades using conventional plant-breeding techniques.

Transgenic Crops Come of Age

By the late 1990s, transgenic crops genetically engineered to resist pests or pesticides or to produce desired products in large quantities have reached commercial production on a large scale and are projected to have a U.S. market of about $6 billion by the year 2005.[3] The most rapidly growing transgenic crop has been Monsanto's Bollgard® cotton, in which a gene from *Bacillus thuringiensis* has been placed in cotton in which it directs the cotton plant to produce natural pesticides effective against cotton bollworm and budworm. In 1997 about 20% of the 14-million-acre U.S. cotton crop was planted to Bollgard, up from 13% planted in 1996, the first year that the new variety was available. Although not entirely effective against the target pests in 1996, transgenic cotton may have saved the need for use of as much as 1 million liters of synthetic chemical pesticides in 1996. Since U.S. cotton growers spend up to $500 million on pesticides each year, the financial savings are substantial.

The natural insecticides produced by the *Bacillus thuringiensis* genes in cotton are proteins, which are generally regarded as the safest of insecticides. This safety is due to several factors, including their production by and confinement to the plants being protected, the natural biodegradability of proteins, and their toxicological specificity for the target pests.

As is the case with synthetic pesticides, insects can develop resistance to the insecticidal proteins generated by the *Bacillus thuringiensis* genes, a process that is accelerated by the fact that the plants produce the insecticide continually so that insects are exposed throughout the growing season. To delay the development of resistance, therefore, growers are required to plant a certain percentage of their crop with nonresistant crop varieties so that insects feeding on these varieties may crossbreed with insect strains that have developed resistance.

An interesting approach to the use of genetic engineering to control pests is the development of transgenic crops that are immune to herbicides. The most successful example of this approach is the development of soybeans, and to a lesser extent cotton, that are not killed by exposure to Roundup® herbicide. This herbicide can be sprayed directly on the resistant crops, leaving them unharmed while killing the competing weeds. About 15% of the 10-million-acre U.S. soybean crop was planted to Roundup-resistant strains in 1997.

As of 1997 transgenic crops were reaching the market for several purposes. These include herbicide-resistant canola, corn, cotton, and soybeans; insect-resistant corn, cotton, and potatoes; and virus-resistant squash. In addition strains of tomatoes had been genetically engineered for slower ripening and increased pectin, and transgenic canola had been developed for increased yields of desired canola oils.

Limits of Technology in Food Production

Despite the enormous potential of the "green revolution," genetic engineering, and more intensive cultivation of land to produce food and fiber, these technologies cannot be relied upon to support an uncontrolled increase in world population and may even simply postpone an inevitable day of reckoning with the consequences of population growth. Changes in climate resulting from global warming (greenhouse effect, Section 12.2), ozone depletion (by chlorofluorocarbons, Section 12.5), or natural disasters, such as massive volcanic eruptions or collisions with large meteorites can, and almost certainly will, result in worldwide famine conditions in the future that no agricultural technology will be able to alleviate.

14.10. AGRICULTURE AND HEALTH

Some authorities hold that soil has an appreciable effect upon health. An obvious way in which such an effect might be manifested is the incorporation into food of micronutrient elements essential for human health. One such nutrient (which is toxic at overdose levels) is selenium. It is definitely known that the health of animals is adversely affected in selenium-deficient areas, as it is in areas of selenium excess. Human health might be similarly affected.

There are some striking geographic correlations with the occurrence of cancer. Some of these correlations may be due to soil type. A high incidence of stomach cancer has been shown to occur in areas with certain types of soil in the Netherlands, the United States, France, Wales, and Scandinavia. These soils are high in organic matter content, are acidic, and frequently are waterlogged. A "stomach cancer-prone life style" has been described,[4] which includes consumption of home-grown food, consumption of water from one's own well, and reliance on native and uncommon foodstuffs.

One possible reason for the existence of "stomach cancer-producing soils" is the production of cancer-causing secondary metabolites by plants and microorganisms. Secondary metabolites are biochemical compounds that are of no apparent use to the organism producing them. It is believed that they are formed from the precursors of primary metabolites when the primary metabolites accumulate to excessive levels.

The role of soil in environmental health is not well known, nor has it been extensively studied. The amount of research on the influence of soil in producing foods that are more nutritious and lower in content of naturally occurring toxic substances is quite small compared to research on higher soil productivity. It is to be hoped that the environmental health aspects of soil and its products will receive much greater emphasis in the future.

Chemical Contamination

Sometimes human activities contaminate food grown on soil. Most often this occurs through contamination by pesticides. An interesting example of such contamination occurred in Hawaii in early 1982. It was found that milk from several sources on Oahu contained very high levels of heptachlor. This pesticide causes cancer and liver disorders in mice; therefore, it is a suspected human carcinogen. Remarkably, in this case it was not until 57 days after the initial discovery that the public was informed of the contamination by the Department of Health. The source of heptachlor was traced to contaminated "green chop," chopped-up pineapple leaves fed to cattle. Although heptachlor was banned for most applications, Hawaiian pineapple growers had obtained special Federal permission to use it to control mealybug wilt. Although it was specified that green chop could not be collected within 1 year of the last application of the pesticide, apparently this regulation was violated, and the result was distribution of contaminated milk to consumers.

In the late 1980s Alar residues on food caused considerable controversy in the marketplace. **Alar**, daminozide, is a growth regulator that was widely used on apples to bring about uniform ripening and to improve firmness and color of the apples. It was discontinued for this purpose after 1988 because of concerns that it might cause cancer, particularly in those children who consume relatively large amounts of apples, apple juice, and other apple products. Dire predictions were made in the industry of crop losses and financial devastation. However, the 1989 apple crop, which was the first without Alar in the U.S. had a value of $1.0 billion, only $0.1 billion less than that of the 1988 crop, and production of apples has continued since then without serious problems from the unavailability of Alar.

CHAPTER SUMMARY

The chapter summary below is presented in a programmed format to review the main points covered in this chapter. It is used most effectively by filling in the blanks, referring back to the chapter as necessary. The correct answers are given at the end of the summary.

Agriculture can be divided into the two main categories of [1]________________ and [2] ____________________. Long before recombinant DNA technology was known, the development of [3]________________ had vastly increased yields and other desired characteristics of a number of important crops. A variable mixture of minerals, organic matter, and water, capable of supporting plant life on the Earth's surface defines [4]__________________. Typical soils exhibit distinctive layers called [5]________________. The process by which water enters the atmosphere from a plant's leaves is called [6]____________________. The liberation of phytotoxic (substance toxic to plants) soluble Fe^{2+} from insoluble $Fe_2O_3(s)$ in soil occurs under [7] ____________________________________ conditions. The soil solution serves to [8] ____

____

____.

The most common elements in the mineral portion of soil are [9]________________ ________________. Among the functions served by organic matter in soil are [10] ____

____

____.

The most significant organic constituent of soil is [11]______________. The reaction Soil}NH_4^+ + $H^+(aq)$ $\leftarrow\rightarrow$ Soil}H^+ + $NH_4^+(aq)$ shows the process of [12]______________ in soil. The reaction Soil}$(H^+)_2$ + $CaCO_3$ $\rightarrow$ Soil}Ca^{2+} + CO_2 + H_2O occurs as part of a process to treat excess [13]______________ in soil. Elements that occur in substantial levels in plant materials or in fluids in the plant are called [14]______________, whereas those that are essential only at very low levels are called [15]______________. Of the former, [16]______________ are the most likely to be lacking and are commonly added to soil as fertilizers. Root nodules on legumes contain [17]______________. In the pH range that is present in most soils, [18]______________ are the phosphorus-containing species generally most utilized by plants. The three primary ways in which pesticides are degraded in or on soil are [19]______________. The [20]______________ is the layer of soil in which plant roots are particularly active. The process associated with drought and loss of fertility by which soil becomes unable to grow significant amounts of plant life is called [21]______________. The loss of soil by the action of wind and water is called [22]______________. No-till agriculture is a process that consists of [23]______________. The land area upon which rainwater falls is called a [24]______________.

Answers

1 crop farming

2 livestock farming

3 hybrids

4 soil

5 horizons

6 transpiration

7 waterlogged reducing

8 provide water to plants, transport chemical species to and from soil particles, provide intimate contact between the solutes and the soil particles, and exchange plant nutrients between roots and solid soil

9 silicon and oxygen

10 source of food for microorganisms, chemical reactions such as ion exchange, and influence on the physical properties of soil

11 soil humus

12 ion exchange

13 acidity

14 macronutrients

15 micronutrients

16 nitrogen, phosphorus, and potassium

17 nitrogen-fixing bacteria

18 $H_2PO_4^-$ and HPO_4^{2-}

19 chemical degradation, photochemical reactions, and biodegradation

20 rhizosphere

21 desertification

22 soil erosion

23 planting a crop among the residue of the previous year's crop, without plowing

24 watershed

QUESTIONS AND PROBLEMS

1. Give two examples of reactions involving manganese and iron compounds that may occur in waterlogged soil.
2. What temperature and moisture conditions favor the buildup of organic matter in soil?
3. What is a minor-use pesticide? What is a specific economic/regulatory problem or concern with minor-use pesticides?
4. Justify the statement that "soil and soil systems are highly complex and variable."
5. Suggest a phenomenon by which heavy crop growth during the summer may have a severe drying effect on soil.
6. List the functions and explain the importance of the soil solution.
7. What is the most significant organic constituent of soil? How is it produced? What does it do in soil?
8. How does cation exchange function in soil? What useful purpose does it serve?
9. Which macronutrients are most likely to be lacking in soil? How may they be replenished?
10. How are the chelating agents that are produced from soil microorganisms involved in soil formation?
11. Some kinds of plants can be "self-fertilizing" with nitrogen. Explain how this works and how a symbiotic relationship with another kind of organism is involved.
12. What is the purpose of treating phosphate minerals with sulfuric or phosphoric acids to make phosphate fertilizers?
13. Under the U.S. FIFRA Act what are some of the factors that must be considered and studied when licensing a new fertilizer?

14. What is no-till agriculture? How are herbicides essential for the practice of this environmentally friendly technique?

15. What was the "green revolution?" How might advances in genetic engineering with recombinant DNA lead to a second, even greater "green revolution?"

16. Suggest how soil might act on pollutants to reduce their harmful effects.

LITERATURE CITED

1 Alcock, R. E., A. E. Johnston, S. P. McGrath, M. L. Berrow, and Kevin C. Jones, "Long-Term Changes in the Polychlorinated Biphenyl Content of United Kingdom Soils," *Environmental Science and Technology*, **27**, 1918-1923 (1993).

2 Pimentel, David, H. Acquay, M. Biltonen, P. Rice, M. Silva, J. Nelson, V. Lipner, S. Giordano, A. Horowitz, and M. D'Amore, "Environmental and Economic Costs of Pesticide Use," *BioScience*, **42**, 750-760 (1992).

3 Thayer, Ann M., "Betting the Transgenic Farm," *Chemical and Engineering News*, April 28, 1997, pp. 15-19.

4 Adams, R. S., Jr., "Soil Variability and Cancer," *Chemical and Engineering News*, June 12, 1978, p. 84.

SUPPLEMENTARY REFERENCES

Alef, Kassem and Paolo Nannipieri, *Methods in Applied Soil Microbiology and Biochemistry*, Academic Press, Orlando, FL, 1995.

Arntzen, Charles J. and Ellen M. Ritter, Eds., *Encyclopedia of Agricultural Science*, Academic Press, Orlando, FL, 1994.

Board on Agriculture, *Ecologically Based Pest Management: New Solutions for a New Century*, National Academy Press, Washington, D.C., 1996.

Brown, Lester R., *Tough Choices: Facing the Challenge of Food Scarcity*, Worldwatch Institute, Washington, D.C., 1996.

Brussaard, Lijbert and Ronald Ferrara-Cerrato, *Soil Ecology in Sustainable Agriculture Systems*, CRC Press/Lewis Publishers, Boca Raton, FL, 1997.

Ellis, Boyd G. and Henry D. Foth, *Soil Fertility*, 2nd ed., CRC Press/Lewis Publishers, Boca Raton, FL, 1997.

Franz, John E., Michael K. Mao, and James A. Sikorski, *Glyphosate: A Unique Global Herbicide*, American Chemical Society, Washington, D.C., 1997.

Hedin, Paul A., Ed., *Phytochemicals for Pest Control* American Chemical Society, Washington, D.C., 1997.

Lal, Rattan, W. H. Blum, and C. Valentin, *Methods for Assessment of Soil Degradation*, CRC Press/Lewis Publishers, Boca Raton, FL, 1997.

McBaride, Murray B., *Environmental Chemistry of Soils*, OUP, New York, NY, 1994.

Marschner, Horst, *Mineral Nutrition of Higher Plants*, 2nd ed., Academic Press, Orlando, FL, 1995.

Montgomery, John H., Ed., *Agrochemicals Desk Reference*, 2nd ed., CRC Press/Lewis Publishers, Boca Raton, FL, 1997.

Paul, Eldor A. and Francis E. Clark, *Soil Microbiology and Chemistry*, 2nd ed., Academic Press Textbooks, San Diego, CA, 1995.

Prakash, Anand and Jagadiswari Rao, *Botanical Pesticides in Agriculture*, CRC Press/Lewis Publishers, Boca Raton, FL, 1997.

Prasad, Rajendra and James F. Power, *Soil Fertility Management for Sustainable Agriculture*, CRC Press/Lewis Publishers, Boca Raton, FL, 1997.

Rechcigl, Jack E. and Nancy A. Rechcigl, *Environmentally Safe Approaches to Crop Disease Control*, CRC Press/Lewis Publishers, Boca Raton, FL, 1997.

Sparks, Donald L., *Environmental Soil Chemistry*, Academic Press, Orlando, FL, 1995.

Tan, Kim H., *Environmental Soil Science*, Marcel Dekker, Inc., New York, NY, 1994.

15 ENVIRONMENTAL GEOLOGY AND GEOSPHERIC POLLUTION

15.1. INTRODUCTION

This chapter deals specifically with the environmental aspects of geology and human interactions with the geosphere. To a large extent it discusses how natural geological phenomena affect the environment through occurrences such as volcanic eruptions that may blast so much particulate matter and acid gas into the atmosphere that it may have a temporary effect on global climate, or massive earthquakes that disrupt surface topography and disturb the flow and distribution of groundwater and surface water. The chapter also covers human influences on the geosphere and the strong connection between the geosphere and the anthrosphere.

Going back several billion years to its formation as a ball of dust particles collected from the universe and held together by gravitational forces, Earth has witnessed constant environmental change and disruption. During its earlier eons Earth was a most inhospitable place for humans and, indeed, for any form of life. Heat generated by gravitational compression of primitive Earth and by radioactive elements in its interior caused much of the mass of the planet to liquify. Relatively high density iron sank into the core, and lighter minerals, primarily silicates, solidified and floated to the surface.

Although in the scale of a human lifetime Earth changes almost imperceptibly, the planet is in fact in a state of constant change and turmoil. It is known that continents have formed, broken apart, and moved around. Rock formations produced in ancient oceans have been thrust up onto continental land and huge masses of volcanic rock exist where volcanic activity is now unknown. Today the angry bowels of Earth unleash enormous forces that push molten rock to the surface and move continents continuously as evidenced from volcanic activity and from earthquakes resulting from the movement of great land masses relative to each other. Earth's surface is constantly changing as new mountain ranges are heaved up and old ones are worn down.

Humans have learned to work with, against, and around natural Earth processes and phenomena to exploit Earth's resources and to make these processes and phenomena work for the benefit of humankind. Human efforts have been moderately successful in mitigating some of the major hazards posed by natural geospheric phenomena. The survival of modern civilization and, indeed, of humankind, will depend upon how intelligently these efforts are applied. That is why it is so important for humans to have a fundamental understanding of the geospheric environment.

An important consideration in human interaction with the geosphere is the application of engineering to geology. Engineering geology takes account of the geological characteristics of soil and rock in designing buildings, dams, highways, and other structures in a manner compatible with the geological strata on which they rest. Engineering geology must consider a large number of geological factors, including type, strength, and fracture characteristics of rock, tendency for landslides to occur, susceptibility to settling, and likelihood of erosion. Engineering geology is an important consideration in land-use planning (see Section 15.14).

Natural Hazards

Earth presents a variety of natural hazards to the creatures that dwell on it. some of these are the result of internal processes that arise from the movement of land masses relative to each other and from heat and intrusions of molten rock from below the surface. The most common such hazards are earthquakes and volcanoes. Whereas internal processes tend to force matter upward, often with detrimental effects, surface processes are those that generally result from the tendency of matter to seek lower levels. Such processes include erosion, landslides, avalanches, mudflows, and subsidence.

A number of natural hazards result from the interaction and conflict between solid Earth and liquid and solid water. Perhaps the most obvious such hazard consists of floods when too much water falls as precipitation and seeks lower levels through streamflow. Wind can team with water to increase destructive effects, such as beach erosion and destruction of beachfront property resulting from wind-driven seawater. Ice, too, can have some major effects on solid Earth. Evidence of such effects from Ice Age times abound with massive glacial moraines left over from deposition of till from melting glaciers, and landscape features carved by advancing sheets of ice.

Anthropogenic Hazards

All too often, attempts to control and reshape the geosphere to human demands have been detrimental to the geosphere and dangerous to human life and well-being. Such attempts may exacerbate damaging natural phenomena. A prime example of this interaction occurs when efforts are made to control the flow of rivers by straightening them and building levees. The initial results can be deceptively favorable in that a modified stream may exist for decades flowing smoothly and staying within the confines imposed by humans. But eventually, the forces of nature are likely to overwhelm the efforts of humans to control them, such as when a record flood breaks levees and destroys structures constructed in flood-prone areas. Landslides of mounds of earthen material piled up from mining can be very destructive. Destruction of wetlands in an effort to provide additional farmland can have some detrimental effects upon wildlife and upon the overall health of ecosystems.

15.2. EARTHQUAKES

As discussed in Section 13.4, earthquakes occur as the result of plate tectonic processes as huge land masses move relative to each other. Adding to the terror of earthquakes is their lack of predictability. An earthquake can strike at any time—

during the calm of late night hours or in the middle of busy rush hour traffic. Although the exact prediction of earthquakes has so far eluded investigators, the locations where earthquakes are most likely to occur are much more well known. These are located in lines corresponding to boundaries along which tectonic plates collide, move relative to each other, and build up stresses that are suddenly released when earthquakes occur. Such interplate boundaries are locations of pre-existing faults and breaks. Occasionally, however, an earthquake will occur within a plate, made more massive and destructive because for it to occur the thick lithosphere has to be ruptured.

The destructive effects of earthquakes are due to the release of energy. The released energy moves from the quakes focus as **seismic waves**. Such waves are of several types. **Surface waves** travel along Earth's surface, whereas **body waves** travel through the interior. The latter are further categorized as **P-waves**, compressional vibrations that result from the alternate compression and expansion of geospheric material, and **S-waves**, consisting of shear waves manifested by sideways oscillations of material. The motions of these waves are detected by a **seismograph**. The two types of waves move at different rates, with P-waves moving faster. By recording arrival times of the two kinds of waves at different seismographic locations, it is possible to quickly and accurately locate the epicenter of an earthquake.

The scale of earthquakes can be estimated by the degree of motion that they cause and by their destructiveness. The former is termed the **magnitude** of an earthquake and is commonly expressed by the **Richter scale**. The Richter scale is open-ended, and each unit increase in the scale reflects a 10-fold increase in magnitude. Several hundred thousand earthquakes with magnitudes from 2 to 3 occur each year; they are detected by seismographs, but are not felt by humans. Minor earthquakes range from 4 to 5 on the Richter scale, and earthquakes cause damage at a magnitude greater than about 5. Great earthquakes, which occur with a frequency of about 1-2 per year, register over 8 on the Richter scale.

The **intensity** of an earthquake is a subjective estimate of its potential destructive effect. On the Mercalli intensity scale, an intensity III earthquake feels like the passage of heavy vehicles; one with an intensity of VII causes difficulty in standing, damage to plaster, and dislodging of loose bricks; whereas a quake with an intensity of XII causes virtually total destruction, throws objects upward, and shifts huge masses of earthen material. Intensity does not correlate exactly with magnitude. Distance from the epicenter, the nature of underlying strata, and the types of structures affected may all result in variations in intensity from the same earthquake. In general, structures built on bedrock will survive with much less damage than those constructed on poorly consolidated material. Displacement of ground along a fault can be substantial, for example, up to 6 meters along the San Andreas fault during the 1906 San Francisco earthquake. Such shifts can break pipelines and destroy roadways. Highly destructive surface waves can shake vulnerable structures apart.

The shaking and movement of ground are the most obvious means by which earthquakes cause damage. In addition to shaking it, earthquakes can cause the ground to rupture, subside, or rise. **Liquefaction** is an important phenomenon that occurs during earthquakes with ground that is poorly consolidated and in which the water table may be high. Liquefaction results from separation of soil particles accompanied by water infiltration such that the ground behaves like a fluid. Another devastating phenomenon consists of **tsunamis**, large ocean waves resulting from earthquake-induced movement of ocean floor. Tsunamis sweeping onshore at speeds up to 1000 km/hr have destroyed many homes and taken many lives, often large distances from the epicenter of the earthquake, itself. This effect occurs when a tsunami approaches land and forms huge breakers, some as high as 10-15 meters.

The loss of life and destruction of property by earthquakes makes them one of nature's more damaging natural phenomena. Literally millions of lives have been lost in past earthquakes, and damage from an earthquake in a developed urban area can easily run into billions of dollars. As examples, a massive earthquake in Egypt and Syria in 1201 A.D. took over 1 million lives, one in Tangshan, China, in 1976 killed about 650,000, and the 1989 Loma Prieta earthquake in California cost about 7 billion dollars. The economic geology (see Section 13.10) associated with earthquakes is very important. Significant progress has been made in designing structures that are earthquake-resistant. As evidence of that, during a 1964 earthquake in Niigata, Japan, some buildings tipped over on their sides due to liquefaction of the underlying soil, but remained structurally intact! Other areas of endeavor that can lessen the impact of earthquakes is the identification of areas susceptible to earthquakes, discouraging development in such areas, and educating the public about earthquake hazards. Accurate prediction would be a tremendous help in lessening the effects of earthquakes, but so far has been generally unsuccessful. Most challenging of all is the possibility of preventing major earthquakes. One unlikely possibility would be to detonate nuclear explosives deep underground along a fault line to release stress before it builds up to an excessive level. Fluid injection to facilitate slippage along a fault has also been considered.

15.3. VOLCANOES

On May 18, 1980, Mount St. Helens, a volcano in Washington State erupted, blowing out about 1 cubic kilometer of material. This massive blast spread ash over half the United States, causing about $1 billion in damages and killing an estimated 62 people, many of whom were never found. Many volcanic disasters have been recorded throughout history. Perhaps the best known of these is the 79 A.D. eruption of Mount Vesuvius, which buried the Roman city of Pompei with volcanic ash.

Temperatures of **lava**, molten rock flowing from a volcano, typically exceed 500°C and may get as high as 1,400°C or more. Lava flows destroy everything in their paths, causing buildings and forests to burn and burying them under rock that cools and becomes solid. Often more dangerous than a lava flow are the **pyroclastics** produced by volcanoes and consisting of fragments of rock and lava. Some of these particles are large and potentially very damaging, but they tend to fall quite close to the vent. Ash and dust may be carried for large distances and, in extreme cases, as was the case in ancient Pompei, may bury large areas to some depth with devastating effects. The explosion of Tambora volcano in Indonesia in 1815 blew out about 30 cubic kilometers of solid material. The ejection of so much solid into the atmosphere had such a devastating effect on global climate that the following year was known as "the year without a summer," causing widespread hardship and hunger because of global crop failures.

A special kind of particularly dangerous pyroclastic consists of **nuée ardente**. This term, French for "glowing cloud," refers to a dense mixture of hot toxic gases and fine ash particles reaching temperatures of 1000°C that can flow down the slopes of a volcano at speeds of up to 100 km/hr. In 1902 a nuée ardente was produced by the eruption of Mont Pelée on Martinique in the Caribbean. Of as many as 40,000 people in the town of St. Pierre, the only survivor was a terrified prisoner shielded from the intense heat by the dungeon in which he was imprisoned.

One of the more spectacular and potentially damaging volcanic phenomena is a **phreatic eruption** that occurs when infiltrating water becomes superheated by hot magma and causes a volcano to literally explode. This happened to uninhabited Krakatoa in Indonesia in 1883, which blew up with an energy release of the order of 100 megatons of TNT. Dust was blown 80 kilometers into the stratosphere, and a perceptible climatic cooling was noted for the next 10 years. As is the case with earthquakes, volcanic eruptions may cause the devastating ocean waves known as tsunamis. Krakatoa produced a tsunami 40 meters high that killed 30 to 40 thousand people on surrounding islands.

Some of the most damaging health and environmental effects of volcanic eruptions are caused by gases released to the atmosphere. Huge quantities of water vapor are often released. The release of dense carbon dioxide gas has the potential to suffocate people near the point of release. Hydrogen sulfide and carbon monoxide are definitely toxic gases that may be released by volcanoes. Volcanoes tend to give off acid gases. One of these is hydrogen chloride, produced by the subduction and heating of sodium chloride entrained in ocean sedimentary material. Sulfur oxides released by volcanoes can have some distinct effects on the atmosphere. In 1982 El Chichón erupted in Mexico, producing comparatively little dust, but huge quantities of sulfur oxides. These gases were converted to sulfuric acid droplets in the atmosphere, which reflected enough sunlight to cause a perceptible cooling in climate. Eventually the sulfuric acid released fell as acidic precipitation, "acid rain."

Volcanic activity has the potential to change the global environment dramatically. Massive volcanic eruptions many millions of years ago were probably responsible for widespread extinctions of organisms on Earth's surface. These effects occur primarily by the ejection of particles and sulfuric acid precursors into the atmosphere causing global cooling and potential harm to the protective stratospheric ozone layer. Although such an extinction event is unlikely in modern times from a volcanic event, an eruption such as the Tambora volcano described above could certainly happen. With humankind "living on the edge" as far as grain supplies are concerned, widespread starvation resulting from a year or two of crop failures would almost certainly occur.

15.4. SURFACE EARTH MOVEMENT

Mass movements are the result of gravity acting upon rock and soil on Earth's surface. This produces a shearing stress on earthen materials located on slopes that can exceed the shear strength of the material and produce landslides and related phenomena involving the downward movement of geological materials. Such phenomena are affected by several factors, including the kinds and therefore strengths of materials, slope steepness, and degree of saturation with water. Usually a specific event initiates mass movement. This can occur when excavation by humans steepens the slopes, by the action of torrential rains, or by earthquakes.

Described in Section 13.4 and illustrated in Figure 13.6, landslides refer to events in which large masses of rock and dirt move downslope rapidly. Such events occur when material resting on a slope at an **angle of repose** is acted upon by gravity to produce a **shearing stress**. This stress may exceed the forces of friction or **shear strength**. Weathering, fracturing, water, and other factors may induce the formation of **slide planes** or **failure planes**, such that a landslide results.

Loss of life and property from landslides can be substantial. In 1970 a devastating avalanche of soil, mud, and rocks initiated by an earthquake slid down Mt.

Huascaran in Peru killing an estimated 20,000 people. Sometimes the effects are indirect. In 1963 a total of 2,600 people were killed near the Vaiont Dam in Italy. A sudden landslide filled the reservoir behind the dam with earthen material and, although the dam held, the displaced water spilled over its abutments as a wave 90 meters high, wiping out structures and lives in its path.

Although often ignored by developers, the tendency toward landslides is predictable, and can be used to determine areas in which homes and other structures should not be built. Slope stability maps based upon the degree of slope, the nature of underlying geological strata, climatic conditions, and other factors can be used to assess the risk of landslides. Evidence of a tendency for land to slide can be observed from effects on existing structures, such as walls that have lost their alignment, cracks in foundations, and poles that tilt. The likelihood of landslides can be minimized by moving material from the upper to the lower part of a slope, avoiding the loading of slopes, and avoiding measures that might change the degree and pathways of water infiltration into slope materials. In cases where the risk is not too severe, retaining walls may be constructed that reduce the effects of landslides.

Several measures can be used to warn of landslides. Simple visual observations of changes in the surface can be indicative of an impending landslide. More sophisticated measures include tilt meters and devices that sense vibrations accompanying the movement of earthen materials.

In addition to landslides, there are several other kinds of mass movements that have the potential to be damaging. **Rockfalls** occur when rocks fall down slopes so steep that at least part of the time the falling material is not in contact with the ground. The fallen material accumulates at the bottom of the fall as a pile of **talus**. A much less spectacular event is **creep**, in which movement is slow and gradual. The action of frost—frost heaving—is a common form of creep. Though usually not life-threatening, over a period of time creep may ruin foundations and cause misalignment of roads and railroads, with significant property damage often resulting.

Special problems are presented by permanently frozen ground in arctic climates such as Alaska or Siberia. In such areas the ground may remain permanently frozen, thawing to only a shallow depth during the summer. This condition is called **permafrost**. Permafrost poses particular problems for construction, particularly where the presence of a structure may result in thawing such that the structure rests in a pool of water-saturated muck resting on a slick surface of frozen water and soil. The construction and maintenance of highways, railroads, and pipelines, such as the Trans-Alaska pipeline in Alaska, can become quite difficult in the presence of permafrost.

Some types of soils, particularly so-called expansive clays, expand and shrink markedly as they become saturated with water and dry out. Although essentially never life-threatening, the movement of structures and the damage caused to them by expansive clays can be very high. Aside from years when catastrophic floods and earthquakes occur, the monetary damage done by the action of expansive soil exceeds that of earthquakes, landslides, floods, and coastal erosion combined!

Sinkholes are a kind of earth movement resulting when surface earth falls into an underground cavity. They rarely injure people but may cause spectacular property damage. Cavities that produce sinkholes may form by the action of water containing dissolved carbon dioxide on limestone (Reaction 6.11.4), loss of underground water during drought or from heavy pumping, thus removing support that previously kept soil and rock from collapsing, heavy underground water flow, and other factors that remove solid material from underground strata.

15.5. STREAM AND RIVER PHENOMENA

A **stream** consists of water flowing through a channel. The area of land from which water is drawn that flows into a stream is the stream's **drainage basin**. The sizes of streams are described by **discharge** defined as the volume of water flowing past a given point on the stream per unit time. Discharge and **gradient**, the steepness of the downward slope of a stream determine the stream **velocity**.

Internal processes raise masses of land and whole mountain ranges, which in turn are shaped by the action of streams. Streams cut down mountain ranges, create valleys, form plains, and produce great deposits of sediment. Thus streams play a key role in shaping the geospheric environment. Streams spontaneously develop bends and curves by cutting away the outer parts of stream banks and depositing materials on the inner parts. These curved features of streams are known as **meanders**. Left undisturbed, a stream forms meanders across a valley in a constantly changing pattern. The cutting away of material by the stream and the deposition of sediment eventually forms a generally flat area. During times of high stream flow, the stream leaves its banks, inundating parts or all of the valley, thus creating a **floodplain**.

A **flood** occurs when a stream develops a high flow such that it leaves its banks and spills out onto the floodplain. Floods are arguably the most common and damaging of surface phenomena in the geosphere. Though natural and in many respects beneficial occurrences, floods cause damage to structures located in their paths, and the severity of their effects is greatly increased by human activities.

A number of factors determine the occurrence and severity of floods. One of these is the tendency of particular geographic areas to receive large amounts of rain within short periods of time. One such area is located in the middle of the continental United States where warm, moisture-laden air from the Gulf of Mexico is carried northward during the spring months to collide with cold air from the north; the resultant cooling of the moist air can cause torrential rains to occur, resulting in severe flooding. In addition to season and geography, geological conditions have a strong effect on flooding potential. Rain falling on a steep surface tends to run off rapidly, creating flooding. A watershed can contain relatively massive quantities of rain if it consists of porous, permeable materials that allow a substantial rate of infiltration, assuming that it is not already saturated. Plants in a watershed tend to slow runoff and loosen soil enabling additional infiltration. Through transpiration (see Section 14.2), plants release moisture to the atmosphere quickly, enabling soil to absorb more moisture.

Several terms are used to describe flooding. When the **stage** of a stream, that is, the elevation of the water surface, exceeds the stream bank level, the stream is said to be at **flood stage**. The highest stage attained defines the flood **crest**. **Upstream** floods occur close to the inflow from the drainage basin, usually the result of intense rainfall. Whereas upstream floods usually affect smaller streams and watersheds, **downstream floods** occur on larger rivers that drain large areas. Widespread spring snowmelt and heavy, prolonged spring rains, often occurring together, cause downstream floods.

Floods are made more intense by higher fractions and higher rates of runoff, both of which may be aggravated by human activities. This can be understood by comparing a vegetated drainage basin to one that has been largely denuded of vegetation and paved over. In the former case, rainfall is retained by vegetation, such as grass cover. Thus the potential flood water is delayed, the time span over which it enters a stream is extended, and a higher proportion of the water infiltrates into the ground. In the latter case, less rainfall infiltrates, and the runoff tends to reach the stream quickly and to be discharged over a shorter time period, thus leading to more severe flooding. These factors are illustrated in Figure 15.1.

The conventional response to the threat of flooding is to control a river, particularly by the construction of raised banks called **levees**. In addition to raising the banks to contain a stream, the stream channel may be straightened and deepened to increase the volume and velocity of water flow, a process called **channelization**. Although effective for common floods, these measures may exacerbate extreme floods by confining and increasing the flow of water upstream such that the capacity to handle water downstream is overwhelmed. Another solution is to construct dams to create reservoirs for flood control upstream. Usually such reservoirs are multipurpose facilities designed for water supply, recreation, and to control river flow for navigation in addition to flood control. The many reservoirs constructed for flood control in recent decades have been reasonably successful. There are, however, conflicts in the goals for their uses. Ideally, a reservoir for flood control should remain largely empty until needed to contain a large volume of floodwater, an approach that is obviously inconsistent with other uses. Another concern is that of exceeding the capacity of the reservoir, or dam failure, the latter of which can lead to catastrophic flooding.

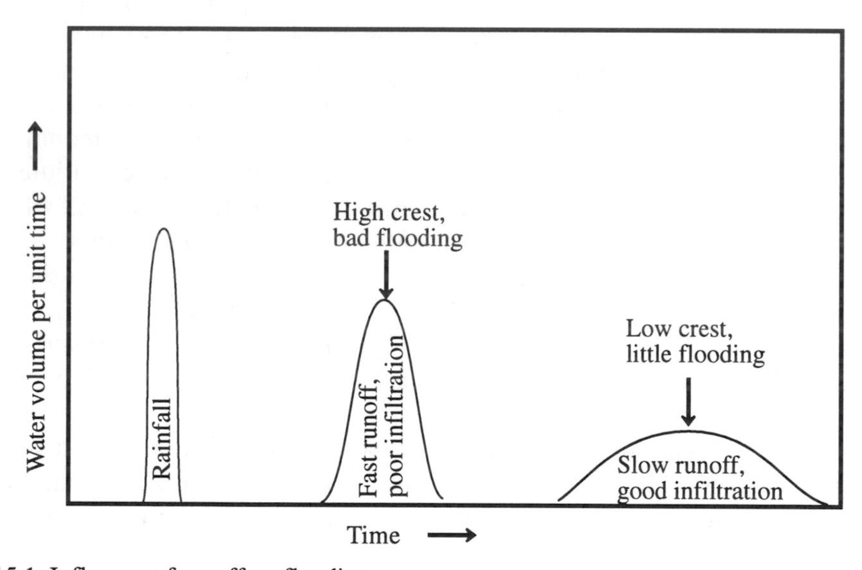

Figure 15.1. Influence of runoff on flooding.

15.6. PHENOMENA AT THE LAND/OCEAN INTERFACE

The coastal interface between land masses and the ocean is an important area of environmental activity. The land along this boundary is under constant attack from the waves and currents from the ocean, so that most coastal areas are always changing. The most common structure of the coast is shown in cross section in Figure 15.2. The beach, consisting of sediment, such as sand formed by wave action on coastal rock, is a sloping area that is periodically inundated by ocean waves. Extending from approximately the high tide mark to the dunes lining the landward edge of the beach is a relatively level area called the **berm**, which is usually not washed over by ocean water. The level of water to which the beach is subjected varies with the tides. Through wind action the surface of the water is in constant motion as undulations called **ocean waves**. As these waves reach the shallow water along the beach, they "touch bottom" and are transformed to **breakers** characterized by crested tops. These breakers crashing onto a beach give it much of its charm, but can also be extremely destructive.

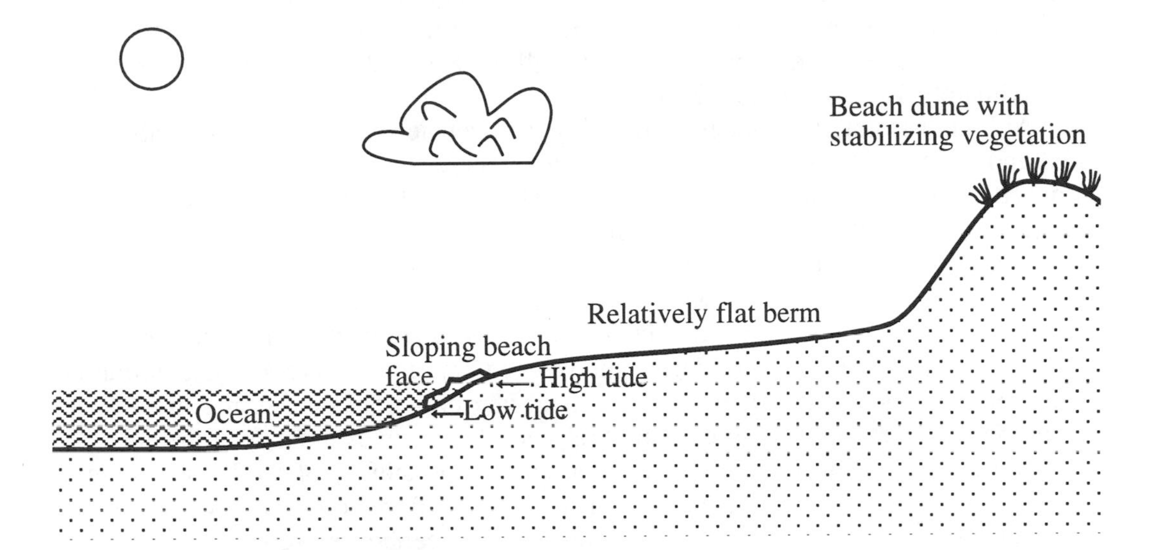

Figure 15.2. Cross section of the ocean/land interface along a beach.

Coastlines exhibit a variety of features. Steep valleys carved by glacial activity, then filled with rising seawater, constitute the fjords along the coast of Norway. Valleys, formerly on land, now filled with seawater, constitute **drowned valleys**. **Estuaries** occur where tidal salt water mixes with inflowing fresh water.

Erosion is a constant feature of a beachfront. Unconsolidated beach sand can be shifted readily, sometimes spectacularly through great distances over short periods of time, by wave action. Sand, pebbles and rock in the form of rounded cobbles constantly wear against the coast by wave action, exerting a constant abrasive action called **milling**. This action is augmented by the chemical weathering effects of seawater, in which the salt content may play a role.

Some of the more striking alterations to coastlines occur during storms, such as hurricanes and typhoons. The low pressure that accompanies severe storms tends to suck ocean water upward. This effect, usually combined with strong winds blowing onshore and coinciding with high tide, can cause ocean water to wash over the berm on a beach to attack dunes or cliffs inland. The result is a storm **surge** that can remove large quantities of beach, damage dune areas, and wash away structures unwisely constructed too close to the shore. Such a surge associated with a hurricane washed away most of the structures in Galveston, Texas, in 1900, claiming 6000 lives.

An especially vulnerable part of the coast consists of barrier islands, long, low strips of land roughly paralleling the coast some distance offshore. High storm surges may wash completely over barrier islands, partially destroying them and shifting them around. Many dwellings unwisely constructed on barrier islands, such as the outer banks of North Carolina, have been destroyed by storm surges during hurricanes.

The Threat of Rising Sea Levels

Large numbers of people live at a level near, or in some cases below, sea level. As a result, any significant temporary or permanent rise in sea level poses significant risks to lives and property. Such an event occurred on February 1, 1953, when high tides and strong winds combined to breach the system of dikes protecting much of the Netherlands from seawater. About one sixth of the country was flooded as far inland as 64 kilometers from the coast, killing about 2000 people and leaving approximately 100,000 without homes.

Although isolated instances of flooding by seawater caused by combinations of tidal and weather phenomena will continue to occur, a much more long-lasting threat is posed by long-term increases in sea level. These could result from global warming due to the greenhouse gas emissions discussed in Chapter 12. Several factors could raise ocean levels to destructive highs, also a result of greenhouse warming. Simple expansion of warmed oceanic water could raise sea levels by about 1/3 meter over the next century. The melting of glaciers, such as those in the Alps, has probably raised ocean levels about 5 cm during the last century, and the process is continuing. The greatest concern, however, is that global warming could cause the great West Antarctic ice sheet to melt, which would raise sea levels by as much as 6 meters.

A great deal of uncertainty exists regarding the possibility of the West Antarctic ice sheet melting with consequent rises in sea level. Current computer models show a compensating effect in that hotter air produced by greenhouse warming could carry much more atmospheric moisture to the Antarctic regions where the moisture would be deposited as snow. The net result could well be an *increase* in solid snow and ice in the Antarctic, and an accompanying *decrease* in sea levels.[1] Some of the uncertainty regarding the status of the West Antarctic ice sheet should be alleviated after the year 2002 when a satellite is to be launched into a polar orbit by the U.S. National Aeronautics and Space Administration that will be capable of measuring surface levels of polar ice caps to the nearest centimeter.

The measurement of sea levels has proven to be a difficult task because the levels of the surface of land keep changing. Land most recently covered with Ice Age glaciers in areas such as Scandinavia is still "springing back" from the immense mass of the glaciers, so that sea levels measured by gauges fixed on land actually appear to be dropping by several millimeters per year in such locations. An opposite situation exists on the east coast of North America where land was pushed outward and raised around the edge of the enormous sheet of ice that covered Canada and the northern U.S. about 20,000 years ago and is now settling back. Factors such as these illustrate the potential advantages of satellite technology in measuring sea levels.

15.7. PHENOMENA AT THE LAND/ATMOSPHERE INTERFACE

The interface between the atmosphere and land is a boundary of intense environmental activity. The combined effects of air and water tend to cause significant changes to the land materials at this interface. The top layer of exposed land is especially susceptible to physical and chemical weathering (see Section 3.10). Here air laden with oxidant oxygen contacts rock, originally formed under reducing conditions, causing oxidation reactions to occur. Acid naturally present in rainwater as dissolved CO_2 or present as pollutant sulfuric, sulfurous, nitric, or hydrochloric acid can dissolve portions of some kinds of rocks. Organisms, such as lichens (Section 3.4), drawing carbon dioxide, oxygen, or nitrogen from air can grow on rock surfaces at the boundary of the atmosphere and geosphere, causing additional weathering to take place.

One of the most significant agents affecting exposed geospheric solids at the atmosphere/geosphere boundary is wind. Consisting of air moving largely in a horizontal fashion, wind both erodes solids and acts as an agent to deposit solids on geospheric surfaces. The influence of wind is especially pronounced in dry areas. A major factor in wind erosion is wind **abrasions** in which solid particles of sand and rock carried by wind tend to wear away exposed rock and soil. Loose, unconsolidated sand and soil may be removed in large volumes by wind, a process called **deflation**.

The potential for wind to move matter is illustrated by the formation of large deposits of **loess** consisting of finely divided soil carried by wind. Loess particles are typically several tens of micrometers in size, small enough to be carried great distances by wind. Especially common are loess deposits that originated with matter composed of rock ground to a fine flour by ice age glaciers. This material was first deposited in river valleys by flood waters issuing from melting glaciers, then blown some distance from the rivers by strong winds after drying out.

One of the more common geospheric features created by wind is a **dune** consisting of a mound of debris, usually sand, dropped when wind slows down. When a dune begins to form, it forms an obstruction that slows wind even more, so that more sediment is dropped. The result is that in the presence of sediment-laden wind, dunes several meters or more high may form rapidly. In forming a dune, heavier, coarser particles settle first so that the matter in dunes is sorted according to size, just like sediments deposited by flowing streams. In areas in which winds are prevalently from one direction, as is usually the case, dunes show a typical shape as illustrated in Figure 15.3. It is seen that the steeply sloping side, called the **slip face**, is downwind.

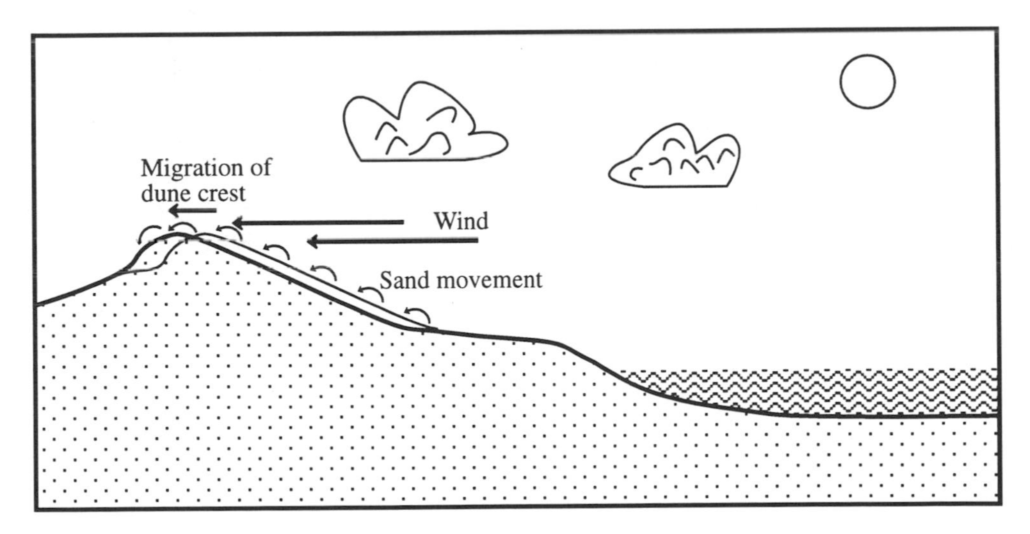

Figure 15.3. Shape and migration of a dune as determined by prevailing wind direction.

Some of the environmental effects of dunes result from their tendency to migrate with the prevailing winds. Migration occurs as matter is blown by the wind up the gently sloping face of the dune and falls down the slip face. Migrating sand dunes have inundated forest trees, and dust dunes in drought-stricken agricultural areas have filled road ditches, causing severely increased maintenance costs.

15.8. EFFECTS OF ICE

The power of ice to alter the geosphere is amply demonstrated by the remains of past glacial activity from the Ice Age. Those large areas of the Earth's surface that were once covered with layers of glacial ice 1 or 2 kilometers in thickness show evidence of how the ice carved the surface, left massive piles of rock and gravel, and left rich deposits of fresh water. The enormous weight of glaciers on Earth's surface compressed it, and in places it is still springing back 10,000 or so years after the glaciers retreated. Today the influence of ice on Earth's surface is minimal, and there is substantial concern that melting of glaciers by greenhouse warming will raise ocean levels so high that many coastal areas will be inundated by seawater.

Glaciers form at sufficiently high latitudes and altitudes such that snow does not melt completely each summer. This occurs when snow becomes compacted over several, to several thousand years such that the frozen water turns to crystals of true ice. Huge masses of ice with areas of several thousand square kilometers or more and often around 1 kilometer thick occur in polar regions and are called **continental glaciers**. Both Greenland and the Antarctic are covered by continental glaciers. **Alpine glaciers** occupy mountain valleys.

Glaciers on a slope flow as a consequence of their mass. This rate of flow is usually only a few meters per year, but may reach several kilometers per year. If a glacier flows into the sea, it may lose masses of ice as icebergs, a process called **calving**. Ice may also be lost by melting along the edges. The processes by which ice is lost is termed **ablation**.

Glacial ice affects the surface of the geosphere by both erosion and deposition. It is easy to imagine that a flowing mass of glacial ice is very efficient in scraping away the surface over which it flows, a process called **abrasion**. Adding to the erosive effect is the presence of rocks frozen into the glaciers, which can act like tools to carve the surface of the underlying rock and soil. Whereas abrasion tends to wear rock surfaces away producing a fine rock powder, larger bits of rock can be dislodged from the surface over which the glacier flows and be carried along with the glacial ice.

When glacial ice melts, the rock that has been incorporated into it is left behind. This material is called **till**, or if it has been carried for some distance by water running off the melting glacier it is called **outwash**. Piles of rock left by melting glaciers produce unique structures called **moraines**.

Although the effects of glaciers described above are the most spectacular manifestations of the action of ice on the geosphere, at a much smaller level ice can have some very substantial effects. These occur through the freezing and expansion of water in pores and small crevices in rock. The expansion of frozen water on and near the surface of rock is a major contributor to physical weathering processes. Freeze/thaw cycles are also very destructive to some kinds of structures, such as stone buildings.

15.9. EFFECTS OF HUMAN ACTIVITIES

Human activities have profound effects on the geosphere. Such effects may be obvious and direct, such as strip mining, or rearranging vast areas for construction projects, such as roads and dams. Or the effects may be indirect, such as pumping so much water from underground aquifers that the ground subsides, or abusing soil such that it no longer supports plant life well and erodes. As the source of minerals and other resources used by humans, the geosphere is dug up, tunnelled, stripped bare, rearranged, and subjected to many other kinds of indignities. The land is often severely disturbed, air can be polluted with dust particles during mining, and water may be polluted. Many of these effects, such as soil erosion caused by human activities are addressed elsewhere in this book.

Extraction of Geospheric Resources: Surface Mining

Many human effects on the geosphere are associated with the extraction of resources from Earth's crust. This is done in a number of ways, the most damaging of which can be surface mining. Surface mining is employed in the United States to extract virtually all of the rock and gravel that is mined, well over half of the coal, and numerous other resources. Properly done, with appropriate restoration practices, surface mining does minimal damage and may even be used to improve surface quality, such as by the construction of surface reservoirs where rock or gravel have been extracted. In earlier times before strict reclamation laws were in effect, surface mining, particularly of coal, left large areas of land scarred, devoid of vegetation, and subject to erosion.

Several approaches are employed in surface mining. Sand and gravel located under water are extracted by **dredging** with draglines or chain buckets attached to large conveyers. In most cases resources are covered with an **overburden** of earthen material that does not contain any of the resource that is being sought. This material must be removed as **spoil**. **Open-pit mining** is, as the name implies, a procedure in which gravel, building stone, iron ore, and other materials are simply dug from a big hole in the ground. Some of these pits, such as several from which copper ore has been taken in the U. S., are truly enormous in size.

The most well known (sometimes infamous) method of surface mining is **strip mining** in which strips of overburden are removed by draglines and other heavy earth-moving equipment to expose seams of coal, phosphate rock, or other materials. Heavy equipment is used to remove a strip of overburden, and the exposed mineral resource is removed and hauled away. Overburden from a parallel strip is then removed and placed over the previously mined strip, and the procedure is repeated numerous times. Older practices left the replaced overburden as relatively steep erosion-prone banks. On highly sloping terrain overburden is removed on progressively higher terraces and placed on the terrace immediately below.

Environmental Effects of Mining and Mineral Extraction

Some of the environmental effects of surface mining have been mentioned above. Although surface mining is most often considered for its environmental effects, subsurface mining may also have a number of effects, some of which are not immediately apparent and may be delayed for decades. Underground mines have a tendency to collapse leading to severe subsidence. Mining disturbs groundwater aquifers. Water seeping through mines and mine tailings may become polluted. One of the more common and damaging effects of mining on water occurs when pyrite, FeS_2, commonly associated with coal, is exposed to air and becomes oxidized to sulfuric acid by bacterial action to produce acid mine water (see Section 7.8). Some of the more damaging environmental effects of mining are the result of the processing of mined materials. Usually ore is only part, often a small part, of the material that must be excavated. Various **beneficiation** processes are employed to separate the useful fraction of ore, leaving a residue of **tailings**. A number of adverse effects can result from environmental exposure of tailings. For example, residues left from the beneficiation of coal are often enriched in acid-producing pyrite. Uranium ore tailings unwisely used as fill material have contaminated buildings with radioactive radon gas.

15.10. AIR POLLUTION AND THE GEOSPHERE

The geosphere can be a significant source of air pollutants. Of these geospheric sources, volcanic activity is one of the most common. Volcanic eruptions, fumaroles, hot springs, and geysers can emit toxic and acidic gases, including carbon monoxide, hydrogen chloride, and hydrogen sulfide. Greenhouse gases that tend to increase global climatic warming — carbon dioxide and methane — can come from volcanic sources. Massive volcanic eruptions may inject huge amounts of particulate matter into the atmosphere. The incredibly enormous 1883 eruption of the East Indies volcano Krakatoa blew about 2.5 cubic kilometers of solid matter into the atmosphere, some of which penetrated well into the stratosphere. This material stayed aloft long enough to circle the Earth several times, causing red sunsets and a measurable lowering of temperature worldwide.

The 1982 eruption of the southern Mexico volcano El Chicón showed the importance of the type of particulate matter in determining effects on climate. The matter given off by this eruption was unusually rich in sulfur, so that an aerosol of sulfuric acid formed and persisted in the atmosphere for about three years, during which time the mean global temperature was lowered by several tenths of a degree due to the presence of atmospheric sulfuric acid. By way of contrast, the eruption of Mt. St. Helens in Washington State in the U.S. two years earlier had little perceptible effect on climate, although the amount of material blasted into the atmosphere was about the same as that from El Chicón. The material from the Mt. St. Helens eruption had comparatively little sulfur in it, so the climatic effects were minimal.

Thermal smelting processes used to convert metal fractions in ore to usable forms have caused a number of severe air pollution problems that have affected the geosphere. Many metals are present in ores as sulfides, and smelting can release large quantities of sulfur dioxide, as well as particles that contain heavy metals, such as arsenic, cadmium, or lead. The resulting acid and heavy metal pollution of surrounding land can cause severe damage to vegetation so that devastating erosion occurs. One such area is around a large nickel smelter in Sudbury, Ontario, Canada, where a large area of land has become denuded of vegetation. Similar dead zones have been produced by copper smelters in Tennessee and in eastern Europe, including the former Soviet Union.

Soil and its cultivation produces significant quantities of atmospheric emissions. Waterlogged soil, particularly that cultivated for rice, generates significant quantities of methane, a greenhouse gas. The microbial reduction of nitrate in soil releases nitrous oxide, N_2O, to the atmosphere. However, soil and rock can also remove atmospheric pollutants. It is believed that microorganisms in soil account for the loss from the atmosphere of some carbon monoxide, which some fungi and bacteria can metabolize. Carbonate rocks, such as calcium carbonate, $CaCO_3$, can neutralize acid from atmospheric sulfuric acid and acid gases.

As discussed in Section 9.6, masses of atmospheric air can become trapped and stagnant under conditions of a temperature inversion in which the vertical circulation of air is limited by the presence of a relatively warm layer of air overlaying a colder layer at ground level. The effects of inversions can be aggravated by topographical conditions that tend to limit circulation of air. Figure 15.4 shows such a condition in which surrounding mountain ridges limit horizontal air movement.

Air pollutants may be forced up a mountain ridge from a polluted area to significantly higher altitudes than they would otherwise reach. Because of this "chimney effect," air pollutants may reach mountain pine forests that are particularly susceptible to damage from air pollutants, such as ozone formed along with photochemical smog.

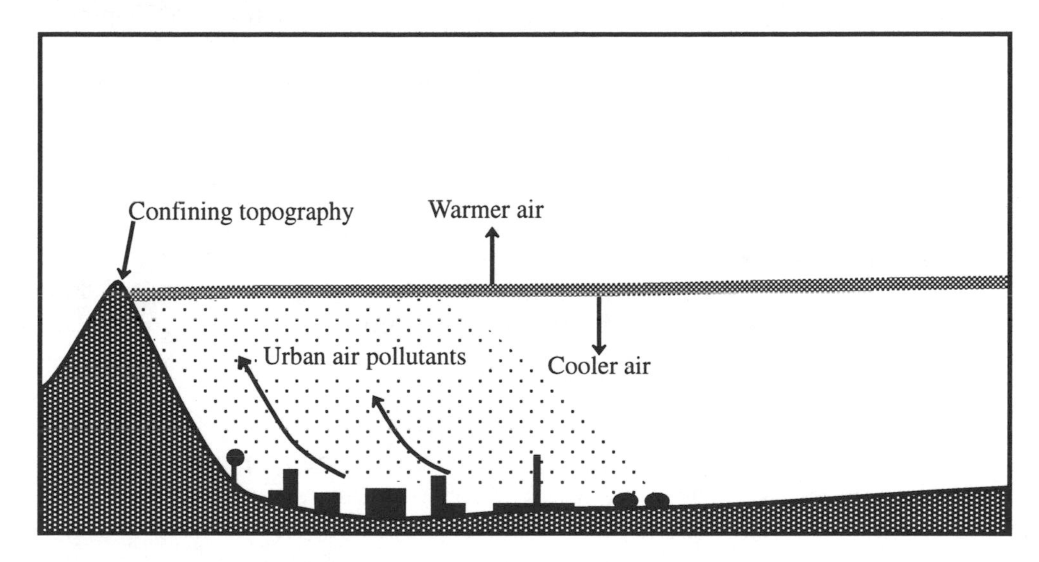

Figure 15.4. Topographical features, such as confining mountain ridges, may work with temperature inversions to increase the effects of air pollution.

15.11. WATER POLLUTION AND THE GEOSPHERE

Water pollution is addressed in detail elsewhere in this book. Much water pollution arises from interactions of groundwater and surface water with the geosphere. These aspects are addressed briefly here.

The relationship between water and the geosphere is twofold. The geosphere may be severely damaged by water pollution. This occurs, for example, when water pollutants produce contaminated sediments, such as those contaminated by heavy metals or PCBs. In some cases the geosphere serves as a source of water pollutants. Examples include acid produced by exposed metal sulfides in the geosphere or synthetic chemicals improperly discarded in landfills.

The sources of water pollution are divided into two main categories. The first of these consists of **point sources**, which enter the environment at a single, readily identified entry point. An example of a point source would be a sewage water outflow. Point sources tend to be those directly identified as to their origins from human activities. **Nonpoint sources** of pollution are those from broader areas. Such a source is water contaminated by fertilizer from fertilized agricultural land, or water contaminated with excess alkali leached from alkaline soils. Nonpoint sources are relatively harder to identify and monitor. Pollutants associated with the geosphere are usually nonpoint sources.

An especially common and damaging geospheric source of water pollutants consists of sediments carried by water from land into the bottoms of bodies of water.

Most such sediments originate with agricultural land that has been disturbed such that soil particles are eroded from land into water. The most common manifestation of sedimentary material in water is opacity, which seriously detracts from the esthetics of the water. Sedimentary material deposited in reservoirs or canals can clog them and eventually make them unsuitable for water supply, flood control, navigation, and recreation. Suspended sediment in water used as a water supply can clog filters and add significantly to the cost of treating the water. Sedimentary material can devastate wildlife habitats by reducing food supplies and ruining nesting sites. Turbidity in water can severely curtail photosynthesis, thus reducing primary productivity necessary to sustain the food chains of aquatic ecosystems (see Chapter 17).

15.12. WASTE DISPOSAL AND THE GEOSPHERE

The geosphere receives many kinds and large amounts of wastes. Its ability to cope with such wastes with minimal damage is one of its most important characteristics and is dependent upon the kinds of wastes disposed on it. A variety of wastes, ranging from large quantities of relatively innocuous municipal refuse to much smaller quantities of potentially lethal radioactive wastes, are deposited on land or in landfills. These are addressed briefly in this section.

Municipal Refuse

The currently favored method for disposing of municipal solid wastes—household garbage—is in **sanitary landfills** consisting of refuse piled on top of the ground or into a depression, such as a valley, compacted, and covered at frequent intervals by soil. This approach permits frequent covering of the refuse so that loss of blowing trash, water contamination, and other undesirable effects are minimized. A completed landfill can be put to beneficial uses, such as a recreational area; because of settling, gas production, and other factors, landfill surfaces are generally not suitable for building construction. Modern sanitary landfills are much preferable to the open dump sites that were once the most common means of municipal refuse disposal.

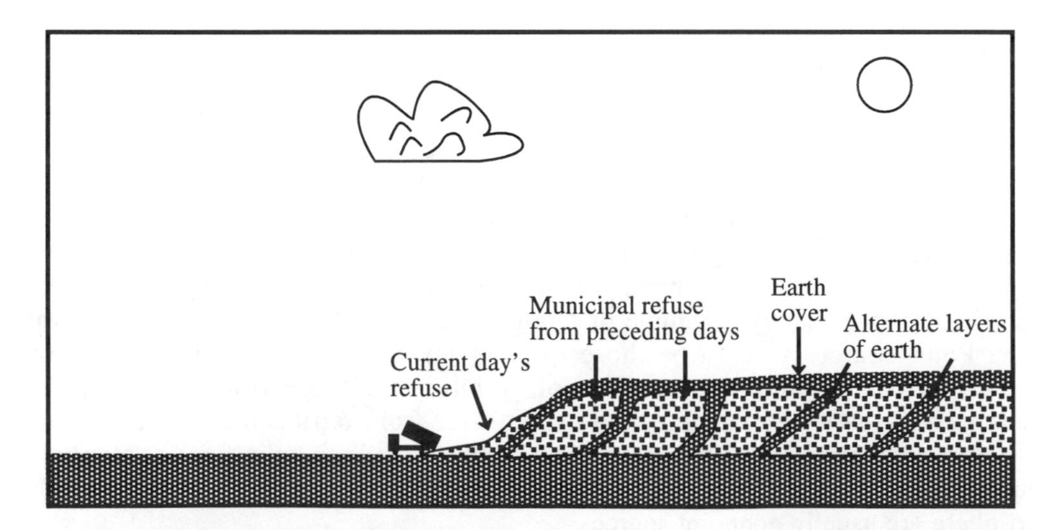

Figure 15.5. Structure of a sanitary landfill.

Although municipal refuse is much less dangerous than hazardous chemical waste, it still poses some hazards. Despite prohibitions against the disposal of cleaners, solvents, lead storage batteries, and other potentially hazardous materials in landfills, materials that pose some environmental hazards do find their way into landfills and can contaminate their surroundings.

Landfills produce both gaseous and aqueous emissions. Biomass in landfills quickly depletes oxygen by aerobic biodegradation of microorganisms in the landfill,

$$\{CH_2O\}(biomass) + O_2 \rightarrow CO_2 + H_2O \quad (15.12.1)$$

emitting carbon dioxide. Over a period of many decades the buried biodegradable materials undergo anaerobic biodegradation,

$$2\{CH_2O\} \rightarrow CO_2 + CH_4 \quad (15.12.2)$$

releasing methane as well as carbon dioxide. Although often impractical and too expensive, it is desirable to reclaim the methane as fuel, and some large sanitary landfills are major sources of methane. Released methane is a greenhouse gas and can pose significant explosion hazards to structures built on landfills. Although produced in much smaller quantities than methane, hydrogen sulfide, H_2S, is also generated by anaerobic biodegradation. This gas is toxic and has a bad odor. In a properly designed sanitary landfill, hydrogen sulfide releases are small and the gas tends to oxidize before it reaches the atmosphere in significant quantities.

Water infiltrating into sanitary landfills dissolves materials from the disposed refuse and runs off as **leachate**. Contaminated leachate is the single greatest potential pollution problem with refuse disposal sites, so it is important to minimize its production by designing landfills in a way that keeps water infiltration as low as possible. The anaerobic degradation of biomass produces organic acids that give the leachate a tendency to dissolve acid-soluble solutes, such as heavy metals. Leachate can infiltrate into groundwater posing severe contamination problems. This is minimized by siting sanitary landfills over formations of poorly permeable clay or depositing layers of clay in the landfill before refuse is put in it. In addition, impermeable synthetic polymer liners may be placed in the bottom of the landfill. In areas of substantial rainfall, infiltration into the landfill exceeds its capacity to hold water so that leachate flows out. In order to prevent water pollution downstream, this leachate should be controlled and treated.

Hazardous chemical wastes are disposed in so-called **secure landfills**, which are designed to prevent leakage and geospheric contamination of toxic chemicals disposed in them. Such a landfill is equipped with a variety of measures to prevent contamination of groundwater and the surrounding geosphere. The base of the landfill is made of compacted clay that is largely impermeable to leachate. An impermeable polymer liner is placed over the clay liner. The surface of the landfill is covered with material designed to reduce water infiltration, and the surface is designed with slopes that also minimize the amount of water running in. Elaborate drainage systems are installed to collect and treat leachate.

The most pressing matter pertaining to geospheric disposal of wastes involves radioactive wastes. Most of these wastes are **low-level** wastes, including discarded radioactive laboratory chemicals and pharmaceuticals, filters used in nuclear reactors, and ion-exchange resins used to remove small quantities of radionuclides from nuclear reactor cooler water. Disposed in properly designed landfills, such wastes pose minimal hazards.

Of greater concern are the **high-level** radioactive wastes, primarily fission products of nuclear power reactors and byproducts of nuclear weapons manufacture. Many of these wastes are currently stored as solutions in tanks at sites such as the Federal nuclear facility at Hanford, Washington, where plutonium was generated in large quantities during post-World War II years. Eventually such wastes must be placed in the geosphere such that they will pose no hazards. Numerous proposals have been advanced for their disposal, including disposal in salt formations, subduction zones in the seafloor, and ice sheets. The most promising sites appear to be those in poorly permeable formations of igneous rock. Basalts, which are strong, glassy igneous types of rock found in the Columbia River plateau, granite, and pyroclastic welded tuffs fused by past high temperature volcanic eruptions are likely possibilities.

15.13. THE GEOSPHERE AND HEALTH

Striking geographic correlations are observed with occurrences of various kinds of diseases. Although most of these relationships are probably due to cultural factors, including lifestyle, choice of diet, and exposure to pollutants, it is an intriguing possibility that some of these correlations are associated with exposure to the geosphere. This is most likely to be the case for plants and wildlife that exist in close association with the geosphere. Formerly, and to a lesser extent perhaps even now, such correlations have been associated with agents in the environment, particularly essential or toxic trace elements.

One of the most well documented associations between geospheric environment and health is the association between iodine deficiency and goiter, which used to be prevalent in some regions, such as the northern inland United States. Iodine is essential for the function of the thyroid gland, and a lack of iodine causes the thyroid to enlarge, causing a condition called goiter. Iodide ion is abundant in seawater, and consumption of seafood and even consumption of food and water exposed to sea salt spray provides sufficient iodine. It is believed that continuous leaching of midcontinent regions by rainwater and, immediately after the last ice age, by glacial melt removed highly soluble iodine from the soil so that foods grown on such soil were deficient in iodine. Happily, the deficiency was readily remedied by placing iodine in salt, and the use of iodized salt has largely resulted in the disappearance of goiter as a human affliction.

Another halide, fluoride, has been associated with human health, in both a positive and a negative manner. On the positive side, fluoride intake helps prevent tooth decay. This was observed as a correlation between low rates of tooth decay and fluoride content of drinking water. There is also some evidence to suggest that a certain level of fluoride may be helpful in preventing degenerative bone loss, osteoporosis, which tends to afflict the aged. Fluoride intake exceeding normal levels by around four-fold or more can cause the formation of dark spots on teeth, so called "mottling" of teeth. At levels around 10 times greater than those that produce tooth mottling, fluoride causes significant problems with bone, including buildup of excess bone tissue and accumulation of calcium in ligaments.

Both excesses and deficiencies of selenium can cause health problems, particularly with livestock. Selenium toxicity is much more common than selenium deficiency. Selenium deficiency has been found to adversely affect plants and animals in some regions. In contrast, some plants, such as those of the genus *Astragalus* occurring in the western United States, concentrate selenium. *Astragalus* plant has been called "loco weed" because cattle that ingest too much selenium from eating it develop "blind staggers," a condition characterized by lack of coordination.

A particularly striking example of the relationship of the geosphere, modified by human activities, to health has been observed in the Kesterson Wildlife Refuge in California. Selenium leached from land used to grow rice in the San Joaquin Valley accumulated in ponds and wetlands in the wildlife refuge. The result was a greatly increased rate of mortality and occurrences of deformities in waterfowl and fish growing in these waters.

The presence of limestone formations, which produce foods and water rich in calcium, may be associated with health effects. Some evidence suggests that livestock grown in limestone-rich areas are healthier in some respects than animals grown in other areas. Some studies also suggest that the consumption of calcium-rich "hard" water from limestone aquifers leads to lower rates of heart disease. Zinc is another essential element that may be associated with relationships between the geosphere and health. The occurrence of radionuclides in the geosphere, particularly radon emitted from uranium deposits, may be associated with elevated levels of cancer.

15.14. LAND USE

At an accelerating pace in recent years, increasing human populations, decreasing availability of land, and increased ability to exploit land, such as through heavy construction projects, have combined to put land in short supply. In addition, disasters resulting from human errors in siting structures on land, such as catastrophic dam failures, have pointed to the importance of using land wisely and of the proper application of geological engineering. The interaction of the geosphere with what humans do with it works both ways. It is essential to understand the nature of the material upon which a structure is placed to ensure that it is stable, safe, and durable. Secondly, the nature and construction of structures placed on land should be consistent with the esthetics, preservation, and quality of the geospheric environment. Modern technology utilizing computers, satellite mapping, and other technologies can be used to great advantage in siting new buildings and other structures.

The considerations outlined above have given rise to the discipline of land use planning. Land use planning is a broad area that must consider not only geological and engineering factors, but social, political, and economic factors as well. Consider, for example, prime farmland located in the path of urban development. For the benefit of humankind in general, the best option by far is for such property to remain as farmland so that it can continue to provide badly needed food. However, owners of the property are justified on purely personal financial grounds in wanting to sell it for a much higher price for development. These are the kinds of issues with which modern land use planning must deal.

An important concept in land planning is that of **multiple use** in which the same land area may be employed for more than one purpose. Forest lands, for example, commonly serve as sources of lumber, recreational areas, and watersheds at the same time. **Sequential use** refers to situations in which land is employed for various purposes in sequence. As an example, land overlaying coal deposits may be strip-mined, then used as sanitary landfill, and finally restored, planted with trees and grass, and employed for recreation.

An important aspect of land use is evaluation of sites with respect to their suitability for various purposes. Such evaluations must take many factors into consideration. It is especially important to consider the characteristics of soil and rock on which structures are to be placed. Large structures may be placed on steep slopes that are based on stable bedrock, but not on unstable formations or those with steeply angled bedding planes prone to slide. The presence of folds and faults may affect building stability. The behavior of soil with water is especially important. A particularly important consideration is the tendency of some clays to swell dramatically when they absorb water.

Land use planning has been largely based on economic and financial factors, with an effort made to maximize economic growth and income. A more comprehensive approach is that of **ecological land use planning**, which integrates ecological, geological, and societal factors along with purely economic considerations. Such land use planning begins with a comprehensive survey of geological, water, economic, societal, and other resources. Goals are identified and ranked. Basic questions must be asked and answered regarding desired population and economic growth, as well as preservation and enhancement of natural features, such as forests, grasslands, wetlands, and wildlife resources. In addition consideration must be given to the preservation of valuable cropland. Resources and proposed development are transferred to maps. With these tools a master plan is developed that is subject to input by experts, the public, and various other interests. Finally, the master plan is implemented through the efforts of governmental and regulatory agencies. Usually the financial stakes are relatively high and as a result the whole process is always very political.

The most important tool for implementing land use plans is **zoning**. Zoning divides land into various categories, including the following:

- Residential (single-family dwellings, duplexes, apartments, condominiums)
- Commercial (shopping malls, strip developments, convenience stores)
- Industrial and manufacturing (heavy industry, light industry)
- Infrastructural (utilities, highways, commuter rail lines)
- Recreational
- Natural wildlife areas
- Watersheds, floodplains, and wetlands
- Agricultural

Zoning is a direct means of controlling development. There are other effective, less direct measures that can be employed. Taxation can be used to encourage desired development and discourage less desirable aspects. Sewer and other utility hookups can be controlled and limited. Measures to control development may occur at several different governmental levels. For example, local and regional zoning regulations must conform to Federal laws pertaining to the preservation of wetlands. Private resources from groups such as local private foundations or the Nature Conservancy are often used to buy and preserve land with particular ecological significance.

CHAPTER SUMMARY

The chapter summary below is presented in a programmed format to review the main points covered in this chapter. It is used most effectively by filling in the blanks, referring back to the chapter as necessary. The correct answers are given at the end of the summary.

Evidence of the high energy processes that occur beneath Earth's surface is provided by [1]____________________. The discipline that takes account of the geological characteristics of soil and rock in designing buildings, dams, highways, and other structures in a manner compatible with the geological strata on which they rest is [2]____________________. Whereas internal processes in Earth tend to force matter [3]__________, surface processes are those that generally result from the tendency of matter to [4]____________________. Earthquakes, occur as the result of [5]______________________________. The Richter scale for the magnitude of earthquakes is [6]______________, and each unit increase in the scale reflects a [7]______________ increase in magnitude. Massive earthquakes can cost in the $ [8]__________ of dollars and the loss of life from a single earthquake has exceeded [9]______________. Liquid rock flowing from volcanoes is called [10]__________. The most recent large volcanic eruption in the continental United States was that of Mount. [11]______________, which occurred on [12]__________. The ejection of huge quantities of sulfur oxides into the atmosphere by a volcano can result in the formation of [13]____________________ in the atmosphere, which can cause a perceptible [14]____________________. Gravity acting upon rock and soil on Earth's surface can cause large [15]______________ that can be very damaging and result in a heavy loss of life. The formation of sinkholes is often the result of the chemical action of dissolved [16]______________ on underground formations of [17]____________________. A [18]______________ consists of water flowing through a channel that has run off from a [19]____________________. A flat valley created by the changing courses of a stream creates a [20]______________. An ocean beach consists of sediment formed by [21]__________ on [22]____________________. The areas where tidal salt water from the ocean mixes with inflowing fresh water are called [23]______________. Topographical features, such as deposits of till and piles of rock forming morains are evidence of [24]______________________________.

The top layer of exposed land at the interface with the atmosphere is especially susceptible to [25]______________________________. An often damaging technique in which strips of overburden are removed by draglines and other heavy earth-moving equipment to expose seams of coal, phosphate rock, or other materials is called [26]____________________. Associated with mining, various [27]______________ processes are employed to separate the useful fraction of ore, leaving a residue of [28]______________. Examples of the influence of "natural air pollution" from the geosphere on the atmosphere are provided by past [29]______________ that have resulted in perceptible cooling of climate. Many metals are present in ores as [30]______________, and smelting can release large quantities of [31]__________________, as well as particles that contain [32]_______________. The two main categories of water pollution sources are [33]____________________________ sources. Municipal landfill surfaces are generally not suitable for building construction because of factors such as [34]______ ______________________________. Water infiltrating into sanitary landfills produces [35]______________. So-called secure chemical landfills are used for the disposal of [36]______________________________________. Striking geographic correlations observed with occurrences of various kinds of diseases may show the relationship between [37]_____________________________________. Strip-mining land for coal followed by use as a sanitary landfill, then restoration with vegetation, and finally designation as a recreational area is an example of [38]______ __________________ of land. The integration of ecological, geological, and societal factors along with purely economic considerations in the use of land is known as [39]__________________________.

Answers

1 volcanic activity and from earthquakes

2 engineering geology

3 upward

4 seek lower levels

5 plate tectonic processes as huge land masses move relative to each other

6 open-ended

7 10-fold

8 billions

9 1 million people

10 lava

11 St. Helens

12 May 18, 1980

13 sulfuric acid droplets

14 cooling in climate

15 mass movements (landslides)

16 carbon dioxide

17 limestone

18 stream

19 drainage basin

20 floodplain

21 wave action

22 coastal rock

23 estuaries

24 the action of ice from past ice ages

25 physical and chemical weathering

26 strip mining

27 beneficiation

28 tailings

29 volcanic eruptions

30 sulfides

31 sulfur dioxide

32 heavy metals

33 point and nonpoint

34 settling and gas production

35 leachate

36 hazardous chemical wastes

37 the geosphere and health

38 sequential use

39 ecological land use planning

QUESTIONS AND PROBLEMS

1. At which locations are earthquakes most likely to occur? Why?
2. What are two scales used to measure the severity of earthquakes?
3. Describe the phenomenon of liquefaction that can occur during earthquakes.
4. What climatic effects may be caused by a major volcanic eruption? What is the historical evidence of such effects?
5. What is the stage of a stream? What is meant by the crest of a flood?
6. What is an estuary? Why are estuaries important?
7. Why have parts of Earth's surface been rising or "springing back" for the last approximately 10,000 years?
8. What is the distinction between continental and alpine glaciers? In which parts of the world are continental glaciers now found? Where were they found many thousands of years ago?
9. In what sense is glacial till analogous to stream load?
10. Describe how surface mining can be among the "most damaging ways" of extracting Earth's resources.

11. What is overburden? How does it become spoil?

12. What is beneficiation? What sort of residue does it leave?

13. In what sense is the type of inorganic matter emitted by volcanoes important in determining climatic effects? Cite a recent example showing such an effect.

14. Distinguish between point sources and nonpoint sources of pollution.

15. What is landfill leachate? What are some of its important potential environmental effects?

16. How is human health related to the distribution of substances in the geosphere? Give a specific example.

17. Distinguish between multiple land use and sequential land use.

18. What is ecological land use planning? What factors does it integrate?

LITERATURE CITED

[1] Schneider, David, "The Rising Seas," *Scientific American*, March 1997, pp. 112-117.

SUPPLEMENTARY REFERENCES

Colley, H., *Introduction to Environmental Geology*, Chapman & Hall, New York, NY, 1997.

Keller, Edward A., *Environmental Geology*, Prentice Hall, Upper Saddle River, NJ, 1996.

Keys, W. Scott, *A Practical Guide to Borehole Geophysics in Environmental Investigations*, CRC Press/Lewis Publishers, Boca Raton, FL, 1997.

Murck, Barbara, *Environmental Geology*, Wiley, New York, NY, 1995.

Oliver, J. E., *Shocks and Rocks: Seismology in the Plate Tectonics Revolution: The Story of Earthquakes and the Great Earth Science Revolution of the 1960s*, American Geophysical Union, New York, NY, 1996.

Pipkin, D. D. and Bernard W. Trent, *Geology and the Environment*, West Publishing Co., Westbury, NY, 1996.

Rydin, Yvonne, *Environmental Impact of Land and Property Management*, Wiley, New York, NY, 1996.

Sutherland, Lin, *The Volcanic Earth: Volcanoes and Plate Tectonics: Past, Present & Future*, New South Wales University Printing Ltd., New South Wales, Australia, 1996.

16 GEOSPHERIC RESOURCES

16.1. INTRODUCTION

Manufacturing requires a steady flow of raw materials—minerals, fuel, wood, and fiber. These can be provided from either **extractive** (nonrenewable) and **renewable** sources. This chapter addresses the extractive industries in which irreplaceable resources are taken from the Earth's crust. Therefore, it deals with a particularly important part of Earth's support system upon which humankind depends—geospheric resources. As such, the chapter emphasizes mineral resources, nonrenewable materials that are extracted from Earth's crust for use in manufacturing, agriculture, and other applications. Addressed in other chapters are related matters pertaining to soil resources (Chapter 14), the use of resources in manufacturing (Chapter 21), and energy utilization (Chapter 22).

The utilization of mineral resources is strongly tied with technology, energy, and the environment. Perturbations in one usually cause perturbations in the others. For example, reductions in automotive exhaust pollutant levels with the use of catalytic devices, discussed in Chapter 11, have resulted in increased demand for platinum metal, a scarce natural resource, and greater gasoline consumption than would be the case if exhaust emissions were not a consideration (a particularly pronounced effect in the earlier years of emissions control). The availability of many metals depends upon the quantity of energy used and the amount of environmental damage tolerated in the extraction of low-grade ores. Many other such examples could be cited. Because of these intimate interrelationships, technology, resources, and energy must all be considered when environmental science is discussed.

The availability of minerals, particularly metals in limited supply, are subject to political influences and financial manipulation. Cartels have been organized to attempt to raise prices. Tin has long been the subject of such a cartel. Governments may limit exports to influence foreign policy, or subsidize prices to discourage competition and increase foreign exchange. Prices fluctuate with costs of labor, materials, property, and pollution controls. During the early 1980s speculators cornered the market for silver, causing panic buying by interests fearing shortages and raising prices several-fold. As often happens, this excursion was followed by a crash in prices. Increased industrial demand, such as accompanies economic development in developing countries, can cause metal prices to increase significantly.

In discussing minerals and fossil fuels in the remainder of this chapter, two terms related to available quantities are used and should be defined. The first of these is **resources**, defined as quantities that are estimated to be *ultimately* available. The second term is **reserves**, which refers to well-identified resources that can be profitably utilized with existing technology.

As is the case of energy supplies—shortages of which were a major driving force leading to the outbreak of World War II, for example—adequate supplies of needed minerals, particularly metals are deemed to be essential for the well-being of a sovereign nation. The United States maintains stockpiles of such materials as insurance against the interruption of supplies and of unreasonable price increases. Those minerals stored for national defense purposes are called **strategic minerals**, and the ones maintained for economic reasons are called **critical minerals**.

16.2. GEOSPHERIC SOURCES OF USEFUL MINERALS

There are numerous kinds of mineral deposits that are used in various ways. The most available deposits have already been exploited. A challenge now is to find ways to utilize available resources in a cost-effective manner consistent with maximum conservation of the resource and environmental protection.

Geological and geochemical factors (see Chapter 13) are crucial in locating deposits of crucial metals. Deposits of metals often occur in masses of igneous rock that have been extruded in a solid or molten state into the surrounding rock strata; such masses are called **batholiths**. Other geological factors to consider include age of rock, fault zones, and rock fractures. The crucial step of finding ore deposits falls in the category of mining geology.

Deposits from Igneous Rocks and Magmatic Activity

Magmatic activity giving rise to the formation of igneous rock formations is responsible for forming numerous kinds of useful mineral deposits. In addition to deposits formed directly from solidifying magma, associated deposits are produced by water in association with magma. Both of these kinds of deposits tend to occur in subduction zones where lithosphere beneath the ocean is being forced under another continental or oceanic plate.

When mineral deposits are formed from cooling magma, they often are produced as coarse-grained material called **pegmatite**, which may be enriched in valuable minerals in the form of relatively large crystals that can be isolated by physical processes to extract the desired mineral. As masses of magma cool and solidify, crystals of minerals often form and, depending upon their densities relative to the magma, either rise or sink in the molten magma, forming enriched layers that can be mined. Such a mineral is magnetite, Fe_3O_4, which is more dense than the comparatively light silicates prevalent in magma; because of its lower density, it settles as a layer when the magma solidifies. Another such dense, settling mineral is chromite, an oxide mineral containing Cr, Mg, Fe, and Al and mined as a source of chromium. Of course, the more valuable a mineral, the more worthwhile it is to separate it from dispersed sources, even if it has not been isolated well within a pegmatite deposit. This is the case with gold, platinum, palladium, and diamond.

Deposits from Hydrothermal Activity

Hot aqueous solutions associated with magma can form rich deposits of minerals. These solutions, either associated with the magma as it comes to the surface or produced by water coming into contact with hot magma, can carry dissolved minerals out of the magma or can pick up minerals from surrounding strata as the magma cools. As the solutions cool, dissolved minerals crystallize from solution, forming **hydrothermal** mineral deposits. Several important metals, including lead, zinc, and copper, are often associated with hydrothermal deposits. The hydrothermal waters are chemically reducing, so that sulfur associated with them is in the reduced sulfide form, S^{2-}. Most metal sulfides are insoluble, and can produce sulfide deposits as shown by the following reaction:

$$Pb + S^{2-} \rightarrow PbS\ (\textit{solid}, \text{galena}) \qquad (16.2.1)$$

In addition to galena two other representatives of the numerous metal sulfide deposits are CuS, HgS (cinnabar), and ZnS (sphalcrite).

The ocean floor in some locations is an especially rich source of hydrothermal deposits consisting of metal sulfides. One such region is the Juan de Fuca ridge in the Pacific Ocean off the northwest coast of the U.S., where significant hydrothermal deposits of silver and zinc sulfides are located on the ocean floor. Similar deposits, including sulfides of copper and lead, are found on the bottom of the Red Sea.

Deposits Formed by Sedimentary or Metamorphic Processes

Some useful mineral deposits are formed as **sedimentary deposits** in association with the formation of sedimentary rocks (see Chapter 13). **Evaporites** are produced when seawater is evaporated. In addition to the obvious example of NaCl deposits of halite, evaporite deposits include sodium carbonates, potassium chloride, gypsum ($CaSO_4 \cdot 2H_2O$), and magnesium salts. Many significant iron deposits consisting of hematite (Fe_2O_3) and magnetite (Fe_3O_4) were formed as sedimentary bands when earth's atmosphere was changed from reducing to oxidizing as photosynthetic organisms produced oxygen. Oxidation reactions, such as

$$4Fe^{2+} + O_2 + 4H_2O \rightarrow 2Fe_2O_3(s) + 8H^+ \qquad (16.2.2)$$

precipitated soluble iron(II) ion in ancient oceans to form insoluble deposits of oxidized iron oxides.

Deposition of suspended rock solids by flowing water can cause segregation of the rocks according to differences in rock size and density. This can result in the formation of useful deposits that are enriched in desired minerals. Originally the minerals were removed from rock deposits upstream by weathering processes and transported by flowing water to the location at which they were deposited as **placer** deposits. During the weathering, transport, and deposition processes that lead to useful placer deposits, the desired minerals are enriched and sorted, whereas impurity constituents are dissolved or washed out by the flowing water. Gravel, sand, and some other minerals, such as gold, often occur in placer deposits.

Some significant placer deposits are now located beneath coastal ocean waters. These were formed during glacial times by streams carrying glacial melt and were deposited on what was then land because of the much shallower ocean depths (much more of Earth's water was tied up as ice on land at the time). Later, as the glacial ice melted and ocean levels raised, these deposits were immersed by seawater, where they remain today as potential mineral sources.

Metamorphic deposits of some minerals have been observed. These occur when minerals, usually of sedimentary origin, become buried and subjected to high pressures and extreme heat. Mineable deposits of graphite, a useful, "slick" form of carbon, have been formed by metamorphic action on fossil carbon, such as coal.

Some mineral deposits are formed by the enrichment of desired constituents when other fractions are weathered or leached away. The most common example of such a deposit is bauxite, Al_2O_3, remaining after silicates and other more soluble constituents have been dissolved by the weathering action of water under the severe conditions of hot tropical climates with very high levels of rainfall. This kind of material is called a **laterite.**

16.3. EVALUATION OF MINERAL RESOURCES

In order to make its extraction worthwhile, a mineral must be enriched at a particular location in Earth's crust relative to the average crustal abundance. Normally applied to metals, such an enriched deposit is called an **ore**. The value of an ore is expressed in terms of a **concentration factor**:

$$\text{Concentration factor} = \frac{\text{Concentration of material in ore}}{\text{Average crustal concentration}} \tag{16.3.1}$$

Obviously, higher concentration factors are always desirable. Required concentration factors decrease with average crustal concentrations and with the value of the commodity extracted. A concentration factor of 4 might be adequate for iron, which makes up a relatively high percentage of Earth's crust. Concentration factors must be several hundred or even several thousand for relatively low-value metals that are not present at very high percentages in Earth's crust. However, for an extremely valuable metal, such as platinum, a relatively low concentration factor is acceptable because of the high financial return obtained from extracting the metal.

Acceptable concentration factors are a sensitive function of the price of a metal. Shifts in price can cause significant changes in which deposits are mined. If the price of a metal increases by, for example, 50%, and the increase appears to be long-term, it becomes profitable to mine deposits that had not been mined previously. The opposite can happen, as is often the case when substitute materials are found, or newly discovered, richer sources go into production.

In addition to large variations in the concentration factors of various ores, there are extremes in the geographic distribution of mineral resources. The United States is perhaps about average for all nations in terms of its mineral resources, possessing significant resources of copper, lead, iron, gold, and molybdenum, but virtually without resources of some important strategic metals, including chromium, tin, and platinum-group metals. For its size and population, South Africa is particularly blessed with some important metal mineral resources.

16.4. EXTRACTION AND MINING

Minerals are usually extracted from Earth's crust by various kinds of mining procedures, but other techniques may be employed as well. The raw materials so obtained include inorganic compounds, such as phosphate rock; sources of metal, such as lead sulfide ore; clay used for firebrick; and structural materials, such as sand and gravel.

Surface mining, which can consist of digging large holes in the ground to remove copper ore, or strip mining (see Section 15.9), is used to extract minerals that occur near the surface. A common example of surface mining is quarrying of rock. Vast areas have been dug up to extract coal.

Because of past mining practices surface mining got a well-deserved bad name. When strip mining (Section 15.9) was employed, the common practice was to dump waste overburden in rather randomly constructed **spoil banks** consisting of poorly compacted, steeply sloped piles of finely divided material. The rock and soil on spoil banks was highly susceptible to erosion and physical and chemical weathering. No effort was made to replace topsoil on the surface of the spoil banks, and that which was there quickly eroded away, so that vegetation on these unsightly piles was sparse. Now, however, with modern reclamation practices, topsoil is first removed and stored to place on top of overburden that is replaced such that it has gentle slopes and proper drainage. Topsoil spread over the top of the replaced spoil, often carefully terraced to prevent erosion, is seeded with indigenous grass and other plants, fertilized, and watered, if necessary, to provide vegetation. The end result of carefully done **mine reclamation** projects is a well-vegetated area suitable for wildlife habitat, recreation, forestry, and other beneficial purposes.

Extraction of minerals from placer deposits formed by deposition from water has obvious environmental implications. Mining of placer deposits can be accomplished by dredging from a boom-equipped barge. Another means that can be used is hydraulic mining with large streams of water. One interesting approach for more coherent deposits is to cut the ore with intense water jets, then suck up the resulting small particles with a pumping system.

For many minerals, underground mining is the only practical means of extraction. An underground mine can be very complex and sophisticated. The structure of the mine depends upon the nature of the deposit. It is of course necessary to have a shaft that reaches to the ore deposit. Horizontal tunnels extend out into the deposit, and provision must be made for sumps to remove water and for ventilation. Factors that must be considered in designing an underground mine include the depth, shape, and orientation of the ore body, as well as the nature and strength of the rock in and around it; thickness of overburden; and depth below the surface.

Usually, significant amounts of processing are required before a mined product is used or even moved from the mine site. Such processing, and the by-products of it, can have significant environmental effects. Even rock to be used for aggregate and for road construction must be crushed and sized, a process that has the potential to emit air-polluting dust particles to the atmosphere. Crushing is also a necessary first step for further processing of ores. Some minerals occur to an extent of a few percent or even less in the rock taken from the mine and must be concentrated on site so that the residue does not have to be hauled far. For metals mining, these processes, as well as roasting, extraction, and similar operations are covered under the category of **extractive metallurgy**.

One of the more environmentally troublesome by-products of mineral refining consists of waste **tailings**. By the nature of the mineral processing operations employed, tailings are usually finely divided, and therefore subject to chemical weathering processes. Heavy metals associated with metal ores can be leached from tailings, producing water runoff contaminated with cadmium, lead, and other pollutants. Adding to the problem are some of the processes used to refine ore. Large quantities of cyanide solution are used to extract low levels of gold from ore, posing obvious toxicological hazards.

16.5. METALS

With an adequate supply of all of the important elements and energy, almost any needed material can be manufactured. Most of the elements, including practically all of those likely to be in short supply, are metals. Some metals are considered especially crucial because of their importance to industrialized societies, uncertain sources of supply, and price volatility in world markets. One of these is antimony, used in auto batteries, fire-resistant fabrics, and rubber. Chromium, another crucial metal, is used to manufacture stainless steel (especially for parts exposed to high temperatures and corrosive gases), jet aircraft, automobiles, hospital equipment, and mining equipment. The platinum-group metals (platinum, palladium, iridium, rhodium) are used as catalysts in the chemical industry, in petroleum refining, and in automobile exhaust antipollution devices. The U.S. imports 87% of these metals and receives the remainder from recycling.

Mining and processing of metal ores involve major environmental concerns, including disturbance of land, air pollution from dust and smelter emissions, and water pollution from disrupted aquifers. This problem is aggravated by the fact that the general trend in mining involves utilization of less rich ores. This is illustrated in Figure 16.1, showing the average percentage of copper in copper ore mined since 1900. The average percentage of copper in ore mined in 1900 was about 4%, but by 1982 it was about 0.6% in domestic ores and 1.4% in richer foreign ores. Ores as low as 0.1% copper may eventually be processed. Increased demand for a particular metal, coupled with the necessity to utilize lower grade ores, has a vicious multiplying effect upon the amount of ore that must be mined and processed, and accompanying environmental consequences.

Figure 16.1. Average percentage of copper in ore that has been mined.

Metals exhibit a wide variety of properties and uses. They come from a number of different compounds; in some cases two or more compounds are significant mineral sources of the same metal. Usually these compounds are oxides or sulfides. However, other kinds of compounds and, in the cases of gold and platinum-group metals, the elemental (native) metals, themselves, serve as metal ores. Table 16.1 lists the important metals, their properties, major uses, and sources.

Table 16.1. Worldwide and Domestic Metal Resources

Metals	Properties[a]	Major uses	Ores, aspects of resources[b]
Aluminum	mp 660°C, bp 2467°C, sg 2.70, malleable, ductile	Metal products, including autos aircraft, electrical equipment. Conducts electricity better than copper per unit mass and is used in electrical transmission lines.	From bauxite ore containing 35-55% Al_2O_3. About 60 million metric tons of bauxite produced worldwide per year. U.S. resources of bauxite are 40 million metric tons, world resources about 15 billion metric tons.
Chromum	mp 1903°C, bp 2642°C, sg 7.14, hard, silvery color	Metal plating, stainless steel, wear-resistant and cutting tool alloys, chromium chemicals, including chromates.	From chromite, an oxide mineral containing Cr, Mg, Fe, Al. Resources of 1 billion metric tons in South Africa and Zimbabwe, large deposits in Russia, virtually none in U.S.
Cobalt	mp 1495°C, bp 2880°C, sg 8.71, bright, silvery	Manufacture of hard, heat-resistant alloys, permanent magnet alloys, driers, pigments, glazes, animal feed additive.	From a variety of minerals, such as linnaeite, Co_3S_4, and as a byproduct of other metals. Abundant global and U.S. resources.
Copper	mp 1083°C, bp 2582°C, sg 8.96, ductile, maleable	Electrical conductors, alloys, chemicals. Many uses.	Occurs in low percentages as sulfides, oxides and carbonates. U.S. consumption 1.5 million metric tons per year. per year. World resources of 344 million metric tons, including 78 million in U.S.
Gold	mp 1063°C, bp 2660°C, sg 19.3	Jewelry, basis of currency, electronics, increasing industrial uses.	In various minerals at only around 10 ppm for ores currently processed in the U.S.; by-product of copper refining. World resources of 1 billion oz, 80 million in U.S.
Iron	mp 1535°C, bp 2885°C, sg 7.86, silvery metal, in (rare) pure form	Most widely produced metal, usually as steel, a high-tensile-strength material containing 0.3-1.7% C. Made into many specialized alloys.	Occurs as hematite (Fe_2O_3), goethite, ($Fe_2O_3 \cdot H_2O$), and magnetite (Fe_3O_4), abundant global and U.S. resources.
Lead	mp 327°C, bp 1750°C, sg 11.35, silvery color	Fifth most widely used metal, storage batteries, chemicals, uses in gasoline, pigments, and ammunition decreasing for environmental reasons.	Major source is galena, PbS. Worldwide consumption about 3.5 million metric tons, 1/3 in U.S. Global reserves about 140 million metric tons, 39 million metric tons U.S.
Manganese	mp 1244°C, bp 2040°C, sg 7.3, hard, brittle, gray-white	Sulfur and oxygen scavenger in steel, manufacture of alloys, dry cells, gasoline additive, chemicals.	Found in several oxide minerals. About 20 million metric tons per year produced globally, 2 million consumed in U.S., no U.S. production, world reserves 6.5 billion metric tons.

Mercury	mp -38°C, bp 357°C, sg 13.6, shiny, liquid metal	Instruments, electronic apparatus, electrodes, chemicals.	From cinnabar, HgS. Annual world production 11,500 metric tons, 1/3 used in U.S. World resources 275,000 metric tons, 6,600 U.S.
Molybdenum	mp 2620°C, bp 4825°C, sg 9.01, ductile, silvery-gray	Alloys, pigments, catalysts, chemicals, lubricants.	Molybdenite (MoS_2) and wulfenite ($PbMoO_4$) are major ores. About 2/3 global Mo production in U.S., large global resources.
Nickel	mp 1455°C, bp 2835°C, sg 8.90, silvery color	Alloys, coins, storage batteries, catalysts (such as for hydrogenation of vegetable oil).	Found in ores associated with iron. U.S. consumes 150,000 metric tons, per year, 10% from domestic production, large domestic reserves of low-grade ore.
Silver	mp 961°C, bp 2193°C, sg 10.5, shiny metal	Photographic film, electronics, sterling ware, jewelry, bearings, dentistry.	Found with sulfide minerals, by-product of Cu, Pb, Zn. Annual U.S. consumption of 150 million troy ounces, short supply.
Tin	mp 232°C, bp 2687°C, sg 7.31	Coatings, solders, bearing alloys, bronze, chemicals, organometallic biocides.	Many forms associated with granitic rocks and chrysolites. Global consumption 190,000 metric tons/year, U.S. 60,000 metric tons/year, world resources 10 million metric tons.
Titanium	mp 1677°C, bp 3277°C, sg 4.5, silvery color	Strong, corrosion-resistant, used in aircraft, valves, pumps, paint pigments.	Commonly as TiO_2, ninth in elemental abundance, no shortages.
Tungsten	mp 3380°C, bp 5530°C, sg 19.3, gray	Very strong, high boiling point, used in alloys, drill bits, turbines, nuclear reactors, to make tungsten carbide.	Found as tungstates, such as scheelite ($CaWO_4$); U.S. has 7% world reserves, China 60%.
Vanadium	mp 1917°C, bp 3375°C, sg 5.87, gray	Used to make strong steel alloys.	In igneous rocks, primarily a by-product of other metals. U.S. consumption of 5,000 metric tons/year equals production.
Zinc	mp 420°C, bp 907°C, sg 7.14, bluish-white	Widely used in alloys (brass), galvanized steel, paint pigments, chemicals. Fourth in world metal production.	Found in many ore minerals. World production is 5 milion metric tons per year (10% from U.S.), U.S. consumption is 1.5 million metric tons. World resources 235 million metric tons, 20% in U.S.

[a] Abbreviations: mp, melting point; bp, boiling point; sg, specific gravity.

[b] All figures are approximate; quantities of minerals considered available depend upon price, technology, recent discoveries, and other factors, so that quantities quoted are subject to fluctuation.

16.6. NONMETAL MINERAL RESOURCES

A number of minerals other than those used to produce metals are important resources. There are so many of these that it is impossible to discuss them all in this chapter; however, mention will be made of the major ones. As with metals, the environmental aspects of mining many of these minerals are quite important. Typically, even the extraction of ordinary rock and gravel can have important environmental effects.

Clays have been discussed as suspended and sedimentary matter in water (Chapter 5) and as secondary minerals in soil (Chapter 14). Various clays are also used for clarifying oils, as catalysts in petroleum processing, as fillers and coatings for paper, and in the manufacture of firebrick, pottery, sewer pipe, and floor tile. The main types of clays that have industrial uses are shown in Table 16.2. U.S. production of clay is about 60 million metric tons per year, and global and domestic resources are abundant.

Fluorine compounds are widely used in industry. Large quantities of fluorspar, CaF_2, are required as a flux in steel manufacture. Synthetic and natural cryolite, Na_3AlF_6, is used as a solvent for aluminum oxide in the electrolytic preparation of aluminum metal. Sodium fluoride is added to water to help prevent tooth decay, a measure commonly called water fluoridation. World reserves of high-grade fluorspar are around 190 million metric tons, about 13% of which is in the United States. This is sufficient for several decades at projected rates of use. A great deal of by-product fluorine is recovered from the processing of fluorapatite, $Ca_5(PO_4)_3F$, used as a source of phosphorus.

Table 16.2. Major Types of Clays and Their Uses in the U.S.

Type of clay	Percent use	Composition	Uses
Miscellaneous	72	Variable	Filler, brick, tile, portland cement, others
Fireclay	12	Variable; can be fired at high temperatures without warping	Refractories, pottery, sewer pipe, tile, brick
Kaolin	8	$Al_2(OH)_4Si_2O_5$; white and can be fired without losing shape or color	Paper filler, refractories, pottery, dinnerware, petroleum-cracking catalyst
Bentonite and fuller's earth	7	Variable	Drilling muds, petroleum catalyst, carriers for pesticides, sealers, clarifying oils
Ball clay	1	Variable, very plastic	Refractories, tile, whiteware

Micas are complex aluminum silicate minerals, which are transparent, tough, flexible, and elastic. Muscovite, $K_2O{\cdot}3Al_2O_3{\cdot}6\ SiO_2{\cdot}2\ H_2O$, is a major type of mica. Better grades of mica are cut into sheets and used in electronic apparatus, capacitors, generators, transformers, and motors. Finely divided mica is widely used in roofing, paint, welding rods, and many other applications. Sheet mica is imported into the United States, and finely divided "scrap" mica is recycled domestically. Shortages of this mineral are unlikely.

In addition to consumption in fertilizer manufacture (Section 14.5), phosphorus is used for supplementation of animal feeds, synthesis of detergent builders, and preparation of chemicals such as pesticides and medicines. The most common phosphate minerals are hydroxyapatite, $Ca_5(PO_4)_3(OH)$, and fluorapatite, $Ca_5(PO_4)_3F$. Ions of Na, Sr, Th, and U are found substituted for calcium in apatite minerals. Small amounts of PO_4^{3-} may be replaced by AsO_4^{3-}, and the arsenic must be removed for food applications. Approximately 17% of world phosphate production is from igneous minerals, primarily fluorapatites. About three-fourths of world phosphate production is from sedimentary deposits, generally of marine origin. Vast deposits of phosphate, accounting for approximately 5% of world phosphate production, are derived from guano droppings of seabirds and bats. Current U.S. production of phosphate rock is around 40 million metric tons per year, most of it from Florida. Tennessee and several of the western states are also major producers of phosphate. Reserves of phosphate minerals in the United States amount to 10.5 billion metric tons, containing approximately 1.4 billion metric tons of phosphorus. Identified world reserves of phosphate rock are approximately 6 billion metric tons.

Pigments and fillers of various kinds are used in large quantities. The only naturally occurring pigments still in wide use are those containing iron. These minerals are colored by limonite, an amorphous brown-yellow compound with the formula $2Fe_2O_3{\cdot}3H_2O$, and hematite, composed of gray-black Fe_2O_3. Along with varying quantities of clay and manganese oxides, these compounds are found in ocher, sienna, and umber. Manufactured pigments include carbon black, titanium dioxide, and zinc pigments. About 1.5 million metric tons of carbon black, manufactured by the partial combustion of natural gas, are used in the U.S. each year, primarily as a reinforcing agent in tire rubber.

Over 7 million metric tons of minerals are used in the U.S. each year as fillers for paper, rubber, roofing, battery boxes, and many other products. Among the minerals used as fillers are, carbon black, diatomite, barite, fuller's earth, kaolin, mica, limestone, pyrophyllite, and wollastonite ($CaSiO_3$).

Although sand and gravel are the cheapest of mineral commodities per ton, the average annual dollar value of these materials is greater than all but a few mineral products because of the huge quantities involved. In tonnage, sand and gravel production is by far the greatest of nonfuel minerals. Almost 1 billion tons of sand and gravel are employed in construction in the U.S. each year, largely to make concrete structures, road paving, and dams. Slightly more than that amount is used to manufacture portland cement and as construction fill. Although ordinary sand is predominantly silica, SiO_2, about 30 million tons of a more pure grade of silica are consumed in the U.S. each year to make glass, high-purity silica, silicon semiconductors, and abrasives.

At present, old river channels and glacial deposits are used as sources of sand and gravel. Many valuable deposits of sand and gravel are covered by construction and lost to development. Transportation and distance from source to use are especially crucial for this resource. Environmental problems involved with defacing land can be severe, although bodies of water used for fishing and other recreational activities frequently are formed by removal of sand and gravel.

The biggest single use for sulfur is in the manufacture of sulfuric acid. However, the element is employed in a wide variety of other industrial and agricultural products. Current consumption of sulfur amounts to approximately 10 million metric tons per year in the United States. The four most important sources of sulfur are (in decreasing order) deposits of elemental sulfur; H_2S recovered from sour natural gas; organic sulfur recovered from petroleum; and pyrite (FeS_2). Supply of sulfur is no problem either in the United States or worldwide. The United

States has abundant deposits of elemental sulfur, and sulfur recovery from fossil fuels as a pollution control measure could even result in surpluses of this element.

Sodium chloride, gypsum, and potassium salts are all important minerals that are recovered as evaporites remaining from the evaporation of seawater. Sodium chloride in the form of mineral halite is used as a raw material for the production of industrially important sodium, chlorine, and their compounds. It is used directly to melt ice on roads, in foods, and in other applications. Potassium salts are, of course, essential ingredients of fertilizers, and have some industrial applications as well. Gypsum, hydrated calcium sulfate, is used to make plaster and wallboard and is an ingredient for the manufacture of portland cement.

16.7. HOW LONG WILL ESSENTIAL MINERALS LAST?

During about a 30-year period following World War II, demand for most important mineral commodities increased at a very rapid rate. Coinciding roughly with the "energy crisis" of the early to mid-1970s, demand slowed. Now, however, with the emergence of newly developing economies, particularly those in the highly populous countries of China and India, it may be assumed that demand for minerals will increase.

To a degree, the economic demand for and price of a resource determine its availability. Higher prices lead to greater exploration, exploitation of less available resources, and often spectacular increases in supply. This phenomenon has led to misinterpretation of the resource supply and demand equation by authorities whose understanding of economics is limited to only conventional monetary supply and demand models, without due consideration of environmental aspects. The fact is that the total available amounts of most resources are limited, painfully so in terms of the time span over which they will be needed by humankind. Although higher prices and improved technologies can increase supplies of critical resources significantly, the ultimate result will be the same—the resource will run out. Furthermore, exploitation of lower and lower grades of resources results in ever-increasing environmental disruption, adding significantly to the cost, considering environmental economics.

Mineral resources may be divided into several categories based upon current production and consumption and known reserves. In the first category are those that are in relatively comfortable supply, with supply of at least 100 to several hundred years. Minerals in this category include bauxite, the source of aluminum, iron ore, platinum-group metals, and potassium salts. In an intermediate category are minerals with a current projected lifetime supply of 25-100 years. These include chromium, cobalt, copper, manganese, nickel, gypsum, phosphate minerals, and sulfur. The most critical group consists of minerals for which the supply based upon current rates of consumption and known reserves is 25 years or less. Among these minerals are sources of lead, tin, zinc, gold, silver, and mercury.

The United States is essentially without economic reserves of a number of essential minerals. These include aluminum, antimony, chromium, cobalt, manganese, tantalum, niobium, platinum, nickel, and tin. Reserves of asbestos, fluorine, and vanadium are quite limited. Larger, but still limited domestic supplies are available of gold, potash, silver, mercury, tungsten, sulfur, and zinc. As far as the United States is concerned, metals of most concern are chromium, manganese, and cobalt. These substances are essential for a modern industrialized economy. Although global supplies are adequate for the immediate future, they are threatened by the potential instability of the countries from which they come—Zaire, Zambia, South Africa, Russia, and other countries in the former Soviet Union.

The world economy will never totally run out of any of the minerals listed above. However, severely constrained supplies of any one or several of them will have some marked effects. For example, world food production now depends on fertilizers, which require phosphorus, of which resources are limited. Within the next century a food crisis related to phosphate shortages may be anticipated.

16.8. WHAT CAN BE DONE ABOUT MINERAL SHORTAGES?

Modern technology and human ingenuity are very effective in alleviating shortages of important minerals. Applications of materials science (see Section 21.5) continue to produce substances made from readily available materials that provide good substitutes for more scarce resources. For example, concrete covered by strong layers of composite materials can readily substitute for iron in construction. Ceramics with special heat- and abrasion-resistant qualities are being used where high-temperature alloys were formerly required.

As minerals become less available, one of the measures to be taken is one that has been taken historically—find more. Modern technology provides a number of useful tools for finding new mineral deposits. Arguably the most useful approach to finding new mineral deposits is through the applications of geology. Recent advances in plate techtonics, for example, have contributed understanding of likely locations of significant mineral deposits.

Slight differences in magnetic field, gravity, and electrical conductivity can be detected very sensitively, and reflect differences in density, magnetic properties, and electrical properties that indicate the presence of remote mineral deposits. These techniques are in the realm of **geophysical prospecting**. Another useful technique for finding mineral deposits is **geochemical prospecting**. As its name implies, this method depends upon detecting the presence of chemical species, usually specific metals. In addition to finding such substances directly in rock, geochemical prospecting can detect evidence of minerals in water some distance from the mineral sources. Even gas analysis can be indicative of some minerals, such as volatile mercury or sulfur compounds from sulfur deposits. Plants, particularly those that concentrate some elements, such as copper, can be analyzed to indicate the presence of minerals. This approach might be termed **biogeochemical prospecting**.

Photography and the measurement of light and infrared radiation from aircraft and from satellites has greatly increased human understanding of Earth and its resources. These techniques, which make it unnecessary to go into remote, poorly accessible, dangerous regions, are in the category of **remote sensing**. Particularly using satellite measurements, it is possible to cover huge areas of Earth's surface in a reasonable period of time. The most ambitious program of remote sensing for mineral exploration is the Landsat satellite system, first launched in 1972 and subsequently followed by other launches. The sensors on these satellites measure visible and infrared radiation. The Landsat images reveal numerous features of Earth's surface including abundance and types of vegetation, soil and rock type, and moisture. Such features in turn may reflect the presence of various kinds of mineral deposits with potential for exploitation.

All the rich mineral ores in readily accessible areas have already been found and exploited. Therefore, any rich deposits will be found in remote locations and hostile environments, such as deep under the ocean. One major possibility is Antarctica, a remote continent noted for its ice, wind, and generally hostile conditions. It is very likely that rich mineral deposits are buried beneath the thick

Antarctic ice sheet. However, the probability of severe environmental damage from extracting minerals there is very high, even if the extreme climate conditions can be overcome. In recognition of that concern 26 nations involved in Antarctic exploration signed a treaty in 1991 banning mineral extraction for 50 years. However, if shortages of crucial minerals become severe, it is likely that efforts will be made to find and extract them in Antarctica.

Exploitation of Lower Grade Ores

Modern technology enables exploitation of lower grade ores, thus significally increasing supplies. A striking example of this phenomenon has been provided by copper. About 100 years ago, the average copper content of ore mined in the United States was around 5%; now it is only about 1/10 that figure for copper ore mined globally. Despite the decline in copper ore quality, during the last 50 years known copper reserves have increased about 5-fold and, adjusted for inflation, the price of copper is now less than it was a century ago.

The ability to exploit much less rich sources of ores has resulted from improved technologies. Of particular importance have been advances in the means of moving huge quantities of rock, essential for the exploitation of lower grade ores. Earth-moving equipment has greatly increased in size and versatility during the last several decades. There has been an environmental cost, of course, for these advances. As an approximation, for each 10-fold decrease in mineral content, it is necessary to move 10 times as much material to obtain the same amount of metal. In addition to disruption of land, disturbed material is more prone to erosion, landslides, and water pollution. Much more energy is required, as is more water for those mining operations that use large quantities of water. Not the least of the factors required for exploiting lower grade resources is the need for additional capital and operating investment, which may be in short supply.

Remote Sources of Minerals

Exciting possibilities exist for the extraction of minerals from remote sources. One possibility is ultra-deep mining under conditions too severe to enable human participation. It may one day be possible to use robots to mine deposits several kilometers deep, where extreme pressures and heat would make it impossible for humans to work.

Another potentially exciting source of minerals is on ocean floors, which remain today largely unexplored. Here, relatively new technologies, such as remote-controlled submarines capable of withstanding crushing pressure have opened new possibilities for exploration and resource utilization. Large areas of the ocean are covered with manganese-rich lumps called **manganese nodules**. Obvious sources of manganese, these lumps also contain other metals, including valuable platinum, copper, and nickel. Extraction of these metals as by-products adds to the economic attractiveness of mining manganese nodules.

Recycling

Recycling should be practiced for all major mineral commodities. Fortunately, both economic and environmental concerns have resulted in vastly increased efforts to recycle materials in recent years. The largest quantity of metal that is recycled consists of ferrous metals (iron). During the last 20 years electric-arc furnaces for iron have become commonplace. Fortunately, these devices require scrap iron as feedstock, and have resulted in a continuing market for recyclable iron scrap. Aluminum ranks next to iron in quantity of metal recycled; about 1/3 of aluminum is recycled in the United States and globally. Particular success has been achieved in the recycling of aluminum beverage cans, with a 4-fold increase during the last 20 years, now approaching 70% of all cans produced. The refining of aluminum metal from bauxite ore is particulary energy consumptive, so a big advantage of aluminum recycling is reduced energy consumption. In addition, recycling produces only about 5% the amount of wastes and potential pollutants as are generated by refining aluminum from ore. Cost savings are huge as well. Other metals that are largely recycled are copper and copper alloys, cadmium, lead, tin, mercury, zinc, silver, and, of course, gold and platinum.

A crucial consideration in recycling is the nature of source material. Copper is relatively easy to recycle because it is often found in a relatively pure form in wire, pipe, and electrical apparatus. Lead in lead storage batteries can simply be melted down and recast into battery electrodes. Large amounts of aluminum are available from waste cans and structural materials. Although iron is largely recycled, it often occurs as specialized alloys containing varying contents of other metals, such as titanium or tungsten. The contents of these elements complicate the utilization of scrap iron.

CHAPTER SUMMARY

The chapter summary below is presented in a programmed format to review the main points covered in this chapter. It is used most effectively by filling in the blanks, referring back to the chapter as necessary. The correct answers are given at the end of the summary.

Manufacturing raw materials from extractive sources are examples of [1]__________ __________ resources. Quantities of raw materials estimated to be ultimately available are called [2]____________, whereas well-identified resources that can be profitably utilized with existing technology are called [3]____________________. Deposits of metals often occur in masses of igneous rock called [4]______________ that have been extruded in a solid or molten state into the surrounding rock strata. Hydrothermal mineral deposits are those formed by [5]______________________ ____.

Solid deposits of normally soluble salts have usually been formed as [6]__________. Minerals left by flowing streams are known as [7]____________________ deposits. Relatively pure deposits of Al_2O_3 called bauxite have been left behind because more [8]_______________ constituents have been [9]______________________ ____ ____________________________. The value of an ore is expressed in terms of a [10]____________________________ defined as [11]______________________ Undesirable mine spoil banks consist of [12]__________________________ ____. Extractive metallurgy consists of [13]__________________________ ______________________________________. Usually metals occur in ores chemically as [14]______________________________. Gold and platinum-group metals may occur as [15] ________________ Kaolin and bentonite are examples of [16] ________________ minerals. Fluorspar, CaF_2, is used as [17] ________________ ________________, whereas cryolite, Na_3AlF_6, is used as [18] ________________ ________________.

Although guano droppings from seabirds and bats are used as a source of an important element, two more common sources of this element are [19] ________________ ________________.

The average total annual monetary value of "cheap" sand and gravel is so high because of [20] ________________. Historically higher prices of minerals have lead to [21] ________________ ________________.

Among the metals with least known reserves are [22] ________________ ________________. Within the next century a food crisis related to shortages of [23] ________________ minerals may be anticipated. In the future mineral shortages can be somewhat alleviated by advanced techniques of [24] ________________ prospecting, [25] ________________ sensing, and exploitation of [26] ________________ ________________. Electric-arc furnaces for iron require [27] ________________ as feedstock and therefore provide a prime example of [28] ________________.

Answers

1 nonrenewable

2 resources

3 reserves

4 batholiths

5 dissolved minerals precipitating from cooled solutions

6 evaporites

7 placer

8 soluble

9 dissolved by the weathering action of water under the severe conditions of hot tropical climates with very high levels of rainfall

10 concentration factor

11 (concentration of material in ore)/(average crustal abundance)

12 waste overburden in poorly compacted, steeply sloped piles of finely divided material

13 concentrating minerals and related operations, such as roasting and extraction

14 oxides or sulfides

15 the elemental metals

16 clay

17 a flux in steel manufacture

18 a solvent for aluminum oxide in the electrolytic preparation of aluminum metal

19 fluorapatite, $Ca_5(PO_4)_3F$ and hydroxyapatite, $Ca_5(PO_4)_3(OH)$

20 the huge quantities involved

21 greater exploration, exploitation of less available resources, and often spectacular increases in supply

22 lead, tin, zinc, gold, silver, and mercury

23 phosphate

24 geophysical, geochemical, and biogeochemical

25 remote

26 lower grade ores

27 scrap iron

28 recycling

QUESTIONS AND PROBLEMS

1. As related to minerals, what is the distinction between resources and reserves?
2. How is the presence of sulfide related to some hydrothermal deposits of metals?
3. What is the role of flowing water in the formation of placer deposits?
4. How are weathering and leaching related to the formation of bauxite deposits? Which metal is extracted from bauxite?
5. What is meant by concentration factor in minerals and how does it relate to their economic recovery?
6. What are mine tailings? What are some of the special environmental problems associated with tailings? Why are they particularly susceptible to weathering?
7. Discuss how technology that enables utilization of less rich ores relates to environmental and energy considerations.
8. Match the following pertaining to metals:

( ) Chromium	A. Used in organometallic biocides
( ) Lead	B. Fourth in world metal production
( ) Mercury	C. Used to make stainless steel
( ) Tin	D. Liquid metal
( ) Zinc	E. From galena

9. What are some of the main uses of clays?
10. How are fluorine compounds used in metal manufacture?
11. What are some of the major applications of phosphate minerals?
12. What is the largest use for sulfur?

13. What is the overall effect of the development of the economies of highly populated countries on the demand for mineral resources?
14. How can modern materials science help alleviate mineral shortages?
15. Explain how modern high technology may assist in finding new mineral sources.
16. How does larger and more sophisticated earth-moving equipment increase the supply of minerals? What are the environmental implications, both good and bad?
17. What are two largely unexplored parts of the Earth where significant mineral resources may yet be found?
18. Explain why the nature of source material is particularly important in recycling.

SUPPLEMENTARY REFERENCES

Annels, Alwyne E., Ed., *Case Histories and Methods in Mineral Resource Evaluation*, American Association of Petroleum Geologists, Tulsa, OK, 1992.

Auty, Richard M., *Sustaining Development in Mineral Economies: The Resource Curse Thesis*, Routledge, New York, NY, 1993.

Common, Michael S., *Environmental and Resource Economics: An Introduction*, 2nd ed., Longman, New York, NY, 1996.

Financial Times Mining International Yearbook 1997 (110th ed.), St. James Press, New York, NY, 1996.

Ripley, Earle A., *Environmental Effects of Mining*, St. Lucie Press, Boca Raton, FL, 1995.

Shaw, T., *Surface Mining and Quarrying: Mechanization, Technology, and Capacity*, Ellis Horwood Ltd., New York, NY, 1996.

LIFE

17 ECOSYSTEMS AND BIOLOGICAL COMMUNITIES

17.1. LIFE AND THE BIOSPHERE

The water-rich boundary region at the interface of Earth's surface with the atmosphere, a paper-thin skin compared to the dimensions of Earth or its atmosphere, is the **biosphere** where life exists. The biosphere includes soil on which plants grow, a small bit of the atmosphere into which trees extend and in which birds fly, the oceans, and various other bodies of water. Although the numbers and kinds of organisms decrease very rapidly with distance above Earth's surface, the atmosphere as a whole, extending many kilometers upward, is essential for life as a source of oxygen, medium for water transport, blanket to retain heat by absorbing outgoing infrared radiation, and protective filter for high-energy ultraviolet radiation. Indeed, were it not for the ultraviolet-absorbing layer of ozone in the stratosphere, life on Earth could not exist in its present form.

This chapter deals with life on Earth. It considers the highly varied locations where life exists and the vastly different conditions of moisture, temperature, sunlight, nutrients, and other factors to which various life forms adapt. Such conditions may be those of the tropics, with abundant moisture, intense sunlight, high temperatures, and relatively little variations in these and other factors. Or they may be characteristics of inland deserts that are hot during the daytime and cold at night, generally very dry, but subject to occasional torrential rainstorms and flash flooding. Life thrives on land surfaces, in bodies of water, and in sediments in water. The extreme variability of environments in which life exists is matched by the remarkable variety, versatility, and adaptability of the communities of organisms that populate these environments. These range from tropical rain forest communities containing thousands of plant, animal, and microbial species in a small area to austere, exposed mountain rocks subjected to extremes of weather and populated by a thin coating of tenacious lichen, a symbiotic combination of fungi and algae that clings as a thin layer to the rock surface. In addition to dealing with organisms and their environment, this chapter also discusses the intricate relationships among organisms that enable them to coexist with each other and their surroundings.

To understand life on Earth it is important to define what life really is. Living organisms are constituted of cells that are bound by a membrane, contain nucleic acid genetic material (DNA) and possess specialized structures that enable the cell to perform its functions. A living organism may consist of only one cell or of billions of cells of many specialized types. All living organisms have two characteristics: they process matter and energy through metabolic processes, and they reproduce. The ability of an organism to process matter and energy is called metabolism. Another important characteristic of living organisms is their ability to maintain an internal environment that is favorable to metabolic processes and that may be quite different from the external environment. Warm-blooded animals, for example maintain internal temperatures that may be much warmer or even cooler than their surroundings. Finally, through succeeding generations living organisms can undergo fundamental changes in their genetic composition that enable them to adapt better to their environment.

Living species are present in the biosphere because they have evolved with the capability to survive and to reproduce. Every single species in the biosphere has become an expert in these two things; otherwise it would not be here. The key factors for existing, at least long enough to reproduce, are the ability to process energy and to process matter. In so doing, life systems and processes are governed by the principles of thermodynamics and the law of conservation of matter. Organisms handle energy and matter in various ways. Plants, for example, process solar energy by photosynthesis and utilize atmospheric carbon dioxide and other simple inorganic nutrients to make their biomass. Herbivores are animals that eat the matter produced by plants, deriving energy and matter for their own bodies from it. Carnivores in turn feed upon the herbivores.

Life forms require several things to exist.[1] The appropriate chemical elements must be present and available. Energy for photosynthesis is required in the form of adequate sunlight. Temperatures must stay within a suitable range and preferably should not be subject to large, sudden fluctuations. Liquid water must be available. And, as noted above, a sheltering atmosphere is required. The atmosphere should be relatively free of toxic substances. This is an area in which human influence can be quite damaging, through release of air pollutants that are directly toxic or which react to form toxic products, such as life-damaging ozone produced through the photochemical smog-forming process.

Individual organisms and groups of organisms must maintain a high degree of stability (**homeostasis**, meaning "same status") through a dynamic balance involving inputs of energy and matter and interaction with other organisms and with their surroundings. This requires a high degree of organization and the ability to make continuous compensating adjustments in response to external conditions. For an individual organism, homeostasis means maintaining temperature, levels of water, inputs of nutrients, and other crucial factors at suitable levels. Arguably the most advantageous evolutionary trait of mammals is their ability to keep their body temperatures within the very narrow limits that are optimum for their biochemical processes. The concept of homeostasis applied to whole ecosystems consisting of groups of organisms and their surroundings is termed **ecosystem stability**. Indeed, homeostasis applies to the entire biosphere. To a large extent environmental science addresses the homeostasis of the biosphere and how the critical factors involved in maintaining the dynamic equilibrium of the biosphere are affected by human activities.

The nature of life is determined by the surroundings in which the life forms must exist. Much of the environment in which organisms live is described by physical factors, including whether or not the surroundings are primarily aquatic or terrestrial. For a terrestrial environment important physical factors are the nature of available soil, availability of water, and availability of nutrients. These are **abiotic factors**. There are also important **biotic factors** relating to the life forms present, their wastes and decomposition products, their availability as food sources, and their tendencies to be predatory or parasitic.

This chapter discusses life on Earth. To understand the nature of life on Earth, it is important to consider what kinds of life are present, how various species fit into specific habitats, how energy and matter are utilized and cycled, and how various species interact with each other and with their environment. These factors are covered by the science of **ecology**, which addresses how organisms interact with their environment and with each other.[2]

17.2. ECOLOGY AND LIFE SYSTEMS

To consider the biosphere and its ecology in their entirety, it is necessary to look at several levels in which life exists. The unimaginably huge numbers of individual organisms in the biosphere belong to **species**, or kinds of organisms. Groups of organisms living together and occupying a specified area over a particular period of time constitute a **population**; and that part of Earth on which they dwell is their **habitat**. In turn, various populations coexist in **biological communities**. Members of a biological community interact with each other and with their atmospheric, aquatic, and terrestrial environments to constitute an **ecosystem**. An ecosystem describes the complex manner in which energy and matter are taken in, cycled, and utilized; the foundation upon which an ecosystem rests is the production of organic matter by photosynthesis. Assemblies of organisms living in generally similar surroundings over a large geographic area constitute a **biome**. Each biome may contain many ecosystems. The following are examples of important kinds of biomes:

- **Tropical rain forests** characterized by warm temperatures throughout the year and having most of their nutrients contained in the bodies of the organisms populating the forest.
- **Warm-climate evergreen forests** found in the southeastern United States.
- **Coniferous forests** in temperate climates that have distinct summer and winter seasons and that are populated by cone-bearing trees with needles, such as cedar, hemlock, and pine.
- **Temperate deciduous forests** growing in regions with hot, wet summers and cold winters populated by trees that grow new leaves and shed them annually.
- **Grasslands** in which grass is anchored in a tough, dense mass of grass roots, and soil called **sod**.
- **Hot deserts** populated by cacti, creosote bush, yucca, and other species adapted to high temperatures and sparse moisture.
- **Cold deserts** in which tough perennial plants, such as sagebrush and some grasses, survive under cool, dry conditions.
- **Tundra** found in arctic regions or at high altitudes. Tundra regions have no trees, cold winters with frost possible even in the summer, permanently frozen subsurface soil called **permafrost**, low productivity, relatively few species, and high vulnerability to environmental insult.

In order to sustain life, an ecosystem must provide energy and nutrients. Energy enters an ecosystem as sunlight. Part of the solar energy is captured by photosynthesis, and part is absorbed to keep organisms warm, which enables their metabolic processes to occur faster. In addition to capturing energy, an ecosystem must provide for recycling essential nutrients, including carbon, oxygen, phosphorus, sulfur, and trace-level metal nutrients, such as iron.

Much of the organization of ecosystems has to do with the acquisition of food by the organisms in it. Virtually all food upon which organisms depend is produced by the fixation of carbon from carbon dioxide and energy from light in the form of energy-rich, carbon-rich biomass through the process of photosynthesis. Photosynthesis can be represented by

$$CO_2 + H_2O + h\nu \rightarrow \{CH_2O\} + O_2 \quad (17.5.1)$$

where $h\nu$ represents light energy absorbed in photosynthesis and $\{CH_2O\}$ represents biomass. Thus the photosynthetic plants in the biosphere are the basic **producers** upon which all other members of the community depend for food and for their existence. The rate of biomass production is called **productivity**. It is conventionally expressed as energy or quantity of biomass per unit area per unit time. The food manufactured by producers is utilized by other organisms generally classified as **consumers**.

The sequence of food utilization, starting with biomass synthesized by photosynthetic producers is called the **food chain**. Numerous food chains exist in ecosystems, and there is crossover and overlap between them. Therefore, food chains are interconnected to form intricate relationships called **food webs**. An example of a food chain would be one in which biomass is produced by unicellular algae in a lake and consumed by small aquatic organisms (copepods), which are eaten by small fish, which are eaten in turn by large fish. Finally, a bald eagle atop the food chain may consume the large fish. As shown by this example, there are several levels of consumption in a food chain called **trophic levels**. In going up the chain, the first through fourth trophic levels are (1) producers, (2) primary consumers (herbivores), (3) secondary consumers (carnivores), and (4) tertiary consumers, sometimes called "top carnivores. In addition to herbivores and carnivores, there are several other classifications of consumers. **Omnivores** eat both plant matter and flesh. **Parasites** draw their nourishment from a living host. **Scavengers**, such as beetles, flies, and vultures, feed on dead animals and plants. **Detritovores**, such as crabs, earthworms, and some kinds of beetles, feed on **detritus** composed of fragments of dead organisms and undigested wastes in feces. Ultimately fungi and microorganisms complete the degradation of food matter to simple inorganic forms that can be recycled through the ecosystem, a process called **mineralization**.

Food webs can be divided into two main categories broadly based upon whether the food is harvested from living populations or from the remains of dead organisms. In a **grazing food web** food, along with the energy and nutrient minerals that it contains, is transferred from plants to herbivores and on to carnivores. Dead organisms become part of a **detritus food web** in which various levels of scavengers degrade the organic matter from the dead organisms.

The energy content of food passing through each trophic level may be consumed by respiratory processes to maintain the metabolic activity and movement of the organisms at that level, eventually to be dissipated as heat. Energy incorporated into the bodies of organisms or excreted from them as waste products can become part of the detritus food web. A relatively small, but very important fraction of the energy is passed on to the next trophic level when the organisms are consumed by predators.

It is instructive to consider the flow of energy utilization through various trophic levels. Typically, based upon 100% of the energy from primary producers, the percentages utilized by various trophic levels are decomposers 25%, herbivores 15%, primary carnivores 1.5%, and top carnivores less than 0.1%.

17.3. WHAT IS A BIOLOGICAL COMMUNITY?

A *biological community* consists of an assembly of organisms that occupy a defined space in the environment. The nature of such a community depends upon the physical and chemical characteristics that influence the life forms in it and upon the interactions of the organisms in the community. A biological community is the biological component of an *ecosystem*, which includes the organisms and their physical environment. The community functions in a manner such that it tends to utilize and convert energy and materials in the most efficient manner that will enable the organisms in it to reproduce and thrive. Therefore, the exchange of matter and energy among the organisms and with their physical environment is a key aspect of an ecosystem. The study of biological communities is called **community ecology**.

There are many interactions of organisms in a community. Many of these interactions are mutually advantageous. For example, grazing animals derive their food from grass growing on grasslands and return nutrients to the grass as manure, urine, and mineralized phosphorus, nitrogen, and potassium from their own bodies after death. Fierce competition exists between species in biological communities. Species compete for the same nutrients, food sources, space, and sunlight. Coexistence often depends upon species finding their own niches. For example, tall trees utilize the sunlight falling directly on a forest canopy for photosynthesis, whereas smaller plants can exist on the more meager light that filters through from above.

Biological communities are subject to constant change. Some of these changes are relatively short term and cyclical, following daily and seasonal patterns. Some desert plants, for example, may lie largely dormant during prolonged periods of dry weather, only to start growing fiercely and burst into bloom when a rare rainstorm occurs. Other changes occur as a habitat undergoes long-term transitions. These may occur, for example, after a forest fire abruptly kills all the tall trees in a community and the plants and the animals in the community undergo several successions until tall trees are eventually re-established. Many transitions are the result of human activities, such as take place when agricultural land is taken out of production for a number of years. In the past, major transitions in biological communities have occurred with changes in climate, such as happened with the retreat of the glaciers after the last Ice Age. Some evidence suggests that global greenhouse warming will cause long-term changes in many biological communities during the next several decades.

Stable biological communities are characterized by a high degree of order and often complex organization. The organisms in a community undergo constant opposing and compensating readjustment of their behavior, feeding, and reproduction in response to each other and to their surroundings. Therefore, such communities are in a state of homeostasis. An established, stable biological community as a whole is homeostatic.

17.4. PHYSICAL CHARACTERISTICS AND CONDITIONS

Biological communities largely develop in response to their physical environment, which includes the three most critical factors of temperature, moisture, and light, as well as other factors, such as quality of soil and nutrient supply. In the case of an aquatic environment, mixing and circulation, which bring nutrients up from sediment deposits are important. Another factor that must be considered is **variability** of the physical environment. Changes in temperature and moisture level can have particularly strong effects on biological communities, forcing them to change in response to the new conditions. Whereas organisms have adapted well to seasonal differences in their environment, sudden and drastic variations can put a lot of stress on a community. With these factors in mind, it is understandable that some of the most thriving, diverse, productive biological communities are found in the tropics where elevated temperatures favor high metabolic activity, intense light enables a high level of photosynthesis, abundant water favors plant growth, and relatively constant climatic conditions make it unnecessary for organisms to waste much of their effort adapting to seasonal change.

Consideration of the crucial factors in maintaining biological communities leads to the following important concepts:

- **Limiting factors** determine whether a species can exist in a biological community and, if so, its abundance, growth, and distribution.
- **Tolerance limits** describe the lower and higher values of limiting factors below and above which a species cannot live.
- The **critical factor** is the one that is closest to a tolerance limit.

Consider a species of plant that under normal circumstances might be the primary producer of biomass in a biological community. The most prominent limiting factors that determine its survival are light, temperature, and water. Typically, during the late summer months, light is abundant and the temperature remains within a range conducive to plant growth, but the supply of moisture becomes marginal due to drought conditions. Therefore, water is the critical factor and in a very severe drought the water supply may fall below the lower tolerance limit such that plants of the particular species wither and die, thus affecting the whole food chain and supply for the biological community.

Topography can have an important influence on a biological community. The population of a community on a south-facing slope (in the Northern Hemisphere) is influenced by the generally warmer temperatures and greater intensity of sunlight compared to that of a north-facing slope. (Advantage has been taken of this phenomenon in reclaimed land on the sites of brown-coal strip mines in Germany. The soil was replaced in terraces having large, gradual south-facing facing slopes where crops were planted and short, steep north-facing slopes that were not cultivated. This has extended the growing season by several days in both spring and fall, thereby increasing crop yields.) Marked contrasts in communities may exist on either side of a mountain range over which moisture-laden air flows, as illustrated in Figure 17.1. As warm moist air flows over the upwind slope, it is forced to rise causing it to cool and release precipitation. On the downwind slope the cool air sinks, becomes warmer, and has a drying effect on the vegetation and terrain that it contacts. The absence of rain in this region is called a "rain shadow."

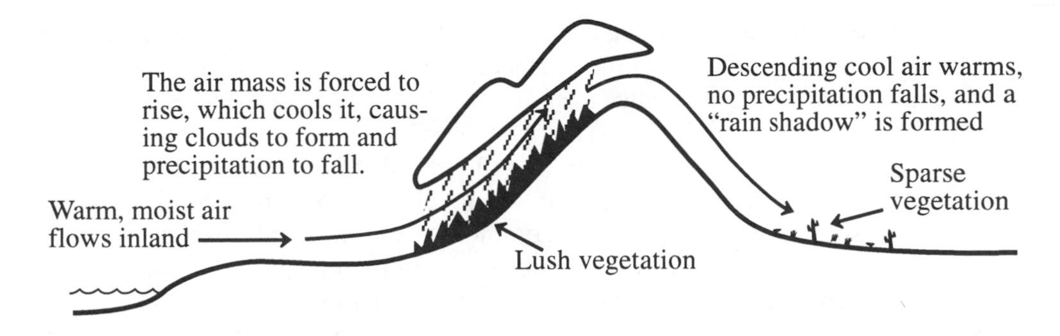

Figure 17.1. Illustration of the effects of topography on biological communities. A mass of warm moist air forced to rise over a mountain range deposits rain and snow as it does so, enabling a productive community to thrive. On the other side of the range a "rain shadow" creates drought conditions in which the productivity is low, and a much different community exists.

17.5. EFFECTS OF CLIMATE

Climate refers to weather conditions that exist over the long term, whereas **weather** describes short-term conditions of temperature, precipitation, humidity, wind, and cloud cover. Organisms and their biological communities are largely determined by climate. This section briefly addresses the effects of various aspects of climate on organisms.

The productivity of ecosystems is a function of climate. The most productive ecosystems are those in the tropics. Productivity is lower in the temperate zones, and much less near the poles. Seasonal cold has a tremendous effect on plants and hence on productivity. Annual plants are killed by freezing, whereas deciduous trees lose their leaves and stop photosynthesis when it freezes. Thus cold weather and freezing temperatures tremendously decrease primary productivity.

Specific species of plants and animals thrive only within certain temperature ranges. Animals are either **poikilothermic**, such that their temperatures remain close to those of their surroundings, or **homeothermic**, using intricate physiological processes to maintain their internal temperatures within a narrow range. There are several important ways in which animals maintain desired temperatures. One of the most interesting of these is through circulatory system "heat exchangers" in which blood flowing in veins from the extremities is warmed by outgoing arterial blood. Fur, feathers, layers of fat, and perspiration are all employed to maintain relatively constant temperatures in warm-blooded animals. Some animals hibernate and thus maintain reduced but adequate body temperatures at a low level of metabolism.

The temperatures of plants are largely those of the surroundings. (An interesting exception is the "warm-blooded" skunk cabbage plant that uses respiratory processes to metabolize energy-producing materials stored in its large root to keep its temperature as much as 30°C higher than the surrounding atmosphere on cold spring days.) However, for most plants temperature extremes cause enzymes to become inactivated and proteins to become denatured. Plant membranes are injured by very high or very low temperatures. Freezing can be very damaging to plants and the first "killing frost" of autumn signals the demise of many plants that have been thriving during the summer. Ice crystals formed when plants freeze destroy tissue so that it can no longer function. Plants do have some mechanisms to cope with freezing. If the temperature decrease is slow enough, plant tissues can lose water so that they become dehydrated and ice crystals form on the plant surfaces. Resistance to freezing is also accomplished by the presence of sugars, sugar alcohols, and amino acids in plant fluids. These substances act as a kind of natural antifreeze. Plants also have mechanisms to resist high temperatures. One of these is for plant leaves to turn at an angle such that they do not receive the full force of mid-day sun.

All life forms require water. Undesirable levels of moisture, usually moisture deficiency, constitute one of the most common environmental stresses faced by organisms. Drought conditions can be very harmful to biological communities. For a terrestrial community a lack of water combined with high temperatures may be devastating to plants and animals. During winter or early spring total levels of water may be adequate, but the water is frozen and not available.

Organisms have various ways to respond to moisture deficiency. Some plants get around the problem by being ephemeral in that their seeds germinate and the plants grow and bloom only when precipitation levels are adequate. Other plant mechanisms for dealing with moisture deficiency include shedding leaves and reducing water loss to the atmosphere by transpiration. This is accomplished by closing stomata and with structures such as thicker leaves or waxy surfaces that slow moisture loss. Succulent plants store large quantities of water in cells, whereas others have extremely deep root systems that reach underground water levels.

Animals have evolved that can tolerate some dehydration. Nocturnal animals avoid the extreme drying conditions of hot desert days. Water generated during metabolic oxidation of food substances is retained in the body. One interesting mechanism of coping with short water supplies is that of producing a very concentrated urine. Such urine, which may have a very pungent odor, is characteristic of the cat family, which is largely adapted to dry conditions.

17.6. SPECIES

The kinds of species present in a biological community are largely determined by their environment and, in turn, influence their surroundings. The roles played by different species can be categorized in several different ways. Before considering populations of organisms, it is instructive to consider the categories and functions of the species that constitute such populations.

It should be noted that individuals in species are not identical, but exhibit slight differences resulting in **genetic diversity**. Living organisms undergo **mutations** arising from changes in their DNA and through recombinations of their genes that occur with reproduction. Mutations that give rise to favorable traits in individuals give them natural advantages so that they are relatively more likely to survive and reproduce. This **natural selection** allows species to better adapt to their environment and, over many generations, results in permanent changes in genetic composition. This is the process of **evolution**, through which new species are formed. Natural selection and evolution are very important in determining the occupants of biological communities. Organisms evolve that are best adapted to climatic conditions, particularly those of temperature and moisture, and to the presence or deficiency of nutrients. They also evolve in reponse to competition from other species, diseases caused by bacteria and viruses, parasites, and predation by other species.

A diverse biological community cannot exist in which all of the species present are competing for exactly the same limited resources. Therefore, in order to thrive, a species must be specialized and have its own **ecological niche**, which basically describes where the organism lives and how it functions in its environment. The ecological niche of a species defines the habitat that it occupies, the food and nutrients that it utilizes, its predators, its interaction with the other biotic and abiotic components of the ecosystem, and the seasons and times that it does various things. In order to coexist with the limited resources available, the members of a biological community must **partition resources**. Organisms vary in the degree to which they require specialized niches.

Niche specialists occupy very well defined niches, tending to require specific kinds of climatic conditions, habitats, and kinds of foods. Specialists are vulnerable to destruction or disturbance of their niches. Such a specialist is the spotted owl, which requires old-growth timber in the U.S. Northwest for its habitat. Tropical rain forests are particularly notable for the wide variety of niche specialists that occupy them. So many niches are possible because of the stratification of the forests and their long-term environmental and biological stability. That is one reason that damage to or destruction of tropical rain forests poses a particular threat of species extinction.

Niche generalists are organisms that are much more adaptable to different environments. Some niche generalists have adapted so well to habitats modified by humans that they have become nuisances. An example is the coyote, which coexists with humans in some suburban areas, occasionally killing an unwary pet cat. Another example is the Canadian goose, which has taken a strong liking to country club lakes and suburban lawns, and may even aggressively object to sharing these niches with the humans that they encounter. The human species, itself, is an excellent, resourceful niche generalist.

It is useful to define several other categories of species in discussing ecosystems. **Native species** are those that are naturally present in an ecosystem, whereas those that have been introduced, usually by human intervention, are called **immigrant species** or **alien species**. In some cases immigrant species have been beneficial, whereas in other cases they have displaced native species with catastrophic results. Some species have such important functions in an ecosystem that they are called **keystone species**. Keystone species include some predators that keep numbers of otherwise destructive species in check, insects that pollinate flowers, and species that modify habitat to provide resources and dwelling space for other species. An example of the last of these is the American alligator, which digs depressions in swampland that provide water habitat for numerous species during droughts.

In considering pollution and damage to ecosystems, **indicator species** are particularly important. These are species whose numbers decline or exhibit symptoms of malaise as a reflection of habitat damage before other major symptoms are observed. The decline of hawks, eagles, and other predatory birds at the top of the food chain was indicative of widespread DDT pesticide pollution during the 1950s. The current loss of frogs and other amphibians worldwide is viewed by some experts as an indication of widespread environmental deterioration and pollution.

17.7. POPULATIONS

Recall that populations consist of groups of organisms living together and occupying a specified area over a particular period of time. Populations have numerous characteristics, including numbers, genetic composition, birth and death rates, and age and sex distribution. This section addresses some of the important factors involved with populations in biological communities.

All existing species are present on Earth because they have developed exceptional abilities to reproduce and to survive long enough to do so. The potential for reproduction always greatly exceeds the ultimate capacity of an environment to support a population. This capacity is known as the **carrying capacity**. When members of a population are newly introduced into an area that is amenable to their growth, their numbers tend to increase very rapidly at a rate approaching exponential growth. Eventually numbers reach a level around the carrying capacity at which point rapid growth ceases abruptly because of some limiting factor, such as limited food, nutrients, water, air, or shelter, or because of stress from crowding. After the very rapid growth phase ceases, the number of organisms tends to fluctuate slightly around the carrying capacity as shown in Figure 17.2.

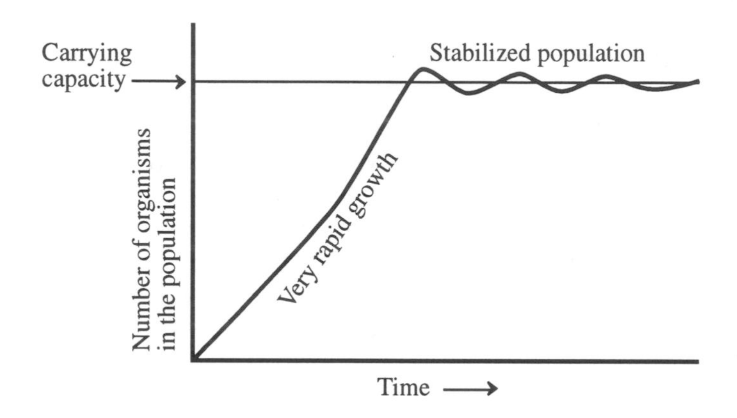

Figure 17.2. Very rapid growth of a population newly introduced into an environment suitable for its survival, followed by stabilization of numbers around the carrying capacity. This is an idealized picture subject to perturbation by many factors, such as predation, disruption of habitat, or disease.

In some cases a large growth in population can cause it to temporarily exceed the carrying capacity by a significant margin, or the carrying capacity can be suddenly reduced by circumstances such as droughts. This can result in a temporary overpopulation until reduced reproduction rates and increased death rates can adjust the population to accord with the carrying capacity. The result can be an abrupt decrease in population, or **population crash**. If the carrying capacity has been altered by overpopulation, such as destruction of grassland by overgrazing, or if it has been reduced by external factors, the new population will stabilize at a figure that reflects the new carrying capacity.

The physical and chemical conditions of a habitat largely determine the organisms that dwell in it. The biological history of a community is also an important factor. Species may dominate simply because they have been there for quite some time, and chance may have played a role in their establishment and survival. The presence and numbers of a species can also be strongly influenced by the other species present. The degree of productivity and the kind of plant biomass produced can largely determine the kinds of organisms in a community. For example, large ruminant animals, such as American Bison, can thrive in areas that are highly productive of prairie grass, and squirrels may thrive in wooded areas that produce large quantities of nuts. Species may modify the environment in ways that assist other species. Damming of streams by beavers, which are keystone species in some habitats, can create pools that provide habitat for certain kinds of plants and other kinds of animals. Predation by other species is also important in determining which species exist in a biological community.

Species divide resources for most effective utilization according to time and space. Some animal species hunt and gather food during the day and others do so at night, thus avoiding needless competition. Some varieties of plants thrive and reproduce in the spring when moisture is abundant, whereas others are adapted to hotter, drier conditions and become dominant during the mid- and late-summer months. One of the major adaptations based upon physical space is that of **vertical stratification**, which is seen, for example, in forests where tall trees get light for photosynthesis from the top layers of the forests and ferns and mosses use the much less intense, diffuse light at ground level. As shown in Figure 17.3, vertical stratification is particularly important in bodies of water that are stratified because of the temperature/density behavior of water.

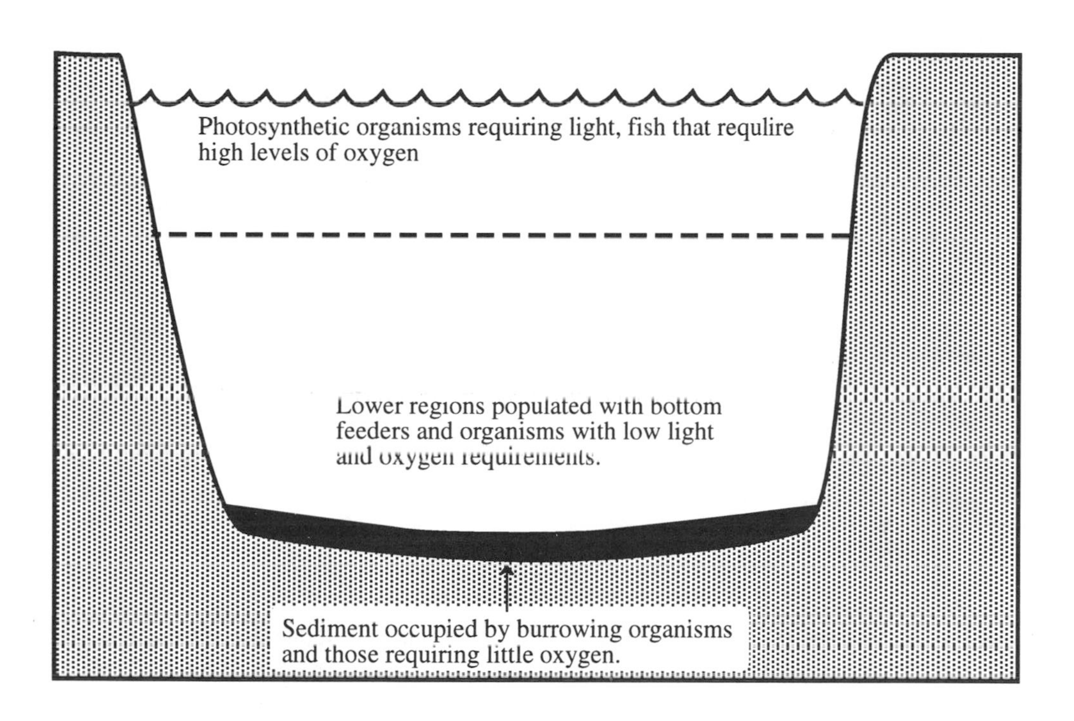

Figure 17.3. Vertical stratification of a biological community is exhibited in bodies of water, which are often physically stratified because of the temperature/density behavior of water.

17.8. SURVIVAL OF LIFE SYSTEMS, PRODUCTIVITY, DIVERSITY, AND RESILIENCE

Biological communities are subject to various stressful changes that affect their populations. These can be divided between the two main categories of catastrophic *vs.* gradual changes, each of which is divided between those caused by natural factors and those resulting from human activities. Examples of catastrophic natural changes include fire, wind damage, flooding, drought, landslide, or volcanic eruption. Catastrophic changes caused by humans include deforestation, cultivation, disruption of surface by strip mining, and some forms of severe pollution. Gradual natural changes include those due to adaptation, evolution, changes in climate, and disease. Gradual changes caused by humans include exploitation of natural, soil, and wildlife resources; salinization of soil due to irrigation; addition or elimination of species; and destruction of habitat.

Ecosystems and biological communities that are suffering from stress exhibit warning signs. One of the most common of these is decreased productivity; if the basic food source for a whole community diminishes, the community as a whole must suffer. Commonly associated with decreased productivity is a loss in nutrients. A well-balanced ecosystem recycles and retains nutrients such as nitrogen. Under stress, such nutrients may be permanently lost from the whole system, whereas levels of pollutants may increase as the ecosystem suffers damage. The loss of indicator species is indicative of problems in an ecosystem, as is a decline in the diversity of species present. Increases in the numbers of some species, particularly predatory insects and disease-causing organisms indicate ecosystem damage.

There are several key parameters that are used to describe the ability of biological communities to survive and thrive. These are the following:

- **Productivity**, the rate at which a biological community generates biomass, almost always by photosynthesis, required to sustain the organisms in it.
- **Diversity**, which describes how many different species are present and their relative abundances.
- **Inertia** describes the resistance of a community to being altered, damaged, or destroyed.
- **Constancy** is the ability to maintain numbers within the optimum limits that can be supported.
- **Resilience**, the ability to recover from perturbations, which may be of a catastrophic nature and kill large numbers of organisms in the community.

The productivity of a biological community by the plant producers in it is influenced by many factors. One such factor for land-based biological communities is the nature of the geosphere to which the community is anchored. A key geospheric component is soil; the quality of soil and its nutrient contents largely determine biomass yield. In some communities soil is virtually absent, and what little productivity there is comes from lichen growing on rock surfaces and hardy plants that anchor their roots between rocks. Topography can influence productivity to an extent. Soil on steep slopes, for example, may be subject to erosion and hence less productive than soil on gently sloping terrain. In general, higher temperatures and higher precipitation favor productivity. Therefore, rain forests tend to be the most productive biological communities, and dry, cold deserts have very little productivity.

Overall, ecosystems in the world's oceans have lower productivity than do those on land. However, biological communities along ocean shores are usually very productive, and estuaries where fresh water and saltwater merge contain the most productive of all biological communities. Warm seawater in tropical regions provides habitat for the most productive marine-based biological communities. On the high seas, cooler temperatures, limited light penetration, and the lack of availability of nutrients tends to keep marine productivity low. Dead biomass and fecal pellets from marine organisms commonly sink to the ocean floor where they are not readily available for recycle of the nutrients, such as phosphorus, that they contain. Areas of particularly high marine productivity occur when nutrient-rich sediments are brought to the surface by convection currents, a phenomenon called **upwelling**.

Species diversity describes an important characteristic of a biological community based upon how many different species are present and their relative abundances. Relatively productive biological communities often show a high level of diversity. Thus, a tropical rain forest provides habitats for a large variety of plants, mammals, reptiles, and lizards; whereas an arctic biological community may have comparatively few. Related to species diversity is the observation that similar biological communities develop under similar conditions in distant geographical locations. **Ecological equivalents** is a term used to describe species that are not closely related to each other genetically, but have evolved similar characteristics and behavior patterns that enable them to occupy similar niches in widely different locations. An example is provided by the African jackal and the North American coyote, which, though entirely different species, fill similar ecological niches on their respective continents and exhibit some similarities in their traits and appearance.

High productivity and species diversity can help a biological community to be resilient and stable. A resilient community subjected to stress, such as from a drought or disease affecting one or more of its members, has the ability to deal with and compensate for change; whereas stress applied to a community with only a few members that is marginally productive may be catastrophic for the community as a whole. One that has a large number of species and a cushion of high productivity will be in a much better position to recover from damaging events and circumstances.

17.9. RELATIONSHIPS AMONG SPECIES

The interactions of species in a biological community have a strong influence on determining its nature. The crucial aspects of such interactions are the following:

- Biomass photosynthesized by plants almost always provides the basic food source upon which the rest of the community depends. Most of this food is usually provided by a **dominant plant species**, such as algae growing in a pond.
- The physical nature of the environment is modified by the species in it, particularly the dominant plant species. For example, sagebrush on arid lands holds soil in place and provides shelter for small animals.
- **Competition** exists between different species for food, sunlight, and space.
- **Beneficial** and **antagonistic** relationships may exist between species.
- **Predation** occurs when an organism feeds upon another.
- **Parasitism** describes the activities of organisms that feed off other live **host organisms**, usually without causing death to the host. **Pathogens** are disease-causing organisms, usually microbial bacteria, protozoa, fungi, or viruses.

As shown in Figure 17.4, a dominant plant species anchors the community as its major producer of biomass. In addition to providing most of the food through photosynthesis that the rest of the community uses, the dominant plant species often acts to modify the physical environment of the community in ways that enable the other species to exist in it. For example, the trees in a forest community provide the physical habitat in which birds can nest, relatively safe from predators. In addition the trees provide shade that significantly modifies the habitat at ground level and prevents the growth of most kinds of low-growing plants.

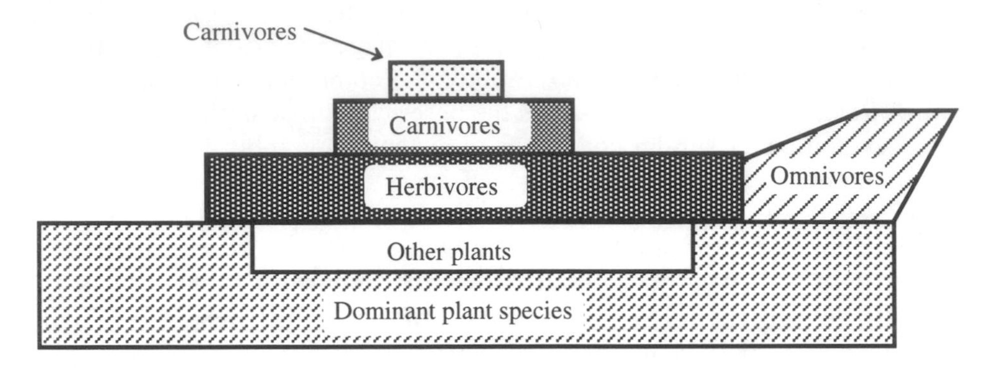

Figure 17.4. A dominant plant species typically provides most of the food for a biological community. The biomass that it produces is consumed primarily by herbivores, which are fed upon by carnivores, of which there may be more than one level. Omnivores feed on both plants and animals.

Competition exists between organisms for energy, matter, and space. Members of the same species may compete, as may members of different species. Adult plants may shade the ground such that seedlings of the same or other plant species cannot get started. Competition can be more overt, such as occurs when plants secrete substances through their leaves or roots that deter the growth of other plants. Animals exhibit **territoriality** in which they define areas from which they try to exclude competing animals, including those of their own species that are not one of a mating pair or part of a larger family unit. When two or more similar species exist in the same biological community on a steady basis, it is because they have evolved in ways that reduce competition for food and space; that is, they do not occupy exactly the same ecological niche. This phenomenon is called the **principle of competitive exclusion**.

Symbiotic relationships are those in which species live in close association with each other. Symbiosis is extremely common and is closely related to the other aspects of biological communities, such as the ways in which organisms utilize energy and nutrients. The classic example of symbiosis is that of lichen in which a species of algae exists within the matrix of filaments produced by a fungus. The fungus provides a hospitable physical environment that is anchored to rock, retaining moisture and extracting nutrients from the rock. The algae fix carbon as biomass that is utilized by the fungi. Other examples include **mycorrhizae**, in which fungal hyphae associate with plant roots, enabling the roots to take up adequate quantities of water and nutrients, and nitrogen-fixing bacteria that exist as nodules on the roots of leguminous plants.

The possible beneficial relationships among organisms can be divided into the three classifications as shown in Figure 17.5. **Mutualism** occurs when two species must have each other to exist; **commensalism** when one species requires the other, but the reverse is not true; and **protocooperation** when two species benefit from each other, but can exist independently.

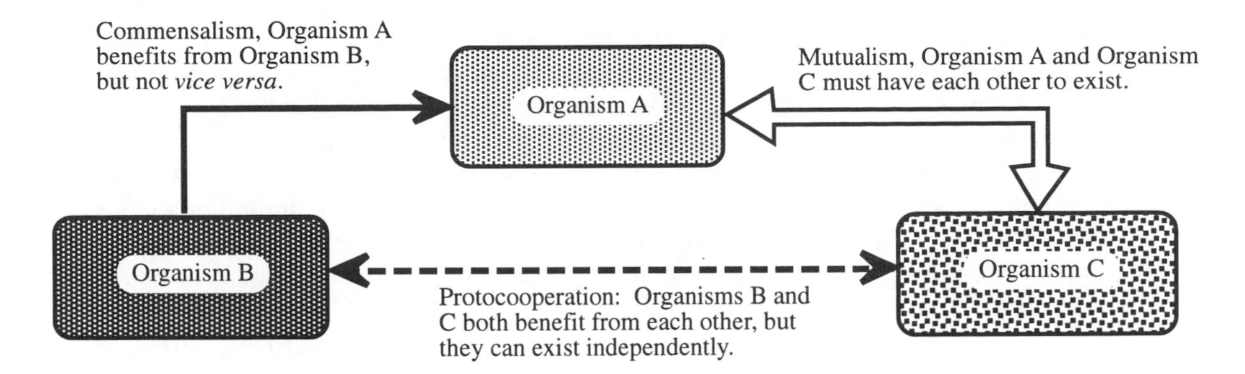

Figure 17.5. Types of beneficial relationships among organisms.

In the broadest definition of the term, **predators** consume the biomass of other organisms. Meat-eating carnivores that pursue and kill their prey are obviously predators. However, so are herbivores that graze on grass, consuming part of the host plant for food, but not killing it. Humans are omnivores that will eat both meat and vegetable matter.

Predation in a biological community can be quite complex. The predators have highly specialized means for getting food, and their potential prey have evolved ways to prevent this from happening. In some cases predation is of benefit to the prey. Through coevolutionary processes predators and prey have evolved such that both may survive. Stages of the life cycle are very important for both predators and prey. Most very young predators are not sufficiently developed to feed on the prey of adults, who care for them until they can feed themselves. Tadpoles are herbivores, but turn into carnivorous frogs. Very young plants provide excellent food sources for small predators, such as cutworms, whereas the fully grown plants are prey for larger animals, such as ruminant animals.

Different levels of feeding exist in a biological community; these are called **trophic levels**. Trophic levels are illustrated clearly in some aquatic ecosystems. In such systems the most abundant primary food source constituting the base of the grazing food web may consist of very small, free-floating photosynthesizing organisms called **phytoplankton**. These are fed upon by very small, free-floating, single-celled and invertebrate animals called **zooplankton**. (The larvae and eggs of many somewhat larger aquatic animals, such as crustaceans, are planktonic.) Small aquatic creatures, such as small fish, feed on the free-floating plankton. Organisms called **filter feeders** efficiently strain water through special structures to remove plankton for food. Organisms that feed on plankton are in turn consumed by larger creatures. This phenomenon gives rise to several levels of carnivores. It can result in the concentration of some kinds of aquatic pollutants that accumulate in progressively higher levels up the food chain, a phenomenon called **biomagnification**.

17.10. CHANGING COMMUNITIES

The nature of biological communities changes over both space and time. The borderline between communities is often a region of unique character and activity. Communities begin, develop, and end, sometimes over very long periods of time, sometimes abruptly. Human intervention is one of the major ways in which biological communities are forced to change.

This section outlines the general way in which a community changes. It should be emphasized, however, that such changes are not as predictable as was once believed. It cannot be known for certain that the succession will proceed in a specific direction, or that it will end up with a readily predicted population of organisms. Random chance plays a part in such uncertainty. Furthermore, human effects, which may be indirect and the result of human activities remote from the community add a note of uncertainty. However, human intervention intelligently applied can actually hasten the succession processes in biological communities and direct them in beneficial ways. Such measures include, as examples, replanting of forest trees after clear cutting or deliberate introduction of prairie grass species on worn-out agricultural land that is being restored to a native state.

The most fundamental way in which a biological community can change is when a new community becomes established on a newly created site where life of the sort that eventually populates the community has been essentially absent. The process of establishing life forms in such an area is called **primary succession**. On land primary succession occurs when a community develops in a location where soil is initially absent. This occurs, for example, with the establishment of a community on a new lava flow or newly exposed rock or gravel removed as overburden in strip mining. The process of colonizing such an area is called **primary succession**. It is one case of the more general process of **ecological succession**, a term used to describe the changes in species abundance as a result of changes in the environment of a biological community. The initial colonization of such a site is accomplished by a **pioneer species** that is hardy enough to exist under the harsh conditions often presented by a new site. Freshly exposed rock is often colonized by colonies of lichen, described above as symbiotic combinations of fungi and algae. The chemical action of lichen dissolves nutrients from the rock and initiates chemical weathering, which is the first step in soil formation. As soil is formed and crevices develop in the rocks, other organisms begin to populate the community, and eventually displace the pioneer species. Newly established aquatic ecosystems may eventually fill in with debris, and a succession of plants follows that can eventually change the system from an aquatic one to a swamp, then to a grassland, and eventually to a forest.

A mature biological community that lasts for very long time periods is called a **climax community**. (Although climax community is used here, some authorities prefer other terms, such as *mature community* or *relatively stable community*, which imply a less permanent state.) Consider the development of a climax community on land. Long before a mature community develops, productivity is low, but goes up gradually as the community acquires increased ability to utilize the resources in it. Soil forms, becomes deeper, and increases in organic content. In general, as the community develops, it becomes richer in organic matter, both living and dead. The number of species usually increases as the total number of organisms rises. The growth of trees and other taller plants enhances the three-dimensional character of the community. The number and variety of habitats become larger. Shading and shelter from larger plants give rise to small microhabitats that are protected from wind, sun, and temperature extremes, thus enabling colonization by small, specialized organisms that thrive in specialized microhabitats. As conditions change, successions of populations of various organisms occupy niches in the community. There is a tendency for successive populations to be both larger and longer lived, such as is the case with trees in an old-growth forest. Finally, a climax community is achieved that has reached a steady-state condition with its terrestrial, atmospheric, and aquatic environment and is in a homeostatic state. It usually has a relatively large amount of organic matter; compare, for example, the high organic content of a rich soil with the negligible amount of organic matter on rock colonized by lichen. A climax community does not exhibit sudden onset of populations of different species or loss of species. The species in the community go through their life cycles in ways that maintain community stability and homeostasis. The following are some important characteristics of a climax community:

- High diversity of life forms
- A high degree of order
- Narrow specialization of species in their appropriate niches with a high degree of stratification
- Conservation and recycling of nutrients

Secondary succession occurs when a less drastic, but still major change has occurred in a biological community. An example of secondary succession would be the establishment of a highly modified community on forest land that has just been subjected to clear-cutting. Annual plants are the first to colonize newly exposed soil. These are followed by small perennials, such as grass, and successively larger perennials, such as shrubs, bushes, small trees, and finally large trees.

A special kind of community is that which is adapted to periodic and dramatic alteration, especially from fire. **Fire-climax communities** are populated by species that quickly become re-established following a fire. Some species of trees even require fire to reproduce.

The borders of different biological communities have special characteristics called **edge effects**. A shoreline where grassland borders on a body of water is such an edge. Such areas provide special habitats for various kinds of plants and animals that can take advantage of both kinds of habitat. Edges often show a high degree of biological diversity. Somewhat similar to edges are **patches** that are present within a community due to some discontinuity in the physical structure of the community. A rock outcropping in the middle of a hilly grassland is an example of a patch. Patches may show a different distribution of species from those present in the main community.

17.11. HUMAN EFFECTS

Human intervention has a large potential effect on biological communities. Sometimes such intervention involves drastic physical alteration of the community, such as by plowing grasslands or cutting forests. Other types of human intervention may be less subtle, but may still have drastic effects.

An important way in which humans may influence biological communities is through the introduction of new species. If it is successful in a community, a new species affects those already there and may significantly modify the physical nature of the habitat. New species may prey upon those already present, or serve as prey that attracts predatory species from outside the community. When forests are cut and grasslands established, larger numbers of herbivores and representatives of species not previously present are attracted. These animals in turn attract carnivores that feed on them. Parasites usually accompany newly introduced species that can serve as their hosts.

Some introduced species are particularly destructive to biological communities and habitat. One of the worst of these is the goat, which has a well-earned reputation for indiscriminate consumption of vegetation, destruction of plant life, and damage to sod with its hard hooves. Rats introduced onto islands have wreaked havoc with indigenous species. Domestic house cats reverting to a wild state have wiped out whole populations of birds. Aggressive bird species, particularly house sparrows and starlings, have displaced more desirable native species.

The human species has become inextricably linked with technology such that in a sense *Homo sapiens* are not "natural animals." Much of what is known about the effects of humans on biological communities is negative — destruction of habitat, emission of pollutants to the environment, a potential permanent change in climate from greenhouse gases. These kinds of influences are unfortunate and very harmful to biological communities. However, humans are linked to technology irreversibly, and it will be necessary for humans to adapt themselves and their technologies to the biological communities upon which ultimately humankind depends for its existence.

17.12. HUMAN ACTIONS TO PRESERVE AND IMPROVE LIFE ON EARTH

Technology can be harnessed to preserve and improve the condition of life on Earth. A prime example of how that is done is provided by agriculture. With the application of plant genetics, herbicides, fertilizers, and advanced cultivation and harvesting techniques, agricultural interests can vastly increase the productivity of a plot of soil. By building terraces and waterways planted to grass that forms a tough, erosion-resistant sod, the productivity of land may be increased while erosion is slowed to a negligible level.

Human intervention can be used to create and enhance habitats that are not maintained for agricultural production. Although not all reservoirs formed by damming streams are desirable, many provide a welcome variety of habitat for species that live in or around bodies of water. Impounding water can cause suspended material to settle from streams, thus improving stream quality below the dams. In a very limited, but encouraging number of cases, human intervention is being applied to reverse damage done to habitats by human activities in the past. Once meandering streams straightened and turned into ugly, erosive ditches by channelization have been restored in some cases to provide the bending channels that make the stream hospitable to life. Productive wetlands are being restored, or even constructed where none existed before, usually as a means to aid wastewater treatment.[3] Badly used, eroded farmland has been converted to forests and grassland. Special structures can be made and sunk in shallow coastal areas to provide shelter and habitat for marine life; even old ship hulls and airplane fuselages have been used for this purpose.

The restoration of ecosystems by human intervention is called **restoration ecology**. Restoration ecology has become a significant area of human endeavor, and it may be hoped that it will increase in importance as technology is used increasingly to benefit the natural environment. The restoration ecologist needs to be familiar with basic ecology, as well as with the kinds of technology used to rebuild ecosystems. A knowledge of related areas, such as geology, hydrology, limnology, and soil science is also required. After catastrophic floods along the Missouri and Mississippi Rivers in the U.S. in 1993, the deliberate decision was made to forego reconstruction of some river dikes destroyed by the flood. Restoration ecology was applied to some limited areas to restore wetlands and river bottom lands for wildlife habitat.

Much of the work that has been done to preserve wildlife and to restore ecosystems in which wild species exist has been the result of efforts to maintain and increase numbers of game animals. Enlightened hunting and fishing laws have reduced the harvest of many species to sustainable levels. In some cases these have brought species back from very low numbers or even the brink of extinction. Important examples in the U.S. are American bison, wood ducks, wild turkeys, snowy egrets, and white-tailed deer. In addition to hunting and fishing restrictions, habitat restoration has been very important in increasing numbers of game animals. Restoration of wetland breeding areas have enabled significant increases in numbers of waterfowl.

Information is essential in order to understand, preserve, and enhance biological systems. The capabilities of technology to gather and process information are enormous. Sophisticated chemical analysis techniques provide detailed profiles of the chemical characteristics of the environment in which organisms live, including both nutrients and pollutants. Sensors for temperature, wind, moisture, and sunlight can be used to give a continuous picture of the physical environment. This and other information can be subjected to sophisticated computer analysis to provide a profile of the life system and to direct human intervention in constructive ways.

One of the more useful relatively recent technologies used to study life systems consists of satellite images of Earth, such as those provided by the Landsat satellite. Such images can be gathered by infrared measurements, digitized, and processed by computer to provide profiles of geological features, water on Earth's surface, and vegetation. By remotely sensing the absorption of electromagnetic radiation at specific wavelengths, instruments mounted on satellites can monitor gases or reactive chemical species in the atmosphere. One example of the latter is ClO, a reactive intermediate produced during the photochemical processes that occur as part of ozone depletion from stratospheric chlorofluorocarbons. The levels of greenhouse gases, including carbon dioxide, nitrous oxide, and methane can also be monitored. This information can be used to predict the effects of atmospheric species on life forms that may occur from global warming or ozone depletion.

Artificial Habitats and Habitat Restoration

To a limited degree plants can be preserved artificially by seed banks in which seeds are stored for long periods of time under appropriate conditions for their preservation. Botanical gardens and arboreta enable growth of plants under artificial conditions that can prevent at least some species from becoming extinct.

The number of animal species that can be maintained in zoos is limited, but this is still of some use for protecting various kinds of animals from extinction. Zoos are being used to a greater degree for wildlife preservation, in some cases with the goal of introducing animal species back into the wild. **Captive breeding** programs have been established to salvage individuals of endangered species from the wild, increase their population by breeding in captivity, and reintroduce them into the wild state. The numbers of endangered bird species have been increased by taking eggs from nests of birds in the wild and hatching them in captivity, sometimes with surrogate parents from other bird species. On a much larger scale, fish hatcheries have been in use for many decades to ensure a steady supply of fingerlings, particularly of trout and salmon species. There have been some tentative successes in captive breeding programs to restore species to the wild. In the U.S., captive peregrine falcon and blackfooted ferret have been reintroduced to some areas. The Arabian oryx (a large species of antelope) has been restored to some of its former habitats in the Middle East. Golden lion tamarins have been reintroduced to rain forests in Brazil. The widely publicized reintroduction of the California condor, a large carrion-eating bird, from individuals bred in captivity has been difficult because of the deaths of many of the specimens released.

A major problem with captive breeding programs has been the vulnerability of limited numbers of any species population to loss. When only a few individuals remain, the sudden onset of disease can be devastating. Not the least of the problems is the limited genetic diversity of a small population and the adverse effects of inbreeding.

17.13. LAWS AND REGULATIONS

Numerous laws and regulations have been applied to the preservation of wildlife and ecosystems. Gaming laws and the regulations fostered by them were mentioned in the preceding section. In the United States such laws have been in effect since the latter 1800s. For the most part these laws restricted the hunting of game and trade in game products, such as pelts.

In 1973 landmark wildlife legislation was passed in the U.S. in the form of the Endangered Species Act. This act was designed to prevent extinction of both animals and plants belonging to threatened species. Because of conflicts of interests, such as that between the lumber industry and groups attempting to preserve Spotted Owl habitat in northwestern U.S. old-growth forests, the enforcement of this act has been controversial. The basic provisions of the Endangered Species Act are the following:

- Requires that a list be compiled of species that are in danger of extinction, thus subject to regulation by the act.
- Prohibits hunting or capture of such species.
- Forbids trade in the products of listed species, such as pelts or feathers.
- Prohibits projects, such as dams, that would threaten listed species.
- Provides for protection of habitat.
- Requires that plans be formulated to restore species listed under the act.

Extinct species are those that no longer exist; examples are the carrier pigeon, the relic leopard frog, and the spiderflower. According to the provisions of the Endangered Species Act, an **endangered species** is one for which the danger of becoming extinct is very high. Over 1000 species are so designated in the U.S., of which almost 40% are mammals. **Threatened species**, such as the gray wolf and grizzly bear, have suffered marked declines in total numbers and extinction or threat thereof in certain localities.

Species recovery plans are mandated under the Endangered Species Act to enable recovery of species to the point that they can be delisted. In a few cases, numbers of some species have increased to a sufficient extent, and conditions for them have improved enough that they have been delisted. One example of such a species is the American alligator, which rebounded from perilously low populations in the late 1960s to relative abundance at present.

CHAPTER SUMMARY

The chapter summary below is presented in a programmed format to review the main points covered in this chapter. It is used most effectively by filling in the blanks, referring back to the chapter as necessary. The correct answers are given at the end of the summary.

Life is found in Earth's [1]________________. The ability of an organism to process matter and energy is called [2]______________, and another important characteristic of living organisms is their ability to [3]______________________ ____. The dynamic balance involving inputs of energy and matter and interaction with other organisms and with the surroundings by which organisms maintain their conditions within acceptable ranges is called [4] __________________. Kinds of organisms are called [5]__________, groups of organisms living together and occupying a specified area over a particular period of time constitute a [6]___________, and that part of Earth on which they dwell is their [7]_____________________. Assemblies of organisms living in generally similar surroundings over a large geographic area constitute a [8]_________, each of which may contain many [9]_______ ________. Photosynthetic plants in the biosphere are the basic [10]____________ and are located at the bottom of the [11]____________. Mineralization is an important process in which food matter is [12]________________________ ______________________________. A biological community is the biological component of an [13]___________, which includes the organisms and [14]_______ ______________________. Stable biological communities are characterized by a high degree of [15] ______________________, the organisms in a community undergo constant [16]____ ____, therefore, such communities as a whole are in a state of [17]________________. [18]________________________ express the boundaries of the [19]____________ ________________________ which determine whether a species can exist in a biological community. Homeothermic animals are those that [20]______________ ____. Individuals in species are not identical, but exhibit slight differences resulting in genetic [21]____________________. Keystone species are those of [22]__________ __________________ in an ecosystem, whereas indicator species are [23]______ ____ ____________. A population crash can result from rapid growth of a population such that the [24]______________________________ is exceeded. Vertical stratification is an example of one of [25]______________________________ ____. Five key parameters that are used to describe the ability of biological communities to survive and thrive are [26]____ ____. Areas of particularly high marine productivity occur when [27]__________________________________ ____. In a biological community most of the food is usually provided by a [28]_________ ___________________________. [29]____________________________ exists between different species for food, sunlight, and space, and [30]________________ ______________________________ relationships may exist between species. As an example of a symbiotic relationship, [31]____________________ associate with plant roots, enabling the roots to [32]____________________________ ______________________________. A requirement for a pioneer species in a community is that it be [33]_______________________________________ ______________________________________. Important characteristics of a climax community are [34]____ ____ ______________________. Much of the work that has been done to preserve wildlife and to restore ecosystems in which wild species exist has been the result of efforts to [35]____ ________________________________. The U.S. Endangered Species Act was designed to [36]____ ________________________________.

Answers

[1] biosphere

[2] metabolism

[3] maintain an internal environment that is favorable to metabolic processes

[4] homeostasis

[5] species

[6] population

[7] habitat

[8] biome

[9] ecosystems

[10] producers

[11] food chain

[12] converted completely to simple inorganic forms

[13] ecosystem

[14] their physical environment

[15] order and organization

[16] opposing and compensating readjustment of their behavior, feeding, and reproduction in response to each other and to their surroundings

[17] homeostasis

[18] Tolerance limits

[19] limiting factors

[20] maintain their internal temperatures within a narrow range

21 diversity

22 particular importance

23 those whose numbers decline or exhibit symptoms of malaise as a reflection of habitat damage before other major symptoms are observed

24 carrying capacity

25 major adaptations based upon physical space

26 productivity, diversity, inertia, constancy, and resilience

27 nutrient-rich sediments are brought to the surface by convection currents

28 dominant plant species

29 Competition

30 beneficial and antagonistic

31 fungal hyphae

32 take up adequate quantities of water and nutrients

33 hardy enough to exist under the harsh conditions often presented by a new site

34 high diversity of life forms, a high degree of order, narrow specialization of species in their appropriate niches, and conservation and recycling of nutrients

35 maintain and increase numbers of game animals

36 prevent extinction of both animals and plants threatened by extinction

QUESTIONS AND PROBLEMS

1. In what sense is the biosphere a particularly thin layer on Earth?
2. Give two characteristics of all living organisms.
3. Define homeostasis and explain why it is important for life.
4. Explain the sense in which a biological community is a subcategory of an ecosystem.
5. What is a food chain and what is the special role of producers in a food chain?
6. As they pertain to biological communities, define and relate limiting factors, tolerance limits, and the critical factor.
7. Based on a climatological/topographical factor, explain why there may be marked differences in the kind and quantity of vegetation on one side of a mountain range as compared to the opposite side.
8. How do plants cope with dry conditions and with subfreezing temperatures?
9. What is an ecological niche? Do you occupy an ecological niche? If so, define and explain it.
10. Explain why niche specialists are more likely than niche generalists to become endangered species.

11. Relate the plot shown in Figure 17.2 to what is happening to Earth's human population. Discuss the ramifications of a stabilized human population and of a possible population crash.
12. How do populations tend to avoid destructive competition in terms of both space and time?
13. Explain how proper balances of the following may enable human populations to exist and thrive: productivity, diversity, inertia, constancy, and resilience.
14. Explain how organisms may be ecological equivalents, even though they are not closely related genetically.
15. What is the special role of a dominant plant species in a biological community?
16. Distinguish among mutualism, commensalism, and protocooperation. Can you cite specific examples of these kinds of relationships from your own life?
17. Relate the phenomenon of biomagnification to the various classes of organisms shown in Figure 17.5.
18. Explain how a pioneer species and a climax community are on opposite ends of the spectrum of geological succession.
19. What are the major characteristics of a climax community?
20. Explain how human actions can be employed to preserve and improve life on Earth?

LITERATURE CITED

[1] Cunningham, William P. and Barbara Woodworth Saigo, *Environmental Science: A Global Concern*, 4th ed., William C Brown Publishers, Dubuque, IA, 1996.

[2] Beeby, A. N. and A. Brennan, *First Ecology*, Chapman & Hall, New York, NY, 1997.

[3] Mulamoottil, George, Barry G. Warner, and Edward A. McBean, *Wetlands*, CRC Press/Lewis Publishers, Boca Raton, FL, 1997.

SUPPLEMENTARY REFERENCES

Arms, Karen, *Environmental Science*, 2nd ed., HBJ College and School Division, Saddle Brook, NJ, 1994.

Atchia, Michael and Shawna Tropp, Eds., *Environmental Management: Issues and Solutions*, Wiley, New York, NY, 1995.

Cheremisinoff, Paul N., Ed., *Ecological Issues and Environmental Impact Assessment*Gulf Publishing, Houston, TX, 1997.

Dennison, Mark A. and James A. Schmid, *Wetland Mitigation: Mitigation Banking and Other Strategies for Development and Compliance*, Government Institutes, Inc., Rockville, MD, 1997.

Grosse, W. Jack, *The Protection and Management of Our Natural Resources, Wildlife, and Habitat*, Oceana, Dobbs Ferry, NY, 1997.

Jackson, Andrew R. and Julie M. Jackson, *Environmental Science: The Natural Environment and Human Impact*, Longman, New York, NY, 1996.

Jørgensen, S. E., B. Halling-Sørensen, and S. N. Nielsen, *Handbook of Environmental and Ecological Modeling*, CRC Press/Lewis Publishers, Boca Raton, FL, 1996.

McKibben, Bill, *Hope, Human and Wild: True Stories of Living Lightly on the Earth*, Little Brown & Company, Boston, MA, 1995.

Miller, G. Tyler, Jr., *Environmental Science: Working with the Earth*, 5th ed., Wadsworth Publishing Co., Belmont, CA, 1997.

Stone, Christopher D., *Should Trees Have Standing?*, Oceana, Dobbs Ferry, NY, 1996.

Trettin, Carl C., Martin F. Jurgensen, Margaret R. Gale, David F. Grigal, and John K. Jeglum, *Northern Forested Wetlands*, CRC Press/Lewis Publishers, Boca Raton, FL, 1997.

18 BIOTRANSFORMATIONS AND BIODEGRADATION

18.1. INTRODUCTION

The water and soil environments receive a variety of xenobiotic compounds that are foreign to living systems. Microorganisms in water and soil act upon these compounds in four different ways: (1) by using them directly as substrates for energy and biomass production, (2) through cometabolism along with primary metabolic processes, (3) by joining them with other chemical species present in the organism through the process of conjugation, or (4) by bioaccumulation. Of these processes, biodegradation, the metabolic breakdown of substances by microorganisms, is the most important.

Detoxication refers to the biological conversion of a toxic substance to a less toxic species, which may still be a relatively complex, or even more complex material. An example of detoxication is illustrated below for the enzymatic conversion of paraoxon (a highly toxic organophosphate insecticide) to *p*-nitrophenol, which has only about 1/200 the toxicity of the parent compound:

$$\underset{\text{Paraoxon}}{(CH_3CH_2O)_2P(=O)-O-C_6H_4-NO_2} \xrightarrow[\text{Enzyme action}]{H_2O,\ \{O\}} HO-C_6H_4-NO_2 + \text{Other products} \qquad (18.1.1)$$

Bioaccumulation is the uptake and concentration of environmental chemicals by living systems. In a general sense the term refers to the process by which substances dissolved and suspended in water or contained in sediments, soil, food, or drinking water are taken into an organism by diffusion from aqueous solution and by ingestion. The term applies especially to aquatic organisms, particularly fish. It may be extended to whole series of organisms in food chains. Uptake of environmental chemicals through food chains can result in much higher levels of the chemicals in organisms than would be expected from simple bioaccumulation, thereby resulting in **biomagnification**. Biomagnification can occur, for example, in a succession of organisms starting with herbivores (which live on plant material), progressing through detritovores (which feed on residues from the herbivores), and terminating with carnivores.

Of all the biologically mediated processes that may operate on environmental xenobiotic species, the uptake by organisms and related phenomena, such as biomagnification, usually changes the substance least. Therefore, biological uptake without any metabolic alteration of xenobiotic substances is addressed in this chapter first.

The process opposite to that in which organisms take up substances from water can be observed as a lowered concentration of xenobiotic in tissue when the organism is placed in an uncontaminated environment. This loss of substance back to the surroundings is called **depuration**. Depuration may occur through passive mechanisms of diffusion or desorption. It may also occur by active excretion or egestion on the part of the organism. Biotransformation that changes the substance to a different form may also occur.

The length to time corresponding to a 50-percent probability that a molecule of a substance will be eliminated from an organism is the **half-time** or **half-life** of the substance. If an organism is placed in uncontaminated water, such as a fish placed in clean water, the half-time is measured as the period required for half of the substance to be eliminated from the organism, or for the tissue concentration to reach half its initial value.

18.2. BIOCONCENTRATION

The tendency of a chemical to leave aqueous solution and enter a food chain is important in determining its environmental effects and is expressed through the concept of bioconcentration. **Bioconcentration** (Figure 18.1) may be viewed as a special case of bioaccumulation in which a *dissolved substance* is selectively taken up from water solution and concentrated in tissue by nondietary routes. Bioconcentration applies especially to the concentration of materials from water into fish. As illustrated in Figure 18.1, the model of bioconcentration is based upon a process by which contaminants in water traverse fish gill epithelium and are transported by the blood through highly vascularized tissues to lipid tissue, which serves as a storage sink for hydrophobic substances. Transport through the blood is affected by several factors, including rate of blood flow and degree and strength of binding to blood plasma protein. Prior to reaching the lipid tissue sink, some of the compound may be

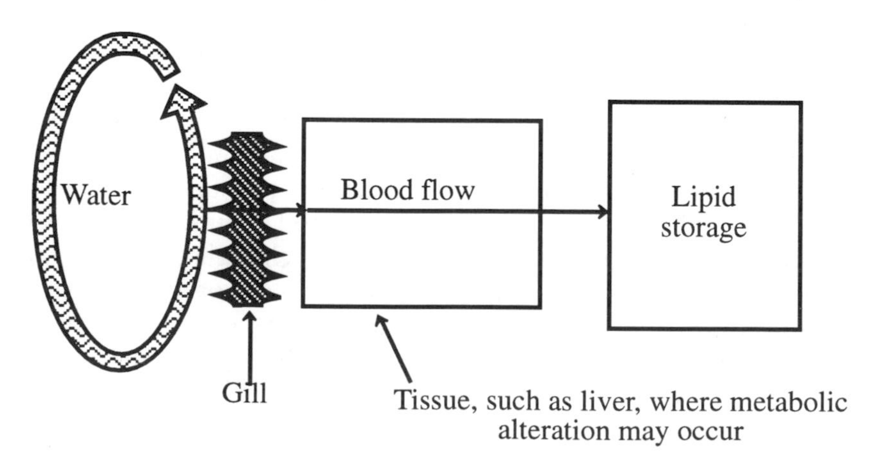

Figure 18.1. Overall pathway of bioconcentration.

metabolized to different forms. The concept of bioconcentration is most applicable under the following conditions:

- The substance is taken up and eliminated *via* passive transport processes.
- The substance is metabolized slowly.
- The substance has a relatively low water solubility.
- The substance has a relatively high lipid solubility.

Substances that undergo bioconcentration are hydrophobic and tend to undergo transfer from water media to fish lipid tissue. The simplest model of bioconcentration views the phenomenon on the basis of the physical properties of the contaminant and does not account for physiologic variables (such as variable blood flow) or metabolism of the substance. Such a simple model forms the basis of the **hydrophobicity model** of bioconcentration in which bioconcentration is regarded from the viewpoint of a dynamic equilibrium between the substance dissolved in aqueous solution and the same substance dissolved in lipid tissue.

Variables in Bioconcentration

There are several important variables in estimating bioconcentration. A basic requirement for uptake of a chemical species from water is whether or not it is in a form that is **bioavailable**. Biouptake may be severely curtailed for substances with extremely low water solubilities or that are bound to particulate matter, rather than being dissolved in water. Dissolved organic matter may also bind to substances and limit their biouptake. A **physiological component** of the process by which a chemical species must traverse membranes in the gills and skin to reach a final lipid sink tends to cause bioconcentration to deviate from predictions based on hydrophobicity alone. Some evidence suggests that the **lipid content** of the subject organism affects bioconcentration. Higher lipid contents in an organism may to a degree be associated with relatively higher bioconcentration. **Molecular shape and size** seem to play a role in bioconcentration. **Distribution** of the chemical species within an organism by blood flow may be relatively slow.

Biotransfer From Sediments

Because of the strong attraction of hydrophobic species for insoluble materials such as humic matter, many organic pollutants in the aquatic environment are held by sediments in bodies of water. Bioaccumulation of these materials must, therefore, consider transfer from sediment to water to organism as illustrated in Figure 18.2.

18.3. BIOCONCENTRATION AND BIOTRANSFER FACTORS

Bioconcentration Factor

Quantitatively, the *hydrophobicity model* of bioconcentration is viewed as an equilibrium between the uptake and elimination of substance "X":

$$X(aq) \leftrightarrow X(lipid) \qquad (18.3.1)$$

Using k_u as the rate constant for uptake and k_e as the rate constant for elimination leads to the following definition of **bioconcentration factor, BCF**:

$$BCF = \frac{k_u}{k_e} = \frac{[X(lipid)]}{[X(aq)]} \qquad (18.3.2)$$

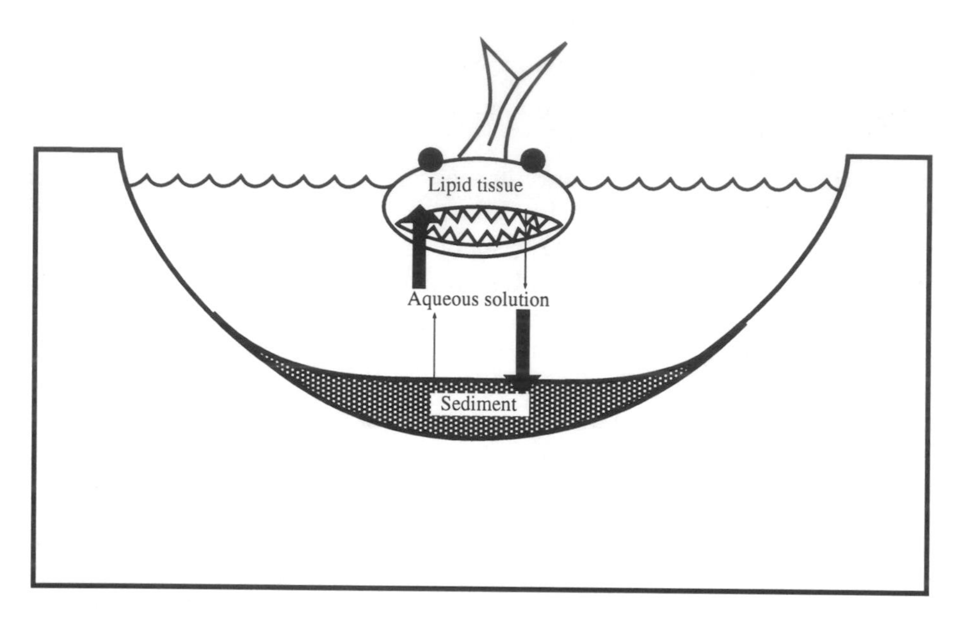

Figure 18.2. Partitioning of a hydrophobic chemical species among sediment, water, and lipid tissue. Heavier arrows denote the preference of the chemical for sediment and lipid tissue compared to aqueous solution.

When $[X(lipid)]/[X(aq)]$ = BCF, the rates of uptake and elimination are equal, the concentration of the xenobiotic substances remains constant (at constant $[X(aq)]$), and the system is in a condition of **dynamic equilibrium** or **steady state**. Values of BCF vary with the nature of the chemical in question, the species of aquatic organism and temperature. The BCF for hexachlorobenzene in rainbow trout at 15°C is 5.5×10^3, a typical value.[1]

Evidence for the validity of the hydrophobicity model of bioconcentration is provided by correlations of it with the **octanol–water partition coefficient**, K_{ow} using *n*-octanol as a surrogate for fish lipid tissue. The measurement of K_{ow} consists of determining the concentration of a hydrophobic contaminant in water-immiscible *n*-octanol relative to water with which it is in equilibrium. Typical K_{ow} values range from 10 to 10^7 corresponding to BCF values of 1 to 10^6. Such K_{ow}/BCF correlations have proven to be reasonably accurate when narrowly defined for a specified class of compounds, most commonly poorly metabolized organohalides. Major inconsistencies appear when attempts are made to extrapolate from one class of contaminants to another.

Biotransfer Factor

A useful measure of bioaccumulation from food and drinking water by land animals is the **biotransfer factor**, BTF, defined as,

$$BTF = \frac{\text{Concentration in tissue}}{\text{Daily intake}} \tag{18.3.3}$$

where the concentration in tissue is usually expressed in mg/kg and daily intake in mg/day. This expression can be modified to express other parameters, such as concentration in milk. As is the case for bioconcentration factors for fish in water, BTF shows a positive correlation with K_{ow} values.

Bioconcentration by Vegetation

Like fish and mammals, vegetation can absorb organic contaminants. In the case of vegetation the bioconcentration factor can be expressed relative to the mass of compound per unit mass of soil. The exact expression for vegetation is,

$$BTF = \frac{\text{Concentration in plant tissue}}{\text{Concentration in soil}} \tag{18.3.4}$$

where the concentration in plant tissue is given in units of mg/kg dry plant tissue and the concentration in soil is in units of mg/kg dry soil. For uptake of hydrophobic substances by plants BCF values are less than 1 and tend to decrease with increasing K_{ow}, the opposite of the trend observed in animals. This is explained by the transport of organic substances by water from soil to plant tissue, which increases with increasing water solubility of the compound and, therefore, with decreasing K_{ow}.

18.4. BIODEGRADATION

Biodegradation, the alteration of chemical species by biochemical processes, may involve relatively small changes in the parent molecule, such as substitution or modification of a functional group. In the most favorable cases, however, the compound is completely destroyed and the end result is conversion of relatively complex organic compounds to CO_2, H_2O, and inorganic salts, a process called **mineralization**. Usually the products of biodegradation are molecular forms that tend to occur in nature and that are in greater thermodynamic equilibrium with their surroundings.

Biochemical Aspects of Biodegradation

Several terms should be reviewed in considering the biochemical aspects of biodegradation. *Biotransformation* is what happens to any substance that is *metabolized* by the biochemical processes in an organism and is altered by these processes. *Metabolism* is divided into the two general categories of *catabolism*, which is the breaking down of more complex molecules, and *anabolism*, which is the building up of life molecules from simpler materials. The substances subjected to biotransformation may be naturally occurring or *anthropogenic* (made by human activities). They may consist of *xenobiotic* molecules that are foreign to living systems.

Biodegradation of an organic compound occurs in a stepwise fashion and is usually not the result of the activity of a single specific organism. Usually several strains of microorganisms, often existing synergistically, are involved. These may utilize different metabolic pathways and a variety of enzyme systems.

Although biodegradation is normally regarded as degradation to simple inorganic species such as carbon dioxide, water, sulfates, and phosphates, the possibility must always be considered of forming more complex or more hazardous chemical species. An example of the latter is the production of volatile, soluble, toxic methylated forms of arsenic and mercury from inorganic species of these elements by bacteria under anaerobic conditions.

It is well known that microbial communities develop the ability to break down xenobiotic compounds metabolically when exposed to them in the environment.[2] This has become particularly obvious from studies of biocidal compounds in the environment. In general, such compounds are readily degraded by bacteria that have been in contact with the compounds for prolonged periods, but not by bacteria from unexposed sites. The development of microbial cultures with the ability to degrade materials to which they are exposed is described as **metabolic adaptation**. In rapidly multiplying microbial cultures enough generations are involved so that metabolic adaptation can include genetic changes that favor microorganisms that have developed the ability to degrade a specific pollutant. Metabolic adaptation may also include increased numbers of microorganisms capable of degrading the substrate in question and enzyme induction.

Cometabolism

Xenobiotic compounds are usually attacked by enzymes whose prime function is to react with other compounds, a process that provides neither carbon nor energy called **cometabolism**. Cometabolism usually involves relatively small modifications of the substance that is cometabolized (the secondary substrate) relative to the primary substrate. The enzymes that carry out cometabolism tend to be nonspecific. As an environmentally significant example of cometabolism, at least one strain of bacteria degrades trichloroethylene with an enzyme system that acts predominantly on phenol. The enzyme activity can be induced by exposure to phenol, after which it acts on trichloroethylene.

In pure cultures of microorganisms, the products of cometabolism tend to accumulate and often do not undergo further degradation. However, in mixed cultures, which are the norm for environmental systems, they may serve as substrates for other organisms so that complete biodegradation results. Therefore, studies of biodegradation in pure cultures are usually of limited utility in predicting what happens in the environment.

An example of cometabolism of pollutants is provided by the white rot fungus, *Phanerochaete chrysosporium*, which degrades a number of kinds of organochlorine compounds, including DDT, PCBs, and chlorodioxins, under the appropriate conditions. The enzyme system responsible for this degradation is one that the fungus uses to break down lignin in plant material under normal conditions.

General Factors in Biodegradation

The rates and efficacy of biodegradation of organic substances depend upon several obvious factors. These include the concentration of the substrate compound; whether or not molecular oxygen, O_2, is available; presence of phosphorus and nitrogen nutrients; availability of trace element nutrients; the presence of a suitable organism; absence of toxic substances; and the presence of appropriate physical conditions (temperature, growth matrix). In addition to their biochemical properties, the physical properties of compounds, including volatility, water-solubility, organophilicity, tendency to be sorbed by solids, and charge play a role in determining the biodegradability of organic compounds.

To a large extent, xenobiotic compounds in the aquatic environment are bound with sediments and suspended solid materials, such as humic acids. This binding plays a large role in biodegradation. Low concentrations of surfactants may affect rates of biodegradation. Competition from other organisms may be a factor in biodegradation of pollutants. “Grazing” by protozoa may result in consumption of bacterial cells responsible for the biodegradation of particular compounds.

Trace amounts of micronutrients are needed to support biological processes and as constituents of enzymes. Important micronutrients are calcium, magnesium, potassium, sodium, chlorine, cobalt, iron, vanadium, and zinc. Sometimes sulfur, phosphorus, and micronutrients must be added to media in which microorganisms are used to degrade hazardous wastes in order for optimum growth to occur.

Biodegradability

The amenability of a compound to biochemical attack by microorganisms is expressed as its **biodegradability**. The biodegradability of a compound is influenced by its physical characteristics, such as solubility in water and vapor pressure, and by its chemical properties, including molecular mass, molecular structure, and presence of various kinds of functional groups, some of which provide a “biochemical handle” for the initiation of biodegradation. With the appropriate organisms and under the right conditions, even substances that are biocidal to most microorganisms can undergo biodegradation. For example, normally bactericidal phenol is readily metabolized by the appropriate bacteria acclimated to its use as a carbon and energy source.

In general, compounds of biological origin readily undergo biodegradation. Thus proteins and carbohydrates are readily metabolized by organisms. Lipids, including fats and oils, are slower to undergo biodegradation because of their generally low solubililities. Initiation of the biodegradation of hydrocarbons tends to be slow. However, if an alcohol (-OH) or carboxylic acid (-CO_2H) group is attached to a hydrocarbon structure, biodegradation is much faster. Hydrolysis generally occurs very readily, so that compounds with ester or amide groups tend to break down quickly.

```
   O  H                        O     |
   ||  |                        ||    |
 —C—N—  Amide group        —C—O—C—  Ester group
                                     |
```

Recalcitrant or **biorefractory** substances are those that resist biodegradation and tend to persist and accumulate in the environment. Such materials are not necessarily toxic to organisms, but simply resist their metabolic attack. Even some compounds regarded as biorefractory may be degraded by microorganisms adapted to their biodegradation. Examples of such compounds and the types of microorganisms that can degrade them include endrin (*Arthrobacter*), DDT (*Hydrogenomonas*), phenylmercuric acetate (*Pseudomonas*), and raw rubber (*Actinomycetes*).

The ultimate in biodegradation is **mineralization** in which all the elements in an organic species are completely converted to simple, thermodynamically stable inorganic forms. The most abundant elements in typical organic compounds are carbon, hydrogen, oxygen, nitrogen, phosphorus, sulfur, and, in the case of many xenobiotic compounds, chlorine. Therefore, the general process of mineralization can be represented by the following:

$$C_cH_hO_oN_nP_pS_sCl_{cl}(\textit{organic compound}) \xrightarrow{\text{Enzymes}}$$
$$cCO_2 + hH_2O + nNH_4^{+} + pHPO_4^{2-} + sSO_4^{2-} + clCl^{-} + \text{energy} \qquad (18.4.1)$$

18.5. ENZYMATIC PROCESSES IN BIODEGRADATION

Most important processes involved in the breakdown of chemical species in the water and soil environments are enzymatic biodegradation of organic matter by microorganisms in the aquatic and terrestrial environments. It occurs by way of a number of stepwise, microbially catalyzed reactions. The major types of these reactions are oxidation; decarboxylation, in which the $-CO_2H$ is replaced with an H atom or −OH group; hydrolysis, which involves the addition of H_2O to a molecule accompanied by cleavage of the molecule into two species; substitution, in which one group of atoms is replaced by another (such as OH for Cl); elimination whereby atoms or groups of atoms are removed from adjacent carbon atoms, which remain joined by a double bond; reduction; dehalogenation; demethylation; deamination (removal of NH_2); condensation, in which two smaller molecules are joined to produce a larger one; conversion of one isomer of a compound to another with the same molecular formula but a different structure; conjugation; and ring cleavage. Examples of these processes are shown in Table 18.1.

Biodegradation can occur under either oxic (aerobic) conditions where molecular O_2 is present, or it may take place under anoxic conditions in the absence of molecular O_2. In the presence of molecular O_2 the ultimate biodegradation product of organic carbon is carbon dioxide, CO_2. Under anoxic conditions a process called **fermentation** occurs in which some of the carbon ends up as reduced organic species. A common example is the production of ethyl alcohol, CH_3CH_2OH, from the fermentation of carbohydrates:

$$C_6H_{12}O_6 \rightarrow 2CO_2 + 2CH_3CH_2OH \qquad (18.5.1)$$

Organic acids are usually products of fermentation, and acidity may increase in a fermentation medium to a high enough level that the fermentation process stops.

Biodegradation of 2,4-D

Herbicidal 2,4-D provides an informative example of ways in which several of the biodegradation processes discussed above are involved in the breakdown of an environmental pollutant. A 2,4-D ester undergoes hydrolysis:

$$\text{(2,4-Cl}_2\text{C}_6\text{H}_3\text{O)–CH}_2\text{–C(=O)–O–CH}_2\text{–CH}_3 + H_2O \xrightarrow{\text{Hydrolysis}} \text{(2,4-Cl}_2\text{C}_6\text{H}_3\text{O)–CH}_2\text{–C(=O)–OH} + \text{HO–CH}_2\text{–CH}_3 \qquad (18.5.2)$$

Table 18.1. Enzymatic Processes in Biodegradation

Process	Example
Oxidation	
Epoxidation, addition of an O atom bridging between two C atoms	benzene $\xrightarrow[\text{epoxidation}]{O_2,\text{ enzyme-mediated}}$ epoxide (O)
Beta oxidation in which straight-chain hydrocarbons are oxidized 2 carbon atoms at a time	$CH_3CH_2CH_2CH_2CO_2H + 3O_2 \rightarrow CH_3CH_2CO_2H + 2CO_2 + 2H_2O$
Ring cleavage (preceded by hydroxylation)	benzene $\xrightarrow{O_2}$ (OH, OH) $\xrightarrow{O_2}$ (CO_2H, CO_2H)
Decarboxylation, replacement of the $-CO_2H$ with an H atom or -OH group	$-\overset{\vert}{\underset{\vert}{C}}-\overset{O}{\overset{\Vert}{C}}-OH \rightarrow -\overset{\vert}{\underset{\vert}{C}}-H + CO_2$
Hydrolysis, addition of H_2O to a molecule accompanied by cleavage	$H_3C-O-\overset{S}{\overset{\Vert}{P}}(-O-CH_3)-S-CH(-\overset{H}{C}H-\overset{O}{\overset{\Vert}{C}}-O-C_2H_5)(-\overset{O}{\overset{\Vert}{C}}-O-C_2H_5) \xrightarrow{H_2O}$ (Malathion) $H_3C-O-\overset{S}{\overset{\Vert}{P}}(-O-CH_3)-S-CH(-\overset{H}{C}H-\overset{O}{\overset{\Vert}{C}}-OH)(-\overset{O}{\overset{\Vert}{C}}-OH) + 2HO-C_2H_5$
Reduction	$C_6H_5-NO_2 \rightarrow C_6H_5-NH_2$
Dehalogenation	$H-\overset{H}{\underset{H}{C}}-\overset{H}{\underset{H}{C}}-Cl \xrightarrow{H_2O} H-\overset{H}{\underset{H}{C}}-\overset{H}{\underset{H}{C}}-Cl + H_2O$
Dealkylation	$R-\overset{H}{N}-CH_3 \xrightarrow{\text{N-dealkylation}} R-N\begin{smallmatrix}H\\H\end{smallmatrix} + H-\overset{O}{\overset{\Vert}{C}}-H$
Conjugation (attachment of a group, such as the methyl group to a xenobiotic compound or metabolite)	$H_3AsO_3 \rightarrow CH_3AsO(OH)_2$

The aliphatic acid portion of the residue may be oxidized:

$$Cl-C_6H_3(Cl)-O-\overset{H}{\underset{H}{C}}-\overset{O}{\overset{\Vert}{C}}-OH + \frac{3}{2}O_2 \rightarrow Cl-C_6H_3(Cl)-OH + 2CO_2 + H_2O \qquad (18.5.3)$$

Chlorine may be removed by hydrolytic dehalogenation in which an OH group replaces Cl:

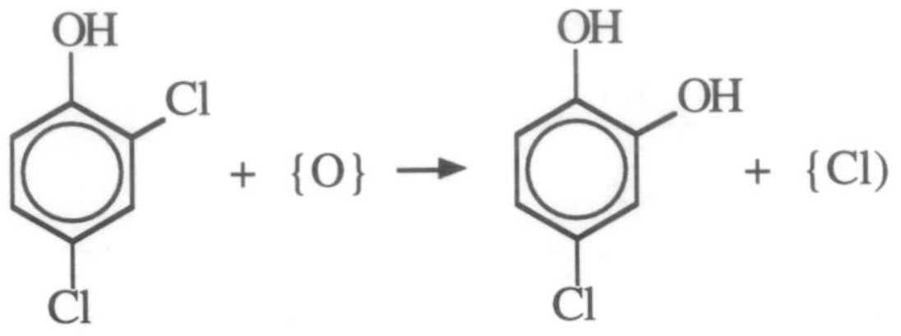

$$\text{(OH, Cl, Cl ring)} + \{O\} \rightarrow \text{(OH, OH, Cl ring)} + \{Cl\} \quad (18.5.4)$$

And the ring may be cleaved:

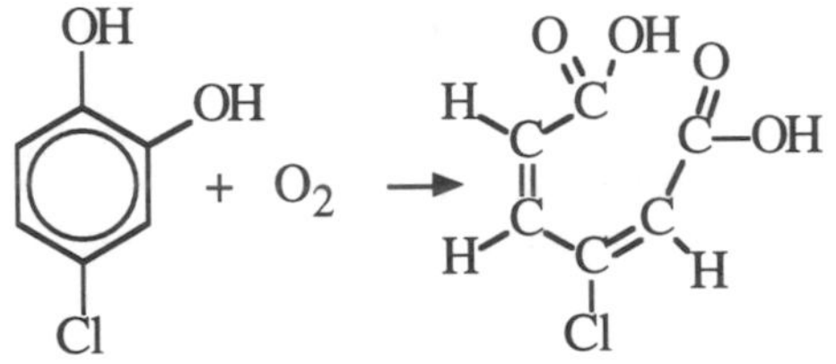

$$\text{(OH, OH, Cl ring)} + O_2 \rightarrow \text{(ring-cleavage product: O, OH, O, C–OH, H, H, H, Cl)} \quad (18.5.5)$$

18.6. BIODEGRADATION OF WASTES IN WATER AND SOIL

The most common deliberate application of biodegradation is for destruction of "nonchemical" sewage, food, and agricultural wastes. Treatment processes for these wastes make use of **bioreactors** that enable contact of the wastes with a high concentration of microorganisms held at a relatively high concentration. As discussed in Section 8.4, bioreactors, such as activated sludge plants, are used to treat municipal sewage; applications of these devices are readily extended to food processing, agricultural, and even some types of chemical wastes.[3]

Biodegradation in bioreactors is used to remove **biochemical oxygen demand** (BOD) from wastewater. If not removed, BOD would consume oxygen in water receiving the wastewater. In a bioreactor in the presence of oxygen, biodegradable organic substances, represented as $\{CH_2O\}$, consume oxygen by microbially mediated aerobic respiration reactions:

$$\{CH_2O\} + O_2 \rightarrow CO_2 + H_2O \quad (18.6.1)$$

This process yields biomass and energy. One of the simplest devices for accomplishing biological waste treatment is the **trickling filter** in which wastewater is sprayed over rocks or other solid support material covered with microorganisms allowing biodegradable wastes in the water to contact both air and a high concentration of microorganisms. Because the microorganisms are contained in a relatively thin layer on a support material, a trickling filter is an example of a **fixed film** bioreactor. A trickling filter is illustrated in Chapter 8, Figure 8.2.

The other major kind of bioreactor in addition to a fixed film bioreactor is the **activated sludge** bioreactor discussed in Section 8.4 and shown in Figure 8.3. In this kind of device a mass of microorganisms is settled from the water and recirculated to incoming water as a "return sludge." Excess sludge built up in the activated sludge bioreactor may be digested in the absence of oxygen by methane-producing anaerobic bacteria to produce methane and carbon dioxide,

$$2\{CH_2O\} \rightarrow CH_4 + CO_2 \quad (18.6.2)$$

reducing the volume of the sludge and producing a methane byproduct, which is used as fuel.

Throughout the 1980s and into the present decade, interest has accelerated in the use of microorganisms for waste treatment. This can be done in several ways, especially **land treatment**, where wastes are put on soil for degradation, or by **composting**, where wastes are mixed with porous, aerated material, such as wood shavings. For more sophisticated treatment bioreactors are used.

A number of hazardous waste compounds are susceptible to destruction by biodegradation. Properties of hazardous wastes can be changed to increase biodegradability. This is especially true of wastes that consist of several constituents, one or more of which inhibit biological processes. Sometimes a waste substance that is toxic to microorganisms at a relatively high concentration is degraded well in more dilute media. Inhibition of biodegradation by extremes of pH may be overcome by neutralizing excess acid or base. Toxic organic and inorganic substances, such as heavy metal ions, can be removed in some cases prior to biodegradation.

Biodegradability of Waste Compounds

Practically all classes of synthetic organic compounds can be at least partially degraded by various microorganisms. For the most part, anthropogenic compounds resist biodegradation much more strongly than do naturally occurring compounds. This is generally due to the absence of enzymes that can bring about an initial attack on the compound. As mentioned earlier in this chapter, a number of physical and chemical characteristics of a compound determine its amenability to biodegradation. Such characteristics include hydrophobicity, solubility, volatility, and affinity for lipids.

Microorganisms in Waste Treatment

Several groups of microorganisms are capable of partial or complete degradation of organic compounds, including those commonly regarded as hazardous. Among the aerobic bacteria, those of the *Pseudomonas* family are the most widespread and most adaptable to the degradation of synthetic compounds, degrading biphenyl, naphthalene, DDT, and many other compounds. Anaerobic bacteria are very fastidious and difficult to study in the laboratory because they require oxygen-free (anoxic) conditions in order to survive. These bacteria catabolize biomass by hydrolytic processes, breaking down proteins, lipids, and saccharides. **Actinomycetes**, microorganisms that are morphologically similar to both bacteria and fungi, are involved in the degradation of a variety of organic compounds, including degradation-resistant alkanes, and lignocellulose. Fungi are particularly noted for their ability to attack long-chain and complex hydrocarbons and are more successful than bacteria in the initial attack on PCB compounds.

Biodegradation in Soil

Microorganisms are much more important than insects, earthworms, and plants in the biodegradation of pesticides and other pollutant organic chemicals in soil. In recent years it has become apparent that the rhizosphere is a particularly important part of soil in respect to biodegradation of wastes.[4] The **rhizosphere** is the layer of soil in which plant roots are particularly active. It is a zone of increased biomass and is strongly influenced by the plant root system and the microorganisms associated with plant roots. The rhizosphere may have more than 10 times the microbial biomass per unit volume compared to nonrhizospheric zones of soil. This population varies with soil characteristics, plant and root characteristics, moisture content, and exposure to oxygen. If this zone is exposed to pollutant compounds, microorganisms adapted to their biodegradation may also be present.

Plants and microorganisms exhibit a strong synergistic relationship in the rhizosphere, which benefits the plant and enables highly elevated populations of rhizospheric microorganisms to exist. Epidermal cells sloughed from the root as it grows and carbohydrates, amino acids, and root-growth-lubricant mucigel secreted from the roots all provide nutrients for microorganism growth. Root hairs provide a hospitable biological surface for colonization by microorganisms.

The biodegradation of many synthetic organic compounds occurs in the rhizosphere. Understandably, studies in this area have focused on herbicides and insecticides that are widely used on crops. Among the organic species for which enhanced biodegradation in the rhizosphere has been demonstrated are the following (associated plant or crop shown in parentheses): 2,4-D herbicide (wheat, African clover, sugarcane, flax), parathion (rice, bush bean), carbofuran (rice), atrazine (corn), diazinon (wheat, corn, peas), volatile aromatic alkyl and aryl hydrocarbons and chlorocarbons (reeds), and surfactants (corn, soybean, cattails). Enhanced biodegradation of polycyclic aromatic hydrocarbons (PAH) was observed in the rhizosperic zones of prairie grasses where PAHs are regularly deposited by grass fires.

CHAPTER SUMMARY

The chapter summary below is presented in a programmed format to review the main points covered in this chapter. It is used most effectively by filling in the blanks, referring back to the chapter as necessary. The correct answers are given at the end of the summary.

Four ways in which microorganims in water and soil act upon xenobiotic compounds are [1] ____.
Bioaccumulation that results in significant concentration of a chemical species is called [2] ____________________. Depuration may occur by means of [3] ________
____.
Bioconcentration applies especially to the [4] ______________________________
____. A basic requirement for uptake of a chemical species from water is whether or not it is in a form that is [5] ___
______________________. In the hydrophobicity model of bioconcentration the equilibrium of the expression, X(*aq*) $\leftrightarrow$ X(*lipid*) is expressed as [6] ___________
__________________. A chemical model of the hydrophobicity model of bioconcentration is provided by correlations with the [7] ______________________
______________. A useful measure of bioaccumulation from food and drinking water by land animals is the [8] ____________________, defined as, [9] __________
____________________. Biotransformation is what happens to any substance that is [10] ______________________, whereas mineralization is the process by which [11] ____

____
__________________________________. Xenobiotic compounds are usually attacked by enzymes whose prime function is [12]______________________
_____________________________________. The rates and efficacy of biodegradation of organic substances depend upon [13]____________________
____
____
____
____. The amenability of a compound to biochemical attack by microorganisms is expressed as its [14]_________________________. The major types of enzymatic processes involved in the breakdown of chemical species in the water and soil environments are [15]____
____
____
____. Biodegradation in bioreactors employed to treat sewage is used to remove [16]_______________
_________________________ from wastewater. Composting is used to treat wastes that are [17]____
_____________________________. Actinomycetes, microorganisms that are morphologically similar to both bacteria and fungi, are involved in the degradation of degradation-resistant [18]____________________________________. The [19]___________________________, the layer of soil in which plant roots are particularly active, may have more than 10 times the microbial biomass per unit volume compared to nonrhizospheric zones of soil.

Answers

1 substrates, cometabolism, conjugation, and bioaccumulation

2 biomagnification

3 passive mechanisms of diffusion or desorption or by active excretion

4 concentration of materials from water into fish

5 bioavailable

6 $BCF = \frac{k_u}{k_e} = \frac{[X(lipid)]}{[X(aq)]}$

7 octanol–water partition coefficient, K_{ow}

8 biotransfer factor, BTF

9 $BTF = \frac{\text{Concentration in tissue}}{\text{Daily intake}}$

10 metabolized

11 all the elements in an organic species are completely converted to simple, thermodynamically stable inorganic forms

12 to react with other compounds

13 concentration of the substrate compound, availability of O_2, presence of phosphorus and nitrogen nutrients, availability of trace element nutrients, presence of a suitable organism, absence of toxic substances, and presence of appropriate physical conditions

14 biodegradability

15 oxidation, decarboxylation, hydrolysis, substitution, reduction, dehalogenation, demethylation, deamination, condensation, conversion of one isomer to another, conjugation, and ring cleavage

16 biochemical oxygen demand (BOD)

17 mixed with porous, aerated material

18 alkanes, and lignocellulose

19 rhizosphere

QUESTIONS AND PROBLEMS

1. Define detoxication and cite an example of the phenomenon.
2. Distinguish among the following: bioaccumulation, biomagnification, and bioconcentration.
3. What is the basis of the hydrophobicity model? Under what circumstances is it most readily applicable?
4. What are the roles of molecular size and shape in bioconcentration? Why may moleular mass alone be insufficient to predict bioconcentration?
5. What is the bioconcentration factor, BCF? How is it calculated mathematically? What are representative values of BCF?
6. How is the octanol-water partition coefficient employed in predicting biouptake? How is this use justified?
7. In which respect do the values and trends of the bioconcentration factors for vegetation differ from those of animals? What is the explanation?
8. Define biodegradation. Define each of the following terms related to biodegradation: mineralization, biotransformation, metabolism, catabolism.
9. How is cometabolism involved in the biodegradation of xenobiotic compounds? What is a secondary substrate in cometabolism? How does *Phanerochaete chrysosporium* illustrate secondary metabolism?
10. List the general factors involved in biodegradation. How might such factors influence biodegradation? What is the influence of binding of xenobiotic compounds to sediments in biodegradation?
11. Some studies suggest that biodegradation rates of substances at relatively higher concentrations are not extrapolatable to very low concentrations. Discuss the effects that this might have upon the long-term persistence of very low levels of pollutants.
12. Nonenzymatic reactions that may be involved in the breakdown of chemical species in the water and soil environments include hydrolysis, oxidation-

reduction, surface-catalyzed, photolytic, and ion-exchange reactions. Which of these may also be biologically mediated enzyme-catalyzed reactions?

13. Name and describe a process involving H_2O that is especially important in the microbial degradation of pesticidal esters, amides, organophosphate esters, and nitriles.

14. List some physical and chemical properties of a compound as well as some biochemical conditions that are favorable for biodegradability to occur.

15. Critique the statement that "recalcitrant or biorefractory substances are those that are toxic to microorganisms that would otherwise degrade them."

16. What is a bioreactor? How are such devices used in the treatment of municipal wastes? Give at least two examples. What do they actually eliminate in municipal wastewater?

LITERATURE CITED

[1] Barron, Mace G., "Bioconcentration," *Environmental Science and Technology*, **24**, 1612-1618 (1990).

[2] Young, Lily Y. and Carl Cernigila, Eds., *Microbial Transformation and Degradation of Toxic Organic Chemicals*, Wiley, New York, NY, 1995.

[3] Asenjo, Juan A. and Jose C. Merchuk, *Bioreactor System Design*, Marcel Dekker, New York, NY, 1995.

[4] Anderson, Todd A., Elizabeth A. Guthrie, and Barbara T. Walton, "Bioremediation in the Rhizosphere," *Environmental Science and Technology*, **27**, 2630-2636 (1993).

SUPPLEMENTARY REFERENCES

Focht, Dennis D., *Principles of Biodegradation*, Chapman & Hall, New York, NY, 1997.

Hinchee, Robert E., Fred J. Brockman, and Catherine M. Vogel, *Microbial Processes for Bioremediation*, Battelle Press, Columbus, OH, 1995.

Iwai, S., *Wastewater Treatment With Microbial Films*, Technomic Publishing Co., Lancaster, PA, 1994.

McDuffie, Norton G., *Bioreactor Design Fundamentals*, Butterworth-Heinemann, Newton, MA, 1991.

NATO, *Microbial Degradation Processes in Radioactive Waste Repository and in Nuclear Fuel Storage Areas*, Kluwer Academic Publishers, Norwell, MA, 1997.

Ratledge, Colin, Ed., *Biochemistry of Microbial Degradation*, Kluwer Academic Publishers, Norwell, MA, 1994.

19 BIOGEOCHEMICAL CYCLES

19.1. MATTER AND CYCLES OF MATTER

Cycles of matter (Figure 19.1), often based on elemental cycles, are very important in the environment. Global geochemical cycles can be regarded from the viewpoint of various reservoirs, such as oceans, sediments, and the atmosphere, connected by conduits through which matter moves continuously. Energy is utilized in the movement of matter, which usually involves a change in chemical and/or physical state. The movement of a specific kind of matter between two particular reservoirs may be reversible or irreversible. The fluxes of movement of various kinds of matter differ greatly as do the contents of such matter in a specified reservoir.

Cycles of matter would occur even in the absence of life on Earth, but are strongly influenced by life forms, particularly plants and microorganisms. Organisms participate in **biogeochemical cycles**, which describe the circulation of matter, particularly plant and animal nutrients, through ecosystems. As part of the carbon cycle, atmospheric carbon in CO_2 is fixed as biomass, and as part of the nitrogen cycle, atmospheric N_2 is fixed in organic matter. The reverse of these processes is **mineralization** in which biologically bound elements are returned to inorganic states. Biogeochemical cycles are ultimately powered by solar energy, fine-tuned and directed by energy expended by organisms. In a sense, the solar-energy-powered hydrologic cycle (Figure 5.1) acts as a continuous conveyer belt to move materials essential for life through ecosystems.

Figure 19.1 shows a general cycle with all five spheres in which matter may be contained and processed. Human activities now have such a strong influence on materials cycles that it is useful to refer to the "anthrosphere" along with the other environmental "spheres" involved in cycles of matter. Using Figure 19.1 as a model, it is possible to arrive at any of the known elemental cycles. Some of the numerous possibilities for materials exchange are summarized in Table 19.1.

Natural cycles have solid, liquid, or gaseous **reservoir stages** in which most of the available material constituting the cycle is contained. Atmospheric nitrogen as elemental N_2 gas is the reservoir stage of the nitrogen cycle. The ocean is a huge liquid reservoir for water. Vast mineral deposits of limestone, $CaCO_3$, or dolomite, $CaCO_3 \cdot MgCO_3$, constitute a reservoir stage for calcium. Such a solid mineral deposit is called a **sedimentary reservoir**.

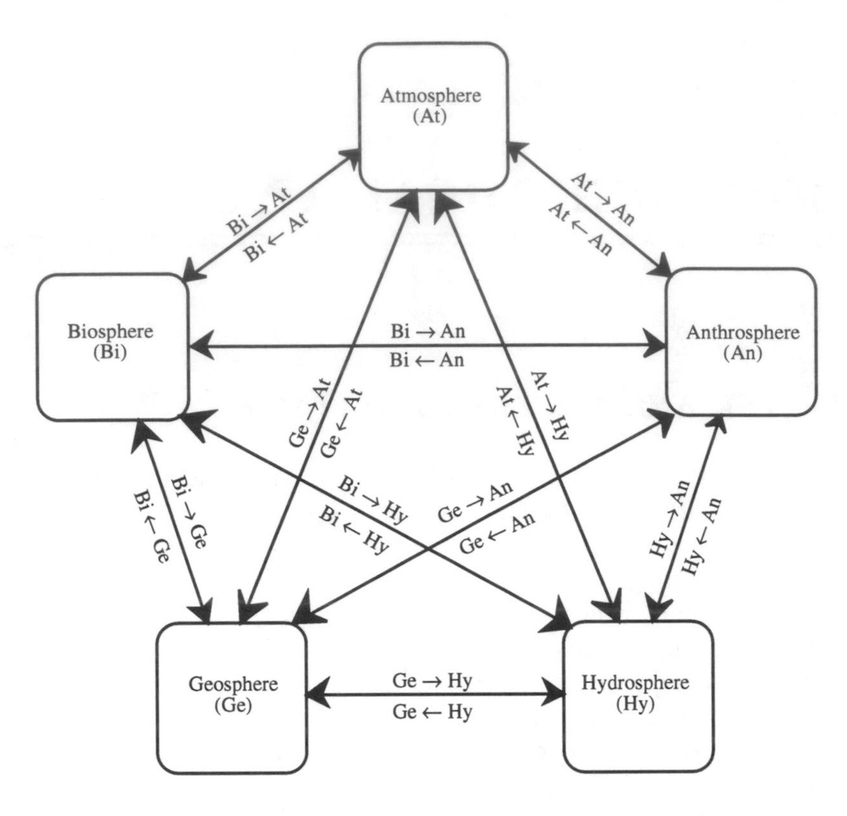

Figure 19.1. General cycle showing interchange of matter among the atmosphere, biosphere, anthrosphere, geosphere, and hydrosphere.

Table 19.1. Interchange of Materials Among the Possible Spheres of the Environment

From: / To:	Atmosphere	Hydrosphere	Biosphere	Geosphere	Anthrosphere
Atmosphere	—	H_2O	O_2	H_2S, particles	SO_2, CO_2
Hydrosphere	H_2O	—	$\{CH_2O\}$	Mineral solutes	Water pollutants
Biosphere	O_2, CO_2	H_2O	—	Mineral nutrients	Fertilizers
Geosphere	H_2O	H_2O	Organic matter	—	Hazardous wastes
Anthrosphere	O_2, N_2	H_2O	Food	Minerals	—

Endogenic and Exogenic Cycles

Materials cycles may be divided broadly between **endogenic cycles**, which predominantly involve subsurface rocks of various kinds, and **exogenic cycles**, which occur largely on Earth's surface and usually have an atmospheric component. These two kinds of cycles are broadly outlined in Figure 19.2. In general, sediment and soil can be viewed as being shared between the two cycles and constitute the predominant interface between them.

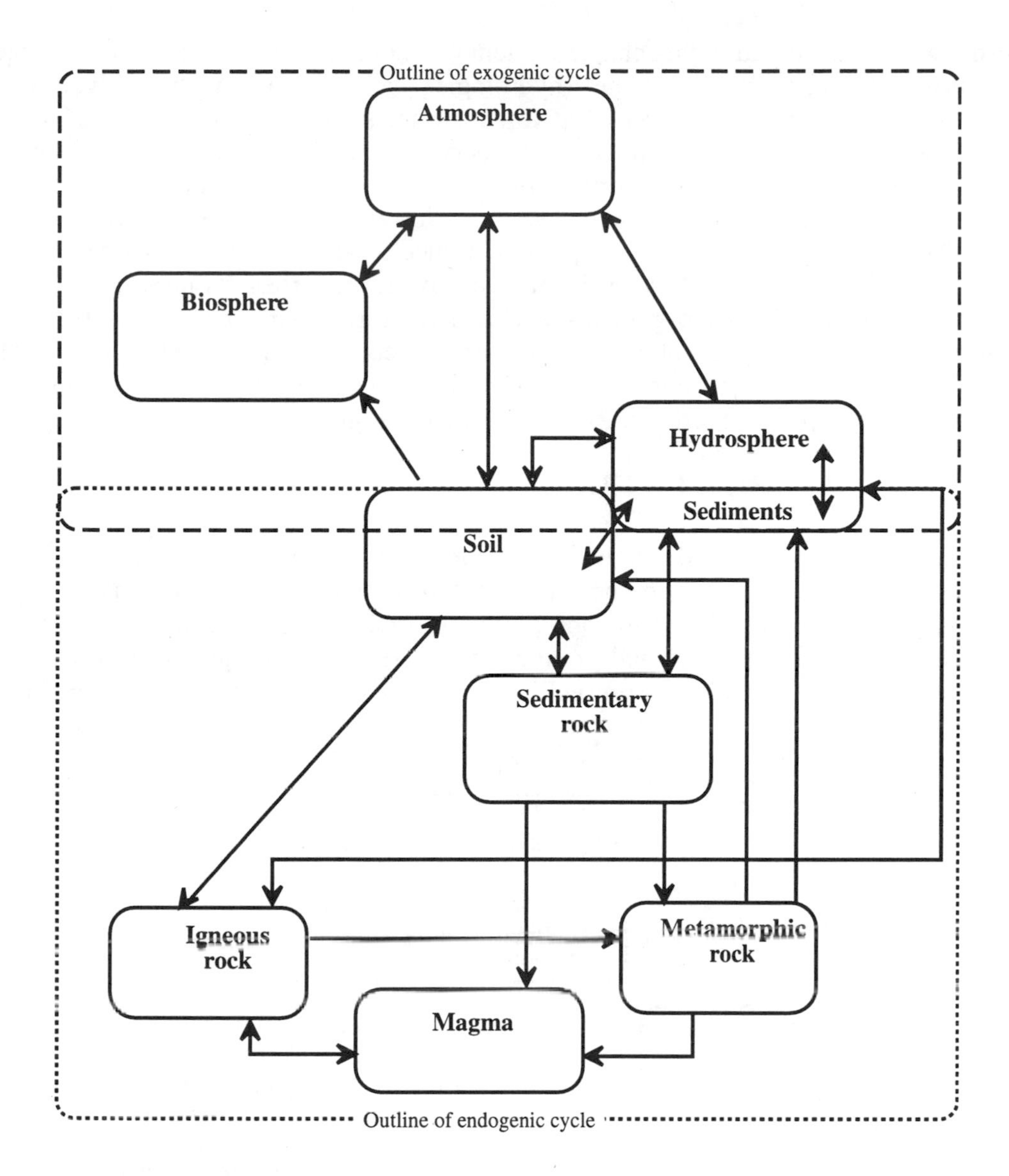

Figure 19.2. General outline of exogenic and endogenic cycles.

Most biogeochemical cycles can be described as **elemental cycles** involving **nutrient elements**, such as carbon, nitrogen, oxygen, phosphorus, and sulfur.[1] Many are exogenic cycles in which the element in question spends part of the cycle in the atmosphere—O_2 for oxygen, N_2 for nitrogen, CO_2 for carbon. Others, notably the phosphorus cycle, do not have a gaseous component and are endogenic cycles. All sedimentary cycles involve **salt solutions** or **soil solutions** (see Section 14.2) that contain dissolved substances leached from weathered minerals; these substances may be deposited as mineral formations, or they may be taken up by organisms as nutrients.

In the remainder of this chapter several of the more important elemental cycles are discussed. Although these are classified according to individual elements, it should be noted that there are strong relationships and interactions between the different cycles. For example, both carbon, considered as part of the carbon cycle, and oxygen, part of the oxygen cycle, are contained in reservoirs of mineral $CaCO_3$.

Although not discussed in this chapter, calcium contained in this mineral has a cycle of its own. Water, part of the water (hydrologic) cycle, is very much involved with the deposition and dissolution of $CaCO_3$. In water solution, this mineral exists as Ca^{2+} and HCO_3^- ions. During photosynthesis carbon is taken from the HCO_3^- ion to incorporate into biomass, another part of the carbon cycle, and elemental oxygen O_2 is released to the atmosphere, which is the main reservoir of the oxygen cycle.

The hydrologic cycle is of unique importance to all the other elemental cycles because of the abundance of water, its unique solvent properties, its importance as an essential nutrient for all organisms and a habitat for many organisms that are involved in various biogeochemical cycles, and its constant movement around Earth that physically transports constituents of other cycles. The hydrologic cycle was discussed in Chapter 5, Section 5.2, and is shown in Figure 5.1.

19.2. THE CARBON CYCLE

Carbon is circulated through the **carbon cycle**, shown in Figure 19.3. This cycle shows that carbon may be present as gaseous atmospheric CO_2, constituting a relatively small, but highly significant, portion of global carbon. Some of the carbon is dissolved in surface water and groundwater as HCO_3^- or molecular $CO_2(aq)$. A very large amount of carbon is present in minerals, particularly calcium and magne-

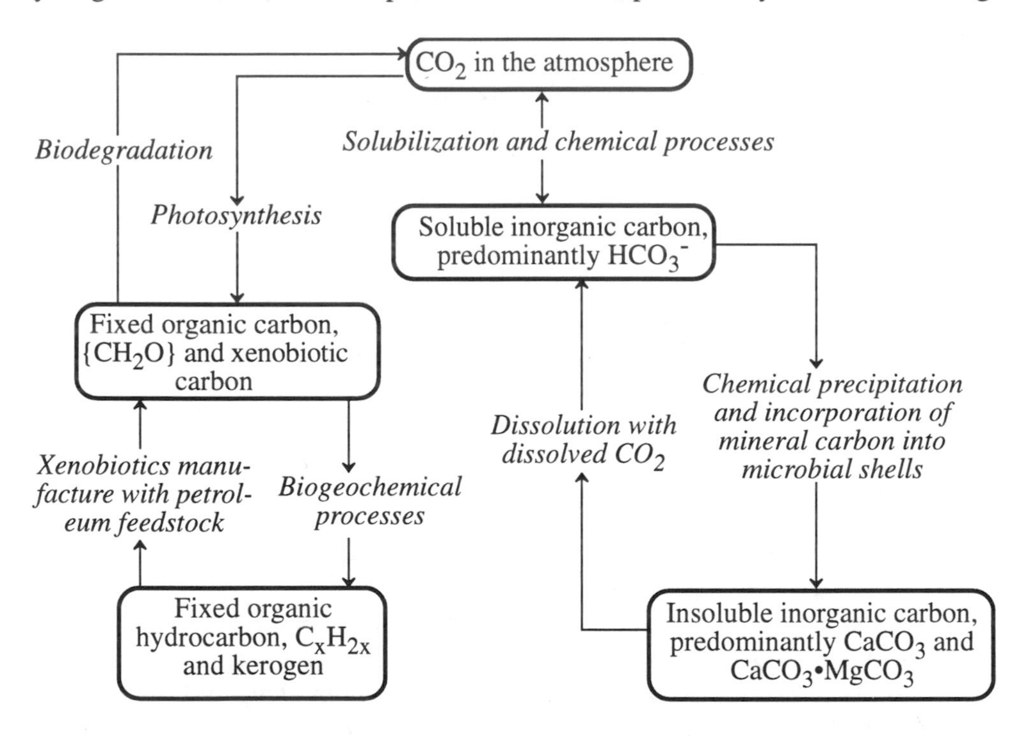

Figure 19.3. Important aspects of the biogeochemical carbon cycle.

sium carbonates, such as $CaCO_3$. Photosynthesis fixes inorganic C as biological carbon, represented as $\{CH_2O\}$, which is a constituent of all life molecules. Therefore, the carbon cycle is strongly linked to all other cycles involving living organisms. Another fraction of carbon is fixed as petroleum and natural gas, with a much larger amount as hydrocarbonaceous kerogen (the organic matter in oil shale), coal, and lignite, represented as C_xH_{2x}. Manufacturing processes are used to convert hydrocarbons to xenobiotic synthetic chemical compounds with functional groups containing halogens, oxygen, nitrogen, phosphorus, or sulfur. Though a very small amount of total environmental carbon, these compounds are particularly significant because of their toxicological chemical effects.

An important aspect of the carbon cycle is that it is the cycle by which solar energy is transferred to biological systems and ultimately to the geosphere and anthrosphere as fossil carbon and as fossil fuels. Organic, or biological carbon, $\{CH_2O\}$, is contained in energy-rich molecules that can react biochemically with molecular oxygen, O_2, to regenerate carbon dioxide and produce energy. This can occur biochemically in an organism through aerobic respiration or it may occur as combustion, such as when wood or fossil fuels are burned.

Microorganisms are strongly involved in the carbon cycle, mediating crucial biochemical reactions discussed later in this section. Photosynthetic algae are the predominant carbon-fixing compounds in water; as they consume CO_2, the pH of the water is raised enabling precipitation of $CaCO_3$ and $CaCO_3$•$MgCO_3$. Organic carbon fixed by microorganisms is transformed by biogeochemical processes to fossil petroleum, kerogen, coal, and lignite. Microorganisms degrade organic carbon from biomass, petroleum, and xenobiotic sources, ultimately returning it to the atmosphere as CO_2.

Photosynthesis and Respiration

Carbon is an essential life element and composes a high percentage of the dry weight of microorganisms. For most microorganisms, the bulk of net energy-yielding or energy-consuming metabolic processes involve changes in the oxidation state of carbon. These chemical transformations of carbon have important environmental implications. When algae and other plants conduct **photosynthesis** to fix CO_2 as carbohydrate, represented as $\{CH_2O\}$,

$$CO_2 + H_2O \xrightarrow{h\nu} \{CH_2O\} + O_2(g) \tag{19.2.1}$$

energy from sunlight is stored as chemical energy in organic compounds. However, when the algae or plants die, bacterial decomposition results in the reverse of the biochemical process represented by the above reaction, energy is released, and oxygen is consumed.

As the reverse of photosynthesis, in the presence of oxygen, the principal energy-yielding reaction of bacteria is the oxidation of organic matter:

$$\{CH_2O\} + O_2 \rightarrow CO_2 + H_2O \tag{19.2.2}$$

This general type of reaction is called **aerobic respiration**, and from it bacteria and other microorganisms extract the energy needed to carry out their metabolic processes, to synthesize new cell material, for reproduction, and for locomotion.

Anaerobic respiration occurs when microorganisms degrade biomass in the absence of oxygen. Certain bacteria can carry out anaerobic respiration using part of the organic matter or biomass as a substitute for molecular oxygen. Methane gas is a common product of anaerobic respiration. The overall reaction by which methane is produced is

$$2\{CH_2O\} \rightarrow CH_4 + CO_2 \tag{19.2.3}$$

Degradation of Biomass by Soil Bacteria and Fungi

One of the most important functions of soil bacteria and fungi, and a crucial link in the carbon cycle, is the biodegradation of dead organic matter consisting predominantly of plant residues. In addition to preventing accumulation of excess waste residue, this composting process converts organic carbon, nitrogen, sulfur, and phosphorus to simple organic forms that can be utilized by plants and is a key part of the biogeochemical cycles of these elements. It also leaves a humus residue that is required for the optimum physical form of soil.

Partial microbial decomposition of organic matter is a key step in producing peat, lignite, coal, oil shale, and petroleum. Under reducing conditions, particularly below water, the oxygen content of the original plant material (approximate empirical formula, $\{CH_2O\}$) is lowered, leaving materials with higher carbon contents.

19.3. THE NITROGEN CYCLE

As shown in Figure 19.4, nitrogen occurs prominently in all the spheres of the environment. The atmosphere is 78% by volume elemental nitrogen, N_2, and is an inexhaustible reservoir of this essential element. Nitrogen, though constituting much less of biomass than carbon or oxygen, is an essential constituent of proteins, enzymes, and nucleic acids.

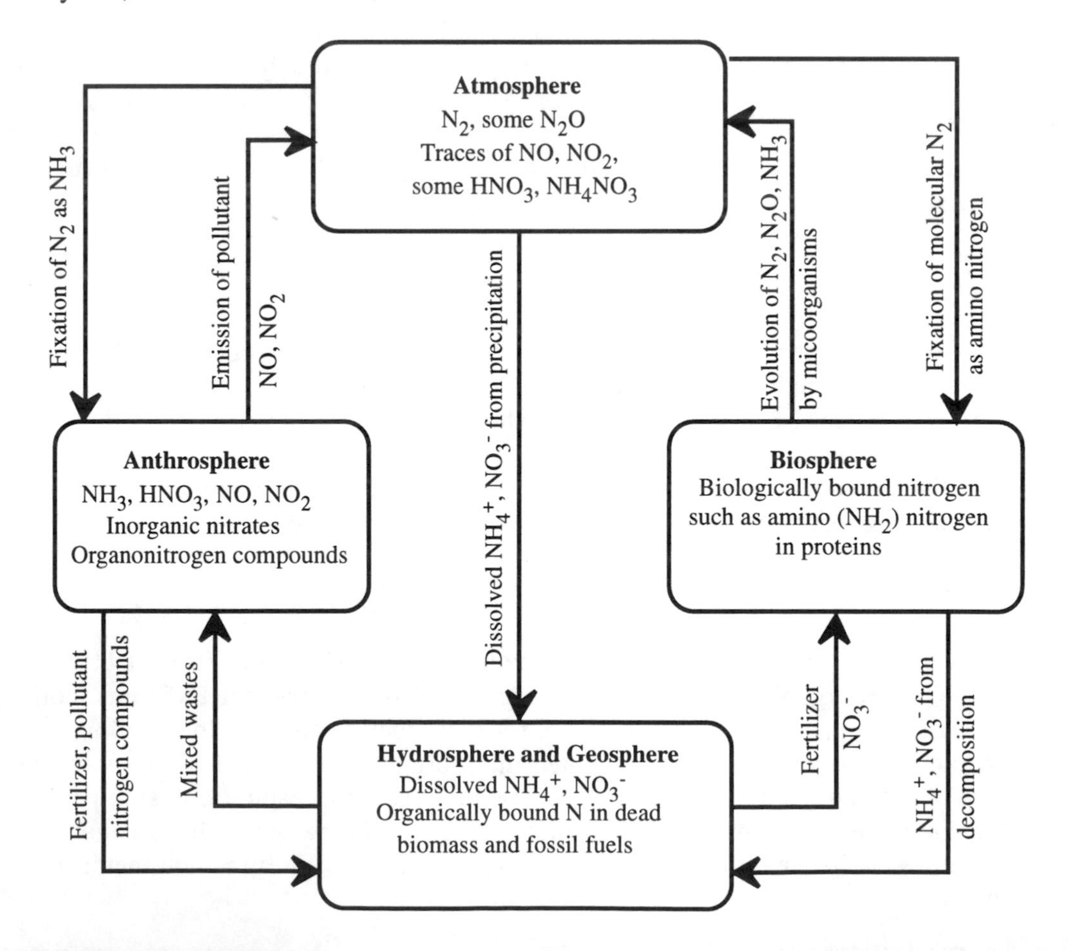

Figure 19.4. The nitrogen cycle.

There is a strong barrier to the exchange of nitrogen between its main reservoir as elemental nitrogen in the atmosphere and other spheres of the nitrogen cycle. This is because the N_2 molecule is so stable that breaking it down to atoms that can be incorporated into inorganic and organic chemical forms of nitrogen is the limiting step in the nitrogen cycle. The breakdown of N_2 does occur by highly energetic processes in lightning discharges that produce nitrogen oxides. Elemental nitrogen is also incorporated into chemically bound forms, or **fixed**, by biochemical processes mediated by microorganisms. The biological nitrogen is mineralized to the inorganic form during the decay of biomass. Large quantities of nitrogen are fixed synthetically under high-temperature and high-pressure conditions according to the following overall reaction:

$$N_2 + 3H_2 \rightarrow 2NH_3 \qquad (19.3.1)$$

Microbial Nitrogen Fixation

The overall microbial process for **nitrogen fixation**, the binding of atmospheric nitrogen in a chemically combined form,

$$3\{CH_2O\} + 2N_2 + 3H_2O + 4H^+ \rightarrow 3CO_2 + 4NH_4^+ \qquad (19.3.2)$$

is actually quite complicated and not completely understood. Biological nitrogen fixation is a key biochemical process in the environment and is essential for plant growth in the absence of synthetic fertilizers.

Only a few species of aquatic microorganisms have the ability to fix atmospheric nitrogen. Among the aquatic bacteria that can do so are photosynthetic bacteria, *Azotobacter*, and several species of *Clostridium*. The best-known and most important form of nitrogen-fixing bacteria is *Rhizobium*, which enjoys a symbiotic (mutually advantageous) relationship with leguminous plants such as clover or alfalfa. The *Rhizobium* bacteria are found in root nodules, special structures attached to the roots of legumes and connected directly to the vascular (circulatory) system of the plant, enabling the bacteria to derive photosynthetically produced energy directly from the plant. Thus, the plant provides the energy required to break the strong triple bonds in the dinitrogen molecule, converting the nitrogen to a reduced form which is directly assimilated by the plant. When the legumes die and decay, NH_4^+ ion is released and is converted by microorganisms to nitrate ion, which is assimilable by other plants.

Some nonlegume angiosperms fix nitrogen through the action of actinomycetes bacteria contained in root nodules. Shrubs and trees in the nitrogen-fixing category are abundant in fields, forests, and wetlands throughout the world. Their rate of nitrogen fixation is comparable to that of legumes.

Nitrification

Nitrification is the conversion of ammoniacal nitrogen (NH_4^+) to nitrate (NO_3^-). It is a very common and extremely important process in water and in soil carried out largely by the action of *Nitrosomonas* and *Nitrobacter* bacteria. The overall nitrification reaction is the following:

$$2O_2 + NH_4^+ \rightarrow NO_3^- + 2H^+ + H_2O \qquad (19.3.3)$$

The reason that nitrification is especially important in nature is because nitrogen is absorbed by plants primarily as nitrate. When fertilizers are applied in the form of ammonium salts or anhydrous ammonia, and when NH_4^+ ion is released by the biodegradation of nitrogen-containing organic matter, a microbial transformation to nitrate enables maximum assimilation of nitrogen by the plants.

Nitrate Reduction and Denitrification

As a general term, **nitrate reduction** refers to microbial processes by which nitrogen in chemical compounds is reduced to lower oxidation states. When free oxygen, O_2, is absent, nitrate, NO_3^-, may be used by some bacteria as an alternate oxygen source. Under these conditions nitrogen in nitrate may be reduced to ammonium ion, NH_4^+.

A special case of nitrate reduction occurs when nitrate is reduced to elemental nitrogen, N_2, which returns to the atmosphere as nitrogen gas. This important link in the nitrogen cycle is called **denitrification**. In addition to returning free nitrogen to the atmosphere, denitrification is useful in advanced water treatment for the removal of nutrient nitrogen. Because of the industrial fixation of nitrogen, human activities have perturbed the nitrogen cycle significantly, in some cases overloading the capacity of natural systems to return nitrogen to the atmosphere by denitrification. Fixed nitrogen can accumulate, particularly in bodies of water, where it contributes to excess algal growth (eutrophication) and in groundwater, which in some cases acquires toxic levels of dissolved nitrate.

19.4. THE OXYGEN CYCLE

The **oxygen cycle** is discussed in Chapter 10 and is illustrated in Figure 10.5. It involves the interchange of oxygen between the elemental form of gaseous O_2, contained in a huge reservoir in the atmosphere and chemically bound O in CO_2, H_2O, and organic matter. It is strongly tied with other elemental cycles, particularly the carbon cycle. Elemental oxygen becomes chemically bound by various energy-yielding processes, particularly combustion and metabolic processes in organisms. It is released in photosynthesis. This element readily combines with and oxidizes other species, such as carbon in aerobic respiration (Reaction 19.2.2), or carbon and hydrogen in the combustion of fossil fuels, such as methane:

$$CH_4 + 2O_2 \rightarrow CO_2 + 2H_2O \qquad (19.4.1)$$

Elemental oxygen also oxidizes inorganic substances, such as iron(II) in minerals:

$$4FeO + O_2 \rightarrow 2Fe_2O_3 \qquad (19.4.2)$$

A particularly important aspect of the oxygen cycle is stratospheric ozone, O_3. As discussed in Chapter 10, Section 10.8, a relatively small concentration of ozone in the stratosphere more than 10 kilometers high in the atmosphere filters out ultraviolet radiation in the wavelength range of 220-330 nm, thus protecting life on Earth from the highly damaging effects of this radiation.

The oxygen cycle is completed when elemental oxygen is returned to the atmosphere. The only significant way in which this is done is through photosynthesis mediated by plants, which liberates elemental oxygen from water. The overall reaction for photosynthesis is given in Equation 19.2.1. Elemental oxygen acting as a biological oxidizing agent is incorporated into water by aerobic respiration as shown in Reaction 19.2.2.

19.5. THE PHOSPHORUS CYCLE

There are no common stable gaseous forms of phosphorus, so the phosphorus cycle is endogenic without an atmospheric component. The main reservoirs of phosphorus are in the geosphere. Geospheric phosphorus is held largely in poorly soluble minerals, such as hydroxyapatite, a calcium salt, deposits of which constitute the major reservoir of environmental phosphate. Soluble phosphorus from phosphate minerals and other sources, such as fertilizers, is taken up by plants as an essential nutrient. Mineralization of biomass by microbial decay returns phosphorus to the salt solution from which it may precipitate as mineral matter.

Figure 19.5 illustrates important aspects of the phosphorus cycle. It involves natural and pollutant sources of phosphorus including biological, organic, and inorganic phosphorus. Phosphorus is essential for life because it is a component of the nucleic acids constituting genetic materials, it is contained in high-energy molecules involved in energy transfer during metabolism, and it is an essential component of bone in animals. Soil and aquatic microbial processes are very important in the phosphorus cycle. Of particular importance is the fact that phosphorus is the most common limiting nutrient in water, particularly for the growth of algae. Bacteria are even more effective than algae in taking up phosphate from water, accumulating it as excess cellular phosphorus that can be released to support additional bacterial growth if the supply of phosphorus becomes limiting. Microorganisms that die

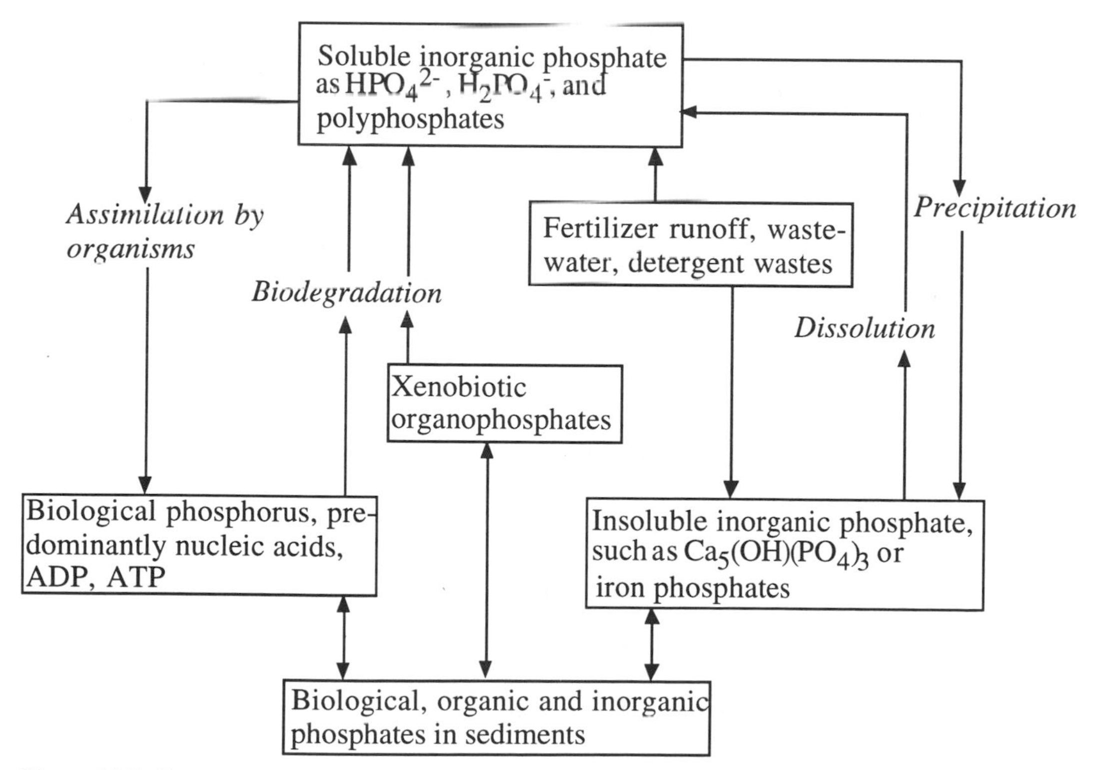

Figure 19.5. Important aspects of the phosphorus cycle.

release phosphorus that can support additional organisms. **Phosphoclastic** bacteria are even "phosphorus-dissolving" in that they produce acidic substances that bring about the release of inorganic phosphorus from poorly soluble phosphorus minerals.

Living organisms constitute an important link in the phosphorus cycle. Birds and bats particularly excrete feces rich in phosphate concentrated by their metabolic processes. In locations where these deposits of feces accumulate, especially some dry coastal areas frequented by sea birds and floors of caves inhabited by bats, deposits of phosphorus-rich **guano** accumulate, which are important commercial sources of phosphate. Another important biological link in the phosphorus cycle occurs when droppings and bodies of organisms that have concentrated phosphorus from sea water are incorporated into sea sediments. These sediments can be brought to the surface along seacoasts by a process called upwelling, resulting in particularly nutrient-rich waters that support productive fisheries.

The anthrosphere is an important reservoir and conduit of phosphorus in the environment. Large quantities of phosphates are extracted from phosphate minerals for fertilizer, industrial chemicals, and food additives. Phosphorus is a constituent of some extremely toxic compounds, especially organophosphate insecticides and military poison nerve gases.

Biodegradation of phosphorus compounds is important in the environment for two reasons. The first of these is that it is a *mineralization* process that releases inorganic phosphorus from the organic form thereby providing a source of algal nutrient orthophosphate and, secondly, biodegradation deactivates highly toxic organophosphate compounds, such as the organophosphate insecticides.

The organophosphorus compounds of greatest environmental concern tend to be sulfur-containing **phosphorothionate** and **phosphorodithioate** ester insecticides with the general formulas illustrated in Figure 19.6, where R represents a hydrocar-

S
‖
RO–P–O-Ar
|
RO

General formula of phosphorothionates

S
‖
C_2H_5O–P–O–(benzene ring)–NO_2
|
C_2H_5O

Parathion

S
‖
RO–P–S–Ar
|
RO

General formula of phosphorodithioates

S H O
‖ | ‖
$(CH_3O)_2P$–S–C–C–O–C_2H_5
|
H–C–C–O–C_2H_5
| ‖
H O

Malathion, a phosphorodithioate insecticide

O
‖
RO–P–O-Ar
|
RO

General formula of phosphate esters

O
‖
C_2H_5O–P–O–(benzene ring)–NO_2
|
C_2H_5O

Paraoxon, a phosphate ester insecticide

Figure 19.6. Phosphorothionate, phosphorodithioate, and phosphate ester insecticides.

carbon or substituted hydrocarbon group. These are used in insecticides because they exhibit higher ratios of insect:mammal toxicity than do their nonsulfur analogs. The metabolic conversion of P=S to P=O (oxidative desulfuration, such as in the conversion of parathion to paraoxon) in organisms is responsible for the insecticidal activity and mammalian toxicity of phosphorothionate and phosphorodithioate insecticides. The biodegradation of these compounds is an important environmental chemical process. Fortunately, unlike the organohalide insecticides that they largely displaced, the organophosphates readily undergo biodegradation and do not bioaccumulate.

19.6. THE SULFUR CYCLE

The sulfur cycle, which is illustrated in Figure 19.7, is relatively complex in that it involves several gaseous species, poorly soluble minerals, and several species in solution. It is tied with the oxygen cycle in that sulfur combines with oxygen to form gaseous sulfur dioxide, SO_2, an atmospheric pollutant, and soluble sulfate ion, SO_4^{2-}. Among the significant species involved in the sulfur cycle are gaseous hydrogen sulfide, H_2S; mineral sulfides, such as FeS; sulfuric acid, H_2SO_4, the main constituent of acid rain; and biologically bound sulfur in sulfur-containing proteins.

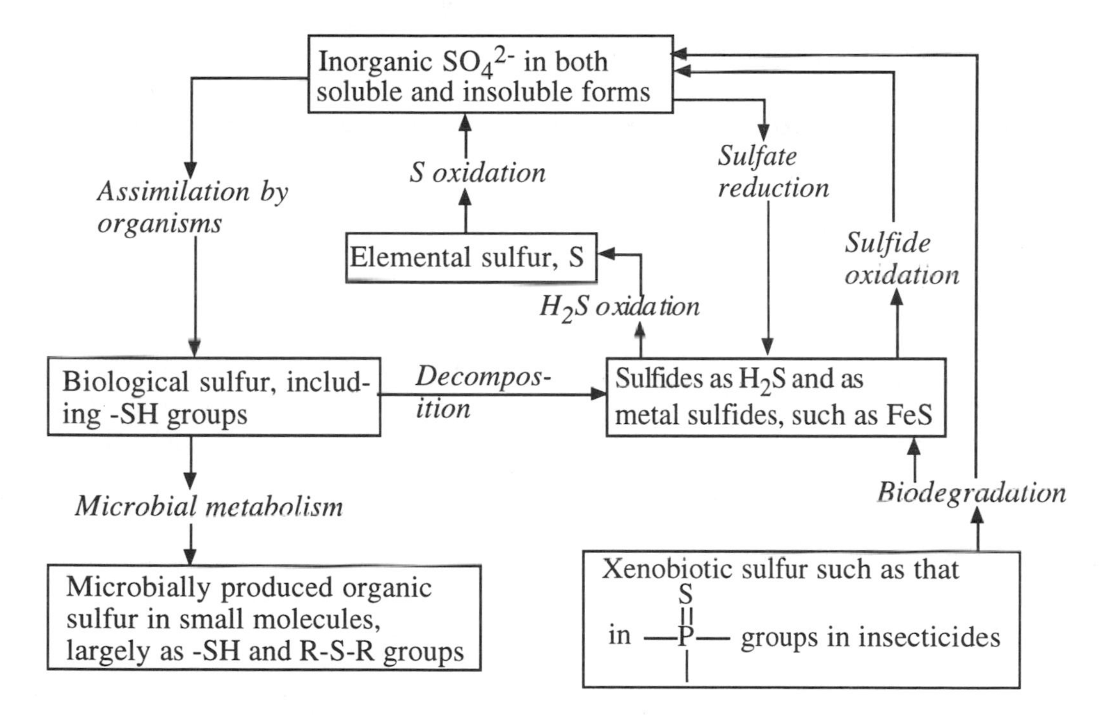

Figure 19.7. Important aspects of the sulfur cycle.

Insofar as pollution is concerned, the most significant part of the sulfur cycle is the presence of pollutant SO_2 gas and H_2SO_4 in the atmosphere. The former is a somewhat toxic gaseous air pollutant evolved in the combustion of sulfur-containing fossil fuels. Sulfur dioxide is discussed further as an air pollutant in Chapter 11. The major detrimental effect of sulfur dioxide in the atmosphere is its tendency to oxidize in the atmosphere to produce sulfuric acid. This species is responsible for acidic precipitation, a major atmospheric pollutant commonly called “acid rain.”

Sulfur compounds are very common in water and soil. Sulfate ion, SO_4^{2-}, is found in varying concentrations in practically all natural waters. Organic sulfur compounds, both those of natural origin and pollutant species, are very common in natural aquatic systems, and the degradation of these compounds is an important microbial process. Sometimes the degradation products, such as the odiferous and toxic H_2S, cause serious problems with water quality.

Although organic sulfur compounds often are the source of H_2S in water, H_2S commonly is produced by the microbial reduction of sulfate. Some bacteria can reduce sulfate ion to H_2S. In so doing, they utilize sulfate as an oxygen source in the oxidation of organic matter. The overall reaction is the following:

$$SO_4^{2-} + 2\{CH_2O\} + 2H^+ \rightarrow H_2S + 2CO_2 + 2H_2O \tag{19.6.1}$$

Because of the high concentration of sulfate ion in seawater, bacterially mediated formation of H_2S causes pollution problems in some coastal areas and is a major source of atmospheric sulfur. In waters where sulfide formation occurs, the sediment is often black in color due to the formation of FeS. Whereas some bacteria can reduce sulfate ion to H_2S, others can oxidize hydrogen sulfide to sulfate.

CHAPTER SUMMARY

The chapter summary below is presented in a programmed format to review the main points covered in this chapter. It is used most effectively by filling in the blanks, referring back to the chapter as necessary. The correct answers are given at the end of the summary.

Biogeochemical cycles describe the circulation of matter, particularly [1]______________________________________, through ecosystems. Natural cycles have [2]_____________________________ in which most of the available material constituting the cycle is contained. [3]_________________ predominantly involve subsurface rocks of various kinds, and [4]_________________ occur largely on Earth's surface, and usually have an atmospheric component. The [5]___________ cycle is of unique importance to all the other elemental cycles. In the carbon cycle, some of the carbon is dissolved in surface water and groundwater as [6]______________________________________, very large amounts of carbon are present in minerals as [7]____, and photosynthesis fixes inorganic C as [8]_____________________. The reaction for photosynthesis is represented as [9]___________________________________, whereas the opposite process called [10]______________________ is represented as [11]______________________________________. The main barrier between incorporation of elemental nitrogen in the atmosphere into other spheres of the nitrogen cycle, a process called [12]______________ arises because of the [13]______________________________. The most important form of nitrogen-fixing bacteria is [14]__________________ which has a [15]_________________ relationship with [16]__________________ plants. The reason that nitrification, represented by the reaction [17]_____________________________, is especially important in nature is because [18]____.

The process that returns free nitrogen to the atmosphere is called [19]____________ _________________. A particularly important aspect of the oxygen cycle is stratospheric [20]_______________. The only significant way in which elemental oxygen is returned to the atmosphere is through [21]___________________________. The main reservoirs of phosphorus are in the [22]______________. Phosphorus is essential for life because [23]____.

Large quantities of phosphates are extracted from phosphate minerals by humans for use as [24]____. Insofar as pollution is concerned, the most significant part of the sulfur cycle is the presence of [25]____.
Although organic sulfur compounds often are the source of H_2S in water, H_2S commonly is produced by [26]____, for which the overall reaction is [27]____.

Answers

[1] plant and animal nutrients

[2] reservoir stages

[3] Endogenic cycles

[4] exogenic cycles

[5] hydrologic

[6] HCO_3^- or molecular $CO_2(aq)$

[7] calcium and magnesium carbonates

[8] biological carbon

[9] $CO_2 + H_2O \xrightarrow{h\nu} \{CH_2O\} + O_2(g)$

[10] aerobic respiration

[11] $\{CH_2O\} + O_2 \rightarrow CO_2 + H_2O$

[12] fixing

[13] extreme stability of the elemental nitrogen molecule

[14] *Rhizobium*

[15] symbiotic

[16] leguminous

[17] $2O_2 + NH_4^+ \rightarrow NO_3^- + 2H^+ + H_2O$

[18] nitrogen is absorbed by plants primarily as nitrate

19 denitrification

20 ozone, O_3

21 photosynthesis mediated by plants

22 geosphere

23 it is a component of the nucleic acids constituting genetic materials; it is contained in high-energy molecules involved in energy transfer during metabolism; and it is an essential component of bone in animals

24 fertilizer, industrial chemicals, and food additives

25 pollutant SO_2 gas and H_2SO_4 in the atmosphere

26 the microbial reduction of sulfate

27 $SO_4^{2-} + 2\{CH_2O\} + 2\,H^+ \rightarrow H_2S + 2CO_2 + 2H_2O$

QUESTIONS AND PROBLEMS

1. What is meant by the reservoir stage of a biogeochemical cycle? Is it plausible that a cycle might have more than one reservoir stage?
2. Why is the phosphorus cycle regarded as being endogenic?
3. Explain how the hydrologic cycle is involved with other important cycles.
4. Although only a very small part of the cycle in terms of mass, why is xenobiotic carbon a particularly important constituent of the carbon cycle?
5. What is the environmental advantage derived from the fact that wood undergoes only partial biodegradation?
6. What are the products of the mineralization of organic carbon, nitrogen, phosphorus, and sulfur?
7. Where are *Rhizobium* bacteria found, and what unique function do they perform?
8. What is the special case of nitrate reduction when molecular N_2 is the nitrogen product?
9. What are three essential life functions served by phosphorus?
10. Considering organophosphorus compounds, is it correct to assume that high toxicity of a compound is always consistent with poor biodegradability?
11. In what sense does "acid rain" provide a link between the sulfur and hydrologic cycles?
12. What purpose is served for some bacteria when they convert sulfate (SO_4^{2-}) to sulfide (H_2S)?

LITERATURE CITED

1 Agren, Goran Ernesto Bosatta, *Theoretical Ecosystem Ecology: Understanding Element Cycles*, Cambridge University Press, 1997.

SUPPLEMENTARY REFERENCES

Lal, Rattan, John H. Kimble, and Ronald F. Follett, *Soil Processes and the Carbon Cycle*, CRC Press/Lewis Publishers, Boca Raton, FL, 1997.

Schlesinger, William H., *Biogeochemistry: An Analysis of Global Change*, 2nd ed., Academic Press, San Diego, CA, 1997.

20 TOXICOLOGY AND TOXICOLOGICAL CHEMISTRY

20.1. INTRODUCTION TO TOXICOLOGY AND TOXICOLOGICAL CHEMISTRY

Ultimately, most pollutants and hazardous substances are of concern because of their toxic effects. The general aspects of these effects and the toxicological chemistry of specific classes of chemical substances are addressed in this chapter. In order to understand toxicological chemistry, it is essential to have some understanding of biochemistry, the science that deals with chemical processes and materials in living systems. Biochemistry was summarized in Chapter 3.

Toxicology

A **poison**, or **toxicant**, is a substance that is harmful to living organisms because of its detrimental effects on tissues, organs, or biological processes. **Toxicology** is the science of poisons. These definitions are subject to a number of qualifications. Whether a substance is poisonous depends upon the type of organism exposed, the amount of the substance, and the route of exposure. In the case of human exposure, the degree of harm done by a poison can depend strongly upon whether the exposure is to the skin, by inhalation, or through ingestion.

Toxicants to which subjects are exposed in the environment or occupationally may be in several different physical forms. This may be illustrated for toxicants that are inhaled. **Gases** are substances such as carbon monoxide in air that are normally in the gaseous state under ambient conditions of temperature and pressure. **Vapors** are gas-phase materials that have evaporated or sublimed from liquids or solids. **Dusts** are respirable solid particles produced by grinding bulk solids, whereas **fumes** are solid particles from the condensation of vapors, often metals or metal oxides. **Mists** are liquid droplets.

Often a toxic substance is in solution or mixed with other substances. A substance with which the toxicant is associated (the solvent in which it is dissolved or the solid medium in which it is dispersed) is called the **matrix**. The matrix may have a strong effect upon the toxicity of the toxicant.

There are numerous variables related to the ways in which organisms are exposed to toxic substances. One of the most crucial of these, **dose,** is discussed in

Section 20.2. Another important factor is the **toxicant concentration**, which may range from the pure substance (100%) down to a very dilute solution of a highly potent poison. Both the **duration** of exposure per exposure incident and the **frequency** of exposure are important. The **rate** of exposure and the total time period over which the organism is exposed are both important situational variables. The exposure **site** and **route** also affect toxicity.

It is possible to classify exposures on the basis of acute *vs.* chronic and local *vs.* systemic exposure, giving four general categories. **Acute local** exposure occurs at a specific location over a time period of a few seconds to a few hours and may affect the exposure site, particularly the skin, eyes or mucous membranes. The same parts of the body can be affected by **chronic local** exposure, for which the time span may be as long as several years. **Acute systemic** exposure is a brief exposure or exposure to a single dose and occurs with toxicants that can enter the body, such as by inhalation or ingestion, and affect organs such as the liver that are remote from the entry site. **Chronic systemic** exposure differs in that the exposure occurs over a prolonged time period.

In discussing exposure sites for toxicants it is useful to consider the major routes and sites of exposure, distribution, and elimination of toxicants in the body as shown in Figure 20.1. The major routes of accidental or intentional exposure to toxicants by humans and other animals are the skin (percutaneous route), the lungs (inhalation, respiration, pulmonary route), and the mouth (oral route); minor routes

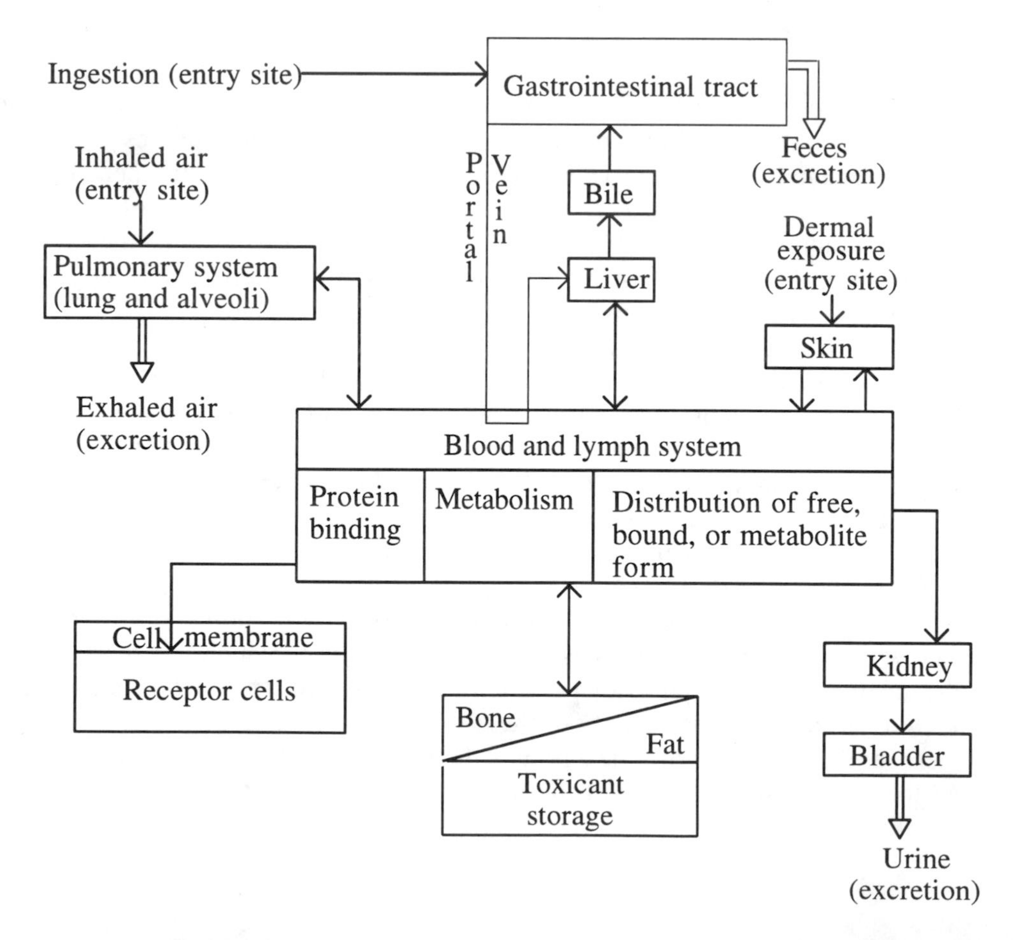

Figure 20.1. Major sites of exposure, metabolism, and storage, routes of distribution and elimination of toxic substances in the body.

of exposure are rectal, vaginal, and parenteral (intravenous or intramuscular, a common means for the administration of drugs or toxic substances in test subjects). The way that a toxic substance is introduced into the complex system of an organism is strongly dependent upon the physical and chemical properties of the substance. The pulmonary system is most likely to take in toxic gases or very fine, respirable solid or liquid particles. In other than a respirable form, a solid usually enters the body orally. Absorption through the skin is most likely for liquids, solutes in solution, and semisolids, such as sludges.

The defensive barriers that a toxicant may encounter vary with the route of exposure. For example, toxic elemental mercury is readily absorbed through the alveoli in the lungs much more readily than through the skin or gastrointestinal tract. Most test exposures to animals are through ingestion or gavage (introduction into the stomach through a tube). Pulmonary exposure is often favored with subjects that may exhibit refractory behavior when noxious chemicals are administered by means requiring a degree of cooperation from the subject. Intravenous injection may be chosen for deliberate exposure when it is necessary to know the concentration and effect of a xenobiotic substance in the blood. However, pathways used experimentally that are almost certain not to be significant in accidental exposures can give misleading results when they avoid the body's natural defense mechanisms.

An interesting historical example of the importance of the route of exposure to toxicants is provided by cancer caused by contact of coal tar with skin. The major barrier to dermal absorption of toxicants is the **stratum corneum** or horny layer. The permeability of skin is inversely proportional to the thickness of this layer, which varies by location on the body in the order soles and palms > abdomen, back, legs, arms > genital (perineal) area. Evidence of the susceptibility of the genital area to absorption of toxic substances is to be found in accounts of the high incidence of cancer of the scrotum among chimney sweeps in London described by Sir Percival Pott, Surgeon General of Britain during the reign of King George III. The cancer-causing agent was coal tar condensed in chimneys. This material was more readily absorbed through the skin in the genital areas than elsewhere leading to a high incidence of scrotal cancer (a condition aggravated by a lack of appreciation of basic hygienic practices, such as bathing and regular changes of underclothing.)

Organisms can serve as indicators of various kinds of pollutants. In this application, organisms are known as biomonitors. Higher plants, fungi, lichens, and mosses can be useful biomonitors for heavy metal pollutants in the environment.[1]

Synergism, Potentiation, and Antagonism

The biological effects of two or more toxic substances can be different in kind and degree from those of one of the substances alone. One of the ways in which this can occur is when one substance affects the way in which another undergoes any of the steps in the kinetic phase as discussed in Section 20.7 and illustrated in Figure 20.9. Chemical interaction between substances may affect their toxicities. Both substances may act upon the same physiologic function or two substances may compete for binding to the same receptor (molecule or other entity acted upon by a toxicant). When both substances have the same physiologic function, their effects may be simply **additive** or they may be **synergistic** (the total effect is greater than the sum of the effects of each separately). **Potentiation** occurs when an inactive substance enhances the action of an active one and **antagonism** when an active substance decreases the effect of another active one.

20.2. DOSE-RESPONSE RELATIONSHIPS

Toxicants have widely varying effects upon organisms. Quantitatively, these variations include minimum levels at which the onset of an effect is observed, the sensitivity of the organism to small increments of toxicant, and levels at which the ultimate effect (particularly death) occurs in most exposed organisms. Some essential substances, such as nutrient minerals, have optimum ranges above and below which detrimental effects are observed (see Section 20.5 and Figure 20.4).

Factors such as those just outlined are taken into account by the **dose-response** relationship, which is one of the key concepts of toxicology. **Dose** is the amount, usually per unit body mass, of a toxicant to which an organism is exposed. **Response** is the effect upon an organism resulting from exposure to a toxicant. In order to define a dose-response relationship, it is necessary to specify a particular response, such as death of the organism, as well as the conditions under which the response is obtained, such as the length of time from administration of the dose. Consider a specific response for a population of the same kinds of organisms. At relatively low doses, none of the organisms exhibits the response (for example, all live) whereas at higher doses, all of the organisms exhibit the response (for example, all die). In between, there is a range of doses over which some of the organisms respond in the specified manner and others do not, thereby defining a dose-response curve. Dose-response relationships differ among different kinds and strains of organisms, types of tissues, and populations of cells.

Figure 20.2 shows a generalized dose-response curve. Such a plot may be obtained, for example, by administering different doses of a poison in a uniform manner to a homogeneous population of test animals and plotting the cumulative per-

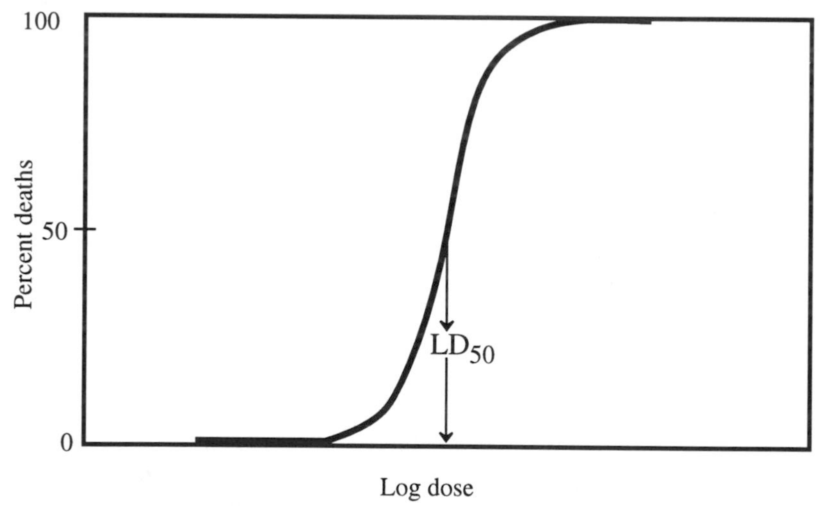

Figure 20.2. Illustration of a dose-response curve in which the response is the death of the organism. The cumulative percentage of deaths of organisms is plotted on the Y axis.

centage of deaths as a function of the log of the dose. The dose corresponding to the mid-point (inflection point) of the resulting S-shaped curve is the statistical estimate of the dose that would kill 50 percent of the subjects and is designated as LD_{50}. The estimated doses at which 5 percent (LD_5) and 95 percent (LD_{95}) of the test subjects die are obtained from the graph by reading the dose levels for 5 percent and 95 percent fatalities, respectively. A relatively small difference between LD_5 and LD_{95}

is reflected by a steeper **S**-shaped curve and vice versa. Statistically, 68 percent of all values on a dose-response curve fall within ± 1 standard deviation of the mean at LD_{50} and encompass the range from LD_{16} to LD_{84}.

20.3. RELATIVE TOXICITIES

Table 20.1 illustrates standard **toxicity ratings** that are used to describe estimated toxicities of various substances to humans. In terms of fatal doses to an adult human of average size, a "taste" of a supertoxic substances (just a few drops or less) is fatal. In June of 1997 a research chemist died as the result of exposure to a minute amount of dimethyl mercury. The exposure had occurred some months before as the result of a few drops of the compound absorbed through the skin after being spilled on latex gloves used by the chemist. Later tests showed that dimethyl mercury penetrates such gloves in a matter of seconds. A teaspoonful of a very toxic substance could be fatal to a human. However, as much as a liter of a slightly toxic substance might be required to kill an adult human.

Table 20.1. Toxicity Scale with Example Substances.[a]

Substance	Approximate LD_{50}	Toxicity rating
	-10^5	1. Practically nontoxic > 1.5×10^4 mg/kg
DEHP[b] →	–	
Ethanol →	-10^4	2. Slightly toxic, 5×10^3 1.5×10^4 mg/kg
Sodium chloride →	–	
Malathion →	-10^3	3. Moderately toxic, 500 – 5000 mg/kg
Chlordane →	–	
Heptachlor →	-10^2	4. Very toxic, 50 – 500 mg/kg
	–	
Parathion →	– 10	5. Extremely toxic, 5 – 50 mg/kg
	–	
TEPP[c] →	– 1	
	–	
Tetrodotoxin[d] →	-10^{-1}	
	–	
	-10^{-2}	6. Supertoxic, <5 mg/kg
	–	
TCDD[e] →	-10^{-3}	
	–	
	-10^{-4}	
	–	
Botulinus toxin →	-10^{-5}	

[a] Doses are in units of mg of toxicant per kg of body mass. Toxicity ratings on the right are given as numbers ranging from 1 (practically nontoxic) through 6 (supertoxic) along with estimated lethal oral doses for humans in mg/kg. Estimated LD50 values for substances on the left have been measured in test animals, usually rats, and apply to oral doses.

[b] Bis(2-ethylhexyl)phthalate

[c] Tetraethylpyrophosphate

[d] toxin from pufferfish

[e] TCDD represents 2,3,7,8,-tetrachlorodibenzodioxin, commonly called "dioxin."

When there is a substantial difference between LD_{50} values of two different substances, the one with the lower value is said to be the more **potent**. Such a comparison must assume that the dose-response curves for the two substances being compared have similar slopes.

Nonlethal Effects

So far, toxicities have been described primarily in terms of the ultimate effect, that is, deaths of organisms, or lethality. This is obviously an irreversible consequence of exposure. In many, and perhaps most, cases, **sublethal** and **reversible** effects are of greater importance. This is obviously true of drugs, where death from exposure to a registered therapeutic agent is rare, but other effects, both detrimental and beneficial, are usually observed. By their very nature, drugs alter biological processes; therefore, the potential for harm is almost always present. The major consideration in establishing drug dose is to find a dose that has an adequate therapeutic effect without undesirable side effects. A dose-response curve can be established for a drug that progresses from noneffective levels through effective, harmful, and even lethal levels. A low slope for this curve indicates a wide range of effective dose and a wide **margin of safety** (see Figure 20.3). This term applies to other substances, such as pesticides, for which a large difference between the dose that kills a target organism and one that harms a desirable species is needed.

20.4. REVERSIBILITY AND SENSITIVITY

Sublethal doses of most toxic substances are eventually eliminated from an organism's system. If there is no lasting effect from the exposure, it is said to be **reversible**. However, if the effect is permanent, it is termed **irreversible**. Irreversible effects of exposure remain after the toxic substance is eliminated from the organism. Figure 20.3 illustrates these two kinds of effects. For various chemicals and different subjects, toxic effects may range from the totally reversible to the totally irreversible.

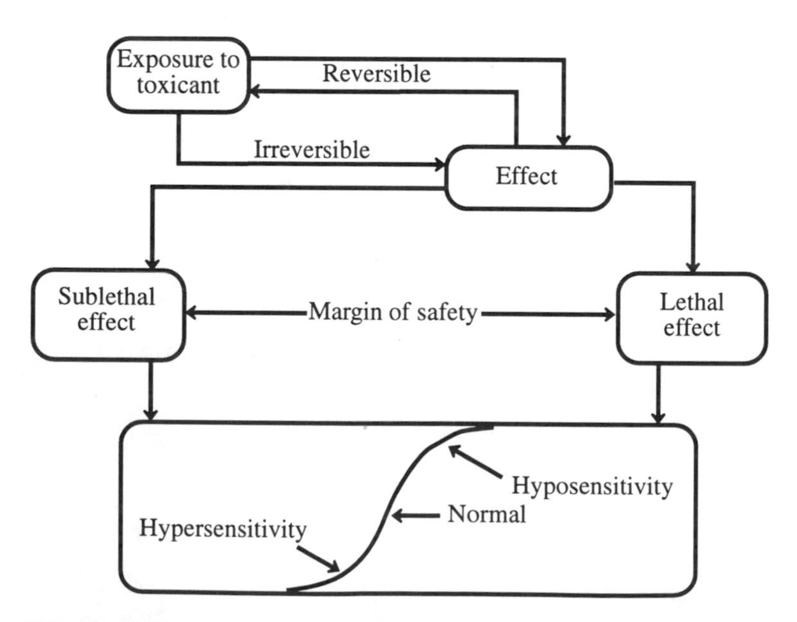

Figure 20.3. Effects of and responses to toxic substances.

Hypersensitivity and Hyposensitivity

Examination of the dose-response curve shown in Figure 20.2 reveals that some subjects are very sensitive to a particular poison (for example, those killed at a dose corresponding to LD_5), whereas others are very resistant to the same substance (for example, those surviving a dose corresponding to LD_{95}). These two kinds of responses illustrate **hypersensitivity** and **hyposensitivity**, respectively; subjects in the mid-range of the dose-response curve are termed **normals**. These variations in response tend to complicate toxicology in that there is not a specific dose guaranteed to yield a particular response, even in a homogeneous population.

In some cases hypersensitivity is induced. After one or more doses of a chemical, a subject may develop an extreme reaction to it. This occurs with penicillin, for example, in cases where people develop such a severe allergic response to the antibiotic that exposure is fatal if countermeasures are not taken.

20.5. XENOBIOTIC AND ENDOGENOUS SUBSTANCES

Xenobiotic substances are those that are foreign to a living system, whereas those that occur naturally in a biologic system are termed **endogenous**. The levels of an endogenous substance must usually fall within a particular concentration range in order for metabolic processes to occur normally. Levels below a normal range may result in a deficiency response or even death, and the same effects may occur above the normal range. This kind of response is illustrated in Figure 20.4.

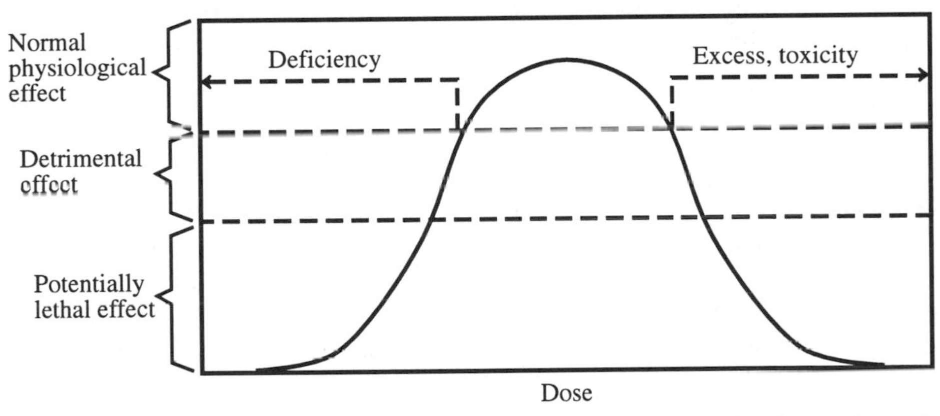

Figure 20.4. Biological effect of an endogenous substance in an organism showing optimum level, deficiency, and excess.

Examples of endogenous substances in organisms include various hormones, blood glucose, and some essential metal ions, including Ca^{2+}, K^+, and Na^+. The optimum level of calcium in human blood serum covers a narrow range of 9 – 9.5 milligrams per deciliter (mg/dL). Below these values a deficiency response known as hypocalcemia occurs, manifested by muscle cramping, and above about 10.5 mg/dL hypercalcemia occurs, the major effect of which is kidney malfunction.

20.6. TOXICOLOGICAL CHEMISTRY

Toxicological chemistry is the science that deals with the chemical nature and reactions of toxic substances, including their origins, uses, and chemical aspects of exposure, fates, and disposal.[2] Toxicological chemistry addresses the relationships between the chemical properties and molecular structures of molecules and their toxicological effects. Figure 20.5 outlines the terms discussed above and the relationships among them.

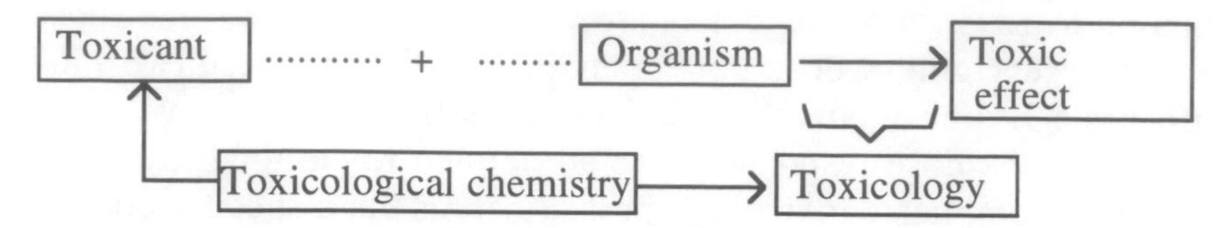

Figure 20.5. Toxicology is the science of poisons. Toxicological chemistry relates toxicology to the chemical nature of toxicants.

Toxicants in the Body

The processes by which organisms metabolize xenobiotic species are enzyme-catalyzed Phase I and Phase II reactions, which are described briefly here.

Phase I Reactions

Lipophilic xenobiotic species in the body tend to undergo **Phase I reactions** that make them more water-soluble and reactive by the attachment of polar functional groups, such as –OH (Figure 20.6). Most Phase I processes are "microsomal mixed-function oxidase" reactions catalyzed by the cytochrome P-450 enzyme system associated with the **endoplasmic reticulum** of the cell and occurring most abundantly in the liver of vertebrates.

Phase II Reactions

The polar functional groups attached to a xenobiotic compound in a Phase I reaction provide reaction sites for **Phase II reactions**. Phase II reactions are **conjugation reactions** in which enzymes attach **conjugating agents** to xenobiotics, their phase I reaction products, and non-xenobiotic compounds (Figure 20.7). The **conjugation product** of such a reaction is usually less toxic than the original xenobiotic compound, less lipid-soluble, more water-soluble, and more readily eliminated

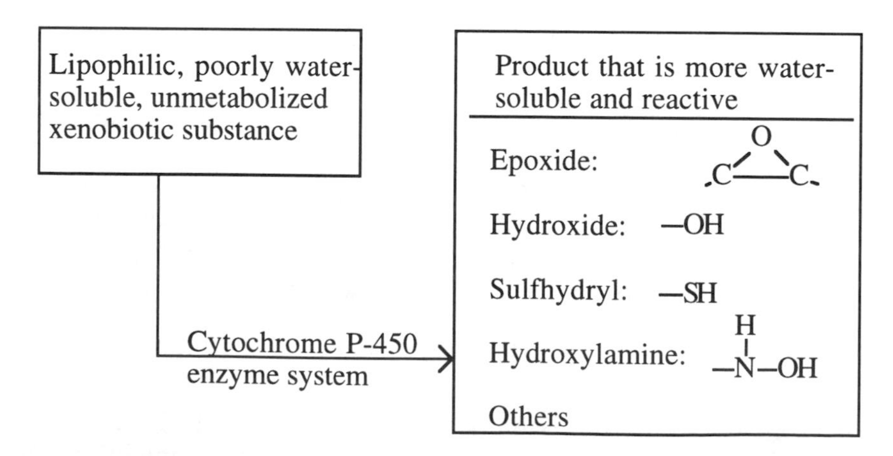

Figure 20.6. Illustration of Phase I reactions.

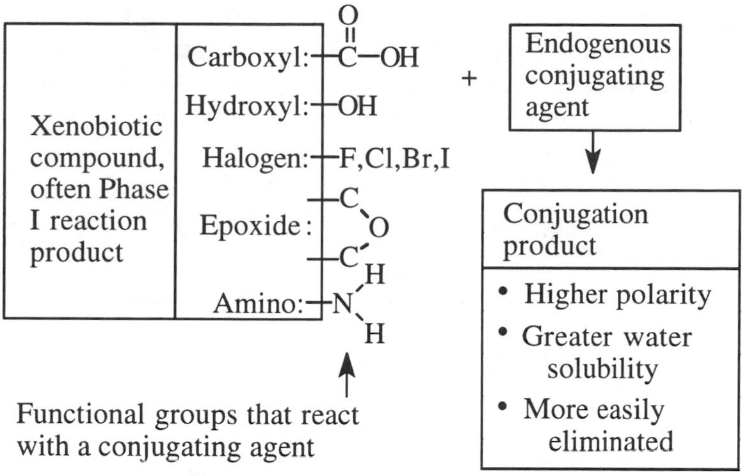

Figurc 20.7. Illustration of Phase II reactions.

eliminated from the body. The major conjugating agents and the enzymes that catalyze their Phase II reactions are glucuronide (UDP glucuronyltransferase enzyme), glutathione (glutathionetransferase enzyme), sulfate (sulfotransferase enzyme), and acetyl (acetylation by acetyltransferase enzymes). The most abundant conjugation products are glucuronides. A glucuronide conjugate is illustrated in Figure 20.8, where -X-R represents a xenobiotic species conjugated to glucuronide and R is an organic moiety. For example, if the xenobiotic compound conjugated is phenol, HXR represents the HOC_6H_5 molecule, X is the O atom, and R represents the phenyl group, C_6H_5.

20.7. KINETIC PHASE AND DYNAMIC PHASE

Kinetic Phase

The major routes and sites of absorption, metabolism, binding, and excretion of toxic substances in the body are illustrated in Figures 20.1. and 20.9. Toxicants in the body are metabolized, transported, and excreted; they have adverse biochemical effects; and they cause manifestations of poisoning. It is convenient to divide these processes into two major phases, a kinetic phase and a dynamic phase.

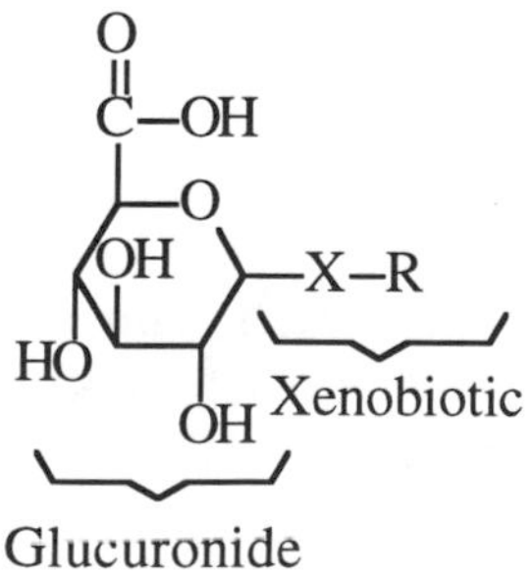

Figure 20.8. Glucuronide conjugate formed from a xenobiotic, HX-R.

In the **kinetic phase**, a toxicant or the metabolic precursor of a toxic substance (**protoxicant**) may undergo absorption, metabolism, temporary storage, distribution, and excretion, as illustrated in Figure 20.9. A toxicant that is absorbed may be passed through the kinetic phase unchanged as an **active parent compound**, metabolized to a **detoxified metabolite** that is excreted, or converted to a toxic **active metabolite**. These processes occur through Phase I and Phase II reactions discussed in the preceding section.

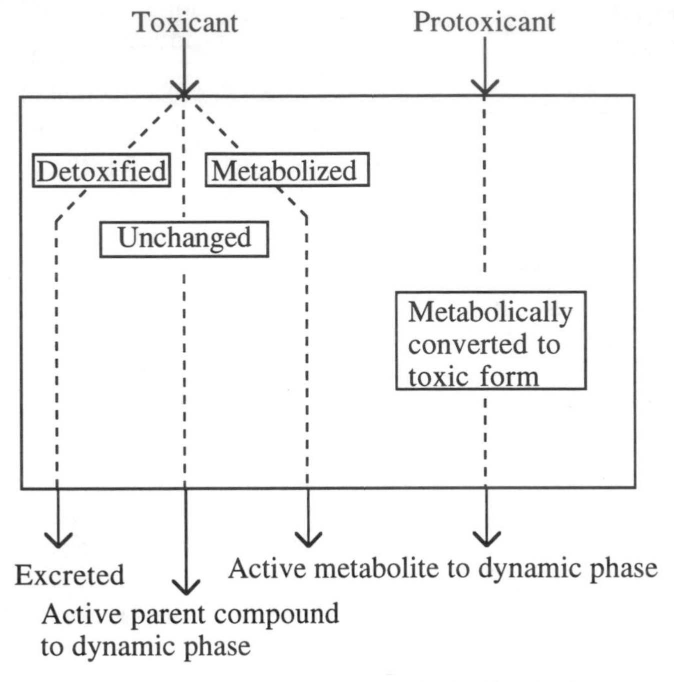

Figure 20.9. Processes involving toxicants or protoxicants in the kinetic phase.

Dynamic Phase

In the **dynamic phase** (Figure 20.10) a toxicant or toxic metabolite interacts with cells, tissues, or organs in the body to cause some toxic response. The three major subdivisions of the dynamic phase are the following:

- **Primary reaction** with a receptor or target organ
- A **biochemical response**
- **Observable effects**

Primary Reaction in the Dynamic Phase

A toxicant or an active metabolite reacts with a receptor. The process leading to a toxic response is initiated when such a reaction occurs. A typical example is when benzene epoxide produced by the metabolic oxidation of benzene,

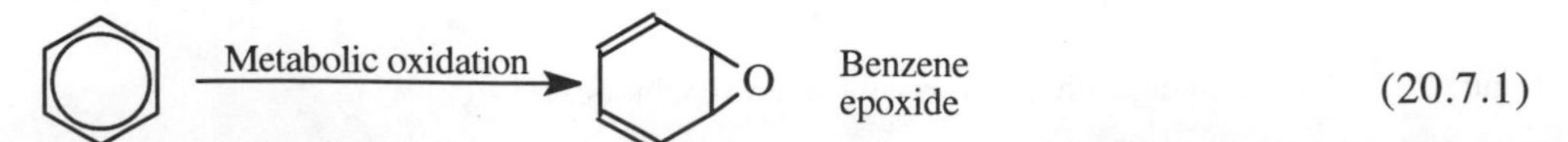

(20.7.1)

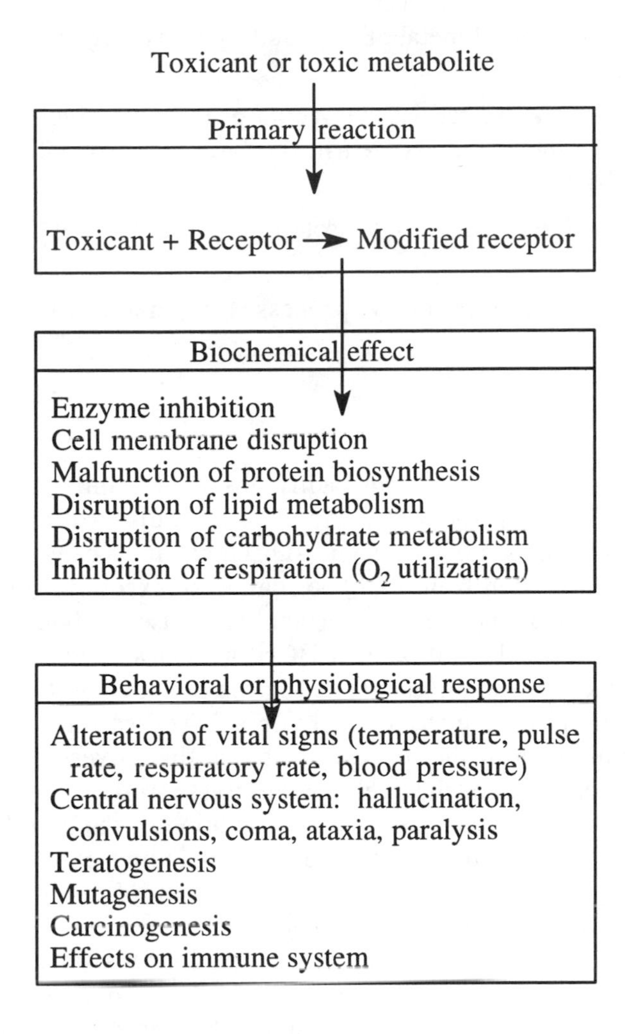

Figure 20.10. The dynamic phase of toxicant action.

forms an adduct with a nucleic acid unit in DNA (receptor) resulting in alteration of the DNA. This reaction is an **irreversible** reaction between a toxicant and a receptor. A **reversible** reaction that can result in a toxic response is illustrated by the binding between carbon monoxide and oxygen-transporting hemoglobin (Hb) in blood:

$$O_2Hb + CO \leftrightarrow COHb + O_2 \quad (20.7.2)$$

Biochemical Effects in the Dynamic Phase

The binding of a toxicant to a receptor may result in some kind of biochemical effect. The major ones of these are the following:

- Impairment of enzyme function by binding to the enzyme, coenzymes, metal activators of enzymes, or enzyme substrates
- Alteration of cell membrane or carriers in cell membranes
- Interference with carbohydrate metabolism

- Interference with lipid metabolism resulting in excess lipid accumulation ("fatty liver")
- Interference with respiration, the overall process by which electrons are transferred to molecular oxygen in the biological oxidation of energy-yielding substrates
- Stopping or interfering with protein biosynthesis by the action of toxicants on DNA
- Interference with regulatory processes mediated by hormones or enzymes.

Responses to Toxicants

Among the more immediate and readily observed manifestations of poisoning are alterations in the **vital signs** of **temperature**, **pulse rate**, **respiratory rate**, and **blood pressure**. Poisoning by some substances may cause an abnormal skin color (jaundiced yellow skin from CCl_4 poisoning) or excessively moist or dry skin. Toxic levels of some materials or their metabolites cause the body to have unnatural **odors**, such as the bitter almond odor of HCN in tissues of victims of cyanide poisoning. Symptoms of poisoning manifested in the eye include **miosis** (excessive or prolonged contraction of the eye pupil), **mydriasis** (excessive pupil dilation), **conjunctivitis** (inflammation of the mucus membrane that covers the front part of the eyeball and the inner lining of the eyelids), and **nystagmus** (involuntary movement of the eyeballs). Some poisons cause a moist condition of the mouth, whereas others cause a dry mouth. Gastrointestinal tract effects including pain, vomiting, or paralytic ileus (stoppage of the normal peristalsis movement of the intestines) occur as a result of poisoning by a number of toxic substances.

Central nervous system poisoning may cause **convulsions**, **paralysis**, **hallucinations**, and **ataxia** (lack of coordination of voluntary movements of the body), as well as abnormal behavior, including agitation, hyperactivity, disorientation, and delirium. Severe poisoning by some substances, including organophosphates and carbamates, causes **coma**, the term used to describe a lowered level of consciousness.

Chronic responses to toxicant exposure include mutations, cancer, birth defects, and effects on the immune system. Other observable effects, some of which may occur soon after exposure, include gastrointestinal illness, cardiovascular disease, hepatic (liver) disease, renal (kidney) malfunction, neurologic symptoms (central and peripheral nervous systems), skin abnormalities (rash, dermatitis).

Often the effects of toxicant exposure are subclinical in nature. These include some kinds of damage to immune system, chromosomal abnormalities, modification of functions of liver enzymes, and slowing of conduction of nerve impulses.

20.8. TERATOGENESIS, MUTAGENESIS, CARCINOGENESIS, IMMUNE SYSTEM EFFECTS, AND REPRODUCTIVE EFFECTS

Teratogenesis

Teratogens are chemical species that cause birth defects. These usually arise from damage to embryonic or fetal cells. However, mutations in germ cells (egg or sperm cells) may cause birth defects, such as Down's syndrome.

The biochemical mechanisms of teratogenesis are varied. These include enzyme inhibition by xenobiotics; deprivation of the fetus of essential substrates, such as vitamins; interference with energy supply; or alteration of the permeability of the placental membrane.

Mutagenesis

Mutagens alter DNA to produce inheritable traits. Although mutation is a natural process that occurs even in the absence of xenobiotic substances, most mutations are harmful. The mechanisms of mutagenicity are similar to those of carcinogenicity, and mutagens often cause birth defects as well. Therefore, mutagenic hazardous substances are of major toxicological concern.

Biochemistry of Mutagenesis

To understand the biochemistry of mutagenesis, it is important to recall from Chapter 3 that DNA contains the nitrogenous bases adenine, guanine, cytosine, and thymine. The order in which these bases occur in DNA determines the nature and structure of newly produced RNA, a substance produced as a step in the synthesis of new proteins and enzymes in cells. Exchange, addition, or deletion of any of the nitrogenous bases in DNA alters the nature of RNA produced and can change vital life processes, such as the synthesis of an important enzyme. This phenomenon, which can be caused by xenobiotic compounds, is a mutation that can be passed on to progeny, usually with detrimental results.

There are several ways in which xenobiotic species may cause mutations. It is beyond the scope of this work to discuss these mechanisms in detail. For the most part, however, mutations due to xenobiotic substances are the result of chemical alterations of DNA, such as those discussed in the two examples below.

Nitrous acid, HNO_2, is an example of a chemical mutagen that is often used to cause mutations in bacteria. To understand the mutagenic activity of nitrous acid it should be noted that three of the nitrogenous bases — adenine, guanine, and cytosine — contain the amino group, $-NH_2$. The action of nitrous acid is to replace amino groups with a hydroxy group. When this occurs, the DNA may not function in the intended manner, causing a mutation to occur.

Alkylation consisting of the attachment of a small alkyl group, such as $-CH_3$ or $-C_2H_5$, to an N atom on one of the nitrogenous bases in DNA is one of the most common mechanisms leading to mutation. The methylation of "7" nitrogen in guanine in DNA to form *N*-Methylguanine is shown in Figure 20.11. *O*-alkylation may also occur by attachment of a methyl or other alkyl group to guanine's oxygen atom.

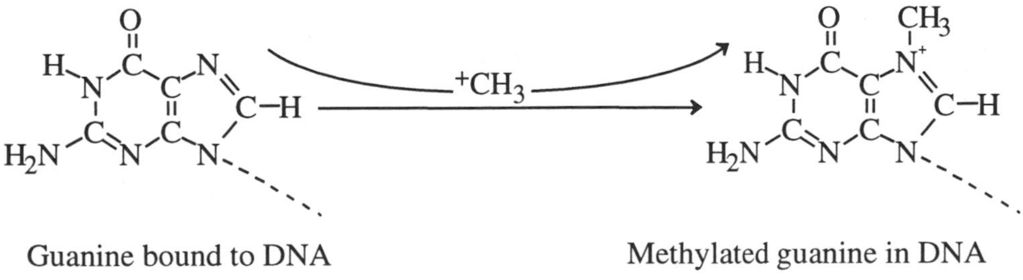

$+CH_3$

Guanine bound to DNA

Methylated guanine in DNA

Figure 20.11. Alkylation of guanine in DNA.

A number of mutagenic substances act as alkylating agents, which usually function by attaching the methyl group to one of the niltrogenous bases in DNA. Prominent among these are the compounds shown in Figure 20.12.

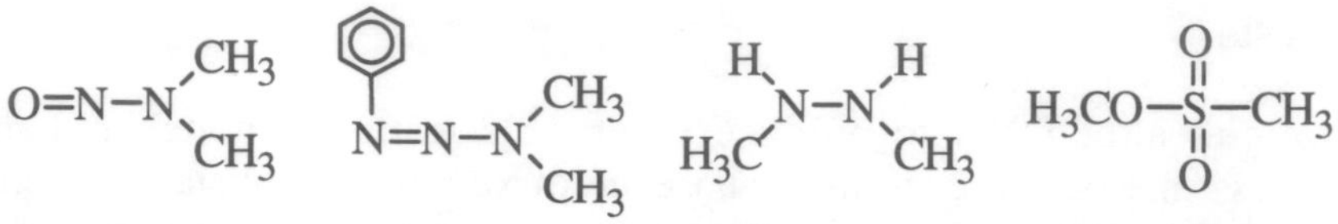

$O{=}N{-}N(CH_3)_2$	$C_6H_5{-}N{=}N{-}N(CH_3)_2$	$H_3C(H)N{-}N(H)CH_3$	$H_3CO{-}S(=O)_2{-}CH_3$
Dimethylnitros-amine	3,3-Dimethyl-1-phenyltriazine	1,2-Dimethylhydra-zine	Methylmethane-sulfonate

Figure 20.12. Examples of simple alkylating agents capable of causing mutations.

Alkylation occurs by way of generation of positively charged electrophilic species that bond to electron-rich nitrogen or oxygen atoms on the nitrogenous bases in DNA. The generation of such species usually occurs by way of biochemical and chemical processes. For example, dimethylnitrosamine (structure in Figure 20.12) is activated by oxidation through cellular NADPH to produce the following highly reactive intermediate:

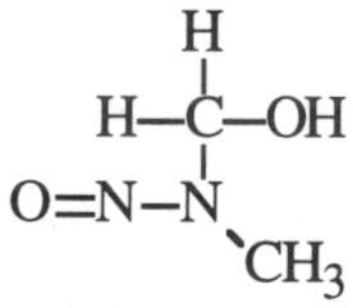

$$O{=}N{-}N(CH_3){-}CH_2OH$$

This product undergoes several nonenzymatic transitions, losing formaldehyde and generating a methyl carbocation, $^+CH_3$, that can methylate nitrogenous bases on DNA:

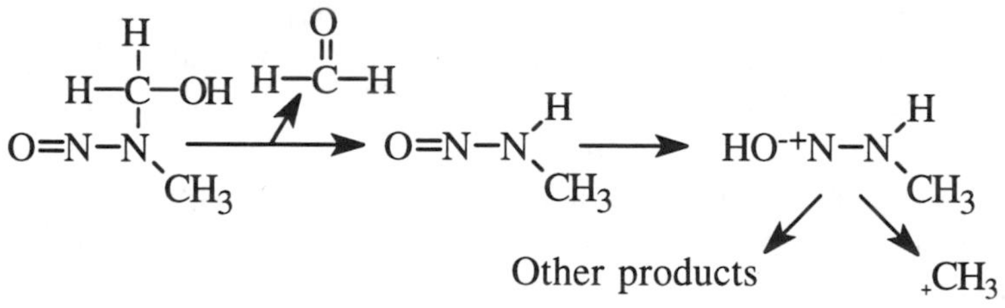

$$O{=}N{-}N(CH_3){-}CH_2OH \xrightarrow{-\,H{-}C(=O){-}H} O{=}N{-}N(H)CH_3 \longrightarrow HO^{-+}N{-}N(H)CH_3 \rightarrow \text{Other products} + {}^{+}CH_3 \quad (20.8.1)$$

One of the more notable mutagens is tris(2,3-dibromopropyl)phosphate, commonly called "tris," that was used as a flame retardant in children's sleepwear. Tris was found to be mutagenic in experimental animals and metabolites of it were found in children wearing the treated sleepwear. This strongly suggested that tris is absorbed through the skin and its uses were discontinued.

Carcinogenesis

Cancer is a condition characterized by the uncontrolled replication and growth of the body's own cells (somatic cells).[3] **Carcinogenic agents** may be categorized as follows:

- Chemical agents, such as nitrosamines and polycyclic aromatic hydrocarbons
- Biological agents, such as hepadnaviruses or retroviruses
- Ionizing radiation, such as X-rays
- Genetic factors, such as selective breeding.

Clearly, in some cases, cancer is the result of the action of synthetic and naturally occurring chemicals. The role of xenobiotic chemicals in causing cancer is called **chemical carcinogenesis**. It is often regarded as the single most important facet of toxicology and clearly the one that receives the most publicity.

Chemical carcinogenesis has a long history. As noted earlier in this chapter, in 1775 Sir Percivall Pott, Surgeon General serving under King George III of England, observed that chimney sweeps in London had a very high incidence of cancer of the scrotum, which he related to their exposure to soot and tar from the burning of bituminous coal. Around 1900 a German surgeon, Ludwig Rehn, reported elevated incidences of bladder cancer in dye workers exposed to chemicals extracted from coal tar; 2-naphthylamine,

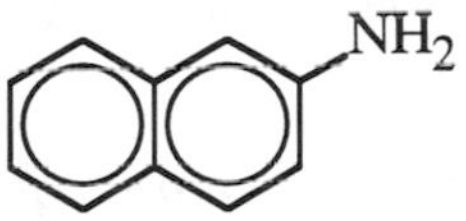

NH_2

was shown to be largely responsible. Other historical examples of carcinogenesis include observations of cancer from tobacco juice (1915), oral exposure to radium from painting luminescent watch dials (1929), tobacco smoke (1939), and asbestos (1960).

Biochemistry of Carcinogenesis

Large expenditures of time and money on the subject in recent years have yielded a much better understanding of the biochemical bases of chemical carcinogenesis. The overall processes for the induction of cancer may be quite complex, involving numerous steps. However, it is generally recognized that there are two major steps in carcinogenesis: an **initiation stage** followed by a **promotional stage**. These steps are further subdivided as shown in Figure 20.13.

Initiation of carcinogenesis may occur by reaction of a **DNA-reactive species** with DNA or by the action of an **epigenetic carcinogen** that does not react with DNA and is carcinogenic by some other mechanism. Most DNA-reactive species are **genotoxic carcinogens** because they are also mutagens. These substances react irreversibly with DNA. They are either electrophilic or, more commonly, metabolically activated to form electrophilic species, as is the case with electrophilic $^{+}CH_3$ generated from dimethylnitrosamine, discussed under mutagenesis above. Cancer-causing substances that require metabolic activation are called **procarcinogens**. The metabolic species actually responsible for carcinogenesis is termed an **ultimate carcinogen**. Some species that are intermediate metabolites between procarcinogens and ultimate carcinogens are called **proximate carcinogens**. Carcinogens that do not require biochemical activation are categorized as **primary** or **direct-acting carcinogens**. Some example procarcinogens and primary carcinogens are shown in Figure 20.14.

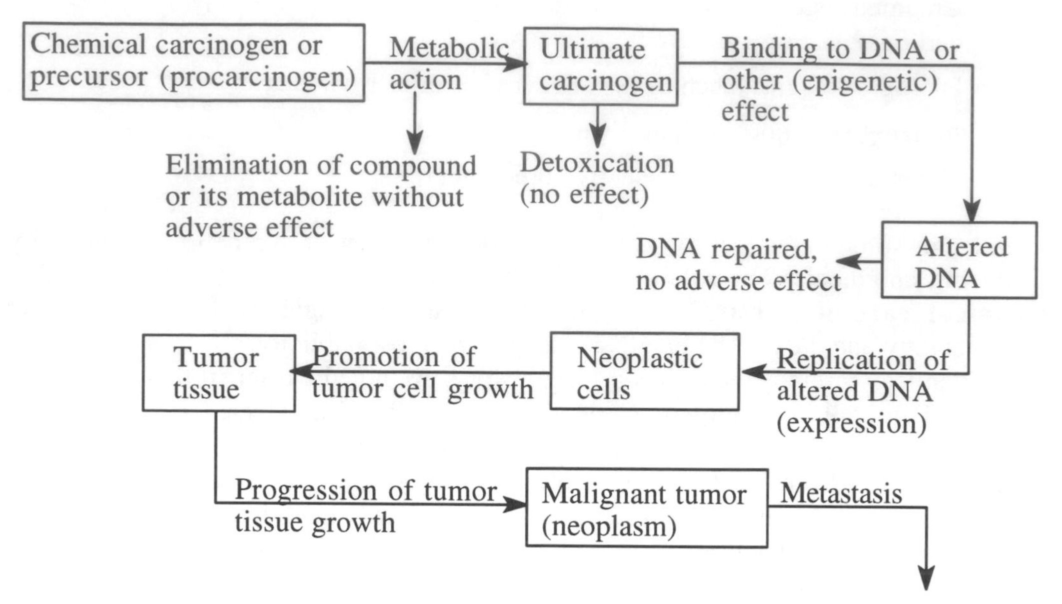

Figure 20.13. Outline of the process by which a carcinogen or procarcinogen may cause cancer.

Naturally occurring carcinogens that require bioactivation

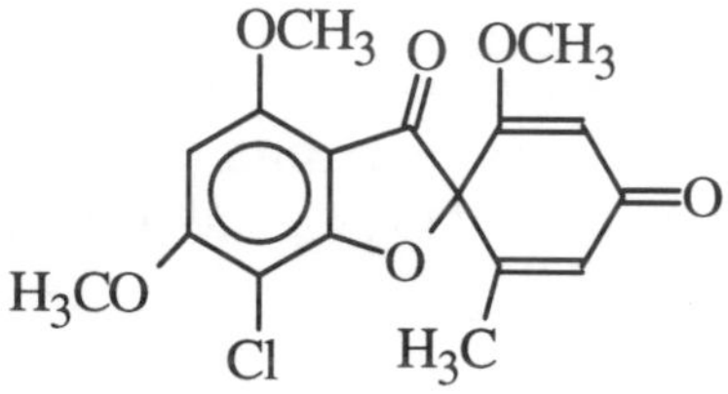

Griseofulvin (produced by *Penicillium griseofulvum*)

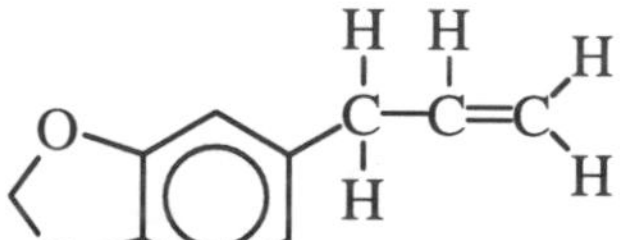

Saffrole (from sassafras)

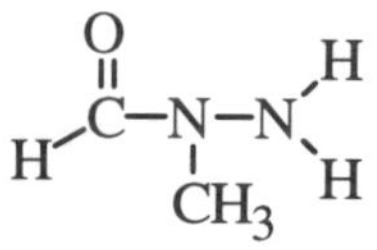

N-methyl-N-formylhydrazine (from edible false morel mushroom)

Synthetic carcinogens that require bioactivation

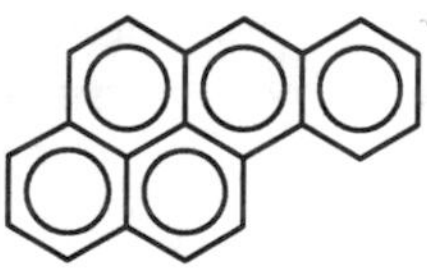

Benzo(a)pyrene

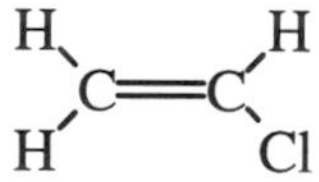

Vinyl chloride

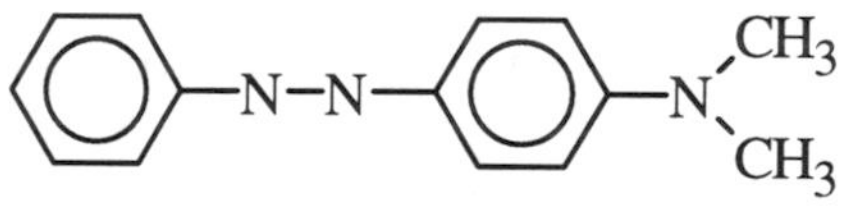

4-Dimethylaminoazobenzene

Primary carcinogens that do not require bioactivation

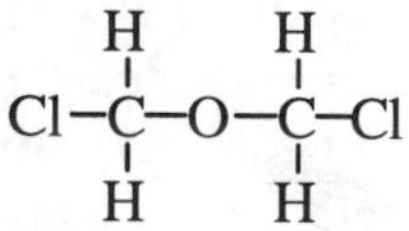

Bis(chloromethyl)-ether

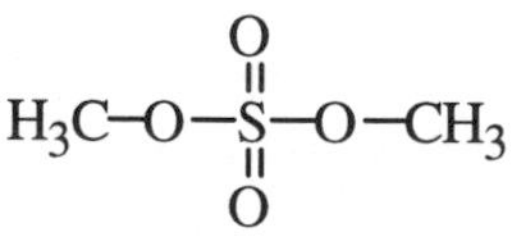

Dimethyl sulfate

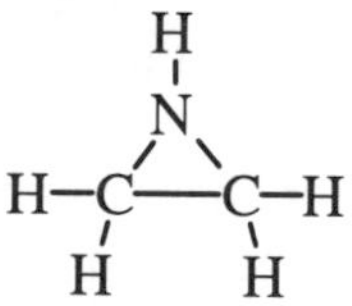

Ethyleneimine

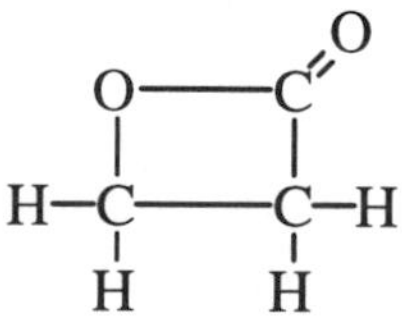

β-Propioacetone

Figure 20.14. Examples of the major classes of naturally occurring and synthetic carcinogens, some of which require bioactivation, and others of which act directly.

Most substances classified as epigenetic carcinogens are **promoters** that act after initiation. Manifestations of promotion include increased numbers of tumor cells and decreased length of time for tumors to develop (shortened latency period). Promoters do not initiate cancer, are not electrophilic, and do not bind with DNA. The classic example of a promoter is a substance known chemically as decanoyl phorbol acetate or phorbol myristate acetate, a substance extracted from croton oil.

Alkylating Agents in Carcinogenesis

Chemical carcinogens usually have the ability to form covalent bonds with macromolecular life molecules. Such covalent bonds can form with proteins, peptides, RNA, and DNA. Although most binding is with other kinds of molecules, which are more abundant, the DNA adducts are the significant ones in initiating cancer. Prominent among the species that bond to DNA in carcinogenesis are the alkylating agents which attach alkyl groups — such as methyl (CH_3) or ethyl (C_2H_5) — to DNA. A similar type of compound, **arylating agents**, act to attach aryl moieties, such as the phenyl group

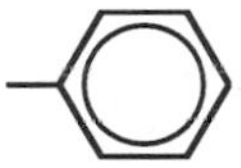

Phenyl group

to DNA. As shown by the examples in Figure 20.15, the alkyl and aryl groups become attached to N and O atoms in the nitrogenous bases that compose DNA. This alteration in the DNA can initiate the sequence of events that results in the growth and replication of neoplastic (cancerous) cells. The reactive species that donate alkyl groups in alkylation are usually formed by metabolic activation as shown for dimethylnitrosamine in the discussion of mutagenesis earlier in this section.

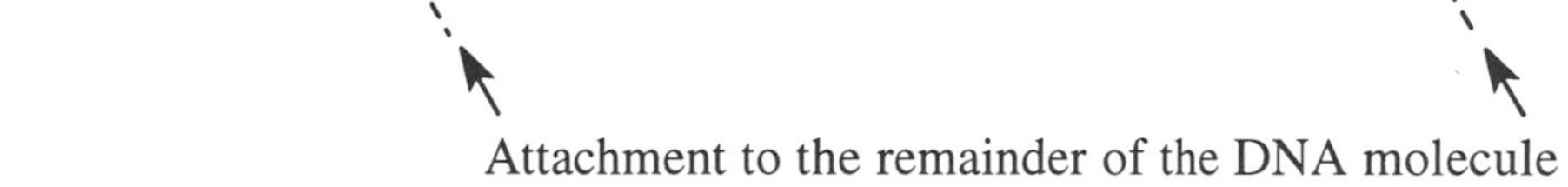

Methyl groups attached to N (left) or O (right) in guanine contained in DNA

Attachment to the remainder of the DNA molecule

Figure 20.15. Alkylated (methylated) forms of the nitrogenous base guanine.

Testing for Carcinogens

Only a few chemicals have definitely been established as human carcinogens. A well-documented example is vinyl chloride, $CH_2{=}CHCl$, which is known to have caused a rare form of liver cancer (angiosarcoma) in individuals who cleaned autoclaves in the poly(vinyl chloride) fabrication industry. In some cases chemicals are known to be carcinogens from epidemiological studies of exposed humans. Animals are used to test for carcinogenicity, and the results can be extrapolated, although with much uncertainty, to humans.

Bruce Ames Test

Mutagenicity used to infer carcinogenicity is the basis of the **Bruce Ames** test, in which observations are made of the reversion of mutant histidine-requiring *Salmonella* bacteria back to a form that can synthesize its own histidine. The test makes use of enzymes in homogenized liver tissue to convert potential procarcinogens to ultimate carcinogens. Histidine-requiring *Salmonella* bacteria are inoculated onto a medium that does not contain histidine, and those that mutate back to a form that can synthesize histidine establish visible colonies that are assayed to indicate mutagenicity.

Animal tests for carcinogens that make use of massive doses of chemicals may give results that cannot be accurately extrapolated to assess cancer risks from smaller doses of chemicals. This is because the huge doses of chemicals used kill large numbers of cells, which the organism's body attempts to replace with new cells. Rapidly dividing cells greatly increase the likelihood of mutations that result in cancer simply as the result of rapid cell proliferation, not genotoxicity.

Immune System Response

The **immune system** acts as the body's natural defense system to protect it from xenobiotic chemicals; infectious agents, such as viruses or bacteria; and neoplastic cells, which give rise to cancerous tissue. Adverse effects on the body's immune system are being increasingly recognized as important consequences of exposure to hazardous substances. Toxicants can cause **immunosuppression**, which is the impairment of the body's natural defense mechanisms. Xenobiotics can also cause the immune system to lose its ability to control cell proliferation, resulting in leukemia or lymphoma.

Another major toxic response of the immune system is **allergy** or **hypersensitivity**. This kind of condition results when the immune system overreacts to the presence of a foreign agent or its metabolites in a self-destructive manner. Among the xenobiotic materials that can cause such reactions are beryllium, chromium, nickel, formaldehyde, pesticides, resins, and plasticizers.

20.9. HEALTH HAZARDS

In recent years attention in toxicology has shifted away from readily recognized, usually severe, acute maladies that developed on a short time scale as a result of brief, intense exposure to toxicants, toward delayed, chronic, often less severe illnesses caused by long-term exposure to low levels of toxicants. Although the total impact of the latter kinds of health effects may be substantial, their assessment is very difficult because of factors such as uncertainties in exposure, low occurrence above background levels of disease, and long latency periods.

Assessment of Potential Exposure

A critical step in assessing exposure to toxic substances, such as those from hazardous waste sites is evaluation of potentially exposed populations. The most direct approach to this is to determine chemicals or their metabolic products in organisms. For inorganic species this is most readily done for heavy metals, radionuclides, and some minerals, such as asbestos. Symptoms associated with exposure to particular chemicals may also be evaluated. Examples of such effects include skin rashes or subclinical effects, such as chromosomal damage.

Epidemiological Evidence

Epidemiological studies applied to toxic environmental pollutants, such as those from hazardous wastes, attempt to correlate observations of particular illnesses with probable exposure to such wastes. There are two major approaches to such studies. One approach is to look for diseases known to be caused by particular agents in areas where exposure is likely from such agents in hazardous wastes. A second approach is to look for **clusters** consisting of an abnormally large number of cases of a particular disease in a limited geographic area, then attempt to locate sources of exposure to hazardous wastes that may be responsible. The most common types of maladies observed in clusters are spontaneous abortions, birth defects, and particular types of cancer.

Epidemiologic studies are complicated by long latency periods from exposure to onset of disease (which, in the case of cancer can be 20 years or more); lack of specificity in the correlation between exposure to a particular waste, pollutant, or substance to which exposure has taken place in the workplace; and the occurrence of a disease, and background levels of a disease in the absence of exposure to a hazardous waste capable of causing the disease.

Estimation of Health Effects Risks

An important part of estimating the risks of adverse health effects from exposure to toxicants involves extrapolation from experimentally observable data. Usually the end result needed is an estimate of a low occurrence of a disease in humans after a long latency period resulting from low-level exposure to a toxicant for a long period of time. The data available are almost always taken from animals exposed to high levels of the substance for a relatively short period of time. Extrapolation is then made using linear or curvilinear projections to estimate the risk to human populations. There are, of course, very substantial uncertainties in this kind of approach.

Risk Assessment

Toxicological considerations are very important in estimating potential dangers of pollutants and hazardous waste chemicals. One of the major ways in which toxicology interfaces with the area of hazardous wastes is in **health risk assessment**, providing guidance for risk management, cleanup, or regulation needed at a hazardous waste site based upon knowledge about the site and the chemical and toxicological properties of wastes in it. Risk assessment includes the factors of site characteristics; substances present, including indicator species; potential receptors; potential exposure pathways; and uncertainty analysis. It may be divided into the following major components:

- Identification of hazard
- Dose-response assessment
- Exposure assessment
- Risk characterization.

CHAPTER SUMMARY

The chapter summary below is presented in a programmed format to review the main points covered in this chapter. It is used most effectively by filling in the blanks, referring back to the chapter as necessary. The correct answers are given at the end of the summary.

A [1]______________________________ is a substance that, above a certain level of exposure or dose, has detrimental effects on tissues, organs, or biological processes. Acute local exposure to a toxicant occurs at a [2]__________ over a relatively short time period whereas systemic exposure occurs with toxicants that [3]____ ________________. Dose is defined as [4]______________________________ ______________________________________, whereas the observed effect of the toxicant is the [5]____________________. A plot of the percentage of organisms that exhibit a particular response as a function of dose is a [6]________________________. When death is the response measured in such a curve, the inflection point is designated as [7]_______. A response to a very low level of toxicant is known as [8]_______________, whereas the requirement of a very high dose to cause a response is called [9]____________. A substance that is normally foreign to an organism is called a [10]__________ substance. Toxicological chemistry is defined as [11]____________________ ____ ____ ____ __________. Lipophilic xenobiotic species in the body tend to undergo Phase I reactions that [12]____ ____.

Phase II reactions are [13]____________________ reactions in which enzymes attach [14]________________________ to xenobiotics, their phase I reaction products, and non-xenobiotic compounds. The product of such a reaction is usually [15]__________ water-soluble than is the original xenobiotic compound. The most abundant conjugation products are [16]______________________. The two major phases that toxicants undergo in the body are [17]________________ ______________________________________. In the first of these a toxicant or a protoxicant may undergo [18]___________________________________ ____. In the dynamic phase a toxicant or toxic metabolite interacts with [19]__________________ _______________ to cause [20]_______________________________. The three major subdivisions of the dynamic phase are [21]________________ ____.

The major kinds of biochemical effects caused by binding of a toxicant to a receptor are [22]____ ____ ____ ____ ____.

Vital signs affected by toxicants are [23]__________________________ ____.

Teratogens are chemical species that cause [24]______________________.

Mutagens alter [25]___________ to produce [26]___________________________. The term that applies to the role of substances foreign to the body in causing the uncontrolled cell replication commonly known as cancer is [27]_______________ ____________________. The two major steps in the overall processes by which xenobiotic chemicals cause cancer are [28]__________________________________ _________________________. Many chemical carcinogens are [29]_____________ __________________, which act to attach alkyl groups. Chemical substances that cause cancer directly are called [30]__________________________________ ____________________, whereas most xenobiotics involved in causing cancer are [31]__________________________________. Mutagenicity used to infer carcinogenicity is the basis of the [32]__________________ test, which makes use of the organism [33]____. Toxicants can cause [34]______________________________, which is the impairment of the body's natural defense mechanisms. Another adverse response of the immune system that can be caused by exposure to xenobiotic compounds is [35]___ ____________________. Epidemiologic studies applied to toxic environmental pollutants attempt to correlate observations of particular illnesses with probable exposure to such wastes, or to look for [36]______________________________ ____________________________.

Answers

1 poison or toxicant

2 specific location

3 enter an organism and are distributed around inside it

4 the degree of exposure of an organism to a toxicant

5 response

6 dose-response curve

7 LD_{50}

8 hypersensitivity

9 hyposensitivity

10 xenobiotic

11 the science that deals with the chemical nature and reactions of toxic substances, including their origins, uses, and chemical aspects of exposure, fates, and disposal

12 make them more water-soluble and reactive by the attachment of polar functional groups, such as –OH

13 conjugation

14 conjugating agents

15 more

16 glucuronides

17 the kinetic phase and the dynamic phase

18 absorption, metabolism, temporary storage, distribution, and excretion

19 cells, tissues, or organs in the body

20 some toxic response

21 primary reaction, biochemical response, and observable effects

22 impairment of enzyme function, alteration of cell membrane, interference with carbohydrate metabolism, interference with lipid metabolism, interference with respiration, stopping or interfering with protein biosynthesis, and interference with regulatory processes mediated by hormones or enzymes

23 temperature, pulse rate, respiratory rate, and blood pressure

24 birth defects

25 DNA

26 inheritable traits

27 chemical carcinogenesis

28 an initiation stage followed by a promotional stage

29 alkylating agents

30 primary or direct-acting carcinogens

31 precarcinogens or procarcinogens

32 Bruce Ames

33 *Salmonella* bacteria

34 immunosuppression

35 hypersensitivity

36 clusters consisting of an abnormally large number of cases of a particular disease in a limited geographic area

QUESTIONS AND PROBLEMS

1. How are conjugating agents and Phase II reactions involved with some toxicants?
2. What are Phase I reactions? What enzyme system carries them out? Where is this enzyme system located in the cell?
3. Name and describe the science that deals with the chemical nature and reactions of toxic substances, including their origins, uses, and chemical aspects of exposure, fates, and disposal.
4. What is a dose-response curve?
5. What is meant by a toxicity rating of 6?
6. What are the three major subdivisions of the *dynamic phase* of toxicity, and what happens in each?
7. Characterize the toxic effect of carbon monoxide in the body. Is its effect reversible or irreversible? Does it act on an enzyme system?

8. Of the following, choose the one that is **not** a biochemical effect of a toxic substance: (a) impairment of enzyme function by binding to the enzyme, (b) alteration of cell membrane or carriers in cell membranes, (c) change in vital signs, (d) interference with lipid metabolism, (e) interference with respiration.

9. Distinguish among teratogenesis, mutagenesis, carcinogenesis, and immune system effects. Are there ways in which they are related?

10. As far as environmental toxicants are concerned, compare the relative importance of acute and chronic toxic effects and discuss the difficulties and uncertainties involved in studying each.

11. What are some of the factors that complicate epidemiologic studies of toxicants?

12. Match the type of exposure on the left, below, with its description or example from the right.

( ) Acute local	1. Cancer of the mouth at age 31 by a person who has been using snuff (chewing tobacco) since age 9
( ) Chronic systemic	2. Bladder cancer developed by coal tar dye workers after many years of work
( ) Acute systemic	3. Fatal dose of cyanide
( ) Chronic local	4. Occurs at a specific location over a time period of a few seconds to a few hours and may affect the exposure site, particularly the skin, eyes or mucous membranes

13. A distinction was made between the kinetic phase and dynamic phase of toxicology. Of the following, the untrue statement pertaining to these is:

( ) In the dynamic phase the toxicant reacts with a receptor or target organ in the primary reaction step.

( ) The kinetic phase involves absorption, metabolism, temporary storage, distribution, and, to a certain extent, excretion of the toxicant or its precursor compound called the protoxicant.

( ) In the dynamic phase there is a biochemical response.

( ) Following the biochemical response, physiological and/or behavioral manifestations of the effect of the toxicant may occur.

( ) Phase 2 reactions commonly occur in the dynamic phase.

14. Pertaining to the major sites of exposure, metabolism, and storage, routes of distribution and elimination of toxic substances from the body as shown in Figure 20.1, the true statement of the following is:

( ) The most likely place to find an unmetabolized, water-insoluble xenobiotic compound several weeks after it was absorbed through the skin is in the liver.

( ) The most likely place for loss from the body of a non-volatile, Phase 2 metabolite of a xenobiotic compound is through urine.

( ) There are no barriers that could prevent a cancer-causing metabolite of a procarcinogen from reaching DNA that it might affect.

( ) An entry site that presents the least barriers, either from a screening organ or from barriers inherent to the entry site, itself, is the lung.

( ) Systemic poisons are most likely to affect the skin at the point where they are absorbed.

15. Why were the genital areas of chimney sweeps in England during the 1700s particularly susceptible to cancer caused by coal tar?
16. Substance "A" and substance "B" taken separately have only minimal toxic effects, whereas taken together at similar dose levels, severe adverse effects occur. What sort of phenomenon is illustrated by this example in relation to toxic substances?
17. In what sense are immunosuppression and hypersensitivity opposites?
18. What is the distinction between epigenetic carcinogens and genotoxic carcinogens?
19. Toxicologically, what is the distinction between a detoxified metabolite and an active metabolite?
20. What are some examples of materials that have been shown directly to produce cancer in humans?

LITERATURE CITED

[1] Butterworth, Frank M., Lynda D. Corkum, and ju Guzman-Rinco, *Biomonitors and Biomarkers As Indicators of Environmental Change: A Handbook*, Plenum Publishing Corporation, Edison, NJ, 1995.

[2] Manahan, Stanley E., *Toxicological Chemistry*, 2nd ed., Lewis Publishers/CRC Press, Boca Raton, FL, 1992.

[3] Lemoine, Nicholas R., John Neoptolemos, and Timothy Cooke, *Cancer: A Molecular Approach*, Blackwell Science Inc., Cambridge, MA, 1994.

SUPPLEMENTARY REFERENCES

Allwood, Michael, Andrew Stanley, and Patricia Wright, Eds., *The Cytotoxics Handbook*, American Pharmaceutical Association, Washington, D.C., 1997.

Ballantyne, Bryan, Timothy Marrs, and Paul Turner, *General and Applied Toxicology: College Edition*, Stockton Press, New York, NY, 1995.

Chang, Louis W., *Toxicology of Metals*, CRC Press/Lewis Publishers, Boca Raton, FL, 1996.

Cockerham, Lorris G. and Barbara S. Shane, *Basic Environmental Toxicology*, CRC Press/Lewis Publishers, Boca Raton, FL, 1994.

Cooper, Andre R., *Cooper's Toxic Exposures Desk Reference*, CRC Press/Lewis Publishers, Boca Raton, FL, 1997.

Draper, William M., Ed., *Environmental Epidemiology*, American Chemical Society, Washington, D.C., 1994.

Hall, Steven K., Joanna Chakraborty, and Randall Ruch, *Chemical Exposure and Toxic Responses*, CRC Press/Lewis Publishers, Boca Raton, FL, 1997.

Landis, Wayne G. and Ming-Ho Yu, *Introduction to Environmental Toxicology*, CRC Press/Lewis Publishers, Boca Raton, FL, 1995.

Lewis, Robert A., *Lewis' Dictionary of Toxicology*, CRC Press/Lewis Publishers, Boca Raton, FL, 1997.

Malachowski, M. J., *Health Effects of Toxic Substances*, Government Institutes, Rockville, MD, 1995.

McClellan, Roger O., Ed., *Critical Reviews in Toxicology* (journal), CRC Press/Lewis Publishers, Boca Raton, FL.

Rea, William H., *Chemical Sensitivity*, Volume 1, *Principles and Mechanisms*, CRC Press/Lewis Publishers, Boca Raton, FL, 1992.

Rea, William H., *Chemical Sensitivity*, Volume 2, *Sources of Total Body Load*, CRC Press/Lewis Publishers, Boca Raton, FL, 1994.

Rea, William H., *Chemical Sensitivity*, Volume 3, *Clinical Manifestations of Pollutant Overload*, CRC Press/Lewis Publishers, Boca Raton, FL, 1995.

Rea, William H., *Chemical Sensitivity*, Volume 4, *Tools for Diagnosis and Methods of Treatment*, CRC Press/Lewis Publishers, Boca Raton, FL, 1997.

Revoir, William H. and Ching-tsen Bien, *Respiratory Protection Handbook*, CRC Press/Lewis Publishers, Boca Raton, FL, 1997.

Revoir, William H. and Ronald W. Michaud, *Principles of Environmental, Health and Safety Management*, Government Institutes, Inc., Rockville, MD, 1995.

Zakrzewski, Sigmund F., *Principles of Environmental Toxicology*, American Chemical Society, Washington, D.C., 1997.

TECHNOLOGY

21 TECHNOLOGY, MANUFACTURING, TRANSPORTATION, AND COMMUNICATION

21.1. INTRODUCTION

Technology, manufacturing, and transportation, which were introduced in Chapter 4, "Technology and Engineering in Environmental Science," are closely interrelated areas that have a tremendous impact upon the environment. This chapter addresses these human activities as a group and as interrelated aspects of the environment in which they function.

As noted in Chapter 4 and discussed in more detail in Section 21.12, **industrial ecology** is an emerging discipline of particular importance for modern society.[1] Industrial ecology operates at the interface of industry with various other spheres of the environment.[2] The practitioner of industrial ecology deals with optimization of the complete cycle of materials from virgin raw material through finished material, components, products, use, obsolescence, and finally disposal or recycle.[3] A primary objective of industrial ecology is to sustain Earth's carrying capacity through the deliberate and rational evolution of culture, economics, and technology. The sucessful application of industrial ecology should enable practices in the anthrosphere to be carried out in a manner that does not put intolerable strain on Earth's support systems. This includes minimization of uses of nonrenewable resources and prevention of release of toxic and pollutant materials.

Life cycle analysis consists of a careful examination of the whole life cycle of a material with a view toward making this cycle as environmentally friendly as possible. Lead-containing materials, such as lead storage batteries, are good candidates for life cycle analysis. Numerous factors are involved in designing materials, processes, and products to be as environmentally friendly as possible.[4] Manufacturing processes can be chosen and designed for lowest possible energy consumption. Materials can be chosen for minimum environmental impact in their extraction, manufacture, and disposal. Special consideration can be given to using materials for optimum recycling. Packaging can receive special consideration with respect to its disposal. Processes can be modified to consume less water and to produce less wastewater, and they can be re-engineered to produce less manufacturing process residue. It is important to reduce the overall production of and risk from pollutants, not simply move them from one medium to another.

Basically, industrial ecology is about setting priorities and making choices pertaining to industrial operations in a manner consistent with environmental protection. An important choice is that of materials; different materials have different kinds and degrees of environmental impacts in their extraction and disposal. Selections must be made among different manufacturing processes. Various combinations of choices of materials and manufacturing technologies have different effects on material sources and energy consumption, as well as different impacts on the water, air, terrestrial, and living environments.

21.2. TECHNOLOGY

As defined in Chapter 4, **technology** refers to the ways in which humans do and make things with materials and energy. During the last two centuries, technology has been driven by engineering, which in turn has been based on an ever increasing knowledge base in the sciences. Technology makes use of tools, techniques, and systems for carrying out specific objectives, such as manufacturing consumer goods; moving people, manufactured goods, and materials; or minimizing environmental impact of human activities. At an accelerated pace during the last several years, technology has made use of computers to accomplish its objectives.

An example of how technology is applied is provided by emission control for automobiles. Since the late 1960s, the objective in automotive pollution control has been to reduce emissions of toxic carbon monoxide, as well as mixtures of nitrogen oxides and hydrocarbons that cause formation of photochemical smog. In earlier years, the technological approach had been to adapt existing hardware to newly manufactured automobiles to reduce emissions. The first measures used were largely plumbing modifications, such as those employed for positive crankcase ventilation, and they provided significantly reduced emissions for relatively low costs. However, as emission standards became more stringent, the costs and complexities of automotive pollution control increased, and a substantial penalty was realized in fuel consumption. More recently, modified approaches have emerged based upon scientific knowledge of fundamental combustion processes. Using sophisticated computer control of such parameters as timing and air/fuel mixtures, these scientifically based measures have enabled much reduced emissions with enhanced fuel economy.

Several aspects of technology should be emphasized here. It is important to realize that technology has a tremendous ability to increase productivity. Therefore, improved technology in areas such as manufacturing and agriculture has been largely responsible for the accelerating production of goods and services that has been characteristic of the last two centuries. In so doing, technology has resulted in the displacement of entire occupations and the creation of entirely new ones. An examination of technology and all of the aspects of the modern world that it influences clearly illustrates its role in the **interconnectedness** of important global and societal issues, such as productivity, resource utilization, environmental impact, employment, and overpopulation.

In general, more intense application of technology has resulted in environmental degradation through release of water and air pollutants, production of hazardous wastes, and extraction of mineral and energy resources. Increasingly, during recent years, however, technology has been directed toward environmental improvement. It is absolutely essential for the well-being of human kind that this trend continue at an accelerating pace.

21.3. AGRICULTURE—THE MOST BASIC INDUSTRY

Agriculture, the production of food by growing crops and livestock, provides for the most basic of human needs. The topic of agriculture was introduced in Chapter 4, and aspects of it were addressed in Chapter 14, "Soil."

No other industry impacts as much as agriculture does on the environment. Agriculture is absolutely essential to the maintenance of the huge human populations now on Earth. The displacement of native plants, destruction of wildlife habitat, erosion, pesticide pollution, and other environmental aspects of agriculture have enormous potential for environmental damage and demand that agricultural practice be as environmentally friendly as possible. On the other hand, growth of domestic crops removes (at least temporarily) greenhouse-gas carbon dioxide from the atmosphere and provides renewable sources of energy and fiber that can substitute for petroleum-derived fuels and materials.

Agriculture can be divided into the two main categories of **crop farming**, in which plant photosynthesis is used to produce grain, fruit, and fiber, and **livestock farming**, in which domesticated animals are grown for meat, milk, and other animal products. The major divisions of crop farming include production of cereals, such as wheat, corn, or rice; animal fodder, such as hay; fruit; vegetables; and specialty crops, such as sugarcane, sugar beets, tea, coffee, tobacco, cotton, and cacao. Livestock farming involves raising of cattle, sheep, goats, swine, asses, mules, camels, buffalo, and various kinds of poultry. In addition to meat, livestock produce dairy products, eggs, wool, and hides. Freshwater fish and even crayfish are raised on "fish farms." Beekeeping provides honey.

Agriculture is based on domestic plants engineered by early farmers from their wild plant ancestors. Without perhaps much of an awareness of what they were doing, early farmers selected plants with desired characteristics for the production of food. This selection of plants for domestic use brought about a very rapid evolutionary change, so profound that the products often barely resemble their wild ancestors. Plant breeding based on scientific principles of heredity is a very recent development dating from early in the present century. One of the major objectives of plant breeding has been to increase yield. An example of success in this area is the selection of dwarf varieties of rice, which yield much better and mature faster than the varieties that they replaced. Such rice was largely responsible for the "green revolution" dating from about the 1950s. Yields can also be increased by selecting for resistance to insects, drought, and cold. In some cases the goal is to increase nutritional value, such as in the development of corn high in lysine, an amino acid essential for human nutrition.

The development of hybrids has vastly increased yields and other desired characteristics of a number of important crops. Basically, **hybrids** are the offspring of crosses between two different **true-breeding** strains. Often quite different from either parent strain, hybrids tend to exhibit "hybrid vigor" and to have significantly higher yields. The most success with hybrid crops has been obtained with corn (maize). Corn is one of the easiest plants to hybridize because of the physical separation of the male flowers, which grow as tassels on top of the corn plant, from female flowers, which are attached to incipient ears on the side of the plant. Despite past successes, application of recombinant DNA technology may eventually overshadow all the advances in plant breeding made to date.

In addition to plant strains and varieties, numerous other factors are involved in crop production. Weather is an obvious factor, and shortages of water, chronic in many areas of the world, are mitigated by irrigation. Here automated techniques and computer control (see Section 21.6) are beginning to play an important, often more environmentally friendly, role by minimizing the quantities of water required. The application of chemical fertilizer has vastly increased crop yields. The judicious application of pesticides, especially herbicides, but including insecticides and fungicides as well, has increased crop yields and reduced losses greatly. Use of herbicides has had an environmental benefit in reducing the degree of mechanical cultivation of soil required. Indeed, "no-till" and "low-till" agriculture are now widely practiced on some crops.

The crops that provide for most of human caloric food intake, as well as much food for animals, are **cereals**, which are harvested for their starch-rich seeds. In addition to corn, mentioned above, wheat used for making bread and related foods, and rice consumed directly, other major cereal crops include barley, oats, rye, sorghum, and millets.

As applied to agriculture and food, **vegetables** are plants or their products that can be eaten directly by humans. A large variety of different parts of plants are consumed as vegetables. These include leaves (lettuce), stems (asparagus), roots (carrots), tubers (potato), bulb (onion), immature flower (broccoli), immature fruit (cucumber), mature fruit (tomato), and seeds (pea). According to this system, fruits, which are bodies of plant tissue containing the seed, are a subclassification of vegetables. A tremendous variety of vegetables are grown and consumed. Some others in addition to those mentioned above include beets, cabbage, celery, leek, pepper, pumpkin, spinach, squash, and watermelon.

For the most part, though not invariably, fruits and nuts (which are fruits, botanically) grow on trees. The ubiquitous peanut is actually a legume. Common fruits include apple, peach, apricot, citrus (orange, lemon, lime, grapefruit), banana, cherry, and various kinds of berries. A truly vast variety of nuts is grown. Among the exotic kinds of nuts are the bambarra groundnut from tropical Africa, the Chile hazel from Chile, the dika nut from West Africa, the pili nut from the Pacific tropics, and the yeheb nut from Somalia. In addition to consumption as food, nuts are also used for oil, ink, varnish, spices, ornamentals, soap substitutes, polishing (ground walnut shells), tanning, and poisons.

21.4. RAW MATERIALS AND MINING

Manufacturing requires a steady flow of raw materials—minerals, fuel, wood, and fiber. These can be provided from either **extractive** (nonrenewable) and **renewable** sources, both of which are discussed briefly in this section.

The extractive industries are those which take irreplaceable resources from the Earth's crust. This is normally done by mining, but may also include pumping petroleum and withdrawal of natural gas. The raw materials extracted may be divided broadly into the categories of inorganic minerals, such as iron ore, clay used for firebrick, and gravel, and materials of organic origin, such as coal, lignite, or petroleum. Inorganic minerals that are mined include aluminum (from bauxite), antimony, asbestos, barite, beryl, bismuth, cadmium, chromite, cobalt, columbium-tantalum, copper, gem and industrial diamond, feldspar, fluorspar, gold, gypsum, iron ore, lead, magnesium, manganese, mercury, molybdenum, nickel, phosphate rock (hydroxyapatite and fluorapatite), platinum group metals, potash, selenium, silver, tellurium, tin, titanium, tungsten, uranium oxide, vanadium, and zinc.

Geological and geochemical factors are crucial in mining, particularly in locating ore deposits (see Chapters 13, 15, and 16). Deposits of metals often occur in masses of igneous rock that have been forced in a solid or molten state into the surrounding rock strata; such masses are called batholiths. Other geological factors to consider include age of rock, fault zones, and rock fractures. The crucial step of finding ore deposits falls in the category of mining geology.

Surface mining, which for coal is usually called strip mining, is used to extract minerals that occur near the surface. A common example of surface mining is quarrying of rock. In some cases where stone is mined for construction, quarrying is done by cutting the stone into slabs and blocks. Most rock is loosened by blasting before being removed as irregular chunks. Although quarry blasting used to be done by expensive and potentially hazardous dynamite or gunpowder, it is now accomplished primarily by granular ammonium nitrate soaked with fuel oil. Though powerful, this explosive is particularly safe for the user because it requires a booster in addition to a simple blasting cap for detonation. Sand and gravel are sometimes surface-mined by simply digging the material from deposits. Coal is commonly strip-mined with giant shovels that are employed primarily for removing overburden, leaving the thinner coal seam to be loaded with smaller equipment.

Gravel, sand, and some other minerals, such as gold, often occur in so-called placer deposits to which they have been carried and deposited by running water. In such cases mining can be accomplished by dredging from a boom-equipped barge. Another approach that can be used is hydraulic mining with large streams of water. One interesting approach for more coherent deposits is to cut the ore with intense water jets, then suck up the resulting small particles with a pumping system.

For many minerals underground mining is the only practical means of extraction. An underground mine can be very complex and sophisticated. The structure of the mine depends upon the nature of the deposit. It is, of course, necessary to have a shaft that reaches to the ore deposit. This is normally a vertical shaft, but for inclined beds it can be inclined. Horizontal tunnels extend out into the deposit, and provision must be made for sumps to remove water and for ventilation. In designing an underground mine many factors must be considered by the mine engineer. These include the depth, shape, and orientation of the ore body, as well as the nature and strength of the rock in and around it; thickness of overburden; and depth below the surface. Of crucial importance is support of the mine roof. In some cases roof support is accomplished with a room and pillar technique in which pillars of ore are left to support the roof. Bolts may have to be put into the roof to secure it, and posts may have to be installed to hold up the roof. In stabilizing mine structures it is important to apply rock mechanics to determine the behavior of rock under stress and how such behavior is altered by mining. The compressive and tensile strengths of the rock must be considered, along with the fact that rock behaves elastically when a load is placed on it.

Usually, significant amounts of processing are required before a mined product is used or even moved from the mine site. Even rock to be used for aggregate and for road construction must be crushed and sized. Crushing is a necessary first step for further processing of ores. Some minerals occur to an extent of a few percent or even less in the rock taken from the mine and must be concentrated on site so that the residue does not have to be hauled far. For metals mining, these processes, as well as roasting, extraction, and similar operations are covered under the category of **extractive metallurgy**.

21.5. MATERIALS SCIENCE

Materials science deals with the composition, properties, and applications of substances used to make devices, machines, and structures. All kinds of materials may be used for various purposes, including metals, plastics, wood, glass, concrete, and ceramics. A vast variety of characteristics of materials must be considered, including strength, weight, fire resistance, and costs. A particularly important consideration in materials science has become the application of industrial ecology in closing materials cycles to eliminate wastes and pollutants.[5]

Polymers

To a very large extent materials consist of polymers made up of smaller monomer molecules (see Chapter 2). There are many kinds of polymers, both natural and synthetic. Among the important natural polymers classified as "materials" are cellulose and lignin in wood, cotton, and rubber. Synthetic polymers include plastics (such as polyethylene), fibers used in textiles, synthetic rubber, and epoxy resins (widely used in composites, see below). One special class of polymers consists of **elastomers** that readily flex and stretch, like rubber. Many polymers are used to make fibers, which can be woven into fabrics. Premium plastics are polymers that have especially outstanding properties, such as high strength or resistance to heat, abrasion, or chemical attack.

Ceramics

One of the fastest growing areas of materials science is ceramics. **Ceramics** are inorganic substances that usually involve ionically bound constituents, silicon, and oxygen, and that are usually formed by high temperature processes. Ceramics may contain aluminum as well as carbides, nitrides, and borides. Ceramics tend to be hard, rigid, and resistant to high temperatures and chemical attack. Because of their desirable properties, they are finding increasing application in a number of areas. The major raw materials used to make ceramics are clays, of which the main types are montmorillonite ($Al_2(OH)_2Si_4O_{10}$), illite ($K_{0\text{-}2}Al_4(Si_{8\text{-}6}Al_{0\text{-}2})O_{20}(OH)_4$,) and kaolinite ($Al_2Si_2O_5(OH)_4$); silica ($SiO_2$); feldspar, such as potassium feldspar, $KAlSi_3O_8$; and various synthetic chemicals.

Ceramics are used for many purposes, including refractory bricks and linings, crucibles, furnace tubes, and as composites with metals. One of the major applications of ceramics is as electrical materials. Some ceramics are excellent insulators, and maintain their insulating properties even at high temperatures. Other ceramics, such as silicon carbide, are semiconductors that are used in transistors, rectifiers, photocells, thermistors, and electrical heating elements. Other materials can be blended with ceramics to give composites that retain the desirable properties of ceramics, such as their thermal resistance, while improving other characteristics, such as resistance to mechanical stress.

Composites

An important class of materials with growing uses consists of **composites** composed of two or more materials to give a material that has better overall properties of strength, density, cost, or electrical resistance/conductivity than their individual constituents. The materials in composites may be organic polymers, inorganic substances, metals, or metal alloys shaped as fibers, rods, particles, porous solids, sheets, or other forms. Composites are usually composed of high-strength reinforcing materials imbedded in some sort of matrix. The matrix can be a metal, plastic, ceramic, or other moldable material. Among the reinforcing materials used are fibers composed of substances such as glass, carbon, or boron nitride, or larger filaments of metal wires, silicon carbide, or other materials.

21.6. AUTOMATION

As noted in Section 4.9, **automation** uses machines working automatically and repetitively in tasks repeated multiple times. Automation is most widely employed on assembly lines. An automated operation is directed by a **control system**, which provides the desired response as a function of time or location. In a relatively simple **open-loop system**, information regarding the desired output is fed to a controller (control actuator) that directs a process to provide the desired response. In a more sophisticated **closed-loop system** the output is measured and compared to the desired response for feedback.

As shown in Figure 21.1, modern automated systems make use of **feedback control** systems containing the following components: (1) input in the form of a **reference value** or **set point** with which output is to be matched, for example, a pH needed for a wastewater stream fed into a bioreactor designed to degrade the wastes in the stream. The pH desired can be that of a standard buffer solution (see Section 2.9); (2) the process being controlled, for example the addition of acid or base as needed for pH adjustment; (3) output, such as the actual pH of the wastewater stream; (4) sensing element, such as a glass electrode that produces an electrical potential indicative of the pH of the stream in which it is immersed; and (5) control mechanism, such as valves attached to tanks of acid and base solution and actuated by small electric motors and servomechanisms.

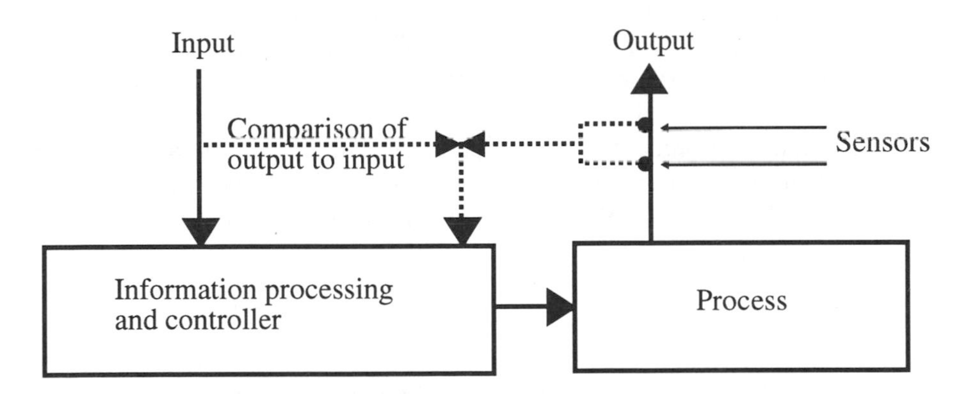

Figure 21.1. Major components of a feedback control system.

In all but the most simple modern automated systems computers are used for control. This enables complex sets of instructions to be given and simultaneous control of a number of interrelated factors. An important aspect of more sophisticated control is the capability for **decision making** based upon computer processing of multiple kinds of information. This can serve a number of purposes, including correction of errors, maintenance of safe conditions, and process optimization. The decision-making process enables human interaction as well by sensing undesirable changes and correcting for them, a process termed *error detection and recovery*.

Automation has been most widely applied in manufacturing and assembly. In this application there are two major approaches. The first of these is **fixed automation** in which the mechanical design of the equipment determines the motions by means of hardware (cams and gears) and wiring, semipermanent features of the equipment that cannot be changed except by rebuilding the device or replacing parts on it. Fixed automation is most applicable to large numbers of repetitive processes that take place over a long time period. **Programmable automation** is used to give a machine instructions that can be changed for different purposes. An advanced form of this type of automation is **flexible automation**, in which the machine is automatically reprogrammed for different tasks on demand, such that it can, for example, do different assembly tasks as different items reach it on an assembly line.

Automation has found a large number of applications in industry. The first of these was in machining, the operations by which metal is shaped by cutting, grinding, and drilling. Chemical synthesis and production is especially amenable to automation because of the relative ease by which flows of materials and heat and parameters such as pressure or pH are controlled or sensed. Electronics manufacturing in its modern form has always taken advantage of automation in functions such as parts placement on circuit boards or wire wrapping, the process in which terminal pins are connected by wires.

Automation has been used for decades in the communications industry, which is now fully automated. This was done through telephone dialing systems, now replaced by "touch tone" telephone systems. Once dominated by somewhat cumbersome, though generally effective, electromechanical switching and control systems, telephone communication is now controlled by sophisticated computer systems (swift, powerful, generally quite reliable, but capable of causing widespread consternation when catastrophic crashes occur).

Transportation makes use of a variety of automation. Simple examples are cruise control and antilock brakes on automobiles. Much more sophisticated automated functions in transportation are automatic pilots for aircraft and complex integrated systems for urban rail and subway systems. A broad range of financial transactions in banking and retail trades are now handled automatically. Another service industry in which automation is playing a major role is in health care, such as in sensing blood glucose levels in diabetics and injecting appropriate amounts of insulin to regulate glucose.

21.7. ROBOTICS

Robotics is an extension of automation in which a machine mimics human activities through physical movement. The most common type of robotic device is the mechanical arm that can pick up objects, move them, and reorient them, or perform other activities, such as welding on an assembly line. A **robot** is a multifunctional device that can perform a variety of tasks by manipulating tools and other objects according to a preprogrammed set of instructions that can be changed as needed. Objects are manipulated by means of mechanical "arms" called **mechanical manipulators** consisting of links fastened by flexible joints that can extend (like a piston in a cylinder), twist, or rotate (Figure 21.2). For grasping parts a robot is equipped with a mechanical "hand," or **gripper**, designed to fit the intended part.

The "brains" of a robot consist of a computer that tells the robot what to do, that is, the motion sequence that it is required to go through. These instructions may be entered into the computer by a skilled programmer using computer language. A more intuitive approach is to actually lead the robot through the desired motion sequence and have it transfer the information to a computer for later repetition. Efforts are now under way to develop robots that can respond to voice instructions. The robot's computer is used for more than just directing its motions, and includes functions such as data processing, communication with other devices and with humans, decision making, and response to sensors. The last of these is a rapidly advancing area as robots become equipped with larger numbers of increasingly sensitive and sophisticated sensors for sensing motion, pressure, temperature, and light. A form of vision that enables robots to recognize shapes, colors, textures, and sizes and locations of objects will become more important in the future.

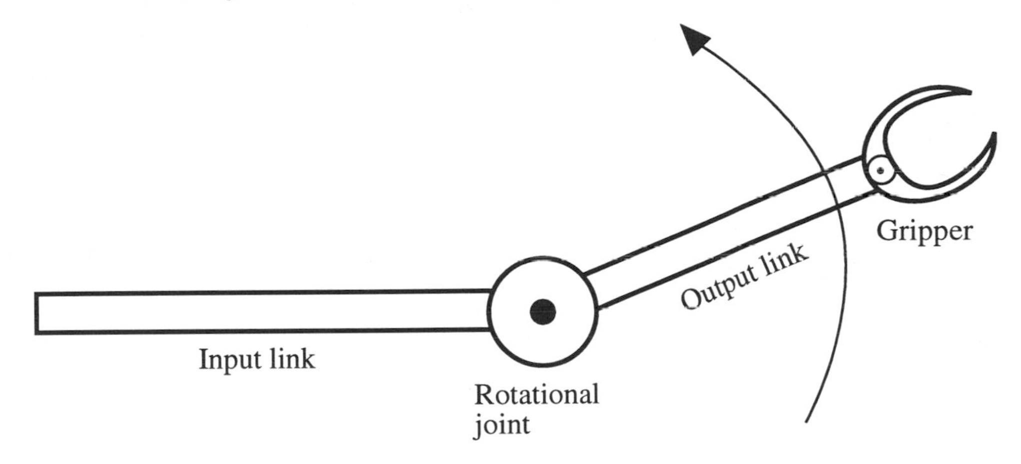

Figure 21.2. Representation of a robot manipulator and gripper with a rotational joint.

The industrial applications of robots can be divided into at least four major areas. The first of these consists of **moving materials and objects**, such as transferring parts from a conveyer to pallets. An important special aspect of moving objects is placing parts in machines in the correct way for further processing. A second major area in which industrial robots are used is in **processing operations**, that is, in performing particular operations or steps in manufacturing. The most common of these is welding metals, such as in automobile assembly; other examples include grinding, smoothing, polishing, and spray painting. Robots are especially useful in **assembly**, where their reliability, speed, and accuracy are of utmost value. Finally, **inspection** for quality control may be accomplished by robots.

21.8. COMPUTERS AND TECHNOLOGY

A major application of computers in industry is for so-called CAD/CAM, **computer-aided design/computer-aided manufacturing**. The tremendous data acquisition and processing capabilities of a computer are used to create and analyze a design, then modify and optimize it. The design can be subjected to analyses, such as those dealing with its heat-transfer or mechanical-strain characteristics. In computer-aided manufacturing, the computer can be used directly to process data regarding production, assembly, and other aspects of manufacturing, and to make adjustments to optimize the manufacturing process.

Applications of CAD/CAM are experiencing explosive growth. For example, CAD/CAM is used to make electric guitars to the specifications of individual musicians.[6] The parts of a guitar, such as its neck, can be simulated for size, strength, and other characteristics, and the whole assembly analyzed for optimum fit. Using known characteristics of the tools to be used, the computer program can deduce the best means of fabricating and fitting each piece. The tools used to cut the individual pieces in turn get their instructions from the CAD/CAM program after the design has been finalized, then perform the cutting and fitting by computer control. The final product is readily visualized in three dimensions by computer and can be viewed from any angle. From three to five months can be saved on producing a guitar using a computerized approach.

The largest project to date using CAD/CAM has been Boeing Aircraft's new 777 jetliner, the first production model plane to be produced ready for flight testing without prior production of a mockup. First flown in 1994, the product was within 0.5 millimeters of perfect alignment, compared to variances of more than 10 mm for previous models. The numbers of changes required during each step of production were only about half those required on previous models. Other examples of CAD/CAM products include newly designed chinaware produced in about a year less than normal production time with 25% less rejected product, new models of steam carpet cleaners brought to market about twice as fast and having much better maintenance records because of computer analysis and redesign of components likely to fail, and miniature joysticks that replace "mice" on laptop computers. It is believed that new automobile designs can move from concept to production about two years quicker with CAD/CAM.

CAD/CAM offers a number of other advantages in addition to better products produced more quickly. It can show, for example, if parts are likely to interfere with each other, which offers obvious safety advantages in areas such as aircraft operation. The interaction of people with controls and machines can be simulated and parts optimized for best interaction and reduction of repetitive motion injury. CAD/CAM has the potential to significantly reduce amounts of materials used and particularly wastage in cutting objects. Therefore, it provides significant environmental benefits.

Computers have found wide application in **modelling**, the construction of mathematically oriented theoretical schemes that mimic complex systems, such as manufacturing processes, ecosystems (see Section 1.7 and Chapter 17), and climate. One caution, not always observed as it should be, is that models are only as good as the information that goes into them and should always be correlated and verified as closely as possible with what actually happens in the systems modelled.

21.9. TRANSPORTATION

Ways of getting around and of moving materials and belongings have always been central to human existence and lifestyle. From the time that humans first started using primitive tools, they have invented devices to increase their ability to move themselves and their burdens. For land transport, the earliest of these were simple poles on which slain animals or other things were carried by two people or dragged by one (devices called the travois or slide car). These developed into skis and sledges. A major advance was the domestication of animals and training them to carry and drag loads. Then came the discovery of the wheel, arguably the greatest single increment in transportation technology of all time.

On water, primitive log dugouts and rafts greatly increased human mobility. These subsequently developed into canoes and boats propelled by human-powered ores and paddles. An advance comparable to the wheel on land was the discovery of the sail and development of means to use it effectively to propel boats and ships on water.

For both land and water transport, the development around 1800 of steam engines light enough to fit on self-propelled vehicles provided an enormous impetus to transportation. Steam-powered ships and boats freed water transport from the vagaries of the wind and enabled movement of boats upstream, though at a fearful price from boiler explosions and fires. The marriage of the steam engine mounted on a locomotive with steel rails enabled the development of railroads, which totally revolutionized land transport and completely changed human economic and social systems. The next huge advance came with the development of successful internal combustion (gasoline) engines in the late 1800s. These relatively light and compact power plants made the automobile a practical reality and made air transport possible. As the 1900s progressed, rail transport gave way (unfortunately, in some respects) to a large extent to airplanes, automobiles, buses, and trucks. Now, at least technically, transport through space is possible.

In more industrialized nations in the modern era, private transport, commuting, and travel over relatively short distances is largely by automobile. The flexibility, convenience, and independence afforded by private automobiles have made them extremely popular. Technical advances have greatly improved automobile efficiency, comfort, and safety (relative to size). The convenience of widespread automobile use has come at a high cost to the environment and, because of automobile accidents, in human lives. Land has been taken over to build highways, and the greater mobility afforded by automobiles has resulted in the conversion of vast areas of agricultural land and wildlife habitat to suburban housing areas. Increased energy demand for private automobiles and pollution from auto exhausts have had marked detrimental effects on the environment.

Air transportation has become the method of choice for long-distance movement of passengers. Movement of people and high-value freight by airplane is now reliable, safe, and relatively inexpensive. Technological advances in aircraft and engine design, as well as operation and control, made possible by better materials and computers, have given a tremendous impetus to air transportation. It is clearly the mode of choice for overseas travel and long-distance travel over land, but definitely needs to be integrated with rail transport (see mixed modes below) for distances of up to a few hundred kilometers.

Advanced technology is being used, and still has a huge unrealized potential, for the improvement of transportation systems in areas such as speed, convenience, and energy efficiency. High-tech computerized control has enabled automobiles to be much more efficient while emitting far smaller amounts of pollutants. Advanced air-traffic-control systems allow the operation of more aircraft in smaller spaces and much closer intervals with much greater safety than was previously possible. **Mixed modes** of transport can be very successful; an example is the movement of truck trailers and their contents over long distances by rail followed by final distribution by tractor trailer truck. A *systems approach* that enables tradeoffs to be made among speed, energy consumption, noise, pollution, convenience, and other factors offers much promise in the transportation area.

21.10. INFORMATION AND COMMUNICATION

The development of the silicon integrated circuit mentioned in Section 4.3 has had a massive impact on capabilities for recording, computing, storing, displaying, and communicating information in vast quantities.

The most rapid means of manipulating massive amounts of information is to combine standard electronics with **photonics**, which deals with information carried by photons of light. Using photonics, information can be acquired by a video camera, stored on compact video disks, transmitted by fiber optics, and displayed with a cathode ray tube or light-emitting diode display. Photonics relates to information transfer and storage in two major ways—optical memory, discussed briefly here, and optical communication, discussed later in this section.

Somewhat misnamed, **optical memory** consists of information recorded on microscopic grooves of a rotating disk, the most familiar example of which is the compact disk (CD) for playing music. The information is recorded on the disk digitally in a binary system represented by series of reflective dots (to represent "1") and nonreflective dots (to represent "0"). The amounts of information that can be so stored are staggering—a single compact video disk 9 cm in diameter can store the contents of a set of encyclopedias!

An emerging aspect of information technology is the **expert system**, which uses computers and appropriate computer programs to do reasoning. Expert systems enable machines to behave as though they possessed intelligence. Therefore, rather than just doing calculations, expert systems employ a form of reasoning to arrive at a conclusion. There are many applications of expert systems. Typical examples would be for quality control in manufactured items or in making decisions regarding the placement of rooms and other features of a house under design. The analogy may be made that whereas computers have vastly increased the speed, quantity, and accuracy of mathematical calculations that are possible, expert systems can be combined with human judgement and reasoning to speed the process by which judgements are made and to increase the probability that the right decisions will be made.

Information Transmission

The acquisition, manipulation, and storage of information are all crucial parts of an information system. No aspect is more important, however, than the transmission of information, quickly, accurately, and over long distances. This constitutes the area of communications, now performed predominantly by telecommunications. The major constituents of a telecommunications system are illustrated in Figure 21.3. These are (1) a means of converting the information in the form in which it is found to a form that can be transmitted, (2) a means and medium for transmission, and (3) a device or system for receiving the transmitted signal and converting it to sound, video display, computer language, or other form understandable by humans or other machines. These three components may be termed, respectively, the **transmitter**, the **medium**, and the **receiver**.

For transmission over a long distance, information is **encoded** to produce a form that the medium employed can handle. The first electrical device capable of virtually instantaneous long-distance communication was the telegraph, which encoded information as dots and dashes, that is, in a **discrete** form. The telephone was able to encode sound waves to continuous electrical waves that could be transmitted over wires, a form of signal called an **analog** signal. Interestingly, modern telecommunications systems convert analog information, such as the human voice, to a discrete form for transmission, then put it back into an analog form at the receiver. This is done by a process called **digitization** in which the analog signal is sampled at regular, closely spaced intervals, and its characteristics converted to numbers which are transmitted. This digital information may then be converted back to analog information.

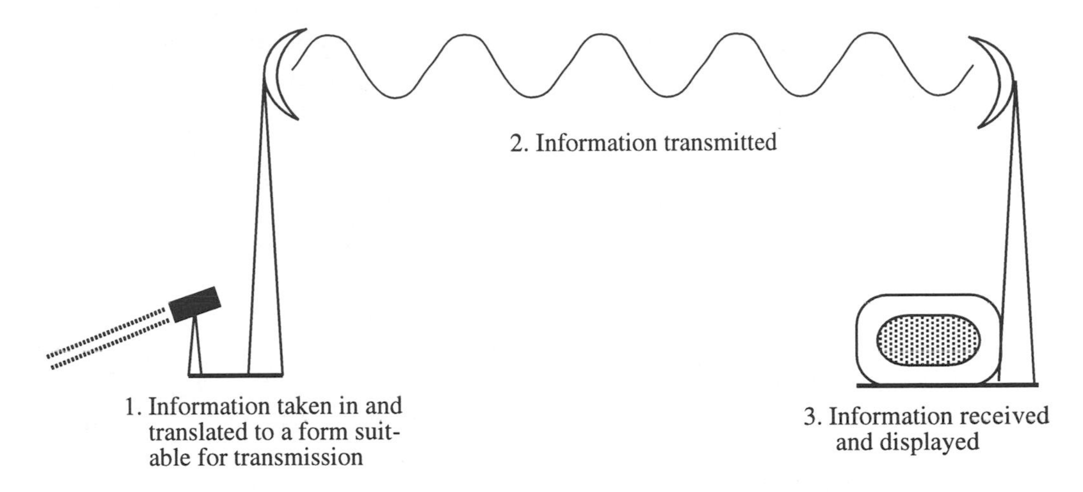

Figure 21.3. Major aspects of communication based on telecommunication.

The medium of photonics (see above) has been adapted to communication with remarkable success through **optical communication** using optical fibers to transmit information. The three major components of an optical communication system are shown in Figure 21.4. One of the keys to its success is the very thin, flexible **optical fiber** through which photons of light can be transmitted. Optical consist of extremely pure glass (essential to prevent unacceptable loss of light intensity during transmission) sheathed with a transparent cladding having a lower index of refraction that prevents loss of light from the fiber. The transmitter in the system takes in a signal, amplifies it, converts it to light with a device called a light emitting diode (LED), and sends the optical signal on its way with an emitter. These light signals lose their power and have to be amplified at regular intervals. This is accomplished by a repeater, which detects the light signal with a photodetector and converts it to an electronic signal that is amplified electronically, then changed back to light and sent on through optical fibers with an emitter. At the receiver the light signal is again converted to an electronic signal by a photodetector, amplified, and put out in the needed form, such as sound in a telephone headset.

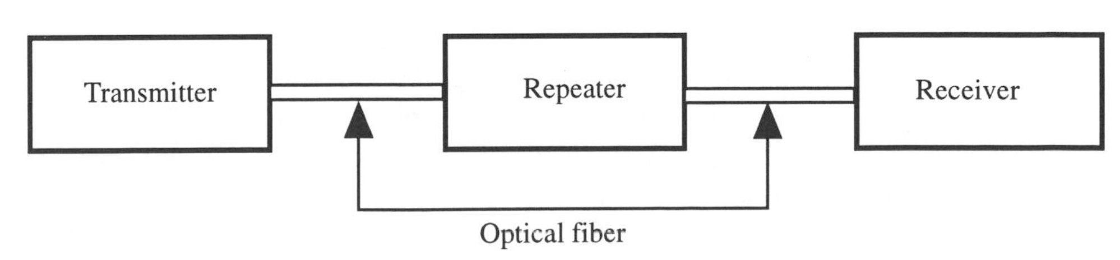

Figure 21.4. Three major components of an optical communications system.

21.11. HIGH TECH

What is high technology, "high tech"? Some have said, "To a caveman, it is the wheel." Certainly, in their day, the telegraph and the steam locomotive were high technology, as was the biplane aircraft of World War I. In present times, however, there is a group of technologies that can be labelled as high tech from the perspective of the modern era. In speaking of high tech a number of terms come to mind, including the following: voice synthesis, bionics, plasma systems, artificial intelligence, computerized language translation, remote sensing, cryogenics, photovoltaics, sonar, computer-controlled prosthetic devices, lasers, composite materials, ceramics, light-emitting diodes, fiber optics, genetic engineering, robotics, cryptography, CAT scanners, MRI imaging, digital audio, digital video, as well as computer-aided design (CAD), instruction (CAI), graphics (CAG), and manufacturing (CAM). Several important aspects of high technology are summarized briefly in this section.

Several areas of high tech have already been addressed in this chapter, including computers, robotics, and certain aspects of materials science. Central to these and all high tech areas are computers and the microchip, integrated circuits, and microprocessors that make them possible. Any other field of modern high technology is dependent upon computers. Computers are central to modern telecommunications, they make robots possible, they are essential to the development of new formulations for exotic materials, and make possible the exacting conditions under which such materials must be made.

Endeavors in **space** certainly can be classified as high tech. The realities of high costs and other more pressing priorities have prevented the development of "factories on the moon" or permanently populated orbitting space stations that many predicted when the "Space Age" became reality in the late 1950s. The exotic space combat visualized in "Star Wars" concepts have given way to a large extent to the decline of major nuclear powers and to the realities of grubby little wars fought by more conventional means. The greatest practical success in space so far has been in the launching of telecommunications satellites that have revolutionized global communications. Space technology is used very successfully in weather forecasting and in studying threats to the global climate, particularly greenhouse warming and atmospheric ozone depletion (see Chapter 12). Additional uses of space technology will undoubtedly develop in the future.

Lasers, devices that generate and transmit a condensed and directed beam of light, probably symbolize high tech more than any device other than the computer. Laser technology is an area that has come of age in areas such as laser printers, laser scanning of purchases, and laser drilling of diamonds. Lasers have also found significant uses in medicine.

Advanced **biotechnology** uses biochemical processes to perform tasks and to make products that would otherwise be impossible to make. Biotechnology directed through bioengineering makes use of enzymes, recombinant DNA, gene splicing, cloning, and other biological phenomena. The most exciting area of biotechnology in recent years is the one dealing with *gene splicing* or *recombinant DNA* wherein DNA material from one organism is inserted into another to give an organism with desired characteristics, such as the ability to make a specific protein. Several significant products have been produced by gene splicing. One such product is Humulin, a form of insulin identical to human insulin; another is human growth hormone. There is a high level of activity in biotechnology as applied to agriculture. Particularly promising are prospects to develop plants that are resistant to insects or to herbicides applied to competing plants, plants that can be made to fix nitrogen (through symbiotic bacteria growing on their roots), and growth promoters for plants and perhaps animals.

High tech is finding increasing uses in medicine and in pharmaceuticals. Examples of high tech medicine include arthroscopic surgery that greatly reduces the invasiveness of surgical procedures, real-time sensors for the control of such things as blood sugar in diabetics, and magnetic resonance imaging techniques. Prosthetic devices controlled by computers have significant promise in aiding the physically handicapped to lead more normal lives. Among the high-tech medical devices in common use are CAT scanners, defibrillators, pacemakers to regulate heartbeat, and heart-lung machines. Medical sonographs are devices that image internal organs to show abnormalities, such as tumor masses, using reflected high frequency sound signals processed by computer. Common organs that are imaged by this technique include the brain (using echoencephalogy) and the heart (using echocardiography). A refined version called doppler sonography is employed to show blood flow. Dialysis machines are devices through which a patient's blood is passed for removal of impurities across a semipermeable membrane as a substitute for nonfunctional kidneys. Another medical area that depends upon high-tech equipment and techniques is nuclear medicine, which uses radioactive substances that emit gamma rays to image body organs to diagnose abnormalities, particularly tumors.

High-tech pharmaceuticals use sophisticated techniques to design and synthesize new drugs that are extremely potent and targeted against specific maladies with minimum side effects. High-tech medicine and pharmaceuticals both make considerable use of biotechnology.

Micromachines

Micromachines consisting of working devices of the order of a millimeter in size have been constructed and may have a number of useful applications in the future. As an example of the capability of micromachining, a model of an automobile with 24 moving parts capable of propelling itself has been built by a subsidiary of Toyota Motor Corporation in Japan (see Figure 21.5).[7] Although there might not be many practical uses for a model of a 1936 Toyota about as large as a grain of rice that can traverse the length of a matchstick in about 1 second, there are numerous potential uses for micromachines. An example of such a device would be a monitor for vital signs, such as blood pressure, attached to a wristwatch and capable of injecting medication if an acute health emergency develops. It may be possible to develop diagnostic devices small enough to propel themselves through blood vessels

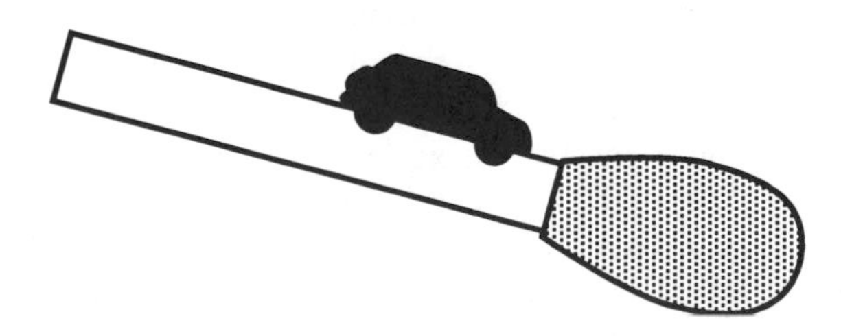

Figure 21.5. Relative sizes of a working miniaturized model of Toyota's first automobile, a 1936 model, compared to a match. The model has 24 parts, including an electric motor that can propel the miniature car at a speed of up to 50 millimeters per second.

and relay pictures outside the body. Such devices might even be equipped to perform microsurgical procedures. Devices to inspect pipes internally are also feasible and would be very useful.

A special category of micromachines consists of those fabricated on semiconductors using technology well developed in making integrated circuits (microchips) for computer and electronics applications. These kinds of devices have been called microelectromechanical systems, or MEMS. An example of such a MEMS device is a miniature pressure sensor containing a silicon diaphragm that flexes in response to pressure.

21.12. CLOSING THE LOOP: INDUSTRIAL ECOLOGY AND SUSTAINABLE DEVELOPMENT

As discussed in Section 17.2, a natural ecosystem is a self-organizing, self-balancing system in which members of a biological community interact with each other and with their atmospheric, aquatic, and terrestrial environments, thus ensuring their own survival and reproduction. To various degrees, species in an ecosystem are dependent upon and take advantage of the products or processes of other species; and the well-being of each individual in the ecosystem is dependent upon the health of the ecosystem as a whole. In a well-balanced natural ecosystem, there is an excellent material balance, and often a virtually complete degree of recycling. The driving force behind natural ecosystems is solar energy, which producer plants in the ecosystem fix as high-grade chemical food energy, which drives the rest of the ecosystem. Organisms in an ecosystem utilize the by-products and waste products of other organisms. Photosynthesis by plants generates by-product molecular oxygen, O_2, that is utilized by animals, fungi, and aerobic bacteria for respiration. Phosphate and nitrogen contained in waste products and biomass of animals is released in an inorganic form when bacteria degrade these substances and in turn are utilized by plants as nutrients. Many other examples can be cited.

Obviously, natural ecosystems have evolved to make the most efficient use possible of energy and materials. It is attractive, therefore, to use natural ecosystems as models for ideal anthropogenic systems in which a diverse assembly of humans, their industries, infrastructures, dwellings, and other things that they use are integrated into an **industrial ecosystem**, or, more broadly defined, an **anthrospheric ecosystem**. As noted in the introduction to this chapter, *industrial ecology* deals with the optimization of the complete cycle of materials from source to ultimate fate, with maximum recycling practiced as part of the entire anthrospheric ecosystem. As with natural ecosystems, anthrospheric ecosystems are driven by an energy source, or sources, and high-grade energy put into the system in the forms of fossil fuels, fissionable uranium, and other energy sources is used and dissipated as the ecosystem operates. Basic tenets of industrial ecology include **sustainable development**, through which living conditions are enhanced in a manner that can be sustained within Earth's carrying capacity, as well as measures to prevent production and release of pollutants and to reduce uses of toxic and otherwise dangerous materials.[8] Just as natural ecosystems are generally quite diversified, the most successful and efficient anthrospheric ecosystems generally exhibit a high degree of versatility.

Without having been recognized as such, anthrospheric ecosystems are as old as civilization and have developed to various degrees of sophistication through the normal activities of humans. Consider as an example a community developed around a single large enterprise, such as a mineral mine. Energy may be used in the community in the forms of coal-generated electricity, gasoline, and diesel fuel for automobiles, trucks, buses, and tractors; and natural gas for heating. The major industry employs many people and provides income that they spend in the private sector and through taxes and fees to maintain a variety of business services, dwellings, homes, schools, and elements of the infrastructure. It is easy to see that such an anthrospheric ecosystem is very dependent upon the single business activity that largely supports it, and any setbacks in that enterprise can have adverse effects on the whole system. Examples abound of communities that have flourished during "boom times" of a central industry, such as a large oil field, then crashed miserably when the business failed. The analogies with poorly diversified natural ecosystems, particularly those largely dependent upon a single producer of biomass, are obvious.

Anthrospheric Ecosystems on an Industry-Wide Basis

Recognition of anthrospheric ecosystems as viable entities provides the opportunity in principle to plan such systems in ways that will optimize their performance. Integrated industries provide excellent opportunities for such planning. The modern petrochemical industry is one of the best examples of a successful and efficient industrial ecosystem. A petrochemical complex is driven by the energy and materials derived from crude oil. This material is distilled to produce a variety of fuel and raw material fractions, including gasoline, diesel fuel, lubricating oil, feedstocks for manufacturing rubber and plastics, and asphalt for paving roads. The value of the products is enhanced by chemically breaking down ("cracking") heavier distillation fractions and synthesizing heavier molecules from lighter constituents (alkylation). Specialized ingredients, such as methyltertiarybutyl ether, used as an antiknock additive in gasoline, are manufactured to enhance quality. Methane can be reacted to make hydrogen used to synthesize ammonia for fertilizer and chemical synthesis. The production of ethene (ethylene) can be enhanced to provide the raw material for making polyethylene plastic. Dozens of different processes and products can be associated with a single petrochemical complex.

The recognition of the huge problems caused by waste products and pollutants has given impetus to industries in which the producers of waste products work in close association with industrial enterprises that can utilize such products as raw materials. Such utilization can be driven by both economics and regulation. Unprofitable by-products of manufacturing that were once discarded as waste materials can become quite valuable as demand changes. In the early days of the petroleum refining industry during the 1800s, gasoline was an undesirable by-product of petroleum, which was refined primarily for kerosene used for cooking and lighting. With the development of the internal combustion engine and the automobile, the production of gasoline became economically viable, and it eventually became the leading product of petroleum refining. In an unregulated environment, the cheapest thing to do with sulfur in fossil fuels is to burn it and discharge it into the atmosphere as sulfur dioxide. However, when sulfur emissions are curtailed by regulation, it may become economically feasible to reclaim sulfur and convert it to a marketable product, such as sulfuric acid.

Large petrochemical complexes are often examples of **vertical integration** in which a single concern owns and controls sources of supply, the manufacturing complex, and the marketing organization. To a large extent vertical integration was practiced by the major U.S. automobile manufacturers, some of which even owned iron and coal mines to produce the steel and energy needed to make automobiles. However, vertical integration can be too unwieldy to be efficient and can stifle competition and innovation. Therefore, modern business enterprises, particularly large retailing concerns, rely on a large number of smaller concerns, independently owned and operated, as sources of supply, manufacturing, and even marketing. This allows for a natural selection and continuous evolution of individual "species" of concerns most fit to contribute efficiently to the overall enterprise.

Anthrospheric Ecosystems on a Community Basis

Advantageous as a planned anthrospheric ecosystem based upon a single industry might be, there are even greater advantages to be had in planning such a system around an entire community with a variety of carefully integrated industrial enterprises providing employment and an economic base. A frequently cited example of such a community is that of Kalundborg, Denmark.[9] Some years ago the electric power plant in Kalundborg began co-production of steam along with electricity to sell steam to the local Statoil petroleum refinery. The availability of steam made it profitable to sell to other local manufacturing plants, greenhouses, homes, and a fish farm. Other mutually advantageous enterprises that have developed in the community are the following:

- Sulfur removed from high-sulfur gas by the Statoil petroleum refinery was sold to a local sulfuric acid manufacturer, and the clean gas product from the petroleum refinery was marketed to a local gypsum wallboard manufacturer.
- A lime scrubber installed to remove sulfur from flue gas at the electric power plant produced calcium sulfate by-product, which was sold to the wallboard manufacturer as a substitute for gypsum.
- Fish processing waste from the fish farm was sold to local farmers as fertilizer.
- A local pharmaceutical concern, which used steam co-produced by the power plant, marketed sterilized fermentation waste sludge to local farmers as fertilizer in quantities amounting to about 1.5 million cubic meters per year.
- Grain from local farmers was used to make fish food for the fish farm.

The possibilities for anthropogenic ecosystems planned around a whole community are many and very promising. The application of computers in planning such communities raises many interesting possibilities for developing integrated anthrospheric ecosystems operating at maximum efficiency and with minimal production of pollutants and wastes. Possible aspects of such a system are shown in Figure 21.6.

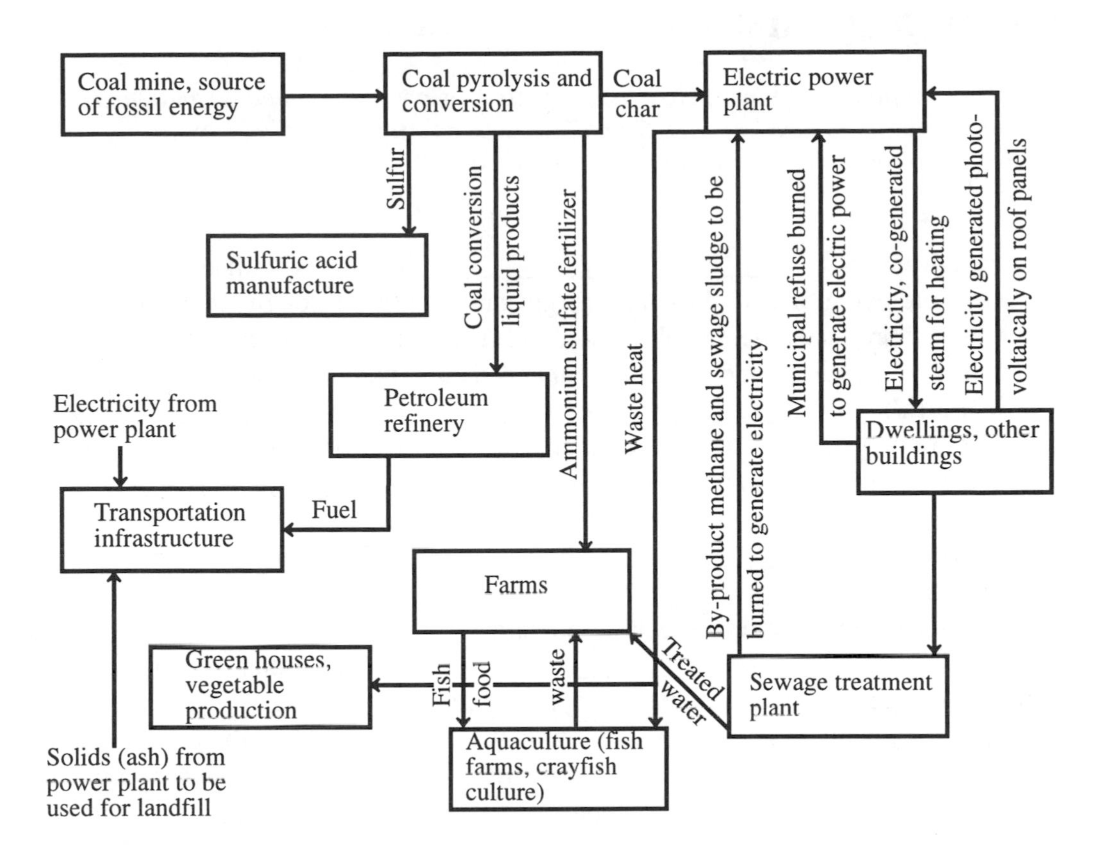

Figure 21.6. Possible components of a community-based anthrospheric ecosystem.

Basic to such a system as the one outlined above are components that provide energy and those that deal with wastes. In the example shown in Figure 21.6, the basis of the energy system is locally mined coal. However, rather than burning the coal directly as fuel, it is first subjected to a coal conversion process that recovers high value hydrocarbon liquids used as petroleum feedstock, potentially polluting sulfur used to manufacture sulfuric acid, and by-product ammonium sulfate sent to farms for fertilizer, leaving only the carbonaceous char (coke) product to be burned. In addition to producing electricity for homes, businesses and a rail-based transportation system, the power plant co-generates steam and produces waste heat utilized to heat buildings and greenhouses and to warm water to maximize aquaculture production. The power plant in turn receives and redistributes electricity generated by photovoltaic panels on roofs. In addition to burning coal for fuel, it burns municipal refuse and sludge and methane gas generated by sewage treatment. In planning the community infrastructure, provision has been made for landfill needed for future construction of road or rail lines, and stabilized waste ash from the power plant is used to make useful landfill.

The community is largely self sufficient in food from nearby farms, greenhouses, and aquaculture operations. Fish are fed with grain from the farms and sediment wastes from the aquaculture operations make excellent fertilizer for farms or for growing vegetables in the greenhouses. Secondary sewage effluent is spread on land for irrigation, thus eliminating the need for advanced treatment of the wastewater. Many other possibilities for a planned community anthrospheric ecosystem may be visualized.

CHAPTER SUMMARY

The chapter summary below is presented in a programmed format to review the main points covered in this chapter. It is used most effectively by filling in the blanks, referring back to the chapter as necessary. The correct answers are given at the end of the summary.

The practitioner of [1]____________________ deals with optimization of the complete cycle of materials from virgin raw material through finished material, components, products, use, obsolescence, and finally disposal or recycle. The process through which living conditions are enhanced in a manner that can be sustained within Earth's carrying capacity defines [2]____________________. Life cycle analysis consists of a careful examination of the whole life cycle of a material with a view toward making this cycle as [3]____________________ __________. The ways in which humans do and make things with materials and energy defines [4]____________________. The two main categories of agriculture are [5]____________________________. The offspring of crosses between two different true-breeding strains are [6]______________ ______________. The [7]____________________ industries are those in which irreplaceable resources are taken from the Earth's crust. [8]____________ __________ deals with the composition, properties, and applications of substances used to make devices, machines, and structures. Cellulose, cotton, rubber, polyethylene, and epoxy resins are all examples of [9]__________________. Among the usually desirable properties of ceramics are that they tend to be [10]____________ ____.
Ceramics usually contain [11]______________________________ and are generally formed by [12]________________________________ processes. Composites are made of [13]________________________________ to give a material that has [14]____________________________________ than any of the constituents singly. Composites are usually composed of [15]__________ ____ imbedded in some sort of [16]__________________. Automation is [17]____________________ ____.
In a [18]________________________ automated system, the output is measured and compared to the desired response for feedback. [19]____________________ is used to give a machine instructions that can be changed for different purposes. A [20]____________________ is a multifunctional device that can perform a variety of tasks by manipulating tools and other objects according to a pre-programmed set of instructions that can be changed as needed. Four major industrial applications of automation are [21]____ ____ ____________________. CAD/CAM stands for [22]____. For both land and water transport, the development around 1800 of [23]____________________ light enough to fit on self-propelled vehicles provided an enormous impetus to transportation. The long-distance movement of truck trailers and their contents by rail followed by final distribution by tractor trailer truck is an example of a [24]________ ______________ of transportation. Photonics is the process by which [25]__________

_______________________. For transmission over a long distance, information is [26]_______________. The process in which an analog signal is sampled at regular, closely spaced intervals, and its characteristics converted to numbers which are transmitted to produce a form that the medium employed can handle is called [27]____ __________. The equivalent to a wire in optical communications is the [28]_______ _____________________. In modern times, voice synthesis, artificial intelligence, remote sensing, and computer-aided design are all examples of [29]______________ _________________.

Answers

[1] industrial ecology

[2] sustainable development

[3] environmentally friendly as possible

[4] tcchnology

[5] crop farming and livestock farming

[6] hybrids

[7] extractive

[8] Materials science

[9] polymers

[10] hard, rigid, and resistant to high temperatures and chemical attack

[11] silicon and oxygen

[12] high temperature

[13] two or more materials

[14] better overall properties

[15] high-strength reinforcing materials

[16] matrix

[17] the use of automatic components in the performance of repetitive tasks

[18] closed loop

[19] Programmable automation

[20] robot

[21] moving materials and objects, processing operations, assembly, and inspection

[22] computer-aided design/computer-aided manufacturing

[23] steam engines

[24] mixed mode

[25] information is carried by light

[26] encoded

[27] digitization

[28] optical fiber

[29] high technology

QUESTIONS AND PROBLEMS

1. Explain how modern industrial ecology contributes to sustainable development, a term defined in Section 4.10.
2. Outline the life cycle analysis of an automobile with the view of minimizing the environmental impact of its manufacture, use, and disposal.
3. Define "interconnectedness" in an environmental context as related to technology.
4. What is the distinction between extractive and renewable sources of raw materials?
5. Explain the relationship of geology and geochemistry to the location and utilization of extractive sources of raw materials.
6. What is extractive metallurgy?
7. What are polymers? How are they important in materials science?
8. What are ceramics? Why are they of growing importance in materials science?
9. What are the advantages of composites as materials?
10. What is the distinction between an open-loop and closed-loop system in automation?
11. What is the function of a set point in a feedback control system used in automation?
12. What is meant by robotics? What is the distinction between robotics and automation?
13. What are the "brains" and "hands," respectively, of a robot?
14. What is meant by CAD/CAM? What advantages does it offer?
15. What are mixed modes of transport? What advantages do they offer?
16. What is the distinction between photonics and electronics?
17. What did the first successful long-distance communication device, the telegraph, have in common with modern digital computers?
18. What are the key components of an optical communication system?
19. What would have been considered "high-tech" devices in the late 1800s.
20. In what respects might it be argued that the biological sciences are the most "high tech" of all modern areas?

LITERATURE CITED

[1] Graedel, T. E. and B. R. Allenby, *Industrial Ecology*, Prentice Hall, Englewood Cliffs, NJ, 1995.

[2] Socolow, R., C. Andrews, F. Berkhout, and V. Thomas, Eds., *Industrial Ecology and Global Change*, Cambridge University Press, Cambridge, UK, 1994.

[3] Allenby, Braden R. and Deanna J. Richards, Eds., *The Greening of Industrial Ecosystems*, National Academy Press, Washington, D.C., 1994.

[4] Jackson, Tim, Ed., *Clean Production Strategies: Developing Preventive Environmental Management*, CRC Press/Lewis Publishers, Boca Raton, FL, 1993.

[5] Ayres, Robert U. and Leslie W. Ayres, *Industrial Ecology: Towards Closing the Materials Cycle*, E. Elgar Pubishers, Cheltenham, U.K., 1996.

[6] Deutsch, Claudia H., "Not Making Them Like They Used To," *New York Times*, March 31, 1997, pp. C1-C2.

[7] Pollack, Andrew, "Japan's Micro-Machine Project Draws Envy and Criticism," *New York Times*, August 20, 1996, p. B5.

[8] Breen, Joseph J., "Designing the Future," *Environmental Science and Technology*, **30**, 258A-259A (1996).

[9] "On Industrial Ecosystems," Chapter 15 in Ayres, Robert U., and Leslie W. Ayres, *Industrial Ecology: Towards Closing the Materials Cycle*, E. Elgar Pubishers, Cheltenham, U.K., 1996, pp. 273-293.

SUPPLEMENTARY REFERENCES

Abdel-Magid, Isam Mohammed, Abdel-Wahid Hago Mohammed, and Donald R. Rowe, *Modelling Methods for Environmental Engineers*, CRC Press/Lewis Publishers, Boca Raton, FL, 1996.

Bregman, Jacob I., Craig A. Kelley, and James R. Melchor, *Environmental Compliance Handbook*, CRC Press/Lewis Publishers, Boca Raton, FL, 1996.

Carberry, Judith, *Environmental and Systems Engineering*, OUP, New York, NY, 1995.

Henry, J. Glenn and Gary W. Heinke, *Environmental Science and Engineering*, 2nd ed., Prentice-Hall, Upper Saddle River, NJ, 1996.

Jeffries, Thomas W. and Liisa Viikari, Eds., *Enzymes for Pulp and Paper Processing*, American Chemical Society, Washington, D.C., 1997.

Kohn, James P., *The Ergonomic Casebook: Real World Solutions*, CRC Press/Lewis Publishers, Boca Raton, FL, 1997.

Koren, Herman, *Illustrated Dictionary of Environmental Health and Occupational Safety*, CRC Press/Lewis Publishers, Boca Raton, FL, 1996.

McKinney, Michael L. and Robert M. Schoch, *Environmental Science: Systems and Solutions*, West Publishing, Westbury, NY, 1996.

Sangeeta, D., *The CRC Handbook of Inorganic Materials Chemistry*, CRC Press/Lewis Publishers, Boca Raton, FL, 1997.

Stern, Martin B. and Zack Mansorf, *Applications of Industrial Hygiene*, Lewis Publishers/CRC Press, Boca Raton, Florida, 1997.

Swaddle, T. W., *Inorganic Chemistry: An Industrial and Environmental Perspective*, Academic Press, San Diego, CA, 1997.

Videla, Hector, J. Fred Wilkes, and Renato Silva, *Manual of Biocorrosion*, CRC Press/Lewis Publishers, Boca Raton, FL, 1997.

22 ENERGY AND RESOURCE UTILIZATION

22.1. INTRODUCTION

The availability of various kinds of resources, such as water, minerals, and soil, has been discussed in other chapters of this book. This chapter, "Energy and Resource Utilization," discusses the availability of one resource that is largely the key to all the rest—energy. To a large extent, if enough energy is available, and if the environmental costs are acceptable, almost any needed resource can be made available. For example, large quantities of energy can be used to extract fresh water from salt water or to concentrate scarce metals from very weak ore sources. The environmental costs can be high, such as increased output of greenhouse gas carbon dioxide from the use of fossil fuels to extract energy, or disruption of land from the mining of huge quantities of low-grade ore.

Resource utilization is intimately tied with environmental concerns and with the welfare of humankind. Humans must use resources—food, wood, hydrocarbons, fuel, water, and minerals—in order to exist. In the modern era, however, a rapidly increasing population combined with higher living standards have put great strains on Earth's resource base. In addition to running short of critical resources it is also a matter of the environmental harm that is done when utilizing such resources.

A crucial aspect of resource utilization is **resource economics** dealing with supply and demand, monetary values, and other economic aspects of resources as they are related to environmental considerations. Perhaps preferably called *environmental* resource economics, this discipline goes beyond the conventional supply/demand relationships, monetary values temporarily assigned to various resources, and other aspects of economics as it has commonly been understood to consider factors such as pollution, general environmental degradation, effects on life-support systems, and other broadly based and environmentally connected economic concerns associated with resource utilization.

Resources

Resources consist of virtually anything of value that can be utilized by humans. **Exhaustible resources** are nonrenewable resources present in a fixed amount on Earth, which cannot be replenished by ordinary means. **Renewable**

resources are those that can in principle be replenished. The distinction between these two categories tends to become blurred. Nonrenewable resources can in fact become renewable if much higher monetary or environmental costs can be borne. Renewable resources, such as vast underground deposits of water left from glacial times, essentially become exhaustible in any sort of reasonable time scale. Some resources do not consist of "things," but are intangibles, such as culture, knowledge, beauty, and satisfaction with life.

There are various categories of resources based upon availability. The total amount of a material may be a poor indicator of its resource availability, which is commonly governed by technology, price, and other factors. Resources that have been accurately identified and are known to be utilizable with current technology at current prices are classified as **proven reserves**. Resources known to exist, but not completely documented or necessarily profitable to recover are **known resources**. Another category of resources consists of those that are likely to exist, but are as yet undiscovered.

22.2. ECONOMICS: THE BROADER VIEW INCLUDING ENVIRONMENT AND RESOURCES

The allocation and use of resources is described by the discipline of **economics**. Economics describes which resources are used, by whom, how much is used, and when they are used. The driving force behind economics is supply and demand. The **law of supply and demand** relates the price of a material or service to its supply. Supply is composed of many complex factors, including the nature and availability of raw materials, costs of converting a raw material to finished goods, availability of alternate materials and goods, and societal/governmental/ regulatory factors. Increasingly, in many cases recyclability, including the availability of the infrastructure to enable recycling, largely determines supply. When the supply of newsprint became tight in the mid-1990s, recycled paper became a factor in making adequate supplies available.

Most goods and services exhibit elasticity in supply and demand in that when the price increases, the supply increases as well, and *vice versa*. In a properly operating economic system the price stabilizes around an equilibrium value that reflects supply and demand. This value can be shifted, sometimes substantially. In modern times technological developments have caused dramatic shifts in price. An especially good example of this is the price of doing mathematical calculations, which has decreased by many orders of magnitude with ongoing developments in computers. Substitution of alternate materials or means of delivering services can drastically reduce effective prices. The use of plastics has reduced the prices of a number of items in inflation-adjusted terms relative to what they would be if synthetic plastics were not available. Substitution of commuter rail transport for the private automobile can greatly reduce the cost of a unit of travel. Recycling tends to limit price increases.

Costs are conventionally viewed as the amount of money that must be exchanged for a particular quantity of goods or services. However, costs may include factors that are not directly monetary, such as environmental degradation or loss of quality of life. A particularly important aspect of environmental economics consists of **external costs** that are borne by those other than individuals receiving a direct, specific benefit. If a waste product is simply pumped into the atmosphere, the cost of producing a specific manufactured item can be quite low to the manufacturer. However, other people have to bear the cost of degraded air quality and health costs from breathing polluted air. External costs are usually diffuse and hard to quantify. Properly applied, environmental economics requires internalization of external costs so that those who gain a benefit pay the full price. For example, a disposal fee attached to the price of a new tire internalizes the cost of disposal that otherwise might have to be borne by the public sector.

The total economic output of a nation is its **gross national product**, commonly used as a measure of well being. The economy of a nation may grow as a whole, but if population growth exceeds total economic growth, the population is not better off. Thus it is a more accurate reflection of human well being to express gross national product per person. This gives an average, however, and large segments of a population may not be more prosperous while the average goes up because of steep increases of income among the top income segment of the population. In most industrialized nations the gross national product per person has increased significantly during the last century. The distribution of gross national product is skewed in that well over half of it is concentrated in North American and European countries, which have only about 1/5 of Earth's population.

Most large public works projects are now subject to **cost/benefit** analyses, which attempt to relate the cost of a project to the benefits that are supposed to accrue from it. Such analyses can become quite elaborate, with complicated flow charts and computer analyses. Inclusion of environmental costs and benefits can significantly complicate cost/benefit analyses, but such analyses are incomplete if environmental factors are not included.

An economy that is "booming" in the conventional sense of increased population, increased industrial production, greater consumption per capita, and rapid depletion of irreplaceable resources puts a heavy burden on life-support systems that maintain humans as well as all other forms of life. Such an economy is oriented toward consumption working in concert with industrial productivity. Eventually, if life as we know it on Earth is to be maintained, it will be necessary to adopt a **steady-state economic system** characterized by complete balance between the demands placed upon Earth's life support system and its ability to meet these demands. The major characteristics of a steady-state economy are the following:

- Essentially complete recycling of nonrenewable materials
- Maximum utilization of renewable energy resources
- Maximum efficiency in energy and materials utilization
- Design and manufacture for maximum durability and recyclability
- Steady level of population with low rates of both births and deaths

A steady-state economy would be one that is beyond the stage of an industrial economy. Whereas an industrial economy is characterized by a high degree of manufacturing, high volumes of materials processed, heavy resource utilization, and a strong tendency to pollute, a postindustrial economy is oriented primarily toward services, information, education, and high technology. Overconsumption is discouraged in a steady-state economy. For example, efforts to reduce electrical power consumption in the U.S. have cut back on the demand for new power plants and have reduced costs. Within the limits imposed by the need to provide adequate food, housing, and other material benefits, a postindustrial economy can provide for the needs of humankind without imposing undue burdens on Earth's material resources.

22.3. MINING AND MINERAL RESOURCES

Manufacturing requires a steady flow of raw materials—minerals, fuel, wood, and fiber. As discussed in Chapter 16, these can be provided from either **extractive** (nonrenewable) and **renewable** sources.[1] The extractive industries are those in which irreplaceable resources are taken from the Earth's crust. This is normally done by mining, but may also include pumping of crude oil and withdrawal of natural gas. The raw materials so obtained may be divided broadly into the two categories of (1) inorganic minerals, such as iron ore, clay used for firebrick, and gravel, and (2) materials of organic origin, such as coal, lignite, or crude oil.

Most of the elements, including practically all of those likely to be in short supply, are metals. Some metals, including antimony, chromium, and the platinum-group metals, are considered especially crucial because of their importance to industrialized societies, uncertain sources of supply, and price volatility in world markets. Mining and processing of metal ores involve major environmental concerns, including disturbance of land, air pollution from dust and smelter emissions, and water pollution from disrupted aquifers. This problem is aggravated by the fact that the general trend in mining involves utilization of less rich ores.

A number of minerals other than those used to produce metals are important resources. These include several important minerals generally considered to be very commonplace, but still essential for a modern industrialized society. Prominent among such minerals are rock, gravel, sand, and clay. Clays, for example, have many uses, including applications for clarifying oils, as catalysts in petroleum processing, as fillers and coatings for paper, and in the manufacture of firebrick, pottery, sewer pipe, and floor tile. Phosphate minerals constitute a particularly crucial nonmetal resource, the most crucial use for which is in fertilizer manufacture. Phosphorus extracted from fluorapatite, $Ca_5(PO_4)_3F$, and hydroxyapatite, $Ca_5(PO_4)_3(OH)$, is also used for supplementation of animal feeds, synthesis of detergent builders, and preparation of chemicals such as pesticides and medicines. Large quantities of carbon black, diatomite, barite, fuller's earth, kaolin, mica, limestone, pyrophyllite, and wollastonite ($CaSiO_3$) are used in the U.S. each year as fillers for paper, rubber, roofing, battery boxes, and many other products. Sulfur from deposits of elemental sulfur, H_2S, recovered from sour natural gas, organic sulfur recovered from petroleum, and pyrite (FeS_2) is used in huge quantities for the manufacture of sulfuric acid and in a wide variety of other industrial and agricultural products.

22.4. WOOD — A MAJOR RENEWABLE RESOURCE

Fortunately, one of the major natural resources in the world, wood, is a renewable resource. Production of wood and wood products is the fifth largest industry in the United States, and forests cover one third of the United States surface area. Wood ranks first worldwide as a raw material for the manufacture of other products, including lumber, plywood, particle board, cellophane, rayon, paper, methanol, plastics, and turpentine.

Chemically, wood is a complicated substance consisting of long cells having thick walls composed of polysaccharides such as cellulose:

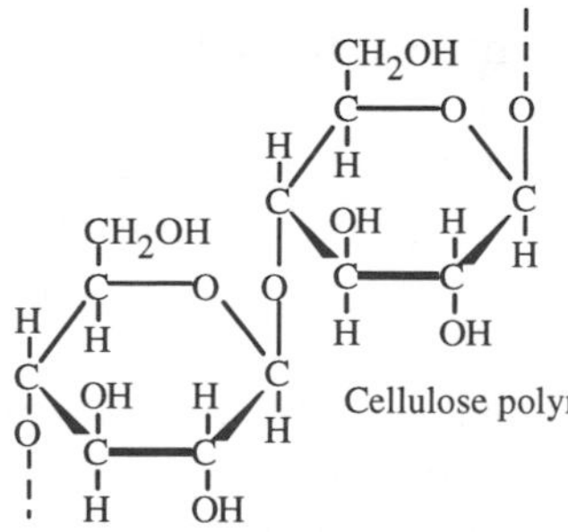

The polysaccharides in cell walls account for approximately three fourths of *solid wood*, wood from which extractable materials have been removed by an alcohol-benzene mixture. Wood typically contains a few tenths of a percent ash (mineral residue left from the combustion of wood).

A wide variety of organic compounds can be extracted from wood by water, alcohol-benzene, ether, and steam distillation. These compounds include tannins, pigments, sugars, starch, cyclitols, gums, mucilages, pectins, galactans, terpenes, hydrocarbons, acids, esters, fats, fatty acids, aldehydes, resins, sterols, and waxes. Substantial amounts of methanol (sometimes called *wood alcohol*) are obtained from wood, particularly when it is pyrolyzed. Methanol, once a major source of liquid fuel, is now being used to a limited extent as an ingredient of some gasoline blends (see gasohol in Section 22.15).

A major use of wood is in paper manufacture. The widespread use of paper is a mark of an industrialized society. The manufacture of paper is a highly advanced technology. Paper consists essentially of cellulosic fibers tightly pressed together. The lignin fraction must first be removed from the wood, leaving the cellulosic fraction. Both the sulfite and alkaline processes for accomplishing this separation have resulted in severe water and air pollution problems, although substantial progress has been made in alleviating these.

Wood fibers and particles can be used for making fiberboard, paper-base laminates (layers of paper held together by a resin and formed into the desired structures at high temperatures and pressures), particle board (consisting of wood particles bonded together by a phenol-formaldehyde or urea-formaldehyde resin) and nonwoven textile substitutes consisting of wood fibers held together by adhesives. Chemical processing of wood enables the manufacture of many useful products, including methanol and sugar. Both of these substances are potential major products from the 60 million metric tons of wood wastes produced in the U.S. each year.

22.5. THE ENERGY PROBLEM

Since the 1973-74 "energy crisis," much has been said and written, many learned predictions have gone awry, and some concrete action has even taken place. Catastrophic economic disruption, people "freezing in the dark," and freeways given over to bicycles (perhaps a good idea) have not occurred. Nevertheless, uncertainties over petroleum availability and price, and disruptions such as the 1990 Gulf War have caused energy to be one of the major problems of modern times.

In the U.S. concern over energy supplies and measures taken to ensure alternate supplies reached a peak in the late 1970s. Significant programs on applied energy research were undertaken in the areas of renewable energy sources, efficiency, and fossil fuels. The financing of these efforts reached a peak around 1980, then dwindled significantly after that date.

The solutions to energy problems are strongly tied to environmental considerations. For example, a massive shift of the energy base to coal in nations that now rely largely on petroleum for energy would involve much more strip mining, potential production of acid mine water, use of scrubbers, and release of greenhouse gases (carbon dioxide from coal combustion and methane from coal mining). Similar examples could be cited for most other energy alternatives.

Dealing with the energy problem requires a heavy reliance on technology, which is discussed in numerous places in this book. Computerized control of transportation and manufacturing processes enables much more efficient utilization of energy. New and improved materials enable higher peak temperatures and therefore greater extraction of usable energy in thermal energy conversion processes. Innovative manufacturing processes have greatly lowered the costs of photovoltaic cells used to convert sunlight directly to energy.

22.6. WORLD ENERGY RESOURCES

At present, most of the energy consumed by humans is produced from fossil fuels. Estimates of the amounts of fossil fuels available differ; those of the quantities of recoverable fossil fuels in the world before 1800 are given in Figure 22.1. By far the greatest recoverable fossil fuel is in the form of coal and lignite. Furthermore, only a small percentage of this energy source has been utilized to date, whereas much of the recoverable petroleum and natural gas has already been consumed. Projected use of these latter resources indicates rapid depletion.

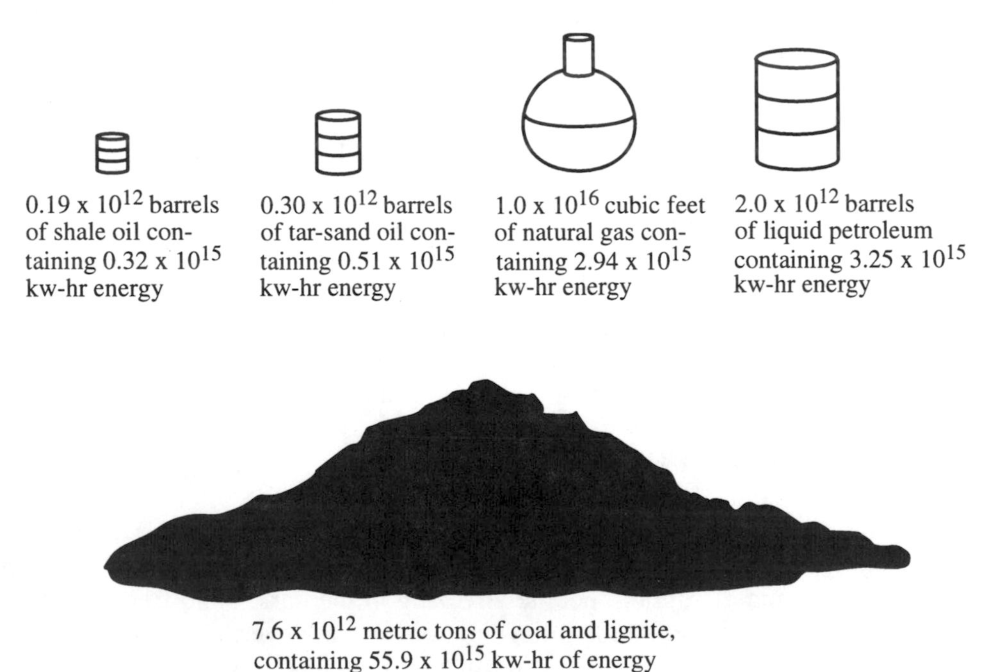

Figure 22.1 Original amounts of the world's recoverable fossil fuels (quantities in thermal kilowatt hours of energy based upon data taken from M. K. Hubbert, "The Energy Resources of the Earth," in *Energy and Power*, W. H. Freeman and Co., San Francisco, 1971).

Although world coal resources are enormous and potentially can fill energy needs for a century or two, their utilization is limited by environmental disruption from mining and emissions of carbon dioxide and sulfur dioxide. These would become intolerable long before coal resources were exhausted. Assuming only uranium-235 as a fission fuel source, total recoverable reserves of nuclear fuel are roughly about the same as fossil fuel reserves. These are many orders of magnitude higher if the use of breeder reactors is assumed. Extraction of only 2% of the deuterium present in the Earth's oceans would yield about a billion times as much energy by controlled nuclear fusion as was originally present in fossil fuels! This prospect is tempered by the lack of success in developing a controlled nuclear fusion reactor. Geothermal power, currently utilized in northern California, Italy, and New Zealand, has the potential for providing a high percentage of energy worldwide. The same limited potential is characteristic of several renewable energy resources, including hydroelectric energy, tidal energy, and wind power. All of these will continue to contribute significant, but relatively small, amounts of energy. Renewable, nonpolluting solar energy comes as close to being an ideal energy source as any available. It almost certainly has a bright future.

22.7. ENERGY CONSERVATION

Any consideration of energy needs and production must take energy conservation into consideration. This does not have to mean cold classrooms with thermostats set at 60°F in mid-winter, nor swelteringly hot homes with no air-conditioning, nor total reliance on the bicycle for transportation, although these, and even more severe, conditions are routine in many countries. The fact remains that the United States has wasted energy at a deplorable rate. For example, U.S. energy consumption is higher per capita than that of some other countries that have equal, or significantly better, living standards. Obviously, a great deal of potential exists for energy conservation that will ease the energy problem.[2]

Transportation is the economic sector with the greatest potential for increased efficiencies. The private auto and airplane are only about one third as efficient as buses or trains for transportation. Transportation of freight by truck requires about 3800 Btu/ton-mile, compared to only 670 Btu/ton-mile for a train. It is terribly inefficient compared to rail transport (as well as dangerous, labor-intensive, and environmentally disruptive). Major shifts in current modes of transportation in the U.S. will not come without anguish, but energy conservation dictates that they be made.

Household and commercial uses of energy are relatively efficient. Here again, appreciable savings can be made. The all-electric home requires much more energy (considering the percentage wasted in generating electricity) than a home heated with fossil fuels. The sprawling ranch-house style home uses much more energy per person than does an apartment unit or row house. Improved insulation, sealing around the windows, and other measures can conserve a great deal of energy. Electric generating plants centrally located in cities can provide waste heat for commercial and residential heating and cooling and, with proper pollution control equipment, can use refuse for a significant fraction of fuel.

As scientists and engineers undertake the crucial task of developing alternative energy sources to replace dwindling petroleum and natural gas supplies, energy conservation must receive proper emphasis. In fact, zero energy-use growth, at least on a per capita basis, is a worthwhile and achievable goal. Such a policy would go a long way toward solving many environmental problems. With ingenuity, planning, and proper management, it could be achieved while increasing the standard of living and quality of life.

22.8. ENERGY CONVERSION PROCESSES

As shown in Figure 22.2, energy occurs in several forms and must be converted to other forms. The efficiencies of conversion vary over a wide range. Conversion of electrical energy to radiant energy by incandescent light bulbs is very inefficient — less than 5% of the energy is converted to visible light and the remainder is wasted as heat. At the other end of the scale, a large electrical generator is around 80% efficient in producing electrical energy from mechanical energy. The once much-publicized Wankel rotary engine converts chemical to mechanical energy with an efficiency of about 18%, compared to 25% for a gasoline-powered piston engine and about 37% for a diesel engine. A modern coal-fired steam-generating power plant converts chemical energy to electrical energy with an overall efficiency of about 40%.

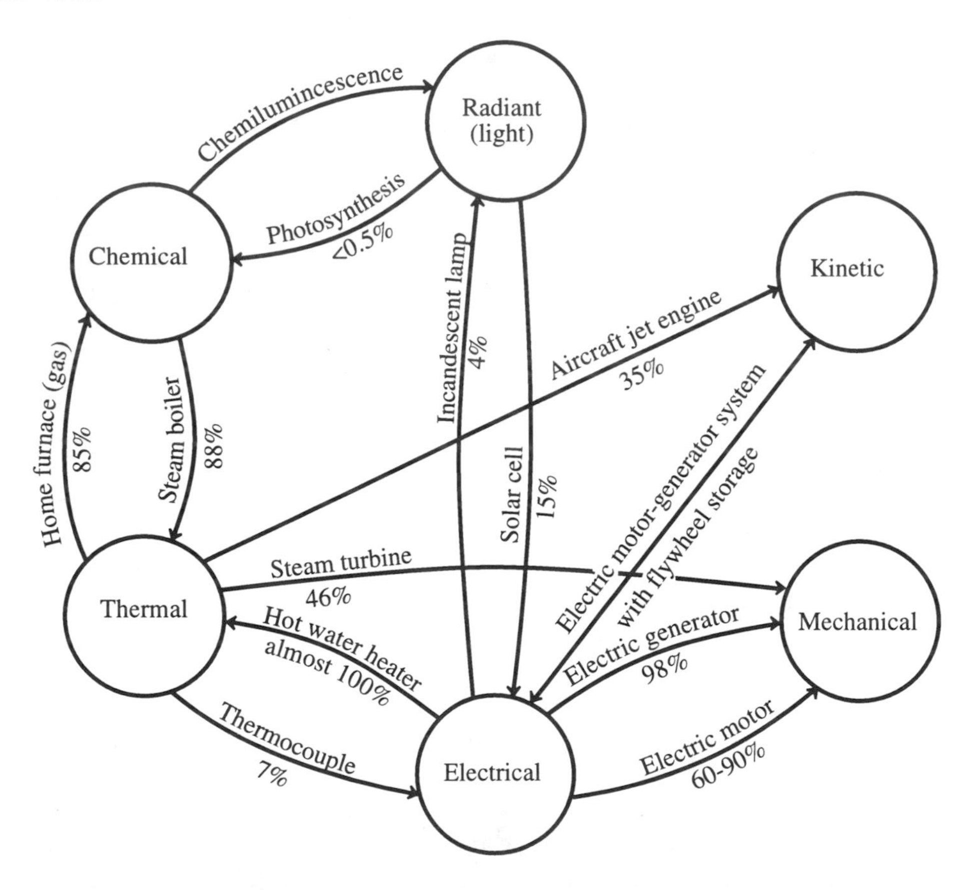

Figure 22.2. Kinds of energy and examples of conversion between them, with conversion efficiency percentages.

One of the most significant energy conversion processes is that of thermal energy to mechanical energy in a heat engine such as a steam turbine. The Carnot equation,

$$\text{Percent efficiency} = \frac{T_1 - T_2}{T_1} \times 100 \tag{22.8.1}$$

states that the percent efficiency is given by a fraction involving the inlet temperature (for example, of steam), T_1, and the outlet temperature, T_2. These temperatures are expressed in Kelvin (°C + 273). Typically, a steam turbine engine operates with approximately 810 K inlet temperature and 330 K outlet temperature. These temperatures substituted into the Carnot equation give a maximum theoretical efficiency of 59%. However, because it is not possible to maintain the incoming steam at the maximum temperature and because mechanical energy losses occur, overall efficiency of conversion of thermal energy to mechanical energy in a modern steam power plant is approximately 47%. Taking into account losses from conversion of chemical to thermal energy in the boiler, the total efficiency is about 40%.

Some of the greatest efficiency advances in the conversion of chemical to mechanical or electrical energy have been made by increasing the peak inlet temperature in heat engines. The use of superheated steam has raised T_1 in a steam power plant from around 550 K in 1900 to about 850 K at present. Improved materials and engineering design, therefore, have resulted in large energy savings.

The efficiency of nuclear power plants is limited by the maximum temperatures attainable. Reactor cores would be damaged by the high temperatures used in fossil-fuel-fired boilers and have a maximum temperature of approximately 620 K. Because of this limitation, the overall efficiency of conversion of nuclear energy to electricity is about 30%.

Most of the 60% of energy from fossil-fuel-fired power plants and 70% of energy from nuclear power plants that is not converted to electricity is dissipated as heat, either to the atmosphere or to bodies of water and streams. The latter is thermal pollution, which may either harm aquatic life or, in some cases, actually increase bioactivity in the water to the benefit of some species. This waste heat is potentially very useful in applications like home heating, water desalination, and aquaculture (growth of plants in water).

Some devices for the conversion of energy are shown in Figure 22.3. Substantial advances have been made in energy conversion technology over many decades and more can be projected for the future. Through the use of higher temperatures and larger generating units, the overall efficiency of fossil-fueled electrical power generation has increased approximately ten-fold since 1900, from less than 4% to a maximum of around 40%. An approximately four-fold increase in the energy-use efficiency of rail transport occurred during the 1940s and 1950s with the replacement of steam locomotives with diesel locomotives. During the coming decades, increased efficiency can be anticipated from such techniques as combined power cycles in connection with generation of electricity. Magnetohydrodynamics (Figure 22.5) probably will be developed as a very efficient energy source used in combination with conventional steam generation. Entirely new devices such as thermonuclear reactors for the direct conversion of nuclear fusion energy to electricity will very likely be developed.

22.9. PETROLEUM AND NATURAL GAS

Since its first commercial oil well in 1859, somewhat more than 100 billion barrels of oil have been produced in the United States, most of it in recent years. In 1994 world petroleum consumption was at a rate of about 65 million barrels per day.

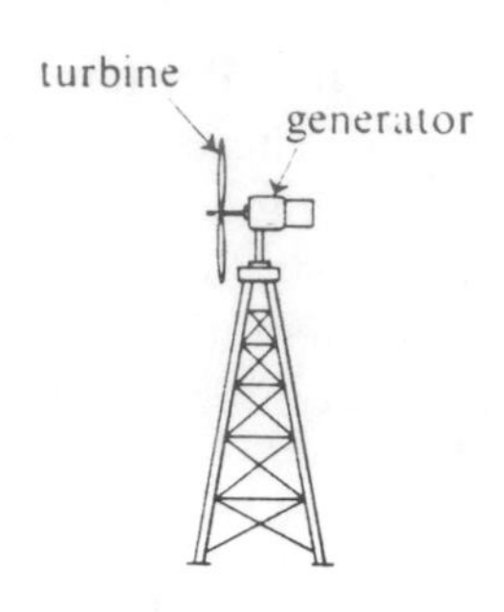

(1) Turbine for conversion of kinetic or potential energy of a fluid to mechanical and electrical energy.

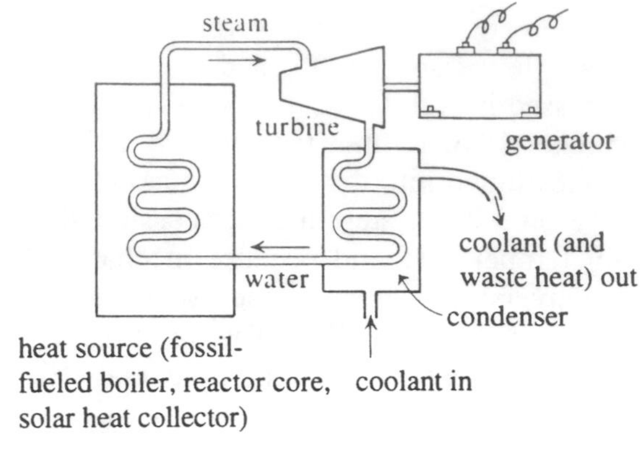

(2) Steam power plant in which high-energy fluid is produced by vaporizing water.

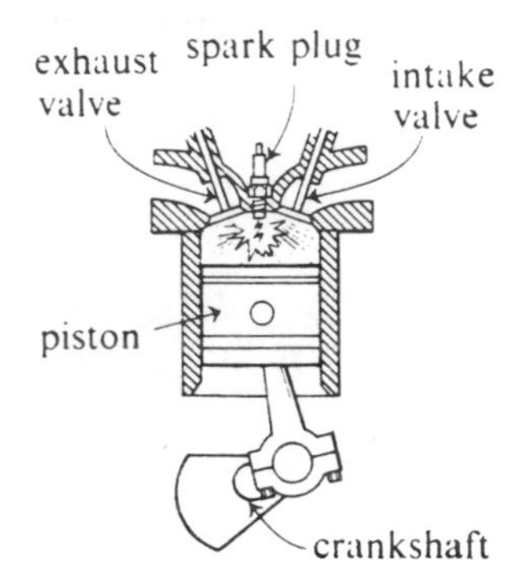

(3) Reciprocating internal combustion engine.

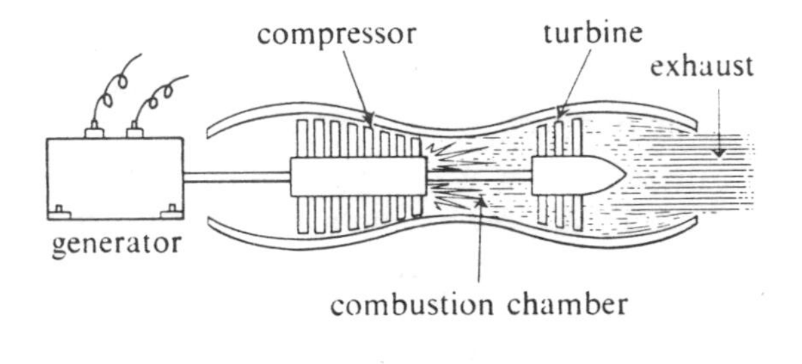

(4) Gas turbine engine. Kinetic energy of hot exhaust gases may propel aircraft.

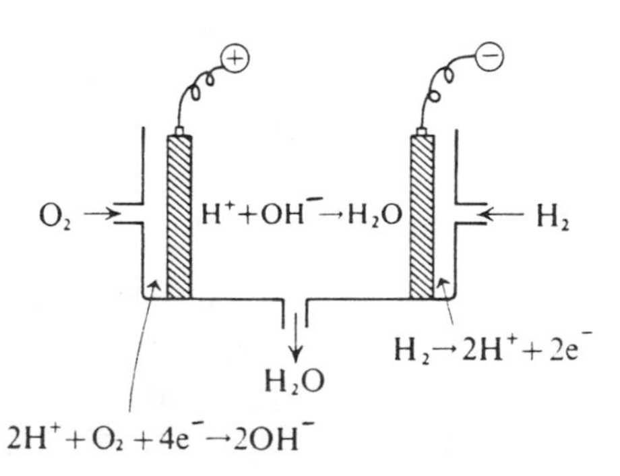

(5) Fuel cell for the direct conversion of chemical energy to electrical energy.

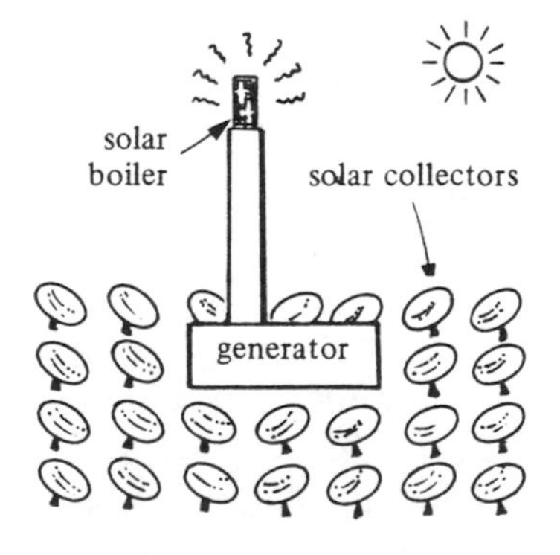

(6) Solar thermal electric conversion.

Figure 22.3. Some energy conversion devices.

Liquid petroleum is found in rock formations ranging in porosity from 10 to 30%. Up to half of the pore space is occupied by water. The oil in these formations must flow over long distances to an approximately 6-inch-diameter well from which it is pumped. The rate of flow depends on the permeability of the rock formation, the viscosity of the oil, the driving pressure behind the oil, and other factors. Because of limitations in these factors **primary recovery** of oil yields an average of about 30% of the oil in the formation, although it is sometimes as little as 15%. More oil can be obtained using **secondary recovery** techniques, which involve forcing water under pressure into the oil-bearing formation to drive the oil out. Primary and secondary recovery together typically extract somewhat less than 50% of the oil from a formation. Finally, **tertiary recovery** can be used to extract even more oil. This normally uses injection of pressurized carbon dioxide, which forms a mobile solution with the oil and allows it to flow more easily to the well. Other chemicals, such as detergents, may be used to aid in tertiary recovery. Currently, about 300 billion barrels of U.S. oil are not available through primary recovery alone. A recovery efficiency of 60% through secondary or tertiary techniques could double the amount of available petroleum. Much of this would come from fields which have already been abandoned or essentially exhausted using primary recovery techniques.

Shale oil is a possible substitute for liquid petroleum. Shale oil is a pyrolysis product of oil shale, a rock containing organic carbon in a complex structure called kerogen. It is believed that approximately 1.8. trillion barrels of shale oil could be recovered from deposits of oil shale in Colorado, Wyoming, and Utah. In the Colorado Piceance Creek basin alone, more than 100 billion barrels of oil could be recovered from prime shale deposits.

Shale oil may be recovered from the parent mineral by retorting the mined shale in a surface retort. This process requires the mining of enormous quantities of mineral and disposal of the spent shale, which has a volume greater than the original mineral. *In situ* retorting limits the control available over infiltration of underground water and resulting water pollution. Water passing through spent shale becomes quite saline, so there is major potential for saltwater pollution.

During the late 1970s and early 1980s, several corporations began building facilities for shale oil extraction in northwestern Colorado. Large investments were made in these operations, and huge expenditures were projected for commercialization. Falling crude oil prices caused all these operations to be cancelled. A large project for the recovery of oil from oil sands in Alberta, Canada, was also cancelled in the 1980s.

Natural gas, consisting almost entirely of methane, has become more attractive as an energy source. This is because of uncertainties regarding natural gas availability, coupled with the potential for the discovery and development of truly enormous new sources of this premium fuel. In 1968, discoveries of natural gas deposits in the U.S. fell below annual consumption for the first time. Price incentives resulting from the passage of the 1978 Gas Policy Act have tended to reverse the adverse trend of natural gas discoveries versus consumption.

In addition to its use as a fuel, natural gas can be converted to many other hydrocarbon materials. It can be used as a raw material for the Fischer-Tropsch synthesis of gasoline. The discovery and development of truly massive sources of natural gas, such as may exist in geopressurized zones, could provide abundant energy reserves for the U.S., though at substantially increased prices.

22.10. COAL

From Civil War times until World War II, coal was the dominant energy source behind industrial expansion in most nations. However, after World War II, the greater convenience of lower-cost petroleum resulted in a decrease in the use of coal for energy in the U.S. and in a number of other countries. Annual coal production in the U.S. fell by about one third, reaching a low of approximately 400 million tons in 1958. Since that time U.S. production has increased substantially and is now around 1 billion tons per year.

Several statistics illustrate the importance of coal as a source of energy by Earth's population. Overall, about 1/3 of the energy used by humankind is provided from coal. The percentage of electricity generated by coal is even higher, around 45%. Almost three fourths of the energy and coke used to make steel, the commodity commonly taken as a measure of industrial development, is provided by coal.

The general term *coal* describes a large range of solid fossil fuels derived from partial degradation of plants. Table 22.1 shows the characteristics of the major classes of coal found in the U.S., differentiated largely by percentage of fixed carbon, percentage of volatile matter, and heating value (*coal rank*). Chemically, coal is a very complex material and is by no means pure carbon. For example, a chemical formula expressing the composition of Illinois No. 6 bituminous coal is $C_{100}H_{85}S_{2.1}N_{1.5}O_{9.5}$.

Table 22.1. Major Types of Coal Found in the United States

	Proximate analysis, percent[a]				
Type of Coal	*Fixed carbon*	*Volatile matter*	*Moisture*	*Ash*	*Range of heating value (Btu/pound)*
Anthracite	82	5	4	9	13,000 - 16,000
Bituminous					
Low-volatile	66	20	2	12	11,000 - 15,000
Medium-volatile	64	23	3	10	11,000 - 15,000
High-volatile	46	44	6	4	11,000 - 15,000
Subbituminous		4032	19	9	8,000 - 12,000
Lignite		3028	37	5	5,500 - 8,000

[a] These values may vary considerably with the source of coal.

Figure 22.4 shows areas in the U.S. with major coal reserves. Anthracite, a hard, clean-burning, low-sulfur coal, is the most desirable of all coals. Approximately half of the anthracite originally present in the United States has been mined. Bituminous coal found in the Appalachian and north central coal fields has been widely used. It is an excellent fuel with a high heating value. Unfortunately, most bituminous coals have a high percentage of sulfur (an average of 2-3%), so the use of this fuel presents environmental problems. Huge reserves of virtually untouched subbituminous and lignite coals are found in the Rocky Mountain states and in the northern plains of Dakotas, Montana, and Wyoming. Despite some disadvantages, the low sulfur content and ease of mining these low-grade fuels are resulting in a rapid increase in their use.

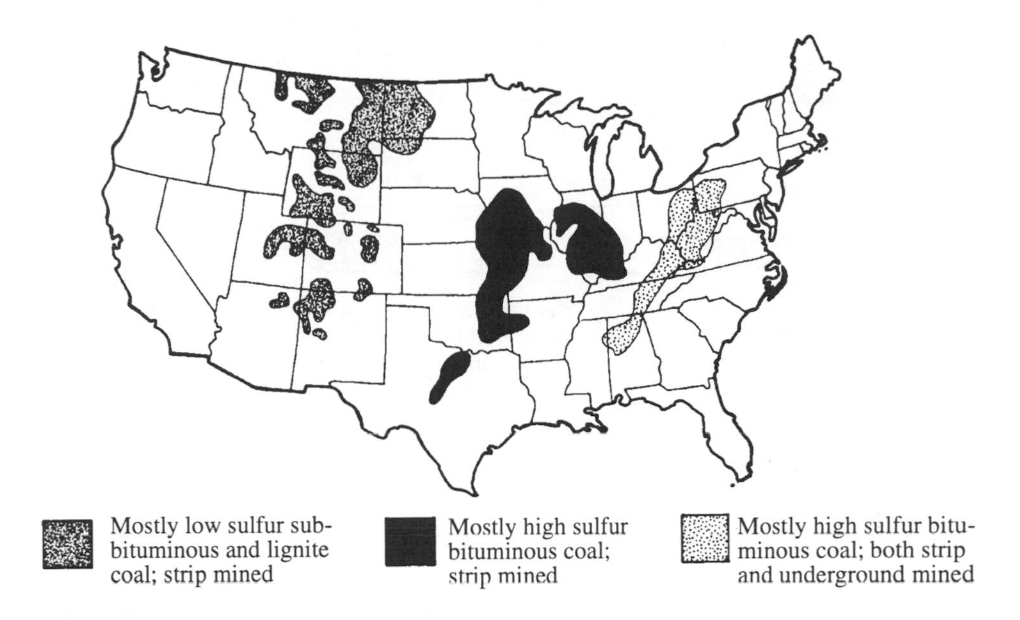

Figure 22.4. Areas with major coal reserves in the coterminous United States.

The extent to which coal can be used as a fuel depends upon solutions to several problems, including (1) minimizing the environmental impact of coal mining; (2) removing ash and sulfur from coal prior to combustion; (3) removing ash and sulfur dioxide from stack gas after combustion; (4) conversion of coal to liquid and gaseous fuels free of ash and sulfur; and, most important, (5) whether or not the impact of increased carbon dioxide emissions upon global climate can be tolerated. Progress is being made on minimizing the environmental impact of mining. As more is learned about the processes by which acid mine water is formed, measures can be taken to minimize the production of this water pollutant. Particularly on flatter lands, strip-mined areas can be reclaimed with relative success. Inevitably, some environmental damage will result from increased coal mining, but the environmental impact can be reduced by various control measures. Washing, flotation, and chemical processes can be used to remove some of the ash and sulfur prior to burning. Approximately half of the sulfur in the average coal occurs as pyrite, FeS_2, and half as organic sulfur. Although little can be done to remove the latter, much of the pyrite can be separated from most coals by physical and chemical processes.

The maintenance of air pollution emission standards requires the removal of sulfur dioxide from stack gas in coal-fired power plants. Stack gas desulfurization presents some economic and technological problems; the major processes available for it are summarized in Section 11.5.

Magnetohydrodynamic power combined with conventional steam generating units has the potential for a major breakthrough in the efficiency of coal utilization.[3] A schematic diagram of magnetohydrodynamic (MHD) generator is shown in Figure 22.5. This device uses a plasma of ionized gas at around 2400°C blasting through a very strong magnetic field of at least 50,000 gauss to generate direct current. The ionization of the gas is accomplished by injecting a "seed" of cesium or potassium salts. In an MHD generator, the ultra-high-temperature gas issuing through a supersonic nozzle contains ash, sulfur dioxide, and nitrogen oxides, which severely erode and corrode the materials used. This hot gas is used to generate steam for a conven-

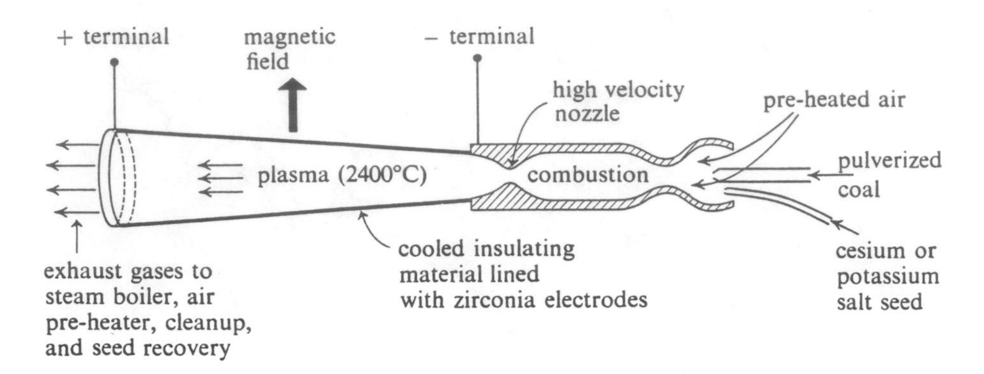

Figure 22.5. A magnetohydrodynamic power generator.

tional steam power plant, thus increasing the overall efficiency of the process. The seed salts combine with sulfur dioxide and are recovered along with ash in the exhaust. Pollutant emissions are low. The overall efficiency of combined MHD-steam power plants should reach 60%, one and one half times the maximum of present steam-only plants. Despite some severe technological difficulties, there is a chance that MHD power could become feasible on a large scale, and an experimental MHD generator was tied to a working power grid in the U.S.S.R. for several years. As of the early 1990s, the U.S. Department of Energy was conducting a proof-of-concept project to help determine the practicability of magnetohydrodynamics.

Coal Conversion

Coal can be converted to gaseous, liquid, or low-sulfur, low-ash solid fuels, such as coal char (coke) or solvent-refined coal (SRC). Coal conversion is an old idea; a house belonging to William Murdock at Redruth, Cornwall, England, was illuminated with coal gas in 1792. The first municipal coal-gas system was employed to light Pall Mall in London in 1807. The coal-gas industry began in the U.S. in 1816. The early coal-gas plants used coal pyrolysis (heating in the absence of air) to produce a hydrocarbon-rich product particularly useful for illumination. Later in the 1800s the water-gas process was developed, in which steam was added to hot coal to produce a mixture consisting primarily of H_2 and CO. It was necessary to add volatile hydrocarbons to this "carbureted" water-gas to bring its illuminating power up to that of gas prepared by coal pyrolysis. The U.S. had 11,000 coal gasifiers operating in the 1920s. At the peak of its use in 1947, the water-gas method accounted for 57% of U.S.-manufactured gas. The gas was made in low-pressure, low-capacity gasifiers that by today's standards would be inefficient and environmentally unacceptable (several locations of these old plants have been designated as hazardous waste sites because of residues of coal tar and other wastes). During World War II, Germany developed a major synthetic petroleum industry based on coal, which reached a peak capacity of 100,000 barrels per day in 1944. A plant now operating in Sasol, South Africa, converts several tens of thousands of tons of coal per day to synthetic petroleum.

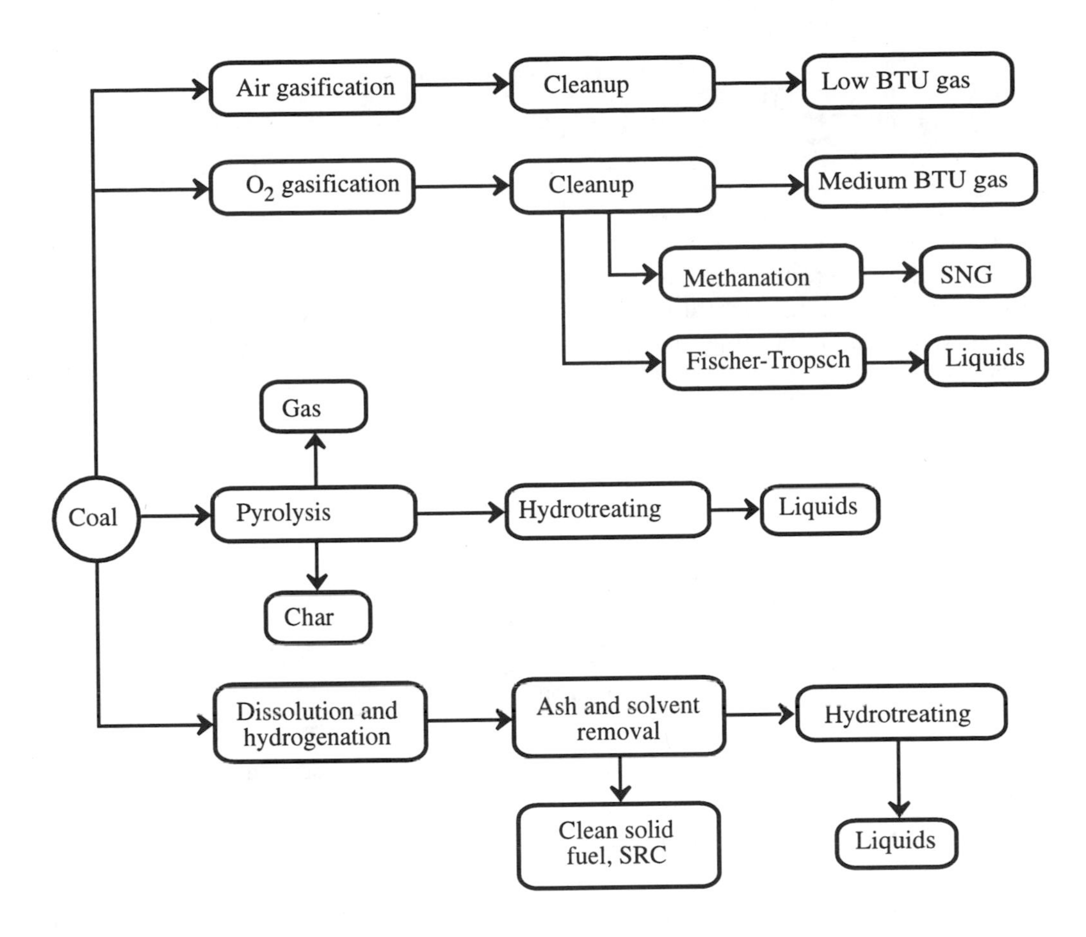

Figure 22.6. Routes to coal conversion.

The two broadest categories of coal conversion are gasification and liquefaction. Arguably the most developed route for coal gasification is the **Texaco process**, which gasifies a water slurry of coal at temperatures of 1250°C to 1500°C and pressures of 350 to 1200 pounds per square inch. Chemical addition of hydrogen to coal can liquefy it and produce a synthetic petroleum product. This can be done with a hydrogen donor solvent, which is recycled and itself hydrogenated with H_2 during part of the cycle. Such a process forms the basis of the successful **Exxon Donor Solvent process**, which has been used in a 250 ton/day pilot plant.

A number of environmental implications are involved in the widespread use of coal conversion. These include strip mining, water consumption in arid regions, lower overall energy conversion compared to direct coal combustion, and increased output of atmospheric carbon dioxide. These plus economic factors have prevented coal conversion from being practiced on a very large scale.

22.11. NUCLEAR FISSION POWER

The awesome power of the atom revealed at the end of World War II held out enormous promise for the production of abundant, cheap energy. This promise has never really come to full fruition, although nuclear energy currently provides a significant percentage of electric energy in many countries.

Nuclear power reactors currently in use depend upon the fission of uranium-235 nuclei by reactions such as

$$^{235}_{92}U + ^{1}_{0}n \rightarrow ^{133}_{51}Sb + ^{99}_{41}Nb + 4^{1}_{0}n \qquad (22.11.1)$$

to produce two radioactive fission products, an average of 2.5 neutrons, and an average of 200 MeV of energy per fission. The neutrons, initially released as fast-moving, highly energetic particles, are slowed to thermal energies in a moderator medium. For a reactor operating at a steady state, exactly one of the neutron products from each fission is used to induce another fission reaction in a chain reaction (Figure 22.8):

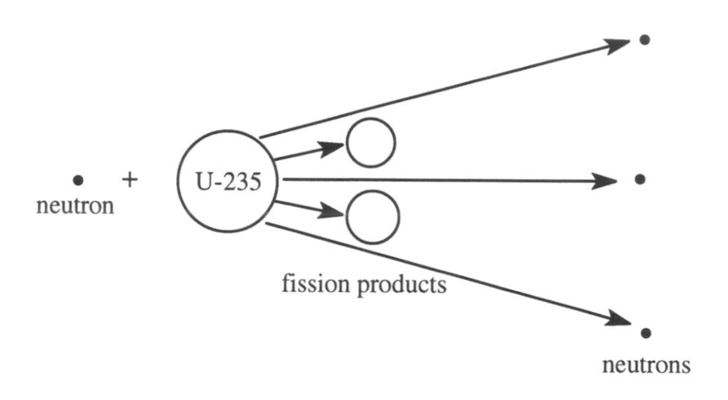

Figure 22.7. Fission of a uranium-235 nucleus.

The energy from these nuclear reactions is used to heat water in the reactor core and produce steam to drive a steam turbine, as shown in Figure 22.8. As noted in Section 22.8, temperature limitations make nuclear power less efficient in converting heat to mechanical energy, and, therefore, to electricity, than fossil energy conversion processes.

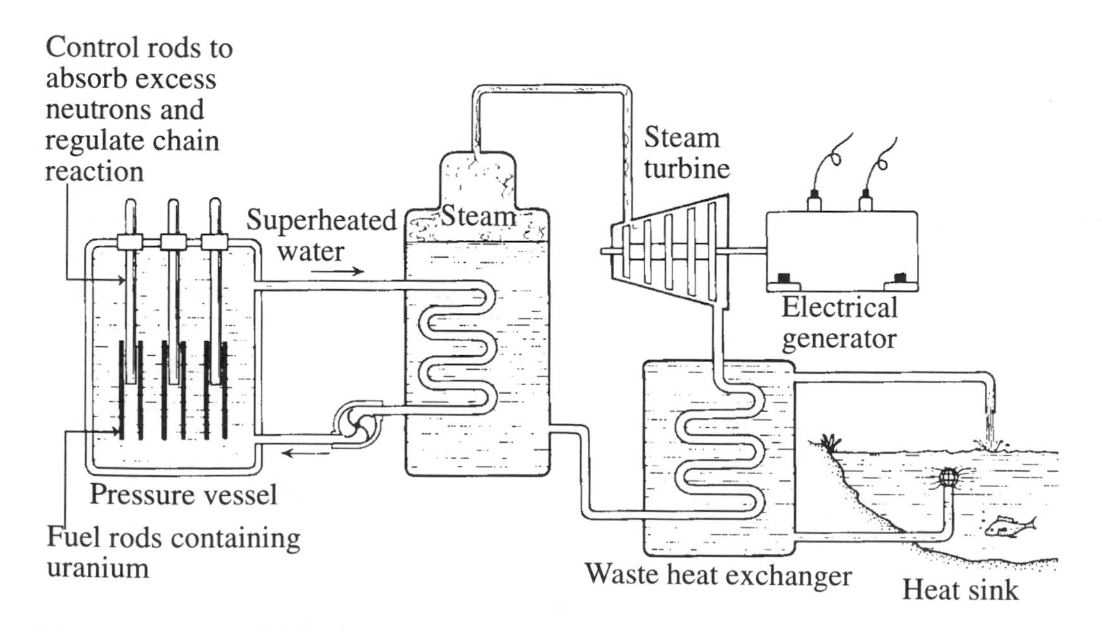

Figure 22.8. A typical nuclear fission power plant.

A limitation of fission reactors is the fact that only 0.71% of natural uranium is fissionable uranium-235. This situation could be improved by the development of **breeder reactors**, which convert uranium-238 (natural abundance 99.28%) to fissionable plutonium-239.

A major consideration in the widespread use of nuclear fission power is the production of large quantities of highly radioactive waste products. These remain lethal for thousands of years. They must either be stored in a safe place or disposed of permanently in a safe manner. At the present time, spent fuel elements are being stored under water at the reactor sites. Eventually, the wastes from this fuel will have to be buried.

Another problem to be faced with nuclear fission reactors is their eventual decommissioning. There are three possible solutions. One is dismantling soon after shutdown, in which the fuel elements are removed, various components are flushed with cleaning fluids, and the reactor is cut up by remote control and buried. "Safe storage" involves letting the reactor stand 30-100 years to allow for radioactive decay, followed by dismantling. The third alternative is entombment, encasing the reactor in a concrete structure.

The course of nuclear power development was altered drastically by two accidents. The first of these occurred on March 28, 1979, with a partial loss of coolant water from the Metropolitan Edison Company's nuclear reactor located on Three Mile Island in the Susquehanna River, 28 miles outside of Harrisburg, Pennsylvania. The result was a loss of control, overheating, and partial disintegration of the reactor core. Some radioactive xenon and krypton gas were released and some radioactive water was dumped into the Susquehanna River. In August of 1993, cleanup workers finished evaporating the water from about 8 million liters of water solution contaminated by the reactor accident, enabling the reactor building to be sealed. A much worse accident occurred at Chernobyl in the Soviet Union in April of 1986 when a reactor blew up spreading radioactive debris over a wide area and killing a number of people (officially 31, but probably many more).[4] Thousands of people were evacuated and the entire reactor structure had to be entombed in concrete. Food was seriously contaminated as far away as northern Scandinavia.

As of 1997, 19 years had passed since a new nuclear electric power plant had been ordered in the U.S. Although this tends to indicate hard times for the nuclear industry, pronouncements of its demise may be premature. Properly designed nuclear fission reactors can generate large quantities of electricity reliably and safely. For example, during the record summer 1993 Mississippi/Missouri River floods, many large fossil-fueled power plants were on the verge of shutting down because of disruptions of fuel supply normally delivered by river barge and train. During that time Union Electric's large Callaway nuclear plant in central Missouri ran continuously at full capacity, immune to the effects of the flood, probably saving a large area from a devastating, long-term power outage. The single most important factor that may lead to renaissance of nuclear energy is the threat to the atmosphere from greenhouse gases produced in large quantities by fossil fuels. It can be argued that nuclear energy is the only proven alternative that can provide the amounts of energy required within acceptable limits of cost, reliability, and environmental effects.

New designs for nuclear power plants should enable construction of power reactors that are much safer and environmentally acceptable than those built with older technologies. The proposed new designs incorporate built-in passive safety features that work automatically in event of problems that could lead to incidents such as TMI or Chernobyl with older reactors. These devices, which depend upon phenomena such as gravity feeding of coolant, evaporation of water, or convection flow of fluids, give the reactor the desirable characteristics of **passive stability**. They have also enabled significant simplification of hardware, with only about half as many pumps, pipes, and heat exchangers as are contained in older power reactors.

22.12. NUCLEAR FUSION POWER

The two main reactions by which energy can be produced from the fusion of two light nuclei into a heavier nucleus are the deuterium-deuterium reaction,

$$ {}^{2}_{1}H + {}^{2}_{1}H \rightarrow {}^{3}_{2}He + {}^{1}_{0}n + 1 \text{ Mev (energy released per fusion)} \qquad (22.12.1)$$

and the deuterium-tritium reaction:

$$ {}^{2}_{1}H + {}^{3}_{1}H \rightarrow {}^{4}_{2}He + {}^{1}_{0}n + 17.6 \text{ Mev (energy released per fusion)} \qquad (22.12.2)$$

The second reaction is more feasible because less energy is required to fuse the two nuclei than to fuse two deuterium nuclei. However, the total energy from deuterium-tritium fusion is limited by the availability of tritium, which is made from nuclear reactions of lithium-6 (natural abundance, 7.4%). The supply of deuterium, however, is essentially unlimited; one out of every 6700 atoms of hydrogen is the deuterium isotope. The ^{3}He by-product of two deuterium nuclei, Reaction 22.12.1, reacts with neutrons, which are abundant in a nuclear fusion reactor, to produce tritium required for Reaction 22.12.2.

The power of nuclear fusion has not yet been harnessed in a sustained, controlled reaction of appreciable duration that produces more power than it consumes. Most approaches have emphasized "squeezing" a plasma (ionized gas) of fusionable nuclei in a strong magnetic field. The two largest such reactors have been the Tokamak Fusion Test Reactor operated by Princeton University for the U.S. Department of Energy and the Joint European Torus located in England. In 1994 a record power level of 10.7 megawatts (MW) was achieved by the Tokamak reactor, though for only about 1 second. This level was only about 20% of the power put into the reactor to achieve fusion, which, of course, would have to be boosted to well over 100% for a self-sustained fusion reactor. In April, 1997, after 15 years of testing, the Tokamak reactor was shut down for the last time because of budget cuts. Limited U.S. funds for the support of fusion energy research may be used for a proposed International Thermonuclear Experimental Reactor sponsored by most European countries, Japan, Russia, and the U.S., although funding for that project is somewhat in doubt.

A great flurry of excitement over the possibility of a cheap, safe, simple fusion power source was generated by an announcement from the University of Utah in 1989 of the attainment of "cold fusion" in the electrolysis of deuterium oxide (heavy water).[5] Funding was appropriated and laboratories around the world were thrown into frenetic activity in an effort to duplicate the reported results. Some investigators reported evidence, particularly the generation of anomalously large amounts of heat, to support the idea of cold fusion, whereas others scoffed at the idea. Since that time, cold fusion has been disproven, and the whole saga of it stands as a classic case of science gone astray.

Controlled nuclear fusion processes would produce almost no radioactive waste products. However, tritium is very difficult to contain, and some release of the isotope would occur. The deuterium-deuterium reaction promises an unlimited source of energy. Either of these reactions would be preferable to fission in terms of environmental considerations. Therefore, despite the possibility of insurmountable technical problems involved in harnessing fusion energy, the promise of this abundant, relatively nonpolluting energy source makes its pursuit well worth a massive effort.

22.13. GEOTHERMAL ENERGY

Underground heat in the form of steam, hot water, or hot rock used to produce steam is already being used as an energy resource. This energy was first harnessed for the generation of electricity at Larderello, Italy, in 1904, and has since been developed in Japan, Russia, New Zealand, the Phillipines, and at the Geysers in northern California.

Underground dry steam is relatively rare, but is the most desirable from the standpoint of power generation. More commonly, energy reaches the surface as superheated water and steam. In some cases, the water is so pure that it can be used for irrigation and livestock; in other cases, it is loaded with corrosive, scale-forming salts. Utilization of the heat from contaminated geothermal water generally requires that the water be reinjected into the hot formation after heat removal to prevent contamination of surface water.

The utilization of hot rocks for energy requires fracturing of the hot formation, followed by injection of water and withdrawal of steam. This technology is still in the experimental state, but promises approximately ten times as much energy production as steam and hot-water sources.

Land subsidence and seismic effects are environmental factors that may hinder the development of geothermal power. However, this energy source holds considerable promise, and its development continues.

22.14. THE SUN: AN IDEAL ENERGY SOURCE

Solar power is an ideal source of energy that is unlimited in supply, widely available, and inexpensive. It does not add to the Earth's total heat burden or produce chemical air and water pollutants. On a global basis, utilization of only a small fraction of solar energy reaching the Earth could provide for all energy needs. In the United States, for example, with conversion efficiencies ranging from 10-30%, it would only require collectors ranging in area from one tenth down to one thirtieth that of the state of Arizona to satisfy present U.S. energy needs. (This is still an enormous amount of land, and there are economic and environmental problems related to the use of even a fraction of this amount of land for solar energy collection. Certainly, many residents of Arizona would not be pleased at having so much of the state devoted to solar collectors, and some environmental groups would protest the resultant shading of rattlesnake habitat.)

Solar power cells (photovoltaic cells) for the direct conversion of sunlight to electricity have been developed and are widely used for energy in space vehicles. With present technology, however, they remain too expensive for large-scale generation of electricity, although the economic gap is narrowing. Most schemes for the utilization of solar power depend upon the collection of thermal energy, followed by conversion to electrical energy. The simplest such approach involves focusing sunlight on a steam-generating boiler. Parabolic reflectors can be used to focus sunlight on pipes containing heat-transporting fluids. Selective coatings on these pipes can be used so that most of the incident energy is absorbed.

The direct conversion of energy in sunlight to electricity is accomplished by special solar voltaic cells.[6] Such devices based on crystalline silicon have operated with a 15% efficiency for experimental cells and 11-12% for commercial units, at a cost of 25-50 cents per kilowatt-hour (kWh), about 5 times the cost of conventionally generated electricity. Part of the high cost results from the fact that the silicon used in the cells must be cut as small wafers from silicon crystals for mounting on the cell surfaces. Significant advances in costs and technology are being made with thin-film photovoltaics, which use an amorphous silicon alloy. A new approach to the design and construction of amorphous silicon film photovoltaic devices uses three layers of amorphous silicon to absorb, successively, short wavelength ("blue"), intermediate wavelength ("green"), and long wavelength ("red") light, as shown in Figure 22.9. Thin-film solar panels constructed with this approach have achieved solar-to-electricity energy conversion efficiencies just over 10%, lower than those using crystalline silicon, but higher than other amorphous film devices. The low cost and relatively high conversion efficiencies of these solar panels should enable production of electricity at only about twice the cost of conventional electrical power, which would be competitive in some situations.

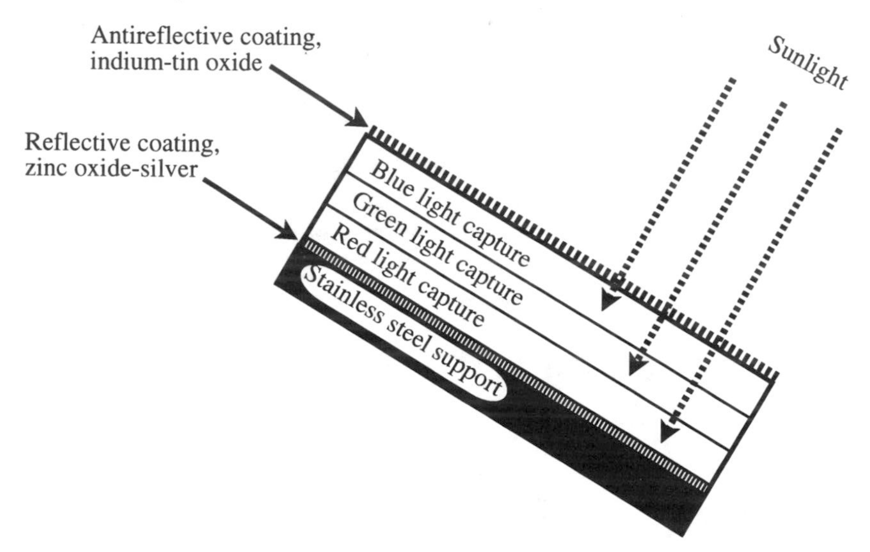

Figure 22.9. High-efficiency thin-film solar photovoltaic cell using amorphous silicon.

A major disadvantage of solar energy is its intermittent nature. However, flexibility inherent in an electric power grid would enable it to accept up to 15% of its total power input from solar energy units without special provision for energy storage. Existing hydroelectric facilities may be used for pumped-water energy storage in conjunction with solar electricity generation. Heat or cold can be stored in water, in a latent form in water (ice) or eutectic salts, or in beds of rock. Enormous amounts of heat can be stored in water as a supercritical fluid contained at high temperatures and very high pressures deep underground. Mechanical energy can be stored with compressed air or flywheels.

Hydrogen gas, H_2, is an ideal chemical fuel that may serve as a storage medium for solar energy. Electricity generated by solar means can be used to electrolyze water:

$$2H_2O + \text{electrical energy} \rightarrow 2H_2(g) + O_2(g) \quad (22.14.1)$$

The hydrogen fuel product, and even oxygen, can be piped some distance and the hydrogen burned without pollution or it may be used in a fuel cell (Figure 22.3). This may, in fact, make possible a "hydrogen economy." Disadvantages of using hydrogen as a fuel include the fact that it has a heating value per unit volume of about one third that of natural gas and that it is explosive over a wide range of mixtures with air.

No really insurmountable barriers exist to block the development of solar energy, such as might be the case with fusion power. In fact, the installation of solar space and water heaters became widespread in the late 1970s, and research on solar energy was well supported in the U.S. until after 1980, when it became fashionable to believe that free-market forces had solved the "energy crisis." With the installation of more heating devices and the probable development of some cheap, direct solar electrical generating capacity, it is likely that during the coming century solar energy will be providing an appreciable percentage of energy needs in areas receiving abundant sunlight.

22.15. ENERGY FROM BIOMASS

All fossil fuels originally came from photosynthetic processes. Photosynthesis does hold some promise of producing combustible chemicals to be used for energy production and could certainly produce all needed organic raw materials. It suffers from the disadvantage of being a very inefficient means of solar energy collection (a collection efficiency of only several hundredths of a percent by photosynthesis is typical of most common plants). However, the overall energy conversion efficiency of several plants, such as sugarcane, is around 0.6%. Furthermore, some plants, such as *Euphorbia lathyrus* (gopher plant), a small bush growing wild in California, produce hydrocarbon emulsions directly. The fruit of the Philippine plant, *Pittsosporum reiniferum*, can be burned for illumination due to its high content of hydrocarbon terpenes (see Section 12.2), primarily α-pinene and myrcene. Conversion of agricultural plant residues to energy could be employed to provide some of the energy required for agricultural production. Indeed, until about 80 years ago, virtually all of the energy required in agriculture—hay and oats for horses, home-grown food for laborers, and wood for home heating—originated from plant materials produced on the land. (An interesting exercise is to calculate the number of horses required to provide the energy used for transportation at the present time in the Los Angeles basin. It can be shown that such a large number of horses would fill the entire basin with manure at a rate of several feet per day.)

Annual world production of biomass is estimated at 146 billion metric tons, mostly from uncontrolled plant growth. Many farm crops and trees can produce 10-20 metric tons per acre per year of dry biomass and some algae and grasses can produce as much as 50 metric tons per acre per year. The heating value of this biomass is 5000-8000 Btu/lb for a fuel having virtually no ash or sulfur (compare heating values of various coals in Table 22.1). Current world demand for oil and gas could be met with about 6% of the global production of biomass. Meeting U.S. demands for oil and gas would require that about 6-8% of the land area of the contiguous 48 states be cultivated intensively for biomass production. Another advantage of this source of energy is that use of biomass for fuel would not add any net carbon dioxide to the atmosphere.

As it has been throughout history, biomass is significant as heating fuel, and in some parts of the world is the fuel most widely used for cooking.[7] For example, as of the early 1990s, about 15% of Finland's energy needs were provided by wood and wood products (including black liquor by-product from pulp and paper manufacture), about 1/3 of which was from solid wood. Despite the charm of a wood fire and the sometimes pleasant odor of wood smoke, air pollution from wood-burning stoves and furnaces is a significant problem in some areas. Currently wood provides about 8% of world energy needs. This percentage could increase through the development of "energy plantations" consisting of trees grown solely for their energy content.

Seed oils show promise as fuels, particularly for use in diesel engines. The most common plants producing seed oils are sunflowers and peanuts. More exotic species include the buffalo gourd, cucurbits, and Chinese tallow tree.

Biomass could be used to replace much of the 100 million metric tons of petroleum and natural gas currently consumed in the manufacture of primary chemicals in the world each year. Among the sources of biomass that could be used for chemical production are grains and sugar crops (for ethanol manufacture), oilseeds, animal by-products, manure, and sewage (the last two for methane generation). The biggest potential source of chemicals is the lignocellulose making up the bulk of most plant material. For example, both phenol and benzene might be produced directly from lignin. Brazil has had a program for the production of chemicals from fermentation-produced ethanol.

Gasohol

A major option for converting photosynthetically produced biochemical energy to forms suitable for internal combustion engines is the production of either methanol or ethanol. Either can be used by itself as fuel in a suitably designed internal combustion engine. More commonly, these alcohols are blended in proportions of up to 20% with gasoline to give **gasohol**, a fuel that can be used in existing internal combustion engines with little or no adjustment.

Gasohol offers some advantages as a fuel. It boosts octane rating and reduces emissions of carbon monoxide. From a resource viewpoint, because of its photosynthetic origin, alcohol may be considered a renewable resource rather than a depletable fossil fuel. The manufacture of alcohol can be accomplished by the fermentation of sugar obtained from the hydrolysis of cellulose in wood wastes and crop wastes. Fermentation of these waste products offers an excellent opportunity for recycling. Cellulose has significant potential for the production of renewable fuels.

Ethanol is most commonly manufactured by fermentation of carbohydrates. Brazil, a country rich in potential to produce biomass, such as sugarcane, has been a leader in the manufacture of ethanol for fuel uses, with 4 billion liters produced in 1982. At one time Brazil had over 450,000 automobiles that could run on pure alcohol, although many of these were converted back to gasoline during the era of relatively low petroleum prices in the 1980s. Significant amounts of gasoline in the United States are supplemented with ethanol, more as an octane-ratings booster, rather than as a fuel supplement.

Methanol, which can be blended with gasoline, can also be produced from biomass by the destructive distillation of wood (Section 22.4) or by converting biomass, such as wood, to CO and H_2, and synthesizing methanol from these gases.

22.16. FUTURE ENERGY SOURCES

As discussed in this chapter, a number of options are available for the supply of energy in the future. The major possibilities are summarized in Table 22.2.

Table 22.2. Possible Future Sources of Energy

Source	Principles
Coal conversion	Manufacture of gas, hydrocarbon liquids, alcohol, or solvent-refined coal (SRC) from coal.
Oil shale	Retorting petroleum-like fuel from oil shale.
Geothermal	Utilization of underground heat.
Gas-turbine topping cycle	Utilization of hot combustion gases in a turbine, followed by steam generation.
MHD	Electricity generated by passing a hot gas plasma through a magnctic field.
Thermionics	Electricity generated across a thermal gradient.
Fuel cells	Conversion of chemical to electrical energy.
Solar heating and cooling	Direct use of solar energy for heating and cooling through the application of solar collectors.
Solar cells	Use of silicon semiconductor sheets for the direct generation of electricity from sunlight.
Solar thermal electric	Conversion of solar energy to heat followed by conversion to electricity.
Wind	Conversion of wind energy to electricity.
Ocean thermal electric	Use of ocean thermal gradients to convert heat energy to electricity.
Nuclear fission	Conversion of energy released from fission of hcavy nuclei to electricity.
Breeder reactors	Nuclear fission combined with conversion of nonfissionable nuclei to fissionable nuclei.
Nuclear fusion	Conversion of energy released by the fusion of light nuclei to electricity.
Bottoming cycles	Utilization of waste heat from power generation for various purposes.
Solid waste	Combustion of trash to produce heat and electricity.
Photosynthesis	Use of plants for the conversion of solar energy to other forms by a biomass intermediate.
Hydrogen	Generation of H_2 by thermochemical means for use as an energy-transporting medium.

CHAPTER SUMMARY

The chapter summary below is presented in a programmed format to review the main points covered in this chapter. It is used most effectively by filling in the blanks, referring back to the chapter as necessary. The correct answers are given at the end of the summary.

In addition to the traditional considerations of economics, resource economics takes account of [1] ____

____

____.

A particularly important aspect of environmental economics consists of external costs that are [2] ____

____.

The major characteristics of a steady-state economy are [3] ____________________

____

____

____

___. Sources of raw materials can be broadly divided between the two categories of [4] ____ sources. The most common raw material for the manufacture of other products worldwide is [5] ______________. Two products that can be obtained by the chemical processing of wood are [6] ________________________________. At present, most of the energy consumed by humans is produced from [7] ____________ __________. Although the most abundant of these is [8] ____________, its utilization is limited by environmental disruption from mining and emissions of [9] __________

____.

Reserves of fissionable nuclear fuels are much higher than those of fossil fuels assuming use of [10] ________________________________. The Carnot equation,

$$\text{Percent efficiency} = \frac{T_1 - T_2}{T_1} \times 100$$

expresses a limitation to [11] ____

____.

About [12] ______________ percent of energy from fossil-fuel-fired power plants and [13] __________ percent of energy from nuclear power plants is not converted to electricity and is [14] ________________________________. Up to half of the pore space in petroleum-bearing formations is occupied by [15] __________. The three levels of petroleum recovery are [16] ____

____________________, the last of which uses [17] ____________________________

________________________________. Other than crude oil pumped directly from underground, two other major sources of liquid petroleum are [18] ______________

______________. The general term coal describes [19] __________________________

____. Major classes of coal are differentiated largely by [20] ____

________________________________ expressed as [21] ____________________.

Coal conversion refers to production of [22] ______________________________.

Five different fuels formed from coal conversion are [23]____. As related to energy, the reaction,

$$ {}^{235}_{92}U + {}^{1}_{0}n \rightarrow {}^{133}_{51}Sb + {}^{99}_{41}Nb + 4{}^{1}_{0}n $$

illustrates [24]________________________________.The thermal efficiency of nuclear reactors is inherently low because of [25]____.

The fissionable isotope of uranium is [26]________________, which is [27]__________ percent of natural uranium. The reaction,

$$ {}^{2}_{1}H + {}^{3}_{1}H \rightarrow {}^{4}_{2}He + {}^{1}_{0}n + 17.6MeV $$

is an example of [28]________________. The most desirable form of geothermal energy is [29]________________________________. An energy source that is "unlimited in supply, widely available, and inexpensive" is [30]________________________. The efficiency of collection of solar energy by photosynthesis is of the order of [31]________________________.

Answers

1 pollution, general environmental degradation, effects on life-support systems, and other broadly based and environmentally connected economic concerns associated with resource utilization

2 borne by those other than individuals receiving a direct, specific benefit

3 essentially complete recycling of nonrenewable materials, maximum utilization of renewable energy resources, maximum efficiency in energy and materials utilization, design and manufacture for maximum durability and recyclability, and steady level of population with low rates of both births and deaths

4 extractive (nonrenewable) and renewable

5 wood

6 methanol and sugar

7 fossil fuels

8 coal

9 carbon dioxide and sulfur dioxide

10 breeder reactors

11 the percentage of energy from a source that can actually be utilized

12 60

13 70

14 dissipated as heat

15 water

16 primary, secondary, and tertiary

17 injection of pressurized carbon dioxide

18 oil shale and tar sands

19 a large range of solid fossil fuels derived from partial degradation of plants

20 percentage of fixed carbon, percentage of volatile matter, and heating value

21 coal rank

22 gaseous, liquid, or low-sulfur, low-ash fuels

23 solvent-refined coal, low-sulfur boiler fuels, liquid hydrocarbon fuels, synthetic natural gas, low-sulfur, low-Btu gas for industrial use

24 production of energy by nuclear fission

25 low operating temperatures that must be maintained because of limitations of structures and materials so that conversion of heat to useful energy as expressed by the Carnot relationship is low

26 uranium-235

27 0.71

28 nuclear fusion

29 underground dry steam

30 solar energy

31 several hundreths of a percent.

QUESTIONS AND PROBLEMS

1. What is meant by resource economics, and why is it so essential for modern societies?
2. What is meant by external costs in environmental economics?
3. Consider the economic system in which you live. To what extent is it a steady-state economic system?
4. Arrange the following energy conversion processes in order from the least to the most efficient: (a) electric hot water heater, (b) photosynthesis, (c) solar cell, (d) electric generator, (e) aircraft jet engine.
5. Considering the Carnot equation and common means for energy conversion, what might be the role of improved materials (metal alloys, ceramics) in increasing energy conversion efficiency?
6. Why is shale oil, a possible substitute for petroleum in some parts of the world, considered to be a pyrolysis product?

7. List some coal ranks and describe what is meant by coal rank.

8. Why was it necessary to add hydrocarbons to gas produced by reacting steam with hot carbon from coal in order to make a useful gas product?

9. What is the principle of the Exxon Donor Solvent process for producing liquid hydrocarbons from coal?

10. As it is now used, what is the principle or basis for the production of energy from uranium by nuclear fission? Is this process actually used for energy production? What are some of its environmental disadavanges? What is one major advantage?

11. What would be at least two highly desirable features of nuclear fusion power if it could ever be achieved in a controllable fashion on a large scale?

12. Justify describing the sun as "an ideal energy source." What are two big disadvantages of solar energy?

13. What are some of the greater implications of the use of biomass for energy? How might such widespread use affect greenhouse warming? How might it affect agricultural production of food?

14. Describe how gasohol is related to energy from biomass.

LITERATURE CITED

[1] Zoebelein, Kans, Ed., *Dictionary of Renewable Resources*, VCH Publishing Co., New York, NY, 1997.

[2] Gottschalk, Charles M., *Industrial Energy Conservation*, John Wiley & Sons, New York, NY, 1996.

[3] Messerle, Hugo K., *Magnetohydrodynamic Electrical Power Generation*, John Wiley & Sons, New York, NY, 1995.

[4] Yaroshinska, Alla, *Chernobyl: The Forbidden Truth*, University of Nebraska Press. Lincoln, NE,1995.

5 Parkin, Lance, *Cold Fusion (Missing Adventures)*, London Bridge Mass Market, London, UK, 1997.

6 Neville, Richard C., *Solar Energy Conversion: The Solar Cell*, 2nd ed., Elsevier Science Ltd, Amsterdam, Netherlands, 1995.

[7] Wereko-Brobby, Charles Y. and Essel B. Hagan, *Biomass Conversion and Technology*, John Wiley & Sons, New York, NY, 1996.

SUPPLEMENTARY REFERENCES

Bisio, Attilio, and Sharon R. Boots, *Energy Technology and the Environment*, John Wiley and Sons, New York, NY, 1995.

Cassedy, Edward S., and Peter Z. Grossman, *Introduction to Energy. Resources, Technology, and Society*, Cambridge University Press, New York, NY, 1990.

Cohen, Bernard L., *The Nuclear Energy Option. An Alternative for the 90s*, Plenum, New York, NY, 1990.

Elliott, David, *Energy, Society, and Environment: Technology for a Sustainable Future*, Routledge, New York, NY, 1997.

Frederick, Kenneth D. and Roger A. Sedjo, *America's Renewable Resources: Historical Trends and Current Challenges*, Resources for the Future, Washington, D.C., 1991.

Institute of Environmental Sciences, *Emerging Environmental Solutions for the Eighties—Energy and the Environment*, Institute of Environmental Sciences, New York, NY, 1997.

Liu, Paul I., *Introduction to Energy and the Environment*, Van Nostrand Reinhold, New York, NY, 1993.

Ogden, Joan M. and Robert H. Williams, *Solar Hydrogen. Moving Beyond Fossil Fuels*, World Resources Institute, Washington, D.C., 1989.

Socolow, Robert H., *Annual Review of Energy and the Environment*, Vol. 22, Annual Reviews, New York, NY, 1997.

Walker, Graham, *The Stirling Alternative: Power Systems, Refrigerants, and Heat Pumps*, Gordon and Breach Publishers, Langhorne, PA, 1994.

23 WASTES FROM THE ANTHROSPHERE

23.1. INTRODUCTION

Human activities produce large quantities of wastes. Some of these are ejected into the atmosphere, where they may reside for some time as pollutants or are transformed by atmospheric chemical processes to other pollutants. Some of the potentially harmful by-products of manufacturing and other human activities are dumped into water and become water pollutants. Other potential pollutants end up on land. A major concern, therefore, with improperly controlled human activities is the production and distribution of hazardous substances and hazardous wastes—wastes from the anthrosphere. These materials and their potential environmental impact are discussed in this chapter.

Hazardous Substances and Hazardous Wastes

A **hazardous substance** is a material that may pose a danger to living organisms, materials, structures, or the environment by explosion or fire hazards, corrosion, toxicity to organisms, or other detrimental effects. What then is a hazardous waste? Although it has has been stated[1] that, "The discussion on this question is as long as it is fruitless," a simple definition of a **hazardous waste** is that it is a hazardous substance that has been discarded, abandoned, neglected, released or designated as a waste material, or one that may interact with other substances to be hazardous. In a simple sense a hazardous waste is a material that has been left somewhere that it may cause harm if encountered.

History of Hazardous Substances

Humans have always been exposed to hazardous substances. Such exposure no doubt occurred in prehistoric times when early humans inhaled noxious volcanic gases or succumbed to carbon monoxide from inadequately vented fires in cave dwellings sealed too well against Ice-Age cold. Slaves in Ancient Greece developed lung disease from weaving mineral asbestos fibers into cloth to make it more degradation-resistant. Some archaeological and historical studies have concluded that lead wine containers were a leading cause of lead poisoning in the more affluent ruling class of the Roman Empire leading to erratic behavior such as fixation on spectacular sporting events, chronic unmanageable budget deficits, poorly regulated financial institutions, and ill-conceived, overly ambitious military ventures in foreign lands. Alchemists who worked during the Middle Ages often suffered debilitating injuries and illnesses resulting from the hazards of their explosive and toxic chemicals. During the 1700s runoff from mine spoils piles began to create serious contamination problems in Europe. As the production of dyes and other organic chemicals developed from the coal-tar industry in Germany during the 1800s, pollution and poisoning from coal-tar by-products was observed. By around 1900 the quantity and variety of chemical wastes produced each year was increasing sharply with the addition of wastes such as spent steel and iron pickling liquor, lead battery wastes, chromic wastes, petroleum refinery wastes, radium wastes, and fluoride wastes from aluminum ore refining. As the century progressed into the World War II era, the wastes and hazardous by-products of manufacturing increased markedly from sources such as chlorinated solvents manufacture, pesticides synthesis, polymers manufacture, plastics, paints, and wood preservatives.

The Love Canal affair of the 1970s and 1980s brought hazardous wastes to the public attention as a major political issue in the U.S. Starting around 1940, this site in Niagara Falls, New York, had received about 20,000 metric tons of chemical wastes containing at least 80 different chemicals. By 1992 state and federal governments had spent well over $100 million to clean up the site and relocate residents.

Other areas containing hazardous wastes that received attention included an industrial site in Woburn, Massachusetts, that had been contaminated by wastes from tanneries, glue-making factories, and chemical companies dating back to about 1850; the Stringfellow Acid Pits near Riverside, California; the Valley of the Drums in Kentucky; and Times Beach, Missouri, an entire town that was abandoned because of contamination by TCDD (dioxin).

Legislation

Governments in a number of nations have passed legislation to deal with hazardous substances and wastes. In the U.S. such legislation has included the following:

- Toxic Substances Control Act of 1976
- Resource, Conservation, and Recovery Act (RCRA) of 1976, amended and strengthened by the Hazardous and Solid Wastes Amendments (HSWA) of 1984
- Comprehensive Environmental Response, Compensation, and Liability Act (CERCLA) of 1980, updated by the Superfund Amendments and Reauthorization Act of 1986 (SARA)

Ideally, enforcement of RCRA enables "cradle-to-grave" control of hazardous waste materials. In so doing, RCRA addresses the generators of hazardous wastes, their transporters, and facilities dealing with their treatment, storage and disposal (TSD facilities). RCRA Subtitle C constitutes a number of important provisions that effectively outline a hazardous waste management program for the U.S. It charges the U.S. Environmental Protection Agency (EPA) with legally defining hazardous wastes by listing them specifically and on the basis of characteristics. Under EPA and state regulations, wastes and their generators must be identified, the generators, transporters, and TSD facilities are subject to inspection and enforcement actions, and appropriate permits are issued, especially for TSD facilities. Several significant kinds of wastes are exempted from the hazardous waste category by legislation. These materials are household wastes; agricultural wastes returned to soil as fertilizer; mining overburden replaced in the mine sites; combustion wastes from coal burned by utilities; drilling waste generated by oil and natural gas exploration; residues remaining from the extraction, beneficiation, and processing of ores; cement kiln dust wastes; arsenic-containing wood wastes generated by end users of treated wood; and some specified chromium-bearing wastes.

To address wastes that had already been disposed, often improperly, the U.S. Congress passed "Superfund" legislation in the form of the Comprehensive Environmental Response, Compensation, and Liability Act of 1980 (CERCLA), updated by the Superfund Amendments and Reauthorization Act of 1986 (SARA). CERCLA legislation deals with actual or potential releases of hazardous materials that have the potential to endanger people or the surrounding environment at uncontrolled or abandoned hazardous waste sites in the U.S. The act requires responsible parties or the U.S. government to clean up waste sites.

Hazardous Wastes and Air and Water Pollution Control

Somewhat paradoxically, measures taken to reduce air and water pollution (Figure 23.1) have had a tendency to increase production of hazardous wastes. Most water treatment processes yield sludges or concentrated liquors that require stabilization and disposal. Hazardous materials, such as heavy metals, removed by the pollution control equipment into the sludges or concentrated liquors may render them hazardous. Air-scrubbing processes likewise produce sludges that may be hazardous. Additives required for efficient scrubbing may increase the quantities of sludges to volumes considerably larger than those of the original hazardous materials. Baghouses and precipitators used to control air pollution all yield significant quantities of solids, some of which are hazardous.

23.2. CLASSIFICATION OF HAZARDOUS SUBSTANCES AND WASTES

Many specific chemicals in widespread use are hazardous because of their chemical reactivities, fire hazards, toxicities, and other properties.[2] There are numerous kinds of hazardous substances, usually consisting of mixtures of specific chemicals. These include the following:

- **Explosives**, such as dynamite, or ammunition
- **Compressed gases**, such as hydrogen and sulfur dioxide
- **Flammable liquids**, such as gasoline and aluminum alkyls
- **Flammable solids**, such as magnesium metal, sodium hydride, and calcium carbide, that burn readily, are water-reactive, or are spontaneously combustible
- **Oxidizing materials**, such as lithium peroxide, that supply oxygen for the combustion of normally nonflammable materials

- **Corrosive materials**, including oleum, sulfuric acid, and caustic soda, which may wound exposed flesh or cause disintegration of metal containers
- **Poisonous materials**, such as hydrocyanic acid or aniline
- **Etiologic agents**, including causative agents of anthrax, botulism, or tetanus
- **Radioactive materials**, including plutonium, cobalt-60, and uranium hexafluoride.

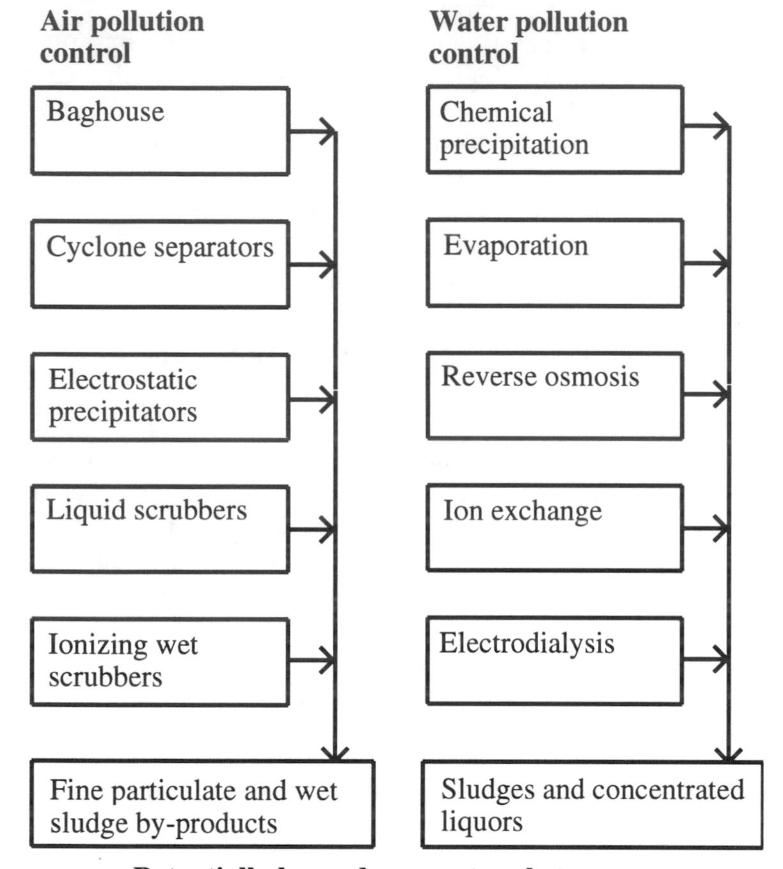

Figure 23.1. Potential contributions of air and water pollution control measures to hazardous wastes production.

Wastes Defined by Characteristics

For regulatory and legal purposes in the U.S. hazardous substances are listed specifically and are defined according to general characteristics. Under the authority of RCRA, the EPA defines hazardous substances in terms of **characteristics**:[3]

- **Ignitability**, characteristic of substances that are liquids whose vapors are likely to ignite in the presence of ignition sources, nonliquids that may catch fire from friction or contact with water and that burn vigorously or persistently, ignitable compressed gases, and oxidizers.

- **Corrosivity**, characteristic of substances that exhibit extremes of acidity or basicity or a tendency to corrode steel.
- **Reactivity**, characteristic of substances that have a tendency to undergo violent chemical change (an explosive substance is an obvious example).
- **Toxicity**, defined in terms of a standard extraction procedure followed by chemical analysis for specific substances.

Listed Wastes

In addition to classification by characteristics, EPA designates more than 450 **listed wastes** that are specific substances or classes of substances known to be hazardous. Each such substance is assigned an EPA **hazardous waste number** in the format of a letter followed by 3 numerals, where a different letter is assigned to substances from each of the four following lists:

- **F-type wastes from nonspecific sources**: For example, quenching waste water treatment sludges from metal heat treating operations where cyanides are used in the process (F012).
- **K-type wastes from specific sources**: For example, heavy ends from the distillation of ethylene dichloride in ethylene dichloride production (K019).
- **P-type acute hazardous wastes**: These are mostly specific chemical species such as fluorine (P056) or 3-chloropropane nitrile (P027).
- **U-Type generally hazardous wastes**: These are predominantly specific compounds such as calcium chromate (U032) or phthalic anhydride (U190).

23.3. FLAMMABLE AND COMBUSTIBLE SUBSTANCES

In a broad sense a **flammable substance** is something that will burn readily, whereas a **combustible substance** requires relatively more persuasion to burn. Before trying to sort out these definitions it is necessary to define several other terms. Most chemicals that are likely to burn accidentally are liquids. Liquids form **vapors**, which are usually more dense than air and thus tend to settle. The tendency of a liquid to ignite is measured by a test in which the liquid is heated and periodically exposed to a flame until the mixture of vapor and air ignites at the liquid's surface. The temperature at which this occurs is called the **flash point**.

With these definitions in mind it is possible to divide ignitable materials into four major classes. A **flammable solid** is one that can ignite from friction or from heat remaining from its manufacture, or which may cause a serious hazard if ignited. Explosive materials are not included in this classification. A **flammable liquid** is one having a flash point below 37.8°C (100°F). A **combustible liquid** has a flash point in excess of 37.8°C, but below 93.3°C. Gases are substances that exist entirely in the gaseous phase at 0°C and 1 atm pressure. A **flammable compressed gas** meets specified criteria for lower flammability limit, flammability range (see below), and flame projection.

In considering the ignition of vapors, two important concepts are those of flammability limit and flammability range. Values of the vapor/air ratio below which ignition cannot occur because of insufficient fuel define the lower **flammability limit**. Similarly, values of the vapor/air ratio above which ignition cannot occur because of insufficient air define the upper flammability limit. The difference between upper and lower flammability limits at a specified temperature is the **flammability range**. Table 23.1 gives some examples of these values for common liquid chemicals.

Table 23.1. Flammabilities of Some Common Organic Liquids

		Volume percent in air	
Liquid	Flash point (°C)[a]	LFL[b]	UFL[b]
Diethyl ether	-43	1.9	36
Pentane	-40	1.5	7.8
Acetone	-20	2.6	13
Toluene	4	1.27	7.1
Methanol	12	6.0	37
Gasoline (2,2,4-trimethylpentane)	---	1.4	7.6
Naphthalene	157	0.9	5.9

[a] Closed-cup flash point test
[b] LFL, lower flammability limit; UFL, upper flammability limit at 25°C.

Combustion of Finely Divided Particles

Finely divided particles of combustible materials are somewhat analogous to vapors in respect to flammability. One such example is a spray or mist of hydrocarbon liquid in which oxygen has the opportunity for intimate contact with the liquid particles; the liquid may ignite at a temperature below its flash point.

Dust explosions can occur with a large variety of solids that have been ground to a finely divided state. Many metal dusts, particularly those of magnesium and its alloys, zirconium, titanium, and aluminum, can burn explosively in air. In the case of aluminum, for example, the highly exothermic (heat-releasing) reaction is the following:

$$4Al(powder) + 3O_2(from\ air) \rightarrow 2Al_2O_3 \tag{23.3.1}$$

Coal dust and grain dusts have caused many fatal fires and explosions in coal mines and grain elevators, respectively. Dusts of polymers such as cellulose acetate, polyethylene, and polystyrene can also be explosive.

Oxidizers

Combustible substances are reducing agents that react with **oxidizers** (oxidizing agents or oxidants) to produce heat. Diatomic oxygen, O_2, from air is the most common oxidizer. Many oxidizers are chemical compounds that contain oxygen in their formulas. The halogens and many of their compounds are oxidizers. Some examples of oxidizers are given in Table 23.2.

Table 23.2. Examples of Some Oxidizers

Name	Formula	State of matter
Ammonium nitrate	NH_4NO_3	Solid
Ammonium perchlorate	NH_4ClO_4	Solid
Chlorine	Cl_2	Gas (stored as liquid)
Fluorine	F_2	Gas
Hydrogen peroxide	H_2O_2	Solution in water
Nitric acid	HNO_3	Concentrated solution
Nitrous oxide	N_2O	Gas (stored as liquid)
Ozone	O_3	Gas
Perchloric acid	$HClO_4$	Concentrated solution
Sodium dichromate	$Na_2Cr_2O_7$	Solid

An example of a reaction of an oxidizer is that of concentrated HNO_3 with copper metal, which gives toxic NO_2 gas as a product:

$$4HNO_3 + Cu \rightarrow Cu(NO_3)_2 + 2H_2O + 2NO_2 \quad (23.3.2)$$

The toxic effects of some oxidizers are due to their ability to oxidize biomolecules.

Whether or not a substance acts as an oxidizer depends upon the reducing strength of the material that it contacts. For example, carbon dioxide is a common fire extinguishing material that can be sprayed onto a burning substance to keep air away. However, aluminum is such a strong reducing agent that carbon dioxide in contact with hot, burning aluminum reacts as an oxidizing agent to give off toxic combustible carbon monoxide gas:

$$2Al + 3CO_2 \rightarrow Al_2O_3 + 3CO \quad (23.3.3)$$

Oxidizers can contribute strongly to fire hazards because fuels may burn explosively in contact with an oxidizer.

Spontaneous Ignition

Pyrophoric substances catch fire spontaneously in air without an ignition source. These include several elements—white phosphorus, the alkali metals (group 1A), and powdered forms of magnesium, calcium, cobalt, manganese, iron, zirconium, and aluminum. Also included are some organometallic compounds, such as lithium ethyl (LiC_2H_4); some metal carbonyl compounds, such as iron pentacarbonyl, $Fe(CO)_5$; and metal and metalloid hydrides, including lithium hydride, LiH; pentaborane, B_5H_9; and arsine, AsH_3. Moisture in air is often a factor in spontaneous ignition. For example, lithium hydride undergoes the following reaction with water from moist air:

$$LiH + H_2O \rightarrow LiOH + H_2 + heat \quad (23.3.4)$$

The heat generated from this reaction can be sufficient to ignite the hydride so that it burns in air:

$$2LiH + O_2 \rightarrow Li_2O + H_2O \quad (23.3.5)$$

Many mixtures of oxidizers and oxidizable chemicals catch fire spontaneously and are called **hypergolic mixtures**. Nitric acid and phenol form such a mixture.

Toxic Products of Combustion

Some of the greater dangers of fires are from toxic products and by-products of combustion. The most obvious of these is toxic carbon monoxide, CO. Toxic SO_2, P_4O_{10}, and HCl are formed by the combustion of sulfur, phosphorus, and organochloride compounds, respectively. A large number of noxious organic compounds such as aldehydes are generated as by-products of combustion. In addition to forming carbon monoxide, combustion under oxygen-deficient conditions produces polycyclic aryl hydrocarbons consisting of fused ring structures. Some of these compounds, such as benzo(a)pyrene, below, are precarcinogens that are acted upon by enzymes in the body to yield cancer-producing metabolites.

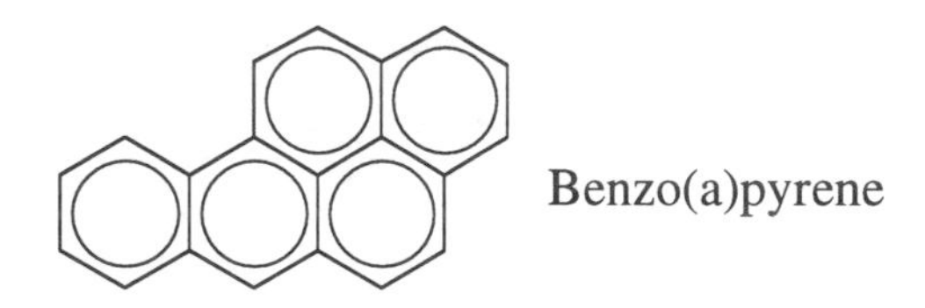

23.4. REACTIVE SUBSTANCES

Reactive substances are those that tend to undergo rapid or violent reactions under certain conditions. Such substances include those that react violently or form potentially explosive mixtures with water. An example is sodium metal, which reacts strongly with water as follows:

$$2Na + 2H_2O \rightarrow 2NaOH + H_2 + heat \quad (23.4.1)$$

This reaction usually generates enough heat to ignite the sodium. Explosives constitute another class of reactive substances. For regulatory purposes substances are also classified as reactive that react with water, acid, or base to produce toxic fumes, particularly those of hydrogen sulfide or hydrogen cyanide.

Many reactions require energy of activation from heat to get them started, the rates of most reactions increase sharply with increasing temperature, and most chemical reactions give off heat. Therefore, once a reaction is started in a reactive mixture, the rate may increase exponentially with time, leading to an uncontrollable event. Other factors that may affect reaction rate include physical form of reactants (for example, a finely divided metal powder that reacts explosively with oxygen, whereas a single mass of metal barely reacts), rate and degree of mixing of reactants, degree of dilution with nonreactive media (solvent), presence of a catalyst, and pressure.

Chemical Structure and Reactivity

As shown in Table 23.3, some chemical structures are associated with high reactivity. High reactivity in some organic compounds results from unsaturated bonds in the carbon skeleton, particularly where multiple bonds are adjacent (allenes, C=C=C) or separated by only one carbon-carbon single bond (dienes, C=C-C=C). Some organic structures involving oxygen are very reactive. Examples are oxiranes, such as ethylene oxide,

$$\mathrm{H{-}\underset{H}{\underset{|}{C}}\overset{\;O\;}{\text{———}}\underset{H}{\underset{|}{C}}{-}H} \quad \text{Ethylene oxide}$$

hydroperoxides (ROOH), and peroxides (ROOR'), where R and R' stand for hydrocarbon moieties such as the methyl group, $-CH_3$. Many organic compounds containing nitrogen along with carbon and hydrogen are very reactive. Included are triazenes (R-N=N-N), some azo compounds (R-N=N-R'), and some nitriles (RC≡N). Functional groups containing both oxygen and nitrogen tend to impart reactivity to an organic compound. Examples are alkyl nitrates ($R-O-NO_2$), alkyl nitrites (R-O-N=O), nitroso compounds (R-N=O), and nitro compounds ($R-NO_2$).

Table 23.3. Examples of Reactive Compounds and Structural Groups

Name	Formula or structural group
Organic	
Allenes	C–C–C
Dienes	C=C-C=C
Azo compounds	C=N-N=C
Triazines	C-N=N-N
Hydroperoxides	R-OOH[a]
Peroxides	R-OO-R'
Alkyl nitrates	$R-O-NO_2$
Nitryl compounds	$R-NO_2$
Inorganic	
Nitrous oxide	N_2O
Nitrogen halides	NCl_3, NI_3
Interhalogen compounds	BrCl
Halogen oxides	ClO_2
Halogen azides	ClN_3
Hypohalites	NaClO

[a] R and R' represent hydrocarbon groups, such as methyl, CH_3.

Many different classes of inorganic compounds are reactive. These include some of the halogen compounds of nitrogen (shock-sensitive nitrogen triiodide, NI_3, is an outstanding example), compounds with metal-nitrogen bonds, halogen oxides (ClO_2), and compounds with oxyanions of the halogens. An example of the last group of compounds is ammonium perchlorate, NH_4ClO_4.

Explosives such as nitroglycerin or TNT that are single compounds containing both oxidizing and reducing functions in the same molecule are called **redox compounds**. Some redox compounds have more oxygen than is needed for a complete reaction and are said to have a positive balance of oxygen, some have exactly the stoichiometric quantity of oxygen required (zero balance, maximum energy release), and others have a negative balance and require oxygen from outside sources to completely oxidize all components.

23.5. CORROSIVE SUBSTANCES

Corrosive substances are regarded as those that dissolve metals or cause oxidized material to form on the surface of metals—rusted iron is a prime example. In a broader sense corrosives are defined as those that cause deterioration of materials, including living tissue, that they contact. Most corrosives belong to at least one of the four following chemical classes: (1) strong acids, (2) strong bases, (3) oxidants, (4) dehydrating agents. Table 23.4 lists some of the major corrosive substances and their effects.

Table 23.4. Examples of Some Corrosive Substances

Name and formula	Properties and effects
Sulfuric acid, H_2SO_4	Strong acid, dehydrating agent, and oxidant, tremendous affinity for water, including water in exposed flesh
Nitric acid, HNO_3	Strong acid and strong oxidizer, corrodes metal, reacts with protein in tissue to form yellow xanthoproteic acid, lesions are slow to heal
Hydrochloric acid, HCl vapor	Strong acid, corrodes metals, gives off HCl gas, which can damage respiratory tract tissue
Hydrofluoric acid, HF	Corrodes metals, dissolves glass, causes bad flesh burns
Alkali metal hydroxides	Strong bases, corrode zinc, lead, and aluminum KOH and NaOH caustic to flesh causing severe burns
Hydrogen peroxide	Oxidizer, all but very dilute solutions of H_2O_2 cause severe burns
Interhalogens, such as ClF, BrF_3	Powerful corrosive irritants that acidify, oxidize, and dehydrate tissue
Halogen oxides such as OF_2, Cl_2O, Cl_2O_7	Powerful corrosive irritants that acidify, oxidize, and dehydrate tissue
Elemental fluorine, chlorine, bromine (F_2, Cl_2, Br_2,)	Very corrosive to mucous membranes and moist tissue, strong irritants

23.6. TOXIC SUBSTANCES

Toxicity is of the utmost concern in dealing with hazardous substances. This includes both long-term chronic effects from continual or periodic exposures to low levels of toxicants and acute effects from a single large exposure. Toxic substances are covered in greater detail in Chapter 20.

Toxicity Characteristic Leaching Procedure

For regulatory and remediation purposes a standard test is needed to measure the likelihood of toxic substances getting into the environment and causing harm to organisms. The test required by the U.S. EPA is the **Toxicity Characteristic Leaching Procedure** (TCLP) designed to determine the mobility of both organic and inorganic contaminants present in liquid, solid, and multiphasic wastes. For analysis of toxic species a solution is leached from the waste or filtered from it and is designated as the TCLP extract. After the TCLP extract is separated from the solids, it is analyzed for a number of specified volatile organic compounds, semi-volatile organic compounds, and metals to determine if the waste exceeds specified levels of these contaminants.

23.7. PHYSICAL FORMS AND SEGREGATION OF WASTES

Three major categories of wastes based upon their physical forms are **organic materials**, **aqueous wastes**, and **sludges**. These forms largely determine the course of action taken in treating and disposing of the wastes. The **level of segregation**, a concept illustrated in Figure 23.2, is very important in treating, storing, and disposing of different kinds of wastes. It is relatively easy to deal with wastes that are not mixed with other kinds of wastes, that is, those that are highly segregated. For example, spent hydrocarbon solvents can be used as fuel in boilers. However, if these solvents are mixed with spent organochloride solvents, the production of contaminant hydrogen chloride during combustion may prevent fuel use and require disposal in special hazardous waste incinerators. Further mixing with inorganic sludges adds mineral matter and water. These impurities complicate the treatment processes required by producing mineral ash in incineration or lowering the heating value of the material incinerated because of the presence of water. Among the most difficult types of wastes to handle and treat are those with the least segregation, of which a “worst-case scenario” would be “dilute sludge consisting of mixed organic and inorganic wastes,” as shown in Figure 23.2.

Concentration of wastes is an important factor in their management. A waste that has been concentrated or preferably never diluted is generally much easier and more economical to handle than one that is dispersed in a large quantity of water or soil. Dealing with hazardous wastes is greatly facilitated when the original quantities of wastes are minimized and the wastes remain separated and concentrated insofar as possible.

23.8. HAZARDOUS WASTES IN THE ENVIRONMENT

Having outlined the nature and sources of hazardous substances and hazardous wastes earlier in this chapter, it is now possible to discuss their environmental behavior according to the following factors:

- Origin
- Transport
- Reactions
- Effects
- Ultimate fate

In addition, consideration must be given to the distribution of hazardous wastes among the geosphere, hydrosphere, atmosphere, and biosphere.

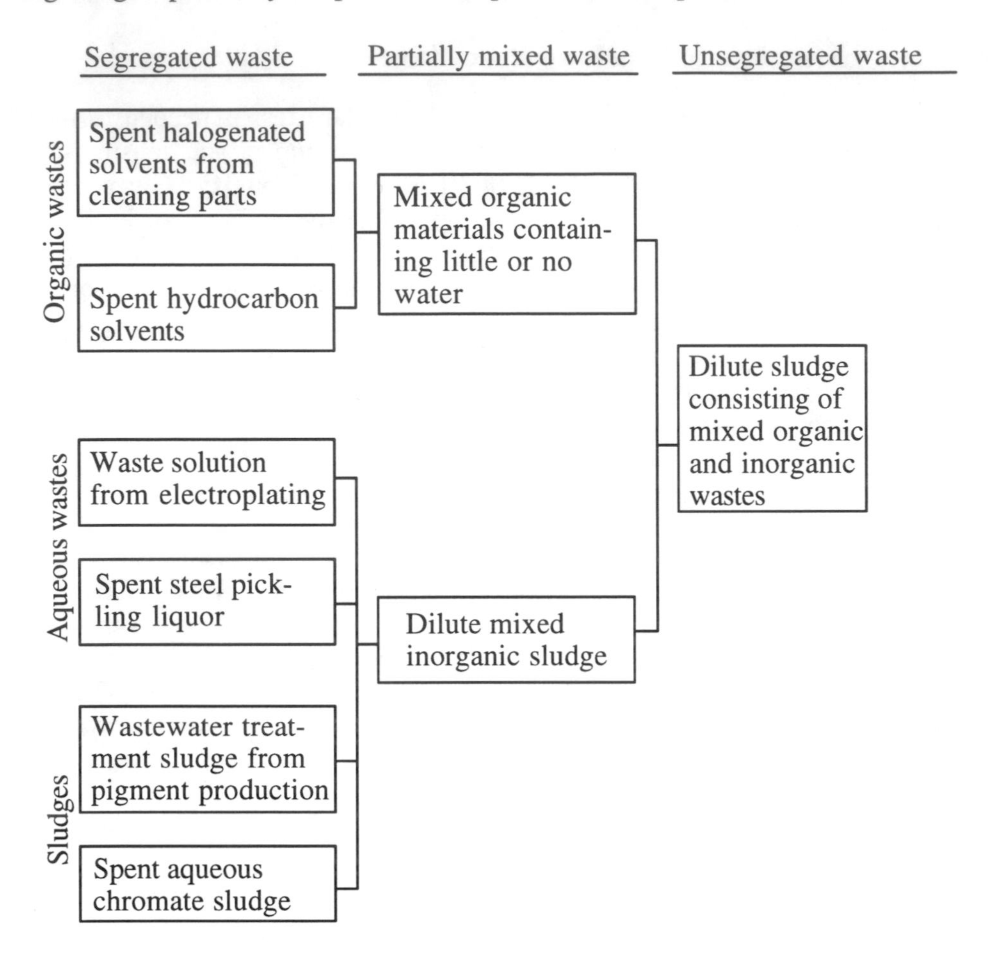

Figure 23.2. Illustration of waste segregation.

Origin of Hazardous Wastes

For purposes of discussion in this chapter, *origin* of hazardous wastes refers to their points of entry into the environment. These may consist of the following:

- Deliberate addition to soil, water, or air by humans
- Evaporation or wind erosion from waste dumps into the atmosphere
- Leaching from waste dumps into groundwater, streams, and bodies of water
- Leakage, such as from underground storage tanks or pipelines
- Evolution and subsequent deposition by accidents, such as fire or explosion
- Release from improperly operated waste treatment or storage facilities

Transport of Hazardous Wastes

The transport of hazardous wastes is largely a function of their physical properties, the physical properties of their surrounding matrix, the physical conditions to which they are subjected, and chemical factors. Highly volatile wastes are obviously more likely to be transported through the atmosphere and more soluble ones to be carried by water. Wastes will move farther and faster in porous, sandy formations than in denser soils. Volatile wastes are more mobile under hot, windy conditions and soluble ones during periods of heavy rainfall. Wastes that are more chemically and biochemically reactive will not move so far as less reactive wastes before breaking down.

Physical Factors

The major physical properties of wastes that determine their amenability to transport are volatility, solubility, and the degree to which they are sorbed to solids, including soil and sediments.

The distribution of hazardous waste compounds between the atmosphere and the geosphere or hydrosphere is largely a function of compound volatility. Compound volatilities are usually measured by vapor pressures, which vary over a wide range.

Usually, in the hydrosphere, and often in soil, hazardous waste compounds are dissolved in water; therefore, the tendency of water to hold the compound is a factor in its mobility. For example, although ethyl alcohol has a higher evaporation rate and lower boiling temperature than toluene, vapor of the latter compound is more readily evolved from soil because of its limited solubility in water compared to ethanol, which is totally miscible with water.

Chemical Factors

As an illustration of chemical factors involved in transport of wastes, consider largely cationic inorganic species. Inorganic species can be divided into three groups based upon their retention by clay minerals. Elements that tend to be highly retained by clay include cadmium, mercury, lead, and zinc. Potassium, magnesium, iron, silicon, and NH_4^+ are moderately retained by clay, whereas sodium, chloride, calcium, manganese, and boron are poorly retained. The apparent retention of the last three elements is probably biased in that they are leached from clay, so that negative retention (elution) is often observed. It should be noted, however, that the retention of iron and manganese is a strong function of oxidation state in that the reduced forms of Mn and Fe are somewhat mobile, whereas the oxidized forms of $Fe_2O_3{\bullet}xH_4O$ and MnO_2 are very insoluble and stay on soil as solids.

Reactions of Hazardous Wastes

Many environmental chemical and environmental biochemical processes operate on hazardous wastes in water, air, and soil. One of the important results of this is that, in addition to primary pollutants added directly to the environment, there are secondary pollutants as well, which may be even more harmful than their precursor species. Reactions that result in the formation of secondary pollutants have been discussed with respect to atmospheric chemistry in Chapters 9-12. An example of the formation of a very damaging secondary pollutant from a relatively less dangerous primary pollutant is the oxidation of primary pollutant sulfur dioxide,

$$2SO_2 + O_2 + 2H_2O \rightarrow 2H_2SO_4 \quad \text{(several steps and intermediates)} \tag{23.8.1}$$

to yield secondary pollutant sulfuric acid, the prime ingredient of acid rain.

Effects of Hazardous Wastes

The effects of hazardous wastes in the environment may be divided among effects on organisms, effects on materials, and effects on the environment. These are addressed briefly here and in greater detail in later sections.

The ultimate concern with wastes has to do with their toxic effects to animals, plants, and microbes. Virtually all hazardous waste substances are poisonous to a degree, some extremely so. The toxicity of a waste is a function of many factors, including the chemical nature of the waste, the matrix in which it is contained, circumstances of exposure, the species exposed, manner of exposure, degree of exposure, and time of exposure. The toxicities of hazardous wastes are discussed in more detail in Chapter 20, "Toxicology and Toxicological Chemistry."

Many hazardous wastes are *corrosive* (Section 23.5) to materials, usually because of extremes of pH or because of dissolved salt content. Oxidant wastes can cause combustible substances to burn uncontrollably. Highly reactive wastes can explode, causing damage to materials and structures. Contamination by wastes, such as by toxic pesticides in grain, can result in substances becoming unfit for use.

In addition to their toxic effects in the biosphere, hazardous wastes can damage air, water, and soil. Wastes that get into air can cause deterioration of air quality, either directly, or by the formation of secondary pollutants. Hazardous waste compounds dissolved in, suspended in, or as surface films on, the surface of water can render it unfit for use and for sustenance of aquatic organisms.

Soil exposed to hazardous wastes can be severely damaged by alteration of its physical and chemical properties and ability to support plants. For example, soil exposed to concentrated brines from petroleum production may become unable to support plant growth so that the soil becomes extremely susceptible to erosion.

Fates of Hazardous Wastes

The fates of hazardous waste substances are addressed in more detail in subsequent sections. As with all environmental pollutants, such substances eventually reach a state of physical and chemical stability, although that may take many centuries to occur. In some cases, the fate of a hazardous waste material is a simple function of its physical properties and surroundings.

The fate of a hazardous waste substance in water is a function of the substance's solubility, density, biodegradability, and chemical reactivity. Dense, water-immiscible liquids may simply sink to the bottoms of bodies of water or aquifers and accumulate there as "blobs" of liquid. This has happened, for example, with hundreds of tons of PCB wastes that have accumulated in sediments in the Hudson River in New York State. Biodegradable substances are broken down by bacteria, a process for which the availability of oxygen is an important variable. Substances that readily undergo bioaccumulation are taken up by organisms, exchangeable cationic materials become bound to sediments, and organophilic materials may be sorbed by organic matter in sediments.

The fates of hazardous waste substances in the atmosphere are often determined by photochemical reactions. Ultimately, such substances may be converted to non-volatile, insoluble matter and precipitate from the atmosphere onto soil or plants.

23.9. HAZARDOUS WASTES IN THE GEOSPHERE

The sources, transport, interactions, and fates of contaminant hazardous wastes in the geosphere involve a complex scheme, some aspects of which are illustrated in Figure 23.3. The primary environmental concern regarding hazardous wastes in the geosphere is the possible contamination of groundwater aquifers by waste leachates and leakage from wastes. As the figure shows, there are a number of possible contamination sources. The most obvious one is leachate from landfills containing hazardous wastes. In some cases, liquid hazardous materials are placed in lagoons, which can leak into aquifers. Leaking sewers can also result in contamination, as can the discharge from septic tanks. Hazardous wastes spread on land can result in aquifer contamination by leachate. Hazardous chemicals are sometimes deliberately disposed of underground in waste disposal wells. This means of disposal can result in interchange of contaminated water between surface water and groundwater at discharge and recharge points.

The transport of contaminants in the geosphere depends largely upon the hydrologic factors governing the movement of water underground and the interactions of hazardous waste constituents with geological strata, particularly unconsolidated earth materials. As shown in Figure 23.4, groundwater contaminated with hazardous wastes tends to flow as a relatively undiluted plug or plume along with the groundwater in an aquifer. The groundwater flow rate depends upon the water gradient and aquifer characteristics, such as permeability and cross-section area. The rate of flow is generally relatively slow; 1 meter per day would be considered fast. Contaminated groundwater can result in contamination of a surface water source. This can occur at a discharge area where the groundwater flows into a lake or stream.

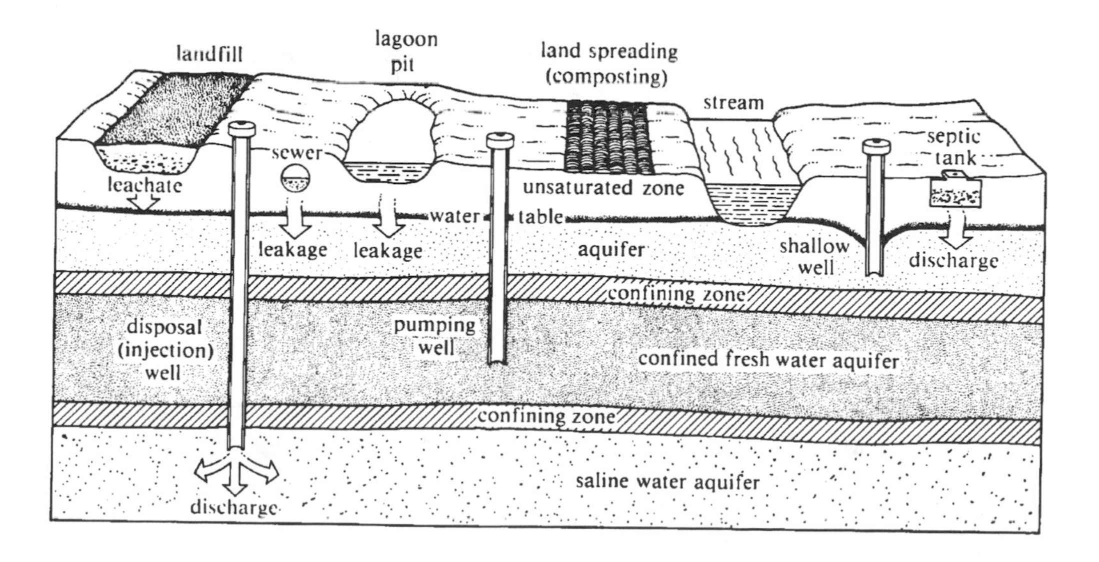

Figure 23.3. Sources, disposal, and movement of hazardous wastes in the geosphere.

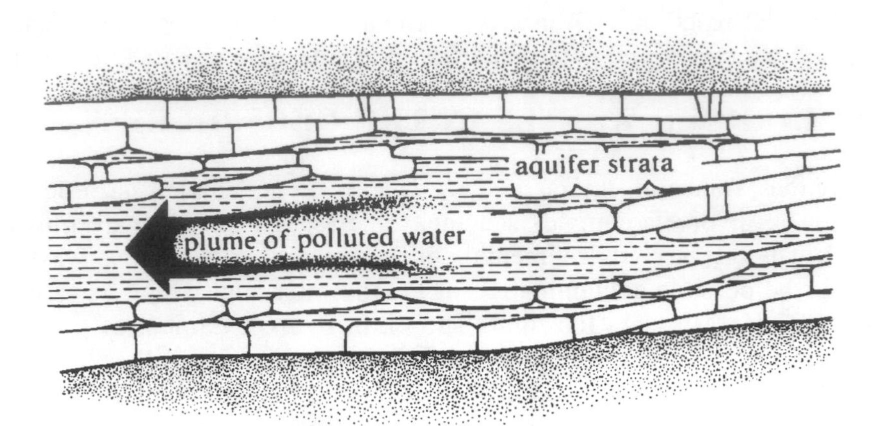

Figure 23.4. Plug-flow of hazardous wastes in groundwater.

As discussed in the preceding section, hazardous waste dissolved in groundwater can be attenuated by soil or rock by means of various sorption mechanisms. Mathematically, the distribution of a solute between groundwater or leachate water and soil is expressed by a **distribution coefficient**, K_d,

$$K_d = \frac{C_S}{C_W} \tag{23.9.1}$$

where C_S is the the concentration of the species in the solid phase and C_W is its concentration in water. This equation assumes that the relative degree of sorption is independent of C_W.

The degree of attenuation depends upon the surface properties of the solid, particularly its surface area. The chemical nature of the attenuating solid is also important because attenuation is a function of the organic matter (humus) content, presence of hydrous metal oxides, and the content and types of clays present. The chemical characteristics of the leachate also affect attenuation greatly. For example, attenuation of metals is very poor in acidic leachate because precipitation reactions, such as,

$$M^{2+} + 2OH^- \rightarrow M(OH)_2(s) \tag{23.9.2}$$

are reversed in acid:

$$M(OH)_2(s) + 2H^+ \rightarrow M^{2+} + 2H_2O \tag{23.9.3}$$

Organic solvents in leachates tend to prevent attenuation of organic hazardous waste constituents.

The degree of attenuation of a pollutant by soil depends upon the water content of the soil. As shown in Chapter 13, Figure 13.9, above the water table there is an unsaturated zone of soil in which attenuation is more highly favored. Normally, soil has a greater surface area at liquid-solid interfaces in this zone so that absorption and ion-exchange processes are favored. Aerobic degradation (see Chapter 18) is possible in the unsaturated zone, enabling more rapid and complete degradation of biodegradable hazardous wastes.

Codisposal of chelating agents with heavy metals can have a strong effect upon the mobility of metal ions in soil. This effect was observed resulting from codisposal of intermediate-level nuclear wastes with chelating agents (generally organic species that bind strongly with metal ions) during the period 1951-1965 at Oak Ridge National Laboratory. The presence of chelating agents resulted from the use of salts of chelating ethylenediaminetetraacetic acid (EDTA) in decontaminating facilities exposed to nuclear wastes. Whereas metal cations are readily held by ion exchange processes and precipitation on soil,

$$2\text{Soil}\}^{-}H^{+} + Co^{2+} \rightarrow (\text{Soil}\}^{-})_2Co^{2+} + 2H^{+} \tag{23.9.4}$$

$$Co^{2+} + 2OH^{-} \rightarrow Co(OH)_2(s) \tag{23.9.5}$$

chelated anionic species, such as CoY^{2-} (where Y^{4-} is the chelating EDTA anion), are not strongly retained by the negatively charged functional groups in soil.

Radionuclides have been buried in shallow trenches on the grounds of Oak Ridge National Laboratory since 1944, so ample time has elapsed to observe the effects of this means of radioactive waste disposal. It has been found that chelating agents used for decontamination, as well as naturally occurring humic substance chelators, are responsible for migration in excess of that expected. Most notably, ^{60}Co has been found outside the disposal trenches.

23.10. HAZARDOUS WASTES IN THE HYDROSPHERE

Figure 23.5 illustrates a typical pathway for the entry of hazardous waste materials into the hydrosphere. Other sources consist of precipitation from the atmosphere with rainfall, deliberate release to streams and bodies of water, runoff from soil, and mobilization from sediments. Once in an aquatic system, hazardous waste species are subject to a number of chemical and biochemical processes, including acid-base, oxidation-reduction, precipitation-dissolution, and hydrolysis reactions, as well as biodegradation.

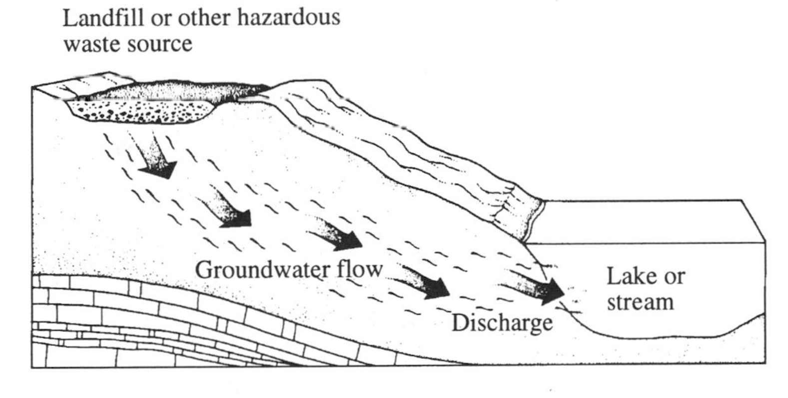

Figure 23.5. Discharge of groundwater contaminated from hazardous waste landfill into a body of water.

The presence of organic matter in water has a tendency to increase the solubility of hazardous organic substances. Typically, the solubility of hexachlorobenzene is 1.8 micrograms per liter in pure water at 25°C, whereas it is 2.3 μg/L in creek water containing organic solutes and 4-4.5 μg/L in landfill leachate.

In considering the processes that hazardous wastes undergo in water, it is important to recall the nature of aquatic systems and the unique properties of water discussed in detail in Chapter 5. Water in the environment is far from pure. Just as the atmosphere is a constantly changing mass of bodies of moving air with different temperatures, pressures, and humidities, bodies of water are highly dynamic systems. Rivers, impoundments, and groundwater aquifers are subject to the input and loss of a variety of materials from both natural and anthropogenic sources. These materials may be gases, liquids, or solids. They interact chemically with each other and with living organisms—particularly bacteria—in the water. They are subject to dispersion and transport by stream flow, convection currents, and other physical phenomena. Hazardous substances or their by-products in water may undergo bioaccumulation through food chains involving aquatic organisms.

Several physical, chemical, and biochemical processes are particularly important in determining the transformations and ultimate fates of hazardous chemical species in the hydrosphere. These include **hydrolysis reactions**, through which a molecule is cleaved with the addition of H_2O; **precipitation reactions**, generally accompanied by **aggregation** of colloidal particles suspended in water; **oxidation-reduction reactions**, generally mediated by microorganisms; **sorption** (adsorption and/or adsorption) of hazardous solutes by sediments and by suspended mineral and organic matter; **biochemical processes**, often involving hydrolysis and oxidation-reduction reactions; **photolysis reactions**; and miscellaneous chemical phenomena.

The hydrolysis of hazardous waste acetic anhydride is illustrated by the following reaction:

$$\mathrm{H_3C{-}C({=}O){-}O{-}C({=}O){-}CH_3} + 2\mathrm{HOH} \rightarrow 2\,\mathrm{H_3C{-}C({=}O){-}OH} \quad (23.10.1)$$

The rates at which compounds hydrolyze in water vary widely. Acetic anhydride hydrolyzes very rapidly. In fact, the great affinity of this compound for water (including water in skin) is one of the reasons that it is hazardous. Once in the aquatic environment, though, acetic anhydride is converted very rapidly to essentially harmless acetic acid. Many ethers, esters, and other compounds formed originally by the joining together of two or more molecules with the loss of water hydrolyze very slowly, although the rate may be greatly increased by the action of enzymes in micoorganisms (biochemical processes).

The formation of precipitates in the form of sludges is one of the most common means of isolating hazardous components from an unsegregated waste. Although solid inorganic ionic compounds are often discussed in terms of very simple formulas, such as $PbCO_3$ for lead carbonate, much more complicated species (for example, $2PbCO_3{\cdot}Pb(OH)_2$) generally result when precipitates are formed in the aquatic environment. For example, a hazardous heavy metal ion in the hydrosphere may be precipitated as a relatively complicated compound, coprecipitated as a minor constituent of some other compound, or sorbed by the surface of another solid.

The major anions present in natural waters and wastewaters are OH^-, CO_3^{2-}, and SO_4^{2-}. Since these anions are all capable of forming precipitates with cationic impurities, such pollutants tend to precipitate as hydroxides, carbonates, and sulfates.

Sorption processes are particularly common methods for the removal of low-level hazardous materials from water. Many heavy metals are sorbed by or coprecipitated with hydrated iron(III) oxide ($Fe_2O_3 \cdot xH_2O$) or manganese (IV) oxide ($MnO_2 \cdot xH_2O$). Oxidation-reduction reactions are very important means of transformation of hazardous wastes in water.

Under many circumstances, biochemical processes largely determine the fates of hazardous chemical species in the hydrosphere. The most important such processes are those mediated by microorganisms, as discussed in Chapters 6 and 18. In particular, the oxidation of biodegradable hazardous organic wastes in water generally occurs by means of microorganism-mediated biochemical reactions. Bacteria produce organic acids and chelating agents, such as citrate, which have the effect of solubilizing hazardous heavy metal ions. Some mobile methylated forms, such as compounds of methylated arsenic and mercury, are produced by bacterial action.

As discussed in Chapters 9-12, photolysis reactions are those initiated by the absorption of light. The effect of photolytic processes on the destruction of hazardous wastes in the hydrosphere is minimal, although some photochemical reactions of hazardous waste compounds can occur when the compounds are present as surface films on water exposed to sunlight.

Groundwater is the part of the hydrosphere most vulnerable to damage from hazardous wastes. Although surface water supplies are subject to contamination, groundwater can become almost irreversibly contaminated by the improper land disposal of hazardous chemicals.

23.11. HAZARDOUS WASTES IN THE ATMOSPHERE

Some chemicals found in hazardous waste sites may enter the atmosphere by evaporation or even as windblown particles. Three major areas of interest in respect to hazardous waste compounds in the atmosphere are their **pollution potential**, **atmospheric fate**, and **residence time**. These strongly interrelated factors are discussed in this section.

Air Pollution Potential of Hazardous Waste Compounds

The pollution potential of hazardous wastes in the atmosphere depends upon whether they are *primary pollutants* that have a direct effect or *secondary pollutants* that are converted to harmful substances by atmospheric chemical processes. Hazardous waste sites do not usually evolve sufficient quantities of air pollutants to give significant amounts of secondary pollutants, so primary air pollutants are the greater concern. Examples of primary air pollutants include toxic organic vapors (vinyl chloride), corrosive acid gases (HCl), and toxic inorganic gases, such as H_2S released by the accidental mixing of waste acid and waste metal sulfides:

$$2HCl + FeS \rightarrow FeCl_2 + H_2S(g) \qquad (23.11.1)$$

Primary air pollutants are most dangerous in the immediate vicinity of a site, usually to workers involved in disposal or cleanup or people living adjacent to the site. Quantities are rarely sufficient to pose any kind of regional air pollution hazard.

The two major kinds of secondary air pollutants from hazardous wastes are those that are oxidized in the atmosphere to corrosive substances and organic materials that undergo photochemical oxidation. Plausible examples of the former are sulfur dioxide released from the action of waste strong acids on sulfites and subsequently oxidized in the atmosphere to corrosive sulfuric acid,

$$SO_2 + {}^1/_2O_2 + H_2O \rightarrow H_2SO_4(aerosol) \quad (23.11.2)$$

and nitrogen dioxide (itself a toxic primary air pollutant) produced by the reaction of waste nitric acid with reducing agents such as metals and oxidized to corrosive nitric acid or converted to corrosive nitrate salts:

$$4HNO_3 + Cu \rightarrow Cu(NO_3)_2 + 2NO_2(g) + 2H_2O \quad (23.11.3)$$

$$2NO_2(g) + {}^1/_2O_2 + H_2O \rightarrow 2HNO_3(aerosol) \quad (23.11.4)$$

$$HNO_3(aerosol) + NH_3(g) \rightarrow NH_4NO_3(aerosol) \quad (23.11.5)$$

Organic species that produce secondary air pollutants are those that form photochemical smog (see Chapter 12). The more reactive of these are unsaturated compounds that react with atomic oxygen or hydroxyl radical in air,

$$R\text{-}CH{=}CH_2 + HO\cdot \rightarrow RCH_2CH_2O\cdot \quad (23.11.6)$$

to yield reactive radicals that participate in chain reactions to eventually yield ozone, organic oxidants, noxious aldehydes, and other products characteristic of photochemical smog.

Fate and Residence Times of Hazardous Waste Compounds in the Atmosphere

An obvious means by which hazardous waste species may be removed from the atmosphere is by **dissolution** in water in the form of cloud or rain droplets. Inorganic acid, base, and salt compounds, such as H_2SO_4, HNO_3, and NH_4NO_3 mentioned above, are readily removed from the atmosphere by dissolution. For vapors of compounds that are not highly soluble in water, solubility information combined with information about rainfall amounts and mixing in the atmosphere can be used to estimate the atmospheric half-life, $\tau_{1/2}$, of the species. Solubility rates may be used to estimate half-lives for substances that are more miscible in water. For poorly water-soluble compounds, such calculations tend to drastically underestimate lifetimes, which indicates that other removal mechanisms must predominate.

The lifetimes of vaporized hazardous waste species removed from the atmosphere through **adsorption by aerosol particles** are limited to that of the sorbing aerosol particles (typically about 7 days) plus the time spent in the vapor phase before adsorption. This mechanism appears to be viable only for highly nonvolatile constituents such as benzo(a)pyrene.

Sorptive removal by soil, water, or plants on the Earth's surface, called **dry deposition**, is another means for physical removal of hazardous substances from the atmosphere. Predictions of dry deposition rate vary greatly with type of compound, type of surface, and weather conditions. For highly volatile organic compounds, such as low-molecular-mass organohalide compounds, predicted rates of dry deposition give atmospheric lifetimes many-fold higher than those actually observed, so, for such compounds, dry deposition is probably not a common removal mechanism.

Predicted rates of physical removal of a number of volatile organic compounds that are not very soluble in water are far too slow to account for the loss of such compounds from the atmosphere, so chemical processes must predominate. As discussed in Chapter 12, Section 12.6, the most important of these processes is reaction with hydroxyl radical, HO•, in the troposphere. Ozone can react with compounds having a double bond. Other oxidant species that might react with hazardous waste compounds in the troposphere and stratosphere are atomic oxygen (O), peroxyl radicals (HOO•), alkylperoxyl radicals (ROO•), and NO_3.

Despite the fact that its concentration in the troposphere is relatively low, HO• is so reactive that it tends to initiate most of the reactions leading to the chemical removal of most refractory organic compounds from the atmosphere. When hydroxyl radical reacts with organic compounds in the atmosphere, new reactive free radicals are formed that undergo further reactions, leading to nonvolatile and/or water-soluble species, which are scavenged from the atmosphere by physical means. These scavengeable species tend to be aldehydes, ketones, or acids. Halogenated organic compounds may lose halogen atoms in the form of halo-oxy radicals and undergo further reactions to form scavengeable species. In general, reactions with species other than HO• or O_3 are not considered significant in the removal of hazardous organic waste compounds from the troposphere.

Photolytic transformations involve direct cleavage (photodissociation) of compounds by reactions with light and ultraviolet radiation:

$$R\text{-}X + h\nu \rightarrow R^\bullet + X^\bullet \qquad (23.11.7)$$

The extent of these reactions varies greatly with light intensity, quantum yields (chemical reactions per quantum of radiation energy absorbed), and other factors. In order for photolysis to be an important process for its removal from the atmosphere, a molecule must have a **chromophore** (light-absorbing group) that absorbs light in a wavelength region of significant intensity in the impinging light spectrum. This requirement limits the importance of photolysis as a removal mechanism to only a few classes of compounds, including conjugated alkenes, carbonyl compounds, some halides, and some nitrogen compounds, particularly nitro compounds. However, these do include a number of the more important hazardous waste compounds.

23.12. HAZARDOUS WASTES IN THE BIOSPHERE

One of the most crucial aspects of fate and toxic effects of environmental chemicals is their accumulation by organisms from their surroundings, including bioaccumulation and biomagnification phenomena (see Chapter 18). **Biodegradation** of wastes is their conversion by biological processes to simple inorganic molecules and, to a certain extent, to biological materials (Figure 23.6). The com-

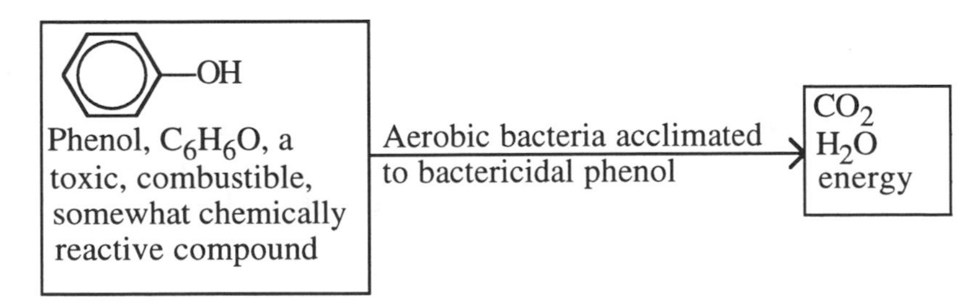

Figure 23.6. Illustration of biological action on a hazardous waste constituent. This example shows mineralization in which the organic compound is degraded completely to simple inorganic species.

plete bioconversion of a substance to inorganic species such as CO_2, NH_3, and phosphate is called **mineralization. Detoxification** refers to the biological conversion of a toxic substance to a less toxic species, which may still be relatively complex, or biological conversion to an even more complex material. An example of detoxification is illustrated below for the enzymatic conversion of paraoxon (a highly toxic organophosphate insecticide) to *p*-nitrophenol, which has only about 1/200 the toxicity of the parent compound:

$$\text{Paraoxon} \xrightarrow[\text{Enzyme action}]{H_2O,\ \{O\}} HO{-}C_6H_4{-}NO_2 + \text{Other products} \qquad (23.12.1)$$

Usually, the products of biodegradation are simpler molecular forms that tend to occur in nature. The definition of biodegradation is illustrated by an example in Figure 23.6. Biodegradation is usually carried out by the action of microorganisms, particularly bacteria and fungi.

Biodegradation Processes

The biotransformations of environmental chemicals, including pesticides and industrial chemicals, in vertebrates (birds, mammals, fish, reptiles) can be of the utmost importance in determining their fates and effects. **Biotransformation** is what happens to any substance that is **metabolized** and thereby altered by biochemical processes in an organism. **Metabolism** is divided into the two general categories of **catabolism**, which is the breaking down of more complex molecules, and **anabolism**, which is the building up of life molecules from simpler materials. The substances subjected to biotransformation may be naturally occurring or *anthropogenic* (made by human activities). They may consist of *xenobiotic* molecules that are foreign to living systems.

An important biochemical process that occurs in the biodegradation of many synthetic and hazardous waste materials is **cometabolism**. Cometabolism does not serve a useful purpose to an organism in terms of providing energy or raw material to build biomass, but occurs concurrently with normal metabolic processes. An example of cometabolism of hazardous wastes is provided by the white rot fungus, *Phanerochaete chrysosporium*. This organism, which has been investigated for its hazardous waste treatment potential, degrades a number of kinds of organochlorine compounds—including DDT, PCBs, and chlorodioxins—under the appropriate conditions. The enzyme system responsible for this degradation is one that the fungus uses to break down lignin in plant material under normal conditions.

Enzymes in Waste Degradation

Enzyme systems hold the key to biodegradation of hazardous wastes. For most biological treatment processes currently in use, enzymes are present in living organisms in contact with the wastes. However, in some cases it is possible to use cell-free extracts of enzymes removed from bacterial or fungal cells to treat hazardous wastes. For this application the enzymes may be present in solution or, more commonly, immobilized in biochemical reactors.

Biodegradation of municipal wastewater and solid wastes in landfills occurs by design. Biodegradation of any kind of waste that can be metabolized takes place whenever the wastes are subjected to conditions conducive to biological processes. The most common type of biodegradation is that of organic compounds in the presence of air, that is, **aerobic processes**. However, in the absence of air, **anaerobic biodegradation** may also take place. Furthermore, inorganic species are subject to both aerobic and anaerobic biological processes.

Although biological treatment of wastes is normally regarded as degradation to simple inorganic species such as carbon dioxide, water, sulfates, and phosphates, the possibility must always be considered of forming more complex or more hazardous chemical species. An example of the latter is the production of volatile, soluble, toxic methylated forms of arsenic and mercury from inorganic species of these elements by bacteria under anaerobic conditions.

For the most part, anthropogenic compounds resist biodegradation much more strongly than do naturally occurring compounds. This is generally due to the absence of enzymes that can bring about an initial attack on the compound. A number of physical and chemical characteristics of a compound are involved in its amenability to biodegradation. Such characteristics include hydrophobicity, solubility, volatility, and affinity for lipids. Some organic structural groups impart particular resistance to biodegradation. These include branched carbon chains, ether linkages, chlorine, amines, methoxy groups, sulfonates, and nitro groups.

Several groups of microorganisms are capable of partial or complete degradation of hazardous organic compounds. Among the aerobic bacteria, those of the *Pseudomonas* family are the most widespread and most adaptable to the degradation of synthetic compounds. Anaerobic bacteria catabolize biomass through hydrolytic processes, breaking down proteins, lipids, and saccharides. They are also known to reduce nitro compounds to amines, degrade nitrosamines, promote reductive dechlorination, reduce epoxide groups to alkenes, and break down aryl structures. **Actinomycetes** are microorganisms that are morphologically similar to both bacteria and fungi. They are involved in the degradation of a variety of organic compounds, including degradation-resistant alkanes, and lignocellulose. Other compounds attacked include pyridines, phenols, nonchlorinated aromatics, and chlorinated aromatics. Fungi are particularly noted for their ability to attack long-chain and complex hydrocarbons and are more successful than bacteria in the initial attack on PCB compounds. Phototrophic microorganisms, which include algae, photosynthetic bacteria, and cyanobacteria (blue-green algae) tend to concentrate organophilic compounds in their lipid stores and induce photochemical degradation of the stored compounds. For example, *Oscillatoria* can initiate the biodegradation of naphthalene by the attachment of –OH groups.

Practically all classes of synthetic organic compounds can be at least partially degraded by various microorganisms. These classes include nonhalogenated alkanes, halogenated alkanes (trichloroethane, dichloromethane), nonhalogenated aryl compounds (benzene, naphthalene, benzo(a)pyrene), halogenated aromatic compounds (hexachlorobenzene, pentachlorophenol), phenols (phenol, cresols), polychlorinated biphenyls, phthalate esters, and pesticides (chlordane, parathion).

CHAPTER SUMMARY

The chapter summary below is presented in a programmed format to review the main points covered in this chapter. It is used most effectively by filling in the blanks, referring back to the chapter as necessary. The correct answers are given at the end of the summary.

Three legislative acts passed in the U.S. to deal with hazardous substances and wastes are 1____
____
____
____
____.

The four characteristics by which hazardous substances are defined are 2________
____.

The four kinds of listed wastes are 3____
____
____
____.

Measurement of the temperature at which a hot liquid ignites is called its 4________
______________________. Values of the vapor/air ratio below which ignition cannot occur because of insufficient fuel define the 5________________________
______________. The difference between upper and lower flammability limits at a specified temperature is the 6____.
Ammonium nitrate, ammonium perchlorate, hydrogen peroxide, and sodium dichromate are all examples of 7____________________________. Such substances can contribute strongly to fire hazards because 8____________________________
____.

White phosphorus; the alkali metals; powdered forms of magnesium, calcium, cobalt, manganese, iron, zirconium and aluminum; some organometallic compounds, such as lithium ethyl (LiC_2H_4) and lithium phenyl (LiC_6H_5); and some metal carbonyl compounds, such as iron pentacarbonyl, $Fe(CO)_5$, are all examples of 9_____
______________________, which 10____
____. The most obvious toxic product of combustion is 11____________________________.
Other products include 12____
____. Reactive substances are those that 13____
____
____ One such substance is sodium metal which undergoes the following reaction with water:
14____.

High reactivity in some organic compounds results from 15__________________
____,
and some organic structures involving oxygen are very reactive. Functional groups containing both oxygen and nitrogen tend to impart 16__________________

to an organic compound. Explosives such as nitroglycerin or TNT are single compounds containing [17]______________________________ in the same molecule. Corrosive substances as the definition is applied to interaction with metals are regarded as [18]____ ______________________________________. The major chemical classes of corrosive substances are [19]____ ____. A measurement of toxicity applied to hazardous wastes for regulatory purposes is [20]_____ __________________________________. Three major categories of wastes based upon their physical forms are [21]__________________________________ ____. In consideration of mixtures and impurities, wastes are best treated that have a high degree of [22]________________________ and [23]_________________________. Major points of origin of hazardous wastes into the environment are [24]_________ ____
____
____
____
____
____.

The major physical properties of wastes that determine their amenability to transport are [25]____
____
____.

Two major subdivisions of hazardous waste pollutants based largely upon their environmental chemical behavior are [26]_______________ pollutants and [27]________ ____________________ pollutants. The effects of hazardous wastes in the environment may be divided among effects on [28]________________________ ____. The fates of hazardous waste substances are largely determined by the tendencies of the substances to attain [29]____. The fate of a hazardous waste substance in water is a function of the substance's [30]_____ ____. The primary environmental concern regarding hazardous wastes in the geosphere is [31]________ ____
____.

The transport of contaminants in the geosphere depends largely upon [32]_________ _________________ that govern the movement of water underground. Mathematically the distribution of a solute between groundwater or leachate water and soil is expressed by a [33]__________________________. Codisposal of [34]_________ __________________ with heavy metals, including radioactive metals, can have a strong effect upon the mobility of metal ions in soil. Among the chemical and biochemical processes to which a hazardous waste species is subjected in an aquatic system are [35]____
____
____. The presence of

organic matter in water has a tendency to increase the [36]________________ of hazardous organic substances. Among the physical, chemical, and biochemical processes that are particularly important in determining the transformations and ultimate fates of hazardous chemical species in the hydrosphere are [37] ________
____
____.
The most important biochemical processes that determine fates of hazardous waste species in water are those that are [38]________________________________.
Three major areas of interest in respect to hazardous waste compounds in the atmosphere are their [39]____
______________________________. The reaction,

$$SO_2 + {}^1/_2O_2 + H_2O \rightarrow H_2SO_4(aerosol)$$

is an example of conversion of a [40]______________ hazardous waste pollutant to a [41]____________________ hazardous waste pollutant. Organic species that produce secondary air pollutants are those that form [42]__________________
________________. The conversion by biological processes of wastes to simple inorganic molecules and, to a certain extent, to biological materials is known as [43]______________, whereas the complete bioconversion of a substance to inorganic species such as CO_2, NH_3, and phosphate is called [44]______________
________________, and [45]____________________________ refers to the biological conversion of a toxic substance to a less toxic species. An important biochemical process that occurs in the biodegradation of many synthetic and hazardous waste materials that does not serve a useful purpose to an organism in terms of providing energy or raw material to build biomass, but occurs concurrently with normal metabolic processes is called [46]________________________.
Microorganisms that are morphologically similar to both bacteria and fungi and that are involved in the degradation of a variety of organic compounds, including degradation-resistant alkanes, and lignocellulose are [47]________________.

Answers

1 Toxic Substances Control Act of 1976, Resource, Recovery and Conservation Act (RCRA) of 1976, and Comprehensive Environmental Response, Compensation, and Liability Act (CERCLA)

2 ignitability, corrosivity, reactivity, toxicity

3 F-type wastes from nonspecific sources, K-type wastes from specific sources, P-type acute hazardous wastes, U-Type generally hazardous wastes

4 flash point

5 lower flammability limit

6 flammability range

7 oxidizers

8 fuels may burn explosively in contact with an oxidizer

9 pyrophoric compounds

10 catch fire spontaneously in air without an ignition source

11 carbon monoxide

12 SO_2, P_4O_{10}, HCl, polycyclic aromatic hydrocarbons

13 tend to undergo rapid or violent reactions under certain conditions

14 $2Na + 2H_2O \rightarrow 2NaOH + H_2 + \text{heat}$

15 unsaturated bonds in the carbon skeleton

16 reactivity

17 both oxidizing and reducing functions

18 those that dissolve metals or cause oxidized material to form on the surface of metals

19 (1) strong acids, (2) strong bases, (3) oxidants, (4) dehydrating agents

20 Toxicity Characteristic Leaching Procedure (TCLP)

21 organic materials, aqueous wastes, and sludges

22 segregation

23 concentration

24 deliberate addition to soil, water, or air by humans; evaporation or wind erosion from waste dumps into the atmosphere; leaching from waste dumps into groundwater, streams, and bodies of water; leakage, such as from underground storage tanks or pipelines; evolution and subsequent deposition by accidents, such as fire or explosion; release from improperly operated waste treatment or storage facilities

25 volatility, solubility, and the degree to which they are sorbed to solids, including soil and sediments

26 primary

27 secondary

28 organisms, materials, and the environment

29 a state of physical and chemical stability

30 solubility, density, biodegradability, and chemical reactivity

31 the possible contamination of groundwater aquifers by waste leachates and leakage from wastes

32 hydrologic factors

33 distribution coefficient

34 chelating agents

35 acid-base, oxidation-reduction, precipitation-dissolution, and hydrolysis reactions, as well as biodegradation

36 solubility

37 hydrolysis reactions, precipitation reactions, oxidation-reduction reactions, sorption, biochemical processes, and photolysis reactions

38 mediated by microorganisms

39 pollution potential, atmospheric fate, and residence time

40 primary

41 secondary

42 photochemical smog

43 biodegradation

44 mineralization

45 detoxification

46 cometabolism

47 actinomycetes

QUESTIONS AND PROBLEMS

1. Match the following kinds of hazardous substances on the left with a specific example of each from the right, below:

1. Explosives	(a) Oleum, sulfuric acid, caustic soda
2. Compressed gases	(b) Magnesium metal, sodium hydride
3. Radioactive materials	(c) Lithium peroxide
4. Flammable solids	(d) Hydrogen, sulfur dioxide
5. Oxidizing materials	(e) Dynamite, ammunition
6. Corrosive materials	(f) Plutonium, cobalt-60

2. Of the following, the property that is **not** a member of the same group as the other properties listed is (a) substances that are liquids whose vapors are likely to ignite in the presence of ignition sources, (b) nonliquids that may catch fire from friction or contact with water and that burn vigorously or persistently, (c) ignitable compressed gases, (d) oxidizers, (e) substances that exhibit extremes of acidity or basicity.

3. In what respects may it be said that measures taken to alleviate air and water pollution tend to aggravate hazardous waste problems?

4. Discuss the significance of LFL, UFL, and flammability range in determining the flammability hazards of organic liquids.

5. Concentrated HNO_3 and its reaction products pose several kinds of hazards. What are these?

6. What are substances called that catch fire spontaneously in air without an ignition source?

7. Name four or five hazardous products of combustion and specify the hazards posed by these materials.

8. What kind of property tends to be imparted to a functional group of an organic compound containing both oxygen and nitrogen?
9. Match the corrosive substance from the column on the left, below, with one of its major properties from the right column:

1. Alkali metal hydroxides	(a) Reacts with protein in tissue to form yellow xanthoproteic acid
2. Hydrogen peroxide	
3. Hydrofluoric acid, HF	(b) Dissolves glass
4. Nitric acid, HNO_3	(c) Strong bases
	(d) Oxidizer

10. Rank the following wastes in increasing order of segregation (a) mixed halogenated and hydrocarbon solvents containing little water, (b) spent steel pickling liquor (c) dilute sludge consisting of mixed organic and inorganic wastes, (d) spent hydrocarbon solvents free of halogenated materials, (e) dilute mixed inorganic sludge.
11. Why is attenuation of metals likely to be very poor in acidic leachate?
12. What are three major properties of wastes that determine their amenability to transport in the environment?
13. What is the influence of organic solvents in leachates upon attenuation of organic hazardous waste constituents?
14. Match the following physical, chemical, and biochemical processes dealing with the transformations and ultimate fates of hazardous chemical species in the hydrosphere on the left with the description of the process on the right, below:

1. Precipitation reactions	(a) Molecule is cleaved with the addition of H_2O
2. Biochemical processes	(b) Often involve hydrolysis and oxidation-reduction
3. Hydrolysis reactions	(c) Generally mediated by microorganisms
4. Sorption	(d) By sediments and by suspended matter
5. Oxidation-reduction	(e) Generally accompanied by aggregation of colloidal particles suspended in water

15. Describe the particular danger posed by codisposal of strong chelating agents with radionuclide wastes. What may be said about the chemical nature of the latter in regard to this danger?
16. Describe a beneficial effect that might result from the precipitation of either $Fe_2O_3{\cdot}xH_2O$ or $MnO_2{\cdot}xH_2O$ from hazardous wastes in water.
17. Why are secondary air pollutants from hazardous waste sites usually of only limited concern as compared to primary air pollutants? What is the distinction between the two?
18. What are the major means by which hazardous waste species may be removed from the atmosphere? What is meant by $\tau_{1/2}$?
19. What may be said about the relative rates of reaction of hazardous waste compounds in the atmosphere with $HO{\cdot}$ and O_3?
20. As applied to hazardous wastes in the biosphere distinguish among biodegradation, biotransformation, detoxification, and mineralization.

21. What is the potential role of *Phanerochaete chrysosporium* in treatment of hazardous waste compounds? For which kinds of compounds might it be most useful?
22. What is a specific example of the formation of relatively more hazardous materials by the action of biological processes on hazardous wastes?
23. Several physical and chemical characteristics are involved in determining the amenability of a hazardous waste compound to biodegradation. These include hydrophobicity, solubility, volatility, and affinity for lipids. Suggest and discuss ways in which each one of these factors might affect biodegradability.
24. List and discuss some of the important processes determining the transformations and ultimate fates of hazardous chemical species in the hydrosphere.
25. Which part of the hydrosphere is most subject to long-term, largely irreversible contamination from the improper disposal of hazardous wastes in the environment?
26. What features or characteristics should a compound possess in order for direct photolyis to be a significant factor in its removal from the atmosphere?
27. List and discuss the significance of major sources for the origin of hazardous wastes, that is, their main modes of entry into the environment. What are the relative dangers posed by each of these? Which part of the environment would each be most likely to contaminate?
28. In what form would a large quantity of hazardous waste PCB likely be found in the hydrosphere?

LITERATURE CITED

[1] Wolbeck, Bernd, "Political Dimensions and Implications of Hazardous Waste Disposal," in *Hazardous Waste Disposal*, Lehman, John P., Ed., Plenum Press, New York, NY, 1982, pp. 7-18.

[2] Karnofsky, Brian, Ed., *Hazardous Waste Management Compliance Handbook*, 2nd ed., Van Nostrand Reinhold, New York, NY, 1997.

[3] Kindschy, Jon W., Marilyn Kraft, and Molly Carpenter, *Guide to Hazardous Materials & Waste Management: Risk, Regulations, and Responsibility*, Solano Press, New York, NY, 1997.

SUPPLEMENTAL REFERENCES

Greenberg, Michael and Dona Schneider, Environmentally Devastated *Neighborhoods: Perceptions, Policies, and Realities*, Rutgers University Press, New Brunswick, NJ, 1996.

Howard, Phillip H., *Handbook of Environmental Fate and Exposure Data for Organic Chemicals*, Volume 5, *Solvents 3*, CRC Press/Lewis Publishers, Boca Raton, FL, 1997.

Kowalski, Kathiann M., *Hazardous Wastesites (Pro/Con)*, Lerner Publications Company, New York, NY, 1996.

MacKay, Donald, Wan Ying Shiu, and Kuo-Ching Ma, *Illustrated Handbook of Physical-Chemical Properties and Environmental Fate for Organic Chemicals*, CRC Press/Lewis Publishers, Boca Raton, FL, 1997.

Manahan, Stanley E., *Hazardous Waste Chemistry, Toxicology and Treatment*, CRC Press/Lewis Publishers, Boca Raton, FL, 1990.

Manahan, Stanley E., *Toxicological Chemistry*, 2nd ed., CRC Press/Lewis Publishers, Boca Raton, FL, 1992.

Selim, Hussein, Michael C. Amacher, and H. Madgi Selim, *Reactivity and Transport of Heavy Metals in Soils*, CRC Press/Lewis Publishers, Boca Raton, FL, 1996.

24 WASTE MINIMIZATION, TREATMENT, AND DISPOSAL

24.1. INTRODUCTION

As discussed in Chapter 23, the manufacturing wastes of most concern are **hazardous wastes**. Until about 1970 in the U.S., such wastes were usually discarded to the environment, often by disposal on land in landfills. With expanded industrial activity following World War II, it became more and more obvious that often discarded wastes did not really "go away," and that they could come back to harm humans, often decades after disposal. Therefore, a concerted effort was made through federal and state legislation to formulate and implement a program to minimize the production of such wastes at their sources, to treat wastes that were produced as effectively as possible, and to ensure that waste disposal was as safe and permanent as possible. The keystone legislation in this effort has been the 1976 Resource Conservation and Recovery Act (RCRA) and its subsequent amendments, most notably the Hazardous and Solid Waste Amendments of 1984. Legislation with similar goals has been implemented in other countries.

This chapter discusses how environmental science and technology can be applied to hazardous waste management to develop measures by which chemical wastes can be minimized, recycled, treated, and disposed. In descending order of desirability, hazardous waste management attempts to accomplish the following:

- Do not produce it
- If making it cannot be avoided, produce only minimum quantities
- Recycle it
- If it is produced and cannot be recycled, treat it, preferably in a way that makes it nonhazardous
- If it cannot be rendered nonhazardous, dispose of it in a safe manner
- Once it is disposed, monitor it for leaching and other adverse effects

24.2. WASTE REDUCTION AND MINIMIZATION

Many hazardous waste problems can be avoided at early stages by **waste reduction** (cutting down quantities of wastes from their sources) and **waste minimization** (utilization of treatment processes that reduce the quantities of wastes requiring ultimate disposal). This section outlines basic approaches to waste minimization and reduction.

There are several ways in which quantities of wastes can be reduced, including source reduction, waste separation and concentration, resource recovery, and waste recycling. The most effective approaches to minimizing wastes center around careful control of manufacturing processes, taking into consideration discharges and the potential for waste minimization at every step of manufacturing. Viewing the process as a whole (as outlined for a generalized manufacturing process in Figure 24.1) often enables crucial identification of the source of a waste, such as a raw material impurity, catalyst, or process solvent. Once a source is identified, it is much easier to take measures to eliminate or reduce the waste.

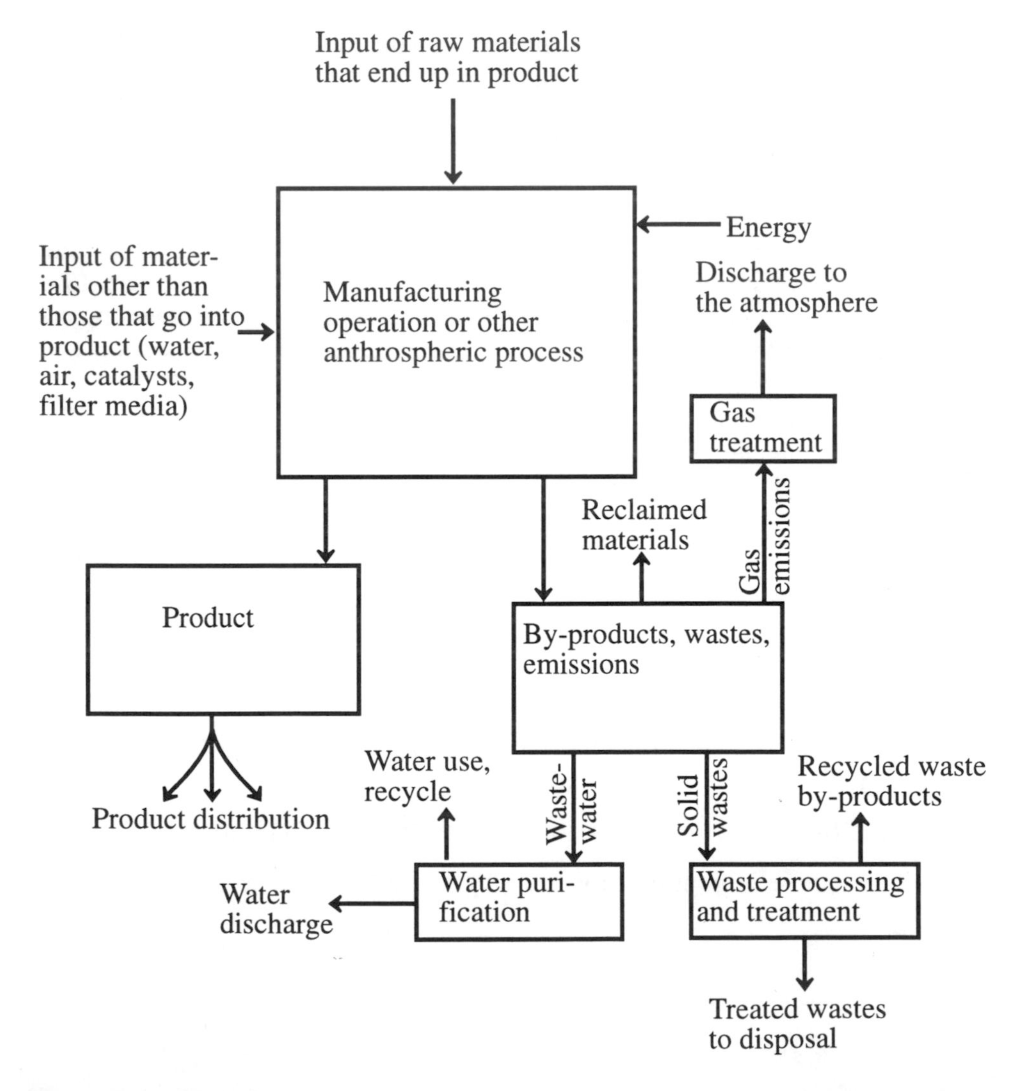

Figure 24.1. Manufacturing or other anthrospheric process from the viewpoint of discharges and waste minimization.

Modifications of the manufacturing process can yield substantial waste reduction. Some such modifications are of a chemical nature. Changes in chemical reaction conditions can minimize production of by-product hazardous substances. In some cases potentially hazardous catalysts, such as those formulated from toxic substances, can be replaced by catalysts that are nonhazardous or that can be recycled rather than discarded. Wastes can be minimized by volume reduction, for example, through dewatering and drying sludge.

24.3. RECYCLING

Wherever possible, recycling and reuse should be accomplished on-site because it avoids having to move wastes and because a process that produces recyclable materials is often the most likely to have use for them. The four broad areas in which something of value may be obtained from wastes are the following:

- Direct recycle as raw material to the generator, as with the return to feedstock of raw materials not completely consumed in a synthesis process
- Transfer as a raw material to another process; a substance that is a waste product from one process may serve as a raw material for another, sometimes in an entirely different industry
- Utilization for pollution control or waste treatment, such as use of waste alkali to neutralize waste acid
- Recovery of energy, for example, from the incineration of combustible hazardous wastes

Examples of Recycling

Recycling of scrap industrial impurities and products occurs on a large scale with a number of different materials. Most of these materials are not hazardous, but, as with most large-scale industrial operations, their recycle may involve the use or production of hazardous substances. Some of the more important examples are the following:

- **Ferrous metals** composed primarily of iron and used largely as feedstock for electric-arc furnaces
- **Nonferrous metals**, including aluminum (which ranks next to iron in terms of quantities recycled), copper and copper alloys, zinc, lead, cadmium, tin, silver, and mercury
- **Metal compounds**, such as metal salts
- **Inorganic substances**, including alkaline compounds (such as sodium hydroxide used to remove sulfur compounds from petroleum products), acids (steel pickling liquor where impurities permit reuse), and salts (for example, ammonium sulfate from coal coking used as fertilizer)
- **Glass**, which makes up about 10 percent of municipal refuse

- **Paper**, commonly recycled from municipal refuse
- **Plastic**, consisting of a variety of moldable polymeric materials and composing a major constituent of municipal wastes
- **Rubber**
- **Organic substances**, especially solvents and oils, such as hydraulic and lubricating oils
- **Catalysts** from chemical synthesis or petroleum processing
- Materials with **agricultural uses**, such as waste lime or phosphate-containing sludges used to treat and fertilize acidic soils

Waste Oil Utilization and Recovery

Waste oil generated from lubricants and hydraulic fluids is one of the more commonly recycled materials. Annual production of waste oil in the U.S. is of the order of 4 billion liters per year. Around half of this amount is burned as fuel and lesser quantities are recycled or disposed as waste. The collection, recycling, treatment, and disposal of waste oil are all complicated by the fact that it comes from diverse, widely dispersed sources and contains several classes of potentially hazardous contaminants. These are divided between organic constituents (polycyclic aromatic hydrocarbons, chlorinated hydrocarbons) and inorganic constituents (aluminum, chromium, and iron from wear of metal parts; barium and zinc from oil additives; lead from leaded gasoline).

Recycling Waste Oil

The processes used to convert waste oil to a feedstock suitable for lubricant formulation are illustrated in Figure 24.2. The first of these uses distillation to remove water and light ends that have come from condensation and contaminant fuel. The second, or processing, step may be a vacuum distillation in which the three products are oil for further processing, a fuel oil cut, and a heavy residue. The processing step may also employ treatment with a mixture of solvents, including isopropyl and butyl alcohols and methylethyl ketone, to dissolve the oil and leave contaminants as a sludge, or contact with sulfuric acid to remove inorganic contaminants followed by treatment with clay to take out acid and contaminants that cause odor and color. The third step shown in Figure 24.2 employs vacuum distillation to separate lubricating oil stocks from a fuel fraction and heavy residue. This phase of treatment may also involve hydrofinishing, treatment with clay, and filtration.

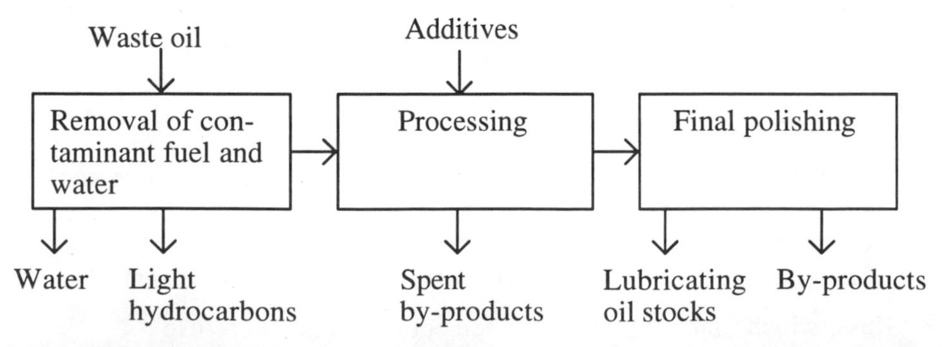

Figure 24.2. Major steps in reprocessing waste oil.

Waste Oil Fuel

For economic reasons, waste oil that is to be used for fuel is given minimal treatment of a physical nature, including settling, removal of water, and filtration. Metals in waste fuel oil become highly concentrated in its fly ash, which may be hazardous.

Waste Solvent Recovery and Recycle

The recovery and recycling of waste solvents is similar in many respects to the recycling of waste oil. Among the many solvents listed as hazardous wastes and recoverable from wastes are dichloromethane, tetrachloroethylene, trichloroethylene, 1,1,1-trichloroethane, benzene, liquid alkanes, 2-nitropropane, methylisobutyl ketone, and cyclohexanone. For reasons of both economics and pollution control, many industrial processes that use solvents are equipped for solvent recycle. The basic scheme for solvent reclamation and reuse is shown in Figure 24.3.

A number of operations are used in solvent purification. Entrained solids are removed by settling, filtration, or centrifugation. Drying agents may be used to remove water from solvents and various adsorption techniques and chemical treatment may be required to free the solvent from specific impurities. Fractional distillation, often requiring several distillation steps, is the most important operation in solvent purification and recycle. It is used to separate solvents from impurities, water, and other solvents.

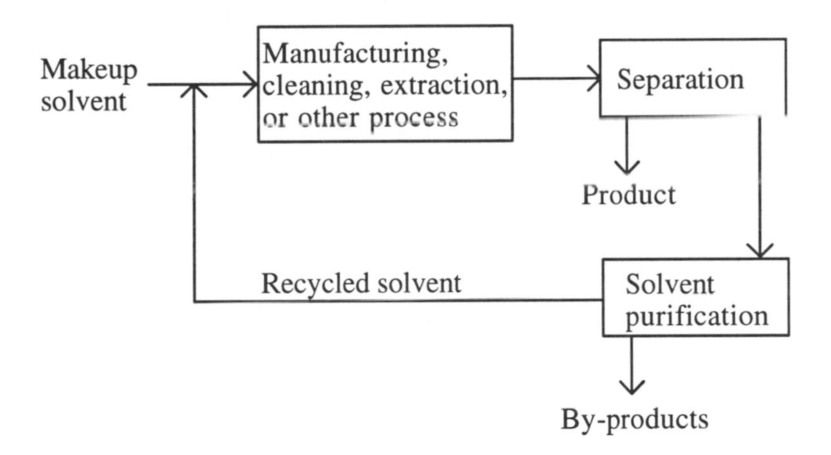

Figure 24.3. Overall process for recycling solvents.

24.4. PHYSICAL METHODS OF WASTE TREATMENT

This section addresses physical methods for waste treatment and the following section addresses methods that utilize chemical processes. In fact, most waste treatment measures have both physical and chemical aspects. The appropriate treatment technology for hazardous wastes obviously depends upon the nature of the wastes. These may consist of volatile wastes (gases, volatile solutes in water, gases or volatile liquids held by solids, such as catalysts), liquid wastes (wastewater, organic solvents), dissolved or soluble wastes (water-soluble inorganic species, water-soluble organic species, compounds soluble in organic solvents) semisolids (sludges, greases), and solids (dry solids, including granular solids with a significant water content, such as dewatered sludges, as well as solids suspended in liquids). The type of physical treatment to be applied to wastes depends strongly upon the physical properties of the material treated, including state of matter, solubility in water and organic solvents, density, volatility, boiling point, and melting point.

As shown in Figure 24.4, waste treatment may occur at three major levels — **primary**, **secondary**, and **polishing** — somewhat analogous to the treatment of wastewater (see Chapter 13). Primary treatment is generally regarded as preparation for further treatment, although it can result in the removal of by-products and reduction of the quantity and hazard of the waste. Secondary treatment detoxifies, destroys, and removes hazardous constituents. Polishing usually refers to treatment of water that is removed from wastes so that it may be safely discharged. However, the term can be broadened to apply to the treatment of other products as well so that they may be safely discharged or recycled.

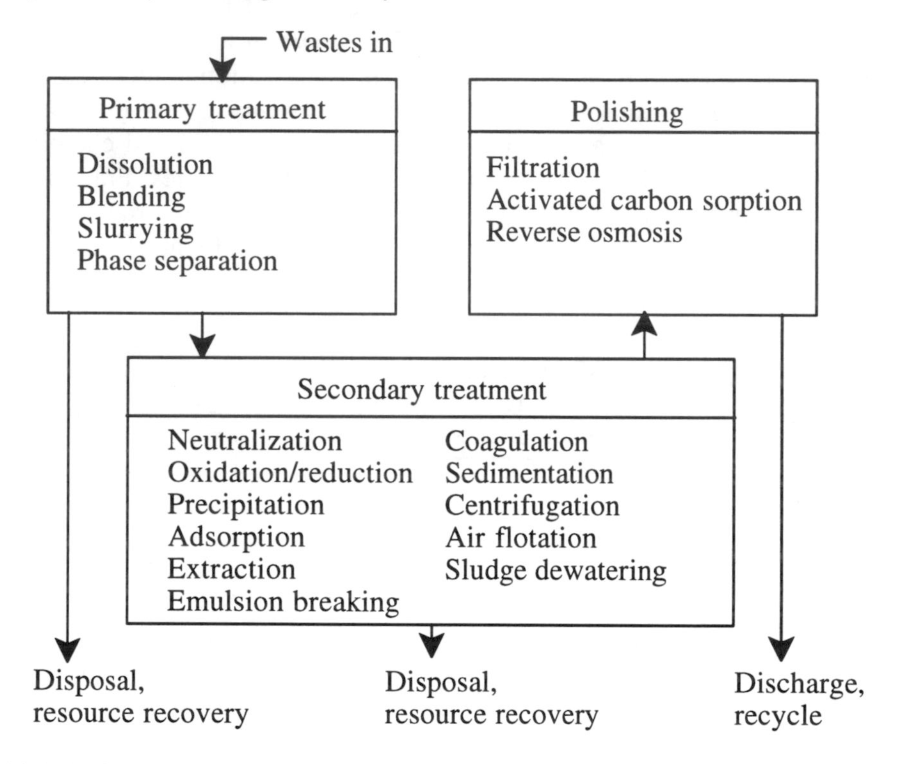

Figure 24.4. Major phases of waste treatment.

Methods of Physical Treatment

Knowledge of the physical behavior of wastes has been used to develop various unit operations for waste treatment that are based upon physical properties. These operations include the following:

- Phase separation
  - Filtration
- Phase transition
  - Distillation
  - Evaporation
  - Physical precipitation
- Phase transfer
  - Extraction
  - Sorption
- Membrane separations
  - Reverse osmosis
  - Hyper- and ultrafiltration

Phase Separations

The most straightforward means of physical treatment involves separation of components of a mixture that are already in two different phases. **Sedimentation** and **decanting** are easily accomplished with simple equipment. In many cases the separation must be aided by mechanical means, particularly **filtration** or **centrifugation**. **Flotation** is used to bring suspended organic matter or finely divided particles to the surface of a suspension. In the process of **dissolved air flotation** (DAF), air is dissolved in the suspending medium under pressure and comes out of solution when the pressure is released as minute air bubbles attached to suspended particles, which causes the particles to float to the surface.

An important and often difficult waste treatment step is **emulsion breaking** in which colloidal-sized **emulsions** are caused to aggregate and settle from suspension. Agitation, heat, acid, and the addition of **coagulants** consisting of organic polyelectrolytes or inorganic substances, such as an aluminum salt, may be used for this purpose. The chemical additive acts as a flocculating agent to cause the particles to stick together and settle out.

Phase Transition

A second major class of physical separation is that of **phase transition** in which a material changes from one physical phase to another. It is best exemplified by **distillation**, which is used in treating and recycling solvents, waste oil, aqueous phenolic wastes, xylene contaminated with paraffin from histological laboratories, and mixtures of ethylbenzene and styrene. Distillation produces **distillation bottoms** (still bottoms), which are often hazardous and polluting. These consist of unevaporated solids, semisolid tars, and sludges from distillation. Specific examples are distillation bottoms from the production of acetaldehyde from ethylene, and still bottoms from toluene reclamation distillation in the production of disulfoton. The landfill disposal of these and other hazardous distillation bottoms used to be widely practiced but is now severely limited:

Evaporation is usually employed to remove water from an aqueous waste to concentrate it. A special case of this technique is **thin-film evaporation** in which volatile constituents are removed by heating a thin layer of liquid or sludge waste spread on a heated surface.

Drying — removal of solvent or water from a solid or semisolid (sludge) or the removal of solvent from a liquid or suspension — is a very important operation because water is often the major constituent of waste products, such as sludges obtained from emulsion breaking. In **freeze drying**, the solvent, usually water, is sublimed from a frozen material. Hazardous waste solids and sludges are dried to reduce the quantity of waste, to remove solvent or water that might interfere with subsequent treatment processes, and to remove hazardous volatile constituents. Dewatering can often be improved with addition of a filter aid, such as diatomaceous earth, during the filtration step.

Stripping is a means of separating volatile components from less volatile ones in a liquid mixture by the partitioning of the more volatile materials to a gas phase of air or steam (steam stripping). The gas phase is introduced into the aqueous solution or suspension containing the waste in a stripping tower that is equipped with trays or packed to provide maximum turbulence and contact between the liquid and gas phases. The two major products are condensed vapor and a stripped bottoms residue. Examples of two volatile components that can be removed from water by air stripping are benzene and dichloromethane. Air stripping can also be used to remove ammonia from water that has been treated with a base to convert ammonium ion to volatile ammonia.

Physical precipitation is used here as a term to describe processes in which a solid forms from a solute in solution as a result of a physical change in the solution, as compared to chemical precipitation (see Section 24.5) in which a chemical reaction in solution produces an insoluble material. The major changes that can cause physical precipitation are cooling the solution, evaporation of solvent, or alteration of solvent composition. The most common type of physical precipitation by alteration of solvent composition occurs when a water-miscible organic solvent is added to an aqueous solution, so that the solubility of a salt is lowered below its concentration in the solution.

Phase Transfer

Phase transfer consists of the transfer of a solute in a mixture from one phase to another. An important type of phase transfer process is **solvent extraction**, a process in which a substance is transferred from solution in one solvent (usually water) to another (usually an organic solvent) without any chemical change taking place. When solvents are used to leach substances from solids or sludges, the process is called **leaching**. Solvent extraction and the major terms applicable to it are summarized in Figure 24.5. The same terms and general principles apply to leaching. The major application of solvent extraction to waste treatment has been in the removal of phenol from by-product water produced in coal coking, petroleum refining, and chemical syntheses that involve phenol.

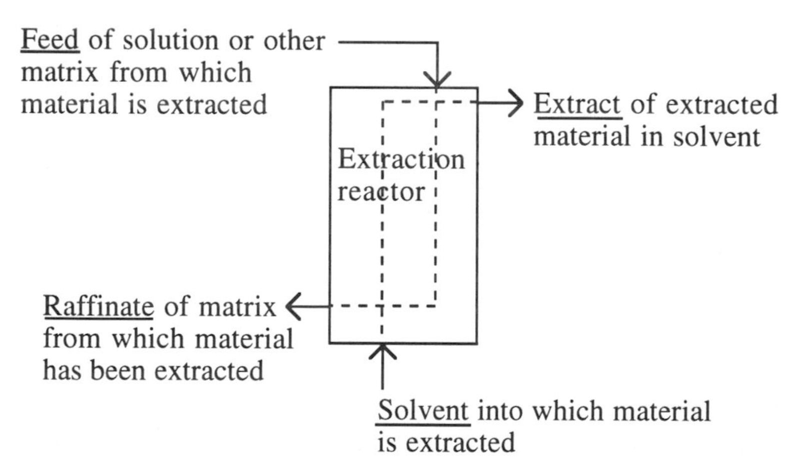

Figure 24.5. Outline of solvent extraction/leaching process with important terms underlined.

One of the more promising approaches to solvent extraction and leaching of hazardous wastes is the use of **supercritical fluids**, most commonly CO_2, as extraction solvents. A supercritical fluid is one that has characteristics of both liquid and gas and consists of a substance above its supercritical temperature and pressure (31.1°C and 73.8 atm, respectively, for CO_2). After a substance has been extracted from a waste into a supercritical fluid at high pressure, the pressure can be released, resulting in separation of the substance extracted. The fluid can then be compressed again and recirculated through the extraction system. Some possibilities for treatment of hazardous wastes by extraction with supercritical CO_2 include removal of organic contaminants from wastewater, extraction of organohalide pesticides from soil, extraction of oil from emulsions used in aluminum and steel processing, and regeneration of spent activated carbon. Waste oils contaminated with PCBs, metals, and water can be purified using supercritical ethane.

Transfer of a substance from a solution to a solid phase is called **sorption**. The most important sorbent is **activated carbon** used for several purposes in waste treatment; in some cases it is adequate for complete treatment. It can also be applied to pretreatment of waste streams going into processes such as reverse osmosis to improve treatment efficiency and reduce fouling. Effluents from other treatment processes, such as biological treatment of degradable organic solutes in water can be polished with activated carbon. Activated carbon sorption is most effective for removing from water those hazardous waste materials that are poorly water-soluble and that have high molecular masses, such as xylene, naphthalene, cyclohexane, chlorinated hydrocarbons, phenol, aniline, dyes, and surfactants. Activated carbon does not work well for organic compounds that are highly water-soluble or polar.

Solids other than activated carbon can be used for sorption of contaminants from liquid wastes. These include synthetic resins composed of organic polymers and mineral substances. Of the latter, clay is employed to remove impurities from waste lubricating oils in some oil recycling processes.

Molecular Separation

A third major class of physical separation is **molecular separation**, often based upon **membrane processes** in which dissolved contaminants or solvent pass through a size-selective membrane under pressure. The products are a relatively pure solvent phase (usually water) and a concentrate enriched in the solute impurities. **Hyperfiltration** allows passage of species with molecular masses of about 100 to 500, whereas **ultrafiltration** is used for the separation of organic solutes with molecular masses of 500 to 1,000,000. With both of these techniques water and lower-molecular-mass solutes under pressure pass through the membrane as a stream of purified **permeate**, leaving behind a stream of **concentrate** containing impurities in solution or suspension. Ultrafiltration and hyperfiltration are especially useful for concentrating suspended oil, grease, and fine solids in water. They also serve to concentrate solutions of large organic molecules and heavy metal ion complexes.

Reverse osmosis is the most widely used of the membrane techniques. Superficially similar to ultrafiltration and hyperfiltration, it operates on a different principle in that the membrane is selectively permeable to water and excludes ionic solutes. Reverse osmosis uses high pressures to force permeate through the membrane, producing a concentrate containing high levels of dissolved salts.

Electrodialysis, sometimes used to concentrate plating wastes, employs membranes alternately permeable to cations and to anions. The driving force for the separation is provided by electrolysis with a direct current between two electrodes. Alternate layers between the membranes contain concentrate (brine) and purified water.

24.5. CHEMICAL TREATMENT: AN OVERVIEW

The applicability of chemical treatment to wastes depends upon the chemical properties of the waste constituents, particularly acid-base, oxidation-reduction, precipitation, and complexation behavior; reactivity; flammability/combustibility;

corrosivity; and compatibility with other wastes. The chemical behavior of wastes translates to various unit operations for waste treatment that are based upon chemical properties and reactions. These include the following:

- Acid/base neutralization
- Chemical extraction and leaching
- Reduction
- Chemical precipitation
- Oxidation/reduction
- Ion exchange

Some of the more sophisticated means available for treatment of wastes have been developed for pesticide disposal.

Acid/Base Neutralization

Waste acids and bases are treated by **neutralization**:

$$H^+ + OH^- \rightarrow H_2O \quad (24.5.1)$$

Although simple in principle, neutralization can present some problems in practice. These include evolution of volatile contaminants, mobilization of soluble substances, excessive heat generated by the neutralization reaction, and corrosion to apparatus. By adding too much or too little of the neutralizing agent, it is possible to get a product that is too acidic or basic.

Lime, $Ca(OH)_2$, is widely used as a base for treating acidic wastes. Because of lime's limited solubility, solutions of excess lime do not reach extremely high pH values. Sulfuric acid, H_2SO_4, is a relatively inexpensive acid for treating alkaline wastes. However, addition of too much sulfuric acid can produce highly acidic products; for some applications, acetic acid, CH_3COOH, is preferable. As noted above, acetic acid is a weak acid and an excess of it does little harm. It is also a natural product and biodegradable.

Neutralization, or pH adjustment, is often required prior to the application of other waste treatment processes. Processes that may require neutralization include oxidation/reduction, activated carbon sorption, wet air oxidation, stripping, and ion exchange. Microorganisms usually require a pH in the range of 6-9, so neutralization may be required prior to biochemical treatment.

Chemical Precipitation

Chemical precipitation is used in hazardous waste treatment primarily for the removal of heavy metal ions from water as shown below for the chemical precipitation of cadmium:

$$Cd^{2+}(aq) + HS^-(aq) \rightarrow CdS(s) + H^+(aq) \quad (24.5.2)$$

Precipitation of Metals

The most widely used means of precipitating metal ions is by the formation of hydroxides such as chromium(III) hydroxide:

$$Cr^{3+} + 3OH^- \rightarrow Cr(OH)_3 \quad (24.5.3)$$

The source of hydroxide ion is a base (alkali), such as lime ($Ca(OH)_2$), sodium hydroxide (NaOH), or sodium carbonate (Na_2CO_3). Most metal ions tend to produce basic salt precipitates, such as basic copper(II) sulfate, $CuSO_4 \cdot 3Cu(OH)_2$, formed as a solid when hydroxide is added to a solution containing Cu^{2+} and SO_4^{2-} ions. The solubilities of many heavy metal hydroxides reach a minimum value, often at a pH in the range of 9-11, then increase with increasing pH values due to the formation of soluble hydroxo complexes, as illustrated by the following reaction:

$$Zn(OH)_2(s) + OH^-(aq) \rightarrow Zn(OH)_3^-(aq) \quad (24.5.4)$$

The chemical precipitation method that is used most is precipitation of metals as hydroxides and basic salts with lime. Sodium carbonate can be used to precipitate hydroxides ($Fe(OH)_3 \cdot xH_2O$), carbonates ($CdCO_3$), or basic carbonate salts ($2PbCO_3 \cdot Pb(OH)_2$). The carbonate anion produces hydroxide by virtue of its hydrolysis reaction with water:

$$CO_3^{2-} + H_2O \rightarrow HCO_3^- + OH^- \quad (24.5.5)$$

Alone, CO_3^{2-} does not give as high a pH as do alkali metal hydroxides, which may be required to precipitate metals that form hydroxides only at relatively high pH values.

The solubilities of some heavy metal sulfides are extremely low, so precipitation by H_2S or other sulfides (see Reaction 24.5.2) can be a very effective means of treatment. Hydrogen sulfide is a toxic gas that is itself considered to be a hazardous waste. Iron(II) sulfide (ferrous sulfide) can be used as a safe source of sulfide ion to produce sulfide precipitates with other metals that are less soluble than FeS. However, toxic H_2S can be produced when metal sulfide wastes contact acid:

$$MS + 2H^+ \rightarrow M^{2+} + H_2S \quad (24.5.6)$$

Some metals can be precipitated from solution in the elemental metal form by the action of a reducing agent, such as sodium borohydride,

$$4Cu^{2+} + NaBH_4 + 2H_2O \rightarrow 4Cu + NaBO_2 + 8H^+ \quad (24.5.7)$$

or with more active metals in a process called **cementation**:

$$Cd^{2+} + Zn \rightarrow Cd + Zn^{2+} \quad (24.5.8)$$

Oxidation/Reduction

As shown by the reactions in Table 24.1, **oxidation** and **reduction** can be used for the treatment and removal of a variety of inorganic and organic wastes. Some waste oxidants can be used to treat oxidizable wastes in water and cyanides.

Ozone, O_3, is a strong oxidant that can be generated on-site by an electrical discharge through dry air or oxygen. Ozone employed as an oxidant gas at levels of 1-2 wt % in air and 2-5 wt % in oxygen has been used to treat a large variety of oxidizable contaminants, effluents, and wastes, including wastewater and sludges containing oxidizable constituents.

Table 24.1. Oxidation/Reduction Reactions Used to Treat Wastes

Waste substance	Reaction with oxidant or reductant
Oxidation of organics	
Organic matter, $\{CH_2O\}$	$\{CH_2O\} + \{O\} \rightarrow CO_2 + H_2O$
Aldehyde	$CH_3CH_2O + \{O\} \rightarrow CH_3COOH$ (carboxylic acid)
Oxidation of inorganics	
Cyanide	$2CN^- + 5OCl^- + H_2O \rightarrow N_2 + HCO_3^- + Cl^-$
Iron(II)	$Fe^{2+} + O_2 + 10H_2O \rightarrow 4Fe(OH)_3 + 8H^+$
Reduction of inorganics	
Chromate	$2CrO_4^{2-} + 3SO_2^- + 4H^+ \rightarrow Cr_2(SO_4)_3 + 2H_2O$
Permanganate	$MnO_4^- + 3Fe^{2+} + 7H_2O \rightarrow MnO_2(s) + 3Fe(OH)_3(s) + 5H^+$

Electrolysis

As shown in Figure 24.6, **electrolysis** is a process in which one species in solution (usually a metal ion) is reduced by electrons at the **cathode** and another gives up electrons to the **anode** and is oxidized there. In hazardous waste applications electrolysis is most widely used in the recovery of cadmium, copper, gold, lead, silver, and zinc.

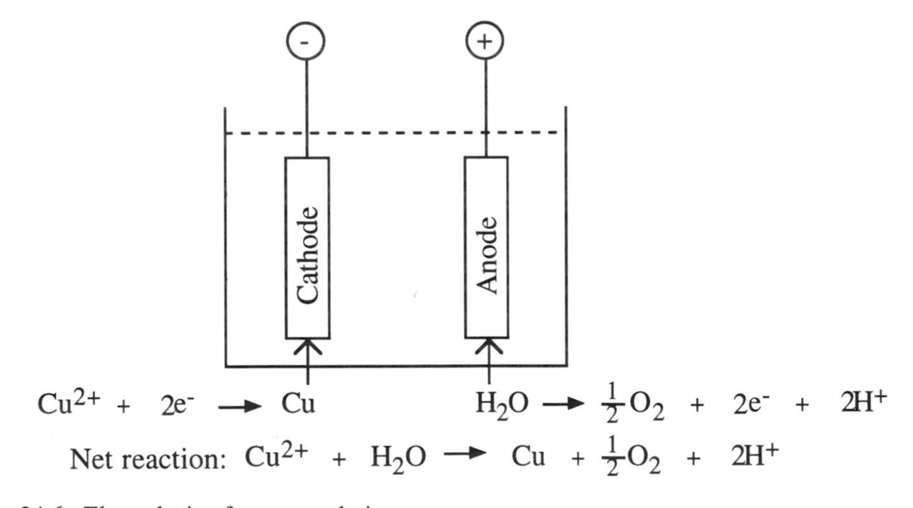

Figure 24.6. Electrolysis of copper solution.

Hydrolysis

One of the ways to dispose of chemicals that are reactive with water is to allow them to react with water under controlled conditions, a process called **hydrolysis**. Inorganic chemicals that can be treated by hydrolysis include metals that react with water; metal carbides, hydrides, amides, alkoxides, and halides; and nonmetal oxyhalides and sulfides. Examples of the treatment of these classes of inorganic species are given in Table 24.2.

Table 24.2. Inorganic Chemicals That May be Treated by Hydrolysis

Class of chemical	Reaction with water
Active metals (calcium)	$Ca + 2H_2O \rightarrow H_2 + Ca(OH)_2$
Hydrides (sodium aluminum hydride	$NaAlH_4 + 4H_2O \rightarrow 4H_2 + NaOH + Al(OH)_3$
Carbides (calcium carbide)	$CaC_2 + 2H_2O \rightarrow Ca(OH)_2 + C_2H_2$
Amides (sodium amide)	$NaNH_2 + H_2O \rightarrow NaOH + NH_3$
Halides (silicon tetrachloride)	$SiCl_4 + 2H_2O \rightarrow SiO_2 + 4HCl$
Alkoxides (sodium ethoxide)	$NaOC_2H_5 + H_2O \rightarrow NaOH + C_2H_5OH$

Organic chemicals may also be treated by hydrolysis. For example, toxic acetic anhydride is hydrolyzed to relatively safe acetic acid:

$$\text{H-}\underset{\text{H}}{\overset{\text{H}}{\text{C}}}\text{-}\overset{\text{O}}{\overset{\|}{\text{C}}}\text{-O-}\overset{\text{O}}{\overset{\|}{\text{C}}}\text{-}\underset{\text{H}}{\overset{\text{H}}{\text{C}}}\text{-H} + H_2O \longrightarrow 2\text{H-}\underset{\text{H}}{\overset{\text{H}}{\text{C}}}\text{-}\overset{\text{O}}{\overset{\|}{\text{C}}}\text{-OH} \quad (24.5.9)$$

Acetic anhydride (an acid anhydride)

Chemical Extraction and Leaching

Chemical extraction or **leaching** in hazardous waste treatment is the removal of a hazardous constituent by chemical reaction with an extractant in solution. Poorly soluble heavy metal salts can be extracted by reaction of the salt anions with H^+ as illustrated by the following:

$$PbCO_3 + H^+ \rightarrow Pb^{2+} + HCO_3^- \quad (24.5.10)$$

Acids also dissolve basic organic compounds such as amines and aniline. Extraction with acids should be avoided if cyanides or sulfides are present to prevent formation of toxic hydrogen cyanide or hydrogen sulfide. Nontoxic weak acids are usually the safest to use. These include acetic acid, CH_3COOH, and the acid salt, NaH_2PO_4.

Chelating agents, such as dissolved ethylenedinitrilotetraacetate (EDTA, HY^{2-}), dissolve insoluble metal salts by forming soluble species with metal ions:

$$FeS + HY^{3-} \rightarrow FeY^{2-} + HS^- \quad (24.5.11)$$

Heavy metal ions in soil contaminated by hazardous wastes may be present in a coprecipitated form with insoluble iron(III) and manganese(IV) oxides, Fe_2O_3 and MnO_2, respectively. These oxides can be dissolved by reducing agents, such as solutions of sodium dithionate/citrate or hydroxylamine. This results in the production of soluble Fe^{2+} and Mn^{2+} and the release of heavy metal ions, such as Cd^{2+} or Ni^{2+}, which are removed with the water.

Ion Exchange

Ion exchange is a means of removing cations or anions from solution onto a solid resin, which can be regenerated by treatment with acids, bases, or salts. The greatest use of ion exchange in hazardous waste treatment is for the removal of low levels of heavy metal ions from wastewater:

$$2H^{+-}\{CatExchr\} + Cd^{2+} \rightarrow Cd^{2+-}\{CatExchr\}_2 + 2H^+ \qquad (24.5.12)$$

Ion exchange is employed in the metal-plating industry to purify rinsewater and spent plating bath solutions. Cation exchangers are used to remove cationic metal species, such as Cu^{2+}, from such solutions. Anion exchangers remove anionic cyanide metal complexes (for example, $Ni(CN)_4{}^{2-}$) and chromium(VI) species, such as $CrO_4{}^{2-}$. Radionuclides may be removed from radioactive wastes and mixed waste by ion exchange resins.

24.6. Photolytic Reactions

Photolytic reactions, or *photolysis*, which use energy from photons of ultraviolet radiation to cause chemical reactions to occur, can be used to destroy a number of kinds of hazardous wastes. In such applications it is most useful in breaking chemical bonds in refractory organic compounds. TCDD, or “dioxin”, one of the most troublesome and refractory of wastes, can be treated by ultraviolet light in the presence of hydrogen atom donors {H} resulting in reactions such as the following:

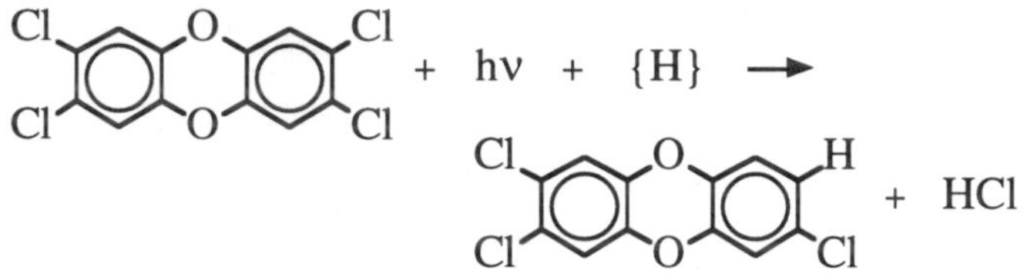

Cl, O, Cl / Cl, O, Cl + hν + {H} →
Cl, O, H / Cl, O, Cl + HCl (24.6.1)

As photolysis proceeds, more H-C bonds are broken, the C-O bonds are broken, and the final product is a harmless organic polymer.

An initial photolysis reaction can result in the generation of reactive intermediates that participate in **chain reactions** that lead to the destruction of a compound. One of the most important reactive intermediates is free radical HO•. In some cases **sensitizers** are added to the reaction mixture to absorb radiation and generate reactive species that destroy wastes.

Hazardous waste substances other than TCDD that have been destroyed by photolysis are herbicides (atrazine), 2,4,6-trinitrotoluene (TNT), and polychlorinated biphenyls (PCBs). The addition of a chemical oxidant, such as potassium peroxidisulfate, $K_2S_2O_8$, enhances destruction by oxidizing active photolytic products.

24.7. THERMAL TREATMENT METHODS

Thermal treatment of hazardous wastes can be used to accomplish most of the common objectives of waste treatment—volume reduction; removal of volatile, combustible, mobile organic matter; and destruction of toxic and pathogenic materials. The most widely applied means of thermal treatment of hazardous wastes is **incineration**. Incineration utilizes high temperatures, an oxidizing atmosphere, and often turbulent combustion conditions to destroy wastes. Methods other than incineration that make use of high temperatures to destroy or neutralize hazardous wastes are discussed briefly at the end of this section.

Incineration

Hazardous waste incineration will be defined here as a process that involves exposure of the waste materials to oxidizing conditions at a high temperature, usually in excess of 900°C. Normally, the heat required for incineration comes from the oxidation of organically bound carbon and hydrogen contained in the waste material or in supplemental fuel:

$$C(\text{organic}) + O_2 \rightarrow CO_2 + \text{heat} \quad (24.7.1)$$

$$4H(\text{organic}) + O_2 \rightarrow 2H_2O + \text{heat} \quad (24.7.2)$$

These reactions destroy organic matter and generate heat required for endothermic reactions, such as the breaking of C-Cl bonds in organochlorine compounds.

Incinerable Wastes

Ideally, incinerable wastes are predominantly organic materials that will burn with a heating value of at least 5,000 Btu/lb and preferably over 8,000 Btu/lb. Such heating values are readily attained with wastes having high contents of the most commonly incinerated waste organic substances, including methanol, acetonitrile, toluene, ethanol, amyl acetate, acetone, xylene, methylethyl ketone, adipic acid, and ethyl acetate. In some cases, however, it is desirable to incinerate wastes that will not burn alone and that require **supplemental fuel,** such as methane and petroleum liquids. Examples of such wastes are nonflammable organochloride wastes, some aqueous wastes, or soil in which the elimination of a particularly troublesome contaminant is worth the expense and trouble of incinerating it. Inorganic matter, water, and organic hetero element contents of liquid wastes are important in determining their incinerability.

Hazardous Waste Fuel

Many industrial wastes, including hazardous wastes, are burned as **hazardous waste fuel** for energy recovery in industrial furnaces and boilers and in incinerators for nonhazardous wastes, such as sewage sludge incinerators. This process is called **coincineration**, and more combustible wastes are utilized by it than are burned solely for the purpose of waste destruction. In addition to heat recovery from combustible wastes, it is a major advantage to use an existing on-site facility for waste disposal rather than a separate hazardous waste incinerator.

Incineration Systems

The four major components of hazardous waste incineration systems are shown in Figure 24.7.

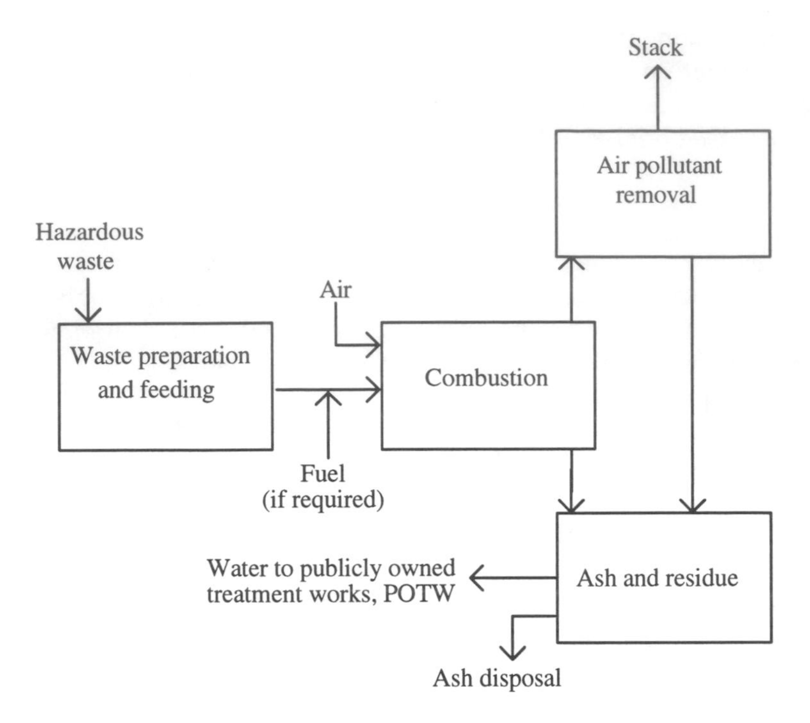

Figure 24.7. Major components of a hazardous waste incinerator system.

Waste preparation for liquid wastes may require filtration, settling to remove solid material and water, blending to obtain the optimum incinerable mixure, or heating to decrease viscosity. Solids may require shredding and screening. Atomization is commonly used to feed liquid wastes. Several mechanical devices, such as rams and augers, are used to introduce solids into the incinerator.

Combustion chambers are usually liquid injection, fixed hearth, rotary kiln, and fluidized bed, types discussed in more detail later in this section.

Often the most complex part of a hazardous waste incineration system is the **air pollution control system**, which involves several operations. The most common operations in air pollution control from hazardous waste incinerators are combustion gas cooling, heat recovery, quenching, particulate matter removal, acid gas removal, and treatment and handling of by-product solids, sludges, and liquids

Hot ash is often quenched in water. Prior to disposal it may require dewatering and chemical stabilization. A major consideration with hazardous waste incinerators and the types of wastes that are incinerated is the disposal problem posed by the ash, especially in respect to potential leaching of heavy metals.

Types of Incinerators

Hazardous waste incinerators may be divided among the following, based upon type of combustion chamber:

- **Rotary kiln** (about 40% of U. S. hazardous waste incinerator capacity) in which the primary combustion chamber is a rotating cylinder lined with refractory materials and an afterburner downstream from the kiln to complete destruction of the wastes

- **Liquid injection incinerators** (also about 40% of U.S. hazardous waste incinerator capacity) that burn pumpable liquid wastes dispersed as small droplets
- **Fixed-hearth incinerators** with single or multiple hearths upon which combustion of liquid or solid wastes occurs
- **Fluidized-bed incinerators** that have a bed of granular solid (such as sand) maintained in a suspended state by injection of air to remove pollutant acid gas and ash products
- **Advanced design incinerators** including **plasma incinerators** that make use of an extremely hot plasma of ionized air injected through an electrical arc; **electric reactors** that use resistance-heated incinerator walls at around 2,200°C to heat and pyrolyze wastes by radiative heat transfer; **infrared systems**, which generate intense infrared radiation by passing electricity through silicon carbide resistance heating elements; **molten salt combustion** that uses a bed of molten sodium carbonate at about 900°C to destroy the wastes and retain gaseous pollutants; and **molten glass processes** that use a pool of molten glass to transfer heat to the waste and to retain products in a poorly leachable glass

Combustion Conditions

The key to effective incineration of hazardous wastes lies in the combustion conditions. These require (1) sufficient free oxygen in the combustion zone; (2) turbulence for thorough mixing of waste, oxidant, and (where used) supplemental fuel; (3) high combustion temperatures above about 900°C to ensure that thermally resistant compounds do react; and (4) sufficient residence time (at least 2 seconds) to allow reactions to occur.

Effectiveness of Incineration

EPA standards for hazardous waste incineration are based upon the effectiveness of destruction of the **principal organic hazardous constituents** (POHC). Measurement of these compounds before and after incineration gives the **destruction removal efficiency** (DRE) according to the formula,

$$DRE = \frac{W_{in} - W_{out}}{W_{in}} \times 100 \qquad (24.7.3)$$

where W_{in} and W_{out} are the mass flow rates of the principal organic hazardous constituent (POHC) input and output (at the stack downstream from emission controls), respectively.

Wet Air Oxidation

Organic compounds and oxidizable inorganic species can be oxidized by oxygen in aqueous solution. The source of oxygen usually is air. Rather extreme conditions of temperature and pressure are required, with a temperature range of 175-327°C and a pressure range of 300-3,000 psig (2070-20,700 kPa). The high pressures allow a high concentration of oxygen to be dissolved in the water and the high temperatures enable the reaction to occur.

Wet air oxidation has been applied to the destruction of cyanides in electroplating wastewaters. The oxidation reaction for sodium cyanide is the following:

$$2Na^+ + 2CN^- + O_2 + 4H_2O \rightarrow 2Na^+ + 2HCO_3^- + 2NH_3 \quad (24.7.4)$$

UV-Enhanced Wet Oxidation

Hydrogen peroxide (H_2O_2) can be used as an oxidant in solution assisted by ultraviolet radiation ($h\nu$). For the oxidation of organic species represented in general as $\{CH_2O\}$, the overall reaction is

$$2H_2O_2 + \{CH_2O\} + h\nu \rightarrow CO_2 + 3H_2O \quad (24.7.5)$$

The ultraviolet radiation breaks chemical bonds and serves to form reactive oxidant species, such as HO•.

24.8. BIODEGRADATION OF WASTES

Biodegradation of wastes is their conversion by biological processes to simple inorganic molecules (mineralization) and, to a certain extent, to biological materials. Usually the products of biodegradation are molecular forms that tend to occur in nature and that are in greater thermodynamic equilibrium with their surroundings than are the starting materials. **Detoxification** refers to the biological conversion of a toxic substance to a less toxic species. Microbial bacteria and fungi possessing enzyme systems required for biodegradation of wastes are usually best obtained from populations of indigenous microorganisms at a hazardous waste site where they have developed the ability to degrade particular kinds of molecules. Although it has some shortcomings in the degradation of complex chemical mixtures, biological treatment offers a number of significant advantages and has considerable potential for the degradation of hazardous wastes, even *in situ*.

Biodegradability

The **biodegradability** of a compound is influenced by its physical characteristics, such as solubility in water and vapor pressure, and by its chemical properties, including molecular mass, molecular structure, and presence of various kinds of functional groups, some of which provide a "biochemical handle" for the initiation of biodegradation. With the appropriate organisms and under the right conditions, even substances such as phenol that are considered to be biocidal to most microorganisms can undergo biodegradation.

Recalcitrant or **biorefractory** substances are those that resist biodegradation and tend to persist and accumulate in the environment. Such materials are not necessarily toxic to organisms, but simply resist their metabolic attack. However, even some compounds regarded as biorefractory may be degraded by microorganisms adapted to their biodegradation; for example DDT is degraded by properly acclimated *Pseudomonas*. Chemical pretreatment, especially by partial oxidation, can make some kinds of recalcitrant wastes much more biodegradable.

Properties of hazardous wastes and their media can be changed to increase biodegradability. This can be accomplished by adjustment of conditions to optimum temperature, pH (usually in the range of 6-9), stirring, oxygen level, and material load. Biodegradation can be aided by removal of toxic organic and inorganic substances, such as heavy metal ions.

Aerobic Treatment

Aerobic waste treatment processes utilize aerobic bacteria and fungi that require molecular oxygen, O_2. These processes are often favored by microorganisms, in part because of the high energy yield obtained when molecular oxygen reacts with organic matter. Aerobic waste treatment is well adapted to the use of an activated sludge process. It can be applied to hazardous wastes such as chemical process wastes and landfill leachates. Some systems use powdered activated carbon as an additive to absorb nonbiodegradable organic wastes.

Contaminated soils can be mixed with water and treated in a bioreactor to eliminate biodegradable contaminants in the soil. It is possible in principle to treat contaminated soils biologically in place by pumping oxygenated, nutrient-enriched water through the soil in a recirculating system.

Anaerobic Treatment

Anaerobic waste treatment in which microorganisms degrade wastes in the absence of oxygen can be practiced on a variety of organic hazardous wastes. Compared to the aerated activated sludge process, anaerobic digestion requires less energy; yields less sludge by-product; generates sulfide (H_2S), which precipitates toxic heavy metal ions; and produces methane gas, CH_4, which can be used as an energy source.

The overall process for anaerobic digestion, is a fermentation process in which organic matter is both oxidized and reduced. The simplified reaction for the anaerobic fermentation of a hypothetical organic substance, "$\{CH_2O\}$", is the following:

$$2\{CH_2O\} \rightarrow CO_2 + CH_4 \tag{24.8.1}$$

In practice, the microbial processes involved are quite complex. Most of the wastes for which anaerobic digestion is suitable consist of oxygenated compounds, such as acetaldehyde or methylethyl ketone.

24.9. LAND TREATMENT AND COMPOSTING

Land Treatment

Soil may be viewed as a natural filter for wastes. Soil has physical, chemical, and biological characteristics that can enable waste detoxification, biodegradation, chemical decomposition, and physical and chemical fixation. Therefore, **land treatment** of wastes may be accomplished by mixing the wastes with soil under appropriate conditions.

Soil is a natural medium for a number of living organisms that may have an effect upon biodegradation of hazardous wastes. Of these, the most important are bacteria, including those from the genera *Agrobacterium*, *Arthrobacteri*, *Bacillus*, *Flavobacterium*, and *Pseudomonas*. In addition, *Actinomycetes* and fungi may be involved in biodegradation of wastes.

Wastes that are amenable to land treatment are biodegradable organic substances. However, in soil contaminated with hazardous wastes, bacterial cultures may develop that are effective in degrading normally recalcitrant compounds through acclimation over a long period of time. Land treatment is most used for petroleum refining wastes and is applicable to the treatment of fuels and wastes from leaking underground storage tanks. It can also be applied to biodegradable organic chemical wastes, including some organohalide compounds. Land treatment is not suitable for the treatment of wastes containing acids, bases, toxic inorganic compounds, salts, heavy metals, and organic compounds that are excessively soluble, volatile, or flammable.

Composting

Composting of hazardous wastes is the biodegradation of solid or solidified materials in a medium other than soil. Bulking material, such as plant residue, paper, municipal refuse, or sawdust may be added to retain water and enable air to penetrate to the waste material. Successful composting of hazardous waste depends upon a number of factors, including those discussed above under land treatment. The first of these is the selection of the appropriate microorganism or **inoculum**. Once a successful composting operation is underway, a good inoculum is maintained by recirculating spent compost to each new batch. Other parameters that must be controlled include oxygen supply, moisture content (which should be maintained at a minimum of about 40%), pH (usually around neutral), and temperature. The composting process generates heat so, if the mass of the compost pile is sufficiently high, it can be self-heating under most conditions. Some wastes are deficient in nutrients, such as nitrogen, which must be supplied from commercial sources or from other wastes.

24.10. PREPARATION OF WASTES FOR DISPOSAL

Immobilization, stabilization, fixation, and solidification are terms that describe techniques whereby hazardous wastes are placed in a form suitable for long-term disposal. These aspects of hazardous waste management are addressed in detail in reference works on the subject.

Immobilization

Immobilization includes physical and chemical processes that reduce surface areas of wastes to minimize leaching. It isolates the wastes from their environment, especially groundwater, so that they have the least possible tendency to migrate. This is accomplished by physically isolating the waste, reducing its solubility, and decreasing its surface area. Immobilization usually improves the handling and physical characteristics of wastes.

Stabilization

Stabilization means the conversion of a waste from its original form to a physically and chemically more stable material. Stabilization may include chemical reactions that produce products that are less volatile, soluble, and reactive.

Solidification, which is discussed below, is one of the most common means of stabilization. Stabilization is required for land disposal of wastes. **Fixation** is a process that binds a hazardous waste in a less mobile and less toxic form; it means much the same thing as stabilization.

Solidification

Solidification may involve chemical reaction of the waste with the solidification agent, mechanical isolation in a protective binding matrix, or a combination of chemical and physical processes. It can be accomplished by evaporation of water from aqueous wastes or sludges, sorption onto solid material, reaction with cement, reaction with silicates, encapsulation, or imbedding in polymers or thermoplastic materials.

In many solidification processes, such as reaction with Portland cement, water is an important ingredient of the hydrated solid matrix. Therefore, the solid should not be heated excessively or exposed to extremely dry conditions, which could result in diminished structural integrity from loss of water. In some cases, however, heating a solidified waste is an essential part of the overall solidification procedure. For example, an iron hydroxide matrix can be converted to highly insoluble, refractory iron oxide by heating, and organics in solidified wastes may be converted to inert carbon by heating. Heating is an integral part of the process of vitrification (see below).

Sorption to a Solid Matrix Material

Hazardous waste liquids, emulsions, sludges, and free liquids in contact with sludges may be solidified and stabilized by fixing onto solid **sorbents**, including activated carbon (for organics), fly ash, kiln dust, clays, vermiculite, and various proprietary materials. Sorption may be done to convert liquids and semisolids to dry solids, improve waste handling, and reduce solubility of waste constituents. Sorption can also be used to improve waste compatibility with substances such as Portland cement used for solidification and setting. Specific sorbents may also be used to stabilize pH and pE (a measure of the tendency of a medium to be oxidizing or reducing, see Chapter 11).

The action of sorbents can include simple mechanical retention of wastes, physical sorption, and chemical reactions. It is important to match the sorbent to the waste. A substance with a strong affinity for water should be employed for wastes containing excess water and one with a strong affinity for organic materials should be used for wastes with excess organic solvents.

Thermoplastics and Organic Polymers

Thermoplastics are solids or semisolids that become liquified at elevated temperatures. Hazardous waste materials may be mixed with hot thermoplastic liquids and solidified in the cooled thermoplastic matrix, which is rigid but deformable. The thermoplastic material most used for this purpose is asphalt bitumen. Other thermoplastics, such as paraffin and polyethylene have also been used to immobilize hazardous wastes.

Among the wastes that can be immobilized with thermoplastics are those containing heavy metals, such as electroplating wastes. Organic thermoplastics repel water and reduce the tendency toward leaching in contact with groundwater. Compared to cement, thermoplastics add relatively less material to the waste.

A technique similar to that described above uses **organic polymers** produced in contact with solid wastes to imbed the wastes in a polymer matrix. Three kinds of polymers that have been used for this purpose include polybutadiene, urea-formaldehyde, and vinyl ester-styrene polymers. This procedure is more complicated than is the use of thermoplastics but, in favorable cases, yields a product in which the waste is held more strongly.

Vitrification

Vitrification or **glassification** consists of imbedding wastes in a glass material. In this application, glass may be regarded as a high-melting-temperature inorganic thermoplastic. Molten glass can be used, or glass can be synthesized in contact with the waste by mixing and heating with glass constituents—silicon dioxide (SiO_2), sodium carbonate (Na_2CO_3), and calcium oxide (CaO). Other constituents may include boric oxide, B_2O_3, which yields a borosilicate glass that is especially resistant to changes in temperature and chemical attack. In some cases glass is used in conjunction with thermal waste destruction processes, serving to immobilize hazardous waste ash consituents. Some wastes are detrimental to the quality of the glass. Aluminum oxide, for example, may prevent glass from fusing.

Vitrification is relatively complicated and expensive, the latter because of the energy consumed in fusing glass. Despite these disadvantages, it is the best immobilization technique for some special wastes and has been promoted for solidification of radionuclear wastes because glass is chemically inert and resistant to leaching. However, high levels of radioactivity can cause deterioration of glass and lower its resistance to leaching.

Solidification With Cement

Portland cement is widely used for solidification of hazardous wastes. In this application, Portland cement provides a solid matrix for isolation of the wastes, chemically binds water from sludge wastes, and may react chemically with wastes (for example, the calcium and base in Portland cement react chemically with inorganic arsenic sulfide wastes to reduce their solubilities). However, most wastes are held physically, not chemically bound, in the rigid Portland cement matrix and are subject to leaching.

As a solidification matrix, Portland cement is most applicable to inorganic sludges containing heavy metal ions that form insoluble hydroxides and carbonates in the basic carbonate medium provided by the cement. The success of solidification with Portland cement strongly depends upon whether or not the waste adversely affects the strength and stability of the concrete product. A number of substances — organic matter such as petroleum or coal; some silts and clays; sodium salts of arsenate, borate, phosphate, iodate, and sulfide; and salts of copper, lead, magnesium, tin, and zinc — are incompatible with Portland cement because they interfere with its set and cure and cause deterioration of the cement matrix with time. However, a reasonably good disposal form can be obtained by absorbing organic wastes with a solid material, which in turn is set in Portland cement. This approach has been used with hydrocarbon wastes sorbed by an activated coal char matrix.

Solidification With Silicate Materials

Water-insoluble **silicates**, (pozzolanic substances) containing oxyanionic silicon such as SiO_3^{2-} are used for waste solidification. These substances include fly ash, flue dust, clay, calcium silicates, and ground-up slag from blast furnaces. Soluble silicates, such as sodium silicate, may also be used. Silicate solidification usually requires a setting agent, which may be Portland cement (see above), gypsum (hydrated $CaSO_4$), lime, or compounds of aluminum, magnesium, or iron. The product may vary from a granular material to a concrete-like solid. In some cases the product is improved by additives, such as emulsifiers, surfactants, activators, calcium chloride, clays, carbon, zeolites, and various proprietary materials.

Success has been reported for the solidification of both inorganic wastes and organic wastes (including oily sludges) with silicates. The advantages and disadvantages of silicate solidification are similar to those of Portland cement discussed above. One consideration that is especially applicable to fly ash is the presence in some silicate materials of leachable hazardous substances, which may include arsenic and selenium.

Encapsulation

As the name implies, **encapsulation** is used to coat wastes with an impervious material so that they do not contact their surroundings. For example, a water-soluble waste salt encapsulated in asphalt would not dissolve, so long as the asphalt layer remains intact. A common means of encapsulation uses heated, molten thermoplastics, asphalt, and waxes that solidify when cooled. A more sophisticated approach to encapsulation is to form polymeric resins from monomeric substances in the presence of the waste.

Chemical Fixation

Chemical fixation is a process that binds a hazardous waste substance in a less mobile, less toxic form by a chemical reaction that alters the waste chemically. Physical and chemical fixation often occur together. Polymeric inorganic silicates containing some calcium and often some aluminum are the inorganic materials most widely used as a fixation matrix. Many kinds of heavy metals are chemically bound in such a matrix, as well as being held physically by it. Similarly, some organic wastes are bound by reactions with matrix constituents.

24.11. ULTIMATE DISPOSAL OF WASTES

Regardless of the destruction, treatment, and immobilization techniques used, there will always remain from hazardous wastes some material that has to be put somewhere. This section briefly addresses the ultimate disposal of ash, salts, liquids, solidified liquids, and other residues that must be placed where their potential to do harm is minimized.

Disposal Aboveground

In some important respects disposal aboveground, essentially in a pile designed to prevent erosion and water infiltration, is the best way to store solid wastes.

Perhaps its most important advantage is that it avoids infiltration by groundwater that can result in leaching and groundwater contamination common to storage in pits and landfills. In a properly designed aboveground disposal facility any leachate that is produced drains quickly by gravity to the leachate collection system, where it can be detected and treated.

Aboveground disposal can be accomplished with a storage mound deposited on a layer of compacted clay covered with impermeable membrane liners laid somewhat above the original soil surface and shaped to allow leachate flow and collection. The slopes around the edges of the storage mound should be sufficiently great to allow good drainage of precipitation, but gentle enough to deter erosion.

Landfill

Landfill historically has been the most common way of disposing of solid hazardous wastes and some liquids, although it is being severely limited in many nations by new regulations and high land costs. Landfill involves disposal that is at least partially underground in excavated cells, quarries, or natural depressions. Usually, fill is continued above ground to most efficiently utilize space and provide a grade for drainage of precipitation.

The greatest environmental concern with landfill of hazardous wastes is the generation of leachate from infiltrating surface water and groundwater with resultant contamination of groundwater supplies. Modern hazardous waste landfills provide elaborate systems to contain, collect, and control such leachate.

There are several components to a modern landfill. A landfill should be placed on a compacted low-permeability medium, preferably clay, which is covered by a flexible-membrane liner consisting of water-tight impermeable material. This liner is covered with granular material in which is installed a secondary drainage system. Next is another flexible-membrane liner above which is installed a primary drainage system for the removal of leachate. This drainage system is covered with a layer of granular filter medium, upon which the wastes are placed. In the landfill, wastes of different kinds are separated by berms consisting of clay or soil covered with liner material. When the fill is complete, the waste is capped to prevent surface water infiltration and covered with compacted soil. In addition to leachate collection, provision may be made for a system to treat evolved gases, particularly when methane-generating biodegradable materials are disposed in the landfill.

The flexible-membrane liner made of rubber (including chlorosulfonated polyethylene) or plastic (including chlorinated polyethylene, high-density polyethylene, and polyvinylchloride), is a key component of approved landfills. It controls seepage out of, and infiltration into the landfill. Liners have to meet stringent standards to serve their intended purpose. In addition to being impermeable, the liner material must be strongly resistant to biodegradation, chemical attack, and tearing.

Capping with a variety of often multilayered material is done to cover the wastes, prevent infiltration of excessive amounts of surface water, and prevent release of wastes to overlying soil and the atmosphere.

Surface Impoundment of Liquids

Many liquid hazardous wastes, slurries, and sludges are placed in **surface impoundments**, which usually serve for treatment and often are designed to be filled in eventually as a landfill disposal site. Most liquid hazardous wastes and a significant fraction of solids are placed in surface impoundments in some stage of treatment, storage, or disposal.

A surface impoundment may consist of an excavated "pit," a structure formed with dikes, or a combination thereof. The construction is similar to that discussed above for landfills in that the bottom and walls should be impermeable to liquids and provision must be made for leachate collection. The chemical and mechanical challenges to liner materials in surface impoundments are severe so that proper geological siting and construction with floors and walls composed of low-permeability soil and clay are important in preventing pollution from these installations.

Deep-Well Disposal of Liquids

Deep-well disposal of liquids consists of their injection under pressure to underground strata isolated by impermeable rock strata from aquifers. Early experience with this method was gained in the petroleum industry where disposal is required of large quantities of saline wastewater coproduced with crude oil. The method was later extended to the chemical industry for the disposal of brines, acids, heavy metal solutions, organic liquids, and other liquids.

A number of factors must be considered in deep-well disposal. Wastes are injected into a region of elevated temperature and pressure, which may cause chemical reactions to occur involving the waste constituents and the mineral strata. Oils, solids, and gases in the liquid wastes can cause problems such as clogging. Corrosion may be severe. Microorganisms may have some effects. Most problems from these causes can be mitigated by proper waste pretreatment.

The most serious consideration involving deep-well disposal is the potential contamination of groundwater. Although injection is made into permeable saltwater aquifers presumably isolated from aquifers that contain potable water, contamination may occur. Major routes of contamination include fractures, faults, and other wells. The disposal well itself can act as a route for contamination if it is not properly constructed and cased or if it is damaged.

Storage of Nuclear Waste

A disposal problem of particular concern is that of nuclear wastes produced by nuclear power generation and production of plutonium for nuclear weapons. The U.S. generates about 6 tons of spent nuclear reactor fuel rods each day, for a current total accumulation of more than 30,000 tons. Most of the spent rods are still stored under water at the reactor sites. In addition there are about 380,000 cubic meters of high-level radioactive wastes accumulated during the production of plutonium for weapons. Although it sounds bad to simply store spent fuel rods on site, such storage is a good interim solution because of the very rapid decrease in radioactivity of the rods during their first several years of storage. The high-level nuclear wastes from weapons production present a much more serious problem because they are stored largely in corrosive solutions and as caked solids contained in vulnerable storage facilities. At the Hanford nuclear site in Washington State where large quantities of military plutonium were produced and processed, high-level nuclear wastes are stored in 177 enormous underground tanks.[1] Most of these tanks have exceeded their life expectancies of around 50 years, and about 1/3 of them are presumed to have leaked part of their contents.

Current U.S. Department of Energy plans call for storing spent fuel and high-level nuclear wastes in chambers excavated in Yucca Mountain in the state of Nevada.[2] Yucca mountain is a long ridge several hundred meters high composed of tuff, a rock derived from volcanic ash around 12 million years ago. The chambers are prepared by excavating horizontally into the mountain about 300 meters below its surface and around 300 meters above the water table. The facility is designed for about 70,000 metric tons of spent nuclear fuel, plus some high-level military wastes. There are numerous concerns about the site, including the possibility of corrosion of waste canisters and formation of radioactive leachate, which may prevent the facility from ever being used. However, the location of the wastes well above the water table, negligible water infiltration due to the low precipitation rate of only about 16 cm of rain per year, the physical and chemical stability of the tuff formations, and other factors make the production of leachate and other pathways of radioactivity release unlikely. The spent fuel rods to be stored in the facility contain large quantities of enriched uranium fuel and plutonium that can be used for reactor fuel, so simply leaving the spent fuel in a disposal site is a wasteful practice, even if it is safe.

24.12. LEACHATE AND GAS EMISSIONS

Leachate

The production of contaminated leachate is a possibility with most disposal sites. Therefore, new hazardous waste landfills require leachate collection/treatment systems and many older sites are required to have such systems retrofitted to them. Modern hazardous waste landfills typically have dual leachate collection systems, one located between the two impermeable liners required for the bottom and sides of the landfill and another just above the top liner of the double-liner system. The upper leachate collection system is called the primary leachate collection system, and the bottom is called the secondary leachate collection system. Leachate is collected in perforated pipes that are imbedded in granular drain material.

Chemical and biochemical processes have the potential to cause some problems for leachate collection systems. One such problem is clogging by insoluble manganese(IV) and iron(III) hydrated oxides upon exposure to air a problem that also afflicts water wells.

Leachate consists of water that has become contaminated by wastes as it passes through a waste disposal site. It contains waste constituents that are soluble, not retained by soil, and not degraded chemically or biochemically. Some potentially harmful leachate constituents are products of chemical or biochemical transformations of wastes.

The best approach to leachate management is to prevent its production by limiting infiltration of water into the site. Rates of leachate production may be very low when sites are selected, designed, and constructed with minimal production of leachate as a major objective. A well-maintained, low-permeability cap over the landfill is very important for leachate minimization.

Hazardous Waste Leachate Treatment

The first step in treating leachate is to characterize it fully, particularly with a thorough chemical analysis of possible waste constituents and their chemical and metabolic products. The biodegradability of leachate constituents should also be determined.

The options available for the treatment of hazardous waste leachate are generally those that can be used for industrial wastewaters. These include biological treatment by an activated sludge, or related process, and sorption by activated carbon usually in columns of granular activated carbon. Hazardous waste leachate can be treated by a variety of chemical processes, including acid/base neutralization, precipitation, and oxidation/reduction. In some cases these treatment steps must precede biological treatment; for example, leachate exhibiting extremes of pH must be neutralized in order for microorganisms to thrive in it. Cyanide in the leachate may be oxidized with chlorine and organics with ozone, hydrogen peroxide promoted with ultraviolet radiation, or dissolved oxygen at high temperatures and pressures. Heavy metals may be precipitated with base, carbonate, or sulfide.

Leachate can be treated by a variety of physical processes. In some cases, simple density separation and sedimentation can be used to remove water-immiscible liquids and solids. Filtration is frequently required and flotation may be useful. Leachate solutes can be concentrated by evaporation, distillation, and membrane processes, including reverse osmosis, hyperfiltration, and ultrafiltration. Organic constituents can be removed by solvent extraction, air stripping, or steam stripping.

Gas Emissions

In the presence of biodegradable wastes, methane and carbon dioxide gases are produced in landfills by anaerobic degradation (see Reaction 24.8.1). Gases may also be produced by chemical processes with improperly pretreated wastes, as would occur in the hydrolysis of calcium carbide to produce acetylene:

$$CaC_2 + 2H_2O \rightarrow 2C_2H_2 + Ca(OH)_2 \qquad (24.12.1)$$

Odorous and toxic hydrogen sulfide, H_2S, may be generated by the chemical reaction of sulfides with acids or by the biochemical reduction of sulfate by anaerobic bacteria (*Desulfovibrio*) in the presence of biodegradable organic matter:

$$SO_4^{2-} + 2\{CH_2O\} + 2H^+ \xrightarrow[\text{bacteria}]{\text{Anaerobic}} H_2S + 2CO_2 + 2H_2O \qquad (24.12.2)$$

Gases such as these may be toxic, they may burn, or they may explode. Furthermore, gases permeating through landfilled hazardous waste may carry along waste vapors, such as those of volatile aryl compounds and low-molecular-mass chlorinated hydrocarbons. Of these, the ones of most concern are benzene, carbon tetrachloride, chloroform, 1,2-dibromoethane, 1,2-dichloroethane, dichloromethane, tetrachloroethane, 1,1,1-trichloroethane, trichloroethylene, and vinyl chloride. Because of the hazards from these and other volatile species, it is important to minimize production of gases and, if significant amounts of gases are produced, they should be vented or treated by activated carbon sorption or flaring.

24.13. *IN SITU* TREATMENT

***In situ* treatment** refers to waste treatment processes that can be applied to wastes in a disposal site by direct application of treatment processes and reagents to the wastes. Where possible, *in situ* treatment is highly desirable as a waste site remediation option. For wastes near the surface, an area of *in situ* treatment that has received a great deal of attention in the last approximately five years is **phyto-remediation**.[3] This process consists of growing plants in contaminated soil or contaminated water. In some cases, the plants metabolically destroy wastes, whereas in others the plants take up heavy metals or radionuclides, which are then removed by harvesting the plant biomass.

In Situ Immobilization

In situ immobilization is used to convert wastes to insoluble forms that will not leach from the disposal site. Heavy metal contaminants including lead, cadmium, zinc, and mercury, can be immobilized by chemical precipitation as the sulfides by treatment with gaseous H_2S or alkaline Na_2S solution. Disadvantages include the high toxicity of H_2S and the contamination potential of soluble sulfide. Although precipitated metal sulfides should remain as solids in the anaerobic conditions of a landfill, unintentional exposure to air can result in oxidation of the sulfide and remobilization of the metals as soluble sulfate salts.

Oxidation and reduction reactions can be used to immobilize heavy metals *in situ*. Oxidation of soluble Fe^{2+} and Mn^{2+} to their insoluble hydrous oxides, $Fe_2O_3 \cdot xH_2O$ and $MnO_2 \cdot xH_2O$, respectively, can precipitate these metal ions and coprecipitate other heavy metal ions. However, subsurface reducing conditions could later result in reformation of soluble reduced species. Reduction can be used *in situ* to convert soluble, toxic chromate to insoluble chromium(III).

Chelation may convert metal ions to less mobile forms, although with most agents chelation has the opposite effect. A chelating agent called Tetran is supposed to form metal chelates that are strongly bound to clay minerals. The humin fraction of soil humic substances likewise immobilizes metal ions.

Solidification *In Situ*

In situ solidification can be used as a remedial measure at hazardous waste sites. One approach is to inject soluble silicates followed by reagents that cause them to solidify. For example, injection of soluble sodium silicate followed by calcium chloride or lime forms solid calcium silicate.

Detoxification *In Situ*

When only one, or a limited number of harmful constituents is present in a waste disposal site, it may be practical to consider detoxification *in situ* as a means of reducing the hazard from the wastes. This approach is most practical for organic contaminants, including pesticides (organophosphate esters and carbamates), amides, and esters. Among the chemical and biochemical processes that can detoxify such materials are chemical and enzymatic oxidation, reduction, and hydrolysis. Chemical oxidants that have been proposed for this purpose include hydrogen peroxide, ozone, and hypochlorite.

Enzyme extracts collected from microbial cultures and purified have been considered for *in situ* detoxification. One cell-free enzyme that has been used for detoxification of organophosphate insecticides is parathion hydrolase. The hostile environment of a chemical waste landfill, including the presence of heavy metal ions and other toxicants that inhibit enzymes, is detrimental to many biochemical approaches to *in situ* treatment. Furthermore, most sites contain a mixture of hazardous constituents, which might require several different enzymes for their detoxification.

Permeable Bed Treatment

Some groundwater plumes contaminated by dissolved wastes can be treated by a permeable bed of material placed in a trench through which the groundwater must flow. For example, limestone contained in a permeable bed neutralizes acid and precipitates some kinds of heavy metals as hydroxides or carbonates. Synthetic ion exchange resins can be used in a permeable bed to retain heavy metals and even some anionic species, although competition with ionic species present naturally in the groundwater can cause some problems with the use of ion exchangers. Activated carbon in a permeable bed will remove some organics, especially less soluble, higher molecular-mass organic compounds.

Permeable bed treatment requires relatively large quantities of reagent, which argues against the use of activated carbon and ion exchange resins. In such an application it is unlikely that either of these materials could be reclaimed and regenerated as is done when they are used in columns to treat wastewater. Furthermore, ions taken up by ion exchangers and organic species retained by activated carbon may be released at a later time, causing subsequent problems. Finally, a permeable bed that has been truly effective in collecting waste materials may, itself, be considered a hazardous waste requiring special treatment and disposal.

In Situ Thermal Processes

Heating of wastes *in situ* can be used to remove or destroy some kinds of hazardous substances. Both steam injection and radio-frequency heating have been proposed for this purpose. Volatile wastes brought to the surface by heating can be collected and held as condensed liquids or by activated carbon.

One approach to immobilizing wastes *in situ* is high temperature vitrification using electrical heating. This process involves placing conducting graphite between two electrodes poured on the surface and passing an electrical current between the electrodes. In principle, the graphite becomes very hot and "melts" into the soil leaving a glassy slag in its path. Volatile species evolved are collected and, if the operation is successful, a nonleachable slag is left in place. It is easy to imagine problems that might occur, including difficulties in getting a uniform melt, problems from groundwater infiltration, and very high consumption of electricity.

Soil Washing and Flushing

Extraction with water containing various additives can be used to cleanse soil contaminated with hazardous wastes. When the soil is left in place and the water pumped into and out of it, the process is called **flushing**; when soil is removed and contacted with liquid the process is referred to as **washing**. Here, washing is used as a term applied to both processes.

The composition of the fluid used for soil washing depends upon the contaminants to be removed. The washing medium may consist of pure water or it may contain acids (to leach out metals or neutralize alkaline soil contaminants), bases (to neutralize contaminant acids), chelating agents (to solubilize heavy metals), surfactants (to enhance the removal of organic contaminants from soil and improve the ability of the water to emulsify insoluble organic species), or reducing agents (to reduce oxidized species). Soil contaminants may dissolve, form emulsions, or react chemically. Inorganic species commonly removed from soil by washing include heavy metals salts; lighter aromatic hydrocarbons, such as toluene and xylenes; lighter organohalides, such as trichloro- or tetrachloroethylene; and light-to-medium molecular mass aldehydes and ketones.

CHAPTER SUMMARY

The chapter summary below is presented in a programmed format to review the main points covered in this chapter. It is used most effectively by filling in the blanks, referring back to the chapter as necessary. The correct answers are given at the end of the summary.

In descending order of desirability, hazardous waste management attempts to accomplish [1]____
____
____
____
____.
Cutting down quantities of wastes from their sources is called [2]______________
______________________ and utilization of treatment processes that reduce the quantities of wastes requiring ultimate disposal is called [3]__________________
_______________. The four broad areas in which something of value may be obtained from recycling wastes are [4]______________________________
____
____
______________________________________. The two main classes of potentially hazardous contaminants in motor oil are [5]_____________________________
____. The three major steps in reprocessing waste lubricating oil are [6]______________________________
____. Ash from waste fuel oil may be hazardous because of [7]______________________________
_______________________________. Among the operations used in solvent purification for recycling are [8]______________________________________
____
____
____.
Primary treatment of wastes is generally regarded as [9]_____________________
____, secondary treatment [10]____,
and polishing as [11]____
____________________________ The main methods of physical treatment can be divided into the categories of [12]______________________________________
____.
A waste treatment process by which colloidal-sized particles are caused to aggregate

and settle from suspension is called [13]____________________and it is aided by [14]____.
A phase transistion process that is used in treating and recycling solvents, waste oil, aqueous phenolic wastes, xylene contaminated with paraffin from histological laboratories, and mixtures of ethylbenzene and styrene is [15]_______________, which produces by-product [16]____________________________, which are often hazardous and polluting. A means of separating volatile components from less volatile ones in a liquid mixture by the partitioning of the more volatile materials to a gas phase of air or steam is called [17]______________________. The major changes in a solution that can cause physical precipitation are [18]_______________
____.
Solvent extraction and leaching are examples of [19]______________________ processes. A solvent extraction process that makes use of CO_2 at very high pressures is [20]______________________________ extraction. Transfer of a substance from a solution to a solid phase is called [21]___________________, for which the most important solid for organics removal is [22]__________________________.
In membrane processes, such as hyperfiltration and ultrafiltration, water and lower molecular-mass solutes under pressure pass through the membrane as a stream of purified [23]__________________, leaving behind a stream of [24]_______________
___________ containing [25]_______________. Reverse osmosis uses a membrane that is [26]____
____.
The major kinds of chemical phenomena used in chemical purification of wastes are [27]____
____
____.
The most widely used means of precipitating metal ions is by [28]_______________
____. The reaction $Cd^{2+} + Zn \rightarrow Cd + Zn^{2+}$ illustrates a process called [29]______________________.
A process that can be used to treat metal carbides, hydrides, amides, alkoxides, and halides; nonmetal oxyhalides and sulfides; and some organic species, such as acetic anhydride is [30]________________________. The most widely applied means of thermal treatment of hazardous wastes is [31]__________________, which utilizes [32]____.
The four main components of hazardous waste incineration systems are [33]________
____
____.
The two most common types of hazardous waste incinerators in the U. S. are [34]____
____.
The type of thermal treatment devices that make use of an extremely hot plasma of ionized air injected through an electrical arc are [35]______________________
____________________. Biodegradation of wastes is [36]____________________
____
____.
Detoxification refers to [37]____
____
____________________________. Recalcitrant or biorefractory substances are

[38]____
____.
Some of the ways in which anaerobic digestion differs from aerobic digestion in the treatment of wastes is [39]____
____
____. Immobilization of wastes
[40]____
whereas stabilization means [41]____________________________________
____.
Some of the sorbents used to solidify and stabilize hazardous wastes are [42]________
____
____.
Vitrification consists of [43]____
____.
Pozzolanic substances used to solidify wastes include [44]____________________
____
__________. Dissolved iron and manganese in hazardous waste leachate can cause problems because of [45]____
____
______________. The best approach to leachate management is [46]____________
____.
Two gases, one explosive and the other highly toxic, produced by anaerobic degradation in a land fill are [47]_____________________________________
____________________. Waste treatment processes that can be applied to wastes in a disposal site by direct application of treatment processes and reagents to the wastes are called [48]______________________ processes. When contaminated soil is left in place and purified by water pumped into and out of it, the process is called [49]__________________________, and when soil is removed and contacted with liquid, the process is referred to as [50]______________________________.

Answers

[1] not produce waste, produce only minimum quantities, recycle, treat in a way that makes the waste nonhazardous, dispose of the waste in a safe manner, monitor disposed wastes for leaching and other adverse effects

[2] waste reduction

[3] waste minimization

[4] direct recycle as raw material to the generator, transfer as a raw material to another process, utilization for pollution control or waste treatment, recovery of energy

[5] organic constituents and inorganic constituents

[6] removal of contaminant fuel and water, processing, and final polishing

[7] its heavy metal content

[8] settling, filtration, centrifugation, application of drying agents, adsorption, chemical treatment, and fractional distillation

9 preparation for further treatment

10 detoxifies, destroys, and removes hazardous constituents

11 usually refers to treatment of water that is removed from wastes

12 phase separation, phase transfer, phase transition, and membrane separations

13 coagulation

14 agitation, heat, acid, and addition of coagulants

15 distillation

16 distillation bottoms

17 stripping

18 cooling the solution, evaporation of solvent, or alteration of solvent composition

19 phase transfer

20 supercritical fluid

21 sorption

22 activated carbon

23 permeate

24 concentrate

25 impurities

26 selectively permeable to water

27 acid/base neutralization, chemical precipitation, chemical extraction and leaching, oxidation/reduction, reduction, and ion exchange

28 the formation of hydroxides

29 cementation

30 hydrolysis

31 incineration

32 high temperatures, an oxidizing atmosphere, and often turbulent combustion conditions

33 waste preparation and feeding, combustion, air pollutant removal , and ash and residue handling

34 rotary kiln and liquid injection incinerators

35 plasma incinerators

36 their conversion by biological processes to simple inorganic molecules and to biological materials

37 the biological conversion of a toxic substance to a less toxic species

38 those that resist biodegradation and tend to persist and accumulate in the environment

39 it requires less energy, yields less sludge by-product, generates sulfide (H_2S), and produces methane gas

40 includes physical and chemical processes that reduce surface areas of wastes to minimize leaching

41 the conversion of a waste from its original form to a physically and chemically more stable material

42 activated carbon, fly ash, kiln dust, clays, vermiculite

43 imbedding wastes in a glass material

44 fly ash, flue dust, clay, calcium silicates, and ground-up slag from blast furnaces

45 clogging by insoluble manganese(IV) and iron(III) hydrated oxides upon exposure to air

46 to prevent its production by limiting infiltration of water into the site

47 methane and carbon dioxide

48 *in situ*

49 flushing

50 washing

QUESTIONS AND PROBLEMS

1. Place the following hazardous waste management options in order of increasing desirability: (a) treat the waste to make it nonhazardous, (b) do not produce waste, (c) minimize quantities of waste produced, (d) dispose of the waste in a safe manner, (e) recycle the waste.

2. Match the waste recycling process or industry from the column on the left with the kind of material that can be recycled from the list on the right, below:

    1. Recycle as raw material to the generator
    2. Utilization for pollution control or waste treatment
    3. Materials with agricultural uses
    4. Organic substances
    5. Energy production

    (a) Waste alkali
    (b) Hydraulic and lubricating oils
    (c) Incinerable materials
    (d) Incompletely consumed feedstock material
    (e) Waste lime or phosphate-containing sludges

3. What material is recycled using hydrofinishing, treatment with clay, and filtration?

4. What is the "most important operation in solvent purification and recycle" that is used to separate solvents from impurities, water, and other solvents?

5. Dissolved air flotation (DAF) is used in the secondary treatment of wastes. What is the principle of this technique? For what kinds of hazardous waste substances is it most applicable?

6. Distillation is used in treating and recycling a variety of wastes, including solvents, waste oil, aqueous phenolic wastes, and mixtures of ethylbenzene and styrene. What is the major hazardous waste problem that arises from the use of distillation for waste treatment?

7. Match the process or industry from the column on the left with its "phase of waste treatment" from the list on the right, below:

1. Activated carbon sorption	(a) Primary treatment
2. Precipitation	(b) Secondary treatment
3. Reverse osmosis	(c) Polishing
4. Emulsion breaking	
5. Slurrying	

8. Supercritical fluid technology has a great deal of potential for the treatment of hazardous wastes. What are the principles involved with the use of supercritical fluids for waste treatment? Why is this technique especially advantageous? Which substance is most likely to be used as a supercritical fluid in this application? For which kinds of wastes are supercritical fluids most useful?

9. What are some advantages of using acetic acid, compared, for example, to sulfuric acid, as a neutralizing agent for treating waste alkaline materials?

10. Which of the following would be **least likely** to be produced by, or used as a reagent for the removal of heavy metals by their precipitation from solution? (a) Na_2CO_3, (b) CdS, (c) $Cr(OH)_3$, (d) KNO_3, (e) $Ca(OH)_2$

11. Both $NaBH_4$ and Zn are used to remove metals from solution. How do these substances remove metals? What are the forms of the metal products?

12. Of the following, thermal treatment of wastes is **not** useful for (a) volume reduction; (b) destruction of heavy metals; (c) removal of volatile, combustible, mobile organic matter; (d) destruction of pathogenic materials; (e) destruction of toxic substances.

13. From the following, choose the waste liquid that is least amenable to incineration and explain why it is not readily incinerated: (a) methanol, (b) tetrachloroethylene, (c) acetonitrile, (d) toluene, (e) ethanol, (f) acetone.

14. Name and give the advantages of the process that is used to destroy more hazardous wastes by thermal means than are burned solely for the purpose of waste destruction.

15. What is the major advantage of fluidized-bed incinerators from the standpoint of controlling pollutant by-products?

16. What is the best way to obtain microorganisms to be used in the treatment of hazardous wastes by biodegradation?

17. What are the principles of composting? How is it used to treat hazardous wastes?

18. How is Portland cement used in the treatment of hazardous wastes for disposal? What might be some disadvantages of such a use?

19. What are the advantages of aboveground disposal of hazardous wastes as opposed to burying wastes in landfills?

20. Describe and explain the best approach to managing leachate from hazardous waste disposal sites.

LITERATURE CITED

[1] Zorpette, Glenn, Hanford's Nuclear Wasteland, *Scientific American*, June, 1996, pp. 88-97.

[2] Whipple, Chris G., Can Nuclear Waste be Stored Safely at Yucca Mountain?, *Scientific American*, June, 1996, pp. 72-79.

[3] Kruger, Ellen L., Todd A. Anderson, and Joel R. Coats, Eds., *Phytoremediation of Soil and Water Contaminants*, American Chemical Society, Washington, DC, 1997.

SUPPLEMENTARY REFERENCES

Brunner, Calvin R., Incineration: Today's Hot Option for Waste Disposal, *Chemical Engineering*, October 12, 1987, pp. 96-106.

Daugherty, Jack, *Industrial Management: A Practical Approach*, Government Institutes, Inc., Rockville, MD, 1996.

Freeman, Harry M., Ed., *Hazardous Waste Minimization*, McGraw-Hill Publishing Co., New York, NY, 1990.

Hinchee, Robert E., *Soil Bioventing*, CRC Press/Lewis Publishers, Boca Raton, FL, 1997.

Horan, N. J., *Environmental Waste Management: A European Perspective*, Wiley, New York, NY, 1996.

Landreth, Robert E. and Paul A. Rebers, *Municipal Solid Wastes* CRC Press/Lewis Publishers, Boca Raton, FL, 1997.

Manahan, Stanley E., *Hazardous Waste Chemistry Toxicology and Treatment*, CRC Press/Lewis Publishers, Boca Raton, FL, 1990.

Manser, A. G. R. and Alan A. Keeling, *Practical Handbook of Processing and Recycling Municipal Waste*, CRC Press/Lewis Publishers, Boca Raton, FL, 1996.

Nyer, E. K., P. L. Palmer, S. Suthersan, S. Fam, F. Johns, D. Kidd, T. L. Crossman, and G. Boettcher, *In Situ Treatment Technology*, CRC Press/Lewis Publishers, Boca Raton, FL, 1996.

Oppelt, E. Timothy, Hazardous Waste Destruction, *Environmental Science and Technology*, **22**, 403–404 (1988).

Riser-Roberts, Eve, *Remediation of Petroleum Contaminated Soils*, CRC Press/Lewis Publishers, Boca Raton, FL, 1997.

Suthersan, Suthan, *Remediation Engineering*, CRC Press/Lewis Publishers, Boca Raton, FL, 1997.

Testa, Stephen M., *The Reuse and Recycling of Contaminated Soil*, CRC Press/Lewis Publishers, Boca Raton, FL, 1997.

INDEX